The Cell Cycle in the Central Nervous System

Contemporary Neuroscience

The Cell Cycle in the Central Nervous System, edited by **Damir Janigro**, 2006

Neural Development and Stem Cells, Second Edition, edited by **Mahendra S. Rao**, 2005

Neurobiology of Aggression: Understanding and Preventing Violence, edited by **Mark P. Mattson**, 2003

Neuroinflammation: Mechanisms and Management, Second Edition, edited by **Paul L. Wood**, 2003

Neural Stem Cells for Brain and Spinal Cord Repair, edited by **Tanja Zigova, Evan Y. Snyder, and Paul R. Sanberg**, 2003

Neurotransmitter Transporters: Structure, Function, and Regulation, Second Edition, edited by **Maarten E. A. Reith**, 2002

The Neuronal Environment: Brain Homeostasis in Health and Disease, edited by **Wolfgang Walz**, 2002

Pathogenesis of Neurodegenerative Disorders, edited by **Mark P. Mattson**, 2001

Stem Cells and CNS Development, edited by **Mahendra S. Rao**, 2001

Neurobiology of Spinal Cord Injury, edited by **Robert G. Kalb** and **Stephen M. Strittmatter**, 2000

Cerebral Signal Transduction: From First to Fourth Messengers, edited by **Maarten E. A. Reith**, 2000

Central Nervous System Diseases: Innovative Animal Models from Lab to Clinic, edited by **Dwaine F. Emerich, Reginald L. Dean, III, and Paul R. Sanberg**, 2000

Mitochondrial Inhibitors and Neurodegenerative Disorders, edited by **Paul R. Sanberg, Hitoo Nishino, and Cesario V. Borlongan**, 2000

Cerebral Ischemia: Molecular and Cellular Pathophysiology, edited by **Wolfgang Walz**, 1999

Cell Transplantation for Neurological Disorders, edited by **Thomas B. Freeman** and **Håkan Widner**, 1998

Gene Therapy for Neurological Disorders and Brain Tumors, edited by **E. Antonio Chiocca** and **Xandra O. Breakefield**, 1998

Highly Selective Neurotoxins: Basic and Clinical Applications, edited by **Richard M. Kostrzewa**, 1998

Neuroinflammation: Mechanisms and Management, edited by **Paul L. Wood**, 1998

Neuroprotective Signal Transduction, edited by **Mark P. Mattson**, 1998

Clinical Pharmacology of Cerebral Ischemia, edited by **Gert J. Ter Horst** and **Jakob Korf**, 1997

Molecular Mechanisms of Dementia, edited by **Wilma Wasco** and **Rudolph E. Tanzi**, 1997

Neurotransmitter Transporters: Structure, Function, and Regulation, edited by **Maarten E. A. Reith**, 1997

Motor Activity and Movement Disorders: Research Issues and Applications, edited by **Paul R. Sanberg, Klaus-Peter Ossenkopp, and Martin Kavaliers**, 1996

Neurotherapeutics: Emerging Strategies, edited by **Linda M. Pullan** and **Jitendra Patel**, 1996

Neuron–Glia Interrelations During Phylogeny: II. Plasticity and Regeneration, edited by **Antonia Vernadakis** and **Betty I. Roots**, 1995

Neuron–Glia Interrelations During Phylogeny: I. Phylogeny and Ontogeny of Glial Cells, edited by **Antonia Vernadakis** and **Betty I. Roots**, 1995

The Biology of Neuropeptide Y and Related Peptides, edited by **William F. Colmers** and **Claes Wahlestedt**, 1993

Psychoactive Drugs: Tolerance and Sensitization, edited by **A. J. Goudie** and **M. W. Emmett-Oglesby**, 1989

The Cell Cycle in the Central Nervous System

Edited by

Damir Janigro, PhD

The Cleveland Clinic Foundation, Cleveland, OH

HUMANA PRESS ✹ TOTOWA, NEW JERSEY

© 2006 Humana Press Inc.
999 Riverview Drive, Suite 208
Totowa, New Jersey 07512

humanapress.com

All rights reserved.

No part of this book may be reproduced, stored in a retrieval system, or transmitted in any form or by any means, electronic, mechanical, photocopying, microfilming, recording, or otherwise without written permission from the Publisher.

All papers, comments, opinions, conclusions, or recommendations are those of the author(s), and do not necessarily reflect the views of the publisher.

For additional copies, pricing for bulk purchases, and/or information about other Humana titles, contact Humana at the above address or at any of the following numbers: Tel.: 973-256-1699; Fax: 973-256-8341; E-mail: orders@humanapr.com, or visit our Website: www.humanapress.com

This publication is printed on acid-free paper. ∞
ANSI Z39.48-1984 (American Standards Institute) Permanence of Paper for Printed Library Materials.

Production Editor: Amy Thau

Cover Illustration: Fig. 3, Chapter 9, "Nonsynaptic GABAergic Communication and Postnatal Neurogenesis," by Xiuxin Liu, Anna J. Bolteus, and Angélique Bordey (background image); Fig. 13, Chapter 29, "Detection of Proliferation in Gliomas by Positron Emission Tomography Imaging," by Alexander M. Spence et al.; Fig. 3, Chapter 24, "Vascular Differentiation and the Cell Cycle," by Luca Cucullo; and Fig. 2, Chapter 32, "Cell Cycle of Encapsulated Cells," by Roberto Dal Toso and Sara Bonisegna (foreground images).

Cover design by Patricia F. Cleary

Photocopy Authorization Policy:
Authorization to photocopy items for internal or personal use, or the internal or personal use of specific clients, is granted by Humana Press Inc., provided that the base fee of US $30 is paid directly to the Copyright Clearance Center at 222 Rosewood Drive, Danvers, MA 01923. For those organizations that have been granted a photocopy license from the CCC, a separate system of payment has been arranged and is acceptable to Humana Press Inc. The fee code for users of the Transactional Reporting Service is: [1-58829-529-X/06 $30].

Printed in the United States of America. 10 9 8 7 6 5 4 3 2 1 5
eISBN: 1-59745-021-9
Library of Congress Cataloging-in-Publication Data

The cell cycle in the central nervous system / edited by Damir Janigro.
    p. cm. -- (Contemporary neuroscience)
  Includes bibliographical references and index.
  ISBN 1-58829-529-X (alk. paper)
  1. Central nervous system--Growth. 2. Cell cycle. 3. Central nervous system--Differentiation. 4. Central nervous system--Diseases.
I. Janigro, Damir. II. Series.
  [DNLM: 1. Central Nervous System--growth & development.
2. Central Nervous System--physiopathology. 3. Cell Cycle. 4. Cell Differentiation.   WL 300 C39305 2006]
QP370.C44 2006
612.8'22--dc22

2005017540

Preface

For many years, it was widely believed that the cell cycle in the central nervous system (CNS) was mostly of a prenatal, developmental nature. The concept of adult neurogenesis remained dormant until recently, while reports of an altered cell cycle in a damaged CNS gained strength. The discovery that the adult mammalian brain creates new neurons from pools of stem cells was a breakthrough in neuroscience. However, cell cycle regulation and disturbances are also a significant event in the life of other, nonneuronal cells of the brain (and spinal cord). *The Cell Cycle in the Central Nervous System* has been assembled with this in mind, and the authorship reflects these concepts.

There is still controversy over how to define a mitotic cell and how to study the relevance of neurogenesis in the CNS. Part I begins with an introduction to some of the tools that neuroscientists have used to determine mitotic propensity in neurons and other CNS cells (Dr. Prayson). The relevance of cell expansion and differentiation, with emphasis on both neuronal and glial cells, is outlined in the chapters by Drs. Taupin and Bradl. The development of blood vessels and their relevance during brain development is discussed in the chapter by Dr. Grant and myself, and Drs. Battaglia and Bassanini describe how impaired cell expansion results in postnatal malformations of cortical structures.

Neurons and glia, brain parenchyma, and cerebral vasculature are regarded today as an integrated system rather than an aggregate of different cell types. The concept of a neurovascular unit is clearly a centerpiece of modern neurobiology. Drs. Walker and Sikorska open Part II with an illustration of how mass screening of genes and gene products can be applied to neurogenesis, and Dr. Lo and colleagues describe how the development of new neurons is counterbalanced by cell death by apoptotic activation. The brief reviews by Drs. Arcangeli and Becchetti and the contributions by Dr. Bordey and colleagues, as well as Dr. Yu, introduce a new and provocative role for ion channels and neurotransmitters expressed in the CNS and apparently involved in the process of cell division and mitotic arrest. The renewal of stem cells in the mammalian brain is introduced by Dr. Arsenijevic.

Part III is devoted specifically to the regulation of cell cycle in glia and how its regulation may fail in pretumor conditions or following a nonneoplastic CNS response to injury (*see* Chapter 12 by Dr. Couldwell and colleagues, and Chapter 13 by Dr. Hallene and myself). In addition to ion channels (Part II, Chapter 8), evidence suggests that electrical field potentials are responsible for the relative quiescence of excitable cells or cells exposed to constant electrical activity (brain, heart, nerve, muscle). This is presented in the chapter by Dr. Dini and colleagues.

The therapeutic success of neurosurgical resections for the treatment of neurological disorders challenges the view that more is necessarily better (Part IV). The chapters by Drs. Taupin and Bengez show that brain injury often translates in cell cycle re-entry. Whether this may be beneficial, and to what extent, is discussed in a cerebrovascular framework by Drs. Kobiler and Glod (Chapter 17), Stanimirovic et al. (Chapter 18), and Moons et al. (Chapter 19).

The possibility that cell cycle re-entry is actually detrimental is presented in Part V. Changes in postmitotic neurons in a variety of pathologies are presented by Drs. York et al., Gustaw et al., Gonzalez-Martinez et al., Eisch and Mandyam, and Casadesus et al. Dr. Cucullo's chapter expands this to the cerebral vasculature.

Cell cycle control fails during tumorigenesis and brain tumors are not an exception. Unfortunately, little progress has been made in the treatment of malignant brain tumors. Part VI focuses on recent advances in the biology and detection of gliomas (Drs. Spence et al., Aeder and Hussaini, Kapoor and O'Rourke, Zhang and Fine), as well as drug resistance (Drs. Teng and Piquette-Miller).

The promises of postnatal neurogenesis and the possible pathological significance of cell cycle re-entry in the central nervous system will greatly influence the neuroscience world in the next several years. There is much hype and controversy surrounding the issue of stem cell research, and also uncertainty concerning the moral and ethical correlates of what we as scientists can do with molecular manipulation of the human genome. In some respects, however, the future is already here, and attempts to treat neurological disorders by gene transfer (Chapter 33), electrical stimulation (Chapter 34), or stem cell introduction (Chapter 35) are presented in Part VII. Drs. DalToso and Bonisegna address the issue of stem cell rejection by the host in Chapter 32, and Chapter 36 by Dr. Aumayr and myself gives a brief overview of how epigenetic modifications may impact CNS development.

Damir Janigro, PhD

Acknowledgment

I would like to thank my wife, Kim A. Conklin, for many years of uninterrupted support and encouragement, and Christine Moore for making all this possible.

Contents

Preface .... *v*
Acknowledgment .... *vii*
Contributors .... *xiii*
Companion CD .... *xvii*

Part I. Cell Cycle During the Development of the Mammalian Central Nervous System

1. Methodological Considerations in the Evaluation of the Cell Cycle in the Central Nervous System 3
   Richard A. Prayson
2. Neural Stem Cells 13
   Philippe Taupin
3. Progenitors and Precursors of Neurons and Glial Cells 23
   Monika Bradl
4. Vasculogenesis and Angiogenesis 31
   Gerald A. Grant and Damir Janigro
5. Neuronal Migration and Malformations of Cortical Development 43
   Giorgio Battaglia and Stefania Bassanini

Part II. Postnatal Development of Neurons and Glia

6. Genome-Wide Expression Profiling of Neurogenesis in Relation to Cell Cycle Exit 59
   P. Roy Walker, Dao Ly, Qing Y. Liu, Brandon Smith, Caroline Sodja, Marilena Ribecco, and Marianna Sikorska
7. Neurogenesis and Apoptotic Cell Death 71
   Klaus van Leyen, Seong-Ryong Lee, Michael A. Moskowitz, and Eng H. Lo
8. Ion Channels and the Cell Cycle 81
   Annarosa Arcangeli and Andrea Becchetti
9. Nonsynaptic GABAergic Communication and Postnatal Neurogenesis 95
   Xiuxin Liu, Anna J. Bolteus, and Angélique Bordey
10. Critical Roles of Ca^{2+} and K^+ Homeostasis in Apoptosis 105
    Shan Ping Yu

11. Mammalian Neural Stem Cell Renewal 119
    Yvan Arsenijevic

Part III. Control of the Cell Cycle and Apoptosis in Glia

12. Methods of Determining Apoptosis in Neuro-Oncology:
    Review of the Literature 143
    **Brian T. Ragel, Bardia Amirlak, Ganesh Rao,
    and William T. Couldwell**

13. Cell Cycle, Neurological Disorders, and Reactive Gliosis 163
    Kerri L. Hallene and Damir Janigro

14. Potassium Channels, Cell Cycle, and Tumorigenesis
    in the Central Nervous System 177
    Gabriele Dini, Erin V. Ilkanich, and Damir Janigro

Part IV. Adult Neurogenesis: A Mechanism for Brain Repair?

15. Enhanced Neurogenesis Following Neurological Disease 195
    Philippe Taupin

16. Endothelial Injury and Cell Cycle Re-Entry 207
    Ljiljana Krizanac-Bengez

17. The Contribution of Bone Marrow-Derived Cells to Cerebrovascular
    Formation and Integrity 221
    David Kobiler and John Glod

18. Microvessel Remodeling in Cerebral Ischemia 233
    Danica B. Stanimirovic, Maria J. Moreno, and Arsalan S. Haqqani

19. Vascular and Neuronal Effects of VEGF in the Nervous System:
    Implications for Neurological Disorders 245
    Lieve Moons, Peter Carmeliet, and Mieke Dewerchin

20. Epidermal Growth Factor Receptor in the Adult Brain 265
    Carmen Estrada and Antonio Villalobo

Part V. Cell Cycle Re-Entry: A Mechanism of Brain Disease?

21. Neurodegeneration and Loss of Cell Cycle Control
    in Postmitotic Neurons 281
    Randall D. York, Samantha A. Cicero, and Karl Herrup

22. Cell Cycle Activation and the Amyloid-β Protein in Alzheimer's Disease 299
    **Katarzyna A. Gustaw, Gemma Casadesus, Robert P. Friedland,
    George Perry, and Mark A. Smith**

23. Neuronal Precursor Proliferation and Epileptic Malformations
    of Cortical Development 309
    **Jorge A. González-Martínez, William E. Bingaman,
    and Imad M. Najm**

24. Vascular Differentiation and the Cell Cycle 319
    Luca Cucullo

Contents

25. Adult Neurogenesis and Central Nervous System
    Cell Cycle Analysis: *Novel Tools for Exploration of the Neural Causes
    and Correlates of Psychiatric Disorders* 331
    Amelia J. Eisch and Chitra D. Mandyam

26. Neurogenesis in Alzheimer's Disease:
    Compensation, Crisis, or Chaos? 359
    **Gemma Casadesus, Xiongwei Zhu, Hyoung-gon Lee,
    Michael W. Marlatt, Robert P. Friedland, Katarzyna A. Gustaw,
    George Perry, and Mark A. Smith**

Part VI. The Biology of Gliomas

27. p53 and Multidrug Resistance Transporters
    in the Central Nervous System 373
    Shirley Teng and Micheline Piquette-Miller

28. Signaling Modules in Glial Tumors and Implications
    for Molecular Therapy 389
    Gurpreet S. Kapoor and Donald M. O'Rourke

29. Detection of Proliferation in Gliomas by Positron Emission
    Tomography Imaging 419
    **Alexander M. Spence, David A. Mankoff, Joanne M. Wells,
    Mark Muzi, John R. Grierson, Janet F. Eary, S. Finbarr O'Sullivan,
    Jeanne M. Link, Daniel L. Silbergeld, and Kenneth A. Krohn**

30. Transition of Normal Astrocytes Into a Tumor Phenotype 433
    Sean E. Aeder and Isa M. Hussaini

31. Mechanisms of Gliomagenesis 449
    Wei Zhang and Howard A. Fine

Part VII. Future Directions

32. Cell Cycle of Encapsulated Cells 465
    Roberto Dal Toso and Sara Bonisegna

33. Viral Vector Delivery to Dividing Cells 477
    Yoshinaga Saeki

34. Electrical Stimulation and Angiogenesis:
    Electrical Signals Have Direct Effects on Endothelial Cells 495
    Min Zhao

35. Development and Potential Therapeutic Aspects of Mammalian
    Neural Stem Cells 511
    L. Bai, S. L. Gerson, and R. H. Miller

36. Mammalian Sir2 Proteins: *A Role in Epilepsy and Ischemia* 525
    Barbara Aumayr and Damir Janigro

Index 541

Contributors

Sean E. Aeder, PhD • *Department of Pathology, University of Virginia, Charlottesville, VA*
Bardia Amirlak, MD • *Department of Neurosurgery, Creighton University, Omaha, NE*
Annarosa Arcangeli, MD, PhD • *Department of Experimental Pathology and Oncology, University of Firenze, Florence, Italy*
Yvan Arsenijevic, PhD • *Unit of Oculogenetics, Department of Ophthalmology, Jules Gonin Eye Hospital, Lausanne, Switzerland*
Barbara Aumayr, BS • *University of Vienna, Vienna, Austria*
L. Bai, MD, PhD • *Department of Neurosciences, Case Western Reserve University, Cleveland, OH*
Stefania Bassanini, PhD • *Department of Experimental Neurophysiology and Epileptology, Instituto Neurologico, Milan, Italy*
Giorgio Battaglia, MD • *Department of Experimental Neurophysiology and Epileptology, Instituto Neurologico, Milan, Italy*
Andrea Becchetti, PhD • *Department of Biotechnology and Bioscience, Università di Milano-Bicocca, Milano, Italy*
William E. Bingaman, MD • *Department of Neurosurgery, The Cleveland Clinic Foundation, Cleveland, OH*
Anna J. Bolteus, PhD • *Department of Neurosurgery, Cellular and Molecular Physiology, Yale University School of Medicine, New Haven, CT*
Sara Bonisegna, PhD • *Biosil-USA, Wilmington, DE*
Angélique Bordey, PhD • *Department of Neurosurgery, Cellular and Molecular Physiology, Yale University School of Medicine, New Haven, CT*
Monika Bradl, PhD • *Center for Brain Research, Department of Neuroimmunology, Medical University Vienna, Vienna, Austria*
Peter Carmeliet, MD, PhD • *Center for Transgene Technology and Gene Therapy, Flanders Interuniversity Institute for Biotechnology, University of Leuven, Leuven, Belgium*
Gemma Casadesus, PhD • *Institute of Pathology, Case Western Reserve University, Cleveland, OH*
Samantha A. Cicero • *Alzheimer Research Lab, Department of Physiology, Case Western Reserve University, Cleveland, OH*
William T. Couldwell, MD, PhD • *Department of Neurosurgery, University of Utah Hospital, Salt Lake City, UT*
Luca Cucullo, PhD • *Cerebrovascular Research Laboratory, Department of Neurosurgery, The Cleveland Clinic Foundation, Cleveland, OH*
Roberto Dal Toso, PhD • *Biosil-USA, Wilmington, DE*
Mieke Dewerchin, PhD • *Center for Transgene Technology and Gene Therapy, Flanders Interuniversity Institute for Biotechnology, University of Leuven, Leuven, Belgium*
Gabriele Dini, PhD • *Cerebrovascular Research Laboratory, Department of Neurosurgery, The Cleveland Clinic Foundation, Cleveland, OH*

JANET F. EARY, MD • *Department of Radiology, University of Washington School of Medicine, Seattle, WA*
AMELIA J. EISCH, PhD • *Department of Psychiatry, University of Texas Southwestern Medical Center, Dallas, TX*
CARMEN ESTRADA, MD, PhD • *Department of Physiology, Facultad de Medicina, University of Cádiz, Cádiz, Spain*
HOWARD A. FINE, MD • *Neuro-Oncology Branch, Center for Cancer Research, National Cancer Institute, Bethesda, MD*
ROBERT P. FRIEDLAND, MD • *Laboratory of Neurogeriatrics, Department of Neurology, Case Western Reserve University, Cleveland, OH*
S. L. GERSON, MD • *Cancer Center, Case Western Reserve University, Cleveland, OH*
JOHN GLOD, MD, PhD • *The Cancer Institute of New Jersey, New Brunswick, NJ*
JORGE A. GONZÁLEZ-MARTÍNEZ, MD, PhD • *Department of Neurosurgery, The Cleveland Clinic Foundation, Cleveland, OH*
GERALD A. GRANT, MD • *Director of Pediatric Neurosurgery, Wilford Hall Medical Center, Lackland Air Force Base, TX*
JOHN R. GRIERSON, PhD • *Department of Radiology, University of Washington School of Medicine, Seattle, WA*
KATARZYNA A. GUSTAW, MD • *Department of Neurodegenerative Diseases, Institute of Agricultural Medicine, Lublin, Poland*
KERRI L. HALLENE, BS • *Cerebrovascular Research Library, Department of Neurological Surgery, The Cleveland Clinic Foundation, Cleveland, OH*
ARSALAN S. HAQQANI, PhD • *Institute for Biological Sciences, National Research Council of Canada, Ottawa, Ontario, Canada*
KARL HERRUP, PhD • *Alzheimer Research Lab, Department of Neuroscience, Case Western Reserve University Medical School, Cleveland, OH*
ISA M. HUSSAINI, PhD • *Department of Pathology, University of Virginia, Charlottesville, VA*
ERIN V. ILKANICH, BS • *Cerebrovascular Research Laboratory, Department of Neurosurgery, The Cleveland Clinic Foundation, Cleveland, OH*
DAMIR JANIGRO, PhD • *Cerebrovascular Research Laboratory, Department of Neurosurgery, The Cleveland Clinic Foundation, Cleveland, OH*
GURPREET S. KAPOOR, PhD • *Department of Neurosurgery, The Hospital of the University of Pennsylvania, Philadelphia, PA*
DAVID KOBILER, PhD • *Department of Infectious Diseases, Israel Institute for Biological Research, Ness-Ziona, Israel*
LJILJANA KRIZANAC-BENGEZ, MD, PhD • *Cerebrovascular Research Laboratory, Department of Neurosurgery, The Cleveland Clinic Foundation, Cleveland, OH*
KENNETH A. KROHN, PhD • *Department of Radiology, University of Washington School of Medicine, Seattle, WA*
SEONG-RYONG LEE, MD, PhD • *Neuroprotection Research Laboratory, Harvard Medical School, Charlestown, MA*
HYOUNG-GON LEE, PhD • *Institute of Pathology, Case Western Reserve University, Cleveland, OH*
JEANNE M. LINK, PhD • *Department of Radiology, University of Washington School of Medicine, Seattle, WA*

Contributors

QING Y. LIU, PhD • *NeuroGenomics Group, Institute for Biological Sciences, National Research Council of Canada, Ottawa, Ontario, Canada*

XIUXIN LIU, PhD • *Department of Neurosurgery, Cellular and Molecular Physiology, Yale University School of Medicine, New Haven, CT*

ENG H. LO, PhD • *Neuroprotection Research Laboratory, Harvard Medical School, Charlestown, MA*

DAO LY, BSc • *NeuroGenomics Group, Institute for Biological Sciences, National Research Council of Canada, Ottawa, Ontario, Canada*

CHITRA D. MANDYAM, PhD • *Department of Psychiatry, University of Texas Southwestern Medical Center, Dallas, TX*

DAVID A. MANKOFF, MD, PhD • *Department of Radiology, University of Washington School of Medicine, Seattle, WA*

MICHAEL W. MARLATT, MS • *Institute of Pathology, Case Western Reserve University, Cleveland, OH*

R. H. MILLER, PhD • *Department of Neurosciences, Case Western Reserve University, Cleveland, OH*

LIEVE MOONS, PhD • *Center for Transgene Technology and Gene Therapy, Flanders Interuniversity Institute for Biotechnology, University of Leuven, Leuven, Belgium*

MARIA J. MORENO, PhD • *Institute for Biological Sciences, National Research Council of Canada, Ottawa, Ontario, Canada*

MICHAEL A. MOSKOWITZ, MD • *Department of Neurology, Massachusetts General Hospital, Charlestown, MA*

MARK MUZI, MS • *Department of Radiology, University of Washington School of Medicine, Seattle, WA*

IMAD M. NAJM, MD • *Department of Neurology, The Cleveland Clinic Foundation, Cleveland, OH*

DONALD M. O'ROURKE, MD • *Department of Neurosurgery, The Hospital of the University of Pennsylvania, Philadelphia, PA*

S. FINBARR O'SULLIVAN, PhD • *Department of Statistics, University of Cork, Cork, Ireland*

GEORGE PERRY, PhD • *Institute of Pathology, Case Western Reserve University, Cleveland, OH*

MICHELINE PIQUETTE-MILLER, PhD • *Faculty of Pharmacy, University of Toronto, Toronto, Ontario, Canada*

RICHARD A. PRAYSON, MD • *Department of Anatomic Pathology, The Cleveland Clinic Foundation, Cleveland, OH*

BRIAN T. RAGEL, MD • *Department of Neurosurgery, University of Utah Hospital, Salt Lake City, UT*

GANESH RAO, MD • *Department of Neurosurgery, University of Utah Hospital, Salt Lake City, UT*

MARILENA RIBECCO, PhD • *NeuroGenomics Group, Institute for Biological Sciences, National Research Council of Canada, Ottawa, Ontario, Canada*

YOSHINAGA SAEKI, MD, PhD • *The Dardinger Laboratory for Neuro-Oncology and Neurosciences, Ohio State University Medical Center, Columbus, OH*

MARIANNA SIKORSKA, PhD • *NeuroGenomics Group, Institute for Biological Sciences, National Research Council of Canada, Ottawa, Ontario, Canada*

DANIEL L. SILBERGELD, MD • *Department of Neurosurgery, University of Washington School of Medicine, Seattle, WA*
MARK A. SMITH, PhD • *Institute of Pathology, Case Western Reserve University, Cleveland, OH*
BRANDON SMITH, MSc • *NeuroGenomics Group, Institute for Biological Sciences, National Research Council of Canada, Ottawa, Ontario, Canada*
CAROLINE SODJA, MSc • *NeuroGenomics Group, Institute for Biological Sciences, National Research Council of Canada, Ottawa, Ontario, Canada*
ALEXANDER M. SPENCE, MD • *Department of Neurology, University of Washington School of Medicine, Seattle, WA*
DANICA B. STANIMIROVIC, MD, PhD • *Institute for Biological Sciences, National Research Council of Canada, Ottawa, Ontario, Canada*
PHILIPPE TAUPIN, PhD • *National Neuroscience Institute, National University of Singapore, Singapore*
SHIRLEY TENG, PhD • *Faculty of Pharmacy, University of Toronto, Toronto, Ontario, Canada*
KLAUS VAN LEYEN, PhD • *Neuroprotection Research Laboratory, Harvard Medical School, Charlestown, MA*
ANTONIO VILLALOBO, MD, PhD • *Institute of Investigational Biomedicine, University of Madrid, Madrid, Spain*
P. ROY WALKER, PhD • *NeuroGenomics Group, Institute for Biological Sciences, National Research Council of Canada, Ottawa, Ontario, Canada*
JOANNE M. WELLS, MS • *Department of Radiology, University of Washington School of Medicine, Seattle, WA*
RANDALL D. YORK, PhD • *Alzheimer Research Lab, Department of Neuroscience, Case Western Reserve University, Cleveland, OH*
SHAN PING YU, MD, PhD • *Department of Pharmaceutical Sciences, School of Pharmacy, Medical University of South Carolina, Charleston, SC*
WEI ZHANG, MD, PhD • *Neuro-Oncology Branch, Center for Cancer Research, National Cancer Institute, Bethesda, MD*
MIN ZHAO, MD, PhD • *Biomedical Sciences, Institute of Medical Sciences, University of Aberdeen, Aberdeen, Scotland, UK*
XIONGWEI ZHU, PhD • *Institute of Pathology, Case Western Reserve University, Cleveland, OH*

Companion CD

Color versions of illustrations listed here are presented on the Companion CD attached to the inside back cover. The image files are organized into folders by chapter number and are viewable in most Web browsers. The number following "f" at the end of the file name identifies the corresponding figure in the text. The Companion CD is compatible with both Mac and PC operating systems.

CHAPTER 4   FIG. 1
CHAPTER 6   FIG. 1
CHAPTER 8   FIG. 1
CHAPTER 10 FIGS. 1 AND 2
CHAPTER 12 FIGS. 1 AND 2
CHAPTER 13 FIG. 1
CHAPTER 18 FIGS. 2 AND 3
CHAPTER 19 FIGS. 1–3
CHAPTER 23 FIGS. 1–3
CHAPTER 24 FIGS. 1, 3–5
CHAPTER 31 FIGS. 1 AND 3
CHAPTER 32 FIGS. 1–3
CHAPTER 36 FIGS. 1 AND 4

I
Cell Cycle During the Development of the Mammalian Central Nervous System

1
Methodological Considerations in the Evaluation of the Cell Cycle in the Central Nervous System

Richard A. Prayson, MD

SUMMARY

A number of modalities have been used in the evaluation of cell cycle proliferation in the central nervous system. The evolution of technology has moved from the routine hematoxylin and eosin stained assessment of mitotic activity to radiolabeling and flow-cytometric methodologies and more recently reliable immunohistochemical approaches. This chapter will review the methodological considerations of each of these modalities and their role in the evaluation of lesions in the central nervous system.

Key Words: Cell proliferation; mitoses; bromodeoxyuridine; flow cytometry; Ki-67; PCNA; thymidine labeling; MIB-1.

1. CELL CYCLE

The cell cycle is the process by which eukaryotic cells undergo cell division. For cell division to be successful, DNA needs to be faithfully replicated and identical chromosomal copies need to be distributed equally among two offspring cells. The process by which this occurs is orderly and remarkably accurate. A whole host of factors are involved in the regulation of the cell cycle. When such regulation becomes aberrant, the end result is often an atypical proliferation of cells resulting in a pathological condition; the prototypical example of this scenario is neoplasia.

There are a variety of stimuli responsible for inducing a cell to undergo cell division or mitosis *(1–5)*. Many of these are protein factors which bind to the surface of the cell receptors and signal the cell that is in the resting phase (G_0) to enter into the cell cycle (i.e., the gap-1 [G_1] phase). Among the more important molecules responsible for this progression are the cyclin-dependent kinases, which are responsible for phosphorylating regulatory proteins. During the G_1 phase, the cell grows and begins production of elements required for DNA synthesis. Cells in the G_1 phase have a diploid number of chromosomes, one set inherited from each parent. Rapidly proliferating cells in humans may progress through the full cell cycle in about 24 h. The G_1 phase may take approx 8–10 h to complete.

The cell then progresses into the synthesis phase (S phase), during which DNA synthesis occurs and chromosomal DNA is replicated. During the S phase, cyclin A-cyclin-dependent kinases 2 complex plays an important role in both the initiation and maintenance of DNA synthesis. The S phase typically takes approx 10 h for completion. The S phase is followed by a 4–5 h gap-2 (G_2) phase. During the G_2 phase, the cell ensures that DNA replication is complete and that DNA damage is repaired prior to the cell entering the mitotic phase of the cell cycle. On entering the mitotic phase, which typically lasts less than 1 h in duration, the cell undergoes a series of events in the process of cell division. The initial portion of the mitotic phase is referred to as prophase. During prophase, the chromosomes become visible as extended double structures.

From: *The Cell Cycle in the Central Nervous System*
Edited by: D. Janigro © Humana Press Inc., Totowa, NJ

By light microscopy, they become shorter and more visible, with each chromosome being composed of two daughter DNA molecules with associated histones and other chromosomal proteins. During metaphase, the kinetochore assembles at each centromere. The kinetochores of sister chromotids then associate with microtubules coming from opposite spindle poles. The chromosomes are aligned at the equator of the cell. The cell then progresses into anaphase, during which the chromosome pairs split and move to opposite poles of the cell. Once chromosome separation has occurred, the mitotic spindles disassemble and the chromosomes decondense during telophase, with the end product being two daughter cells, each containing a complete set of the chromosomal material. Cells may either go on to G_1 phase or can enter the G_0 phase.

2. MITOSIS COUNTS

Possibly, the oldest method for assessing cell proliferation and active division of cells has been the evaluation of mitosis counts. The methodology has long been the gold standard for evaluating proliferative activity in a lesion. The diagnosis of many neoplasms is often dependent on an evaluation of mitotic activity. The method has the advantage of being cheap and can be performed with relative ease on routinely processed, hematoxylin and eosin stained histological sections. Obviously, evaluation of mitotic figures in tissue sections only captures those cells that are in the M phase of the cell cycle, which is a relatively short portion of the entire cycle.

A number of factors affecting mitosis counts that are important to consider have been described. Delays in fixation time, temperature at which the specimen is stored prior to fixation, and the quality or type of fixative used may all potentially impact on mitosis counts *(6–13)*. The rate of penetration of the fixative and the size of the tissue specimen being fixed can result in degenerative changes, particularly in unfixed areas of the tissue, which may make identification of mitotic figures more difficult. There is some debate in the literature regarding the relative importance of delays in fixation and their effect on mitosis counts. It appears that cells can either enter or exit the cell cycle after removal from the body. The apparent decrease in mitotic activity that has been described by some as being attributable to delays in fixation are most likely related to the inability to identify mitotic figures in cells that are undergoing degenerative changes. Decreased temperature (e.g., as a result of refrigeration) may slow these degenerative changes that make identification of mitotic figures less difficult. The type of fixative used may also affect one's ability to recognize mitotic figures. Fixatives with low pH and including mercury-containing compounds, such as Bouin's fixative, tend to increase tissue and cell shrinkage, resulting in smaller sized nuclei *(14)*. Recognition of mitotic figures in smaller sized cells may be more difficult. Formalin fixative that is inadequately buffered may also have a lower pH and induce similar morphological alterations.

In addition to the issues of fixation, tissue staining and sectioning may also influence one's assessment of mitotic activity *(15–18)*. The presence of nuclear pyknosis, nuclear folding, or nuclear hyperchromasia may result in morphological changes resembling mitotic figures. With thicker tissue sectioning, increased numbers of mitotic figures in a given microscopic field may be generated and may also create a challenge in terms of identifying figures in multiple planes of focus.

Different methodologies for actually reporting mitotic activity have also been promulgated. The different methodologies may result in significantly different mitosis counts. One of the more common approaches is to evaluate the number of mitotic figures present in a certain number of high-power fields. Interestingly, the area of a high-power field can vary considerably depending on the make of microscope *(19)*. In 1981, Ellis and Whitehead noted high-power field areas ranging from 0.071 to 0.414 mm^2 in their survey of 26 microscopes of different makes and specifications. Therefore, more precise reporting of mitotic activity per certain number of high-power fields should include a calculated area of the high-power field. In the evaluation of particularly cellular specimens, such as tumors, or specimens with mixed populations of cells, the identification of a mitotic figure with a particular cell type may be difficult. The number of high-power fields that are assessed can also vary, depending on the methodology. Some advocate scanning the whole slide to find an area with the highest mitotic activity and report the single highest count per contiguous

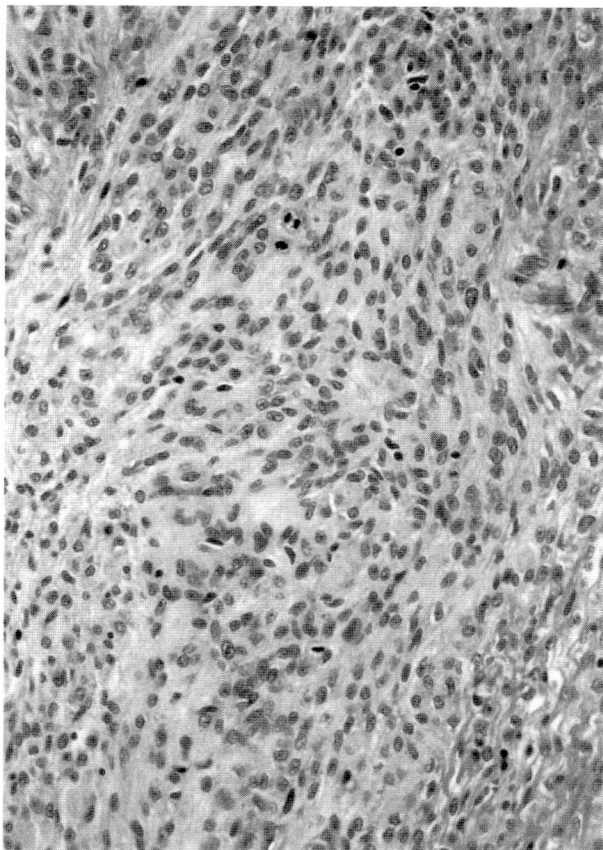

Fig. 1. Anaplastic meningioma with readily identifiable mitotic activity. The current World Health Organization guidelines for the diagnosis of anaplastic (malignant) meningioma grade III includes a high mitotic index—20 or more mitoses per 10 high-power fields, defined as 0.16 mm^2 (hematoxylin and eosin, original magnification ×400).

10 high-power fields (20); others advocate randomly selecting a field to commence counting and to evaluate 40 or 50 consecutive high-power fields. The differences in the final results obtained by using these two different approaches may be strikingly different (20).

The experience of the individual assessing mitotic activity is also an important factor, albeit frequently unrecognized (15,16,21). The number of artifacts that can be generated associated with fixation and staining can be problematic at times even for the most experienced morphologist. Apoptotic cells and inflammatory cells may also mimic mitotic figures. One is advised to count certain mitotic figures only.

The assessment of central nervous system lesions for mitotic activity is generally an exercise reserved for the evaluation of certain tumors. Mitotic activity can be observed in association with inflammatory and reactive conditions, particularly in areas of granulation tissue with small vessel and fibroblastic proliferation. In general, mitotic figures are not seen in association with gliosis, in which the histological changes are predominantly that of hypertrophy of cells rather than marked hyperplasia. With certain tumors, criteria have been developed in which the evaluation of mitotic activity is important. This has been particularly well defined in the setting of meningiomas, in which the current World Health Organization grading system incorporates mitosis counts in defining atypical and anaplastic meningiomas (22) (Fig. 1). Identification of morphologically atypical mitotic figures has classically been viewed as a feature of neoplasia

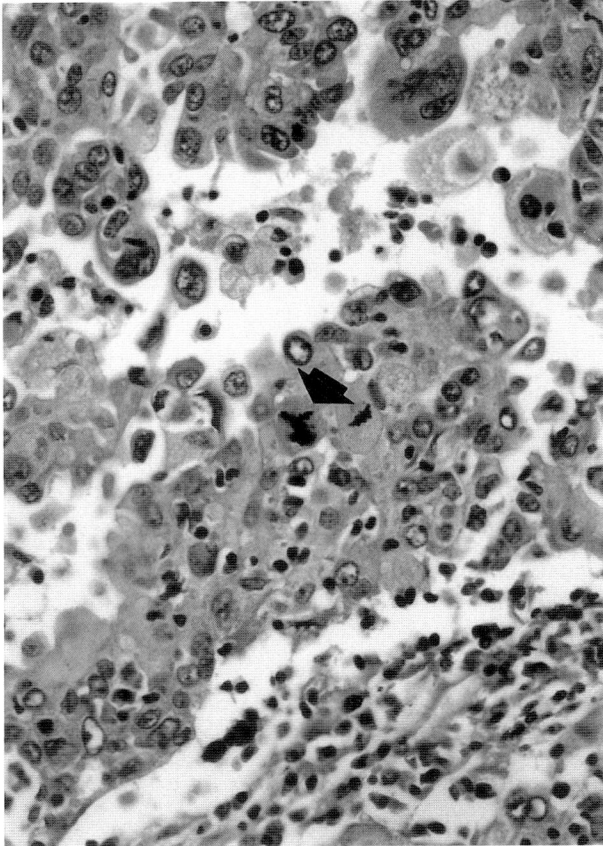

Fig. 2. An atypical mitotic figure (arrow) in a metastatic breast carcinoma in the central nervous system (hematoxylin and eosin, original magnification ×525).

(Fig. 2). Unfortunately, differentiation of abnormal from normal mitotic figures is not always straightforward, and in some cases, tends to be quite observer-dependent.

3. THYMIDINE LABELING

Among some of the earliest approaches for alternatively evaluating cell cycle were thymidine labeling and bromodeoxyuridine labeling. In contrast to mitotic activity, evaluation of tritiated thymidine labeling is an assessment of predominantly the S phase of the cell cycle (Fig. 3). This approach is predicated on the incorporation of a radioactively labeled DNA precursor (thymidine) into cells during the S phase of the cell cycle. This approach may also be used to measure the duration of the cell cycle *(23,24)*.

In the evolution of the methodology, earlier, patients were infused with a radioactive agent prior to surgery. The tissue harvested at surgery was processed routinely and an evaluation of the incorporation of radioactive labeling was assessed *(25,26)*. In vitro approaches using tissue samples that have already been removed were later developed *(27,28)*. These involved utilization of a freshly excised tissue and assessing incorporation of the tritiated thymidine. Tissue sections were incubated with the tritiated thymidine prior to fixation and processing. Various conditions including incubation in a hyperbaric environment or use of 5-fluorouridine 2´deoxy-*S*-fluorouridine as potentiating agents can facilitate the uptake of thymidine. Sections are generated and developed

Methodological Considerations

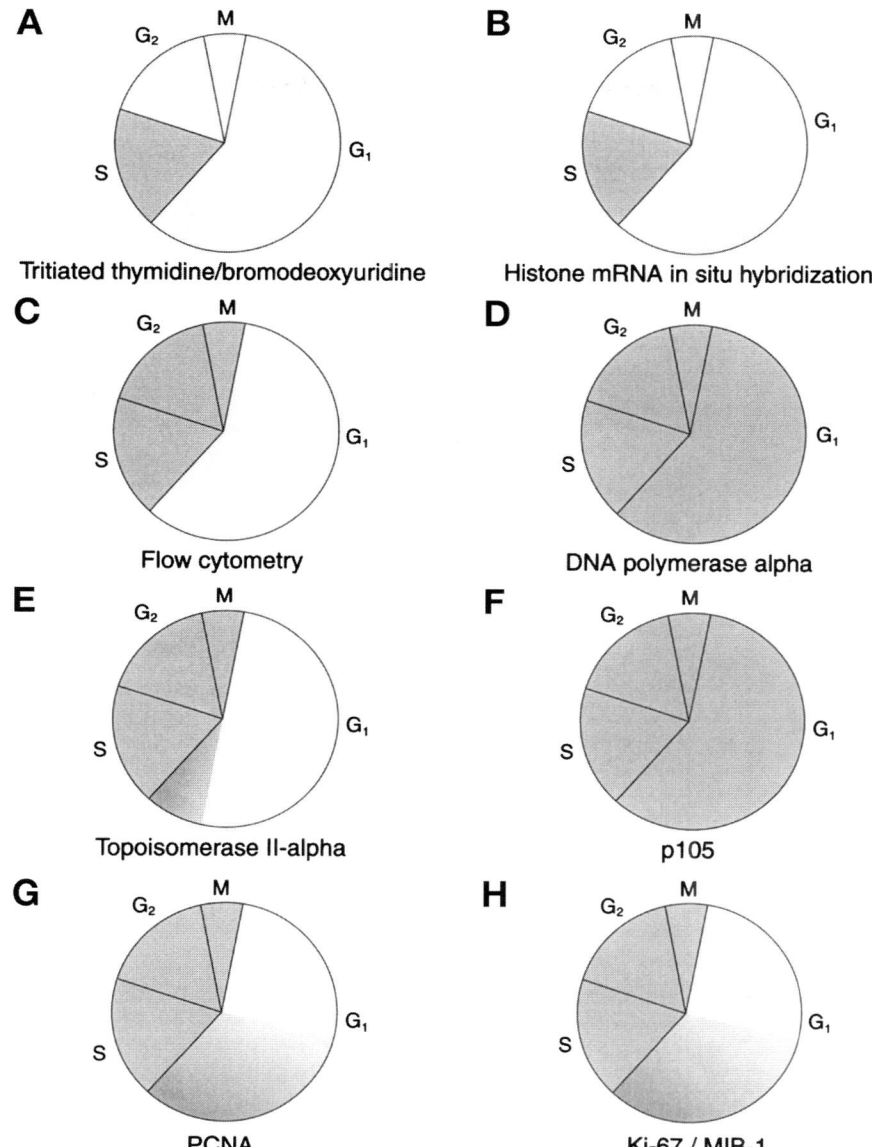

Fig. 3. The figure summarizes the portions of the cell cycle that are assessed (shaded areas) with various methodologies.

using autoradiographic methodologies. Labeling indexes can be calculated by determining a percentage of positive-staining cells.

There are several known potential drawbacks to this approach. The radioactive procedure requires either infusion of the patient prior to the removal of tissue or a fresh tissue. The utilization of the radioactive isotope and all of the concomitant issues that entails needs to be considered. The methodology is also somewhat drawn out, and frequently takes around 7–10 d before results are obtained. There are also issues regarding the interpretation of results and deciding what exactly represents a positive finding.

Bromodeoxyuridine is a thymidine analog which similarly allows for the assessment of the S phase of the cell cycle. Initially developed as a radioactive methodology, more recent alternative ways of evaluating bromodeoxyuridine using nonradioactive methodologies including immunohistochemistry, flow cytometry, and immunofluorescence methodologies have been devised *(29,30)*. Both in vivo and in vitro approaches have been used and, in general, results are somewhat comparable, although labeling indices in the in vivo method tend to be slightly higher *(31)*.

4. HISTONE EVALUATION

Another method for evaluating the S phase of the cell cycle is the *in situ* histone hybridization *(32–34)*. Histones are a group of nuclear proteins whose production is generally restricted to the S phase of the cell cycle. Because histone mRNA has a half-life of approx 10 min, evaluation of histone-3 or histone-4 mRNA allows for a somewhat specific evaluation of the S phase *(35)*. Fresh or formalin-fixed materials can be evaluated, allowing for retrospective evaluation of tissues. Problems related to the loss of mRNA or inability of the probe to reach the target mRNA when fixed tissues are used may limit its utility in certain circumstances, owing to underestimating the true rate of proliferation.

5. FLOW CYTOMETRY

Evaluation of tissues using flow cytometry and image cytometric analysis systems allows for a broader assessment of cell proliferation. The methodology is based on an evaluation of cell DNA content by staining the DNA *(36)*. Either a Feulgen staining procedure or the use of fluorescent markers, such as propidium iodide, can stoichiometrically bind to the DNA, with the intensity of staining being directly related to the amount of DNA present in the cell. This methodology is particularly useful in identifying cells which have increased DNA content, corresponding to the S, G_2, and M phases of the cell cycle. Results may be reported numerically or in a histogram formation.

Both fresh and fixed tissue samples may be used, although better results are usually obtained when fresh tissue is used *(37)*. Use of archived tissue may result in the generation of increased cellular debris or fragmentation, which may confound the results. The procedure requires a tissue sample of sufficient size in order to obtain enough cells for proper analysis. The approach does provide a fairly rapid evaluation of a large number of cells in a relatively short period of time. In contrast to manually assessed, immunohistochemical methodologies, thresholds for what represents a positive result can be set, thereby minimizing problems related to intraobserver variability. In contrast to immunohistochemical approaches which allow for the visual localization of staining, the triage of tissues for flow-cytometric evaluation is often somewhat blinded. Specimens may contain tissue elements that are not the target of evaluation, which may skew one's results. For example, vascular proliferative areas in a high-grade glioma may falsely increase the apparent rate of proliferation in the tumor itself. Similarly, incorporation of a non-neoplastic tissue in the evaluation of a diploid tumor may affect one's results. Use of microdissection techniques or histological evaluation of tissues prior to triage can eliminate or minimize some of these problems. The presence of overlapping aneuploid peaks may make accurate determination of the S phase difficult. Finally, the cost of the procedure, particularly the equipment, is considerably more than the cost associated with immunohistochemical methodologies. Results obtained using flow-cytometric approaches generally correlate well with immunohistochemical methodologies.

6. IMMUNOHISTOCHEMICAL METHODOLOGIES

A variety of antibody markers have been developed over the past two decades, which allow for a ready evaluation of cell proliferation. These approaches have the advantage of being relatively easy to perform, are relatively cost-effective, and provide quick results.

DNA polymerase α is a cell cycle-related enzyme that is expressed during the G_1, S, G_2, and M phases of the cell cycle *(38–40)*. The protein does not survive formalin fixation and routine histological processing and the staining requires a fresh or a frozen tissue. Similar to DNA polymerase α, p105 is expressed during similar phases of the cell cycle. p105 is a nuclear associated protein which may play a role in the production of mature RNA transcripts involved with cell cycle progression *(41–44)*. The expression of this protein may vary during different phases of the cell cycle, resulting in a gradation of staining intensity which may be difficult to interpret and limit its utility as a marker yielding reproducible results. Either fresh or fixed tissues can be evaluated with this antibody.

DNA topoisomerase-II α is a protein involved with untangling DNA strands prior to the cell entering mitosis *(45)*. The enzyme is produced in the late G_1 and S phases of the cycle and may be present during the G_2 and M phases, during which it begins to undergo degradation *(46–48)*. Archival materials may be used for evaluation.

Proliferating cell nuclear antigen (PCNA) is a nonhistone nuclear protein that is associated with the function of DNA polymerase δ. Numerous monoclonal antibodies to PCNA have been developed, which recognize different forms of the protein and are localized to different regions of the cell nucleus *(49)*. PCNA expression varies during different phases of the cell cycle; its production increases during the G_1 phase, remains increased during the S phase, and diminishes during the G_2 and M phases. Similar to p105, this variability of expression results in variable staining intensity that may be difficult to interpret. The presence of low levels of PCNA in non-cycling G_0 cells and the long half-life of PCNA (approx 20 h) may also further complicate the interpretation. Either fresh or fixed tissues may be evaluated. The variability of PCNA antibodies that are currently available, each recognizing a slightly different epitope, can result in some differences in staining and be reflective of slightly different cell cycle distributions.

Probably, the most reliable and widely used marker of cell proliferation is Ki-67 or MIB-1 antibody. The Ki-67 antibody was generated by immunizing mice with the nuclei of a Hodgkin's lymphoma cell line *(50)*. The specific isotope associated with Ki-67 remains unknown, although the gene associated with it is situated on chromosome 10. The antigen is expressed during a portion of the G_1 phase and in the S, G_2, and M phases of the cell cycle *(51)*. When initially manufactured, it was restricted for use with fresh or frozen tissue. In early 1990s, the monoclonal antibody MIB-1 was developed to the Ki-67 antigen that was able to be used with formalin-fixed materials *(52)* and benefited from microwave processing to enhance antigen retrieval *(53)*. More recently, Ki-67 antibodies which work on fixed tissue have been developed.

Several methodological considerations are important to recognize when using the Ki-67 antibody. Selection of the tissue sample to evaluate is an important consideration. Not all lesions demonstrate uniform cell proliferation. This is particularly true for many gliomas which are well-known to be heterogeneous in terms of cell proliferation (Fig. 4). A variety of technical aspects can affect staining. The source or type of antibody used, and dilution of antibody and staining conditions including buffers used can affect staining results *(54)*. Delays in formalin fixation do not appear to affect Ki-67 staining.

The approach to evaluating and reporting the staining results can be variable. By convention, similar to assessing mitotic activity, the most proliferative area (region with the highest staining) is assessed. Only nuclear staining is interpreted as positive. An attempt is made to evaluate only the cells of interest; double-immunolabeling can be sometimes useful in this endeavor. For example, if one is interested in evaluating cell proliferation among microglial cells, double-labeling with Ki-67 and CD68 antibodies might be an useful approach. A determination of what degree of staining will be interpreted as positive must also be made and will obviously vary from individual to individual. Some have advocated the use of image analysis systems to address this issue. Such systems can allow for the ready assessment of large numbers of cells and can set a uniform staining threshold of positivity. The results are typically reported as a labeling index, reflecting a percentage of positive-staining cells per total number of cells of interest that are being evaluated.

Studies that have evaluated interobserver variability in the determination of labeling indices have noted significant differences among observers, reflective of many of the previously

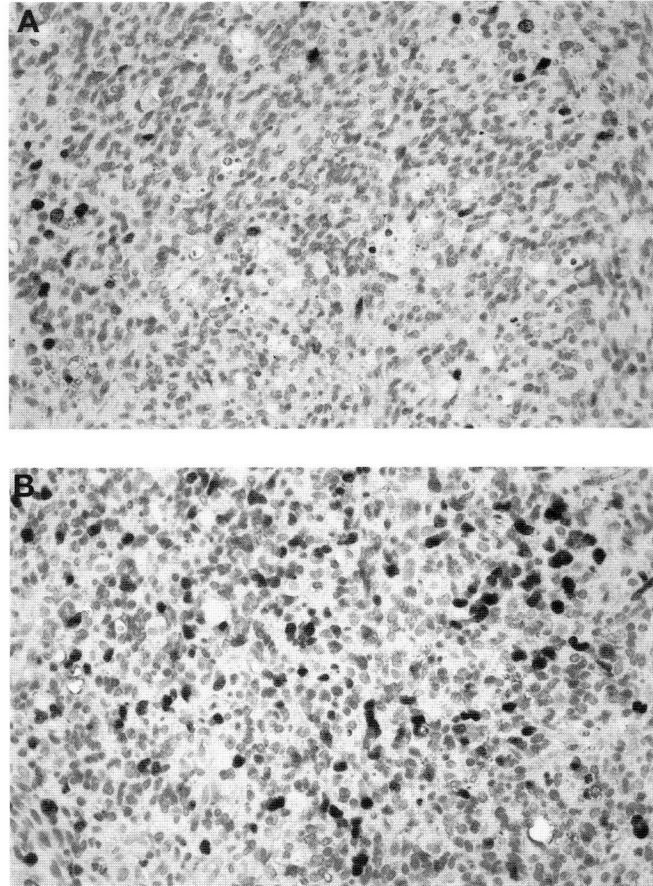

Fig. 4. Two contiguous high magnification fields of glioblastoma multiform immunostained with MIB-1 antibody to highlight the regional variability in cell proliferation that marks many gliomas (MIB-1 immunostained, original magnification ×400).

enumerated issues *(55,56)*. Given this variability, one should be cautioned against the establishment of specific cutoff values for the purpose of clinical diagnosis or prognostication.

ACKNOWLEDGMENT

Special thanks are given to Denise Egleton for her help in the preparation of this manuscript.

REFERENCES

1. Dirks PB, Rutka JT. Current concepts in neuro-oncology: The cell cycle—a review. Neurosurgery 1997;40:1000–1015.
2. Heichman KA, Roberts JM. Rules to replicate by. Cell 1997;79:557–562.
3. Laskey RA, Fairman MP, Blow JJ. S phase of the cell cycle. Science 1989;246:609–614.
4. Nurse P. Regulation of the eukaryotic cell cycle. Eur J Cancer 1997;7:1002–1004.
5. Iliakis G. Cell cycle regulation in irradiated and nonirradiated cells. Semin Oncol 1997;24:602–615.
6. Baak JPA. Mitosis counting in tumors. Hum Pathol 1990;21:683–685.
7. Bergers E, Jannink I, van Diest PI, et al. The influence of fixation delay on mitotic activity and flow cytometric cell cycle variables. Hum Pathol 1997;28:95–100.

8. Donhuijsen K, Schmidt U, Hirche H, van Beuningin D, Budach V. Changes in mitotic rate and cell cycle fractions caused by delayed fixation. Hum Pathol 1990;21:709–714.
9. Bullough WS. Mitotic activity in the tissues of dead mice, and in tissues kept in physiologic salt solution. Exper Cell Res 1950;1:410–420.
10. Evans N. Mitotic figures in malignant tumors as affected by time before fixation of tissues. Arch Pathol 1986;17:1122–1125.
11. Graem N, Helweg-Larsen K. Mitotic activity and delay in fixation of tumor tissue: The influence of delay in fixation on mitotic activity in human osteogenic sarcoma grown in athymic nude mice. Acta Pathol Microbiol Scand Sec A 1979;87:375–378.
12. Cross SS, Start RD, Smith JHF. Does delay in fixation affect the number of mitotic figures in processed tissue? J Clin Pathol 1990;43:597–599.
13. Edwards JL, Donaldson JT. The time of fixation and the mitotic index. Am J Clin Pathol 1964;41:155–162.
14. Baak JPA, Noteboom E, Koevoets JJM. The influence of fixatives and other variations in tissue processing on nuclear morphometric features. Anal Quant Cytol Histol 1989;11:219.
15. Norris HJ. Mitosis counting III. Hum Pathol 1976;7:482,483.
16. Donhuijsen K. Mitosis counts: Reproducibility and significance in grading malignancy. Hum Pathol 1986;17:1122–1125.
17. Cross SS, Start RD. Estimating mitotic activity in tumors. Histopathology 1996;29:485–488.
18. Kempson RL. Mitosis counting II. Hum Pathol 1976;7:482,483.
19. Ellis PSJ, Whitehead R. Mitosis counting—a need for reappraisal. Hum Pathol 1981;12:3,4.
20. Prayson RA, Hart WA. Mitotically active leiomyomas of the uterus. Am J Clin Pathol 1992;97:14–20.
21. Linden MD, Torres FX, Kubus J, Zarbo RJ. Clinical application of morphologic and immunocytochemical assessments of cell prolifertaior.. Am J Clin Pathol 1992;97(Suppl 1):S4–S13.
22. Louis DN, Scheithauer BW, Budka H, von Diemling A, Kepes JJ. Meningiomas. In: Kleihues P, Cavenee WK, eds. Tumours of the Nervous System. Lyon, France: IARC Press; 2000, pp 176–184.
23. Meyer JS. Cell kinetic measurements in human tumors. Pathol Ann 1981;16:53–81.
24. Schultze B, Maurer W, Hagenbusch H. A two emulsion autoradiographic technique and the discrimination of the three different types of labeling after double labeling with ^{3}H- and ^{14}C-thymidine. Cell Tissue Kinet 1976;9:245–255.
25. Hoshino T, Townsend J, Muraoka I, Wilson C. An autoradiographic study of human gliomas: Growth kinetics of anaplastic astrocytoma and glioblastoma multiforme. Brain 1980;103:967–984.
26. Hoshino T. A commentary on the biopsy and growth kinetics of low-grade and high-grade gliomas. J Neurosurg 1984;61:895–900.
27. Broggi G, Franzini A, Costa A, Melcarne A, Allegranza A. Cell kinetics of neuroepithelial tumors in serial stereotactic biopsies. A new combined approach. Appl Neurophysiol 1985;48:472–476.
28. Broggi G, Franzini A, Ferraresi S, et al. Cell kinetics and multimodal prognostic evaluation in glial tumours investigated by serial stereotactic biopsy. Acta Neurochir 1988;94:53–56.
29. Gratzner HG. Monoclonal antibody to 5-bromo- and 5-iododeoxyuridine: A new reagent for detection of DNA replication. Science 1982;218:474,475.
30. Meyer JS, Nauert J, Koehm S, et al. Cell kinetics of human tumors by in vitro bromodeoxyuridine labeling. J Histochem Cytochem 1989;37:1449–1954.
31. Meyer JS, Marchosky A, Hickey WF. Cell kinetic classification of tumors of the nervous system by DNA precursor labeling in vitro. Hum Pathol 1993;24:1357–1364.
32. Stein GS, Stein JL, van Wijnen AJ, Lian JB. Histone gene transcription: A model for responsiveness to an integrated series of regulatory signals mediating cell cycle control and proliferation/differentiation interrelationships. J Cell Biochem 1994;54:393–404.
33. Chou MY, Chang ALC, McBride J, Donoff B, Gallagher GT, Wong DTW. A rapid method to determine proliferation patterns of normal and malignant tissues by H3 mRNA *in situ* hybridization. Am J Pathol 1990;136:729–733.
34. Bosch FX, Udvarhelyi N, Venter E, et al. Expression of the histone H3 gene in benign semi-malignant lesions of the head and neck: A reliable proliferation marker. Eur J Cancer 1993;29A:1454–1461.
35. Heintz N, Sive HL, Roeder RG. Regulation of human histone gene expression: Kinetics of accumulation and changes in the rate of synthesis and in the half-lives of individual histone mRNAs during the HeLa cell cycle. Mol Cell Biol 1993;3:539–550.
36. Quirke P, Dyson JED. Flow cytometry: Methodology and applications in pathology. J Pathol 1986;149:79–87.

37. Hedley DW. Flow cytometry using paraffin-embedded tissue: Five years on. Cytometry 1989;10:229–241.
38. Bensch KG, Tanaka S, Hu S. Intracellular localization of human DNA polymerase α with monoclonal antibodies. J Biol Chem 1982;257:8391–8396.
39. Tanaka S, Shi-Zhen H, Shu-Fong Wang T, et al. Preparation and preliminary characterization of monoclonal antibodies against human DNA polymerase α. J Biol Chem 1982;257:8386–8390.
40. Shibuya M, Miwa T, Hoshino T. Embedding and fixation techniques for immunohistochemical staining with anti-DNA polymerase α and Ki-67 monoclonal antibodies to analyze the proliferative potential of tumors. Biotechnic Histochem 1992;67:161–164.
41. Clevenger CV, Epstein AL. Identification of a nuclear protein component of interchromatin granules using a monoclonal antibody and immunogold electron microscopy. Exp Cell Res 1984;151:194–207.
42. Clevenger CV, Epstein AL, Bauer KD. Modulation of the nuclear antigen p105 as a function of cell cycle progression. J Cell Physiol 1987;130:336–343.
43. Swanson SA, Brooks JJ. Proliferation markers Ki-67 and p105 in soft tissue lesions. Am J Pathol 1990;137:1491–1500.
44. Appley AJ, Fitzgibbons PL, Chandrasoma PT, Hinton DR, Apuzzo ML. Multiparameter flow cytometric analysis of neoplasms of the central nervous system: Correlation of nuclear antigen p105 and DNA content with clinical behavior. Neurosurgery 1990;27:83–96.
45. Holm C, Stearns T, Botstein D. DNA topoisomerase II muscle act at mitosis to prevent nondisjunction and chromosome breakage. Mol Cell Biol 1989;9:159–168.
46. Woessner RD, Mattern MR, Mirabelli CK, Johnson RK, Drake FH. Proliferation- and cell cycle-dependent differences in expression of the 170 kilodalton and 180 kilodalton forms of topoisomerase II in NIH-3T3 cells. Cell Growth Differ 1991;103:2569–2581.
47. Kellner U, Heidebrecht H, Rudolph P, et al. Detection of human topoisomerase II-alpha in cell lines and tissues: Characterization of five novel monoclonal antibodies. J Histochem Cytochem 1997;45:251–263.
48. Holden JA, Townsend JJ. DNA topoisomerase II-alpha as a proliferation marker in astrocytic neoplasms of the central nervous system: Correlation with MIB-1 expression and patient survival. Mod Pathol 1999;12:1094–1100.
49. Waseem NH, Lane DP. Monoclonal antibody analysis of proliferating cell nuclear antigen (PCNA) structural conversion and the detection of a nucleolar form. J Cell Sci 1990;96:121–129.
50. Gerdes J, Schwab U, Lemke H, et al. Production of a mouse monoclonal antibody reactive with human nuclear antigen associated with cell proliferation. Int J Cancer 1983;31:13–20.
51. Scholzen T, Gerdes J. The Ki-67 protein: From the known and the unknown. J Cell Physiol 2000;182:311–322.
52. Cattoretti G, Becker MHG, Kay G, et al. Monoclonal antibodies against recombinant parts of the Ki-67 antigen (MIB1 and MIB3) detect proliferating cells in microwave processed formalin-fixed paraffin sections. J Pathol 1992;168:357–364.
53. McCormick D, Chong H, Hobbs C, Datta C, Hall PA. Detection of the Ki-67 antigen in fixed and wax-embedded sections with the monoclonal antibody MIB1. Histopathology 1993;22:355–360.
54. Torp SH. Proliferative activity in human glioblastomas: evaluation of different Ki-67 equivalent antibodies. J Clin Pathol: Mol Pathol 1997;50:198–200.
55. Grzybicki DM, Liu Y, Moore SA, et al. Interobserver variability associated with the MIB-1 labeling index. High levels suggest limited prognostic usefulness for patients with primary brain tumors. Cancer 2001;92:2720–2726.
56. Prayson RA, Castilla EA, Hembury TA, Liu W, Noga CM, Prok AL. Interobserver variability in determining MIB-1 labeling indices in oligodendrogliomas. Ann Diagn Pathol 2003;7:9–13.

2
Neural Stem Cells

Philippe Taupin, PhD

SUMMARY

Neural stem cells (NSCs) are the self-renewing, multipotent cells that generate neurons, astrocytes, and oligodendrocytes in the nervous system. In the fetus, NSCs participate to the development of the nervous system. Stem cells are present in many tissues of adult mammals where they contribute to cellular homeostasis and regeneration after injury. The central nervous system (CNS), unlike other adult tissues, elicits limited capacity to recover from injury. It was believed contrary to other adult tissues that the CNS lacks stem cells, and thus the capacity to generate new nerve cells. In the 1960s, preliminary studies by Altman and Das gave the first evidence that new neuronal cells were being generated in the adult brain. In the following decades, with the emergence of new technologies for identifying and characterizing neural progenitor and stem cells in vivo, and in vitro, new studies have contributed to confirm that neurogenesis occurs in the adult brain, and that NSCs reside in the adult CNS. Thus overturning the long-held dogma that we are born with a certain number of nerve cells and that the brain cannot generate new neurons and renew itself. In this chapter, we will review the evidences that neurogenesis occurs throughout adulthood in discrete regions of the adult brain and that NSCs reside in the CNS of mammals, including human beings. We will review and discuss the different theories regarding the origin of NSCs in the adult in the brain and spinal cord.

Key Words: Mammals; multipotential; self-renewal; progenitor cell; spinal cord.

1. ADULT NEUROGENESIS AND NEURAL STEM CELLS OF THE MAMMALIAN CENTRAL NERVOUS SYSTEM

The first evidence that neurogenesis occurs in the adult mammalian brain came from studies conducted by Altman and Das *(1)*. Altman and Das reported, using [^{3}H]-thymidine autoradiographic labeling, evidences that new neuronal cells are generated in the adult rat dentate gyrus (DG) of the hippocampus. In a second study, Altman reported evidence of cell proliferation in the ventricular zone, migration, and persisting neurogenesis in the olfactory bulb (OB) *(2)*. Similarly, in a previous study, Adrian and Walker *(3)* failed to observe neurogenesis, but reported cell genesis of glial cells and inflammatory cells in the mouse spinal cord. Until the early 1990s, these studies were marginal, though a few reports supported the seminal work of Altman and Das that neurogenesis occurs in the adult mammalian brain *(4)*. Two major advances in the early 1990s contributed to the emergence of adult neurogenesis and neural stem cells (NSCs) as a major field for biological research and cellular therapy: the validation and wide use of bromodeoxyuridine (BrdU), a marker for dividing cells, as a tool for studying adult neurogenesis and the isolation and characterization of neural progenitor and stem cells in vitro from adult mouse brain by Reynolds and Weiss *(5)*.

From: *The Cell Cycle in the Central Nervous System*
Edited by: D. Janigro © Humana Press Inc., Totowa, NJ

1.1. Labeling of Dividing Cells in the Central Nervous System

The currently used protocol for characterizing neurogenesis in vivo consists of administering a marker of cellular division: BrdU, to perform histological labeling with antibodies against BrdU and other markers of nerve cells, and to perform analysis by confocal microscopy. BrdU is a thymidine analog that incorporates into DNA during S phase of the cell cycle that can be used to visualize cell proliferation *(6)*. BrdU is administered intraperitoneally in the animals and inserted into the DNA of dividing cells, including in the central nervous system (CNS) *(7)*. Histological studies allow the characterization of the newly generated neuronal cells and their fates, by multiple labeling with antibodies against BrdU and markers of interest, such as nestin *(8–12)*, sox-2 *(13–17)*, and oct-3/4 *(18)*, markers of neural progenitor and stem cells, β-tubulin type III *(19–21)*, neuronal nuclear antigen *(22)*, microtubule-associated protein-2 *(23)* and calbindin *(24)*, markers of neuronal cells, and O4 *(25)*, NG2 *(26–29)*, and glial fibrillary acidic protein *(30)*, markers of glial cells. BrdU has some limitations as a marker for dividing cells. Indeed, BrdU can also label DNA undergoing repair and cells that are initiating cell death by apoptosis *(31)*. Thus, when using BrdU for characterizing cell division additional controls must be performed to confirm the specificity of the labeling.

To this aim, other markers of the cell cycle, such as Ki-67, and proliferating cell nuclear antigen (PCNA), are being used to further confirm that cells are dividing, rather than in the process of DNA repair. Ki-67 is a nuclear protein expressed in all phases of the cell cycle except the resting phase *(32,33)*. PCNA is a cell cycle-dependent nuclear protein that serves as a cofactor of DNA polymerase and has a role in ensuring the fidelity of DNA replication *(34)*. PCNA is expressed in the S phase of cell division *(35)*. Ki-67 is consistently absent in quiescent cells and is not detectable during DNA repair processes. Thus, Ki-67 offers a reliable marker for cell division. Other markers of the cell cycle, such as PCNA, are also detected in cells undergoing DNA repair *(36)* and have the same limitation as BrdU regarding its specificity as a marker for studying cell division. The technique known as "terminal deoxynucleotidyltransferasemediated dUTP-biotin nick-end labeling" allows determining cells undergoing apoptosis *(37)*. Cell death can also be characterized by the identification of proteases, such as caspases *(38–40)*. All this labeling can be performed simultaneously with BrdU labeling and allow to confirm cell-division analysis. Lastly, one of the most convincing techniques for identifying newly generated cells involves administering retrovirus-carrying genes such as the gene of the green fluorescent protein. Retrovirus infect only dividing cells *(41,42)*, and thus not only offer a strategy for identifying newly generated cells' origin and fate, but also for tracking cell migration and physiological studies. Such protocols have been applied to characterize neurogenesis in the CNS and have confirmed that BrdU is a valid marker for studying neurogenesis *in situ (43–48)*.

1.2. Neurogenesis in the Adult CNS

Neurogenesis occurs mainly in two areas of the adult brain: the subgranular zone (SGZ) of the DG of the hippocampus *(49)*, and the anterior part of the subventricular zone (SVZ) *(44)*, along the ventricle, of several species including human *(50–52)*. Newly generated neuronal cells in the SGZ migrate to the granular layer of the DG, where they extend axonal projections to the CA3 area. Newly generated neuronal cells in the SVZ migrate to the OB, through the rostromigratory stream, where they differentiate into interneurons of the OB *(53–55)*. Particularly, the BrdU labeling paradigm has been used to label newly generated neuronal cells in the adult human CNS. Eriksson et al. *(50)* reported neurogenesis in the human DG from postmortem patients who had been treated with BrdU during the course of cancer treatment. More recently, Sanai et al. *(51)* reported neurogenesis in the human SVZ from organotypic slices of human adult SVZ. Newly generated neuronal cells in the DG and OB establish synaptic contacts and functional connections with neighboring cells *(56–61)*. Thus, it is a functional neurogenesis. It is estimated that as many as 9000 new cells are generated per day in a young adult rat DG *(62)*, though a significant proportion of these newly generated neuronal cells are lost within 2 wk in the DG *(62)*. Hippocampal neurogenesis contributes about 3.3% per month or about 0.1% per day of the granule cell population *(63)*. As in the DG, a significant fraction of newly generated neuronal cells in the adult SVZ are believed to undergo programmed cell death rather than achieving maturity *(64)*. It is estimated that 65.3–76.9% of the bulbar neurons are replaced during a 6-wk period in the adult rodent *(65)*. It has been reported that in the adult DG, newborn granule cells survive for at least 8 mo in rodents *(1)*, 12 wk in the macaque *(66)*, and 2 yr in humans *(50)*. Thus, the granule cells born during adulthood that become integrated into circuits and survive to maturity are very stable and may permanently replace granule cells born during development *(67)*.

More recent studies have reported that neurogenesis occurs in other areas of the adult mammalian brain, albeit at lower levels. Rietze et al. *(68)* reported low level of neurogenesis in the Ammon's horn CA1 of the adult mouse hippocampus. Gould et al. *(66,69)* reported that neurogenesis occurs in the neocortex of adult primates. Bedard et al. *(70)* and Bernier et al. *(71)* reported that neurogenesis occurs in the adult monkey striatum and amygdala, respectively. More recently, Zhao et al. *(72)* reported that neurogenesis occurs also in the adult mice substantia nigra. Some of these results have been contradicted by other studies. Kornack and Rakic *(73)* reported cell proliferation without neurogenesis in adult primate neocortex, whereas Lie et al. *(74)* and Frielingsdorf et al. *(75)* did not report evidence for new dopaminergic neurons in the adult mammalian substantia nigra. Thus, the confirmation of neurogenesis in these two later areas of the adult mammalian brain remains questionable.

In the adult spinal cord, cell genesis occurs without neurogenesis *(3)*. Horner et al. *(76)* reinvestigated neurogenesis in the adult spinal cord by BrdU labeling and confocal microscopy. Cell division occurs throughout the adult spinal cord and is not restricted to the lining of the central canal, with the majority of dividing cells residing in the outer circumference of the spinal cord. Horner et al. confirmed that newly generated cells in the spinal cord express markers of both immature and mature glial cells, astrocytes and oligodendrocytes, but not of neurons. It is estimated that 0.75% of all astrocytes and 0.82% of all oligodendrocytes are derived from a dividing population over a 4-wk period. These data confirmed that gliogenesis, but not neurogenesis, occurs in the adult spinal cord.

Thus, neurogenesis occurs in the adult mammalian brain, and it is hypothesized that neurogenesis arises from residual stem cells in the brain *(77,78)*. In contrast to the adult brain, newly generated cells in the adult spinal cord give rise to new cells restricted to the glial phenotype. Two hypotheses can be formulated to explain such discrepancies. First, the adult spinal cord, as opposed to the adult brain, does not contain NSCs, but restricted glial progenitor cells. Alternatively, the adult spinal cord would contain NSCs, but the environment would prevent these cells to differentiate into neuronal lineage. Thus, the presence of NSCs in the adult CNS remains to be resolved and the mechanisms of NSCs' fate determination remains to be characterized.

1.3. Neural Stem and Progenitor Cells of the CNS

The demonstration that NSCs exist in the adult CNS lie on two main criteria: self-renewal and multipotentiality. The demonstration that putative NSCs are multipotent relies on showing that the three main phenotypes of the CNS, neurons, astrocytes, and oligodendrocytes can be generated from single cells. The demonstration that putative NSCs can self-renew relies on showing that cells maintain their multipotentiality over time. These two criteria have not been established yet in vivo; however, cells with self-renewing and multipotential properties have been isolated from the adult brain and characterized in vitro. In 1992, Reynolds and Weiss *(5)* were the first to isolate and characterize in vitro, a population of undifferentiated cells, from adult mouse striatal tissue including the SVZ, capable of generating the three main phenotypes of the CNS. This population of cell was termed neural progenitor cells (NPCs) because their stem cell properties had yet to be demonstrated. The NPCs were found to be immunoreactive for the intermediate filament protein nestin, a marker of neural progenitor and stem cells *(4,8–12)*. In 1995, Gage et al. *(79)* isolated a population of cells with similar properties from the adult rat hippocampus, the second neurogenic area of the adult CNS. In both models, NPCs were isolated and cultured in vitro, in defined medium in the presence of trophic factors. The two models of NPCs differ by the trophic factors used to isolate and expand them and by the growth characteristics of the cells. Whereas NPCs isolated by Reynolds and Weiss *(5)* grow as neurospheres in the presence of epidermal growth factor, NPCs isolated by Gage et al. *(79)* grow as monolayer in the presence of fibroblast growth factor (FGF-2). In 1996, Gritti et al. *(80)* isolated and characterized in vitro self-renewing, multipotent NSCs from adult mouse striatal tissue including the SVZ, and in 1997, Palmer et al. *(81)* reported the isolation and characterization of self-renewing, multipotent NSCs from adult rat hippocampus. Because NPCs and self-renewing multipotent NSCs have been isolated and characterized in vitro from different areas of the adult CNS, including the spinal cord, and from different species, including human *(4,74,82–90)*.

One of the limitations in characterizing self-renewal, multipotential properties of putative NSCs in vitro is the difficulty of culturing isolated single cells. Epidermal growth factor and FGF-2, regulate the proliferation of NPCs in vitro *(5,79)*. Other studies have shown that unknown factors, particularly derived from conditioned medium, are required to stimulate NSC proliferation in vitro from single cells *(81,86,91–95)*. Taupin et al. *(96)* purified and characterized a factor derived from the conditioned medium of adult hippocampal-derived NPCs, and required with FGF-2 for the proliferation of NSCs in vitro, from a single cell, and to stimulate adult neurogenesis in vivo. The isolated factor is the glycosylated form of the protease inhibitor cystatin C *(97)*, whose *N*-glycosylation is required for its activity as a cofactor of FGF-2 (CCg) *(96)*. Thus, FGF-2 requires an autocrine/paracrine cofactor, CCg, for its mitogenic activity on NSCs. It has been a result of the isolation and characterization of CCg that we have been able to isolate adult human progenitor cells from autopsy and biopsy brain *(88)*.

In vivo data show that gliogenesis, but not neurogenesis, occurs in the adult spinal cord. It is hypothesized that the adult spinal cord, as opposed to the adult brain, does not contain NSCs, but restricted glial progenitor cells *(76)*. Alternatively, the adult spinal cord would contain NSCs, but the environment would prevent these cells to differentiate into neuronal lineage. The isolation and characterization of self-renewing, multipotent NSCs from the adult spinal cord suggest that the adult spinal cord contains putative NSCs, and that the environment would prevent these cells to differentiate into neuronal lineage *(84,87)*. In support of this contention, Shihabuddin et al. *(87)* reported that with transplantation in the adult spinal cord, adult spinal-cord-derived neural progenitor and stem cells elicited only glial phenotypes, whereas when transplanted into the DG, neuronal phenotypes were also observed. Thus, the clonally expanded spinal-cord-derived neural progenitor and stem cells, when transplanted in the adult spinal cord, behave like endogenous proliferating spinal-cord cells, by differentiating into glia only *(76)*. The ability of the cells to differentiate into neuronal phenotype in heterotypic transplantation studies suggest that adult spinal-cord-derived neural progenitor and stem cells are induced to express mature neuronal phenotype by environmental signals.

Thus, putative NSCs reside in the adult brain not exclusively in the neurogenic areas in the adult brain, but also in nonneurogenic areas where they will remain quiescent. However, the criteria used to characterize self-renewing, multipotent NSCs, although are well accepted to show that a single cell is a NSC in vitro, are not absolute. The main criticism resides in the number of subcloning steps that one must show to qualify a cell as self-renewing in vitro *(84)*. Recent reports have challenged the isolation and characterization of self-renewing, multipotent NSCs from the adult DG, claiming the DG contains restricted progenitors, highlighting the limitation of in vitro studies to identify putative NSCs, but also differences in isolation procedures *(98,99)*.

2. ORIGIN OF NSCs IN THE ADULT CNS

The fact that a cell can be labeled in vivo by administration of [^{3}H]-thymidine, BrdU, or retroviral labeling does not mean that it is a stem cell. Self-renewing, multipotent NSCs can be isolated from the adult brain and expanded in vitro, hence NSC research has aimed at identifying the origin of the newly generated neuronal cells in the adult mammalian brain. It is currently hypothesized that neurogenesis arises from residual stem cells in the adult brain. There are several hypotheses and theories regarding the identity and origin of NSCs in the adult brain. One theory contends that the NSCs of the adult SVZ are differentiated ependymal cells that express the intermediate filament protein nestin *(96)*. The other theory identifies them as astrocyte-like cells expressing glial fibrillary acidic protein and nestin in the SVZ *(100–106)* and that would originate from a pool of slowly dividing cells *(107)*. An astroglial origin for NSCs in the hippocampus has also received much of support *(108)*.

In the adult spinal cord, it has been hypothesized that the central canal is the presumed location of the putative NSCs, because cells in the corresponding region of the brain, that is, the SVZ, can proliferate and differentiate to neurons and glia postnatally *(95,100–106,109)*. Regarding the location of the progenitor cells in the spinal cord, the data from Horner et al. *(76)* predicts otherwise. Horner et al. reported that cell division occurs throughout the adult spinal cord, and is not restricted to the lining of the central canal, with the majority of dividing cells residing in the outer circumference of the spinal cord. Thus, glial progenitor cells exist also in the outer circumference of the spinal cord. Horner et al. *(76)* proposed two models regarding the origin of glial progenitor cell in the adult spinal cord. One model contends that a stem cell exists at the ependymal layer, and divides asymmetrically. A daughter cell then migrates to the outer circumference of the spinal cord where it exists as a bipotent or glial progenitor cell and begins to divide more rapidly. The other model predicts that a glial progenitor and stem cell population may exist in the outer circumference of the spinal cord where cell division is more common. This model functionally separates ependymal cell division from the proliferative zone of the outer annuli. Yamamoto et al. *(110)* isolated and characterized neural progenitor and stem cells from the periventricular area, but also from other regions of the parenchyma, supporting previous evidence by Horner et al. *(76)* that NSCs in the adult spinal cord are not restricted to the periventricular area, although putative NSCs in the adult spinal cord remain to be identified.

3. CONCLUSIONS

Neurogenesis occurs in the adult brain and NSCs reside in the adult CNS. The identification of the putative NSCs in the adult CNS remains the source of intense debate. The identification of molecular markers will ultimately define such cells. Several teams have attempted to identify specific markers of NSCs by gene profiling *(111–114)*. However, the reported results are questionable owing to the heterogeneity of such culture; they contain NSCs, and more mature, yet undifferentiated, cells termed NPCs *(4,115–121)*. Homogeneous populations of NSC/NPCs have been isolated using cell surface markers from human fetal spinal cord and brain tissues *(122)*, by promoter-targeted selection from adult rat *(123)* and human SVZ *(124)*, and from the lateral ventricle wall of adult mice by negative selection *(125)*. These studies, by providing homogeneous populations of NSCs, will allow us to further study the origin and molecular identity of the NSCs.

ACKNOWLEDGMENT

PT was supported by grants from the NMRC, BMRC, and the Juvenile Diabetes Research Foundation.

REFERENCES

1. Altman J, Das GD. Autoradiographic and histological evidence of postnatal hippocampal neurogenesis in rats J Comp Neurol 1965;124:319–336.
2. Altman J. Autoradiographic and histological studies of postnatal neurogenesis. IV. Cell proliferation and migration in the anterior forebrain, with special reference to persisting neurogenesis in the olfactory bulb. J Comp Neurol 1969;137:433–458.
3. Adrian EK Jr, Walker BE. Incorporation of thymidine-H3 by cells in normal and injured mouse spinal cord. J Neuropathol Exp Neurol 1962;21:597–609.
4. Taupin P, Gage FH. Adult neurogenesis and neural stem cells of the central nervous system in mammals. J Neurosci Res 2002;69:745–749.

5. Reynolds BA, Weiss S. Generation of neurons and astrocytes from isolated cells of the adult mammalian central nervous system. Science 1992;255:1707–1710.
6. Dolbeare F. Bromodeoxyuridine: a diagnostic tool in biology and medicine, Part II: Oncology, chemotherapy and carcinogenesis. Histochem J 1995;27:923–964.
7. del Rio JA, Soriano E. Immunocytochemical detection of 5′-bromodeoxyuridine incorporation in the central nervous system of the mouse. Brain Res Dev Brain Res 1989;49:311–317.
8. Hockfield S, McKay RD. Identification of major cell classes in the developing mammalian nervous system. J Neurosci 1985;5:3310–3328.
9. Frederiksen K, McKay RD. Proliferation and differentiation of rat neuroepithelial precursor cells in vivo. J Neurosci 1988;8:1144–1151.
10. Lendahl U, Zimmerman LB, McKay RD. CNS stem cells express a new class of intermediate filament protein. Cell 1990;60:585–595.
11. Cattaneo E, McKay R. Proliferation and differentiation of neuronal stem cells regulated by nerve growth factor. Nature 1990;347:762–765.
12. Dahlstrand J, Lardelli M, Lendahl U. Nestin mRNA expression correlates with the central nervous system progenitor cell state in many, but not all, regions of developing central nervous system. Brain Res Dev Brain Res 1995;84:109–129.
13. Wood HB, Episkopou V. Comparative expression of the mouse Sox1, Sox2 and Sox3 genes from pre-gastrulation to early somite stages. Mech Dev 1999;86:197–201.
14. Zappone MV, Galli R, Catena R, et al. Sox2 regulatory sequences direct expression of a (beta)-geo transgene to telencephalic neural stem cells and precursors of the mouse embryo, revealing regionalization of gene expression in CNS stem cells. Development 2000;127:2367–2382.
15. Graham V, Khudyakov J, Ellis P, Pevny L. SOX2 functions to maintain neural progenitor identity. Neuron 2003;39:749–765.
16. D'Amour KA, Gage FH. Genetic and functional differences between multipotent neural and pluripotent embryonic stem cells. Proc Natl Acad Sci USA 2003;100 (Suppl 1):11,866–11,872.
17. Komitova M, Eriksson PS. Sox-2 is expressed by neural progenitors and astroglia in the adult rat brain. Neurosci Lett 2004;369:24–27.
18. Okuda T, Tagawa K, Qi ML, et al. Oct-3/4 repression accelerates differentiation of neural progenitor cells in vitro and in vivo. Brain Res Mol Brain Res 2004;132:18–30.
19. Lee MK, Tuttle JB, Rebhun LI, Cleveland DW, Frankfurter A. The expression and posttranslational modification of a neuron-specific beta-tubulin isotype during chick embryogenesis. Cell Motil Cytoskeleton 1990;17:118–132.
20. Menezes JR, Luskin MB. Expression of neuron-specific tubulin defines a novel population in the proliferative layers of the developing telencephalon. J Neurosci 1994;14:5399–5416.
21. Fanarraga ML, Avila J, Zabala JC. Expression of unphosphorylated class III beta-tubulin isotype in neuroepithelial cells demonstrates neuroblast commitment and differentiation. Eur J Neurosci 1999; 11:516–527.
22. Mullen RJ, Buck CR, Smith A. NeuN, a neuronal specific nuclear protein in vertebrates. Development 1992;116:201–211.
23. Bernhardt R, Matus A. Light and electron microscopic studies of the distribution of microtubule-associated protein 2 in rat brain: a difference between dendritic and axonal cytoskeletons. J Comp Neurol 1984;226:203–221.
24. Sloviter RS. Calcium-binding protein (calbindin-D28k) and parvalbumin immunocytochemistry: localization in the rat hippocampus with specific reference to the selective vulnerability of hippocampal neurons to seizure activity. J Comp Neurol 1989;280:183–196.
25. Reynolds R, Hardy R. Oligodendroglial progenitors labeled with the O4 antibody persist in the adult rat cerebral cortex in vivo. J Neurosci Res 1997;47:455–470.
26. Levine JM, Stallcup WB. Plasticity of developing cerebellar cells in vitro studied with antibodies against the NG2 antigen. J Neurosci 1987;7:2721–2731.
27. Nishiyama A, Lin XH, Giese N, Heldin CH, Stallcup WB. Co-localization of NG2 proteoglycan and PDGF alpha-receptor on O2A progenitor cells in the developing rat brain. J Neurosci Res 1996;43:299–314.
28. Keirstead HS, Levine JM, Blakemore WF. Response of the oligodendrocyte progenitor cell population (defined by NG2 labelling) to demyelination of the adult spinal cord. Glia 1998;22:161–170.
29. Horner PJ, Thallmair M, Gage FH. Defining the NG2-expressing cell of the adult CNS. J Neurocytol 2002;31:469–480.

30. Debus E, Weber K, Osborn M. Monoclonal antibodies specific for glial fibrillary acidic (GFA) protein and for each of the neurofilament triplet polypeptides. Differentiation 1983;25:193–203.
31. Cooper-Kuhn CM, Kuhn HG. Is it all DNA repair? Methodological considerations for detecting neurogenesis in the adult brain. Brain Res Dev Brain Res 2002;134:13–21.
32. Scholzen T, Gerdes J. The Ki-67 protein: from the known and the unknown. J Cell Physiol 2000;182:311–322.
33. Zacchetti A, van Garderen E, Teske E, Nederbragt H, Dierendonck JH, Rutteman GR. Validation of the use of proliferation markers in canine neoplastic and non-neoplastic tissues: comparison of KI-67 and proliferating cell nuclear antigen (PCNA) expression versus in vivo bromodeoxyuridine labelling by immunohistochemistry. APMIS 2003;111:430–438.
34. Kurki P, Vanderlaan M, Dolbeare F, Gray J, Tan EM. Expression of proliferating cell nuclear antigen (PCNA)/cyclin during the cell cycle. Exp Cell Res 1986;166:209–219.
35. Takahashi T, Caviness VS JR. PCNA-binding to DNA at the G1/S transition in proliferating cells of the developing cerebral wall. J Neurocytol 1993;22:1096–1102.
36. Hall PA, McKee PH, Menage HD, Dover R, Lane DP. High levels of p53 protein in UV-irradiated normal human skin. Oncogene 1993;8:203–207.
37. Gavrieli Y, Sherman Y, Ben-Sasson SA. Identification of programmed cell death in situ via specific labeling of nuclear DNA fragmentation. J Cell Biol 1992;119:493–501.
38. Namura S, Zhu J, Fink K, et al. Activation and cleavage of caspase-3 in apoptosis induced by experimental cerebral ischemia. J Neurosci 1998;18:3659–3668.
39. Pompeiano M, Blaschke AJ, Flavell RA, Srinivasan A, Chun J. Decreased apoptosis in proliferative and postmitotic regions of the Caspase 3-deficient embryonic central nervous system. J Comp Neurol 2000;423:1–12.
40. Ekdahl CT, Mohapel P, Elmer E, Lindvall O. Caspase inhibitors increase short-term survival of progenitor-cell progeny in the adult rat dentate gyrus following status epilepticus. Eur J Neurosci 2001;14:937–945.
41. Mozdziak P, Schultz E. Retroviral labeling is an appropriate marker for dividing cells. Biotech Histochem 2000;75:141–146.
42. Kay MA, Glorioso JC, Naldini L. Viral vectors for gene therapy: the art of turning infectious agents into vehicles of therapeutics. Nat Med 2001;7:33–40.
43. Corotto FS, Henegar JA, Maruniak JA. Neurogenesis persists in the subependymal layer of the adult mouse brain. Neurosci Lett 1993;149:111–114.
44. Luskin MB. Restricted proliferation and migration of postnatally generated neurons derived from the forebrain subventricular zone. Neuron 1993;11:173–189.
45. Seki T, Arai Y. Highly polysialylated neural cell adhesion molecule (NCAM-H) is expressed by newly generated granule cells in the dentate gyrus of the adult rat. J Neurosci 1993;13:2351–2358.
46. Kuhn HG, Dickinson-Anson H, Gage FH. Neurogenesis in the dentate gyrus of the adult rat: age-related decrease of neuronal progenitor proliferation. J Neurosci 1996;16:2027–2033.
47. Liu J, Solway K, Messing RO, Sharp FR. Increased neurogenesis in the dentate gyrus after transient global ischemia in gerbils. J Neurosci 1998;18:7768–7778.
48. Jin K, Sun Y, Xie L, et al. Directed migration of neuronal precursors into the ischemic cerebral cortex and striatum. Mol Cell Neurosci 2003;24:171–189.
49. Cameron HA, Wolley CS, McEwen BS, Gould E. Differentiation of newly born neurons and glia in the dentate gyrus of the adult rat. Neurosci 1993;56:337–344.
50. Eriksson PS, Perfilieva E, Bjork-Eriksson T, et al. Neurogenesis in the adult human hippocampus. Nat Med 1998;4:1313–1317.
51. Sanai N, Tramontin AD, Quinones-Hinojosa A, et al. Unique astrocyte ribbon in adult human brain contains neural stem cells but lacks chain migration. Nature 2004;427:740–744.
52. Bedard A, Parent A. Evidence of newly generated neurons in the human olfactory bulb. Brain Res Dev Brain Res 2004;151:159–168.
53. Lois C, Alvarez-Buylla A. Long-distance neuronal migration in the adult mammalian brain. Science 1994;264:1145–1148.
54. Doetsch F, Garcia-Verdugo JM, Alvarez-Buylla A. Cellular composition and three-dimensional organization of the subventricular germinal zone in the adult mammalian brain. J Neurosci 1997;17:5046–5061.
55. Bonfanti L, Peretto P, Merighi A, Fasolo A. Newly-generated cells from the rostral migratory stream in the accessory olfactory bulb of the adult rat. Neuroscience 1997;81:489–502.

56. Kaplan MS, Bell DH. Neuronal proliferation in the 9-month-old rodent-radioautographic study of granule cells in the hippocampus. Exp Brain Res 1983;52:1–5.
57. Markakis EA, Gage FH. Adult-generated neurons in the dentate gyrus send axonal projections to field CA3 and are surrounded by synaptic vesicles. J Comp Neurol 1999;406:449–460.
58. Van Praag H, Schinder AF, Christie BR, Toni N, Palmer TD, Gage FH. Functional neurogenesis in the adult hippocampus. Nature 2002;415:1030–1034.
59. Carlen M, Cassidy RM, Brismar H, Smith GA, Enquist LW, Frisen J. Functional integration of adult-born neurons. Curr Biol 2002;12:606–608.
60. Belluzzi O, Benedusi M, Ackman J, LoTurco JJ. Electrophysiological differentiation of new neurons in the olfactory bulb. J Neurosci 2003;23:10,411–10,418.
61. Yamada M, Onodera M, Mizuno Y, Mochizuki H. Neurogenesis in olfactory bulb identified by retroviral labeling in normal and 1-methyl-4-phenyl-1,2,3,6-tetrahydropyridine-treated adult mice. Neuroscience 2004;124:173–181.
62. Cameron HA, McKay RD. Adult neurogenesis produces a large pool of new granule cells in the dentate gyrus. J Comp Neurol 2001;435:406–417.
63. Kempermann G, Kuhn HG, Gage FH. More hippocampal neurons in adult mice living in an enriched environment. Nature 1997;386:493–495.
64. Morshead CM, van der Kooy D. Postmitotic death is the fate of constitutively proliferating cells in the subependymal layer of the adult mouse brain. J Neurosci 1992;12:249–256.
65. Kato T, Yokouchi K, Fukushima N, Kawagishi K, Li Z, Moriizumi T. Continual replacement of newly-generated olfactory neurons in adult rats. Neurosci Lett 2001;307:17–20.
66. Gould E, Vail N, Wagers M, Gross CG. Adult-generated hippocampal and neocortical neurons in macaques have a transient existence. Proc Natl Acad Sci USA 2001;98:10,910–10,917.
67. Dayer AG, Ford AA, Cleaver KM, Yassaee M, Cameron HA. Short-term and long-term survival of new neurons in the rat dentate gyrus. J Comp Neurol 2003;460:563–572.
68. Rietze R, Poulin P, Weiss S. Mitotically active cells that generate neurons and astrocytes are present in multiple regions of the adult mouse hippocampus. J Comp Neurol 2000;424:397–408.
69. Gould E, Reeves AJ, Graziano MS, Gross CG. Neurogenesis in the neocortex of adult primates. Science 1999;286:548–552.
70. Bedard A, Cossette M, Levesque M, Parent A. Proliferating cells can differentiate into neurons in the striatum of normal adult monkey. Neurosci Lett 2002;328:213–216.
71. Bernier PJ, Bedard A, Vinet J, Levesque M, Parent A. Newly generated neurons in the amygdala and adjoining cortex of adult primates. Proc Natl Acad Sci USA 2002;99:11,464–11,469.
72. Zhao M, Momma S, Delfani K, et al. Evidence for neurogenesis in the adult mammalian substantia nigra. Proc Natl Acad Sci USA 2003;100:7925–7930.
73. Kornack DR, Rakic P. Cell proliferation without neurogenesis in adult primate neocortex. Science 2001;294:2127–2130.
74. Lie DC, Dziewczapolski G, Willhoite AR, Kaspar BK, Shults CW, Gage FH. The adult substantia nigra contains progenitor cells with neurogenic potential. J Neurosci 2002;22:6639–6649.
75. Frielingsdorf H, Schwarz K, Brundin P, Mohapel P. No evidence for new dopaminergic neurons in the adult mammalian substantia nigra. Proc Natl Acad Sci USA 2004;101:10,177–10,182.
76. Horner PJ, Power AE, Kempermann G, et al. Proliferation and differentiation of progenitor cells throughout the intact adult rat spinal cord. J Neurosci 2000;20:2218–2228.
77. Gross CG. Neurogenesis in the adult brain: death of a dogma. Nat Rev Neurosci 2000;1:67–73.
78. Gage FH. Mammalian neural stem cells. Science 2000;287:1433–1438.
79. Gage FH, Coates PW, Palmer TD, et al. Survival and differentiation of adult neuronal progenitor cells transplanted to the adult brain. Proc Natl Acad Sci USA 1995;92:11,879–11,883.
80. Gritti A, Parati EA, Cova L, et al. Multipotential stem cells from the adult mouse brain proliferate and self-renew in response to basic fibroblast growth factor. J Neurosci 1996;16:1091–1100.
81. Palmer TD, Takahashi J, Gage FH. The adult rat hippocampus contains primordial neural stem cells. Mol Cell Neurosci 1997;8:389–404.
82. Palmer TD, Ray J, Gage FH. FGF-2-responsive neuronal progenitors reside in proliferative and quiescent regions of the adult rodent brain. Mol Cell Neurosci 1995;6:474–486.
83. Reynolds BA, Weiss S. Clonal and population analyses demonstrate that an EGF-responsive mammalian embryonic CNS precursor is a stem cell. Dev Biol 1996;175:1–13.
84. Weiss S, Dunne C, Hewson J, et al. Multipotent CNS stem cells are present in the adult mammalian spinal cord and ventricular neuroaxis. J Neurosci 1996;16:7599–7609.

85. Palmer TD, Markakis EA, Willhoite AR, Safar F, Gage FH. Fibroblast growth factor-2 activates a latent neurogenic program in neural stem cells from diverse regions of the adult CNS. J Neurosci 1999;19:8487–8497.
86. Gritti A, Frolichsthal-Schoeller P, Galli R, et al. Epidermal and fibroblast growth factors behave as mitogenic regulators for a single multipotent stem cell-like population from the subventricular region of the adult mouse forebrain. J Neurosci 1999;19:3287–3297.
87. Shihabuddin LS, Horner PJ, Ray J, Gage FH. Adult spinal cord stem cells generate neurons after transplantation in the adult dentate gyrus. J Neurosci 2000;20:8727–8735.
88. Palmer TD, Schwartz PH, Taupin P, Kaspar B, Stein SA, Gage FH. Cell culture. Progenitor cells from human brain after death. Nature 2001:411:42,43.
89. Schwartz PH, Bryant PJ, Fuja TJ, Su H, O'Dowd DK, Klassen H. Isolation and characterization of neural progenitor cells from post-mortem human cortex. J Neurosci Res 2003;74:838–851.
90. Markakis EA, Palmer TD, Randolph-Moore L, Rakic P, Gage FH. Novel neuronal phenotypes from neural progenitor cells. J Neurosci 2004;24:2886–2897.
91. Temple S. Division and differentiation of isolated CNS blast cells in microculture. Nature 1989;340:471–473.
92. Kilpatrick TJ, Bartlett PF. Cloning and growth of multipotential neural precursors: requirements for proliferation and differentiation. Neuron 1993;10:255–265.
93. Davis AA, Temple S. A self-renewing multipotential stem cell in embryonic rat cerebral cortex. Nature 1994;372:263–266.
94. Qian X, Davis AA, Goderie SK, Temple S. FGF2 concentration regulates the generation of neurons and glia from multipotent cortical stem cells. Neuron 1997;18:81–93.
95. Johansson CB, Momma S, Clarke DL, Risling M, Lendahl U, Frisen J. Identification of a neural stem cell in the adult mammalian central nervous system. Cell 1999;96:25–34.
96. Taupin P, Ray J, Fischer WH, et al. FGF-2-responsive neural stem cell proliferation requires CCg, a novel autocrine/paracrine cofactor. Neuron 2000;28:385–397.
97. Turk V, Bode W. The cystatins: protein inhibitors of cysteine proteinases. FEBS lett 1991;285:213–219.
98. Seaberg RM, van der Kooy D. Adult rodent neurogenic regions: the ventricular subependyma contains neural stem cells, but the dentate gyrus contains restricted progenitors. J Neurosci 2002;22:1784–1793.
99. Becq H, Jorquera I, Ben-Ari Y, Weiss S, Represa A. Differential properties of dentate gyrus and CA1 neural precursors. J Neurobiol 2005;62:243–261.
100. Doetsch F, Caille I, Lim DA, Garcia-Verdugo JM, Alvarez-Buylla A. Subventricular zone astrocytes are neural stem cells in the adult mammalian brain. Cell 1999;97:703–716.
101. Chiasson BJ, Tropepe V, Morshead CM, van der Kooy D. Adult mammalian forebrain ependymal and subependymal cells demonstrate proliferative potential, but only subependymal cells have neural stem cell characteristics. J Neurosci 1999;19:4462–4471.
102. Laywell ED, Rakic P, Kukekov VG, Holland EC, Steindler DA. Identification of a multipotent astrocytic stem cell in the immature and adult mouse brain. Proc Natl Acad Sci USA 2000;97:13,883–13,888.
103. Seri B, Garcia-Verdugo JM, McEwen BS, Alvarez-Buylla A. Astrocytes give rise to new neurons in the adult mammalian hippocampus. J Neurosci 2001;21:7153–7160.
104. Morshead CM, Garcia AD, Sofroniew MV, van Der Kooy D. The ablation of glial fibrillary acidic protein-positive cells from the adult central nervous system results in the loss of forebrain neural stem cells but not retinal stem cells. Eur J Neurosci 2003;18:76–84.
105. Imura T, Kornblum HI, Sofroniew MV. The predominant neural stem cell isolated from postnatal and adult forebrain but not early embryonic forebrain expresses GFAP. J Neurosci 2003;23:2824–2832.
106. Garcia AD, Doan NB, Imura T, Bush TG, Sofroniew MV. GFAP-expressing progenitors are the principal source of constitutive neurogenesis in adult mouse forebrain. Nat Neurosci 2004;7:1233–1241.
107. Morshead CM, Reynolds BA, Craig CG, et al. Neural stem cells in the adult mammalian forebrain: a relatively quiescent subpopulation of subependymal cells. Neuron 1994;13:1071–1082.
108. Seri B, Garcia-Verdugo JM, McEwen BS, Alvarez-Buylla A. Astrocytes give rise to new neurons in the adult mammalian hippocampus. J Neurosci 2001;21:7153–7160.
109. Alonso G. Neuronal progenitor-like cells expressing polysialylated neural cell adhesion molecule are present on the ventricular surface of the adult rat brain and spinal cord. J Comp Neurol 1999;414:149–166.

110. Yamamoto S, Yamamoto N, Kitamura T, Nakamura K, Nakafuku M. Proliferation of parenchymal neural progenitors in response to injury in the adult rat spinal cord. Exp Neurol 2001;172:115–127.
111. Geschwind DH, Ou J, Easterday MC, et al. A genetic analysis of neural progenitor differentiation. Neuron 2001;29:325–339.
112. Terskikh AV, Easterday MC, Li L, et al. From hematopoiesis to neuropoiesis: evidence of overlapping genetic programs. Proc Natl Acad Sci USA 2001;98:7934–7939.
113. Ramalho-Santos M, Yoon S, Matsuzaki Y, Mulligan RC, Melton DA. "Stemness": transcriptional profiling of embryonic and adult stem cells. Science 2002;298:597–600.
114. Ivanova NB, Dimos JT, Schaniel C, Hackney JA, Moore KA, Lemischka IR. A stem cell molecular signature. Science 2002;298:601–604.
115. Yaworsky PJ, Kappen C. Heterogeneity of neural progenitor cells revealed by enhancers in the nestin gene. Dev Biol 1999;205:309–321.
116. Kornblum HI, Geschwind DH. Molecular markers in CNS stem cell research: hitting a moving target. Nat Rev Neurosci 2001;2:843–846.
117. Chu VT, Gage FH. Chipping away at stem cells. Proc Natl Acad Sci USA 2001;98:7652,7653.
118. Suslov ON, Kukekov VG, Ignatova TN, Steindler DA. Neural stem cell heterogeneity demonstrated by molecular phenotyping of clonal neurospheres. Proc Natl Acad Sci USA 2002;99:14,506–14,511.
119. Evsikov AV, Solter D. Comment on " 'Stemness':transcriptional profiling of embryonic and adult stem cells" and "a stem cell molecular signature." Science 2003;302:393.
120. Fortunel NO, Otu HH, Ng HH, et al. Comment on " 'Stemness':transcriptional profiling of embryonic and adult stem cells" and "a stem cell molecular signature." Science 2003;302:393.
121. Chin VI, Taupin P, Sanga S, Scheel J, Gage FH, Bhatia SN. Microfabricated platform for studying stem cell fates. Biotechnol Bioeng 2004;88:399–415.
122. Uchida N, Buck DW, He D, et al. Direct isolation of human central nervous system stem cells. Proc Natl Acad Sci USA 2000;97:14,720–14,725.
123. Wang S, Roy NS, Benraiss A, Goldman SA. Promoter-based isolation and fluorescence-activated sorting of mitotic neuronal progenitor cells from the adult mammalian ependymal/subependymal zone. Dev Neurosci 2000;22:167–176.
124. Roy NS, Benraiss A, Wang S, et al. Promoter-targeted selection and isolation of neural progenitor cells from the adult human ventricular zone. J Neurosci Res 2000;59:321–331.
125. Rietze RL, Valcanis H, Brooker GF, Thomas T, Voss AK, Bartlett PF. Purification of a pluripotent neural stem cell from the adult mouse brain. Nature 2001;412:736–739.

3
Progenitors and Precursors of Neurons and Glial Cells

Monika Bradl, PhD

SUMMARY

The central nervous system is an orderly, highly complicated structure comprising neurons and glia. These cells trace back to neuroepithelial stem cells of the ventricular zone. The creation of differentiated neurons, astrocytes, and oligodendrocytes and their progenitors/precursors proceeds through extensive phases of proliferation, lineage specification, and long distance migration. This chapter briefly summarizes the current knowledge about these milestones in the developing central nervous system.

Key Words: Neurons; astrocytes; oligodendrocytes; radial glial cells; microglia; progenitors; precursors.

1. STEM CELLS IN THE CENTRAL NERVOUS SYSTEM

Neurons and glial cells of the central nervous system (CNS) derive from stem cells of the neural plate. It is not quite clear whether all cells found in the neural plate are stem cells, whether stem cells represent a minor, but evenly distributed population of cells, or whether stem cells are only found in the midline and the lateral edges of the neural plate *(1)*. In any case, these stem cells are mitotically active and change over time. First, they expand and undergo many rounds of symmetric divisions. As development proceeds, these stem cells then start with asymmetrical divisions and give rise to neurons (in the neurogenic phase) and then to astrocytes and oligodendrocytes (in the gliogenic phase). The developmental cues responsible for the completion of this program involves the action of inductive signals, which are either produced from tissues outside the CNS, or from signaling centers within the CNS. Depending on the interpretation of these signals, transcription factors are activated which mediate the acquisition of different cell fates, at the correct time, the correct position, and in correct numbers *(2)*.

2. NEURONS

All neurons of the CNS derive from neuroepithelial cells of the ventricular zone. From the onset of neurogenesis, neurons are generated by two types of divisions: asymmetric divisions of cells at the apical side of the neuroepithelium, which gives rise to one neuron and one neuroepithelial cell which is able to undergo another neuron-generating division, and symmetric divisions of progenitor cells located at the basal side of the ventricular zone which give rise to two neurons *(3)*. Hence, there are two different types of neuronal progenitor cells in the mammalian CNS that differ from each other in the numbers of neurons born: basal progenitors appear to produce twice as many neurons than apical progenitors and seem to be the major source of cortical neurons *(3)*.

From: *The Cell Cycle in the Central Nervous System*
Edited by: D. Janigro © Humana Press Inc., Totowa, NJ

Once born, cortical neurons have to migrate long distances guided by processes of radial glial cells to finally reach their proper destination and to give rise to the six-layered mammalian cortex. The earliest neurons arriving at the presumptive cortex form the preplate, which is then further subdivided by later-arriving neurons *(4)*.

The formation of neurons in the spinal cord is influenced by two different external signaling centers: First, by the epidermis expressing bone morphogenetic proteins (BMPs) 4 and 7. These proteins induce the roof plate cells to produce and secrete BMP-4, which in turn leads to a cascade of transforming growth factor-β-related factors spreading ventrally and second, by the notochord secreting sonic hedgehog (SHH). This factor induces the floor plate cells to produce SHH, which then diffuses dorsally. The dorsal → ventral gradient in the concentration of transforming growth factor-β related factors and the ventral → dorsal gradient of SHH are then read and interpreted. Depending on the concentrations of these factors, different transcription factors are expressed, and neuronal identity is specified *(5)*.

3. THE NEUROGENIC → GLIOGENIC SWITCH POINT

The basic mechanisms underlying the production of neurons and their specification are well understood. Much less is known about the signals necessary to induce the formation of glial cells. Around midgestation, the ventricular zone stem cells stop to produce neurons, and start to produce glia. The factors responsible for this switch from neurogenesis to gliogenesis are largely undefined. Recently, however, two factors have been discovered which might help to regulate the neurogenic → gliogenic switch.

The first factor is the transcription factor Sox9, which seems to determine the glial fate choice in the developing spinal cord. Transgenic mice with a CNS-specific ablation of Sox9 are unable to produce astrocyte and oligodendrocyte progenitors. Instead, they transiently produce increased numbers of motoneurons *(6)*. Murine oligodendrocyte progenitors also express Sox8 and Sox10 and these factors eventually compensate for the loss of Sox9 in the oligodendrocyte lineage. This leads to a recovery of oligodendrocyte numbers at later stages of development. Astrocytes, however, do not have this means of compensation. Consequently, their numbers do not rebound *(6)*.

The second factor possibly involved in the neurogenic → gliogenic fate decision is Notch. It was observed that the conditional ablation of Notch caused the premature generation of neuronal cells, a loss of glia cells expressing the astrocyte marker glial acidic fibrillary protein (in mice, *[7]*), and a loss of oligodendrocytes (in zebrafish, *[8,9]*). Moreover, overexpression of Notch blocked CNS neurogenesis and caused an excess of oligodendrocyte progenitors in the ventral spinal cord *(8,10)*. This effect is probably mediated by reducing the activity of neurogenin 1, a factor that normally promotes neurogenesis and inhibits glial differentiation *(11,12)*.

4. OLIGODENDROCYTES

Oligodendrocyte development has been extensively studied both in vitro and in vivo, with somewhat different results.

In vitro, oligodendrocyte development follows a predictable pathway. The first step in this pathway is the generation of tripotential glial-restricted precursor cells, possibly derived from neuroepithelial stem cells. At the end of the developmental program in vitro, cultures of glia-restricted cells give rise to two distinct astrocyte populations or oligodendrocytes. The oligodendrocyte development from these glial-restricted precursor cells proceeds through another type of intermediate cell: a bipotential cell alternatively referred to as oligodendrocyte/type-2 astrocyte (O2A) progenitor or as oligodendrocyte precursor cell (OPC) *(13)*, which can give rise to just one astrocyte population (the type-2 astrocytes) and to oligodendrocytes *(14)*. Glial-restricted precursor cells can be isolated from both ventral and dorsal areas of the murine embryonic spinal cord, although ventral-derived glial-restricted precursor cells were more likely to generate O2A/OPC cells and oligodendrocytes than were their dorsal counterparts *(13)*.

There is no doubt that oligodendrocytes differentiate through these different steps in vitro; the question still remains whether they do so in vivo. In vivo, oligodendrocyte precursors in spinal cord, hindbrain, midbrain and caudal forebrain originate from two ventral domains of neuroepithelium at either site of the floor plate *(2,15)*. OPCs are characterized by the presence of transcription factors olig1 and olig2, by expression of proteolipid protein and its smaller splice variant, DM20, and by expression of the receptor for platelet-derived growth factor α. It is not quite clear yet, whether these markers define just one or several different oligodendrocyte lineages *(16,17)*. However, there is no dispute about the initial specification and production of these cells. This depends on SHH, and it occurs in the ventral neuroepithelial motor neuron progenitor (pMN) domains at a timepoint in development when the capacity to produce somatic motor neurons has already been lost *(9,16)*. The oligodendrocyte precursors then leave the ventricular surface and start to disseminate throughout the gray and white matter, mainly in the ventral half of the spinal cord, but later also in the dorsal regions *(16)*. In contrast to the data obtained in vitro, there is no evidence to date for an additional, dorsally located spinal cord region, which does give rise to oligodendrocytes in vivo. There is also no evidence that the OPCs in vivo produce oligodendrocytes and astrocytes. Because cells of the pMN domain respond to SHH with the expression of olig2, a transcription factor needed to specify motoneurons and oligodendrocytes, it seems that oligodendrocytes in vivo originate from a progenitor that is not glial-restricted *(9)*.

How to reconcile these different findings? It seems likely that in vitro data reveal the potential of the progenitor cells to develop along a certain lineage, but that this program is much more restricted in vivo, possibly through the action of factors such as platelet-derived growth factor, bone morphogenetic protein-4, thyroid hormone, or others *(13)*. Recently a second source of oligodendrocytes was identified. Oligodendrocytes in the telencephalon originate in the anterior entopeduncular area and migrate then tangentially into more dorsal regions, spreading through the mean ganglionic eminence, the lateral ganglionic eminence and eventually the cerebral cortex. This site of origin is found in chicken and mammals and is highly conserved during evolution *(16,18,19)*. Several lines of evidence suggest that SHH signaling is also necessary for the specification of oligodendrocyte progenitors in the telencephalon. First, there is a tight temporal and spatial correlation between SHH expression and oligodendrogenesis and second, loss or inhibition of SHH blocks the differentiation of telencephalic oligodendrocytes in vitro and in vivo *(20–22)*.

Which factors control the numbers of oligodendrocytes developing in a certain region? Answers to this question come from the work in the developing rat optic nerve. Here, it was observed that the proliferation of oligodendrocyte precursors crucially depends on the electrical activity in neighboring axons. If the electrical activity of retinal ganglion cells and their axons were silenced by an intraocular injection of tetrodotoxin, the number of oligodendrocyte precursors dropped by approx 80%. This effect could be circumvented by experimentally increasing the concentration of platelet-derived growth factor, which is present in the optic nerve and stimulates the proliferation of oligodendrocyte precursors in culture. These data also suggested that the axonal electric activity helps to control the number of oligodendrocytes developing in a defined region, and that this effect is mediated by the production and/or release of growth factors *(23,24)*. Hence, axons control the rate of proliferation and/or survival of developing oligodendrocyte precursors. However, they are not required for the migration and/or differentiation of these cells.

Most of our current knowledge about factors guiding the migration of oligodendrocyte precursors derives from a recent study in the newborn rat optic nerve. Using a special labeling approach for migrating cells, it was shown that cells with "features of the oligodendrocyte lineage" do not move randomly *(25)*. Instead, they migrate directly toward the optic nerve head and away from the chiasm. It was also observed that a molecule needed to guide axons, netrin-1, is produced in the lateral edges of the optic chiasma and functions as a repellent for these cells *(25)*. Based on these findings, a very attractive model was proposed describing the migration of oligodendrocyte precursors in the optic nerve *(26)*. According to this model, the migration is guided by a gradient of netrin-1 that emanates from the optic chiasm. Because netrin-1 acts as a repellent, the precursor cells migrate away from the chiasm toward the retina. At the nerve head, the junction between the retina and the optic nerve, the oligodendrocyte precursors could encounter again a netrin-1 producing environment *(27)*, which could effectively prevent precursor cells from entering the retina. This model is consistent with the spatial restriction of myelination seen in the optic nerve.

5. ASTROCYTES

As mentioned earlier, in vitro data suggest that astrocytes derive from a tripotential glia-restricted precursor, giving rise to two different astrocyte and one oligodendrocyte population *(13)*, or from a bipotential O2A/OPC *(14)*. However, just as seen with oligodendrocytes before, the situation is much more complex in vivo.

In the postnatal brain, astrocytes are formed from multipotent precursors in the subventricular zone. Other brain astrocytes derive from radial glia cells, once they have fulfilled their role as cellular substrates for the migration of neuronal progenitors *(28)*, and last, in the embryonic spinal cord, astrocytes originate from the p2 domain, a region found in the ventral half of the spinal cord ventricular zone. This region lies adjacent to the pMN domain, which gives rise first to motoneurons, and then to oligodendrocytes *(6)*. Hence, at least in the spinal cord, astrocytes and oligodendrocytes seem to develop independently from each other, in complementary domains of the ventricular zone *(29)*. The development of astrocytes is not induced by SHH signaling *(29)*. This is in marked contrast to the situation of spinal cord neurons and oligodendrocytes, which largely depend on the expression of SHH signaling. In addition, SHH seems to restrict the ventral extension of the astrocyte progenitor domain and to reduce astrocyte development *(30)*. Instead of relying on SHH, the formation of astrocytes seems to depend on BMP-4. Overexpression of BMP-4 results in a 40% increase in the density of astrocytes, accompanied by a decrease in the numbers of oligodendrocytes *(31)*.

Current knowledge about the migration of astrocyte progenitors derives from a recent study in the rat forebrain. Here it was shown that progenitors which develop into astrocytes are produced in the subventricular zone and migrate into the overlying white matter along a plane perpendicular to the rostrocaudal axis of the subventricular zone *(32)*. The migration patterns of these cells were associated with the processes of radial glial cells indicating that radial glial cells do not only provide the migratory scaffold for neuronal, but also for glial progenitor cells. Some of the astrocyte progenitors then remain in the white matter, whereas others continue their movement and end up in the cortex *(32)*.

Which factors guide the migration of astrocytes? Sugimoto and colleagues *(25)* described two different cell populations migrating within the newborn rat optic nerve. One cell population had small nuclei and expressed some markers of the oligodendrocyte lineage, whereas the other cell population had large nuclei and did not express these oligodendrocyte markers. This latter population of cells showed a directed movement away from the optic chiasm toward the nerve head, because it was repelled by semaphorin-3a, a molecule produced in the optic chiasm. It was then speculated that the cells with the large nuclei represent a subset of astrocyte precursors in the optic nerve and that semaphorin-3a forms a guidance cue gradient for these cells along the optic nerve *(26)*.

6. RADIAL GLIAL CELLS

Radial glial appear at the onset of neurogenesis and are the first cells, which can be clearly distinguished from the neuroepithelial stem cells *(33)*. They are mitotically active and have a typical bipolar morphology. The nuclei of these cells are located within the ventricular zone and undergo interkinetic nuclear movements with cell cycle progression *(34)*. The processes of these cells, which extend over long distances and reach the pial surface of the developing brain, serve as the migratory framework for neuronal cells. Astrotactin, neuregulin and β1-integrin are neuronal ligands for glia-guided migrations in the developing cerebellar cortex, and α3-integrin in the developing cerebral cortex *(35)*.

For a long time, radial glial cells were considered to be astroglial progenitors. This assumption was based on several observations. First, radial glial cells express proteins typical of astrocytes, such as glial fibrillary acidic protein (in primates) or the astrocyte-specific glutamate transporter and the brain lipid binding protein, respectively (in rodents and chicken) *(36)*. These molecules are expressed when radial glia cells differentiate from neuroepithelial cells, but are then down regulated when neurogenesis is completed. Second, with the exception of Bergmann glia cells in the cerebellum, the Müller glia in the retina, and the radial glia in the dentate gyrus of the hippocampus *(37)* found in the adult CNS, all other mammalian radial glial cells disappear or transform into multipolar astrocytes at the end of the neuronal generation and migration phase *(38,39)*.

Recent experiments, however, demonstrate that the situation is much more complex than previously anticipated. Radial glial cells are not a homogeneous cell type. The expression of glia markers by radial glial cells differs between brain regions and vertebrate species and these cells do not only express astrocytic markers, but also vimentin, nestin and the RC2 antigen, which are molecules characteristic for neural precursor cells *(36)*. The expression of transcription factors and the fate of radial glial cells depend on their position along the dorsoventral axis of the telencephalon:

- In the murine dorsal telencephalon, radial glial cells express the transcription factor Pax6 and will give rise to glutamatergic projection neurons.
- In the lateral ganglionic eminence, they express the transcription factor Gsh2 and will produce olfactory bulb interneurons and some striatal projection neurons I.
- In the medial ganglionic eminence, they contain the transcription factor olig2 and will form oligodendrocytes and subtypes of γ-aminobutyric acid ergic interneurons *(37)*.

Hence, radial glial cells contribute to different cell lineages in distinct regions of the developing CNS (to oligodendrocytes, neurons, and astrocytes). Current experimental evidence suggests that these fate differences are mediated by cell-autonomous differences *(37)*.

7. MICROGLIA CELLS

For a long time, the origin of microglia cells was a matter of debate. At the beginning of the 20th century, Pio del Rio-Hortega recognized microglia as separate glia cells of the CNS and described their intimate involvement in brain pathology and their origin from monocytic/mesodermal cells that enter the embryonic brain. Then, some in vitro studies suggested a neuroectodermal origin of microglia. Today, overwhelming experimental evidence indicates that microglia cells derive from infiltrating hematopoietic cells during the early development of the CNS *(40)*.

Several lines of evidence suggest that microglia cells are myeloid progenitors not yet committed to a defined macrophage or dendritic cells (DCs) phenotype *(41)*. The microglia cell pool in the CNS is rather stable. This points to a capacity for self-renewal, which is not a characteristic feature of terminally differentiated cells. In vitro microglia cells express receptors for the stem cell factors c-kit-ligand and flt-3-ligand, which are only found on myeloid progenitor cells, but not on fully mature myeloid DC or resident macrophages, and they respond to lineage growth factors such as macrophage colony-stimulating factor and granulocyte/monocyte colony-stimulating factor *(41)*. Moreover, macrophages and dendritic cells can be generated from mixed glial cultures in vitro *(41)* and are also found in the CNS in vivo (i.e., in rats with experimental autoimmune encephalitis) *(42)*.

Hence, microglia cells are not terminally differentiated along the myeloid lineage, and can be skewed toward an immature DC-like or macrophage-like phenotype by granulocyte/monocyte colony-stimulating factor and macrophage colony stimulating factor. This undifferentiated state of microglia cells is thought to contribute to the immunoprivileged state of the CNS parenchyma *(40)*.

8. CONCLUSION

In the past few years, an unexpected and new role has been attributed to glia cells. That of stem cells/progenitors in the adult and embryonic brain *(33)*. This assignment was based on the notion that cells expressing the glial fibrillary acidic protein, a marker frequently associated with astrocytes, can turn into neurons *(43–45)*. However, the expression of one single molecule does not necessarily imply the presence of other characteristic features of a certain cell type and is certainly not sufficient to clearly define a cell. Therefore, for the time being, the attribution of stem cell/progenitor features to glial cells should be taken with some grain of salt *(46,47)*.

REFERENCES

1. Temple S. The development of neural stem cells. Nature 2001;414:112–117.
2. Marti E, Bovalenta P. Sonic hedgehog in CNS development: one signal, multiple outputs. Trends Neurosci 2002;25:89–96.
3. Haubensak W, Attardo A, Denk W, Huttner WB. Neurons arise in the basal neuroepithelium of the early mammalian telencephalon: a major site of neurogenesis. Proc Natl Acad Sci USA 2004;101: 3196–3201.
4. Gleeson JG, Walsh CA. Neuronal migration disorder: from genetic diseases to developmental mechanisms. Trends Neurosci 2000;23:352–359.
5. Gilbert SF. Developmental Biology, 6th ed. Sunderland MA: Sinauer Associates, Inc. 2000.
6. Stolt CC, Lommes P, Sock E, Chaboissier M-C, Schedl A, Wegner M. The Sox9 transcription factor determines glial fate choice in the developing spinal cord. Genes Dev 2003;17:1677–1689.
7. Grandbarbe L, Bouissac J, Rand M, Hrabe de Angelis M, Artavanis-Tsakonas S, Mohier E. Delta-Notch signaling controls the generation of neurons/glia from neural stem cells in a stepwise process. Development 2003;130:1391–1402.
8. Park HC, Appel B. Delta-Notch signaling regulates oligodendrocyte specification. Development 2003;130:3747–3755.
9. Colognato H, ffrench-Constant C. Mechanisms of glial development. Curr Opin Neurobiol 2004;14:37–44.
10. Wang S, Sdrulla AD, diSibio G, et al. Notch receptor activation inhibits oligodendrocyte differentiation. Neuron 1998;21:63–75.
11. Sun Y, Nadal-Vicens M, Misono S, et al. Neurogenin promotes neurogenesis and inhibits glial differentiation by independent mechanisms. Cell 2001;104:365–376.
12. Lundkvist J, Lendahl U. Notch and the birth of glial cells. Trends Neurosci 2001;24:492–494.
13. Gregori N, Pröschel C, Noble M, Mayer-Pröschel M. The tripotential glial-restricted precursor (GRP) cell and glial development in the spinal cord: generation of bipotential oligodendrocyte-type-2 astrocyte progenitor cells and dorsal–ventral differences in GRP cell function. J Neurosci 2002;22: 248–256.
14. Raff MC, Miller RH, Noble M. A glial progenitor cell that develops in vitro into an astrocyte or an oligodendrocyte depending on the culture medium. Nature 1983;303:390–396.
15. Miller RH. Oligodendrocyte origins. Trends Neurosci 1996;19:92–96.
16. Perez Villegas EM, Olivier C, Spassky N, et al. Early specification of oligodendrocytes in the chick embryonic brain. Dev Biol 1999;216:98–113.
17. Ivanova A, Nakahira E, Kagawa T, et al. Evidence for a second wave of oligodendrogenesis in the postnatal cerebral cortex of the mouse. J Neurosci Res 2003;73:581–592.
18. Olivier C, Cobos I, Perez Villegas EM, et al. Monofocal origin of telencephalic oligodendrocytes in the anterior entopeduncular area of the chick embryo. Development 2001;128:1757–1769.
19. Qi Y, Stapp D, Qiu M. Origin and molecular specification of oligodendrocytes in the telencephalon. Trends Neurosci 2002;25:223–225.
20. Tekki-Kessaris N, Woodruff R, Hall AC, et al. Hedgehog-dependent oligodendrocyte lineage specification in the telencephalon. Development 2001;128:2545–2554.
21. Nery S, Wichterle H, Fishell DG. Sonic hedgehog contributes to oligodendrocyte specification in the mammalian forebrain. Development 2001;128:527–540.
22. Sussel L, Marin O, Kimura S, Rubenstein JL. Loss of Nkx2.1 homeobox gene function results in a ventral to dorsal molecular respecification within the basal telencephalon: evidence for a transformation of the pallidum into the striatum. Development 1999;126:3359–3370.

23. Barres BA, Raff MC. Proliferation of oligodendrocyte precursor cells depends on electrical activity in axons. Nature 1993;361:258–260.
24. Barres BA, Raff MC. Axonal control of oligodendrocyte development. J Cell Biol 1999; 147:1123–1128.
25. Sugimoto Y, Taniguchi M, Yagi T, Akagi Y, Nojyo Y, Tamamaki N. Guidance of glial precursor cell migration by secreted cues in the developing optic nerve. Development 2001;128:3321–3330.
26. Tsai H-H, Miller RH. Glial cell migration directed by axon guidance cues. Trends Neurosci 2002;25:173–175.
27. Deiner MS, Kennedy TE, Fazeli A, Serafini T, Tessier-Lavigne M, Sretavan DW. Netrin-1 and DCC mediate axon guidance locally at the optic disc: loss of function leads to optic nerve hypoplasia. Neuron 1997;19:575–589.
28. Culican SM, Baumrind NL, Yamamoto M, Pearlman AL. Cortical radial glia: identification in tissue culture and evidence for their transformation to astrocytes. J Neurosci 1990;10:684–692.
29. Pringle NP, Yu W-P, Howell M, Colvin JS, Ornitz DM, Richardson WD. Fgfr3 expression by astrocytes and their precursors: evidence that astrocytes and oligodendrocytes originate in distinct neuroepithelial domains. Development 2003;130:93–102.
30. Agius E, Soukkarieh C, Danesin C, et al. Converse control of oligodendrocyte and astrocyte lineage development by Sonic hedgehog in the chick spinal cord. Dev Biol 2004;270:308–321.
31. Gomes WA, Mehler MF, Kessler JA. Transgenic overexpression of BMP-4 increases astroglial and decreases oligodendroglial lineage commitment. Dev Biol 2003;255:164–177.
32. Suzuki SO, Goldman JE. Multiple cell populations in the early postnatal subventricular zone take distinct migratory pathways: a dynamic study of glial and neuronal progenitor migration. J Neurosci 2003;23:4240–4250.
33. Doetsch F. The glial identity of neural stem cells. Nat Neurosci 2003;6:1127–1134.
34. Misson JP, Edwards MA, Yamamoto M, Caviness VS. Mitotic cycling of radial glial cells of the fetal murine cerebral wall: a combined autoradiographic and immunohistochemical study. Brain Res 1988;466:183–190.
35. Adams NC, Tomoda T, Cooper M, Dietz G, Hatten ME. Mice that lack astrotactin have slowed neuronal migration. Development 2002;129:965–972.
36. Campbell K, Götz M. Radial glia: multi-purpose cells for vertebrate brain development. Trends Neurosci 2002;25:235–238.
37. Kriegstein AR, Götz M. Radial glia diversity: a matter of cell fate. Glia 2003;43:37–43.
38. Levitt P, Rakic P. Immunoperoxidase localization of glial fibrillary acidic protein in radial glial cells and astrocytes of the developing rhesus monkey brain. J Comp Neurol 1980;193:815–840.
39. Voigt T. Development of glial cells in the cerebral wall of ferrets: direct tracing of their transformation from radial glia into astrocytes. J Comp Neurol 1989;289:74–88.
40. Raivich G, Bohatschek M, Kloss CU. Werner A, Jones LL, Kreutzberg GW. Neuroglial activation repertoire in the injured brain: graded response, molecular mechanisms and cues to physiological function. Brain Res Brain Res Rev 1999;30:77–105.
41. Santambrogio L, Belyanskaya SL, Fischer FR, et al. Developmental plasticity of CNS microglia. Proc Natl Acad Sci USA 2001;98:6295–6300.
42. Fischer HG, Reichmann G. Brain dendritic cells and macrophages/microglia in central nervous system inflammation. J Immunol 2001;166:2717–2726.
43. Doetsch F, Caille I, Lim DA, Garcia-Verdugo JM, Alvarez-Buylla A. Subventricular zone astrocytes are neural stem cells in the adult mammalian brain. Cell 1999;97:703–716.
44. Imura T, Kornblum HI, Sofroniew MV. The predominant neural stem cell isolated from postnatal and adult forebrain but not early embryonic forebrain expresses GFAP. J Neurosci 2003;23:2824–2832.
45. Zhuo L, Theis M, Alvarez-Maya I, Brenner M, Willecke K, Messing A. hGFAP-cre transgenic mice for manipulation of glial and neuronal function in vivo. Genesis 2001;31:85–94.
46. Barres BA. What is a glial cell. Glia 2003;43:4,5.
47. Goldman S. Glia as neural progenitor cells. Trends Neurosci 2003;26:590–596.

4
Vasculogenesis and Angiogenesis

Gerald A. Grant, MD and Damir Janigro, PhD

SUMMARY

Vasculogenesis, angiogenesis, and arteriogenesis are terms used to describe the formation of blood vessels. Embryonic neovascularization by vasculogenesis refers to the formation of primitive blood vessels inside the embryo and its surrounding membranes and involves the *in situ* differentiation of mesoderm-derived angioblasts, which aggregate and form *de novo* blood vessels. Vascularization of the brain occurs primarily through angiogenesis. Angiogenesis involves the formation of new blood vessels via sprouting or splitting from pre-existing vessels and occurs both pre- and postnatally. Arteriogenesis refers to the enlargement of pre-existing collateral arterioles to form larger arteries. Specific angiogenic factor signaling systems choreograph each step of blood vessel formation. Vasculogenesis and angiogenesis are not exclusive processes but instead constitute complementary mechanisms for postnatal neovascularization.

Key Words: Angiogenesis; vasculogenesis; endothelial cells; central nervous system.

1. PRENATAL VASCULOGENESIS

Vasculogenesis gives rise to the heart and the first primitive vascular plexus inside the embryo and in its surrounding membranes, as the yolk sac circulation. The angioblasts are endothelial precursor cells (EPCs) without a lumen and are induced by fibroblast growth factors (FGFs) to differentiate from mesoderm and then organize to form primitive blood vessels. Early in embryogenesis, angioblasts migrate from the splanchnopleuric mesoderm into the head region to cover the developing brain. The differentiation of these angioblasts then depends on a paracrine signaling system in part owing to vascular endothelial growth factor (VEGF) and its high-affinity tyrosine kinase receptor (VEGFR-1).

Endothelial progenitor cells arise from migrating mesodermal cells (*see* Fig. 1). These progenitor cells have the capacity to proliferate, migrate, and differentiate into endothelial lineage cells but have not yet acquired characteristic mature endothelial cell markers. Hematopoietic stem cells (HSCs) and EPCs are thought to derive from a common precursor cell called the hemangioblast *(1)*. The cells destined to generate hematopoietic cells are situated in the center of blood islands in the yolk sac of the embryo. The EPCs, in contrast, are located at the periphery of the blood islands and fuse together to form the extraembryonic vascular network, which grows toward the embryo.

Key molecular players, which determine the fate of the hemangioblast, have not been fully elucidated. However, several key factors have been identified to play a role in the molecular regulation of this event *(2)*. Studies in quail/chick chimeras showed that FGF-2 mediates the induction of EPCs from the mesoderm *(3)*. These embryonic EPCs express Flk-1, the type-2 receptor for VEGF, and respond to VEGF for proliferation and migration. The Flk-1-expressing mesodermal cell has also been defined as an embryonic common vascular progenitor that

From: *The Cell Cycle in the Central Nervous System*
Edited by: D. Janigro © Humana Press Inc., Totowa, NJ

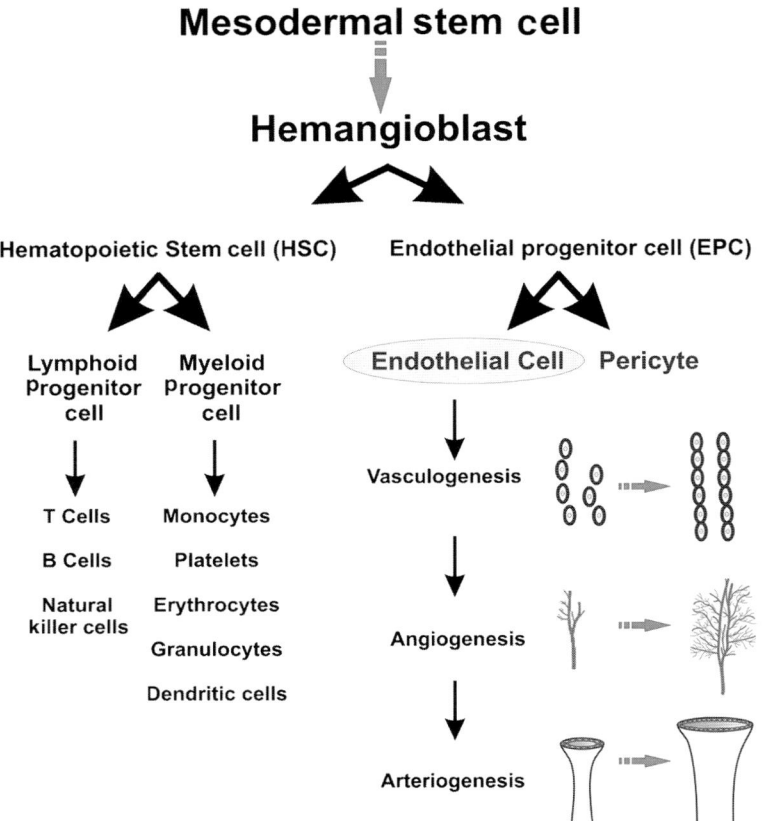

Fig. 1. Schematic representation of vascular development. (Please *see* companion CD for color version of this figure.)

differentiates into endothelial and smooth muscle cells in response to VEGF and platelet-derived growth factor (PDGF), respectively *(4)*. A second tyrosine kinase receptor for VEGFR-1, also mediates the biological effects of VEGF along with Flk-2 (VEGFR-2). Along with VEGF, and Flk-1, Flt-1 is highly expressed during early embryonic development and active vasculogenesis *(5)*. After circulation is initiated, vascular smooth muscle cells and pericytes are recruited to the endothelial tubes.

Both the neural tube and somites are essential for providing key molecular signals for vascular development. The first intra-embryonic angioblasts appear at the single somite stage and interconnection with their extra-embryonic counterparts is established at the two-somite stage *(6,7)*. Neural crest-derived cells provide smooth muscle cells of the blood vessel wall. Somite-derived angioblasts migrate through the embryo and differentiate into endothelial cells, which line the interior of blood vessels. Some endothelial cells also stem from the yolk sac. Therefore, the origin and assembly of embryonic blood vessels not only involve multiple sources for precursor cells but also are influenced by various combinations of proliferation, migration, differentiation, and of cell–cell and cell–matrix interactions.

A primary vascular plexus surrounds the neural tube at day 3 of embryonic chick development. This vascular plexus is derived from migratory angioblasts that have invaded the head at an early stage. From the perineural vascular plexus, capillary sprouts invade the neuroectoderm at day 3. The endothelial cells first locally degrade the perineural basement membrane and then migrate deeper into the neuroectoderm and branch in the subependymal layer of the brain. These steps are highly reproducible and are similar to tumor angiogenesis, a process that recapitulates embryonic brain angiogenesis. Early signs of remodeling in the first vascular plexus become apparent and are defined as rearrangement of vascular segments to give rise to larger vessels. Further expansion of the primary and secondary vascular plexus occurs by the process of angiogenesis. Vascularization of organs derived from mesoderm and endoderm, such as the lung and spleen, occurs primarily by vasculogenesis. In contrast, vascularization of the brain is derived predominantly from angiogenesis.

2. PRENATAL ANGIOGENESIS

Organs, such as the brain, derived from the ectoderm–mesoderm are vascularized by angiogenesis *(7)*. Angiogenesis is characterized by a series of steps including degradation of the basement membrane, endothelial cell proliferation, invasion of the surrounding stroma, and structural reorganization into a novel functional vascular network through the recruitment of perivascular supporting cells, such as pericytes. This complex process involves the involvement of multiple regulatory factors such as growth factors, adhesion molecules, and extracellular matrix proteolytic degrading enzymes matrix metalloproteinases (MMPs) that degrade the basement membrane. Angiogenesis is the process of remodeling and expansion of pre-existing vessels (primary capillary plexus) during vasculogenesis. The process of vasculogenesis and angiogenesis not only includes the differentiation of cells but the morphogenesis of an intricate multicellular arrangement of cells. The cellular basis of the morphogenesis is differential cell affinity. Cell adhesion specificity occurs in response to paracrine signaling (growth factors) during vessel maintenance.

The most significant steps in angiogenesis include the following:

1. New capillaries originating from small venules.
2. Local degradation of basement membrane (type IV collagen and laminin) on the side of the venule closest to the angiogenic stimulus (collagenase, plasminogen activators, etc.).
3. Migration of endothelial cells toward the angiogenic stimulus.
4. Alignment of endothelial cells.
5. Formation of a lumen.
6. Establishment of flow.
7. Alignment of pericytes and smooth muscle cells along the endothelial cells.
8. Formation of a new basement membrane.

Growth and remodeling rely on similar branching processes, which are in part directed by flow conditions. Branching occurs by splitting of the vessel lumen by intussusceptive microvascular growth and by capillary sprouting through endothelial cell migration from preexisting vessels, proliferation, and tube formation. Remodeling also includes pruning, which occurs by endothelial cell division. Recent evidence has indicated that endothelial cells express specific molecular markers, ephrin-B2 and Eph-B4, which identify arterial and venous endothelial cell fate, respectively *(8)*. Remodeling is then necessary to optimize the functional adaptation of the newly formed network and implies the addition of new vessels and the deletion of previous ones.

3. TELENCEPHALIC ANGIOGENESIS

Organs such as the brain and neuroectoderm that are derived from ectoderm–mesoderm are vascularized by angiogenesis *(9)*. The brain grows to a certain size before any blood supply is present in the leptomeninges or in the choroid plexus. The microvasculature in the telencephalic wall develops with the development of the cerebral wall, presumably responding to the need of local regional constituents of the cerebral wall as they appear, migrate, and mature. Therefore, angiogenesis is driven by the metabolic demands of the expanding neuroectoderm. Angioblastic islands coalesce to form an epiparenchymal leptomeningeal plexus in the primitive meninges by embryonic day 4.5 in the chick and E 11.5 in the rat *(10)*. These channels spread from the base of the brain to the convexity with later maturation of arteries and veins. Vascular branching is accompanied by vessel maturation, with recruitment of pericytes and formation of contacts between vessels and astrocytic processes. Endothelial cell invasion of the brain proceeds from rhombencephalic to more rostral levels in contrast to the spinal cord which is characterized by rostrocaudal progression *(11)*. At the end of the second month of gestation, endothelial channels sprout from both leptomeningeal arteries and veins, enter the forebrain at right angles, and reach the subventricular germinal tissue at approx 12 wk gestation *(12,13)*. These endothelial channels, in turn, give rise to the continuous capillary network of the human forebrain. At this stage, the complex vascular unit is not only defined by the endothelial cell but also the astrocytes which envelop the endothelial cells along with pericytes, smooth muscle cells, macrophages, and microglia. The striatal channels are mature before the extrastriatal channels. The extrastriatal channels are governed by the large increase in cortical surface area and the increase in distance from the cortex to the ventricle. Capillaries branch from channels of all sizes, creating a continuous bed. Thereafter, endothelial cell proliferation is dramatically downregulated. In the mouse neuroectoderm, for example, endothelial cell turnover is approx 3 yr. The resting endothelium of a normal adult is in the quiescent state with only 0.01% of endothelial cells undergoing cell division at any given time *(14)*.

4. MOLECULAR REGULATION OF ANGIOGENESIS

A number of growth factors have been identified which play a role in the regulation of angiogenesis. These include FGF-1 and -2, VEGF-A, -B, and -C, transforming growth factor (TGF)-α, interleukin-8, angiopoietins (Angs), PDGF, MMPs, integrins, as well as the Eph-B/ephrin-B system of tyrosine kinase receptors and their ligands.

4.1. Vascular Endothelial Growth Factor

VEGF is an early positive regulator of vasculogenesis, acting as a mitogen with a primary specificity for endothelial cells. VEGF promotes migration, proliferation, and tube formation in endothelial cells. The VEGF family of 34–45 kDa dimeric glycosylated proteins consists of five known isoforms produced by alternative splicing from a single gene mapped to chromosome 6p21.3 *(15)*. The actions of VEGF are mediated through two receptors, VEGFR-2 and VEGFR-1 *(16,17)*. All five isoforms of VEGF bind Flt-1 and Flk-2, inducing dimerization and triggering kinase activation and cytoplasmic signal transduction cascades related to vasculogenesis and angiogenesis control. Whereas the Flk-1 receptor is dominant during early vasculogenesis, Flt-1 is prominent during remodeling of the primary vascular plexus and subsequent angiogenesis and endothelial tube formation. Flk-1 is integral to the generation of hemangioblasts and endothelial cells. VEGF is a relatively specific mitogen and chemotactic factor for endothelial cells. VEGF is downregulated in the adult normal brain but is upregulated in hypoxic/ischemic conditions or in tumors (i.e., gliomas) *(18,19)*. The effects are dose dependent, because overexpression or exogenous administration increases blood vessel density *(20)*. VEGF stimulates endothelial nitric oxide synthase resulting in nitric oxide generation and activation of the angiogenic cascade, which induces endothelial cell production of proteases necessary for degradation of basement membrane during angiogenesis, promotes monocyte and neutrophil migration, as well as increases microvascular permeability. Hypoxia has been shown to induce VEGF-A transcription as well as promote stabilization of VEGF-A mRNA *(21–24)*. Factors that modulate angiogenesis indirectly include PDGF and TGF-β, both of which can also potentiate VEGF-A expression *(25,26)*.

VEGF receptors which regulates angiogenesis are located at the abluminal, basal surface of endothelial cells. VEGF induces proliferation of the endothelial cell toward the extracellular matrix and stimulates endothelial cells to expose tissue factor. Tissue factor initiates the plasma coagulation protease cascade through the extrinsic pathway and generates thrombin. Thrombin forms an extracellular fibrin barrier from the VEGF-dependent fibrinogen extravasation, activates progelatinase-A (pro-MMP2) which destroys the basal membrane and allows proliferation of endothelial cells, and finally induces endothelial cell proliferation which potentiates the VEGF effect *(27)*. VEGF contributes to blood–brain barrier (BBB) breakdown by exacerbating degradation of the junctional proteins occludin and zonula occludens-I *(28)*. Increased microvascular permeability and protein extravasation are additional essential steps in angiogenesis. The increased vessel permeability is the result of direct effects of VEGF on endothelial cells, through mobilization of endothelial cytosolic calcium and increases in fenestrae and pinocytic vesicles. The increased vessel permeability is thought largely responsible for the increased vasogenic edema around brain tumors.

4.2. Angiopoietins

Despite their crucial role in the formation of blood vessels, members of the VEGF family act in concert with other endothelial growth factors such as the Angs. Although four members of the Ang family have been cloned, Ang-1 and Ang-2 are the best characterized. Angs constitute a family of endothelial growth factors that are ligands for the tyrosine kinase receptor family Tie (Tie-1 and Tie-2). Tie receptors form another family of receptor tyrosine kinases and like the VEGFRs are selectively expressed on endothelial cells. The Tie-1 receptor tyrosine kinase has been implicated with a survival function for endothelial cells and a stabilization effect on the vessel wall. Ang-1 is a ligand that specifically activates the Tie-2 receptor, whereas Ang-2 shares 60% sequence identity with Ang-1 and binds Tie-2 with similar affinity but acts as an Ang-1 antagonist and induces autophosphorylation of Tie-2 *(29)*. Ang-1 is also expressed on cells adjacent to endothelial cells, suggesting a paracrine regulation. Although Ang-1 is chemotactic for endothelial cells it does not cause endothelial cell proliferation or tube formation in vitro. The Ang/Tie system is thought to be activated after VEGF and cooperates with VEGF-A to promote endothelial cell migration, proliferation, and tube formation. Unlike VEGF, Ang-1 and Ang-2 and the receptor Tie-2 appear to act in later stages of vascular development, during vascular remodeling and maturation, probably by playing a role in the interaction of endothelial cells with other vascular elements, such as smooth muscle cells and pericytes. VEGF is often coexpressed with Ang-2 at the tumor margin in which angiogenesis is dominant and Ang-2 facilitates endothelial cell proliferation and migration, and prevents the recruitment of smooth muscle cells. Ang-2 also causes regression of newly formed vessels through endothelial cell apoptosis as an antagonist to Ang-1 *(30)* and provides a key destabilizing signal to allow neovascularization to occur *(31)*. In contrast, Ang-1 appears to promote remodeling and stabilization of VEGF-A-induced vessels by the recruitment of smooth muscle cells and pericytes to the developing vascular network and therefore functions as an agonist during vascular development *(32)*. Tie receptors in combination with Ang-1 are critical for development of vascular polarity during angiogenesis *(33)* and targeted disruption results in mice lacking endothelial cell structural integrity *(34)*. Hypoxia and VEGF have been reported to increase Ang-2 expression in endothelial cells in vitro and coincide with endothelial proliferation in vivo *(18,35)*. Ang-2 expression in astrocytes, a component of the BBB, may also contribute to this process.

4.3. Transforming Growth Factor

TGF-β dimers are important to signal mesenchymal cells to differentiate into pericytes and smooth muscle cells, whereas inhibiting endothelial cell proliferation. However, TGF-β at high concentrations has been observed to stimulate endothelial cell growth. TGF-β seems to contribute to angiogenesis by stabilizing newly formed capillary sprouts. TGF-β is also implicated in brain-tumor-associated angiogenesis and upregulated by hypoxia, ischemia, and traumatic brain injury *(36)*.

4.4. Fibroblast Growth Factor

FGFs are involved in a variety of cellular processes and include FGF acidic (FGFa), FGF basic (FGFb), and 21 related FGFs. FGF-2 works synergistically with PDGF-B to promote angiogenesis and regulate extracellular matrix molecules to form new capillary cord structures. FGF-2 initiates vasculogenesis with VEGF stimulation and is detected in the basal lamina of blood capillaries, primarily at sites of vessel branching and in the endothelium of the capillaries of some tumors *(37)*. FGFa and FGFb are potent stimulators of endothelial cell migration, proliferation, sprouting, and tube formation. The FGFs have prominent roles in cerebral ischemia, arteriovenous malformations (AVMs), and glioma-associated angiogenesis *(20)*.

4.5. Platelet-Derived Growth Factor

PDGF, an endothelial cell mitogen, is also a major mediator of angiogenic activity. PDGF promotes angiogenesis, mediates proliferation and migration of vascular smooth muscle cells and pericytes, and enhances the expression of VEGF.

4.6. Ephrins

Another receptor tyrosine kinase family, the Eph-B receptors and their ligands, the ephrins, have been shown to promote embryonic angiogenesis. The Eph receptor tyrosine kinases regulate neural crest migration and axon guidance in the developing brain and are involved in vascular development during embryogenesis and angiogenesis in adults. The ephrins are membrane attached and their binding to the receptor requires cell–cell contacts. Ephrin-B2 in particular mediates angiogenic sprouting into the brain during embryogenesis *(38)*.

4.7. Matrix Metalloproteinases

Endothelial cells use several molecular mechanisms to break down components of the extracellular matrix. Of these, is the expression of MMPs, which include the collagenases, stromelysins, and gelatinases, and serine proteases (i.e., plasminogen activator and its receptor). The activities of these zinc-dependent proteases facilitate new vessel formation at the tumor perimeter by breaking down the extracellular matrix and allowing new vessel sprouting.

4.8. Integrins

Proliferating vascular cells require adhesion molecules to establish contact with the extracellular matrix and to generate the traction required for changes in morphology. Four classes of transmembrane adhesion molecules have been identified in vascular cells: (1) integrins, (2) selectins (E-selectin), (3) cadherins (VE-cadherin), and (4) immunoglobulins. The activity of the integrins, $\alpha_V\beta_3$ and $\alpha_V\beta_5$ in particular, are upregulated during tumor neoangiogenesis, repair, and retinal neovascularization *(39)*.

5. POSTNATAL ANGIOGENESIS AND VASCULOGENESIS

When does blood vessel development end? Never. However, in contrast to the developing brain, only 0.3% of endothelial cells in the adult rat brain demonstrate ^{3}H incorporation *(40)*. Endothelial proliferation processes after childhood are restricted to a few physiological processes such as the menstrual cycle in females and wound-healing *(41)*. In the normal adult central nervous system, inhibitors of angiogenesis predominate and vascular quiescence is maintained. The complex process of adult angiogenesis starts with the local degradation of the basement membrane surrounding the endothelial cells *(10,42)*. Postnatal neovascularization recapitulates embryogenesis and occurs through the following:

1. Capillary sprouting of resident endothelial cells (angiogenesis).
2. Proliferation of pre-existing arteriolar connections (arteriogenesis).
3. *De novo* vascularization from endothelial cell precursors (vasculogenesis), which is also operative in the adult *(43,44)*.

Self-renewing HSCs in the bone marrow have full hemangioblast activity in the adult state and can clonally differentiate into all hematopoietic lineages including endothelial cells and contribute to neovascularization *(45)*. The identification of putative HSCs in peripheral blood and bone marrow has constituted inferential evidence for HSCs in adult tissues *(46–48)*. In vitro, these cells differentiate into endothelial lineage cells. Vascularization does, in fact, make a significant contribution to postnatal neovascularization. Bone marrow transplant experiments have demonstrated the incorporation of bone marrow-derived EPCs isolated from peripheral blood CD34, Flk-1, or AC133-antigen-positive cells into new foci of physiological and pathological neovascularization *(2)*. VEGF is clearly the most critical factor for vasculogenesis and angiogenesis and the mobilization of EPCs from bone marrow. Human EPCs have also been isolated from peripheral blood of adults, expanded in vitro, and committed into an endothelial lineage *(49)*. The role of EPCs in tumor angiogenesis has been demonstrated by several groups. Davidoff et al. showed that bone marrow-derived EPCs contribute to tumor neovasculature and that bone marrow cells transduced with an antiangiogenic gene can restrict tumor growth in mice *(50)*. Lyden et al. also recently demonstrated the critical role of bone marrow-derived EPCs in tumor neovascularization *(51)*. EPCs isolated from adult species have characteristics similar to those of embryonic angioblasts and have the capacity to proliferate, migrate, and differentiate into endothelial lineage cells but have not yet acquired mature endothelial markers. EPCs are mobilized from bone marrow into the circulation and then home to the site of neovascularization in response to physiological and pathological stimuli, thereby contributing to postnatal neovascularization.

Paracrine control of neural regeneration and angiogenesis has also been demonstrated. Therefore, angiogenic growth factors released by EPCs modulate nerve regeneration, whereas neurotrophins, most often associated with the promotion of neuronal growth and survival, stimulate neovascularization *(52)*.

Many of the important regulators of embryonic angiogenesis play very similar roles in angiogenesis in the adult under pathological conditions. Embryonic mechanisms become reactivated in neoplasia or wound-healing, chronic inflammation, restenosis, and atherosclerosis. In experimental models of stroke, upregulation of VEGF, Ang-1, and Ang-2 coincides with endothelial cell proliferation *(35,53)*. Tumors often "switch" to an angiogenic phenotype by starting to release angiogenic growth factors to simulate normal angiogenesis, that is, by endothelial sprouting. The term angiogenic switch is somewhat misleading because it implied that angiogenesis is either toggled on or off, whereas in fact, neovascularization is a continuum and becomes increasingly prevalent during tumor progression. There is a net dynamic balance of promoters and inhibitors that regulate blood vessel growth in a tightly regulated fashion. However, there is a distinct difference in wound-healing and tumor angiogenesis. Hypoxia-induced angiogenesis subsequently ceases in wound-healing but in tumors the pathological structure of the vascular network perpetuates the hypoxia, which drives the angiogenesis. From an angiogenic point of view, tumor angiogenesis is an unlimited process with unbalanced vessel growth and is characterized by a "wound that does not heal" *(5)*. In contrast, during normal vessel development and stabilization, cell replication stops, the basement membrane is reformed, and the new vessel is invested with pericytes, which inhibit endothelial cell proliferation *(54)*.

High-grade gliomas are characterized by extensive microvascular proliferation and a higher degree of vascularity than low-grade gliomas and normal brain. More significantly, the degree of neovascularization in high-grade gliomas are histological indicators of the degree of malignancy and correlate well with tumor grade and prognosis of patients. It is postulated that the progression of a low-grade astrocytoma to a highly vascularized glioblastoma includes an "angiogenic switch" *(55)*. Local hypoxia adjacent to necrotic areas within the tumor or decreased glucose levels within the tumor trigger VEGF expression and induce neovascularization *(56)*. VEGF-A released by tumors promotes the recruitment of EPCs and hematopoietic cells from the basement membrane and induces these cells to migrate to the tumor and become incorporated into the developing neovasculature *(51)*. Soluble VEGF expressed by glioma cells functions as a ligand for the VEGFRs, Flt-1 and Flk-1, which belong to the family of endothelial-specific receptor tyrosine kinases. VEGF is the primary mediator of glioma angiogenesis and is

Table 1
Positive and Negative Regulators of Angiogenesis

Positive regulators of angiogenesis	Negative regulators of angiogenesis
VEGF-A, -B, -C, and -D	Platelet factor 4
Ang-1/Tie-2	Ang-2
FGFa and FGFb	Cytokines (interferon-α, interleukin-12)
PDGF	Thrombospondin-1
TGF-α and -β	TIMP-1, TIMP-2, TIMP-3
Integrins, selectins, and MMPs	Nitric oxide
EGF	Angiostatin
Immunoglobulin superfamily	Endostatin (e.g., PECAM-1)
EGF	p53 and VHL
Interleukin-8	Plasminogen activator/inhibitor

also necessary for tumor vessel maintenance. VEGF has also been shown to act as a vascular permeability factor and contributes not only to the microvascular proliferation but also the formation of peritumoral edema associated with high-grade gliomas *(57)*. Furthermore, dexamethasone, a synthetic glucocorticoid, has been shown to decrease peritumoral brain edema by a reduction in VEGF expression in tumor cells *(55)*. Downregulation of VEGF expression, however, is attenuated in hypoxic tumor cells, suggesting that the in vivo efficacy of dexamethasone might be limited in glioblastoma because of the hypoxic microenvironment *(58)*. An increase in VEGF expression and a decrease in Ang-1 expression may function in a reciprocal manner to open the BBB and cause vasogenic edema. Ang-1 expression remains constant during progression from low to high-grade astrocytomas *(59)*. Ang-2 and Tie-2; however, become expressed only in tumor and vascular endothelial cells *(60)*. Other neuroepithelial tumors such as pilocytic astrocytomas, oligodendrogliomas, mixed oligoastrocytomas, medulloblastomas, meningiomas, and hemangioblastomas have also been shown to upregulate VEGF mRNA.

The relationship of common cerebrovascular malformations, such as AVMs and cavernous malformations (CCMs) to basic mechanisms of vasculogenesis and angiogenesis are also being explored *(6)*. AVMs are dynamic lesions characterized by increased endothelial cell turnover and express elevated levels of VEGF, FGFb, TFG-α, TGF-β, and Ang-2. AVMs develop *in utero* and are thought to be related to a defect in the ephrin signaling system, which is involved in the differentiation of arteries and veins and the formation of capillary beds *(20)*. Although CCMs do not demonstrate high-flow conditions that lead to hemodynamic stress, they also express angiogenic factors. In hereditary hemorrhagic telangiectasia characterized by the presence of AVMs, there is an autosomal dominant pattern of inheritance similar to CCMs, which suggests that underlying genes might regulate critical aspects of vascular morphogenesis *(61)*.

6. INHIBITORS OF ANGIOGENESIS

The precise mechanism by which angiogenesis is suppressed in the central nervous system is not clear although offers potential therapeutic targets at multiple levels *(62,63)*. The regulation of angiogenesis is governed by a balance of positive and negative regulators (*see* Table 1). However, a number of endogenous molecules with antiangiogenic properties have been identified including angiostatin, a cleavage product of plasminogen, and endostatin, a carboxy-terminal fragment of collagen XVIII. Exogenous angiogenic inhibitors have also been identified and act at distinct steps in the angiogenic response. After the initial formation of malignancy, the continued growth of a glioma is critically dependent on its angiogenic potential to sustain metabolic needs of tumor cells. Because angiogenesis is virtually absent in normal adults, therapies aimed at specifically interrupting angiogenesis within tumors should be well tolerated. Therefore, the detection of potent angiogenic inhibitors in tumors enhances the chances of imbalance between promoters and inhibitors of angiogenesis. The problem of drug resistance is limited because endothelial cells do not have as much genetic heterogeneity or instability characteristic of tumor cells. Furthermore, endothelial cells lie in direct contact with the blood and therefore, agents used to inhibit their proliferation or signaling can reach their targets without being affected by the BBB unlike chemotherapeutic agents *(63)*. A number of clinical trials have been initiated using anti-angiogeneic therapy in patients with brain tumors with mixed results thus far *(62,63)*, because the inhibition of individual modulators of angiogenesis may not yield the expected impact on prognosis *(64)*. Furthermore, unlike the cytotoxic agents, angiogenesis inhibitors would be expected to have a cytostatic effect may more impact progression free survival instead of tumor size control. Several studies have looked into the in vitro and in vivo effects of targeting VEGF or its receptors on angiogenesis *(63)*. Other strategies have been employed to use anti-VEGF-A antibodies *(65)* adenoviral vectors for gene delivery for expression of angiostatin, MMP inhibitors (tissue inhibitors of metalloproteinase 1 and 2) and protein kinase inhibitors. Thalidomide, which is known to inhibit endothelial cell proliferation, has also been studied in a phase II trial in patients with recurrent glioma with a limited response in a trial that combined radiotherapy with escalating doses of thalidomide *(66)*.

7. CONCLUSIONS

Angiogenesis is a complex process regulated by well-coordinated steps including production and release of angiogenic factors, proteolytic degradation of extracellular matrix to allow formation of capillary sprouts and the directional migration of microvascular cells. Mounting evidence demonstrates the critical role for continued angiogenesis and even vasculogenesis in the adult state to recapitulate embryonic processes to maintain homeostasis.

REFERENCES

1. Flamme IRW. Induction of vasculogenesis and hematopoiesis in vitro. Development 1992;116: 435–439.
2. Asahara TAK. Endothelial progenitor cells for postnatal vasculogenesis. Am J Physiol Cell Physiol 2004;287:C575–C579.
3. Poole TFEB, Cox CM. The role of FGF and VEGF in angioblast induction and migration during vascular development. Dev Dyn 2001;220:1–17.
4. Yamashita JIH, Hirashima M, Ogawa M, et al. Flk-1 positive cells derived from embryonic stem cells serve as vascular progenitors. Nature 2000;408:92–96.
5. Patan S. Vasculogenesis and angiogenesis as mechanisms of vascular network formation growth and remodeling. J Neuro-Oncol 2000;50:1–15.
6. Gault JS, Sarin H, Nabil A, Shenkar R, Awad I. Pathobiology of Human Cerebrovascular Malformations: Basic Mechanisms and Clinical Relevance. Neurosurgery 2004;55(1):1–20.
7. Yamada S, ed. Arteriovenous Malformations in Functional Areas of the Brain. Armonk, NY: Futura Publishing, 1999.
8. Wang HC, Chen ZF, Anderson DJ. Molecular distinction and angiogenic interaction between embryonic arteries and veins revealed by ephrin-B2 and its receptor Eph-B4. Cell 1998;93:741–753.
9. Pardanaud LY, Yassine F, Dieterlen-Lievre F. Relationship between vasculogenesis, angiogenesis, and haematopoiesis during avian ontogeny. Development 1989;105:473–485.
10. D'Angelo MA, Afanasieva T, Aguzzi A. Angiogenesis in transgenic models of multistep carcinogenesis. J Neuro-Oncol 2000;50:89–98.
11. Kurz H. Physiology of angiogenesis. J Neuro-Oncol 2000;50:17–35.
12. Marin-Padilla M. Early vascularization of the embryonic cerebral cortex: Colgi and electron microscopic studies. J Comp Neurol 1985;241:237–249.
13. Strong L. The early embryonic pattern of internal vascularization of the mammalian cerebral cortex. J Comp Neurol 1964;123:121–138.
14. Webb CVWG. Genes that regulate metastasis and angiogenesis. J Neuro-Oncol 2000;50:71–87.
15. Ferrara N. Vascular endothelial growth factor and the regulation of vascular angiogenesis. Recent Prog Horm Res 2000;55:15–36.

16. Ferrara N. Role of vascular endothelial growth factor in regulation of physiological angiogenesis. Am J Physiol Cell Physiol 2001;280:C1358–C1366.
17. Yancopoulus GS, Davis S, Gale NW, Rudge JS. Vascular specific growth factors and blood vessel formation. Nature 2000;407:242–248.
18. Lopes M. Angiogenesis in brain tumors. Microsc Res Tech 2003;60:225–230.
19. Damert AM, Machein M, Breier G, et al. Upregulation of vascular endothelial growth factor (VEGF) in the vasculature of oligodendrogliomas. Neuropathol Appl Neurobiol 1997;24:29–35.
20. Harrigan M. Angiogenic factors in the central nervous system. Neurosurgery 2003;53(3):1–23.
21. Jansen M, de Witt Hamer PC, Witmer AN, Troost D, van Noorden CJ. Current perspectives on antiangiogenesis strategies in the treatment of malignant gliomas. Brain Res Rev 2004;45:143–163.
22. Damert AM, Machein M, Breier G, Fujita MQ, Hanahan D, Risau W, Plate KH. Upregulation of vascular endothelial growth factor expression in a rat glioma is conferred by two distinct hypoxia-driven mechanisms. Cancer Res 1997;57:3860–3864.
23. Ikeda E, Achen MG, Brier G, Risau W. Hypoxia-induced transcriptional activation and increased mRNA stability of vascular endothelial growth factor in C6 glioma cells. J Biol Chem 1995;270:19,761–19,766.
24. Plate KB, Brier G, Welch H, Mennel H, Risau W. Vascular endothelial growth factor and glioma angiogenesis: Coordinate induction of VEGF receptors, distribution of VEGF protein and possible in vivo regulatory mechanisms. Int J Cancer 1994;59:520–529.
25. Wang DH, Su Huang HJ, Kazlauskas A, Cavence WK. Induction of vascular endothelial growth factor expression in endothelial cells by platelet-derived growth factor through the activation of phyosphatidylinositol 3-kinase. Cancer Res 1999;59:11,464–11,472.
26. Sanchea-Elsner T, Botella LM, Velasco B, Corbi A, Attisano L, Bernabeu C. Synergistic cooperation between hypoxia and transforming growth factor-beta pathways on human endothelial growth factor gene expression. J Biol Chem 2001;276:38,527–38,535.
27. Chiarugi V. Molecular polarity in endothelial cells and tumor-induced angiogenesis. Oncol Res 2000;12:1–4.
28. Antonetti DA, Barber AJ, Hollinger LA, Wolpert EB, Gardner TW. Vascular endothelial growth factor induces rapid phosphorylation of tight junction proteins occludin and zonula occludens. J Biol Chem 1999;274:23,463–23,467.
29. Jones N, Iljin K, Dumont DJ, Alitalo K. Tie receptors: New modulators of angiogenic and lymphangiogenic responses. Nat Rev Mol Cell Biol 2001;2:257–267.
30. Maisonpierre PS, Suri C, Jones PF, et al. Angiopoietin-2, a natural antagonist for Tie2 that disrupts in vivo angiogenesis. Science 1997;277:55–60.
31. Lobov IB, Brooks PC, Lang RA. Angiopoietin-2 displays VEGF-dependent modulation of capillary structure and endothelial cell survival in vivo. Proc Natl Acad Sci USA 2002;99(17):11,205–11,210.
32. Visconti RR, Richardson CD, Sato TN. Orchestration of angiogenesis and arteriovenous contribution by angiopoietins and vascular endothelial growth factor. Proc Natl Acad Sci 2002;99:8219–8224.
33. Loughna S, Sato TN. A combinatorial role of angiopoietin-1 and orphan receptor TIE1 pathways in establishing vascular polarity during angiogenesis. Mol Cell 2001;7:233–239.
34. Sato TN, Tozawa Y, Deutsch U. Distinct roles of the receptor tyrosine kinase Tie-1 and Tie-2 in blood vessel formation. Nature 1995;376:70–74.
35. Beck H, Acker T, Wiessner C, Allegrini P, Plate K. Expression of angiopoietin-1, angiopoietin-2, and tie receptors after middle cerebral artery occlusion in the rat. Am J Pathol 2000;157:1473–1483.
36. Breier G, Blum S, Peli J. Transforming growth factor-beta and Ras regulate the VEGF/VEGF-receptor system during tumor angiogenesis. Int J Cancer 1976;97:142–148.
37. Ingber D. Extracellular matrix and cell shape: Potential control points for inhibition of angiogenesis. J Cell Biochem 1991;47:236–241.
38. Wang HU, Chen ZF, Anderson DJ. Molecular distinction and angiogenic interaction betwteen embryonic arteries and veins revealed by ephrin-B2 and its receptor Eph-B4. Cell 1998;93:741–753.
39. Friedlander MB, Brooks PC, Shaffer RW, Kincaid CM, Varner JA, Cheresh DA. Definition of two angiogenic pathways by distinct alpha V integrins. Scin 1995;270:1500–1502.
40. Robertson PL, Du Bois M, Bowman PD, Goldstein GW. An in vivo and in vitro study. Brain Res 1985;355:219–223.
41. Hobson B, Denekamp J. Endothelial proliferation in tumors and normal tissues: continuous labeling studies. Br J Cancer 1984;49:405–413.
42. D' Angelo MG, Afanasieva T, Aguzzi A. Angiogenesis in transgenic models of multistep carcinogenesis. J Neuro-Oncol 2000;50:89–98.

43. Takahashi T, Kalka C, Masuda H, et al. Ischemia and cytokine-induced mobilization of bone marrow-derived endothelial progenitor cells for neovascularization. Nat Med 1999;5:434–438.
44. Carmeliet P. Mechanisms of angiogenesis and arteriogenesis. Nat Med 2000;6:389–395.
45. Grant MB, May WS, Caballero S, et al. Adult hematopoietic stem cells provide financial hemangioblast activity during retinal neovascularization. Nat Med 2002;8:607–612.
46. Brugger W, Heimfeld S, Berenson RJ, Mertelsmann R, Kanz L. Reconstitution of hematopoiesis after high-dose chemotherapy by autologous progenitor cells generated ex vivo. N Engl J Med 1995;333:283–287.
47. Crosby JR, Kaminski WE, Schatteman G, et al. Endothelial cells of hematopoietic origin make a significant contricution to adult blood vessel formation. Circ Res 2000;87:728–730.
48. Ashara T, Murohara T, Sullivan A, et al. Isolation of putative progenitor endothelial cells for angiogenesis. Science 1997;275:964–967.
49. Ashara T, Masuda H, Takahashi T. Bone marrow origin of endothelial progenitor cells responsible for postnatal vasculogenesis in physiological and pathological neovascularization. Circ Res 1999;85:221–228.
50. Davidoff AM, Ng CY, Brown P, et al. Bone marrow-derived cells contribute to tumor neovasculature and, when modified to express an angiogenesis inhibitor, can restrict tumor growth in mice. Clin Cancer Res 2001;7:2870–2879.
51. Lyden D, Hattori K, Dias S. Impaired recruitment of bone-marrow derived endothelial and hematopoietic precursor cells blocks tumor angiogenesis and growth. Nat Med 2001;7:1194–1201.
52. Chao M. Neurotrophins and their receptors: a convergence point for many signaling pathways. Nat Rev Neurosci 2003;4:299–309.
53. Lin TN, Wang CK, Cheung WM, Hsu CY. Induction of angiopoietin and Tie receptor mRNA expression after cerebral ischemia-reperfusion. J Cereb Blood Flow Metab 2000;20:387–395.
54. Folkman J, Klagsbrun M, Sasse J, Wadzinski M, Ingber D, Vlodavsky I. A herparin-binding angiogenic protein-basic fibroblast growth factor is stored with basement membrane. Am J Pathol 1998;130:393–400.
55. Machein MR, Plate KH. VEGF in brain tumors. J Neuro-Oncol 2000;50:109–150.
56. Plate KH, Breir G, Weich HA, Risau W. Vascular endothelial growth factor is a potential tumor angiogenesis factor in human gliomas in vivo. Nature 1992;359:845–847.
57. Berkman RA, Merrill MJ, Reinhold WC. Expression of the vascular permeability factor/vascular endothelial growth factor gene in central nervous system neoplasms. J Clin Invest 1993;91:153–159.
58. Criscuolo GR, Balledux JP. Clinical neurosciences in the decade of the brain: Hypotheses in neuro-oncology-VEG/PF acts upon the actin cytoskeleton and is inhibited by dexamethasone: Relevance to tumor angiogenesis and vasogenic edema. Yale J Biol Med 1996;69:337–355.
59. Ding H, Roncari L, Wu X, Shannon P, Naggy A, Guha A. Expression and hypoxic regulation of angiopoietins in human astrocytomas. Neurooncology 2001;3:1–10.
60. Audero E, Cascone I, Zanon I, et al. Expression of angiopoietin-1 in human glioblastomas regulates tumor-induced angiogenesis in vivo and in vitro studies. Arterioscler Thromb Vasc Biol 2001;21: 536–541.
61. Marchuk DA, Srinivasan S, Squire TL, Zawistowki JS. Vascular morphogenesis: tales of two syndromes. Hum Mol Genet R 2003;12:97–112.
62. Kirsch M, Schackert G, Black PM. Angiogenesis, metastasis, and endogenous inhibition. J Neuro-Oncol 2000;41(2)173–180.
63. Puduvalli VK, Sawaya R. Antiangiogenesis—therapeutic strategies and clinical implications for brain tumors. J Neuro-Oncol 2000;50(1,2):189–200.
64. Puduvalli VK. Inhibition of angiogenesis as a therapeutic strategy against brain tumors. Cancer Treat Res 2004;117:307–336.
65. Millauer BS, Shawver LK, Plate KH, Risau W. Glioblastoma growth inhibited in vivo by a dominant-negative Flk-1 mutant. Nature 1994;367:576–579.
66. Short SC, Traish D, Dowe A, Hines F, Gore M, Brada M. Thalidomide as an anti-angiogenic agent in relapsed gliomas. J Neuro-Oncol 2001;51:41–45.

5
Neuronal Migration and Malformations of Cortical Development

Giorgio Battaglia, MD and Stefania Bassanini, PhD

SUMMARY

Any impairment of the complex processes underlying brain ontogenesis will eventually determine developmental brain abnormalities, which are now recognized as important cause of developmental disabilities and focal epilepsy. In this chapter, we will review the so far established pathways of neuronal migration and describe the clinical and pathogenetic aspects of the more relevant types of developmental brain malformations encountered in clinical practice, such as focal cortical dysplasia, periventricular nodular heterotopia, polymicrogyria, schizencephaly, lissencephaly, and band heterotopia.

Key Words: Brain dysgenesis; cerebral cortex; focal cortical dysplasia; lissencephaly; periventricular nodular heterotopia; polymicrogyria; schizencephaly; subcortical band heterotopia.

1. INTRODUCTION

The development of the human brain is a complex process based on a precise sequence of temporally and spatially regulated cellular events. In the last 30 yr the integrated cellular processes involved in brain morphogenesis have been extensively explored, and particular attention has been devoted to the process of neuronal migration, which is the most relevant mechanism responsible for the final assembly of the brain. Impairments of the processes underlying brain ontogenesis determine developmental brain abnormalities, which is an important cause of developmental disabilities and focal epilepsy, as clearly revealed by the large-scale use of magnetic resonance imaging (MRI) because the mid-1980s.

Brain developmental abnormalities have been first dubbed as neuronal migration disorders, cerebral dysgeneses, or cortical dysplasia, and then, more recently, they have been termed malformations of cortical development (MCDs) *(1)*. Regardless of the names, which are still frequently used as synonyms, they are all abnormalities of the normal cerebral structure originating during ontogenesis from either genetic or acquired impairments of the complex processes by which the brain is eventually built. This chapter describes the so far established pathways of neuronal migration, and then focuses on the more relevant types of MCDs encountered in clinical practice, their causative genes, and underlying pathogenetic mechanisms.

2. MECHANISMS OF NEURONAL MIGRATION IN CORTICAL DEVELOPMENT

A general rule in the vertebrate central nervous system is that neurons migrate from the sites in which they are generated to distant sites in which they permanently reside after development.

From: *The Cell Cycle in the Central Nervous System*
Edited by: D. Janigro © Humana Press Inc., Totowa, NJ

In the specific case of the cerebral cortex, neurons are produced by sequential division of progenitor cells in a specialized germinal zone deep in the brain called the ventricular zone; from the germinative neuroepithelium in the ventricular zone they must migrate into the regions of the future cortex. The neuronal migration is highly specific, as neurons navigate to predetermined locations that define the six functional layers of the adult cerebral cortex.

As the first cortical neurons are formed, they shortly migrate out of the germinal zone and then differentiate to form the first cortical structure termed "preplate" *(2)*. After preplate formation, neurons later born and destined to the cortex migrate from the ventricular zone and settle into the middle of the preplate, thus forming the cortical plate. The progressive accumulation of immature neurons divides the preplate into an outer layer, the marginal zone, largely made up of Cajal-Retzius neurons, which progressively differentiates into layer I of the mature cortex, and the subplate, a transient cortical structure important in determining the connectivity pattern of the future cortex *(2)*. Within the cortical plate proper, early generated neurons will reside in layer VI, whereas later born cells will migrate past the existing cells to reside in progressively more superficial layers (future layers V–II), thus producing an "inside-out" pattern of cortical generation.

The main pattern of cell migration within the cortex is the radial migration. By reconstructing serial electron microscopy sections, Rakic first demonstrated that cortical neurons migrate along radial glial fibers, which provides the scaffold for migrating neurons to reach their final target layers *(3)*. Radial glia are bipolar cells with a nucleus in the ventricular zone, one short process extended with a large end-foot to the adjacent ventricular surface, and a second process projecting to the pial surface. A two-way signaling process occurs between the migrating neuron and the radial glial fiber that permits the neuroblast to migrate and provides a signal to maintain the structure of the radial glial fiber. Radial glia are mitotically active throughout neurogenesis, but after the migration of cortical neurons has been completed, radial glia scaffolding disappears as some of the glial cells degenerate, whereas others re-enter the mitotic cycle and transform into protoplasmic and fibrous astrocytes *(4)*.

It has been also shown later that two modes of radial migration take place in the developing cortex: somal translocation and locomotion *(5)*. Somal translocation consists of movement of the soma and nucleus with shortening of a radially oriented leading process that terminates at the pial surface. In locomotion, migrating neurons display a saltatory pattern of movement maintaining the leading process attached to the scaffold of radial glia as the soma moves forward. The migration speed of somal translocation is faster than that of locomoting cells. Early generated neurons (e.g., preplate neurons) migrate through somal translocation, whereas later generated neurons may use locomotion to migrate to the cortical plate, and then somal translocation to complete migration within the cortical plate.

In addition to cell locomotion and somal translocation, another type of migration takes place in the subcortical zone of the developing cortex. It has been recently shown that the intermediate zone contains an abundant population of multipolar cells *(6)*. Multipolar cells express neuronal markers and extend and retract dynamically multiple thin process in various directions, independent of the radial glial fibers, as their cell bodies slowly move. This kind of migration has been termed *multipolar migration*: multipolar cells, before entering the cortical plate as locomotion cells, may subserve in the intermediate zone functions related to axon growth and possibly migration of other neurons *(6)*.

Radial glia may not provide the exclusive route for neuronal migration. A significant proportion of cortical neurons do not migrate to their destinations along radially oriented glial fascicles but rather along nonradial pathways. The lateral ganglionic eminence (LGE) and medial ganglionic eminence (MGE) are the source of neurons in the subventricular/intermediate zones and in the marginal zone at different stages of corticogenesis, which follow a tangential migration to their final positions in the developing cortex *(7,8)*. The cellular substratum for tangential migration is uncertain, but the large number of axons interconnecting the thalamus and cortical plate and tangentially crossing the intermediate zone may represent the guidance for these migrating neurons. Tangentially migrating neurons express the neurotransmitter γ-aminobutyric acid (GABA) *(8)*, and double mutations of the homeobox genes regulating LGE/MGE differentiation (*Dlx1*, *Dlx2*, *Mash1*) determine dramatic reduction in hippocampal and cortical GABA-ergic interneurons *(9,10)*. The present evidence indicates that most cortical GABA-ergic interneurons originate in LGE and MGE, even it is not yet clear whether this is true for all GABA-ergic cortical neurons *(11)*. By contrast, glutamatergic projection neurons, both in the cortex and in the subpallial telencephalon, reach their final position by radial migration *(11)*.

Neuronal migration does not necessarily end during brain ontogenesis. Indeed, neurogenesis also continues in the hippocampal dentate gyrus and the olfactory bulb of adult rodents. It has been demonstrated that in the mature brain of mammals newly generated neurons are derived from precursor cells in the subventricular zone of the lateral ventricles. These neurons are capable to migrate for considerable distances before differentiate in the olfactory bulb *(12)*. The migration of neurons to the olfactory bulb in adult vertebrates occurs through both radial and nonradial modes, and it has been referred to as rostral migration.

Finally, it has been recently shown that radial glia may function not only as monorails for guiding migrating neurons but also as neuronal precursors. By means of time-lapse images of radial glial cells transfected with retroviral vectors encoding enhanced green fluorescent proteins, the group of Arnold Kriegstein was able to label small clones of mitotic radial glia and postmitotic neurons, demonstrating that neurons were asymmetrically generated by radial glia and migrate along clonally related radial glial cells *(13)*.

3. DEVELOPMENTAL BRAIN MALFORMATIONS IN HUMANS

Neuropathologists discovered the existence of developmental brain malformations more than a century ago, but only in the past two decades the widespread use of modern imaging techniques in medical practice has allowed the in vivo diagnosis of an increasing number of patients affected by cerebral malformations. Brain malformations are now recognized as a major cause of developmental disabilities and focal epilepsy. These abnormalities may involve most parts of the brain or they may be restricted to localized brain areas. Accordingly, the spectrum of the clinical presentation may be greatly different, ranging from patients with severe developmental delay and therefore reproductive disadvantage, to patients with normal neurological and intellectual abilities who draw medical attention for the onset of focal epilepsy in the second decade of life or even later.

Capitalizing on MRI improvement and the consequent collection of large series of homogeneous cases, clinical neurologists attempted to subdivide and clarify MCDs. In the more comprehensive classification system *(1)*, brain malformations were subdivided in those as a result of abnormal cell proliferation and apoptosis, to abnormal neuronal migration, and to abnormal cortical organization. This classification system tried to include the various embryological, histopathological, and genetic aspects of MCDs, even if it was mainly based on imaging data. Other classification schemes were mainly based on molecular genetics *(14)*, whereas others were specifically aimed at classifying specific types of MCDs *(15)*. The main value of the different classifications is that they all try to provide a useful framework for further classifying affected patients. It should be however borne in mind that our knowledge of the etiology and pathogenesis of MCDs is still largely incomplete; indeed, both etiological factors and pathogenetic mechanisms are yet unknown in the majority of affected patients. Therefore, every classification should be considered as provisional and subjected to rapid changes as new genetic and pathogenetic data are made available. On the other hand, the analysis of larger series of homogeneous patients, both familial and sporadic, is the better way to improve our understanding of why malformations are formed in the developing brain, and it will likely provide the better way to choose the treatment for the associated epilepsy and to give a genetic counseling, when appropriate.

Regardless of the different classifications, we will hereafter focus on some types of MCDs which are more relevant in the clinical practice for their frequency: focal cortical dysplasia (FCD), periventricular nodular heterotopia (PNH), polymicrogyria (PMG), and schizencephaly, and, finally, lissencephaly and band heterotopia (Table 1).

Table 1
Main Types of MCDs

	Neuropathology features	MRI features	Clinical features	Etiology	Involved genes	Available animal models
Focal cortical dysplasia (FCD)	Focal architectural abnormalities Focal cytoarchitectural abnormalities Focal cytoarchitectural abnormalities + dysplastic cells (cytomegalic pyramidal neurons and balloon cells)	Normal Focal cortical thickening Blurring of the gray/white matter interface Increased signal of the underlying white matter	Focal epilepsy, with high-frequency seizures Repeated episodes of epileptic status MR and neurological deficits in some cases	Not known	Not known Possible somatic mutations in clonally related dysplastic cells	None
Periventricular nodular heterotopia (PNH)	Nodular masses of neurons and glia close to the germinal matrix with no laminar organization Nodules may extend to the overlying cortex	Nodules isointense to cortical gray matter bulging into the lateral ventricular walls MCM, cerebellar hypoplasia, hippocampal, and midline abnormalities may be associated	Focal epilepsy, frequently drug-resistant Mild MR and neurological deficits in some cases	Genetically determined Early lesions of selected regions of the PV neurogerminative epithelium	*Filamin 1* on Xq28 *ARFGEF2* (in rare families)	Rats prenatally treated with methylazoxy methanol (MAM)
Polymicrogyria (PMG)	Multiple small gyri separated by shallow sulci Unlayered PMG, radial arrangement of neurons, no laminar organization Four-layered PMG, with cell sparse layer instead of layer V	Deep cortical infoldings Focal cortical thickenings for the packing of multiple microgyri Symmetrical location in bilateral cases	Focal epilepsy Modest impairments of cognitive functions Severe MR and neurological deficits in bilateral cases	Genetically determined in few cases Ischemic midcortical damage in late migrational stages	*MECP2* and *PAX6* (in rare cases)	Freeze-lesion rat model

Schizencephaly	Clefts of the cerebral hemispheres Polymicrogyric cortex surrounding the cleft	Clefts extending from the cerebral surface to the lateral ventricles Ventricular enlargement PMG located symmetrically to the cleft acc or sod may be associated	MR Motor deficits focal epilepsy	Genetically determined in few cases Ischemic midcortical damage in late-migrational stages	EMX2 (in rare patients)	Hamsters infected with Kilham strain of mumps virus
Lissencephaly and Subcortical Band Heterotopia (liss/SBH)	Absence or reduction of the normal gyral pattern Abnormally thick cerebral cortex, with four primitive layers Columnar organization of neurons with the subcortical heterotopia	Thick cortical ribbon with fewer gyri, particularly on the posterior regions Continuous symmetrical bands of gray matter beneath the cortex more evident rostrally Ventricular enlargement	Severe MR Motor delay with axial hypotonia Early onset focal epilepsy Dysmorphic features and visceral abnormalities may be associated	Genetically determined (possibly in all cases)	LIS1 XLIS or CX Reln ARX Fukutin	*tish* Rat Null heterozygous LIS1 mice

acc, corpus callosum agenesis; mcm, mega cisterna magna, MR, mental retardation; PV, periventricular; sod, septo-optic dysplasia.

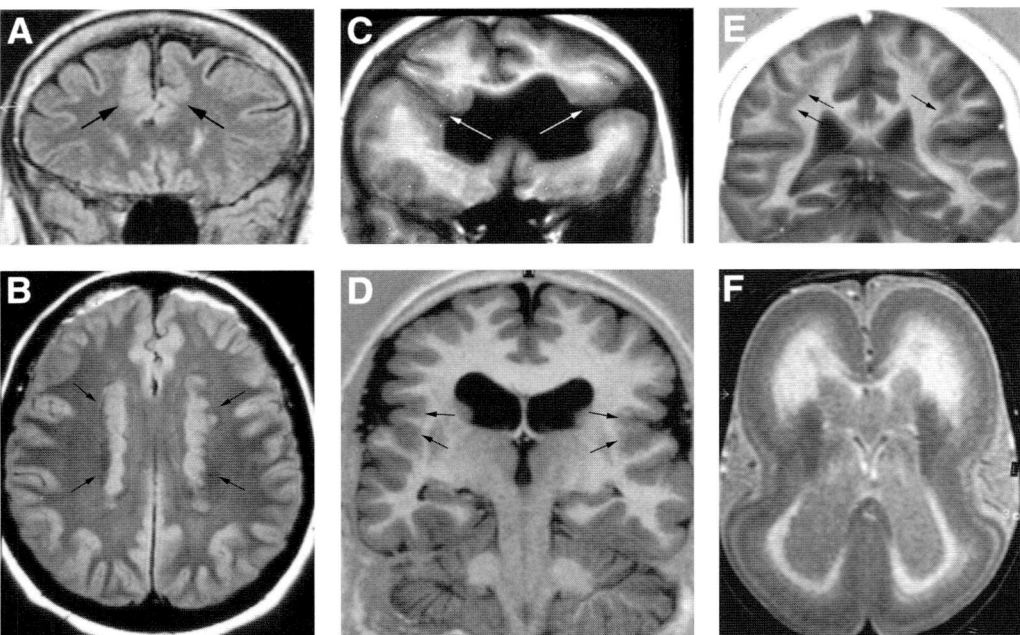

Fig. 1. MRI features of different types of human MCDs. (**A**) Coronal MRI image from a patient with FCD in the mesial frontal lobes (arrows), showing signal hyperintensity and loss of the normal gray/white matter interface. (**B**) Axial MRI image from a patient with bilateral and symmetrical PNH, showing multiple nodules (arrows) lining the entire extension of the lateral wall of the lateral ventricles. (**C**) Coronal MRI section from a patient with bilateral schizencephaly. Left open lip (arrow on the right) and right closed lip (arrow on the left) clefts are lined by polymicrogyric cortex. (**D**) Coronal MRI image from a patient with bilateral perysylvian PMG, demonstrating the bilateral abnormality of the opercular regions (arrows), characterized by multiple gyri and shallow sulci. (**E**) Coronal MRI image from a patient with subcortical band heterotopia, demonstrating thick subcortical heterotopic bands (arrows) bilaterally located beneath the cortex, more evident on the right hemisphere (left side of the MRI image). (**F**) Axial MRI image from a patient with lissencephaly: note the bilaterally thick cortical mantle and the almost complete absence of the normal gyral pattern.

3.1. Focal Cortical Dysplasia

FCD is the most frequently reported MCD in surgical series, that is, in series of patients in whom specimens are available for the neuropathological analysis, accounting for up to 20–50% of all cases subjected to epilepsy surgery *(16,17)*. They are most likely nonhomogeneous malformations, because the term was first used in 1971 in a pioneer work by Taylor and co-workers *(18)*, it has been subsequently used to indicate a wide range of abnormalities of the cortical structure. Some FCD cases are characterized by modest architectural abnormalities of the neocortex, revealed by postsurgical neuropathological analysis but not detectable by MRI. On the other hand, some FCD cases, including those described by Taylor and co-workers in the original report, are characterized by severe cytoarchitectural abnormalities of the neocortex associated with clearly dysplastic cells such as cytomegalic pyramidal neurons and balloon cells (Table 1). In most cases they are revealed by MRI as focal cortical thickenings (Fig. 1) with blurring of the gray matter/white matter interface and increased signal of the underlying white matter *(18,19)*.

Clinically, affected patients are frequently characterized by focal epilepsy, with onset in the first decade of life, which may be associated with motor and language deficits and mental retardation *(19)*. High-frequency focal seizures, refractory to drug treatment, are frequently reported, and they are not infrequently associated with repeated episodes of epileptic status *(20)*. In Taylor-type FCD patients the interictal intralesional electroencephalogram (EEG) activity is characterized by a distinctive pattern of repetitive, high amplitude, fast spikes followed by slow waves, and intermingled by flattening of the EEG tracings *(19,20)*. This peculiar intralesional EEG pattern is selectively observed in Taylor-type FCD, and not reported in other FCD patients, thus suggesting that epileptogenetic phenomena are different in the diverse types of this brain malformation. Regarding surgical outcome, about 40–75% of patients achieve Engel class Ia (i.e., no seizure after surgery) after a follow-up period of 1 yr *(19,21)*. Recent data indicate that surgical outcome is better if presurgical evaluation by means of stereo-EEG, rather than subdural grids or electrocorticography, guide the surgery *(19,20)*.

The pathogenesis of the different types of FCD is still unclear, there is no good evidence that FCD may be inherited, and no appropriate animal models are available for studying this particular type of MCD. However, recent data based on genetic analysis at the single cell level have suggested that dysplastic neurons and balloon cells in Taylor-type FCD could derive from a subpopulation of progenitor cells carrying somatic mutations affecting as yet unidentified genes involved in neuronal differentiation *(22)*. Regarding mechanisms leading to epileptogenesis, recent studies on human surgical samples have demonstrated different GABA-ergic inputs and differential expression of GABA and glutamate receptor subunits in dysplastic vs nondysplastic neurons, thus indicating that epileptogenic mechanisms may differ in the different FCD types *(23)*.

3.2. Periventricular Nodular Heterotopia

PNH is the most frequent MCDs in most clinical series based on MRI *(24)*. They are made up by round nodular masses of normal neurons and glial cells with no laminar organization, located close to the periventricular germinal matrix (Table 1). For this particular location within the brain and the normal features of the heterotopic cells, they have been considered as the result of a primary failure of neuronal migration *(1)*. As FCD, PNHs also are most likely not homogeneous in terms of etiology. Bilateral and symmetrical PNHs occur mostly in females, and most of the reported familial cases are causally related to point mutations of the *filamin* A (filamin 1 [*FLN1*]) gene on Xq28 *(25,26)*. Few familial cases are etiologically linked to gene(s) different from filamin A but most likely acting in the same molecular pathways *(27)*. On the other hand, unilateral PNHs are probably determined by ischemic events impairing *in utero* the perfusion of a limited part of the developing brain *(28)*. In addition, large unilateral PNHs extending from the subependymal region to grossly malformed overlying cortices have been termed subcortical heterotopia *(29)*. But periventricular and subcortical heterotopia are most likely different extensions of the same type of brain dysgenesis. On MRI, heterotopic nodules appear as nodular masses, isointense to cortical gray matter, and bulging into the walls of the lateral ventricles (Fig. 1); mega cisterna magna or cerebellar hypoplasia may be associated findings.

Clinically, epilepsy is the main clinical feature (in 80–100% of reported cases), and the only symptom in most patients, with age of onset in the second decade of life *(28,30)*. However, the clinical picture is related to the extension of the heterotopia, and, accordingly, mild mental retardation and neurological deficits have been described in some patients, most likely related to the involvement of the neocortical areas overlying the heterotopic nodules *(28,32)*. In rare young male patients with bilateral PNH and diffuse gyral abnormalities, the clinical picture is characterized by mental retardation, neurological deficits, and dysmorphic features *(32)*. In addition, PNH may be associated in rare cases with Ehlers-Danlos syndrome *(33)*.

Mutations of the *FLN1* gene have been reported in the majority of familial cases and in 8–19% of sporadic cases with bilateral and symmetrical PNH. The *FLN1* produces a 250 kDa actin-crosslinking phosphoprotein with three major functional domains: the N-terminal actin-binding domain is structurally similar to actin-binding domains of dystrophin and α-actinin, whereas the C-terminal domain is a site of dimerization and linkage to several membrane proteins, particularly integrin. It has been hypothesized that the disruption of the actin cross-linking domain may alter the beginning of neuronal migration because the migrating neurons cannot extend their growth cones along radial glia cells. Interestingly, it has been recently demonstrated that *FLN1* mutations determining gain-of-function and not reducing the level of full-length *FLN1*, do not cause PNH, but complex congenital malformations termed otopalatodigital spectrum disorders *(34)*.

Regarding the pathogenesis of epileptic discharges, alterations of the α-subunit of the Ca^{2+}/calmodulin-dependent kinase II and the *N*-methyl-D-asparate receptor complex have been recently found in the epileptogenic cortex and nodules of human patients with unilateral nodules. These abnormalities may at least contribute to the intrinsic hyperexcitability associated with this brain malformation *(35)*. These data have been recently confirmed *(36)* in an animal model for human PNH, the rat prenatally treated with methylazoxymethanol *(37)*. Recent studies on this rat model have demonstrated that the methylazoxymethanol-induced ablation of an early wave of cortical neurons is sufficient to alter the migration and differentiation of subsequently generated neurons, which in turn set the base for the formation of the heterotopic nodules *(38)*. This cellular mechanism may explain the genesis of the malformation also in human patients *(39)*.

3.3. PMG and Schizencephaly

The term PMG indicates a brain malformation made by an excessive number of abnormally small gyri, separated by shallow and enlarged sulci, or fused together through layers I. Microscopically, two histological types of PMG have been described: in unlayered PMG the cortical neurons display radial arrangement but no laminar organization, whereas the four-layered PMG is thought to be the result of ischemic damage to middle cortical layers in the late stage of neuronal migration, leading to neuronal loss particularly evident in layer V (Table 1). Even if the term is obviously derived from early neuropathology studies, PMG may be diagnosed through MRI for the presence of cortical infoldings and focal cortical thickenings because of the packing of multiple microgyri (Fig. 1). Patients with bilateral as well as unilateral PMG have been described. The bilaterally affected patients have been further subdivided as affected by frontal *(41)*, perysylvian *(40)*, the most frequent form (Fig. 1), and parieto-occipital PMG *(42)*. In unilaterally affected patients, PMG may involve a single lobe or be diffuse to an entire hemisphere. In keeping with the anatomical extension of the malformation, the clinical presentation of PMG patients may vary, ranging from patients with focal epilepsy or selective impairment of cognitive functions *(43)*, to patients with severe epileptic encephalopathy, mental retardation, and neurological deficits. In particular, patients with bilateral perysylvian PMG, in addition to mental retardation and epilepsy, are affected by dysarthria and facio-glosso-masticatory diplegia owing to the bilateral opercular involvement.

The existence of familial cases suggests a genetic cause for PMG. However, even if mutations of *MECP2* and *PAX6* genes in single PMG cases *(44,45)* and linkages to different chromosomal regions have been reported, no gene has yet been convincingly linked to any of the different types of bilateral PMG.

Schizencephaly is a rare MCD characterized by a gray matter-lined cleft spanning the cerebral hemisphere from the pial surface to the lateral ventricle (Fig. 1). It was first described by Yakovlev and Wadsworth almost 60 yr ago *(46,47)*. In the original report of five autoptic cases, the authors distinguished open- or close-lipped (Fig. 1) clefts (with unapposed or apposed cortical lips, respectively), and defined schizencephaly as a true brain malformation determined by agenesis of the cerebral mantle in the early stages of development. Patients affected by schizencephaly show motor deficits and mental retardation of varying severity, in relation to type, location, and size of the clefts, and to the presence of associated brain abnormalities (Table 1); in addition, they are frequently affected by focal epilepsy *(48)*. The occurrence of familial cases has been supported by the original assumption by Yakovlev and Wadsworth suggesting a genetically determined origin of the malformation. Almost 10 yr ago, this hypothesis was supported by the report of heterozygous mutations in the homeobox gene *EMX2* in seven out of eight sporadic schizencephaly patients *(49)*.

In the classification of MCD by Barkovich and co-workers, schizencephaly and PMG are grouped together as abnormalities of cortical organization *(1)*, and indeed the two malformations are frequently associated in the same patient. PMG is not only located within or adjacent to the schizencephalic cleft, but it is also frequently found in different brain regions of the same patient, for instance, symmetrically located in the hemisphere contralateral to the cleft *(50)*. These data suggest that they both share common pathogenetic mechanisms. However, no common genetic factors have been described, and data proving a genetic origin of PMG are scarce. In addition, also the causal relationship between schizencephaly and *EMX2* mutations is still a matter of debate. After the original reports, no further schizencephaly cases have been ever because associated with mutations in the *EMX2* gene or other genes. In addition, data from mutant *Emx2* knockout mice do not support the etiological role of this gene: first, the occurrence of dominant mutations would not be consistent with the observation that heterozygous *Emx2*$^{+/-}$ knockout mice display a normal phenotype *(51)*; second, homozygous *Emx2*-null mice die soon after birth because of the absence of the urogenital apparatus, with cerebral hypoplasia and hippocampal abnormalities but no evidence of brain malformations resembling human schizencephaly *(51)*. Therefore, it is more likely at the present point that both PMG and schizencephaly are late-migrational abnormalities owing to the clastic effect of either endogenous or exogenous factors during brain development *(52)*.

Indeed, the two animal models more closely resembling PMG and schizencephaly are both determined by exogenous lesions during the late migrational period: the freeze lesion model of PMG, in which a small region of the rat cerebral cortex is supercooled in the immediate postnatal period *(53)*, and the virus-induced model of schizencephaly, obtained in hamsters by inoculating the Kilham strain of mumps virus in the late gestational period *(54)*. It should be mentioned that some data from human patients, such as the presence of cytomegalovirus DNA and antibodies in affected children, indeed support the possibility that schizencephaly may be because of the clastic effect of viral agents *(55)*.

3.4. Lissencephaly and Band Heterotopia

This group encompasses severe forms of MCD, all characterized by the smooth appearance of the brain surface owing to the scarcity of the normal gyral pattern of the neocortex (Fig. 1). The terms lissencephaly (smooth brain), agyria (smooth cortex with no sulci), and pachygyria (thickened hard cortex with widened gyri and few sulci), all derived from neuropathology, are still used in clinical practice, sometimes interchangeably, to define different cases of this group of MCD. Clinically, the lissencephalies are a heterogeneous group of cerebral malformations, and different types of lissencephaly have been described (Table 1), such as the autosomal recessive lissencephaly with cerebellar hypoplasia, because of the disruption of the *reelin* gene *(56)*; the X-linked lissencephaly with abnormal genitalia, as a result of mutations of the homeobox gene *ARX (57,58)*; and the so-called "cobblestone lissencephaly," or Fukuyama-type congenital muscular dystrophy, related to mutations of the Fukutin gene *(59,60)*. However, the most frequent forms are lissencephaly caused by mutations of the *LIS1* gene on the short arm of chromosome 17 *(61)*, and lissencephaly/band heterotopia as a result of mutations of the *XLIS* or *DCX* gene on chromosome Xq22 *(62)*.

LIS1 lissencephaly may be isolated, and clinically characterized by severe mental retardation, hypotonia, motor delay, and epilepsy; or associated with dysmorphic facial features, heart and kidney abnormalities, in the Miller-Dieker syndrome *(63)*. The latter is the consequence of contiguous gene large deletions (including *LIS1*) on chromosome 17, whereas the former is owing to smaller deletions, or even intragenic mutations, of the *LIS1* gene, the severity of the clinical picture reflecting the severity of the loss-of-function of the *LIS1* gene product. On the other hand, mutations of the *DCX* gene determine classical lissencephaly in hemizygous males and band heterotopia in heterozygous females. Subcortical band heterotopia is made up by symmetric bands of heterotopic neurons located in the white matter just beneath a simplified or even a normal neocortex (Fig. 1). It was previously recognized as "double cortex" syndrome, and it is mainly characterized by the association of epilepsy and mental retardation of varying severity *(64)*.

The similar spectrum of phenotypes in human patients with either *LIS1* or *DCX* mutations suggest that the two proteins may act in the same molecular pathway. The protein product of the *LIS1* gene was originally identified as the β-subunit of the intracellular brain form of platelet-activating factor acetylhydrolase, but it was later recognized as a protein of the microtubule network regulating microtubule dynamics *(65)*. According to more recent studies, *LIS1* may play a role in the extension of the leading process of migrating cells by interacting with tubulin, and it also regulates the function and distribution along microtubules of dynein, which acts as a microtubule-associated motor protein in the retrograde transport of vesicles and organelles *(66,67)*. These combined data suggest not only a function for *LIS1* in migration but also a dynein-related function in axon growth and differentiation. On the other hand, the *DCX* gene product, or doublecortin, is also a microtubule-associated protein *(68)*. However, in contrast to *LIS1*, it is only expressed in postmitotic immature neurons, it is concentrated at the extremities of growing neurites, and it may therefore play a role in the growth of axonal processes *(69)*. Thus, both genes may influence neuronal migration through different mechanisms in related molecular pathways involving the association with the neuronal cytoskeleton.

If one wishes to study the formation and the epileptogenesis in lissencephaly, two different animal models may be of value: the seizure-prone mutant rat called *tish (70)*, and the null heterozygous *LIS1* mice *(71)*. The former is a spontaneously occurring mutant rat, with an autosomal recessive mode of inheritance, characterized by the presence of large collections of neurons symmetrically located beneath the normal cortex of the two hemispheres; it is particularly interesting to study the contribution of the heterotopic band of neurons to the genesis and maintenance of epilepsy *(70)*. The latter are transgenic mice with graded reductions of the LIS1 protein, particularly relevant in demonstrating how subplate formation, interkinetic nuclear migration, and cell death are dependent on the cellular levels of the LIS1 protein *(71)*.

4. CONCLUDING REMARKS

The basic science studies dedicated to neuronal migration and brain ontogenesis contribute to progressively uncover the complexity of mechanisms underlying the development of the human brain. Advancements in our knowledge of these complex processes are of great value in understanding the pathogenesis of developmental brain malformations in humans. In turn, the analysis of patients with developmental brain malformations have revealed mutations in newly discovered or even in already known proteins, which play important roles in the migration of neurons or more generally in the proper patterning of the developing brain. As in many other instances in neuroscience, the possibility of combining basic science and clinical data offer an extraordinary powerful tool for unravelling the secrets of brain function and malfunction.

ACKNOWLEDGMENTS

The authors wish to thank Dr. Luisa Chiapparini for the MRI images, and Dr. Tiziana Granata and Dr. Franco Taroni for the critical reading of the manuscript. This study was partially supported by Grant No. R-03-29 from the "Pierfranco and Luisa Mariani Foundation for Pediatric Neurology."

REFERENCES

1. Barkovich AJ, Kuzniecky RI, Jackson GD, Guerrini R. Dobyns WB. Classification system for malformations of cortical development: update 2001. Neurology 2001;57:2168–2178.
2. Allendoerfer KL, Shatz CJ. The subplate, a transient neocortical structure: its role in the development of connections between thalamus and cortex. Annu Rev Neurosci 1994;17:185–218.
3. Rakic P. Mode of cell migration to the superficial layers of fetal monkey neocortex. J Comp Neurol 1972;145:61–84.

4. Schmechel DE, Rakic PA. Golgi study of radial glial cells in developing monkey telencephalon: morphogenesis and transformation into astrocytes. Anat Embryol 1979;156:115–152.
5. Nadarajah B, Brunstrom JE, Grutzendler J, Wong RO, Pearlman AL. Two modes of radial migration in early development of the cerebral cortex. Nat Neurosci 2001;4:143–150.
6. Tabata H, Nakajima K. Multipolar migration: the third mode of radial neuronal migration in the developing cerebral cortex. J Neurosci 2003;23:9996–10,001.
7. Tamamaki N, Fujimori KE, Takauji R. Origin and route of tangentially migrating neurons in the developing neocortical intermediate zone. J Neurosci 1997;17:8313–8323.
8. Lavdas AA, Grigoriou M, Pachnis V, Parnavelas JG. The medial ganglionic eminence gives rise to a population of early neurons in the developing cerebral cortex. J Neurosci 1999;19:7881–7888.
9. Anderson SA, Qiu M, Bulfone A, et al. Mutations of the homeobox genes Dlx-1 and Dlx-2 disrupt the striatal subventricular zone and differentiation of late born striatal neurons. Neuron 1997;19:27–37.
10. Pleasure SJ, Anderson S, Hevner R, et al. Cell migration from the ganglionic eminences is required for the development of hippocampal GABAergic interneurons. Neuron 2000;28:727–740.
11. Marin O, Rubenstein JL. A long, remarkable journey: tangential migration in the telencephalon. Nat Rev Neurosci 2001;2:780–790.
12. Lois C, Alvarez-Buylla A. Long-distance neuronal migration in the adult mammalian brain. Science 1994;264:1145–1148.
13. Noctor SC, Flint AC, Weissman TA, Dammerman RS, Kriegstein AR. Neurons derived from radial glial cells establish radial units in neocortex. Nature 2001;409:714–720.
14. Sarnat HB. Molecular genetic classification of central nervous system malformations. J Child Neurol 2000;15:675–687.
15. Palmini A, Najim I, Avanzini G, et al. Terminology and classification of the cortical dysplasias. Neurology 2004;62:S2–S8.
16. Duchowny M, Levin B, Jayakar P, et al. Temporal lobectomy in early childhood. Epilepsia 1992;33:298–303.
17. Wolf HK, Campos MG, Zentner J, et al. Surgical pathology of temporal lobe epilepsy. Experience with 216 cases. J Neuropathol Exp Neurol 1993;52:499–506.
18. Taylor DC, Falconer MA, Bruton CJ, Corsellis JA. Focal dysplasia of the cerebral cortex in epilepsy. J Neurol Neurosurg Psychiatry 1971;34:369–387.
19. Tassi L, Colombo N, Garbelli R, et al. Focal cortical dysplasia: neuropathological subtypes, EEG, neuroimaging and surgical outcome. Brain 2002;125:1719–1732.
20. Palmini A, Gambardella A, Andermann F, et al. Intrinsic epileptogenicity of human dysplastic cortex as suggested by corticography and surgical results. Ann Neurol 1995;37:476–487.
21. Sisodiya SM. Surgery for malformations of cortical development causing epilepsy. Brain 2000;123:1075–1091.
22. Hua Y, Crino PB. Single cell lineage analysis in human focal cortical dysplasia. Cereb Cortex 2003;13:693–699.
23. Crino PB, Duhaime AC, Baltuch G, White R. Differential expression of glutamate and GABA-A receptor subunit mRNA in cortical dysplasia. Neurology 2001;56:906–913.
24. Raymond AA, Fish DR, Sisodiya SM, Alsanjari N, Stevens JM, Shorvon SD. Abnormalities of gyration, heterotopias, tuberous sclerosis, focal cortical dysplasia, microdysgenesis, dysembryoplastic neuroepithelial tumour and dysgenesis of the archicortex in epilepsy. Clinical, EEG and neuroimaging features in 100 adult patients. Brain 1995;118:629–660.
25. Eksioglu YZ, Scheffer IE, Cardenas P, Knoll J, DiMario F, Ramsby G. Periventricular heterotopia: an X-linked dominant epilepsy locus causing aberrant cerebral cortical development. Neuron 1996;16: 77–87.
26. Fox JW, Lamperti ED, Eksioglu YZ, Hong SE, Feng Y, Graham DA. Mutations in filamin 1 prevent migration of cerebral cortical neurons in human periventricular heterotopia. Neuron 1998;21:1315–1325.
27. Sheen VL, Ganesh VS, Topcu M, et al. Mutations in ARFGEF2 implicate vesicle trafficking in neural progenitor proliferation and migration in the human cerebral cortex. Nat Genet 2004;36:69–76.
28. Battaglia G, Granata T, D'Incerti L, et al. Periventricular nodular heterotopia: epileptogenic findings. Epilepsia 1997;38:1173–1182.
29. Barkovich AJ. Morphologic characteristics of subcortical heterotopia: MR imaging study. Am J Neuroradiol 2000;21:290–295.

30. Dubeau F, Tampieri D, Lee N, et al. Periventricular and subcortical nodular heterotopia. A study of 33 patients. Brain 1995;118:1273–1287.
31. Battaglia G, Arcelli P, Granata T, et al. Neuronal migration disorders and epilepsy: a morphological analysis of three surgically treated patients. Epilepsy Res 1996;26:49–58.
32. Guerrini R, Dobyns WB. Bilateral periventricular nodular heterotopia with mental retardation and frontonasal malformation. Neurology 1998;51:499–503.
33. Thomas P, Bossan A, Lacour JP, Chanalet S, Ortonne JP, Chatel M. Ehlers-Danlos syndrome with subependymal periventricular heterotopias. Neurology 1996;46:1165–1167.
34. Robertson SP, Twigg SR, Sutherland-Smith AJ, et al. Localized mutations in the gene encoding the cytoskeletal protein filamin A cause diverse malformations in humans. Nat Genet 2003;33:487–491.
35. Battaglia G, Pagliardini S, Ferrario A, et al. AlphaCaMKII and NMDA-receptor subunit expression in epileptogenic cortex from human periventricular nodular heterotopia. Epilepsia 2002;43:209–216.
36. Gardoni F, Pagliardini S, Setola V, et al. The NMDA receptor complex is altered in an animal model of human cerebral heterotopia. J Neuropathol Exp Neurol 2003;62:662–675.
37. Colacitti C, Sancini G, DeBiasi S, et al. Prenatal methylazoxymethanol treatment in rats produces brain abnormalities with morphological similarities to human developmental brain dysgeneses. J Neuropathol Exp Neurol 1999;58:92–106.
38. Battaglia G, Bassanini S, Granata T, Setola V, Giavazzi A, Pagliardini S. The genesis of epileptogenic cerebral heterotopia: clues from experimental models. Epileptic Disord 2003;5(Suppl 2):S51–S58.
39. Battaglia G, Pagliardini S, Saglietti L, et al. Neurogenesis in cerebral heterotopia induced in rats by prenatal methylazoxymethanol treatment. Cereb Cortex 2003;13:736–748.
40. Guerrini R, Barkovich AJ, Sztriha L, Dobyns WB. Bilateral frontal polymicrogyria: a newly recognized brain malformation syndrome. Neurology 2000;54:909–913.
41. Kuzniecky R, Andermann F, Guerrini R. Congenital bilateral perisylvian syndrome: study of 31 patients. The CBPS Multicenter Collaborative Study. Lancet 1993;341:608–612.
42. Guerrini R, Dubeau F, Dulac O, et al. Bilateral parasagittal parietooccipital polymicrogyria and epilepsy. Ann Neurol 1997;41:65–73.
43. Galaburda AM, Sherman GF, Rosen GD, Aboitiz F, Geschwind N. Developmental dyslexia: four consecutive patients with cortical anomalies. Ann Neurol 1985:18:222–233.
44. Geerdink N, Rotteveel JJ, Lammens M, et al. MECP2 mutation in a boy with severe neonatal encephalopathy:clinical, neuropathological and molecular findings. Neuropediatrics 2002;33:33–36.
45. Mitchell TN, Free SL, Williamson KA, et al. Polymicrogyria and absence of pineal gland due to PAX6 mutation. Ann Neurol 2003;53:658–663.
46. Yakovlev P, Wadsworth RC. Schizencephalies: a study of the congenital clefts in the cerebral mantle, I: clefts with fused lips. J Neuropath Exp Neurol 1946;5:116–130.
47. Yakovlev P, Wadsworth RC. Schizencephalies: a study of the congenital clefts in the cerebral mantle, II: clefts with hydrocephalus and lips separated. J Neuropath Exp Neurol 1946;5:169–206.
48. Granata T, Battaglia G, D'Incerti L, et al. Schizencephaly: neuroradiologic and epileptologic findings. Epilepsia 1996;37:1185–1193.
49. Brunelli S, Faiella A, Capra V, et al. Germline mutations in the homeobox gene EMX2 in patients with severe schizencephaly. Nat Genet 1996;12:94–96.
50. Barkovich AJ, Kjos BO. Schizencephaly: correlation of clinical findings with MR characteristics. Am J Neuroradiol 1992;13:85–94.
51. Pellegrini M, Mansouri A, Simeone A, Boncinelli E, Gruss P. Dentate gyrus formation requires Emx2. Development 1996;122:3893–3898.
52. Barkovich AJ, Gressens P, Evrard P. Formation, maturation, and disorders of brain neocortex. Am J Neuroradiol 1992;13:423–446.
53. Dvorak K, Feit J. Migration of neuroblasts through partial necrosis of the cerebral cortex in newborn rats-contribution to the problems of morphological development and developmental period of cerebral microgyria. Histological and autoradiographical study. Acta Neuropathol (Berl) 1977;38(3):203–212.
54. Takano T, Takikita S, Shimada M. Experimental schizencephaly induced by Kilham strain of mumps virus: pathogenesis of cleft formation. Neuroreport 1999;10:3149–3154.
55. Iannetti P, Nigro G, Spalice A, Faiella A, Boncinelli E. Cytomegalovirus infection and schizencephaly: case reports. Ann Neurol 1998;43:123–127.
56. Hong SE, Shugart YY, Huang DT, et al. Autosomal recessive lissencephaly with cerebellar hypoplasia is associated with human RELN mutations. Nat Genet 2000;26:93–96.

57. Dobyns WB, Berry-Kravis E, Havernick NJ, Holden KR, Viskochil D. X-linked lissencephaly with absent corpus callosum and ambiguous genitalia. Am J Med Genet 1999;86:331–337.
58. Stromme P, Mangelsdorf ME, Shaw MA, et al. Mutations in the human ortholog of Aristaless cause X-linked mental retardation and epilepsy. Nat Genet 2002;30:441–445.
59. Fukuyama Y, Osawa M, Saito K. Congenital muscular dystrophies: an overview. In: Arzimanoglou A, Goutieres F (eds), Trends in Child Neurology. Paris: John Libbey Eurotext, 1996, 107–135.
60. Kobayashi K, Sasaki J, Kondo-Iida E, et al. Structural organization, complete genomic sequences and mutational analyses of the Fukuyama-type congenital muscular dystrophy gene, fukutin. FEBS Lett 2001;489:192–196.
61. Reiner O, Carrozzo R, Shen Y, et al. Isolation of a Miller-Dieker lissencephaly gene containing G protein beta-subunit-like repeats. Nature 1993;364:717–721.
62. des Portes V, Pinard JM, Billuart P, et al. A novel CNS gene required for neuronal migration and involved in X-linked subcortical laminar heterotopia and lissencephaly syndrome. Cell 1998;92:51–61.
63. Dobyns WB, Stratton RF, Greenberg F. Syndromes with lissencephaly. I: Miller-Dieker and Norman-Roberts syndromes and isolated lissencephaly. Am J Med Genet 1984;18:509–526.
64. Barkovich AJ, Guerrini R, Battaglia G, et al. Band heterotopia: correlation of outcome with magnetic resonance imaging parameters. Ann Neurol 1994;36:609–617.
65. Sapir T, Elbaum M, Reiner O. Reduction of microtubule catastrophe events by LIS1, platelet-activating factor acetylhydrolase subunit. EMBO J 1997;16:6977–6984.
66. Smith DS, Niethammer M, Ayala R, et al. Regulation of cytoplasmic dynein behaviour and microtubule organization by mammalian Lis1. Nat Cell Biol 2000;2:767–775.
67. Gupta A, Tsai LH, Wynshaw-Boris A. Life is a journey: a genetic look at neocortical development. Nat Rev Genet 2002;3:342–355.
68. Francis F, Koulakoff A, Boucher D, et al. Doublecortin is a developmentally regulated, microtubule-associated protein expressed in migrating and differentiating neurons. Neuron 1999;23:247–256.
69. Friocourt G, Koulakoff A, Chafey P, et al. Doublecortin functions at the extremities of growing neuronal processes. Cereb Cortex 2003;13:620–626.
70. Lee KS, Schottler F, Collins JL, et al. A genetic animal model of human neocortical heterotopia associated with seizures. J Neurosci 1997;17:6236–6242.
71. Gambello MJ, Darling DL, Yingling J, Tanaka T, Gleeson JG, Wynshaw-Boris A. Multiple dose-dependent effects of Lis1 on cerebral cortical development. J Neurosci 2003;23:1719–1729.

II

Postnatal Development of Neurons and Glia

6
Genome-Wide Expression Profiling of Neurogenesis in Relation to Cell Cycle Exit

P. Roy Walker, PhD, Dao Ly, BSc, Qing Y. Liu, PhD, Brandon Smith, MSc, Caroline Sodja, MSc, Marilena Ribecco, PhD, and Marianna Sikorska, PhD

1. INTRODUCTION

Neurogenesis is the process by which new brain cells are produced either during development or in the adult brain. More specifically, it is "the proliferation of neuronal precursor cells to produce neurons." Both definitions embody a key role for the cell cycle in the process particularly because the brain is an architecturally complex, multicompartmented tissue and the correct numbers of neurons (and glial cells) must be placed into each compartment. The process is made more complicated by the fact that neurons within each compartment are highly specialized, mandating that the new neurons also have the correct phenotype. Therefore, a mechanistic understanding of neurogenesis requires an understanding of several processes—control of the cell cycle to generate neurons in sufficient numbers, spatial mechanisms that ensure the correct number of cells in each compartment, the differentiation process that transforms a progenitor cell into a neuron, and an explanation of how so many neuronal subtypes can be readily created. Equally important is an understanding of the temporal coordination of these four processes, particularly regarding cell cycle exit.

As a result of studies carried out at many stages of development, in many regions of the brain, numerous genes have been associated with each of these processes [1]. More recently, a number of microarray studies have added, or implicated, additional genes [2–9] and we are now essentially in a position to catalog all of the genes contributing to complex phenotypic changes such as neurogenesis. The challenge is to synthesize all of these data into a comprehensive model by which the genome orchestrates the process in terms of the activity of transcription factors and ultimately an understanding of the *cis*-regulatory code embodied in the sequence of DNA and the contribution of epigenetic mechanisms. Mechanisms by which transcription factor networks regulate the spatial organization of the brain have been described by Edlund and Jessell [10] and Jessell [11]. A more explicit gene-regulatory network controlling embryonic specification elucidates by Oliveri and Davidson [12] serves as an essential goal toward a complete understanding of transcriptional control of neurogenesis. In this chapter we review contributions to our understanding of the regulation of proliferation and differentiation in neurogenesis using a data set obtained from genome-wide expression profiling and quantitative polymerase chain reaction analysis of the human NT2 embryonal carcinoma line stimulated by retinoic acid (RA) to differentiate along the neural lineage into postmitotic neurons and astrocytes [13] together with supporting information from the literature. Genome-wide expression profiling provides unprecedented comprehensive insight into the extent to which the genome is transformed as cells shift from a germ cell state to that of a fully differentiated neuron. Because germ

From: *The Cell Cycle in the Central Nervous System*
Edited by: D. Janigro © Humana Press Inc., Totowa, NJ

cells are replicating pluripotent cells, whereas neurons are truly postmitotic with a complex and unique cellular structure, they provide an excellent model system to study the complexity and dynamics of the differentiation process. Importantly, the NT2 cell line is one of the few in which the entire program of human neuronal differentiation can be studied on a genome-wide scale.

Expression profiling, coupled to appropriate data mining, provides not only a comprehensive catalog of genes involved, but also identifies all the cellular processes and ultimately gene-regulatory networks that contribute to the change in cellular phenotype. Previously unknown genes and expressed sequence tags (ESTs) can be associated with specific cellular states or gene networks based on coclustering. Time series expression profiling offers a temporal view of when major changes in processes such as epigenetic DNA modification and chromatin remodeling are occurring relative to more readily observable processes, such as extracellular matrix and cytoskeleton remodeling. It highlights the role that the various levels of control of the activity of the genome play in directing commitment, patterning, and differentiation and, more specifically, the role of extrinsic and intrinsic factors.

2. PHASES OF GENE EXPRESSION IN NEUROGENESIS

2.1. Exit From the Germ Cell State

The undifferentiated NT2 cells are embryonal carcinoma cells with a germ cell phenotype *(14)*. Cells are not in a germ cell or stem cell state by default. They actively express genes that maintain that phenotype. A catalog of all genes contributing to the germ cell phenotype can be readily obtained from expression profiling because a number of studies have now been completed *(15–17)*. The genes that are downregulated in the NT2 cells immediately following RA treatment can be considered key regulators of the germ cell state (Fig. 1) *(13)*. They include the octamer transcription factors *OCT4 (POU5F1)* and *NANOG*, and growth factors and their receptors—fibroblast growth factor (*FGF*) *4*, transforming growth factor-β (*TGF-β*), *FGFR1*, *FGFR4*, growth and differentiation factor (*GDF*) *3* (a member of the *TGF*-β superfamily of growth and differentiation factors *[18]*) and *Stella/pgc7/dppa3*, a gene of unknown function. Downregulation of these genes results in exit from the germ cell phenotype and the cells become irreversibly committed to either differentiate or to die by apoptosis. Thus, if RA is removed after 4 d of treatment the cells cannot re-express any of these germ cell-specific genes. Significantly, even though they are rapidly downregulated they are not all direct targets of RA. For example, the rapid decrease in *FGF4* levels is caused by disruption of the Oct4-Sox2 transcription factor heterodimer on the *FGF4* promoter *(19)* and it is *OCT4* that is the direct target of RA *(20)*. Other than for *OCT4*, relatively little is known about downstream targets of the other transcription factor (TF) genes that are downregulated as the cells exit the germ cell state.

These observations illustrate that maintenance of the stem cell phenotype and even the early stages of exit from this state and commitment to differentiation is highly dependent on extrinsic factors—*TGF-β, GDF3, FGF8, FGF4*, and possibly Stella *(9)*. On exposure to RA, the expression of these extrinsic signals and/or their receptors is attenuated, with the exception of *FGF8*, which remains elevated to at least day 20 (Fig. 1). Notably, downregulation of these genes *per se* does not result in differentiation or even cell cycle arrest, because the genes are inactivated by day 4, but many cells continue to proliferate for at least several more days and, in the case of cells committed to become astrocytes, several more weeks.

2.2. Data Mining Reveals the Major Phases of Gene Expression in Neurogenesis

Many genes have been shown to be involved in different aspects of neurogenesis such as patterning, commitment, neurogenesis, and maturation of neurons. Although these studies have been pivotal in identifying the cast of characters, essentially one gene at a time, they leave a great deal of uncertainty of exactly how such a multitude of quite disparate proteins can be involved and the exact nature of the hierarchy of control, if any, and their temporal relationship to cell cycle exit. Global time series gene expression profiling permits one to establish a temporal basis for their involvement and appropriate clustering can identify which gene expressions precede

Genome-Wide Expression Profiling

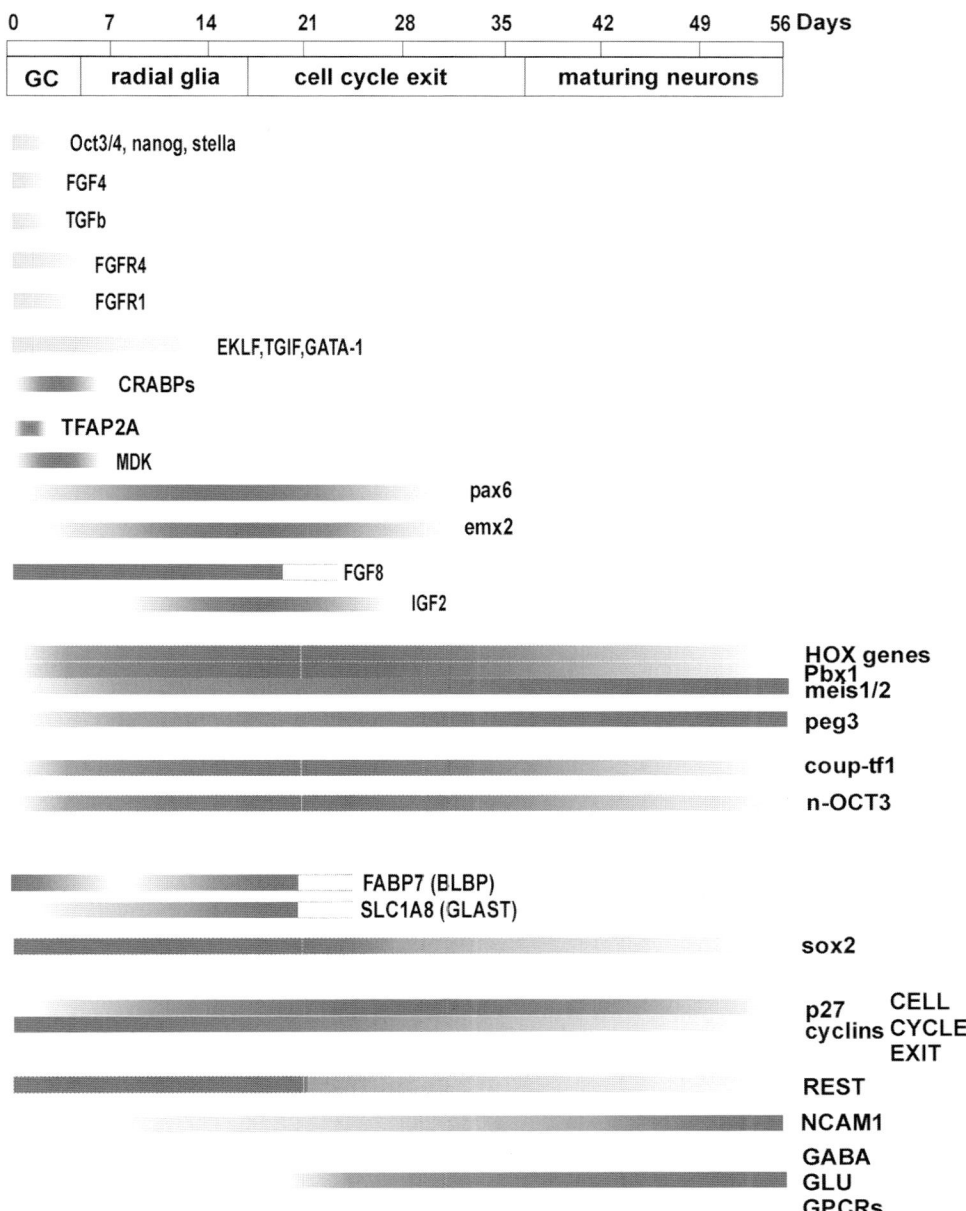

Fig. 1. Changes in the expression of selected genes during neurogenesis. (Please *see* companion CD for color version of this figure.)

cell cycle exit, which are tightly coupled to it and which follow cell cycle exit (Fig. 1). Using a combination of clustering techniques we have been able to distinguish specific subsets of genes that are expressed or repressed in temporally distinct patterns before cell cycle exit from those genes that are expressed concurrently with cell cycle exit and those that appear to be expressed in the postmitotic maturing neurons. There are three major phases of *increased* gene expression and two major phases of *decreased* gene expression (Fig. 2). The phases of increased gene expression include those genes that are expressed transiently before cell cycle exit, genes that are associated with cell cycle exit and genes that are induced post–cell cycle exit. The two major

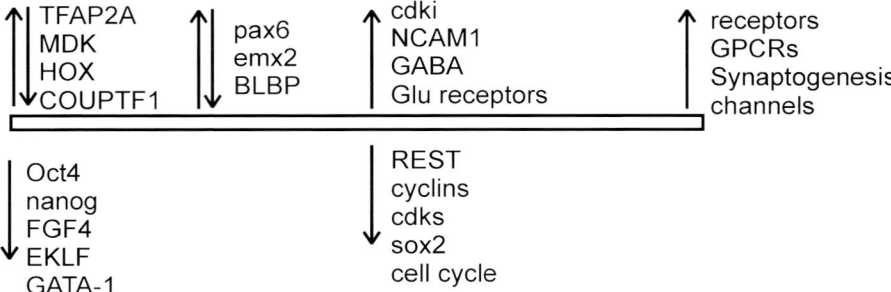

Fig. 2. Major phases of differential gene expression during neurogenesis.

phases of decreased expression are those occurring immediately following *RA* addition and those that occur at the time the cells exit the cell cycle (Fig. 2).

The vast majority of genes differentially expressed during RA-induced neurogenesis are tightly coupled to cell cycle exit, falling into two clusters showing essentially reciprocal changes (Table 1). These two clusters account for 94% (2606) of the differentially expressed genes indicating the fundamental importance of cell cycle exit. The downregulated cluster contains all of the cell cycle-related genes (cyclins, genes associated with biogenesis and replication, etc.), whereas the upregulated genes are typical of the differentiated neuronal phenotype *(13)*. This program of gene expression commences as early as 7 d after RA addition and continues throughout the time-course with the majority of the cells exiting the cell cycle from 14 d of RA treatment onwards. Because the cells exit the cell cycle over a period of days it is not possible to deduce the exact kinetics of change in each individual cell, but clustering easily associates all the genes involved in and coupled to this process.

2.3. Genes Expressed Transiently Prior to Cell Cycle Exit

Committed cells transition through several states prior to becoming fully functional neurons or astrocytes. Most recently, radial glia have emerged as a definable population of neuronal progenitors *(21)*. In RA-treated NT2 cells, *BLBP (FABP7)* and *GLAST (SCL1A8),* which are reliable markers for radial glia, are expressed starting at day 7. Thus, the cells have exited the germ cell state by day 2 and become definably radial glial by day 7 Data mining clusters the genes that are differentially expressed during this time period based on their temporal profiles. Because it is now believed that the fate of the cell population is determined primarily before cell cycle exit, these clusters should contain the genes and ESTs associated with commitment and spatial patterning. There are three clusters of transiently expressed genes and there is also a cluster of genes that are repressed immediately by RA. Interestingly, there are no genes transiently downregulated (i.e., genes whose expression levels decrease and are subsequently restored) at any stage of the entire time-course. A total of about 180 genes and ESTs are distributed among these clusters representing only 6% of the total differentially expressed genes (Table 1). Of these, 72% are known genes and 28% are unknown (ESTs) or poorly annotated. Assuming that the 19,200 genes profiled in this study represent approximately half of all genes in the human genome, the number of extrapolated genes involved in these key stages of neuronal differentiation is remarkably small.

Genes coding for transcription factors, growth factors, and receptors selected from these clusters were validated by quantitative polymerase chain reaction and their overall patterns of change are represented in Fig. 1. As expected, the genes responding most rapidly to *RA* include genes associated with RA transport and processing (*CRABP1, CRABP2, CYP26A1 [22]*). The fact that their expression is transient demonstrates that *RA* is no longer needed to drive differentiation after day 7 or to directly effect cell cycle exit and to direct maturation in postmitotic cells. Moreover, it renders the mature neurons and even their radial glia precursors unable to

Table 1
Categories of Genes Differentially Expressed in Relation to the Cell Cycle

Gene categories	Differentially expressed before cell cycle exit	Downregulated at cell cycle exit	Upregulated at cell cycle exit	Upregulated in maturation
Number of genes	180	1053	1553	187
Known genes (%)	72	70	34	15
Energetics/metabolism (%)	13	26	22	12
Transcription/translation (%)	5	18	5	6
Chromatin remodeling (%)	0	5	5	1
ECM/cytoskeleton (%)	21	7	16	25
Signaling (%)	53	29	37	42
Others (%)	83	15	15	14

respond to RA as an extrinsic factor. It is clear from this study that as cells move along the differentiation path, they downregulate their ability to exist in the previous cell state. This forms the basis of restriction because the cells lose potency and this occurs before cell cycle exit. A key gene that responds immediately to RA, which is also the gene whose expression is the most transient, is that of *TFAP2A*. It is expressed for no more than 1–2 d and is downregulated by day 4. Recent studies have shown that *TFAP2A* is essential for morphogenesis of the lens vesicle *(23)* and trophoblast differentiation *(24)*. Significantly, *TFAP2A* is a regulator of the expression of some *HOX* genes and of midkine (*MDK*), which is also in the cluster of early responding genes. Midkine, a secretory factor, is a member of the *TGF-β* superfamily, is also directly induced by RA *(25–27)*, and known to play a role in neuronal cell survival and differentiation. It has been reported to be able to replace RA as a differentiating factor for P19 cells *(28)*. The relationship between RA and *TFAP2A* in the control of this gene is not clear. The cells begin to express and secrete midkine almost immediately and expression persists until day 7. It is believed to mediate its neurotrophic effects in an autocrine or paracrine fashion via the activation of the ERK, AKT kinase and PI-3 kinase signal transduction pathways *(29,30)*, but its direct downstream targets are not known.

The initiation of *HOX* gene expression, along with *PBX1* parallels that of *TFAP2A* and is already evident by day 2. The pattern of *HOX* gene expression, however, is not transient and is maintained past cell cycle exit through to the mature neuronal stage. *HOX* genes are critical for correct anterior–posterior patterning of the neural tube and thereby confer high level positional information to the cells. Some, but not all, of the *HOX* genes are under direct control of RA *(31)* because RA response elements have been found in some promoters. Moreover, both the *HOXA7* and *HOXA4* promoters have *TFAP2A*-binding sites *(32,33)*, suggesting a role for both inducers in establishing the repertoire of *HOX* gene expression that underlies patterning.

Induction of the *HOX* genes is followed closely by the expression of *PAX6* and *EMX2*. These two genes, which are essential for corticogenesis *(34)*, also impart positional information and are responsible for the arealization of the cortex *(35,36)*. Emx2 protein is restricted to the ventricular zone *(37)* of the developing cortex consistent with its observed transient expression before cell cycle exit. Because *PAX6* is not expressed in the developing lens vesicle in *TFAP2A*-null mice it is likely that it is induced indirectly by RA via *TFAP2A*. *PAX6* expression is a major determinant of cortical radial glia differentiation *(38,39)* and the appearance of *BLBP* coincides with the peak of *PAX6* expression showing that RA directs the differentiation of radial glia before being downregulated. It is also noteworthy that *PAX6* and *HOX* DNA-binding motifs are located in the promoter of the *NCAM1* gene that is expressed in those radial glia cells destined to become neurons *(40)*, thereby establishing a route by which transient RA expression also directs the neurogenic fate of radial glia. *NCAM1* is expressed as one of the genes tightly associated with cell cycle exit and remains expressed on the maturing neurons (Fig. 1). The expression of *EMX2* and *PAX6* are only transient, compared with the *HOX* genes, suggesting that the *HOX* genes may impart additional information on postmitotic cells, for example, directing the cells to express the correct neuronal-subtype phenotype. Relatively little is known about potential interactions between *PAX6* or *EMX2* and the *HOX* genes in terms of how they may coordinate the spatial organization of the cortex.

Also commencing expression at day 4, in parallel with *PAX6* and *EMX2*, is the expression of the *COUP-TF1(NR2F1)*, *MEIS1/2*, *PEG1,3,10*, and *n-OCT3* (*POU3F2*, brn2). Three of the genes expressed in this cluster, *MEIS1,2*, *COUP-TF1*, and *n-OCT3*, are specifically involved in the differentiation of neural cells. *COUP-TF1*, in particular, plays a wide-ranging role at several stages of neural development. Although inducible by RA, *COUP-TF1* is a negative regulator of RA-mediated changes in gene expression *(41)* and may participate in a feedback loop that down-regulates the ability of radial glial cells to respond to RA. Of these genes, *n-OCT3* is also directly inducible by *TFAP2A (42)*. Unlike *EMX2* and *PAX6*, the expression of this cluster of genes remains elevated throughout the differentiation process. Although generally known to be involved in neural differentiation, their roles, with the exception of *COUP-TF1*, are not well defined. The expression of these genes is not confined to the nervous system, suggesting that they play more generic roles in the gene regulatory networks that control differentiation gaining specificity by interacting with tissue-specific genes. The nature of such interaction between the transiently expressed genes is not clear. *EMX2*, *PAX6*, *COUP-TF1*, and *n-OCT3* are part of an intrinsic program that patterns cortical areas *(34,36,37,43,44)*. *PAX6* and *EMX2* interact together to initiate regionalization and *COUP-TF1*, ostensibly acting independently, maintains the identity beyond cell cycle exit. It is possible, therefore, that *COUP-TF1* (and possibly the *MEIS* and *PEG* genes) participate in a regulatory network that carries positional information into postmitotic cells. *n-OCT3* appears to be essential for the migration of postmitotic cortical neurons *(44)*.

These studies also highlight key roles for four imprinted genes that have received relatively little attention. They are paternally expressed *MEST (PEG1)*, *PEG3*, and *PEG10*, and maternally expressed *H19*. All are known to be involved in embryonic development, but their roles remain undefined. *PEG1* may be involved in growth control *(45)*. *PEG3* is important for TNF-NFκ-β signaling and *p53*-mediated apoptosis *(46)*. The *H19* gene produces a transcript that is not translated and there is evidence that it is also involved in the negative control of cell growth *(47)*. *H19* may also be involved in the expression of the closely linked gene *IGF-2 (48)*, which increases transiently after day 7 (Fig. 1). Generally, these genes are not cell-type-specific and are involved in the development of a number of tissues.

Finally, there is a group of genes that are rapidly downregulated immediately upon RA treatment. In this cluster are a number of transcription factors known to be also involved in other cell fates, including *TGIF*, *EKLF* and *GATA-1*. For example, in neurogenesis *TGIF* competes with *MEIS2* for its target genes, such as the dopamine receptor *D(1A)* gene, thereby interfering with neuronal differentiation. Additionally, TGIF binds to RA response elements and prevents the retinoid X receptor RXR from binding and activating transcription *(51)*. It also interacts with Sma- and Mad-related protein (SMAD) to override TGF-β's growth inhibitory signals. TGIF is also involved in spermatogenesis *(49,50)*. GATA-1, a master regulator or erythroid development *(52)*, is also downregulated as well as the amino-terminal enhancer of split transcriptional repressor. EKLF is another key gene involved in erythroid development along with *NFE2L3* that is also downregulated in neurogenesis. It appears, therefore, that RA directly represses commitment to the hematopoietic cell fate.

2.4. Gene Expression Changes Associated With Cell Cycle Exit

The second major phase of gene expression occurs at cell cycle exit and involves the vast majority of the genes that are differentially expressed during Neurogenesis. It includes the genes that effect cell cycle exit, as well has differentiation-associated genes whose expression is tightly coupled to cell cycle exit (Table 1). The majority of the cells destined to become neurons exit the cell cycle commencing around day 7–14, although a few cells exit the cycle much earlier, whereas cells destined to become astrocytes continue to proliferate for several weeks before finally becoming postmitotic. In addition, most early genes, not already downregulated, are downregulated at this time thereby preventing the cells from being able to re-enter any of the previous states.

The cell cycle is a fundamental intrinsic program of all cells and cell cycle exit is a highly orchestrated process *(53)*. In the case of differentiating neuronal precursors, it involves the coordinated expression of thousands of genes. For any given cell, cell cycle exit appears to be rapid despite the massive changes in gene expression that accompany the process. Most notably, the neuron-restrictive silencing factor (NRSF), REST *(54–56)*, is downregulated at this time (Fig. 2) and the expression of *NCAM1*, which is a marker for committed neurons, increases in parallel with cell cycle exit demonstrating that cell cycle exit is a fundamental step in neuronal differentiation.

What actually determines the timing of cell cycle exit is still not clear. Before RA is added no cells exit the cycle, but after transient exposure to RA all the cells will eventually exit the cell cycle and become differentiated without the addition of any other factors implying that RA alone can initiate a complete generic program of neuronal differentiation, although contributions from serum factors cannot be ruled out. However, RA is not required at the actual time of cell cycle exit or at the time when specific neuronal subtypes assume their final identity. Significantly, cell cycle exit parallels the accumulation of Cdk inhibitors (Fig. 1) and it has been proposed that these inhibitors gradually accumulate and individual cells exit the cycle as a threshold is reached *(10)*. For example, p27/Kip1 and p18/INK4c have been shown to time cell cycle exit in differentiating oligodendrocytes *(57)*. p27/Kip1 is one of the Cdks that accumulate in the differentiating NT2 cells (Fig. 1). There is also some evidence *(10,58)* that micro-RNAs, the products of the heterochronic genes, play a key post-transcriptional role in the timing of cell cycle exit.

A brief summary of the main categories of genes, whose expression is tightly associated with cell cycle exit is given in Table 1. The downregulated genes contain a high proportion of genes associated with anabolic metabolism, energy production, protein synthesis, and the transcription/replication machinery. The upregulated genes contain high proportions of genes associated with the cytoskeleton and signaling. A very large proportion of the upregulated genes are unknown or poorly characterized, especially during maturation in which 85% are not characterized. Clearly, there is much to be learned about the phenotype of mature neurons. Thus, transient exposure of the undifferentiated embryonal germ cells to RA leads to the activation of an intrinsic, global program of genome activity, which appears to drive the cells to become postmitotic neurons (or astrocytes).

2.5. Genes Expressed After Cell Cycle Exit

The final major phase of gene expression is the induction of the genes that characterize the neuronal phenotype. This includes the expression of cytoskeleton-associated genes, the population of membranes with neurotransmitter receptors and genes associated with synaptogenesis. One of the genes that is downregulated at the time of cell cycle exit is the NRSF/REST repressor which has been shown to be a key regulatory gene in determining the neuronal phenotype as well as ensuring that these genes are not expressed in nonneuronal cells, including germ cells and radial glia. Thus many of the genes that are upregulated at cell cycle exit and possibly some of the later-induced genes are a consequence of relief from this repression. Little is known about the regulation of genes induced at later stages of maturation, primarily because the genes are so poorly characterized. Of the known genes that are upregulated, there are genes associated with path-finding, axonogenesis, and synaptogenesis.

3. MECHANISMS OF GENOME ACTIVATION

3.1. Intrinsic and Extrinsic Contributions

Extrinsic factors are essential for the survival of precursor cells *(59)* and the undifferentiated NT2 embryonal cells, for example, are absolutely dependent on such factors present in serum. The serum withdrawal experiments demonstrate that undifferentiated NT2 germ cells are completely dependent on extrinsic factors for survival and possibly for proliferation even though the cells themselves secrete a number of factors that may either contribute to survival/proliferation or promote maintenance of the embryonic phenotype. On RA addition, there is a rapid shift away from dependence on extrinsic factors at the onset of differentiation as predicted by Edlund and Jessell *(10)*. Serum can be removed from the cultures after 4–8 d of retinoic treatment and the cells survive and continue to differentiate. Following RA addition, a number of growth factor receptor genes are rapidly repressed as well as the expression of *FGF4* and *TGF-*β and other cytokines (Fig. 1, *see also* ref. *3*). Significantly, the downregulated expression of secreted growth factors, most notably *FGF4*, which also renders the cells nontumorigenic, does not lead to cell cycle exit. It is also important to note that the cells still use autocrine/paracrine mechanisms during differentiation. For example, there is transient expression of midkine followed by the transient expression of *IGF-2* before cell cycle exit. Thus, a fundamental early effect of the hydrophobic nuclear hormone is to initiate a switch from reliance on extrinsic polypeptide factors toward an intrinsic program with limited proliferation potential followed by neuronal differentiation.

Thus, RA alone triggers a program of gene expression in vitro that recapitulates that occurring in vivo without the need for additional factors. It is not clear at this stage whether serum supplies additional differentiation factors or whether it supplies only survival factors. RA serves as a differentiation cue and directs the expression of both intrinsic factors, such as transcription factors, and extrinsic factors such as growth factors. The use of autocrine and/or paracrine mechanisms blurs the distinction between extrinsic and intrinsic factors.

3.2. Transient States of the Genome

It is evident, therefore, that RA-stimulated differentiation of NT2 cells reflects a largely intrinsic program. The intrinsic program is affected by the sequential, but predominantly transient expression or repression of a number of transcription factors. These TFs fall into four major functional categories:

1. Repression of TFs such as *OCT4* and *NANOG,* which maintain the stem cell phenotype.
2. Repression of TFs such as *EKLF* and *GATA-1* to restrict cell fate to the neural lineage.
3. Longer-term expression of TFs associated with AP patterning, such as *HOX* genes and *PBX1*.
4. Transient expression of TFs, such as *EMX2* and *PAX6,* associated with arealization and radial glia formation and differentiation.

Changes in transcription factor expression occur sequentially and mirror the appearance of markers of intermediate cell state(s) such as *BLBP, GLAST*, and finally *NCAM1*. Of the TFs (as well as signaling molecules and receptors) differentially expressed before cell cycle exit very few are persistent—most of them are either expressed transiently before cell cycle exit or are downregulated in parallel with cell cycle exit. In addition, a number of TFs that were already expressed in the undifferentiated cells are also downregulated at this time. A key consequence of this is the prevention of the committed neuronal cells to exist in any previous cell state. Notable exceptions are the *HOX* genes and *COUP-TF1*, which play additional roles in corticogenesis.

In contrast, the TF changes that occur at the time the cells exit the cell cycle, for example, the downregulation of REST, are persistent, indicative of the emergence of a stable TF state in the postmitotic cells that characterizes the neuronal phenotype. Thus, cell cycle exit plays a key role in establishing this stable state. In a recent review, Oliveri and Davidson *(12)* have shown that the onset of embryonic development is characterized by the existence of a number of transient TF states, which gradually evolve to stable TF states that define specific cell fates or differentiated phenotypes. They also provide direct evidence to show how transient TF expression can influence gene expression networks, irreversibly directing them into stable states.

3.3. General Mechanisms of Changes in Cellular Phenotype

Epigenetic mechanisms, such as DNA methylation and histone modification, are involved in the process of making commitment irreversible by silencing key genes, rendering those areas of the genome unavailable for transcription. Transient gene expression followed by epigenetic modification progressively and irreversibly commits the cells to a specific phenotype. The cells then become no longer dependent on earlier extrinsic/autocrine–paracrine factors. Restriction can therefore be viewed as the sequential elimination of a cell's ability to exist in the previous cell state. Several mechanisms contribute to this, including modulation of the extracellular matrix, changes in the expression and secretion of growth and differentiating factors, withdrawal of cell surface receptors (primarily peptide hormone) and downregulation of associated signaling pathways, downregulation of TF complexes, and epigenetic modification of DNA and chromatin. Indeed, all changes in gene expression take place against a backdrop of changes in DNA and chromatin structure, which physically alters the availability of genes for transcription. Numerous genes, including histone deacetylase and DNA methyltransferase are involved in this process and many of these genes are downregulated at cell cycle exit thereby locking the genome into a final stable TF state (Table 1). DNA replication may be an essential component of this mechanism.

3.4. How Does Transient Exposure to RA Effect a Major Differentiation Process?

RA is the sole extrinsic factor supplied to the NT2 cells and it unleashes a massive reprogramming of the genome, ultimately involving the differential expression of more than 3000 genes. Because RA is only required transiently what does it really do? Clearly, it does not direct the whole program of differentiation and it also does not directly direct cell cycle exit which does not occur in most cells until after day 8. In reality, it likely only directly alters the expression of a handful of genes. In a comprehensive analysis of the literature Balmer and Blomhoff *(31)* identified 532 genes modulated by RA, but concluded that as few as 27 of these genes may be direct targets. The latter included *CRABP1, RARA, RARB, HOXA1, A4, B1, B4*, and *D4*. A second category of additional candidate direct targets included *MEIS1/2, CRABP1, cyp26, MDK*, and *TFAP2*. All of these genes are involved at the earliest stages of neurogenesis.

One of the clearest examples of a direct target in the NT2 cells is the effect of RA on *OCT4* expression. RA has a direct negative effect on the promoter activity of this gene, leading to its repression *(60)*. Interestingly there is also a repressor-binding site for *COUP-TF1*, which is also induced by RA (Fig. 1). Repression of *OCT4*, in turn, leads to downregulation of the expression of *FGF4*, which is one of the secreted polypeptide factors responsible for maintaining the cells in an embryonic state. Thus, *FGF4* downregulation is indirect, resulting from the dissociation of the oct4–sox2 transactivating heterodimer from the *FGF4* promoter. Following downregulation, the *OCT4* promoter is then DNA-methylated, leading to irreversible silencing of the gene *(61)*. *TFAP2A* has also been shown to be a direct target of RA *(62)* and to play a role in the induction of many downstream genes.

Although RA directly modulates the transient expression of only a few genes, it is capable of directing a major change in cellular phenotype. Because RA induces *TFAP2* and *TFAP2* induces *PAX6*, the generation of the neurogenic radial glia phenotype is an intrinsic program driven by RA. Moreover, *n-OCT3* induces expression of *BLBP (62)* and BLBP is essential for creation of the radial glia scaffold to support the migration of postmitotic cortical neurons. Furthermore, because *TFAP2A* is an inducer of *PAX6*, as well as some of the *HOX* genes, and *PAX6* and *HOX* genes can collaborate to induce *NCAM1* there is a direct pathway between *RA* and induction of the neuronal phenotype via the intermediate radial glia phenotype with *PAX6*, thereby influencing the neurogenic potential of the radial glia *(39)*.

Both *OCT4* and *PAX6* are key genes that define particular cell states—the germ cell and radial glia phenotypes, respectively. Interestingly, both of these genes, which are major determinants of cell identity, interact with *SOX2* and control genes as heterodimers (Fig. 1). *SOX2* is highly expressed in germ cells and stem cells and remains high throughout differentiation until the cells exit the cell cycle. There is evidence that *SOX2* modulates the transcriptional activity of *PAX6 (64)* and possibly *OCT4 (65)* by tightly controlling their DNA-binding activity.

3.5. Final Comments

Differentiation of neuronal cells involves two major irreversible steps. The first is exit from the germ or stem cell phenotype and the second is exit from the cell cycle. The bulk of gene expression changes occur at the latter stage. Exit from the germ-cell state is driven by the repression of a small number genes commencing within 1–2 d of RA treatment. Commitment to the neural cell fate is also driven by the expression of a small number of genes together with the repression of genes associated with alternative fate(s). The induced genes reach their maximum levels of expression by 6–8 d and after this time RA can be removed and differentiation will continue in serum-free medium. Because the vast majority of cells are still cycling at this time commitment to differentiation occurs while the cells are still replicating, but is only manifested after cell cycle exit.

One can view the differentiation process unleashed by RA as one consisting of a master intrinsic program that runs to produce a "generic" neuron (or astrocyte) in a defined temporal order with no formal requirement for extrinsic input *(66)*. Extrinsic cues, which are primarily positional, may have temporally restricted opportunities to impinge on this program to create region-specific neuronal subtypes. Thus the overall timing of the program and its final outcome could be impacted by extrinsic factors secreted locally which impinge on the master program to dictate cell identity. These extrinsic cues likely impact on the repertoire of *HOX* gene expression, for example, and on the levels of *PAX6* and *EMX2*, thereby influencing positional identity and the creation of neuronal subtypes. The timing of cell cycle exit and finally maturation may be intrinsic or may depend on the exact spatial context of each particular cell, or group of cells, within developing brain regions. It is evident that most of these extrinsic factors must act early and transiently on cells that are still proliferating and it is the final cell cycle that locks in the neuronal identity. Additional extrinsic factors can however impact on maturation, but they cannot fundamentally change the phenotype of the neuron; for example, during axonal path-finding and synaptogenesis to guide the extending processes to their correct destination.

REFERENCES

1. Gangemi RM, Perera M, Corte G. Regulatory genes controlling cell fate choice in embryonic and adult neural stem cells. J Neurochem 2004;89:286–306.
2. Bani-Yaghoub M, Felker JM, Ozog MA, Bechberger JF, Naus CC. Array analysis of the genes regulated during neuronal differentiation of human embryonal cells. Biochem Cell Biol 2001;79:387–398.
3. Luo Y, Cai J, Liu Y, et al. Microarray analysis of selected genes in neural stem and progenitor cells. J Neurochem 2002;83:1481–1497.
4. Freemantle SJ, Kerley JS, Olsen SL, Gross RH, Spinella MJ. Developmentally-related candidate retinoic acid target genes regulated early during neuronal differentiation of human embryonal carcinoma. Oncogene 2002;21:2880–2889.
5. Przyborski SA, Smith S, Wood A. Transcriptional profiling of neuronal differentiation by Human Embryonal Carcinoma stem cells in vitro. Stem Cells 2003;21:459–471.
6. Li A, Zhu X, Brown B, Craft CM. Gene expression networks underlying RA-induced differentiation of human retinoblastoma cells. Invest Ophthalmol Vis Sci 2003;44:996–1007.
7. Evans SJ, Choudary PV, Vawter MP, et al. DNA microarray analysis of functionally discrete human brain regions reveals divergent transcriptional profiles. Neurobiol Dis 2003;14:240–250.
8. Ahn JI, Lee KH, Shin DM, et al. Comprehensive transcriptome analysis of differentiation of embryonic stem cells into midbrain and hindbrain neurons. Dev Biol 2004;265:491–501.
9. Brandenberger R, Wei H, Zhang S, et al. Transcriptome characterization elucidates signaling networks that control human ES cell growth and differentiation. Nat Biotechnol 2004;22:707–716.
10. Edlund T, Jessell TM. Progression from extrinsic to intrinsic signaling in cell fate specification, a view from the nervous system. Cell 1999;96:211–224.
11. Jessell TM. Neuronal specification in the spinal cord, inductive signals and transcriptional codes. Nat Rev Genet 2000;1:20–29.
12. Oliveri P, Davidson EH. Gene regulatory network controlling embryonic specification in the sea urchin. Curr Opin Genet Dev 2004;14:351–360.
13. Walker PR, Ly D, Liu QY, et al. Genome wide transcriptional analysis of neurogenesis in human NT2 cells, unpublished.
14. Andrews PW. Human teratocarcinoma stem cells, glycolipid antigen expression and modulation during differentiation. J Cell Biochem 1987;35:321–332.
15. Loring JF, Porter JG, Seilhammer J, Kaser MR, Wesselschmidt R. A gene expression profile of embryonic stem cells and embryonic stem cell-derived neurons. Restor Neurol Neurosci 2001;18:81–88.

16. Rao RR, Stice SL. Gene expression profiling of embryonic stem cells leads to greater understanding of pluripotency and early developmental events. Biol Reprod, 2005;71:1772–1778.
17. Bhattacharya B, Miura T, Brandenberger R, et al. Gene expression in human embryonic stem cell lines, unique molecular signature. Blood 2004;103:2956–2964.
18. Jones CM, Simon-Chazottes D, Guenet JL, Hogan BL. Isolation of Vgr-2, a novel member of the transforming growth factor-beta-related gene family. Mol Endocrinol 1992;6:1961–1968.
19. Fraidenraich D, Lang R, Basilico C. Distinct regulatory elements govern Fgf4 gene expression in the mouse blastocyst, myotomes, and developing limb. Dev Biol 1998;204:197–209.
20. Sylvester I, Scholer HR. Regulation of the Oct-4 gene by nuclear receptors. Nucleic Acids Res 1994;22:901–911.
21. Anthony TE, Klein C, Fishell G, Heintz N. Radial glia serve as neuronal progenitors in all regions of the central nervous system. Neuron 2004;41:881–890.
22. Maden M. Retinoid signalling in the development of the central nervous system. Nat Rev Neurosci 2002;3:843–853.
23. West-Mays JA, Zhang J, Nottoli T, et al. AP-2 alpha transcription factor is required for early morphogenesis of the lens vesicle. Dev Biol 1999;206:46–62.
24. Cheng YH, Aronow BJ, Hossain S, Trapnell B, Kong S, Handwerger S. Critical role for transcription factor AP-2 alpha in human trophoblast differentiation. Physiol Genomics 2004;18:99–107.
25. Maruta H, Bartlett PF, Nurcombe V, et al. Midkine (MK), a RA (RA)-inducible gene product, produced in *E. coli* acts on neuronal and HL60 leukemia cells. Growth Factors 1993;8:119–134.
26. Pedraza C, Matsubara S, Muramatsu T. A RA-responsive element in human midkine gene. J Biochem (Tokyo) 1995;117:845–849.
27. Zhuang Y, Faria TN, Chambon P, Gudas LJ. Identification and characterization of RA receptor beta2 target genes in F9 teratocarcinoma cells. Mol Cancer Res 2003;1:619–630.
28. Michikawa M, Xu RY, Muramatsu H, Muramatsu T, Kim SU. Midkine is a mediator of RA induced neuronal differentiation of embryonal carcinoma cells. Biochem Biophys Res Commun 1993; 192:1312–1318.
29. Owada K, Sanjyo N, Kobayashi T, et al. Midkine inhibits apoptosis via extracellular signal regulated kinase (ERK) activation in PC12 cells. J Med Dent Sci 1999;46:45–51.
30. Owada K, Sanjo N, Kobayashi T, et al. Midkine inhibits caspase-dependent apoptosis via the activation of mitogen-activated protein kinase and phosphatidylinositol 3-kinase in cultured neurons. Neurochemistry 1999;73:2084–2092.
31. Balmer JE, Blomhoff R. Gene expression regulation by RA. J Lipid Res 2002;43:1773–1808.
32. Kim MH, Cho M, Park D. Sequence analysis of the 5′-flanking region of the gene encoding human HOXA-7. Somat Cell Mol Genet 1998;24:371–374.
33. Doerksen LF, Bhattacharya A, Kannan P, Pratt D, Tainsky MA. Functional interaction between a RARE and an AP-2 binding site in the regulation of the human HOX A4 gene promoter. Nucleic Acids Res 1996;24:2849–2856.
34. Job C, Tan SS. Constructing the mammalian neocortex, the role of intrinsic factors. Dev Biol 2003;257:221–232. Erratum in Dev Biol 259, 188-91.
35. Bishop KM, Rubenstein JL, O'Leary DD. Distinct actions of Emx1, EMX2, and PAX6 in regulating the specification of areas in the developing neocortex. J Neurosci 2002;22:7627–7638.
36. Hamasaki T, Leingartner A, Ringstedt T, O'Leary DD. EMX2 regulates sizes and positioning of the primary sensory and motor areas in neocortex by direct specification of cortical progenitors. Neuron 2004;43:359–372.
37. Cecchi C. Emx2, a gene responsible for cortical development, regionalization and area specification. Gene 2002;291:1–9.
38. Gotz M, Stoykova A, Gruss P. PAX6 controls radial glia differentiation in the cerebral cortex. Neuron 1998;21:1031–1044.
39. Heins N, Malatesta P, Cecconi F, et al. Glial cells generate neurons, the role of the transcription factor PAX6. Nat Neurosci 2002;5:308–315.
40. Meech R, Kallunki P, Edelman GM, Jones FS. A binding site for homeodomain and Pax proteins is necessary for L1 cell adhesion molecule gene expression by Pax-6 and bone morphogenetic proteins. Proc Natl Acad Sci USA 1999;96:2420–2425.
41. Neuman K, Soosaar A, Nornes HO, Neuman T. Orphan receptor COUP-TF I antagonizes RA-induced neuronal differentiation. J Neurosci Res 1995;41:39–48.
42. Atanasoski S, Toldo SS, Malipiero U, Schreiber E, Fries R, Fontana A. Isolation of the human genomic brain-2/N-Oct 3 gene (POUF3) and assignment to chromosome 6q16. Genomics 1995;26:272–280.

43. Zhou C, Tsai SY, Tsai MJ. COUP-TFI, an intrinsic factor for early regionalization of the neocortex. Genes Dev 2001;15:2054–2059.
44. McEvilly RJ, de Diaz MO, Schonemann MD, Hooshmand F, Rosenfeld MG. Transcriptional regulation of cortical neuron migration by POU domain factors. Science 2002;295:1528–1532.
45. Shi W, Lefebvre L, Yu Y, et al. Loss-of-imprinting of Peg1 in mouse interspecies hybrids is correlated with altered growth. Genesis 2004;39:65–72.
46. Johnson MD, Wu X, Aithmitti N, Morrison RS. Peg3/Pw1 is a mediator between p53 and Bax in DNA damage-induced neuronal death. J Biol Chem 2002;277:23,000–23,007.
47. Yamamoto Y, Nishikawa Y, Tokairin T, Omori Y, Enomoto K. Increased expression of H19 non-coding mRNA follows hepatocyte proliferation in the rat and mouse. J Hepatol 2004;40:808–814.
48. Vernucci M, Cerrato F, Pedone PV, Dandolo L, Bruni CB, Riccio A. Developmentally regulated functions of the H19 differentially methylated domain. Hum Mol Genet 2004;13:353–361.
49. Ayyar S, Jiang J, Collu A, White-Cooper H, White RA. Drosophila TGIF is essential for developmentally regulated transcription in spermatogenesis. Development 2003;130:2841–2852.
50. Yang Y, Hwang CK, D'Souza UM, Lee SH, Junn E, Mouradian MM. Three-amino acid extension loop homeodomain proteins MEIS2 and TGIF differentially regulate transcription. J Biol Chem 2000;275:20,734–20,741.
51. Bertolino E, Reimund B, Wildt-Perinic D, Clerc RG. A novel homeobox protein which recognizes a TGT core and functionally interferes with a retinoid-responsive motif. J Biol Chem 1995;270:31,178–31,188.
52. Welch JJ, Watts JA, Vakoc CR, et al. Global regulation of erythroid gene expression by transcription factor GATA-1. Blood, in press.
53. Cremisi F, Philpott A, Ohnuma S. Cell cycle and cell fate interactions in neural development. Curr Opin Neurobiol 2003;13:26–33.
54. Lunyak VV, Prefontaine GG, Rosenfeld MG. REST and peace for the neuronal-specific transcriptional program. Ann NY Acad Sci 2004;1014:110–120.
55. Lunyak VV, Burgess R, Prefontaine GG, et al. Corepressor-dependent silencing of chromosomal regions encoding neuronal genes. Science 2002;298:1747–1752.
56. Bruce AW, Donaldson IJ, Wood IC, et al. Genome-wide analysis of repressor element 1 silencing transcription factor/neuron-restrictive silencing factor (REST/NRSF) target genes. Proc Natl Acad Sci USA 2004;101:10458–10463.
57. Tokumoto YM, Apperly JA, Gao FB, Raff MC. Posttranscriptional regulation of p18 and p27 Cdk inhibitor proteins and the timing of oligodendrocyte differentiation. Dev Biol 2002;245:224–234.
58. Ohnuma S, Harris WA. Neurogenesis and the cell cycle. Neuron 2003;40:199–208.
59. Calof AL. Intrinsic and extrinsic factors regulating vertebrate neurogenesis. Curr Opin Neurobiol 1995;5:19–27.
60. Schoorlemmer J, van Puijenbroek A, van Den Eijnden M, Jonk L, Pals C, Kruijer W. Characterization of a negative retinoic acid response element in the murine Oct4 promoter. Mol Cell Biol 1994;14:1122–1136.
61. Deb-Rinker P, Ly D, Jezierski A, Sikorska M, Walker PR. Sequential DNA methylation of the Nanog and Oct-4 upstream regions in human NTZ cells during neuronal differentiation. J Biol Chem 2005;280:6257–6260.
62. Luscher B, Mitchell PJ, Williams T, Tjian R. Regulation of transcription factor AP-2 by the morphogen retinoic acid and by second messengers. Genes Dev 1989;3:1507–1517.
63. Josephson R, Muller T, Pickel J, et al. POU transcription factors control expression of CNS stem cell-specific genes. Development 1998;125:3087–3100.
64. Aota S, Nakajima N, Sakamoto R, Watanabe S, Ibaraki N, Okazaki K. Pax6 autoregulation mediated by direct interaction of Pax6 protein with the head surface ectoderm-specific enhancer of the mouse Pax6 gene. Dev Biol 2003;257:1–13.
65. Berndt CT, Nowling T, Rizzino A. Transcription factor sox-2 inhibits co-activator stimulated transcription. Mol Reprod Dev 2004;69:260–267.
66. Cayouette M, Barres BA, Raff M. Importance of intrinsic mechanisms in cell fate decisions in the developing rat retina. Neuron 2003;40:897–904.

7
Neurogenesis and Apoptotic Cell Death

Klaus van Leyen, PhD, Seong-Ryong Lee, MD, PhD, Michael A. Moskowitz, MD, and Eng H. Lo, PhD

SUMMARY

Traditionally, the adult brain has been thought of as a very static structure. Essentially, the neurons you are born with are all you have; once they are lost, they cannot be regenerated. This view left very little room for neurogenesis, the birth of new neurons; or apoptosis, neuronal-programmed cell death, for that matter. In the past several years, this belief has been thoroughly upended. In this chapter, we will try to give a perspective on the exciting recent developments and present the new views of life and death in the adult brain that have arisen from them. We have divided this chapter into two segments. The first part deals with postnatal functions of neurogenesis and apoptotic cell death in the healthy brain, whereas the second part focuses on neurogenesis and apoptotic processes following stroke and other neurodegenerative diseases.

Key Words: Apoptosis; neurogenesis; brain development; stroke; Alzheimer's disease; subventricular zone; subgranular zone; BrDu; double cortin; caspase.

1. INTRODUCTION

Whereas prenatal neural development is characterized by rampant yet tightly coordinated neurogenesis and apoptosis, postnatal neural development in the adult mammalian brain presents a much more sedate picture. Only two areas of the brain, the subventricular zone (SVZ) and the subgranular zone (SGZ), are known to exhibit appreciable levels of neurogenesis after birth. The former supplies fresh interneurons for the olfactory bulb, contributing to increased diversity of the olfactory system. In contrast, the latter is assumed to be involved in short-term memory formation in the hippocampus. In addition, neuronal precursor cells can be isolated from various brain regions, but it is unclear that how much they contribute to neurogenesis under normal conditions.

Whereas adult neurogenesis has recently come into the spotlight for its therapeutic potential in neurodegenerative diseases, apoptosis in the postnatal brain is generally seen as a detrimental process. Still, it may also have a balancing function, to weed out those newly formed neurons that have not functionally integrated in the already existing neural network.

The detection of both apoptotic events and increased neurogenesis in the adult brain following injury, finally, seems to provide multiple opportunities for therapeutic intervention to limit brain damage and improve recovery after injury.

2. NEUROGENESIS IN THE ADULT MAMMALIAN BRAIN

Ever since the days of Ramón y Cajal, the adult mammalian brain has been seen as a very static structure. Specifically, neurons were believed to be formed before or around birth, with subsequent increases in brain volume stemming from increased gliogenesis. The pioneering work of Altman in the early 1960s demonstrating neurogenesis in the adult rat brain *(1–3)* did

From: *The Cell Cycle in the Central Nervous System*
Edited by: D. Janigro © Humana Press Inc., Totowa, NJ

not gain mainstream acceptance at that time, similar to subsequent studies that relied heavily on ^{3}H-thymidine labeling and electron microscopy *(4,5)*. The situation changed in the mid-1990s, in part because of the introduction of additional markers for neural precursor cells *(6–8)*. Since then, the field has expanded rapidly.

3. FUNCTIONS OF ADULT MAMMALIAN NEUROGENESIS

To date, two functions of neurogenesis in the healthy adult mammalian brain are known. The rodent SVZ harbors neuronal precursors, principally in its anterior part, and generates neuroblasts that migrate tangentially into the olfactory bulb, thus forming the so-called rostral migratory stream. After arrival, they differentiate into new interneurons that help to determine olfactory specification. Whereas in humans the olfactory sense is not as well developed as in the rodent, an SVZ featuring markers for neuronal precursors can also be defined for the adult human brain *(9)*.

In addition, neuronal precursor cells in the SGZ are now assumed to contribute to memory formation in the adult hippocampus *(10,11)*. This may be one cause for the detrimental effects of radiation treatment on memory formation in the hippocampus *(12)*. Radiation damages neuronal precursors at much lower doses than mature neurons. Imaging studies have also shown increased posterior hippocampal gray matter volume in experienced London taxi drivers *(13,14)*. Whether this increase is caused by neurogenesis is unclear, however, because the phenomenon was not analyzed at the single-cell level.

Other areas of the brain besides SVZ and SGZ have also been found to harbor neuronal precursor cells. Among these are the Macaque cortex *(15)* and the visual cortex in the rodent *(4)*. Additional findings include the striatal parenchyma *(16)* and the primate amygdala *(17)*. Neural stem cell cultures have been isolated from adult spinal cord, cerebral cortex and white-matter tracts, suggesting a fairly widespread occurrence. Whether these neuronal precursors contribute functionally in the healthy brain is at present unknown.

4. FACTORS INFLUENCING NEUROGENESIS

Environmental factors seem to play a major role in mammalian adult neurogenesis. Running increases cell proliferation and neurogenesis, as well as long-term potentiation, in the adult mouse dentate gyrus *(18,19)*. Learning is known to increase the number of granule cells *(20)*. In the visual cortex neurogenesis is stimulated by an enriched environment *(4)*. In addition, as discussed below, various forms of injury are known to stimulate neurogenesis and neuronal migration.

On a molecular level, vascular endothelial growth factor (VEGF) is known to be necessary for exercise-induced adult hippocampal neurogenesis *(21)*. Brain-derived neurotrophic factor (BDNF) and serotonin cooperate in enhancing neurogenesis *(22)*. *See* Section 6 for trophic factors involved in songbird adult neurogenesis *(23)*.

In addition, the newly formed neuron must integrate into an already existing network. Factors involved in neuronal pathfinding and axon guidance during development, of which semaphorin-3A and its receptors neuropilin and plexin are the best-studied examples, are also expressed in the adult, preferentially in regions in which neurogenesis occurs *(24)*. Thus, these guidance molecules may also have their role in rendering neurogenesis functional *(25)*. Downregulation of semaphorin-3A may contribute to aberrant mossy fiber sprouting in the hippocampal dentate gyrus during epilepsy *(26)*.

5. MARKERS FOR NEURONAL DIFFERENTIATION

Incorporation of ^{3}H-thymidine and later bromodeoxyuridine (BrdU) as indicators for dividing cells, along with neuronal morphology, were initially taken as markers for newly generated neurons, generating much controversy *(27)*. In recent years, a number of additional markers have been introduced to define a complete differentiation sequence *(28)*, so most of these issues have now been resolved. It appears that newly generated adult neurons are derived from radial glial cells *(29)*. Widely used molecular markers for neuronally differentiating cells include BrDU, Hu, doublecortin for migrating neurons *(30,31)*, nestin, β-tubulin-III, and polysialylated neural cell adhesion molecule.

6. NEUROGENESIS AND APOPTOSIS IN SONGBIRDS

Some of the clearest indications of postnatal neurogenesis have been found in songbirds. In several species, the ability to form songs requires neurogenesis and arises in males at the juvenile stage *(32,33)*. The acquisition of song leads to massive structural changes in the male brain. On a molecular level, in the canary high-vocal-center region, testosterone upregulates VEGF and its receptor, VEGFR2. Later, BDNF is also increased, presumably secreted by endothelial cells. Here, increased angiogenesis goes along with neurogenesis *(23)*.

Songbirds also provide a clear example of a physiological function for apoptosis well after birth. In the zebra finch, females do not sing. During development, the high-vocal-center region and robust nucleus of the archistriatum robustus archistriatalis areas of the forebrain, which in the male are involved in song learning, are eliminated in young females, at least partially through apoptosis. This removal is paradoxically inhibited by treating the females with estrogen *(34,35)*.

7. NEURONAL APOPTOSIS IN THE ADULT BRAIN

The first is to limit the life-span of new neurons that have not proven to be productively integrated into the existing neural network. Many adult-generated neurons in the hippocampus, indeed the majority, die within weeks of their birth and do not survive to establish connections with other cells or brain regions *(20,36)*. Presumably, this is because of the lack of trophic factors such as BDNF necessary for neuronal survival. This function of apoptosis is akin to that in the developing brain and amounts to a "pruning" of nonproductive neurons.

The removal of apoptotic cells presumably occurs mainly by phagocytosis through primary microglial cells *(37)*. Whether or not other mechanisms also contribute to the clearance of apoptotic neurons is at present not clear. Markers for neuronal apoptosis are discussed in Section 9.

8. AGE EFFECTS

During brain development in the embryonic stage, growth occurs through rampant neurogenesis and production of glial cells. In addition, neuronal apoptosis contributes to brain development by limiting growth when no longer needed, approximately one-half of the newly formed neurons apoptose during maturation of the nervous system *(38)*. A classical example is closure of the neural tube *(39)*, which is defective in (e.g., caspase-9) knockout mice *(40,41)*. A variety of other apoptotic gene knockouts are also embryonic lethal *(42,43)*, indicating the crucial importance of apoptosis during development. In humans, there is still a substantial increase in the number of neurons after birth; the number has been estimated to double between the ages of 15 mo and 6 yr *(44)*. The capacity for apoptosis appears to be higher in neonates than in the adult organism (*see* Fig. 1), although the molecular machinery for apoptosis is expressed throughout life. There also appears to be a decline in neurogenesis with age, although in rodents this varies for different species *(45,46)*.

9. APOPTOSIS IN DISEASED BRAIN

Whereas the physiological role of apoptosis in the sculpting of a developing brain is well accepted, the idea that this process contributes to the pathology of a diseased brain is a more recent one. Over the last decade there has been an explosion of data that define the molecular details of how adult brain cells undergo apoptotic-like cell death in a wide range of central nervous system (CNS) disorders including Alzheimer's, Parkinson's, Huntington's, amyotrophic lateral sclerosis, and stroke. For each disorder, there are not only unique features but also many

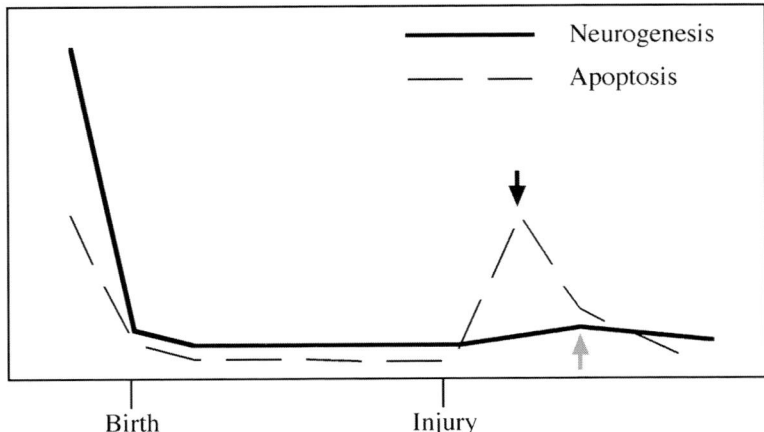

Fig. 1. Schematic diagram of apoptosis and neurogenesis in the brain. Neuroprotective strategies would lead to a ↓ reduction of apoptotic processes or an ↑ increase in neurogenesis to replace dying neurons.

common mechanisms. A full delineation of the specific molecules and unique pathways for each disorder are beyond the scope of this short chapter, and the reader is referred to many excellent prior reviews (38,47–52). Here, we will only briefly summarize the fundamental principles involved.

In essence, energetic and oxidative stress triggered either by genetic defects or external insults trigger a cascade of neuronal death that can be broadly assembled at a mitochondrial locus. The central executioner enzymes in apoptosis comprise the family of cysteine aspartate proteases called caspases, and these are expressed in neurons and become activated during cellular stress and neurodegeneration (53). Prodeath molecules such as Fas ligand and tumor necrosis factor-α bind onto their respective receptors and activate the extrinsic death pathway involving upstream caspase-8. Internally, mitochondrial perturbations trigger the release of additional prodeath signals such as cytochrome-c and Apaf-1, which assemble to form the apoptosome with activated caspase-9 to constitute the intrinsic pathway. Ultimately, downstream caspases such as the prototypical caspase-3 kill cells by cleaving critical cell repair enzymes and homeostatic as well as cytoskeletal proteins. Neurons cultured from mice genetically deficient in caspase-3 are relatively resistant to cell death caused by oxygen-glucose deprivation (54). Similar results can be documented in vivo, with the strongest evidence coming from models of stroke. Chemical inhibitors of caspase or genetic deletion of caspase genes render mice resistant to cerebral ischemia (54,55). The role of caspases in neuronal death may be especially prominent under conditions of milder injury because adenosine triphosphate is required for apoptosome assembly. This suggests that caspase-mediated apoptosis may be emphasized in slower processes of neurodegeneration in which injury may be more subtle over longer periods of time.

Whereas caspases may form the central machinery for executing apoptosis, it is increasingly recognized that a complex web of cell signaling molecules play critical regulatory roles for controlling the execution of cell death. Neuronal apoptosis can be suppressed by prolife kinases such as extracellular signal-regulated kinase (ERK) and Akt. For example, NMDA-receptor activation engages Ras-growth hormone-releasing factor (GRF) exchange factors and upregulates ERK and cyclic AMP-response element-binding protein pathways that mediate endogenous prosurvival responses to cellular stress. Double-knockout mice lacking both Ras-GRF1 and Ras-GRF2 are less resistant to ischemic injury and show increased infarction after focal cerebral ischemia (56). The net effect of ERK activation, however, may also depend on the severity of ischemia. Whereas ERK may be beneficial after moderate ischemia, sustained overactivation of ERK after severe ischemia may paradoxically promote cell death (57). Another important mediator that has emerged is Akt, a serine–threonine kinase that phosphorylates and inactivates cell death mediators such as BAD, caspase-9, and glycogen synthase kinase-3. In cultured neurons, Akt protects against a wide array of excitotoxic, oxidative, and apoptotic insults *(58)*. In a mouse model of focal cerebral ischemia, Akt is rapidly downregulated in the dying ischemic core, whereas increased Akt and neuronal survival in the cortex accompanies superoxide dismutase 1 overexpression, a superoxide radical scavenging protein *(59)*.

In addition to prosurvival responses, neuronal stress also triggers deleterious intracellular signals. The two stress-activated protein kinases, *c*-jun-N-terminal kinase (JNK) and p38, have been intensely investigated in recent years *(60)*. JNK phosphorylates Bax and enhances its mitochondrial translocation, in which it then augments proapoptotic caspase activation. Once again, drawing from the experimental stroke literature is informative. Elevated phospho-JNK colocalizes with terminal deoxynucleotidyltransferasemediated deoxyuridine 5'-triphosphate-biotin nick-end labeling-positive apoptotic neurons in mouse focal cerebral ischemia and inhibition of JNK protects injured brain *(61)*. Consistent with these pharmacological data, ischemic brain injury is reduced in knockout mice lacking the neuron-specific JNK3 isoform *(62)*. Because these stress kinase signals play key roles in the crosstalk between multiple cell death pathways *(63)*, they are attractive targets in the context of neurodegeneration where excitotoxicity, oxidative stress, and apoptosis may converge. Ultimately, multiple pathways are involved so targeting the overall imbalance between prodeath and prolife signaling mechanisms may be required. Finally, in addition to kinase signaling, heat shock proteins also play critical roles by binding and regulating key proteins in the apoptosis cascade *(64–67)*. Not surprisingly, then mechanisms of cell death in neurons are highly redundant, tightly regulated, and very complex.

In recent years, caspase-independent apoptotic pathways have also been dissected. One of these pathways involves the release of apoptosis-inducing factor (AIF), a 67-kDa flavoprotein stored within the same mitochondrial compartment as cytochrome-*c*. The role of AIF in mitochondrial function may involve endogenous antioxidant homeostasis. DNA damage and oxidative or excitotoxic stress lead to release of AIF. The nuclear enzyme poly-adenosine diphosphate-ribose polymerase was recently implicated as a trigger *(68,69)*. On its relocation from the mitochondrial intermembrane space, AIF translocates to the nucleus in which it binds to DNA, promotes chromatin condensation, and kills cells by a complex series of events exhibiting an apoptosis phenotype. Cell death by AIF seems resistant to treatment with pan-caspase inhibitors but is blocked partially by Bcl-2. In addition to AIF, multiple other mediators will surely to emerge as the complex web of apoptosis in neuron death continues to be investigated.

Although apoptosis in adult brain is typically seen as a detrimental process to be blocked in order to rescue neurons in CNS disorders, it may be important to recognize that alternate scenarios may also exist. If selective and controlled dropout of irreversibly damaged neurons can be accomplished, this may still be preferable to large-scale tissue damage through necrosis with its uncontrolled release of intracellular contents and neuroinflammation. Whether this is good or bad ultimately depends on the balance between the number of cells that are sacrificed and those that can be saved from secondary injury. This provocative idea remains an unproven hypotheses; ameliorating apoptosis in diseased brain is still a reasonable target. At least in animal models, inhibition of apoptosis has proven to be neuroprotective.

10. NEUROGENESIS IN DISEASED BRAIN

One of the most exciting frontiers in neuroscience today involves the idea of regrowing brain tissue after acute injury and neurodegeneration. Neurogenesis clearly continues in adult brain *(70)*, and recent findings indicate that these processes can be used to rescue damaged brain tissue. The implications are broad. Here, we will use stroke and Alzheimer's disease (AD) as two "case study" examples.

After transient global ischemia in rat brain, neurogenesis is increased in progenitor cell compartments within the SVZ and SGZ *(71)*, and this endogenous response can be significantly amplified by treatment with growth factors *(72)*. A combination of epidermal growth factor and fibroblast growth factor (FGF)-2 stimulated neurogenesis and re-established up to

50% of cortical area 1 neurons in the damaged hippocampus *(73)*. Most importantly, these neurons were functional; hippocampal slice long-term potentiation was restored, and neurological function was improved on Morris water-maze tests. In addition to the two major sites of neurogenesis in the SVZ and SGZ, more limited progenitors may also be present in other areas of adult brain. In a rat model of cortical apoptosis triggered by focal photochemical injury, dormant progenitor cells within cortex were induced to generate new neurons *(74)*. Thus, it is conceivable that many brain areas can be rescued after stroke.

Neurogenesis in adult brain is mostly restricted to selected regions. Under normal conditions, progenitor cells in the SVZ migrate via the rostral migratory stream toward the olfactory bulb, and neuroblasts from the SGZ mostly serve the hippocampus. An important question is whether damaged cells can be replaced in brain areas not normally targeted by endogenous neurogenesis. Two studies show that this is possible *(16,75)*. After transient focal cerebral ischemia in rats, BrdU labeling of cell proliferation was significantly amplified in the SVZ. By approx 2 wk after stroke, chains of doublecortin-positive neuroblasts become diverted from the rostral migratory stream and migrate toward the damaged striatum. One to three months later, these new cells have assumed neuronal markers such as DARPP-32, consistent with their striatal location. More recent studies have built on these findings to find ways of amplifying this endogenous response in damaged brain. Administration of many pleiotropic molecules has been demonstrated to elevate the neurogenic response and significantly improve outcomes in rodent models of stroke. These include statins *(76)*, VEGF *(77)*, erythropoietin *(78)*, and CD34 cells derived from umbilical cord blood *(79)*. As we better understand how neurogenesis in normal and damaged brain is regulated, we may one day be able to control and precisely target these processes for therapeutic benefit.

Another CNS disorder in which neuronal loss could potentially be replaced via neurogenesis is AD. It is well established that Aβ can be neurotoxic. Surprisingly, however, it was recently demonstrated that soluble Aβ-42 oligomers could induce neurogenesis in isolated neural stem cell cultures *(80)*, suggesting that endogenous beneficial responses may also exist Alzheimer's brain. Others have found that expression of neurogenic markers such as doublecortin and polysialylated adhesion molecules are indeed elevated in AD brain samples *(81)* and a transgenic mouse model that express the Swedish and Iowa amyloid precursor protein mutations *(82)*. In contrast, the presenilin-1 mutant mouse showed impaired neurogenesis in the hippocampus *(83)*. The overall neurogenic response in AD may depend on model systems; the clinical implications of these intriguing data warrant deeper investigation.

In conclusion, both apoptosis and neurogenesis persist in adult brains and may be differentially regulated in disease conditions. As we begin to dissect the signals responsible for mediating these endogenous responses, it is hoped that therapeutic opportunities may be revealed for ameliorating the devastating consequences of CNS injury and neurodegeneration.

REFERENCES

1. Altman J. Are new neurons formed in the brains of adult mammals? Science 1962;135:1127–1128.
2. Altman J, Das GD. Post-natal origin of microneurones in the rat brain. Nature 1965;207:953–956.
3. Altman J, Das GD. Autoradiographic and histological evidence of postnatal hippocampal neurogenesis in rats. J Comp Neurol 1965;124:319–335.
4. Kaplan MS. Environment complexity stimulates visual cortex neurogenesis: death of a dogma and a research career. Trends Neurosci 2001;24:617–620.
5. Gross CG. Neurogenesis in the adult brain: death of a dogma. Nat Rev Neurosci 2000;1:67–73.
6. Kempermann G, Kuhn HG, Gage FH. More hippocampal neurons in adult mice living in an enriched environment. Nature 1997;386:493–495.
7. Palmer TD, Ray J, Gage FH. FGF-2-responsive neuronal progenitors reside in proliferative and quiescent regions of the adult rodent brain. Mol Cell Neurosci 1995;6:474–486.
8. Gage FH, Ray J, Fisher LJ. Isolation, characterization, and use of stem cells from the CNS. Annu Rev Neurosci 1995;18:159–192.

9. Bernier PJ, Vinet J, Cossette M, Parent A. Characterization of the subventricular zone of the adult human brain: evidence for the involvement of Bcl-2. Neurosci Res 2000;37:67–78.
10. Kempermann G, Wiskott L, Gage FH. Functional significance of adult neurogenesis. Curr Opin Neurobiol 2004;14:186–191.
11. Cameron HA, McKay RD. Adult neurogenesis produces a large pool of new granule cells in the dentate gyrus. J Comp Neurol 2001;435:406–417.
12. Monje ML, Palmer T. Radiation injury and neurogenesis. Curr Opin Neurol 2003;16:129–134.
13. Maguire EA, Spiers HJ, Good CD, Hartley T, Frackowiak RS, Burgess N. Navigation expertise and the human hippocampus: a structural brain imaging analysis. Hippocampus 2003;13:250–259.
14. Maguire EA, Gadian DG, Johnsrude IS, et al. Navigation-related structural change in the hippocampi of taxi drivers. Proc Natl Acad Sci USA 2000;97:4398–4403.
15. Gould E, Reeves AJ, Graziano MS, Gross CG. Neurogenesis in the neocortex of adult primates. Science 1999;286:548–552.
16. Arvidsson A, Collin T, Kirik D, Kokaia Z, Lindvall O. Neuronal replacement from endogenous precursors in the adult brain after stroke. Nat Med 2002;8:963–970.
17. Bernier PJ, Bedard A, Vinet J, Levesque M, Parent A. Newly generated neurons in the amygdala and adjoining cortex of adult primates. Proc Natl Acad Sci USA 2002;99:11,464–11,469.
18. van Praag H, Kempermann G, Gage FH. Running increases cell proliferation and neurogenesis in the adult mouse dentate gyrus. Nat Neurosci 1999;2:266–270.
19. van Praag H, Christie BR, Sejnowski TJ, Gage FH. Running enhances neurogenesis, learning, and long-term potentiation in mice. Proc Natl Acad Sci USA 1999;96:13,427–13,431.
20. Gould E, Beylin A, Tanapat P, Reeves A, Shors TJ. Learning enhances adult neurogenesis in the hippocampal formation. Nat Neurosci 1999;2:260–265.
21. Fabel K, Tam B, Kaufer D, et al. VEGF is necessary for exercise-induced adult hippocampal neurogenesis. Eur J Neurosci 2003;18:2803–2812.
22. Mattson MP, Maudsley S, Martin B. BDNF and 5-HT: a dynamic duo in age-related neuronal plasticity and neurodegenerative disorders. Trends Neurosci 2004;27:589–594.
23. Louissaint A, Jr, Rao S, Leventhal C, Goldman SA. Coordinated interaction of neurogenesis and angiogenesis in the adult songbird brain. Neuron 2002;34:945–960.
24. Giger RJ, Pasterkamp RJ, Heijnen S, Holtmaat AJ, Verhaagen J. Anatomical distribution of the chemorepellent semaphorin III/collapsin-1 in the adult rat and human brain: predominant expression in structures of the olfactory–hippocampal pathway and the motor system. J Neurosci Res 1998;52:27–42.
25. de Wit J, Verhaagen J. Role of semaphorins in the adult nervous system. Prog Neurobiol 2003;71:249–267.
26. Holtmaat AJ, Gorter JA, De Wit J, et al. Transient downregulation of Sema3A mRNA in a rat model for temporal lobe epilepsy. A novel molecular event potentially contributing to mossy fiber sprouting. Exp Neurol 2003;182:142–150.
27. Cooper-Kuhn CM, Kuhn HG. Is it all DNA repair? Methodological considerations for detecting neurogenesis in the adult brain. Brain Res Dev Brain Res 2002;134:13–21.
28. Kempermann G, Jessberger S, Steiner B, Kronenberg G. Milestones of neuronal development in the adult hippocampus. Trends Neurosci 2004;27:447–452.
29. Alvarez-Buylla A, Lim DA. For the long run: maintaining germinal niches in the adult brain. Neuron 2004;41:683–686.
30. Francis F, Koulakoff A, Boucher D, et al. Doublecortin is a developmentally regulated, microtubule-associated protein expressed in migrating and differentiating neurons. Neuron 1999;23:247–256.
31. Rao MS, Shetty AK. Efficacy of doublecortin as a marker to analyse the absolute number and dendritic growth of newly generated neurons in the adult dentate gyrus. Eur J Neurosci 2004;19:234–246.
32. Goldman SA, Nottebohm F. Neuronal production, migration, and differentiation in a vocal control nucleus of the adult female canary brain. Proc Natl Acad Sci USA 1983;80:2390–2394.
33. Nottebohm F. Neuronal replacement in adult brain. Brain Res Bull 2002;57:737–749.
34. Konishi M, Akutagawa E. Growth and atrophy of neurons labeled at their birth in a song nucleus of the zebra finch. Proc Natl Acad Sci USA 1990;87:3538–3541.
35. Konishi M, Akutagawa E. Neuronal growth, atrophy and death in a sexually dimorphic song nucleus in the zebra finch brain. Nature 1985;315:145–147.
36. Hastings NB, Gould E. Rapid extension of axons into the CA3 region by adult-generated granule cells. J Comp Neurol 1999;413:146–154.

37. Stolzing A, Grune T. Neuronal apoptotic bodies: phagocytosis and degradation by primary microglial cells. FASEB J 2004;18:743–745.
38. Ekshyyan O, Aw TY. Apoptosis in acute and chronic neurological disorders. Front Biosci 2004;9:1567–1576.
39. Weil M, Jacobson MD, Raff MC. Is programmed cell death required for neural tube closure? Curr Biol 1997;7:281–284.
40. Kuida K, Haydar TF, Kuan CY, et al. Reduced apoptosis and cytochrome c-mediated caspase activation in mice lacking caspase 9. Cell 1998;94:325–337.
41. Hakem R, Hakem A, Duncan GS, et al. Differential requirement for caspase 9 in apoptotic pathways in vivo. Cell 1998;94:339–352.
42. Lomaga MA, Henderson JT, Elia AJ, et al. Tumor necrosis factor receptor-associated factor 6 (TRAF6) deficiency results in exencephaly and is required for apoptosis within the developing CNS. J Neurosci 2000;20:7384–7393.
43. Honarpour N, Gilbert SL, Lahn BT, Wang X, Herz J. Apaf-1 deficiency and neural tube closure defects are found in fog mice. Proc Natl Acad Sci USA 2001;98:9683–9687.
44. Shankle WR, Landing BH, Rafii MS, Schiano A, Chen JM, Hara J. Evidence for a postnatal doubling of neuron number in the developing human cerebral cortex between 15 months and 6 years. J Theor Biol 1998;191:115–140.
45. Kuhn HG, Dickinson-Anson H, Gage FH. Neurogenesis in the dentate gyrus of the adult rat: age-related decrease of neuronal progenitor proliferation. J Neurosci 1996;16:2027–2033.
46. Amrein I, Slomianka L, Poletaeva II, Bologova NV, Lipp HP. Marked species and age-dependent differences in cell proliferation and neurogenesis in the hippocampus of wild-living rodents. Hippocampus. 2004;14(8):1000–1010.
47. Bredesen DE. Apoptosis: overview and signal transduction pathways. J Neurotrauma 2000;17: 801–810.
48. Jellinger KA. Cell death mechanisms in neurodegeneration. J Cell Mol Med 2001;5:1–17.
49. Vila M, Przedborski S. Targeting programmed cell death in neurodegenerative diseases. Nat Rev Neurosci 2003;4:365–375.
50. Mattson MP. Apoptosis in neurodegenerative disorders. Nat Rev Mol Cell Biol 2000;1:120–129.
51. Graham SH, Chen J. Programmed cell death in cerebral ischemia. J Cereb Blood Flow Metab 2001;21:99–109.
52. Moskowitz MA, Lo EH. Neurogenesis and apoptotic cell death. Stroke 2003;34:324–326.
53. Troy CM, Salvesen GS. Caspases on the brain. J Neurosci Res 2002;69:145–150.
54. Le DA, Wu Y, Huang Z, et al. Caspase activation and neuroprotection in caspase-3-deficient mice after in vivo cerebral ischemia and in vitro oxygen glucose deprivation. Proc Natl Acad Sci USA 2002;99:15,188–15,193.
55. Endres M, Namura S, Shimizu-Sasamata M, et al. Attenuation of delayed neuronal death after mild focal ischemia in mice by inhibition of the caspase family. J Cereb Blood Flow Metab 1998;18: 238–247.
56. Tian X, Gotoh T, Tsuji K, Lo EH, Huang S, Feig LA. Developmentally regulated role for Ras-GRFs in coupling NMDA glutamate receptors to Ras, Erk and CREB. EMBO J 2004;23:1567–1575.
57. Chu CT, Levinthal DJ, Kulich SM, Chalovich EM, DeFranco DB. Oxidative neuronal injury. The dark side of ERK1/2. Eur J Biochem 2004;271:2060–2066.
58. Luo HR, Hattori H, Hossain MA, et al. Akt as a mediator of cell death. Proc Natl Acad Sci USA 2003;100:11,712–11,717.
59. Noshita N, Sugawara T, Lewen A, Hayashi T, Chan PH. Copper-zinc superoxide dismutase affects Akt activation after transient focal cerebral ischemia in mice. Stroke 2003;34:1513–1518.
60. Wada T, Penninger JM. Mitogen-activated protein kinases in apoptosis regulation. Oncogene 2004;23:2838–2849.
61. Okuno S, Saito A, Hayashi T, Chan PH. The c-Jun N-terminal protein kinase signaling pathway mediates bax activation and subsequent neuronal apoptosis through interaction with bim after transient focal cerebral ischemia. J Neurosci 2004;24:7879–7887.
62. Kuan CY, Whitmarsh AJ, Yang DD, et al. A critical role of neural-specific JNK3 for ischemic apoptosis. Proc Natl Acad Sci USA 2003;100:15,184–15,189.
63. Bossy-Wetzel E, Talantova MV, Lee WD, et al. Crosstalk between nitric oxide and zinc pathways to neuronal cell death involving mitochondrial dysfunction and p38-activated K^+ channels. Neuron 2004;41:351–365.

64. Klettner A. The induction of heat shock proteins as a potential strategy to treat neurodegenerative disorders. Drug News Perspect 2004;17:299–306.
65. Paschen W. Shutdown of translation: lethal or protective? Unfolded protein response versus apoptosis. J Cereb Blood Flow Metab 2003;23:773–779.
66. Kroemer G. Heat shock protein 70 neutralizes apoptosis-inducing factor. Sci World J 2001;1: 590–592.
67. Yenari MA. Heat shock proteins and neuroprotection. Adv Exp Med Biol 2002;513:281–299.
68. Yu SW, Wang H, Dawson TM, Dawson VL. Poly (ADP-ribose) polymerase-1 and apoptosis inducing factor in neurotoxicity. Neurobiol Dis 2003;14:303–317.
69. Yu SW, Wang H, Poitras MF, et al. Mediation of poly (ADP-ribose) polymerase-1-dependent cell death by apoptosis-inducing factor. Science 2002;297:259–263.
70. Taupin P, Gage FH. Adult neurogenesis and neural stem cells of the central nervous system in mammals. J Neurosci Res 2002;69: 745–749.
71. Sharp FR, Liu J, Bernabeu R. Neurogenesis following brain ischemia. Brain Res Dev Brain Res 2002;134:23–30.
72. Yoshimura S, Takagi Y, Harada J, et al. FGF-2 regulation of neurogenesis in adult hippocampus after brain injury. Proc Natl Acad Sci USA 2001;98:5874–5879.
73. Nakatomi H, Kuriu T, Okabe S, et al. Regeneration of hippocampal pyramidal neurons after ischemic brain injury by recruitment of endogenous neural progenitors. Cell 2002;110:429–441.
74. Magavi SS, Macklis JD. Induction of neuronal type-specific neurogenesis in the cerebral cortex of adult mice:manipulation of neural precursors in situ. Brain Res Dev Brain Res 2002;134:57–76.
75. Parent JM, Vexler ZS, Gong C, Derugin N, Ferriero DM. Rat forebrain neurogenesis and striatal neuron replacement after focal stroke. Ann Neurol 2002;52:802–813.
76. Chen J, Zhang ZG, Li Y, et al. Statins induce angiogenesis, neurogenesis, and synaptogenesis after stroke. Ann Neurol 2003;53:743–751.
77. Sun Y, Jin K, Xie L, et al. VEGF-induced neuroprotection, neurogenesis, and angiogenesis after focal cerebral ischemia. J Clin Invest 2003;111:1843–1851.
78. Wang L, Zhang Z, Wang Y, Zhang R, Chopp M. Treatment of stroke with erythropoietin enhances neurogenesis and angiogenesis and improves neurological function in rats. Stroke 2004;35: 1732–1737.
79. Taguchi A, Soma T, Tanaka H, et al. Administration of CD34+ cells after stroke enhances neurogenesis via angiogenesis in a mouse model. J Clin Invest 2004;114:330–338.
80. Lopez-Toledano MA, Shelanski ML. Neurogenic effect of beta-amyloid peptide in the development of neural stem cells. J Neurosci 2004;24:5439–5444.
81. Jin K, Peel AL, Mao XO, et al. Increased hippocampal neurogenesis in Alzheimer's disease. Proc Natl Acad Sci USA 2004;101:343–347.
82. Jin K, Galvan V, Xie L, et al. Enhanced neurogenesis in Alzheimer's disease transgenic (PDGF-APPSw,Ind) mice. Proc Natl Acad Sci USA 2004;101:13,363–13,367.
83. Wen PH, Hof PR, Chen X, et al. The presenilin-1 familial Alzheimer disease mutant P117L impairs neurogenesis in the hippocampus of adult mice. Exp Neurol 2004;188:224–237.

8
Ion Channels and the Cell Cycle

Annarosa Arcangeli, MD, PhD and Andrea Becchetti, PhD

SUMMARY

There has been a recent resurgence of studies on the relation between ion fluxes and cell proliferation. It is now clear that the activity and the expression of several classes of ion channels not only depend on the proliferative or quiescent state of cells, but in turn, regulate these states. Thus, ion channel function is not limited to excitability and ion homeostasis, but is intimately linked to the regulation of all aspects of cellular life, from fertilization to proliferation, from wound healing to the establishment of embryonic axes. Channel activity affects these processes by participating in the control of cell functions as diverse as migration, proliferation, apoptosis, neurite extension, and pathfinding. The precise mechanisms of these actions are still largely unknown, but the convergence of modern cell physiology with the application of molecular biology to mammalian cells appears very promising. We briefly review some of the recent advances in this field, from the standpoint of the cell cycle. We summarize what is known on the relationship between ion channels and cell cycle in normal and neoplastic cells, and focus on some regulatory pathways which modulate the transition between proliferative and nonproliferative states, with special attention to those triggered by integrin-mediated cell adhesion to the extracellular matrix.

Key Words: Ion channels; cell cycle; integrin receptors; tumors; glioma; cell cycle checkpoints; cell adhesion; ERG.

1. INTRODUCTION

Many evidences suggest that ion channels expressed onto the plasma membrane participate to the regulation of the cell proliferation. In many cell types, proper functionality of different types of K^+, Cl^-, and Ca^{2+} channels appears to be necessary for normal cell cycling. For example, the effect of Ca^{2+} channels is often assumed to be related to the production of transient increases in intracellular Ca^{2+} ($[Ca^{2+}]_i$) during cell cycle progression. Regarding other ion channels, inhibition of K^+ and/or Cl^- channels often blocks the cell cycle with no other toxic effect. The evidence is particularly abundant for K^+ channels, actually too abundant to detail here (for review, *see* ref. *1–4*). Most often, the effect on cell cycle seems to depend on the activity of voltage-gated K^+ channels, such as Kv-type delayed rectifiers and ether-a-gó-gó (EAG) channels. However, cell cycle progression has been found to be affected by many other types of K^+ channels, such as K_{IR}-type inward rectifiers *(5)*, adenosine triphosphate (ATP)-sensitive K^+ currents *(6,7)* and Ca^{2+}-activated K^+ currents *(8,9)*. Cl^- channel function has also been implicated in cell cycle progression. Inhibition of Cl^- currents blocks proliferation of endothelial cells *(10)*, rat microglia *(11)*, mouse liver cells *(12)*, and smooth muscle *(13)*. For review of earlier literature on chloride channels, *see* ref. *14*.

Most of the above results have been obtained in normal or transformed cell lines, but increasing evidence points to a role of ion channels in disparate organic processes that depend on cell

From: *The Cell Cycle in the Central Nervous System*
Edited by: D. Janigro © Humana Press Inc., Totowa, NJ

proliferation, such as gliosis, neural development, and wound healing. After injury of the central nervous system (CNS), astrocytes (or Muller cells in retina) dedifferentiate and proliferate until an astrocytic scar is formed ("reactive gliosis"). This reaction is common in neurological diseases, and particularly prominent in epilepsy and stroke. Interestingly, fully differentiated astrocytes and Muller cells are characterized by a large K_{IR} conductance, which maintains the cells hyperpolarized (15). This current is downregulated in proliferative gliotic cells in vitro (16) and in vivo (17–19), a phenomenon associated to cell depolarization, in agreement with the evidence obtained in cell lines. Other ion channels, such as delayed rectifying K^+ channels, are instead normally expressed, whereas others are upregulated, i.e., Ca-activated K^+ channels in Muller cells (15,19) and references therein.

Cell division, and particularly the orientation of its axis, is also a major determinant of developmental processes. From the standpoint of the present review, we notice that recent evidence suggests that the transepithelial electric fields regulate such processes. Electric fields across neuroepithelia in amphibia are necessary for proper CNS development (20–22), and electric fields also affect orientation and rate of mitosis during wound-healing in mammalian cornea (23). These observations underline the importance of understanding the role of different ion channels in cell cycle regulation and may have general implications for the CNS biology, for example, during asymmetrical divisions of cortical neurons. However, the ionic basis of these interesting processes is still poorly understood (24,25).

What function may ion currents fulfill during cell cycle? Alternatively is channel activation necessary for reasons independent of current flow or membrane potential (V_m)? These points are still matter of debate and several possibilities exist, not mutually exclusive. Table 1 summarizes results from a few particularly detailed studies. In general, the cycling state is linked to a depolarized state, whereas hyperpolarization is associated to the differentiated state (26). However, it is difficult to distinguish whether ion channel activity regulates the progression of cell cycle or viceversa. Experiments with channel blockers tend to suggest that, in some cases at least, the former is true. On this assumption, it still remains to be understood whether the putative causative effect of channel activity depends on ion flow (leading to changes in V_m and/or alteration of ion concentration), or on direct triggering of intracellular pathways by conformational transitions of channel proteins. This point is difficult to resolve in the absence of clear ways of distinguishing the effects produced by ion flow to those generated by conformational change, for example, by using drugs that distinguish permeation from gating. In the following sections, we briefly review some possible reasons that may make ion channel involvement necessary in the modulation of cell cycle, giving special emphasis to mammalian cells. In particular, we discuss the possible roles of ion channels during the different cell cycle phases, during tumor proliferation, and the functional link between integrin receptors and intracellular signaling cascades.

2. ION CHANNELS AND CELL CYCLE PHASES

If ion channels regulate proliferation, their activity must be somehow connected to the cell cycle machinery. Different hypotheses have been proposed to account for this connection. First, a number of studies report that K^+ and Cl^- channels modify their activity and/or expression during the mammalian cell cycle. Thus, these proteins are not only involved in the regulating an on–off switch at G_0/G_1, as might be surmised based on the evidence summarized in the previous section, but participate to the modulation of the cell cycle checkpoints, and especially the transition between G_1 and S. Day et al. showed that a large conductance K^+ channel is active after fertilization in murine zygotes (27). This channel inactivates during the S phase and G_2, and activates again at mitosis. It is controlled by a cytoplasmic oscillator, which does not require the functioning of the nuclear cycle, although it interacts with it (28). A similar dependence on the cell cycle phases has been observed in other experimental systems. In HeLa cells, an inwardly rectifying K^+ current was found to be upregulated in G_1 and mitosis, and downregulated in S (29). In

Table 1
Relation Between Ion Channel Expression/Activity and Cell Cycle

Channel type	Cell type	Is channel activity necessary for cell cycle?	Do cell cycle phases regulate the channel?	Channel expression during cell cycle	Does channel activity change?	ΔV_m in cycling cells	References
$I_K(V)$	Human T-lymphocytes	Yes	ND	ND	PHA[a] activates	PHA-induced hyperpolar	(118)
240 pS K^+ channel	Murine zygote	No	Yes	Unchanged	↓ in G_1-S ↑ in G_2-M	Depolarization in S	(27,28)
I_K(IR)	HeLa cells	ND	Yes (highest I in G_1)	ND	No alteration in single-channel properties	ND	(29)
Herg1 Herg1b[b] (Kv11.1)	Mammalian neuroblastoma and leukemia cells	Yes	Yes	Herg1/Herg1b ratio decreases in S	Activation[c]-shift at G_1-S	Depolarization in S	(38,40)
Ca^{2+} channels (L- and T-type)	Rat smooth muscle	ND	Yes (higher I in G_1 and S)	ND	No alteration in V-dependence	ND	(62)
r-EAG (Kv10.1)	Xenopus oocyte CHO	Yes	Yes	ND	Altered permeation	Depolarization	(33,34,102)
$I_{Cl(vol)}$	Mouse liver cell	Yes	Yes (higher I in cycling cells)	ND	ND	ND	(12)
Kv1.3 Kv1.5	Oligodendrocyte progenitors	Yes (Kv1.3)	Yes	Synthesis in G_1	No	ND	(30)
$I_{Cl(vol)}$	Nasopharyngeal carcinoma	Yes	Yes	ND	Low I in M	ND	(43)

Data from a number of particularly detailed studies are summarized here and listed in the leftmost column. When the papers refer to a specific channel clone, the modern nomenclature is also given.

[a]PHA is a potent mitogen for T-lymphocytes.

[b]HERG1b is an N-truncated isoform of HERG1. The two subunit can coassemble. HERG1b tends to favor more depolarized V_m at the steady state (117,118).

[c]We adopt here the terminology that is now common for HERG's voltage-dependent parameters. Originally (38), we had described our results in terms of an inactivation-shift, because we were using a different model of HERG's properties.

CHO, chinese hamster ovary cells; hyperpolar, hyperpolarization; I, whole-cell current density; I_{Cl}, chloride current; $I_{Cl(vol)}$, volume-activated chloride current; I_K(IR), inwardly-rectifying potassium current; I_K(V), voltage-dependent potassium current; ND, not determined; PHA, phytohemagglutinin; r-EAG, rat ether-a-gó-gó.

oligodendrocyte progenitor cells, upregulation of Kv1.3 and Kv1.5 subunits during G_1 leads to K^+ current increase, and inhibition of Kv1.3 prevents G_1/S transition *(30)*.

In other cell types, the connection between V_m and cell cycle occurs through regulation of voltage-dependent K^+ channels belonging to the EAG family *(31)*. At least three subtypes are known: EAG, EAG-related (ERG), and EAG-like. EAG and ERG, in particular, appear to be involved in cell proliferation.

EAG channels resemble typical delayed rectifiers, with activation induced by depolarization and scarce inactivation *(32)*. Brüggemann et al. *(33)* showed that rat-*eag* clones transiently expressed in *Xenopus* oocytes are downregulated by mitosis-promoting factor. Subsequently, in chinese hamster ovary cells cells stably expressing r-EAG, it was found that the permeability of this channel-type shifts after activation of the mitosis-promoting factor, because of block by intracellular Na^+ *(34)*.

ERG channels, on the other hand, activate slowly on depolarizing from a negative V_m. Superimposed to the activation, however, inactivation also occurs, with quicker kinetics *(35,36)*. This inactivation is quickly removed on repolarizing and produces functional inward rectification, before deactivation takes place. This phenomenon is important during the repolarization of spiking cells, such as cardiac myocytes. On the other hand, the steady state properties of ERG seem more relevant in the context of cell cycle progression, which is a relatively slow process. Three *erg* clones are known to date: *erg*1, *erg*2, and *erg*3 *(37)*, with similar biophysical properties. The crossover of the steady state activation and inactivation curves produces appreciable current at V_m around –40 mV, a crucial range for cycling cells. We focus our discussion on the human (h)ERG1 product, the one preferentially expressed in cycling cells. In human and murine neuroblastoma cells, hERG was shown to be regulated by the cell cycle clock *(38)*. In these cells, as in many other cell lines *(39)*, the resting V_m depends on the activation state of hERG channels. Moreover, the midpoint of channel activation (*see* footnote "c" on Table 1) also depends on the cell cycle phase. It shifts during the G_1/S phase, producing a membrane depolarization during the S phase as compared with G_1. The scatter of this parameter decrease in synchronized cells *(38)*. More recently, we have found that, besides the gating properties, the expression of the hERG1 product onto the membrane oscillates during the cell cycle, being upregulated in G_1 and downregulated in S *(40)*. Moreover, the N-truncated isoform hERG1b is instead upregulated during S. Interestingly, the biophysical properties of hERG1b tend to facilitate cell depolarization compared with those of hERG1. Thus, a complex regulatory network seems to concentrate on hERG1 channels to modulate V_m during cell cycle.

In analogy with K^+ currents, for Cl^- currents also it was found that a cycle of channel activity is synchronized with the cell cycle *(41,42)*. Recent work tends to support this idea. In nasopharingeal carcinoma cells, for example, the activity of volume-activated Cl^- channels is high during G_1 phase, decreases in S and increased again in M *(43)*.

Overall, most of these observations come down to the fact that different types of K^+ and Cl^- channels are upregulated in G_1 and M and downregulated during the S phase. This may be achieved through an increase of channel expression or activity, or both. Regardless of the mechanism, which appears to be different in different cell types, the biological significance of these observations is still unclear. Why do K^+ and Cl^- channels show cell cycle-dependent oscillations in activity and expression?

2.1. Relation to Transmembrane Ca^{2+} Flux

One possibility is that V_m oscillations in phase with the cell cycle are necessary to control Ca^{2+} influx at precise points through the cycle *(44)*. Studies in several species have shown that transient increases of $[Ca^{2+}]_i$ are required for cells to proceed through the cell cycle checkpoints at G_1/S, G_2/M and mitosis exit *(45,46)*. The general idea is that Ca^{2+} operates as a second messenger to modulate calmodulin (CaM) *(46,47)*. In mammalian cells, CaM has been shown to be associated to the centrosome, in the mitotic apparatus *(48,49)*. Moreover, its expression increases during G_1 and its activation is necessary for mitotic progression *(50–53)*. The targets of

Ca^{2+}/CaM are unknown, but there is evidence that in some cell types the effect on the cell cycle machinery is mediated by activation of CaM kinase II *(47,54,55)*. In HeLa cells, in particular, a Ca^{2+}/CaM-dependent phosphorylation activates the tyrosine phosphatase p54$^{cdc25-C}$, which triggers mitosis through activation of the p34-cyclinB protein kinase *(47)*. The evidence on calcium and cell cycle comes largely from experiments performed in oocytes, which owing to their dimensions, generate large calcium signals and are relatively easy to manipulate. Technical difficulties have been instead encountered in measuring calcium transients of low amplitude in mammalian cells, although the studies available suggest that what has been found in oocytes has general significance *(56,57)*. In oocytes, the calcium transients are usually produced through calcium release from intracellular stores, whereas in small mammalian cells the contribution of extracellular calcium is probably crucial, because the intracellular reserve is insufficient to sustain prolonged calcium oscillations *(58)*. In this case, calcium channels onto the plasma membrane must play a major role, either to refill intracellular stores or directly contributing to the dynamics of [Ca^{2+}]$_i$ oscillations. In fact, a number of studies have shown a relation between plasma membrane Ca^{2+} channels and proliferation, especially in dependency of mitogenic stimuli and in lymphocyte activation *(58–61)*. If Ca^{2+} influx is required for proper cell division, K$^+$ and/or Cl$^-$ channel-mediated changes in V_m may control this influx in at least two ways. First, membrane hyperpolarization increases the electrochemical gradient for Ca^{2+}. This is thought to be important, at least, during lymphocyte proliferation, in which a K$^+$ channel-mediated hyperpolarization potentiates Ca^{2+} currents flowing through store-operated channels on the plasma membrane *(58,59)*. Second, V_m may modulate the activity of voltage-gated Ca^{2+} channels. Many types of these channels are known, but for our purposes it is useful to distinguish two functional classes: low threshold ("T," activating around –60 mV) and high-threshold ("L," activating around –20 mV). The former have been shown to be expressed frequently in the proliferative phases, whereas the expression and activity of the latter seem to prevail in nonproliferative states (reviewed in ref. 62). In particular, in smooth muscle cells, T-type currents were found to be strongly upregulated in G$_1$ and especially in S, but were absent in G$_0$ and M, whereas the L-type currents had more uniform properties during all phases *(62)*. In these cells, therefore, the pattern of T channel activity parallels the one found in other cell types for K$^+$ or Cl$^-$ channels. It is possible that the moderate cell cycle-dependent hyperpolarization often observed in many cell lines may bring V_m within a range permissive for T-channel steady state currents, that is, around –40 mV. It is intriguing to note that this is also the typical range of hERG channel activation at rest. The homeostasis of [Ca^{2+}]$_i$ and its relations with cell physiology is, however, an issue far from being settled. It is likely that a full understanding of the putative interactions between K$^+$ and Ca^{2+} channels at cell cycle will have to consider the intervention of the other possible ways of calcium entry from the bath, such as the recently discovered transient receptor potential channels, nonvoltage-dependent cationic channels permeable to calcium *(63,64)*.

2.2. Relation to Volume Homeostasis

Besides the putative effects through modulation of Ca^{2+} influx, there are of course other possibilities to explain the relation between ion channels and cell cycle. Indeed, in murine embryos, the correlation between K$^+$ channel activation and [Ca^{2+}]$_i$ increase has been questioned *(28)*. Several authors have suggested that mitosis is sensitive to blockers of Cl$^-$ and K$^+$ channels because the latter are necessary for regulatory volume decrease (RVD) *(1,65)*. RVD is a necessary compensatory response during the early phases of proliferation. In fact, a number of cell types tend to swell during the initial phases of cell division and this might interphere with the cell cycle clock, for example, may dilute some cell cycle regulator. The most suggestive recent evidence comes from observations on a nontransformed line of mouse hepatocytes *(12)*. In these cells, the Cl$^-$ currents activated by cell swelling are absent in quiescent cells but can be measured in cycling cells. Moreover, inhibition of Cl$^-$ channels blocks the hepatocyte proliferation, as it is the case in other cells types mentioned previously. A full discussion of the possible interplay of cell cycle and RVD can be found in the reviews by Nilius *(14,44,65)*.

3. ION CHANNELS AND INTRACELLULAR SIGNALING PATHWAYS

The link between ion channels and the control of cell cycle progression must depend on a relation between ion channel proteins and the intracellular signal transduction pathway converging onto the activation of the growth-controlling genes. We focus our discussion on two such pathways: the signaling cascade triggered by growth factor receptors and the cellular pathways switched on by cell adhesion-mediated by integrin receptors.

3.1. Ion Channels and Growth Factor Receptors

Cell proliferation in mammalian cells is often triggered by growth factor binding to specific receptors. These receptors are usually protein tyrosine kinases that autophosphorylate on binding of the ligand, thus recruiting adaptor protein(s) that in turn activate other downstream proteins. In this way, a kinase cascade is switched on. A common final step of these cascades is the phosphorylation of the extracellular signal-regulated protein kinase 2 (ERK2) mitogen-activated protein (MAP) kinase. On activation, this protein translocates to the nucleus to phosphorylate specific transcription factors. Analogous processes are often stimulated by cytokines, whose receptors dimerize after ligand binding, thus recruiting and activating a Janus kinase, which targets a signal transducer and activator of transcription (STAT). Again, activated STATs translocate to the nucleus, and form a DNA-binding complex that ultimately activates transcription of a set of proliferation-related genes.

A special mention goes to another crucial player in intracellular signaling: Src and its family. Src is a nonreceptor intracellular protein tyrosine kinase that participates in a multitude of cell processes, including cell adhesion, growth, migration and differentiation. Src transduces signals from a variety of receptors, including integrins (*see* Subheading 3.2.), connecting them to the downstream signaling cascades.

Ion channels are common targets for protein phosphorylation and dephosphorylation; both are important regulatory mechanisms in excitable and nonexcitable cells. In addition to the extensive literature about the regulation of ion channels by serine–threonine kinases (reviewed in ref. *66*), increasing evidence suggests that ion channels are also regulated by phosphorylation on tyrosine residues through either receptor or nonreceptor tyrosine kinases *(67)*. It is thus conceivable that growth factors regulate ion channel expression and function through this mechanism. We will briefly review some of the recent evidence, especially as related to cell proliferation.

Many studies have concentrated on the Ca^{2+} channel regulation. Brain-derived neurotrophic factor, for example, upregulates Ca^{2+} channels in rat embryo motoneurons, through the activation of TrkB receptor *(68)*. Platelet-derived growth factor receptor and nerve growth factor exert a similar action in PC12 rat pheochromocytoma cells *(69)*. In both cases an increase in channel expression was detected. On the other hand, a direct tyrosine phosphorylation of the α_1-subunit of the neuronal L-type Ca^{2+} channel occurs after insulin-like growth factor-1 receptor activation *(70)*. The tyrosine kinase activity of the epidermal growth factor (EGF) was also shown to directly modulate L-type Ca^{2+} channel activity in transformed endothelial cells *(71)*, as well regarding activate Cl^- conductance in murine mammary cells *(72)*.

Regarding K^+ channels, initial evidence for voltage-dependent K^+ (Kv) channel regulation by tyrosine phosphorylation was based on the effects of pharmacological inhibitors of tyrosine kinases and phosphatases. Subsequently, biomolecular and biochemical methods allowed the identification of the mechanisms whereby tyrosine kinases modulate K^+ channel activity, at least in some experimental systems. In T-lymphocytes, Kv1.3 turned out to be tyrosine-phosphorylated after stimulation of Fas receptor, and the phosphorylation was produced by the Src-type tyrosine kinase p56^{Lck} *(73)*. When Kv1.3 is coexpressed in HEK293 cells with the EGF receptor or v-Src, the channels becomes phosphorylated on tyrosine and the current is inhibited *(74)*. A similar inhibition was observed in K_{IR} channels, on activation of nerve growth factor or EGF receptors *(75)*. In contrast, tyrosine phosphorylation of native Kv currents in mouse Schwann cells through the Src-family kinase p55^{Fyn} leads to current activation; this effect is mediated by a direct association between the kinase and the channel protein *(76)*. Particularly relevant to the present review is that in a human myeloblastic leukemia-1 cell line, K$^+$ channels are activated by EGF. Such activation is required for G$_1$/S phase transition *(77)*. The suppression of K$^+$ channels prevents the activation of ERK2 in response to EGF and serum. Hence, K$^+$ channels appear to be part of the EGF-mediated MAP kinase pathway that often controls cell proliferation *(78)*.

On the cytokine receptor side of the picture, the same authors recently demonstrated that the expression of hERG channels facilitates tumor cell proliferation caused by tumor necrosis factor (TNF)-α Moreover, hERG coprecipitates with the TNF receptor-1 on the plasma membrane of HEK-transfected cells *(79)*. The growth-promoting effect of hERG may result from the increased activity of the adaptor protein NF-κB, which mediates TNF-α-induced cell proliferation. Cytokines also stimulated the *herg* messenger expression in human hemopoietic progenitor cells and the concomitant entry into the S phase *(80)*. Finally, hERG channels turned out to be positively regulated by the nonreceptor tyrosine kinase Src in the ret microglia MLS-9 cell line. Interestingly Src co-immunoprecipitates with the channel protein, in these cells *(81)*.

3.2. Ion Channels and Integrin Receptors

Adhesive receptors of the integrin family mediate cell adhesion to the extracellular matrix (ECM). Integrins anchor the ECM to the actin cytoskeleton and trigger multiple signaling pathways that regulate cell migration, proliferation, differentiation, and prevent apoptosis *(82)*. One important pathway is centred on activation of the focal adhesion kinase (FAK), which, through recruitment of a number of structural and signaling components like Src, leads to the activation of MAP kinases. Another interesting pathway is the integrin-mediated activation of the ERK cascade involving the transmembrane protein caveolin-1, the kinase Fyn, and the adaptor protein Shc (but not FAK). Recent studies indicate a link between signaling through integrins and activation of small guanosine triphosphatases (GTPases), particularly Rho-A, Rac-1 and CDC42 *(82)*. It is evident that adhesion signals greatly overlap with those activated by growth factor or cytokines receptors *(83)*.

Integrins form physical complexes at the cell membrane with cell surface receptors, giving rise to signaling platforms at the adhesive sites *(84)*. In particular, integrins are often recovered with Caveolin-1, a protein, which characterizes special plasma membrane microdomain, known as caveolae *(85)*. Caveolae are detergent-resistant membrane areas belonging to the so-called lipid rafts, which not only represent an alternative endocytic pathway, but also seem to act as organized transducing centres that concentrate key signaling molecules *(86)*. Integrin association in lipid rafts/caveolae modulates integrin signaling, although the function of this process is unclear *(86)*.

Evidence has recently emerged concerning the functional and sometimes molecular–physical association of ion channels with integrins; such interaction might have a prominent role in the regulation of different cellular activities. Most studies indicate that integrins can regulate ion channel activity. For example integrin-dependent adhesion initiates Ca^{2+} influx in various types of cells: endothelial cells, fibroblasts, osteoclasts, leukocytes, hepatocytes, smooth muscle cells, and epithelial cells *(87)*. In vascular muscle cells, Ca^{2+} channels are acutely regulated by integrin-activation and this leads to regulate cell tone and constriction *(87)*. A long lasting activation of hERG K$^+$ channels has been reported to occur in neuronal and hemopoietic tumor cells after integrin-mediated adhesion *(88,38,89)*. In these cell lines, hERG activation is associated with the induction of neurite extension and osteoclastic differentiation, respectively. Finally, in human neutrophils, activation of the β$_2$-subunit of the integrin receptor triggers Cl$^-$ efflux, necessary to regulate both spreading and respiratory burst *(90)*.

Some of the above integrin-mediated effects on ion channels have been shown to depend on tyrosine phosphorylation in vascular smooth muscle cells *(87)*. In other cases, it is instead the ion channel activation subsequent to integrin-mediated adhesion that regulates the phosphorylated state of intracellular proteins, such as FAK *(91)*. K$^+$ channels can also regulate integrin expression: in human preosteoclastic cells, hERG channel activation is apparently responsible for an up regulation of α$_v$β$_3$ expression on the plasma membrane *(89)*.

Another important aspect of integrin/ion channel interaction is that integrins may contribute to localize ion channels onto the membrane. Kv1.3 channels are necessary for activation of

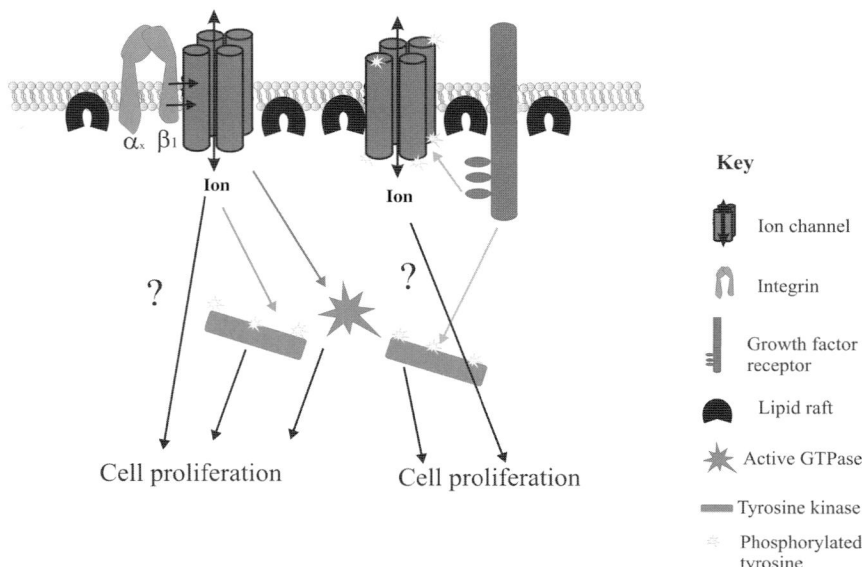

Fig. 1. Mechanisms used to signaling cell proliferation by interactions between growth factors receptors/integrins and ion channels. Ion channels can regulate cell proliferation through intracellular pathways that are integrin-dependent (illustrated on the left) and/or growth factor-dependent (illustrated on the right). Integrins can directly activate ion channels, which in turn contribute to modulate integrin-dependent signaling (i.e., tyrosine kinase phosphorylation and GTPase activation). Growth factor receptors activate ion channels, possibly through phosphorylation of the channel protein (on the right). Alternatively, and more speculatively, ion channel activity can be more directly linked to the control of cell proliferation, through variations of the membrane potential or of the ionic homeostasis. (Please *see* companion CD for color version of this figure.)

β_1-integrins and subsequent integrin-dependent adhesion and migration in T-lymphocytes. Such functional activation relies on the physical association between the two molecules in T-lymphocytes *(92)* and melanoma cells *(93)*. In the latter, Kv1.3/β_1-integrin interaction is promoted by cell adhesion and inhibited by channel blockers *(93)*. β_1-integrins also assemble with GIRK channels in reconstituted systems (oocytes), and thus modulate channel activity *(94)*. A Ca^{2+}-activated Cl^- channel, CLCA2, serves as a β_4-integrin binding partner for adhesion between endothelial cells and breast cancer cells: this interaction seems to be involved in the regulation of tumor metastasis *(95)*. A macromolecular complex between β_1-integrin subunit and hERG channels takes place in both neuroblastoma *(96)* and HEK293 cells stably transfected with the *herg1* gene. The β_1/hERG complex turned out to be localized in adhesion sites in conjunction with caveolin-1, thus suggesting that it is located within specific membrane microdomains, that is caveolae/lipid rafts *(97)*. These results point to an intriguing aspect of integrin–ion channel interaction, that is the possibility that ion channels form macromolecular complexes with integrins, and that such complex, after translocation into caveolae/lipid rafts, recruit and activate diverse signaling proteins. In fact caveolae/lipid rafts can act as organized transducing centers that concentrate key signaling molecules (*see* Subheading 3.2.). Different types of ion channels have been demonstrated to localize in lipid rafts *(98,99)* and such localization can determine the signaling properties of ion channels in excitable cells *(100)*. In HEK cells stably transfected with hERG1, the β_1/hERG complex in fact can also recruit the tyrosine kinase FAK as well as the small GTPase, Rac-1. Interestingly, both FAK phosphorylation and Rac-1 activity appear to be dependent on hERG currents *(97)*.

Figure 1 summarizes the main regulative link between ion channels and integrin or growth factor receptors.

4. ION CHANNELS AND NEOPLASTIC GROWTH

Although we are still a long way from cataloguing cancer as a channelopathy (i.e., a disorder arising directly from ion channel dysfunction), considerable evidence suggests that ion channels are involved in cancer progression and pathology (*see also* Chapter 14 by Dini et al.). Different laboratories have demonstrated that the expression of certain oncogenes directly affects Na^+, K^+, and Ca^{2+} channel function *(101)*. Conversely, an oncogenic potential has been recently ascribed to genes encoding K^+ channel proteins belonging to the EAG family *(102)*.

The contributions of ion channels to the neoplastic phenotype are as diverse as the ion channel families themselves. Most studies concern the involvement of ion channels, especially K^+ channels, in cell cycle regulation (*see* Heading 2.). Not surprisingly, the same ion channel mechanisms that regulate cell proliferation are involved in the control of its flip side, apoptosis *(103)*: Indeed cell shrinkage is one of the early events marking the onset of apoptotic cell death. On the other hand, the potential role of ion channels in tumor-induced angiogenesis and cell invasion has not been much attended to. Invasive growth and cell migration are highly regulated process. The migrating cells must secrete matrix proteases that disrupt the ECM to permit easier transit through the surrounding environment. In addition the migrating cells must profoundly reshape their structure, through massive cytoskeletal rearrangements. There is increasing information concerning ion channel involvement in cytoskeleton reshaping and cell–cell interaction and also some evidence that tumor cell invasion can be halted by the use of channel blockers *(104,105)*.

Finally, the specific pattern of ion channel expression is altered in neoplastic cells as compared with normal counterparts. Evidences about brain tumors will be reported in the next subheading. We recall here the data obtained in tumor cell lines *(39)* and primary tumors. In primary colorectal cancers an aberrant overexpression of hERG channels was recently reported *(105)*, whereas prostate tumor cells express voltage-activated Na^+ currents *(106)* that are one of the hallmarks of excitable cells.

4.1. Ion Channels and Tumors of the Nervous System

Over the past two decades, a number of studies have suggested a role for ion channel activity in the pathogenesis and malignancy of brain tumors. In gliomas, among the most prevalent and aggressive brain tumors, the ion channel expression profile is profoundly altered, and some of these alterations determine many of the features strictly related to tumor establishment and progression, such as relentless growth and invasiveness. Some reports suggest that glioma cells acquire neuron-like biophysical properties, such as a robust increase in Na^+ current density and the reduction of glial-specific inward rectifier currents. This implies a marked depolarization of resting membrane voltage, low K^+ resting conductance and, subsequently, the acquired ability to generate action potential-like responses whose pathological effects are still unclear *(107–109)*. Disruption of glutamate reuptake mechanism in glioma cells is owing to the downregulation of Na^+-dependent glutamate transporters and has also been linked to the ability of glial tumors to destroy peritumoral tissue, and thus expand and invade central nervous system tissue, by causing glutamate-induced excitotoxic neuronal cell death *(110)*. Glutamate also appears to serve as an autocrine signal stimulating migration in glioma cells by inducing $[Ca^{2+}]_i$ oscillations, essential to trigger cell motility, through activation of glutamate receptors, such as GLUR1 and GLUR4 *(111)*.

Many authors have reported a direct role for K^+ and Cl^- channels in cell-volume control and, secondarily, in proliferation and motility (*see* Heading 4). Specific Cl^- channels are indeed expressed in glioma cell lines, and the blockade of tumor-specific Cl^- channels prevents cell shrinkage and reduces invasiveness of glioma cells *(112)*. A model has been proposed in which Cl^- efflux, accompanied by K^+ efflux, causes an osmotically induced water loss and subsequent cell-shrinkage, which enables the cell to infiltrate the narrow extracellular brain spaces.

More recently, large conductance Ca^{2+}-activated K^+ currents have been described in glioblastoma cell lines *(113)*, whereas a mislocalization of K_{IR} channels occurs in malignant glioma cell lines *(114)*. Moreover, a preferential expression of the EAG family currents has been recently detected in human gliomas (refs. *115–117*). A similar preferential expression of an inward, amiloride-sensitive, Na^+ conductance in high-grade gliomas had been previously reported *(101)*. The meaning of these findings is still unclear.

ACKNOWLEDGMENTS

We thank Profs. Massimo Olivotto and Enzo Wanke for advice and encouragement. We apologize to the many authors of valuable work that we were not able to cite because of space limitations. This paper was supported by the Telethon Fondazione Onlus (Project No GGP02208 to A.A. and GTF03007 to A.B.), and the Italian Association for Cancer Research (AIRC) dell'Università e Ricerca Scientifica e Tecnologica (Cofin 2003 to A.A and FAR 2004 to A.B.)

REFERENCES

1. Dubois JM, Rouzaire-Dubois B. Role of potassium channels in mitogenesis. Prog Biophys Mol Biol 1993;59:1–21.
2. Wonderlin WF, Strobl JS. Potassium channels and G1 progression. J Memb Biol 1996;154:91–107.
3. Wang Z. Roles of K^+ channels in regulating tumour cell proliferation and apoptosis. Pflugers Archiv. 2004;448:274–286.
4. Pardo LA. Voltage-gated potassium channels in cell proliferation. Physiology 2004;19:285–292.
5. Lepple-Wienhues A, Berweck S, Bohmig M, et al. K^+ channels and the intracellular calcium signal in human melanoma cell proliferation. J Membr Biol 1996;151:146–157.
6. Lee YS, Sayeed MM, Wurster RD. In vitro antitumor activity of cromakalim in human brain tumor cells. Pharmacology 1994;49:69–74.
7. Malhi H, Irani AN, Rajvanshi P, et al. K_{ATP} channels regulate mitogenically induced proliferation in primary rat hepatocytes and human liver cell lines. Implications for liver growth control and potential therapeutic targeting. J Biol Chem 2000;275:26,050–26,057.
8. Huang MH, Wu SN, Chen CP, Shen AY. Inhibition of Ca^{2+}-activated and voltage-dependent K^+ currents by 2-mercaptophenyl-1, 4-naphtoquinone in pituitary GH3 cells: contribution to its antiproliferative effect. Life Sci 2002;70:1185–1203.
9. Liu X, Chang Y, Reinhart PH, Sontheimer H, Chang Y. Cloning and characterization of glioma BK, a novel BK channel isoform highly expressed in human glioma cells. J Neurosci 2002;22:1840–1849.
10. Voets T, Szucs G, Droogmans G, Nilius B. Blockers of volume-activated Cl^- currents inhibit endothelial cell proliferation. Pflugers Archiv 1995;431:132–134.
11. Schlichter LC, Sakellaropoulos G, Ballyk B, Pennefather PS, Phipps DJ. Properties of K^+ and Cl^- channels and their involvement in proliferation of rat microglial cells. Glia 1996;17:225–236.
12. Wondergem R, Gong W, Monen SH, et al. Blocking swelling-activated chloride currents inhibits mouse liver cell proliferation. J Physiol 2001;532:661–672.
13. Xiao GN, Guan YY, He H. Effects of Cl^- channels blockers on endothelin-1-induced proliferation of rat vascular smooth muscle cells. Life Sci 2002;70:2233–2241.
14. Nilius B, Eggermont J, Voets T, Buyse G, Manolopoulos VG, Droogmans G. Properties of volume-regulated anion channels in mammalian cells. Prog Biophys Mol Biol 1997;68:69–119.
15. Bringmann A, Francke M, Pannicke T, et al. Role of glial K^+ channels in ontogeny and gliosis: a hypothesis based upon studies on Muller cells. Glia 2000;29:35–44.
16. MacFarlane SN, Sontheimer H. Electrophysiological changes that accompany reactive gliosis in vitro. J Neurosci 1997;17:7316–7329.
17. D'Ambrosio R, Maris DO, Grady MS, Winn HR, Janigro D. Impaired K^+ homeostasis and altered electrophysiological properties of post-traumatic hippocampal glia. J Neurosci 1999;19:8152–8162.
18. Schröder W, Hager G, Kouprijanova E, et al. Lesion-induced changes of electrophysiological properties in astrocytes of the rat dentate gyrus. Glia 1999;28:166–174.
19. Bordey A, Lyons SA, Hablitz JJ, Sontheimer H. Electrophysiological characteristics of reactive astrocytes in experimental cortical dysplasia. J Neurophysiol 2001;85:1719–1731.
20. Hotary KB, Robinson KR. Endogenous electrical currents and voltage gradients in *Xenopus* embryos and the consequences of their disruption. Dev Biol 1994;166:279, 800.

21. Shi R, Borgens RB. Three-dimensional gradients of voltage during development of nervous system as invisible coordinates for the establishment of embryonic patterns. Dev Dyn 1995;202:101–114.
22. Borgens RB, Shi R. Uncoupling histogenesis from morphogenesis in the vertebrate by collapse of the transneural tube potential. Dev Dyn 1995;203:456–467.
23. Song B, Zhao M, Forrester JV, McCaig CD. Electrical cues regulate the orientation and frequency of cell division and the rate of wound healing in vivo. Proc Natl Acad Sci USA 2002;99:13577–13582.
24. McCaig CD, Rajnicek AM, Song B, Zhao M. Has electrical growth cone guidance found its potential. Trends Neurosci 2002;25:354–359.
25. Nuccitelli R. A role for endogenous electric fields in wound healing. Curr Top Dev Biol 2003;58:1–26.
26. Binggeli R, Weinstein RC. Membrane potentials and sodium channels: hypotheses for growth regulation and cancer formation based on changes in sodium channels and gap junctions. J Theor Biol 1986;123:377–401.
27. Day ML, Pickering SJ, Johnson MH, Cook DI. cell cycle control of a large conductance K^+ channel in mouse early embryos. Nature 1993;365:560–562.
28. Day ML, Johnson MH, Cook DI. A cytoplasmic cell cycle controls the activity of a K^+ channel in pre-implantation mouse embryos. EMBO J 1998;17:1952–1960.
29. Takahashi A, Yamaguchi H, Miyamoto H. Change in K^+ current of HeLa cells with progression of the cell cycle studied by patch-clamp technique. Am J Physiol 1993;265:C328–C336.
30. Chittajallu R, Chen Y, Wang H, et al. Regulation of Kv1 subunit expression in oligodendrocyte progenitor cells and their role in G_1/S phase progression of the cell cycle. Proc Natl Acad Sci USA 2002;99:2350–2355.
31. Warmke JW, Ganetzky B. A family of potassium channel genes related to eag in Drosophila and mammals. Proc Natl Acad Sci USA 1994;91:3438–3442.
32. Ludwig J, Terlau H, Wunder F, et al. Functional expression of a rat homologue of the voltage-gated ether-a-gó-gó potassium channel reveals differences in selectivity and activation kinetics between the Drosophila channel and its mammalian counterpart. EMBO J 1994;13: 4451–4458.
33. Brüggemann A, Stühmer W, Pardo LA. Mitosis-promoting factor-mediated suppression of a cloned delayed rectifier potassium channel expressed in *Xenopus* oocytes. Proc Natl Acad Sci USA 1997;94:537–542.
34. Pardo LA, Bruggemann A, Camacho J, Stühmer W. Cell cycle-related changes in the conducting properties of r-eag K^+ channels. J Cell Biol 1998;143:767–775.
35. Smith PL, Baukrowitz T, Yellen G. The inward rectification mechanism of the HERG cardiac potassium channel. Nature 1996;379:833–836.
36. Spector PS, Curran ME, Zou A, Keating MT, Sanguinetti MC. Fast inactivation causes rectification of the Ik_r channel. J Gen Physiol 1996;107:611–619.
37. Shi W, Wymore RS, Wang HS, et al. Identification of two nervous system-specific members of the erg potassium channel gene family. J Neurosci 1997;17:9423–9432.
38. Arcangeli A, Bianchi L, Becchetti A, et al. A novel inward-rectifying K^+ current with a cell cycle dependence governs the resting potential of mammalian neuroblastoma cells. J Physiol 1995;489: 455–471.
39. Bianchi L, Wible B, Arcangeli A, et al. Herg encodes a K^+ current highly conserved in tumors of different histogenesis: a selective advantage for cancer cells? Cancer Res 1998;58:815–822.
40. Crociani O, Guasti L, Balzi E, et al. Cell cycle-dependent expression of HERG1 and HERG1B isoforms in tumor cells. J Biol Chem 2003;278:2947–2955.
41. Block ML, Moody WJ. A voltage-dependent chloride current linked to the cell cycle in Ascidian embryos. Science 1990;247:1090–1092.
42. Bubien JK, Kirk KL, Rado TA, Frizzell RA. Cell cycle dependence of chloride permeability in normal and cystic fibrosis lymphocytes. Science 1990;248:1416–1419.
43. Chen L, Wang L, Zhu L, et al. Cell cycle dependent expression of volume-activated chloride currents in nasopharyngeal carcinoma cells. Am J Physiol 2002;283:C1313–C1323.
44. Nilius B, Schwarz G, Droogmans G. Control of intracellular calcium by membrane potential in human melanoma cells. Am J Physiol 1993;265:C1501–C1510.
45. Whitaker M, Larman MG. Calcium and mitosis. Semin Cell Dev Biol 2001;12:53–58.
46. Kahl CR, Means AR. Regulation of cell cycle progression by calcium/calmodulin-dependent pathways. Endocr Rev 2003;24:719–736.
47. Patel R, Holt M, Philipova R, et al. Calcium/calmodulin-dependent phosphorilation and activation of human cdc25-C at the G_2/M phase transition in HeLa cells. J Biol Chem 1999;274:7958–7968.

48. Welsh MJ, Dedman JR, Brinkley BR, Means AR. Calcium-dependent regulator protein: localization in mitotic apparatus of eukaryotic cells. Proc Natl Acad Sci USA 1978;75:1867–1871.
49. Welsh MJ, Dedman JR, Brinkley BR, Means AR. Tubulin and calmodulin. Effects of microtubule and microfilament inhibitors on localization in the mitotic apparatus. J Cell Biol 1979;81:624–634.
50. Sasaki Y, Hidaka H Calmodulin and cell proliferation. Biochem Biophys Res Commun 1982;104:451–456.
51. Rasmussen CD, Means AR. Calmodulin is required for cell cycle progression during G1 and mitosis. EMBO J 1989;8:73–82.
52. Török K, Wilding M, Groigno L, Patel R, Whitaker M. Imaging the spatial dynamics of calmodulin activation during mitosis. Curr Biol 1998;8:692–699.
53. Li C-J, Heim R, Lu P, Pu Y, Tsien RY, Chang DC. Dynamic redistribution of calmodulin in HeLa cells during cell division as revealed by a GFP-calmodulin fusion protein technique. J Cell Sci 1999;112:1567–1577.
54. Baitinger C, Alderton J, Poenie M, Schulman H, Steinhardt RA. Multifunctional Ca^{2+}/calmodulin-dependent kinase is necessary for nuclear envelope breakdown. J Cell Biol 1990;111:1763–1773.
55. Waldmann R, Hanson PI, Schulman H. Multifunctional Ca^{2+}/calmodulin-dependent protein kinase made Ca^{2+} independent for functional studies. Biochemistry 1990;29:1679–1684.
56. Poenie M, Alderton JM, Steinhardt RA, Tsien RY. Calcium rises abruptly and briefly throughout the cell at the onset of anaphase. Science 1986;233:886–889.
57. Kao JP, Alderton JM, Tsien RY, Steinhardt RA. Active involvement of Ca^{2+} in mitotic progression of Swiss 3T3 fibroblasts. J Cell Biol 1990;111:183–196.
58. Berridge MJ, Lipp P, Bootman MD. The versatility and universality of calcium signalling. Nat Rev Mol Cell Biol 2000;1:11–21.
59. Lewis RS, Cahalan MD. Potassium and calcium channels in lymphocytes. Annu Rev Immunol 1995;13:623–653.
60. Lovisolo D, Distasi C, Antoniotti S, Munaron L. Mitogens and calcium channels. News Physiol Sci 1997;12:279–285.
61. Munaron L. Calcium signalling and control of cell proliferation by tyrosine kinase receptors. Int J Mol Med 2002;10:671–676.
62. Kuga T, Kobayashi S, Hirakawa Y, Kanaide H, Takeshita A. Cell cycle-dependent expression of L- and T-type Ca^{2+} currents in rat aortic smooth muscle cells in primary culture. Circ Res 1996;79:14–19.
63. Clapham DE, Runnels LW, Strubing C. The TRP ion channel family. Nat Rev Neurosci 2001;2:387–396.
64. Inoue R, Hanano T, Shi J, Mori Y, Ito Y. Transient receptor potential protein as a novel non-voltage-gated Ca^{2+} entry channel involved in diverse pathophysiological functions. J Pharmacol Sci 2003;91:271–276.
65. Nilius B. Chloride channels go cell cycling. J Physiol 2001;532:581.
66. Ismailov II, Benos DJ. Effects of phosphorylation on ion channel function. Kidney Int 1995;48:1167–1179.
67. Davis MJ, Wu X, Nurkiewicz TR, et al. Regulation of ion channels by protein tyrosine phosphorylation. Am J Physiol 2001;281:H1835–H1862.
68. Baldelli P, Magnelli V, Carbone E. Selective up-regulation of P- and R-type Ca^{2+} channels in rat embryo motoneurons by BDNF. Eur J Neurosci 1999;11:1127–1133.
69. Black MJ, Woo Y, Rane SG. Calcium channel upregulation in response to activation of neurotrophin and surrogate neurotrophin receptor tyrosine kinases. J Neurosci Res 2003;74:23–36.
70. Bence-Hanulec KK, Marshall J, Blair LAC. Potentiation of neuronal L Calcium channels by IGF-1 requires phosphorylation of the $\alpha 1$ subunit on a specific tyrosine residue. Neuron 2000;27:121–131.
71. Mergler S, Dannowski H, Bednarz J, Engelmann K, Hartmann C, Pleyer U. Calcium influx induced by activation of receptor tyrosine kinases in SV40-transfected human corneal endothelial cells. Exp Eye Res 2003;77:485–495.
72. Abdullaev IF, Sabirov RZ, Okada Y. Upregulation of swelling-activated Cl- channel sensitivity to cell volume by activation of EGF receptors in murine mammary cells. J Physiol 2003;549(3):749–758.
73. Szabo I, Gulbins E, Apfel H, et al. Tyrosine phosphorylation-dependent suppression of a voltage-gated K^+ channel in T lymphocytes upon Fas stimulation. J Biol Chem 1996;34:20,465–20,469.

74. Bowlby MR, Fadool DA, Holmes TC, Levitan IB. Modulation of the Kv 1.3 potassium channel by receptor tyrosine kinases. J Gen Physiol 1997;110:601–610.
75. Wischmeyer E, Doring F, Karschin A. Acute suppression of inwardly rectifying Kir 2.1 channels by direct tyrosine kinase phosphorylation. J Biol Chem 1998;273:34,063–34,068.
76. Sobko A, Peretz A, Attali B. Constitutive activation of delayed-rectifier potassium channels by a Src family tyrosine kinase in Schwann cells. EMBO J 1998;17:4723–4734.
77. Wang L, Xu B, White RE, Lu L. Growth factor-mediated K^+ channel activity associated with human myeloblastic ML-1 cell proliferation. Am J Physiol 1997;273:C1657–C1665.
78. Xu D, Wang L, Dai W, Lu L. A requirement for K^+-channel activity in growth factor-mediated extracellular signal-regulated kinase activation in human myeloblastic leukemia ML-1 cells. Blood 1999;94:139–145.
79. Wang H, Zhang Y, Cao L, et al. Herg K^+ channel, a regulator of tumor cell apoptosis and proliferation. Cancer Res 2003;62:4843–4848.
80. Pillozzi S, Brizzi MF, Balzi M, et al. HERG potassium channels are constitutively expressed in primary human acute myeloid leukemias and regulate cell proliferation of normal and leukemic hemopoietic progenitors. Leukemia 2002;16:1791–1798.
81. Cayabyab FS, Schlichter LC. Regulation of an ERG K^+ current by Src tyrosine kinase. J Biol Chem 2002;277:13,673–13,681.
82. Juliano RL. Signal transduction by cell adhesion receptors and the cytoskeleton: functions of integrins, cadherins, selectins, and immunoglobulin-superfamily members. Annu Rev Pharmacol Toxicol 2002;42:283–323.
83. Giancotti FG, Tarone G. Positional control of cell fate through joint integrin/receptor protein kinase signaling. Annu Rev Cell Dev Biol 2003;19:173–206.
84. Brown EJ. Integrin-associated proteins. Curr Opin Cell Biol 2002;14:603–607.
85. Baron W, Decker L, Colognato H, ffrench-Constant C. Regulation of integrin growth factor interactions in oligodendrocytes by lipid raft microdomains. Curr Biol 2003;13:151–155.
86. Lai EC. Lipid rafts make for slippery platforms. J Cell Biol 2003;162:365–370.
87. Davis MJ, Wu X, Nurkiewicz TR, et al. Regulation of ion channels by integrins. Cell Biochem Biophys 2002;36:41–66.
88. Arcangeli A, Becchetti A, Mannini A, et al. Integrin-mediated neurite outgrowth in neuroblastoma cells depends on the activation of potassium channels. J Cell Biol 1993;122:1131–1143.
89. Hofmann G, Bernabei PA, Crociani O, et al. HERG K^+ channels activation during beta(1) integrin-mediated adhesion to fibronectin induces an up-regulation of alpha(v)beta(3) integrin in the preosteoclastic leukemia cell line FLG 29.1. J Biol Chem 2001;276:4923–4931.
90. Menegazzi R, Busetto S, Decleva E, Cramer R, Dri P, Patriarca P. Triggering of chloride ion efflux from human neutrophils as a novel function of leukocyte beta 2 integrins: relationship with spreading and activation of the respiratory burst. J Immunol 1999;162:423–434.
91. Bianchi L, Arcangeli A, Bartolini P, Mugnai G, Wanke E, Olivotto M. An inward rectifier K^+ current modulates in neuroblastoma cells the tyrosine phosphorylation of the pp125FAK and associated proteins: role in neuritogenesis. Biochem Biophys Res Commun 1995;210:823–829.
92. Levite M, Cahalon L, Peretz A, et al. Extracellular K^+ and opening of voltage-gated potassium channels activate T cell integrin function: physical and functional association between Kv1.3 channels and beta1 integrins. J Exp Med 2000;191:1167–1176.
93. Artym VV, Petty HR. Molecular proximity of Kv1.3 voltage-gated potassium channels and beta(1)-integrins on the plasma membrane of melanoma cells: effects of cell adherence and channel blockers. J Gen Physiol 2002;120:29–37.
94. McPhee JC, Dang YL, Davidson N, Lester HA. Evidence for a functional interaction between integrins and G protein-activated inward rectifier K^+ channels. J Biol Chem 1998;273:34,696–34,702.
95. Abdel-Ghany M, Cheng H-C, Elble RC, Lin H, DiBiasio J, Pauli BU. The interating binding domains of the β_4 integrin and calcium-activated chloride channels (CLCAs) in metastasis. J Biol Chem 2003;278:49,406–49,416.
96. Cherubini A, Pillozzi S, Hofmann G, et al. HERG K^+ channels and beta1 integrins interact through the assembly of a macromolecular complex. Ann NY Acad Sci 2002;973:559–561.
97. Cherubini A, Hofmann G, Pillozzi S, et al. hERG1 Channels are physically linked to beta1 integrins and modulate adhesion-dependent signalling. Mol Biol Cell 2005;16:2972–2983.
98. Martens JR, O'Connell K, Tamkun M. Targeting of ion channels to membrane microdomains: localization of KV channels to lipid rafts. Trends Pharmacol Sci 2004;25:16–21.

99. O'Connell KM, Martens JR, Tamkun MM. Localization of ion channels to lipid Raft domains within the cardiovascular system. Trends Cardiovasc Med 2004;14:37–42.
100. Brady JD, Rich TC, Le X, et al. Functional role of lipid raft microdomains in cyclic nucleotide-gated channel activation. Mol Pharmacol 2004;65:503–511.
101. Bubien JK, Keeton DA, Fuller CM, et al. Malignant human gliomas express an amiloride-sensitive Na^+ conductance. Am J Physiol 1999;276:C1405–C1410.
102. Pardo LA, del Camino D, Sanchez A, et al. Oncogenic potential of EAG K^+ channels. EMBO J 1999;18:5540–5547.
103. Yu SP. Regulation and critical role of potassium homeostasis in apoptosis. Prog Neurobiol 2003;70:363–386.
104. Fraser SP, Salvador V, Manning EA, et al. Contribution of functional voltage-gated Na^+ channel expression to cell behaviours involved in the metastatic cascade in rat prostate cancer: I. Leteral motility. J Cell Physiol 2003;195:479–487.
105. Lastraioli E, Guasti L, Crociani O, et al. Herg1 gene and HERG1 protein are overexpressed in colorectal cancers and regulate cell invasion of tumor cells. Cancer Res 2004;64:606–611.
106. Grimes JA, Djamgoz MBA. Electrophysiological characterization of voltage-gated Na^+ current expressed in highly metastatic Mat-LyLu cell line of rat prostate cancer. J Cell Physiol 1995;175:50–58.
107. Bordey A, Sontheimer H. Electrophysiological properties of human astrocytic tumor cells in situ: enigma of spiking glial cells. J Neurophysiol 1998;79:2782–2793.
108. Labrakakis C, Patt S, Hartmann J, Kettenmann H. Glutamate receptor activation can trigger electrical activity in human glioma cells. Eur J Neurosci 1998;10:2153–2162.
109. Gritti A, Rosati B, Lecchi M, Vescovi AL, Wanke E. Excitable properties in astrocytes derived from human embryonic CNS stem cells. Eur J Neurosci 2000;12:3549–3559.
110. Ye ZC, Rothstein JD, Sontheimer H. Compromised glutamate transport in human glioma cells: reduction–mislocalization of sodium-dependent glutamate transporters and enhanced activity of cystine-glutamate exchange. J Neurosci 1999;19:10,767–10,777.
111. Ishiuchi S, Tsuzuki K, Yoshida Y, et al. Blockage of Ca^{2+}-permeable AMPA receptors suppresses migration and induces apoptosis in human glioblastoma cells. Nat Med 2002;8:971–978.
112. Soroceanu L, Manning TJ, Jr, Sontheimer H. Modulation of glioma cell migration and invasion using Cl^- and K^+ ion channel blockers. J Neurosci 1999;19:5942–5954.
113. Ranson CB, Sontheimer H. BK channels in human glioma cells. J Neurophysiol 2001;85:790–803.
114. Olsen ML, Sontheimer H. Mislocalization of Kir channels in malignant glia. Glia 2004;46:63–73.
115. Becchetti A, De Fusco M, Crociani O, et al. The human Ether a gò-gò like (HELK2) K^+ channel: functional expression in heterologous systems and properties in astrocytoma cells. Eur J Neurosci 2002;16:415–428.
116. Patt S, Preubat K, Beetz C, et al. Expression of ether a go-go potassium channels in human gliomas. Neurosci Letts 2004;368:249–253.
117. Masi A, Bechetti A, Restano-Cassulini R, et al. HERG1 channels are overexpressed in glioblastoma multiforme and modulate VEGF secretion in glioblastoma-cells. Br J Cancer, in press.
118. DeCoursey TE, Chandy GK, Gupta S, Cahalan MD. Voltage-gated K^+ channels in human lymphocytes: a role in mitogenesis? Nature 1984;307:465–468.
119. Lees-Miller JP, Kondo C, Wang L, Duff HJ. Electrophysiological characterization of an alternatively processed ERG K^+ channel in mouse and human hearts. Circ Res 1997;81:719–726.
120. London B, Trudeau MC, Newton KP, et al. Two isoforms of the mouse ether-a-gò-gò-relatedbgene coassemble to form channels with properties similar to the rapidly activating component of the cardiac delayed rectifier K^+ current. Circ Res 1997;81:870–878.

9
Nonsynaptic GABAergic Communication and Postnatal Neurogenesis

Xiuxin Liu, PhD, Anna J. Bolteus, PhD, and Angélique Bordey, PhD

SUMMARY

This chapter argues for the view that intercellular γ-aminobutyric acid (GABA)ergic signaling between neural stem cells and neuronal precursors in the subventricular zone (SVZ), also called subependymal zone *(1)*, is critical for controlling postnatal neurogenesis, and that nonsynaptic communication between SVZ cells displays characteristics of that between cells of neuron–glial networks elsewhere in the brain. GABA plays an important signaling role in developmental processes, such as embryonic cell proliferation *(2,3)*, migration *(4–8)*, and differentiation *(9–12)*. The function of GABA on the various developmental stages has been the subject of many reviews *(13–17)*. This chapter will focus on the signaling function of GABA in postnatal neurogenesis in the SVZ *(18–20)*. The cellular architecture of the SVZ and the GABAergic system will be described. Although no anatomically defined synaptic contacts have been observed, local GABAergic signaling between SVZ stem cells and neuronal precursors, or between neuronal precursors, exists and is thought to maintain a balance between the production and mobilization of SVZ precursors.

Key Words: Stem cells; proliferation; migration; neurotransmitter.

1. CELLULAR ARCHITECTURE AND THE γ-AMINOBUTYRIC ACID-ERGIC SYSTEM IN THE SUBVENTRICULAR ZONE

The postnatal subventricular zone (SVZ) contains the largest pool of dividing neural precursors in the adult brain. The cellular architecture of the SVZ has been extensively reviewed elsewhere *(21–25)*. Briefly, the SVZ consists of a network of interconnected channels including the three major cell types *(26)*: class III β-tubulin (TuJ1)-immunopositive neuronal precursors (approx 65–75%) closely ensheathed by glial fibrillary acidic protein (GFAP)-immunopositive cells (approx 20–30%), and a small number of immature, highly proliferative precursors. In the channels, neuronal precursors are densely packed, and migrate throughout the SVZ and along the rostral migratory stream (RMS) toward the olfactory bulb in which they differentiate into interneurons *(27–30)*. The GFAP+ cells of the SVZ display characteristics of stem cells *(21,26)* and will be called astrocyte-like cells in this chapter because they share some, but not all, properties of astrocytes. Because of the proximity of astrocyte-like cells to neuronal precursors, astrocyte-like cells are in an optimal location to send and receive signals from neuronal precursors. Elimination of neuronal precursors and highly proliferative precursors stimulates the proliferation of astrocyte-like cells that regenerate the entire SVZ *(31)*, suggesting that local cues or signaling molecules released from neuronal precursors regulate the proliferation of astrocyte-like cells.

The recent finding of γ-aminobutyric acid (GABA) and its synthetic enzyme (glutamic acid decarboxylase 67 [GAD-67]) in neuronal precursors both in vitro *(32)* and in acute slices *(18,33)*

From: *The Cell Cycle in the Central Nervous System*
Edited by: D. Janigro © Humana Press Inc., Totowa, NJ

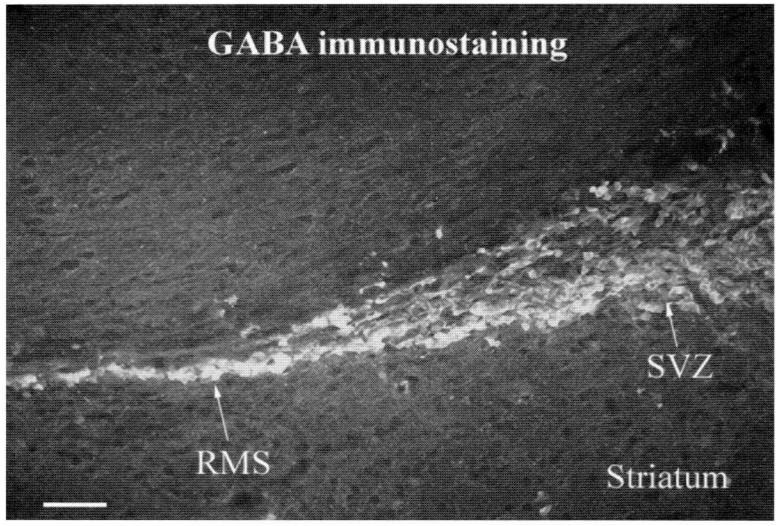

Fig. 1. GABA immunostaining (green) within the SVZ and RMS in a sagittal slice. Scale bar: 100 μm.

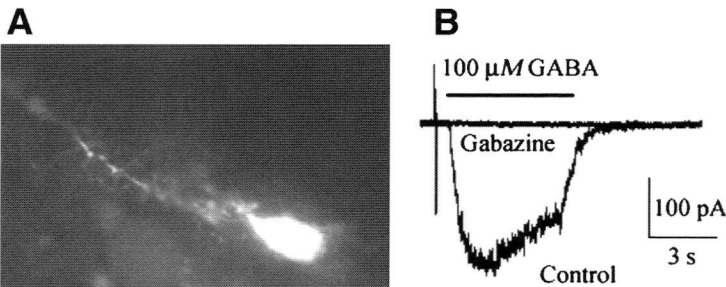

Fig. 2. Astrocyte-like cells express functional $GABA_A$ receptors. **(A)** Photograph of a lucifer yellow-filled astrocyte-like cell that was recorded in the SVZ. Astrocyte-like cells were recorded in slices from transgenic mice expressing GFAP on the promoter of GFP. Recorded astrocyte-like cells have an elongated cell body and a main process bearing some branches. Scale bar: 20 μm. **(B)** 100 μM GABA-induced currents before and during application of the $GABA_A R$ antagonist gabazine (50 μM). Astrocyte-like cells were held at –70 mV and recorded with a 145 mM intracellular KCl solution.

once again attracted attention to the GABAergic system. GABA is present in the cytoplasm of most cells in the SVZ (Fig. 1) and essentially in all TuJ1-positive cells but not astrocyte-like cells *(18–20,32)*. The presence of GABA in neuronal precursors suggested that these cells could release GABA and that surrounding SVZ precursors express GABA receptors and transporters to clear GABA from the extracellular space. The action of GABA can be mediated by at least three types of GABA receptors called $GABA_A$, $GABA_B$, and $GABA_C$ receptors *(34)*. $GABA_B$ and $GABA_C$ receptors have not been detected in neuronal progenitors and agonists of $GABA_B$ receptors do not affect the migration or proliferation of neuronal precursors *(18,20,33)*. However, the expression of $GABA_A$ receptor subunit transcripts and proteins was analyzed using polymerase chain reaction and immunocytochemistry in neuronal precursors in isolated cultures *(32)* and neurospheres *(20)*. Furthermore, using electrophysiological techniques, $GABA_A$ responses were identified and characterized in both neuronal precursors *(33)* and astrocyte-like cells *(19)* (Fig. 2) in acute slices. Interestingly, $GABA_A$ responses in neuronal precursors and astrocyte-like cells display different sensitivities to zinc and benzodiazepine *(19,33)*, suggesting that these cells

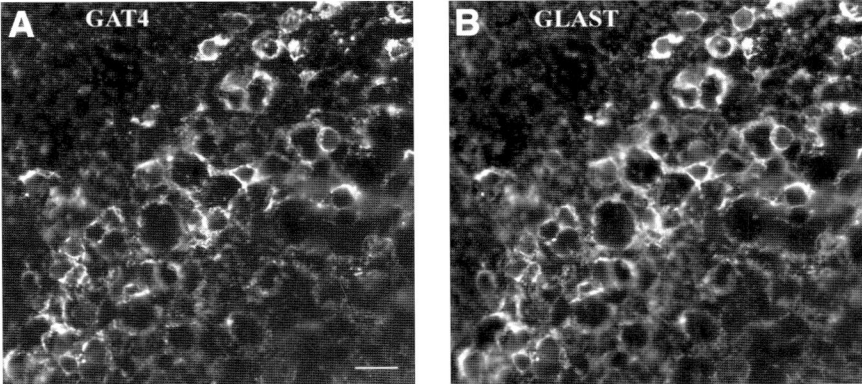

Fig. 3. Immunostainings for GAT4 (**A**) and GLAST (**B**) in the rostral migratory stream of a coronal slice. Scale bar: 30 μm.

express $GABA_A$ receptors with different subunit compositions. The exact cellular distribution and subunit composition of $GABA_A$ receptors in SVZ precursors remain to be determined. Elsewhere in the brain, GABA levels are tightly regulated by high-affinity GABA transporters *(35–37)*. Four types of GABA transporters belonging to the family of Na^+- and Cl^--coupled transporters *(36)* have been cloned, namely GAT1, GAT2, GAT3, and GAT4 in mice *(38,39)*. Among the four subtypes of GABA transporters expressed in the mouse brain, GAT1 and GAT4 are highly expressed in neurons and astrocytes, respectively *(36)*. GAT1 immunostaining was not detected in the SVZ or RMS *(18)*. However, GAT4 immunostaining was observed in astrocyte-like cells but not in neuronal precursors *(18)* (Fig. 3, astrocyte-like cells were identified with glutamate–aspartate transporter immunostaining *[40]*). Importantly, GABA transporters are expressed on the processes of astrocyte-like cells that ensheath neuronal precursors *(18)* and may thus create a microenvironment around clusters of neuronal precursors. It is interesting to draw parallels between the GABAergic system in the SVZ and elsewhere in the brain in which astrocytes ensheath GABAergic synapses. Astrocytes express both $GABA_A$ receptors and GABA transporters that are located to detect synaptically released GABA *(41,42)*. Figure 4 illustrates a simplified diagram of the SVZ where GABA, $GABA_A$ receptors, and GABA transporters are displayed and support the notion of a tightly regulated communication between neuronal precursors and astrocyte-like cells.

2. PARACRINE GABA SIGNALING BETWEEN SVZ PRECURSORS

This section describes that GABA provides nonsynaptic (also called paracrine) communication signals between astrocyte-like cells and neuronal precursors, and between neuronal precursors themselves. To consider GABA as an intercellular signaling molecule, GABA needs to be released, activate GABA receptors, and be removed from the extracellular space because GABA is not degraded. It was recently reported that $GABA_A$ receptors in neuronal precursors and astrocyte-like cells are tonically activated by endogenous GABA *(18,20)*, suggesting that there is enough ambient GABA to activate $GABA_A$ receptors in SVZ precursors. Furthermore, using mass spectrometry or patch-clamp recordings of astrocyte-like cells, it was reported that SVZ precursors spontaneously release GABA, which was enhanced by depolarizing cells with high KCl *(18–20)*. Neuronal precursors were found to display spontaneous 10–20 mV depolarizations, which may account for the spontaneous release of GABA *(19,43)*. The nature of theses spontaneous depolarizations and the mechanisms of GABA release remain unknown. In synaptically silent neurons in the neonatal hippocampus, the release of GABA was nonsynaptic and independent of synaptosome-associated protein 25, a SNARE protein, which is necessary for vesicular exocytosis in neurons *(44)*. Similarly, Maric et al. *(11)* found no evidence of GABA compartmentalization in vesicles in embryonic neocortical neurons in vitro *(11)*. In SVZ cells,

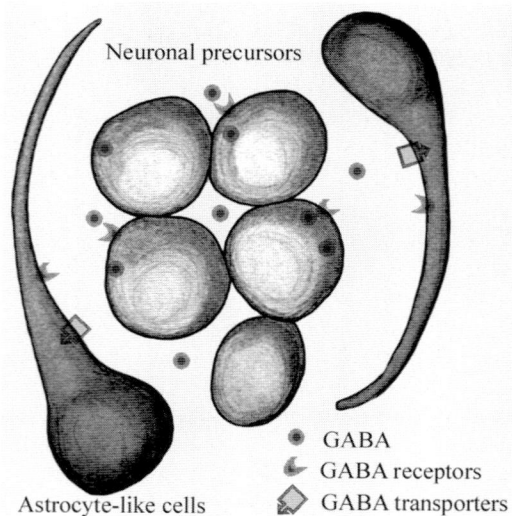

Fig. 4. Cellular architecture and local GABA signaling in the SVZ. In a coronal section, neuronal precursors are surrounded by astrocyte-like cells that express GABA transporters. Most of the neuronal precursors contain GABA and all of the neuronal precursors express $GABA_A$ receptors.

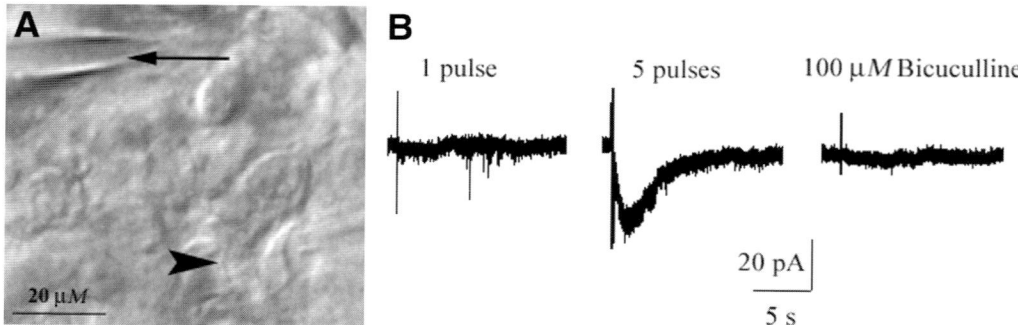

Fig. 5. Focal electrical stimulation of SVZ precursors evoked nonsynaptic activation of $GABA_A$ receptors in astrocyte-like cells. **(A)** Photograph of a recorded astrocyte-like cell (arrowhead) and a stimulating electrode (arrow) placed in the SVZ. **(B)** Whereas single electrical stimulation in the SVZ is not sufficient to induce inward currents in astrocyte-like cells, tetanic electrical stimulation (five pulses of 200 µs at 50 Hz) induced slow, bicuculline-sensitive inward currents in astrocyte-like cells.

no clusters of vesicles have been reported; however, neuronal precursors contain individual vesicles *(26)* and synapsin III, a protein-associated with synaptic vesicles *(45,46)*, suggesting that GABA may be released from neuronal precursors via vesicular exocytosis. Independent of the mode of GABA release, there is direct evidence that nonsynaptic GABA released from SVZ precursors following extracellular KCl increase or tetanic stimulation activates $GABA_A$ receptors in astrocyte-like cells (Fig. 5 for electrically induced GABA release). GABA release was nonsynaptic because it was independent of action potentials, extracellular calcium, and displayed a slow time-course. These data were consistent with the lack of anatomically defined synaptic contacts onto astrocyte-like cells *(26)*. Synaptically released GABA elsewhere in the brain is removed from the extracellular space by high-affinity transporters. In the SVZ, pharmacological inhibition of GABA transporters altered the kinetics of evoked $GABA_A$ currents in astrocyte-like cells *(19)* and reduced the speed of neuronal precursor migration via $GABA_A$ receptor activation

(18). Therefore, in the SVZ GABA transporters play an important role in clearing GABA that is released from neuronal precursors, thereby controlling the degree of $GABA_A$ receptor activation in astrocyte-like cells. The SVZ can thus be viewed as a local GABAergic network in which GABA provides communication signals between precursors despite the absence of synapses.

Thus far, we described the SVZ as an independent network from surrounding brain regions. However, SVZ cells are adjacent to the striatum, which is rich in GABAergic cells and terminals. It is unknown whether synaptically released GABA from GABAergic striatal terminals could diffuse from the striatum to the SVZ and activate $GABA_A$ receptors in neuronal precursors and/or astrocyte-like cells. Although Liu et al. *(19)* found no $GABA_A$ responses in astrocyte-like cells following electrical stimulation in the striatum, this study was intentionally designed to avoid $GABA_A$ receptor activation following stimulation in the striatum. Conceptually, it is conceivable that GABA or other diffusible factors released from striatal cells or terminals limit neurogenesis, in particular cell migration into the striatum or cell proliferation. As an example, the diffusible factor Slit, which is secreted from the septum, is a chemorepellent that directs the migration of SVZ neuronal precursors toward the olfactory bulb *(47–49)*.

It is once again interesting to draw parallels between the neuronal precursor and astrocyte-like cell communication, and that between neurons and astrocytes. At GABAergic synapses, synaptically released GABA diffuses outside the synaptic cleft and activates GABA transporters in astrocytes *(41)*. Furthermore, inhibition of GABA transporters affects the degree of postsynaptic $GABA_A$ receptor activation and thus synaptic transmission *(50–54)*. Astrocytic processes encapsulate either one synapse or a group of GABAergic synapses such as glomeruli in the cerebellum *(55)*. This latter configuration resembles that of the SVZ in which processes of astrocyte-like cells encapsulate clusters of migrating neuronal precursors (*see* Fig. 3). Whereas the role of synaptically released GABA on the biology of astrocytes remains unclear, the next section will describe an important regulation of astrocyte-like cell proliferation, and also neuronal precursor proliferation and migration by nonsynaptically released GABA from SVZ neuronal precursors.

3. FUNCTIONAL CONSEQUENCES OF $GABA_A$ RECEPTOR ACTIVATION IN ASTROCYTE-LIKE CELLS AND NEURONAL PRECURSORS

It was recently reported that activation of $GABA_A$ receptors by endogenous GABA decreases the number of both proliferative astrocyte-like cells *(19)* and neuronal precursors *(20)*, as treatment of slices with the $GABA_A$ receptor antagonist, bicuculline, resulted in a significant increase in the number of bromodeoxyuridine-immunopositive astrocyte-like cells and neuronal precursors in the SVZ *(19,20)*. As more neuronal precursors are generated, it is expected that more GABA is released in the extracellular space resulting in increased ambient GABA levels and $GABA_A$ receptor activation in SVZ precursors. Because astrocyte-like cells generate neuronal precursors *(31,56)*, increase in the number of neuronal precursors seems to serve as a negative feedback to decrease astrocyte-like cell proliferation and neuronal precursor production by activating $GABA_A$ receptors. This negative feedback fits well with the constant migration of neuronal precursors away from the SVZ to the olfactory bulb *(30,57,58)*, which limits ambient GABA accumulation, and with the increased proliferation of astrocyte-like cells following elimination of neuronal precursors *(31)*. A similar feedback exerted on precursor proliferation by GABA may also occur in the embryonic ventricular zone in which tonic $GABA_A$ receptor activation by ambient GABA limits the proliferation of precursors *(2,3)*. Interestingly, the proliferative precursors in the ventricular zone, which are thought to be radial glial cells *(59,60)*, transform astrocyte-like cells in the postnatal SVZ *(61)*. The postnatal SVZ can thus be viewed as an interface between the embryonic ventricular zone and the adult brain, and share properties of both systems. GABA is also known to regulate the speed of embryonic neuronal precursor migration *(14)*. Similarly, in the postnatal SVZ, GABA exerts a tonic influence on the speed of neuronal precursor migration in the SVZ and RMS *(18)*. In particular, it was found that GABA reduces the speed of neuronal precursor migration via $GABA_A$ receptor activation. Both GABA uptake into astrocyte-like cells and GABA release from neuronal precursors influence the speed of cell migration. As GABA release is enhanced by cell depolarization and GABA depolarizes neuronal precursors *(33)*, increases in ambient GABA levels are expected to promote GABA release. However, accumulation of ambient GABA is prevented by uptake into astrocyte-like cells, which ensheath chains of migrating precursors *(26,27,62)* and thus maintain a proper balance between the levels of GABA and GABA release. It is conceivable that astrocyte-like cells create a microgradient of GABA concentration inside the chains of neuronal precursors; precursors in contact with astrocyte-like cells may migrate faster than those in the middle of the chains in which GABA concentration is the highest. Cells inside the clusters may migrate at a lower speed and may be searching for directional cues before migrating at higher speed toward the olfactory bulb. Furthermore, it is possible that GABA reduces the speed of migration to promote cell differentiation, as GABA is known to promote neurite outgrowth and maturation of GABAergic interneurons *(11,13,63,64)*.

4. MECHANISMS OF ACTION OF GABA VIA $GABA_A$ RECEPTOR ACTIVATION

$GABA_A$ receptor activation is usually thought to regulate cell development by triggering membrane depolarization. Furthermore, activation of $GABA_A$ receptors depolarized both embryonic cells *(13,65)* and postnatal SVZ neuronal precursors *(32,33)*. GABA-induced depolarization has been shown to be sufficient to increase intracellular Ca^{2+} concentration by opening voltage-gated Ca^{2+} channels, thereby activating Ca^{2+}-dependent second messenger pathways *(3,20,66–68)*. In the postnatal SVZ, application of a $GABA_A$ receptor agonist, muscimol at 100 μM, also depolarized neuronal precursors sufficiently to activate voltage-gated Ca^{2+} channels *(20)*. Furthermore, the action of muscimol on cell proliferation was occluded with a blocker of voltage-gated Ca^{2+} channels or a blocker of mitogen-activated protein kinase, suggesting that $GABA_A$ receptor activation reduces the cell cycle progression by promoting Ca^{2+} influx via voltage-gated Ca^{2+} channels and inhibiting the mitogen-activated protein kinase pathway. Although it is possible that ambient GABA concentration reaches 100 μM around SVZ neuronal precursors, such an increase is likely transient and baseline ambient GABA concentration should presumably be less than 10 μM. It would thus be interesting to examine whether a lower concentration of muscimol or repeated transient applications of 100 μM muscimol also involve voltage-gated Ca^{2+} channels in the modulation of cell proliferation. The action of GABA (10 μM) on the speed of neuronal precursor migration was independent of cell depolarization and voltage-gated Ca^{2+} channels but interfered with intracellular Ca^{2+} dynamics *(18)*. The intracellular pathway involved in the action of GABA on cell migration remains unknown. It is important to mention that neuronal precursors display spontaneous Ca^{2+} transients whose frequency is modulated by GABA applications (Grosmaitre and Bordey, unpublished observations). Changes in the frequency of Ca^{2+} transients could control the speed of migration and progression through the cell cycle. Interestingly, 10 μM GABA reduced the speed of cell migration whereas 100 μM GABA had an opposite effect, which was similar to that of bicuculline. The authors showed that $GABA_A$ receptor desensitized in approx 15 min during 100 μM GABA application, resulting in a blockade of $GABA_A$ receptors. However, 100 μM muscimol and bicuculline had opposite effects on cell proliferation. This discrepancy is difficult to explain at the present time but likely involves different interactions with intracellular Ca^{2+} dynamics and Ca^{2+}-dependent second messengers that regulate the speed of migration and the progression through the cell cycle. Regarding astrocyte-like cells, it is unknown how $GABA_A$ receptor activation reduced cell cycle progression *(19)*. GABA is expected to depolarize astrocyte-like cells, as previously reported for radial glial cells *(3)*, the ancestors of astrocyte-like cells *(69)*, and astrocytes *(70,71)*. It was also reported that GABA-induced depolarization of radial glial cells was sufficient to open voltage-gated Ca^{2+} channels. These results are somewhat unexpected because all of these glial cells have a very low input resistance and a hyperpolarized resting potential (between –80 and –90 mV), which would limit the amplitude of GABA-induced depolarization. However, it is possible that $GABA_A$ receptors are located in the fine glial processes (*see* Fig. 2 for astrocyte-like cells in the SVZ), which likely have a high-input resistance. GABA may also indirectly affect cell migration and proliferation by inducing the release of diffusible factors such as brain-derived neurotrophic factor (BDNF), as previously reported for the maturation of hippocampal interneurons *(12)*. Furthermore, high-affinity receptors for BDNF are expressed in SVZ cells *(72)*, and intraventricular infusion of BDNF promoted neurogenesis from SVZ cells *(73)*.

5. FUTURE DIRECTIONS

Overall, local GABA signaling between SVZ precursors contributes to the maintenance of the cellular architecture of the SVZ by orchestrating the production and migration of SVZ precursors. However, it is unclear how GABA is released from neuronal precursors and which intracellular messengers are activated following $GABA_A$ receptor activation in either astrocyte-like cells or neuronal precursors. It is also unknown whether GABA reduces both the speed of migration and progression through the cell cycle to promote differentiation of neuronal precursors during their migration along the SVZ and RMS. Furthermore, it remains to be determined whether drugs regulating $GABA_A$ receptor or GABA transporter function affects neurogenesis in the SVZ in vivo. It is interesting to determine whether SVZ neurogenesis is affected not only by a local GABAergic system but also by GABAergic inputs from the surrounding brain regions. Although astrocyte-like cells express GABA transporters and may thus prevent GABA spillover from the striatum to the SVZ, it is unknown whether GABA uptake into astrocyte-like cells can influence their proliferation or fate. Finally, interactions between GABA and growth factors or other neurotransmitters remain to be explored in the postnatal SVZ.

REFERENCES

1. Boulder Committee. Embryonic vertebrate central nervous system:revised terminology. Anat Rec 1970;66:257–261.
2. Haydar TF, Wang F, Schwartz ML, Rakic P. Differential modulation of proliferation in the neocortical ventricular and subventricular zones. J Neurosci 2000;20:5764–5774.
3. LoTurco JJ, Owens DF, Heath MJ, Davis MB, Kriegstein AR. GABA and glutamate depolarize cortical progenitor cells and inhibit DNA synthesis. Neuron 1995;15:1287–1298.
4. Behar TN, Li YX, Tran HT, et al. GABA stimulates chemotaxis and chemokinesis of embryonic cortical neurons via calcium-dependent mechanisms. J Neurosci 1996;16:1808–1818.
5. Behar TN, Schaffner AE, Scott CA, O'Connell C, Barker JL. Differential response of cortical plate and ventricular zone cells to GABA as a migration stimulus. J Neurosci 1998;18:6378–6387.
6. Behar TN, Schaffner AE, Tran HT, Barker JL. GABA-induced motility of spinal neuroblasts develops along a ventrodorsal gradient and can be mimicked by agonists of GABAA and GABAB receptors. J Neurosci Res 1995;42:97–108.
7. Behar TN, Schaffner AE, Scott CA, Greene CL, Barker JL. GABA receptor antagonists modulate postmitotic cell migration in slice cultures of embryonic rat cortex. Cereb Cortex 2000;10:899–909.
8. Fueshko SM, Key S, Wray S. GABA inhibits migration of luteinizing hormone-releasing hormone neurons in embryonic olfactory explants. J Neurosci 1998;18:2560–2569.
9. Spoerri PE. Neurotrophic effects of GABA in cultures of embryonic chick brain and retina. Synapse 1988;2:11–22.
10. Spoerri PE, Wolff JR. Effect of GABA-administration on murine neuroblastoma cells in culture. I. Increased membrane dynamics and formation of specialized contacts. Cell Tissue Res 1981;218: 567–579.
11. Maric D, Liu QY, Maric I, et al. GABA expression dominates neuronal lineage progression in the embryonic rat neocortex and facilitates neurite outgrowth via GABA(A) autoreceptor/Cl-channels. J Neurosci 2001;21:2343–2360.
12. Marty S, Berninger B, Carroll P, Thoenen H. GABAergic stimulation regulates the phenotype of hippocampal interneurons through the regulation of brain-derived neurotrophic factor. Neuron 1996;16:565–570.

13. Owens DF, Kriegstein AR. Is there more to GABA than synaptic inhibition? Nat Rev Neurosci 2002;3:715–727.
14. Barker JL, Behar T, Li YX, et al. GABAergic cells and signals in CNS development. Perspect Dev Neurobiol 1998;5:305–322.
15. Cherubini E, Gaiarsa JL, Ben Ari Y. GABA:an excitatory transmitter in early postnatal life. Trends Neurosci 1991;14:515–519.
16. Nguyen L, Rigo JM, Rocher V, et al. Neurotransmitters as early signals for central nervous system development. Cell Tissue Res 2001;305:187–202.
17. Meier E, Hertz L, Schousboe A. Neurotransmitters as developmental signals. Neurochem Int 1991;19:1–15.
18. Bolteus A, Bordey A. GABA release and uptake orchestrate neuronal precursor migration in the postnatal subventricular zone. J Neurosci 2004;24:7623–7631.
19. Liu X, Wang Q, Haydar TF, Bordey A. Nonsynaptic GABAergic signaling in the postnatal subventricular zone controls GFAP-expressing cell proliferation. Nat Neurosci 2005, in press.
20. Nguyen L, Malgrange B, Breuskin I, et al. Autocrine/paracrine activation of the GABA(A) receptor inhibits the proliferation of neurogenic polysialylated neural cell adhesion molecule-positive (PSA-NCAM+) precursor cells from postnatal striatum. J Neurosci 2003;23:3278–3294.
21. Garcia-Verdugo JM, Doetsch F, Wichterle H, Lim DA, Alvarez-Buylla A. Architecture and cell types of the adult subventricular zone:in search of the stem cells. J Neurobiol 1998;36:234–248.
22. Peretto P, Merighi A, Fasolo A, Bonfanti L. The subependymal layer in rodents:a site of structural plasticity and cell migration in the adult mammalian brain. Brain Res Bull 1999;49:221–243.
23. Doetsch F. The glial identity of neural stem cells. Nat Neurosci 2003;6:1127–1134.
24. Temple S, Alvarez-Buylla A. Stem cells in the adult mammalian central nervous system. Curr Opin Neurobiol 1999;9:135–141.
25. Conover JC, Allen RL. The subventricular zone:new molecular and cellular developments. Cell Mol Life Sci 2002;59:2128–2135.
26. Doetsch F, Garcia-Verdugo JM, Alvarez-Buylla A. Cellular composition and three-dimensional organization of the subventricular germinal zone in the adult mammalian brain. J Neurosci 1997;17:5046–5061.
27. Lois C, Garcia-Verdugo JM, Alvarez-Buylla A. Chain migration of neuronal precursors. Science 1996;271:978–981.
28. Luskin MB. Neuroblasts of the postnatal mammalian forebrain:their phenotype and fate. J Neurobiol 1998;36:221–233.
29. Belluzzi O, Benedusi M, Ackman J, LoTurco JJ. Electrophysiological differentiation of new neurons in the olfactory bulb. J Neurosci 2003;23:10,411–10,418.
30. Carleton A, Petreanu LT, Lansford R, Alvarez-Buylla A, Lledo PM. Becoming a new neuron in the adult olfactory bulb. Nat Neurosci 2003;6:507–518.
31. Doetsch F, Garcia-Verdugo JM, Alvarez-Buylla A. Regeneration of a germinal layer in the adult mammalian brain. Proc Natl Acad Sci USA 1999;96:11,619–11,624.
32. Stewart RR, Hoge GJ, Zigova T, Luskin MB. Neural progenitor cells of the neonatal rat anterior subventricular zone express functional GABA(A) receptors. J Neurobiol 2002;50:305–322.
33. Wang DD, Krueger DD, Bordey A. GABA depolarizes neuronal progenitors of the postnatal subventricular zone via $GABA_A$ receptor activation. J Physiol (Lond) 2003;550:785–800.
34. Chebib M, Johnston GA. The 'ABC' of GABA receptors:a brief review. Clin Exp Pharmacol Physiol 1999;26:937–940.
35. Larsson OM, Griffiths R, Allen IC, Schousboe A. Mutual inhibition kinetic analysis of gamma-aminobutyric acid, taurine, and beta-alanine high-affinity transport into neurons and astrocytes:evidence for similarity between the taurine and beta-alanine carriers in both cell types. J Neurochem 1986;47:426–432.
36. Borden LA. GABA transporter heterogeneity: pharmacology and cellular localization. Neurochem Int 1996;29:335–356.
37. Breckenridge RJ, Nicholson SH, Nicol AJ, Suckling CJ, Leigh B, Iversen L. Inhibition of neuronal GABA uptake and glial beta-alanine uptake by synthetic GABA analogues. Biochem Pharmacol 1981;30:3045–3049.
38. Liu QR, Lopez-Corcuera B, Mandiyan S, Nelson H, Nelson N. Molecular characterization of four pharmacologically distinct gamma-aminobutyric acid transporters in mouse brain [corrected] [published erratum appears in J Biol Chem 1993;268(12):9156]. J Biol Chem 1993;268: 2106–2112.

39. Lopez-Corcuera B, Liu QR, Mandiyan S, Nelson H, Nelson N. Expression of a mouse brain cDNA encoding novel gamma-aminobutyric acid transporter. J Biol Chem 1992;267:17,491–17,493.
40. Braun N, Sevigny J, Mishra SK, et al. Expression of the ecto-ATPase NTPDase2 in the germinal zones of the developing and adult rat brain. Eur J Neurosci 2003;17:1355–1364.
41. Kinney GA, Spain WJ. Synaptically evoked GABA transporter currents in neocortical glia. J Neurophysiol 2002;88:2899–2908.
42. de Blas AL. Monoclonal antibodies to specific astroglial and neuronal antigens reveal the cytoarchitecture of the Bergmann glia fibers in the cerebellum. J Neurosci 1984;4:265–273.
43. Wang DD, Bordey A. Cell-attached measurement of the resting potential of neuronal progenitors in the subventricular zone of postnatal mice. Soc Neurosci Abstr 2003;564:17.
44. Demarque M, Represa A, Becq H, Khalilov I, Ben Ari Y, Aniksztejn L. Paracrine intercellular communication by a Ca^{2+}- and SNARE-independent release of GABA and glutamate prior to synapse formation. Neuron 2002;36:1051–1061.
45. Pieribone VA, Porton B, Rendon B, Feng J, Greengard P, Kao HT. Expression of synapsin III in nerve terminals and neurogenic regions of the adult brain. J Comp Neurol 2002;454:105–114.
46. Kao HT, Porton B, Czernik AJ, et al. A third member of the synapsin gene family. Proc Natl Acad Sci USA 1998;95:4667–4672.
47. Hu H, Rutishauser U. A septum-derived chemorepulsive factor for migrating olfactory interneuron precursors. Neuron 1996;16:933–940.
48. Hu H. Chemorepulsion of neuronal migration by Slit2 in the developing mammalian forebrain. Neuron 1999;23:703–711.
49. Nguyen-Ba-Charvet KT, Picard-Riera N, Tessier-Lavigne M, Baron-Van Evercooren A, Sotelo C, Chedotal A. Multiple roles for slits in the control of cell migration in the rostral migratory stream. J Neurosci 2004;24:1497–1506.
50. Wall MJ, Usowicz MM. Development of action potential-dependent and independent spontaneous GABAA receptor-mediated currents in granule cells of postnatal rat cerebellum. Eur J Neurosci 1997;9:533–548.
51. Dingledine R, Korn SJ. Gamma-aminobutyric acid uptake and the termination of inhibitory synaptic potentials in the rat hippocampal slice. J Physiol (Lond) 1985;366:387–409.
52. Hablitz JJ, Lebeda FJ. Role of uptake in gamma-aminobutyric acid (GABA)-mediated responses in guinea pig hippocampal neurons. Cell Mol Neurobiol 1985;5:353–371.
53. Roepstorff A, Lambert JD. Factors contributing to the decay of the stimulus-evoked IPSC in rat hippocampal CA1 neurons. J Neurophysiol 1994;72:2911–2926.
54. Rekling JC, Jahnsen H, Mosfeldt Laursen A. The effect of two lipophilic gamma-aminobutyric acid uptake blockers in CA1 of the rat hippocampal slice. Br J Pharmacol 1990;99:103–106.
55. Palay SL, Chan-Palay V. Cerebellar Cortex, cytology and Organization. New York: Springer-Verlag;1974, p 236.
56. Doetsch F, Caille I, Lim DA, Garcia-Verdugo JM, Alvarez-Buylla A. Subventricular zone astrocytes are neural stem cells in the adult mammalian brain. Cell 1999;97:703–716.
57. Lois C, Alvarez-Buylla A. Long-distance neuronal migration in the adult mammalian brain. Science 1994;264:1145–1148.
58. Luskin MB, Boone MS. Rate and pattern of migration of lineally-related olfactory bulb interneurons generated postnatally in the subventricular zone of the rat. Chem Senses 1994;19:695–714.
59. Noctor SC, Flint AC, Weissman TA, Wong WS, Clinton BK, Kriegstein AR. Dividing precursor cells of the embryonic cortical ventricular zone have morphological and molecular characteristics of radial glia. J Neurosci 2002;22:3161–3173.
60. Noctor SC, Flint AC, Weissman TA, Dammerman RS, Kriegstein AR. Neurons derived from radial glial cells establish radial units in neocortex. Nature 2001;409:714–720.
61. Tramontin AD, Garcia-Verdugo JM, Lim DA, Alvarez-Buylla A. Postnatal development of radial glia and the ventricular zone (VZ): a continuum of the neural stem cell compartment. Cereb Cortex 2003;13:580–587.
62. Peretto P, Merighi A, Fasolo A, Bonfanti L. Glial tubes in the rostral migratory stream of the adult rat. Brain Res Bull 1997;42:9–21.
63. Barbin G, Pollard H, Gaiarsa JL, Ben Ari Y. Involvement of GABAA receptors in the outgrowth of cultured hippocampal neurons. Neurosci Lett 1993;152:150–154.
64. Marty S, Berninger B, Carroll P, Thoenen H. GABAergic stimulation regulates the phenotype of hippocampal interneurons through the regulation of brain-derived neurotrophic factor. Neuron 1996;16:565–570.

65. Ben Ari Y, Tseeb V, Raggozzino D, Khazipov R, Gaiarsa JL. Gamma-aminobutyric acid (GABA): a fast excitatory transmitter which may regulate the development of hippocampal neurones in early postnatal life. Prog Brain Res 1994;102:261–273.
66. Owens DF, Boyce LH, Davis MB, Kriegstein AR. Excitatory GABA responses in embryonic and neonatal cortical slices demonstrated by gramicidin perforated-patch recordings and calcium imaging. J Neurosci 1996;16:6414–6423.
67. Yuste R, Katz LC. Control of postsynaptic Ca^{2+} influx in developing neocortex by excitatory and inhibitory neurotransmitters. Neuron 1991;6:333–344.
68. Lin MH, Takahashi MP, Takahashi Y, Tsumoto T. Intracellular calcium increase induced by GABA in visual cortex of fetal and neonatal rats and its disappearance with development. Neurosci Res 1994;20:85–94.
69. Tramontin AD, Garcia-Verdugo JM, Lim DA, Alvarez-Buylla A. Postnatal development of radial glia and the ventricular zone (VZ): a continuum of the neural stem cell compartment. Cereb Cortex 2003;13:580–587.
70. Bekar LK, Walz W. Intracellular chloride modulates A-type potassium currents in astrocytes. Glia 2002;39:207–216.
71. Kettenmann H, Backus KH, Schachner M. Gamma-aminobutyric acid opens Cl^- channels in cultured astrocytes. Brain Res 1987;404:1–9.
72. Tirassa P, Triaca V, Amendola T, Fiore M, Aloe L. EGF and NGF injected into the brain of old mice enhance BDNF and ChAT in proliferating subventricular zone. J Neurosci Res 2003;72:557–564.
73. Pencea V, Bingaman KD, Wiegand SJ, Luskin MB. Infusion of brain-derived neurotrophic factor into the lateral ventricle of the adult rat leads to new neurons in the parenchyma of the striatum, septum, thalamus, and hypothalamus. J Neurosci 2001;21:6706–6717.

10
Critical Roles of Ca^{2+} and K^+ Homeostasis in Apoptosis

Shan Ping Yu, MD, PhD

SUMMARY

Apoptosis occurs during the development, aging, and in various disease states. The apoptotic biochemical cascade involves activation of caspases, release of mitochondrial apoptotic factors, and nuclear DNA fragmentation, subjecting to regulation by pro- and antiapoptotic Bcl-2 genes. Emerging evidence supports that apoptosis is controlled by ionic mechanisms involving changes in Ca^{2+} and K^+ homeostasis. It is proposed that Ca^{2+} changes in cell organelles mainly the endoplasmic reticulum (ER) and mitochondria, but not the cytosolic Ca^{2+}, play critical roles in regulating and mediating apoptosis events. The Bcl-2 family members control the ER-mitochondria amplification loop of apoptosis. Overexpression of Bcl-2 or deficient for Bax and Bak lowers ER resting Ca^{2+} concentration ($[Ca^{2+}]_{ER}$) and secondarily decreased mitochondrial Ca^{2+} uptake. This fine-tuning of Ca^{2+} compartmentalization provides a mechanism for temporal and spatial regulation of Ca^{2+} signaling in apoptosis. Compelling evidence also reveals that excessive K^+ efflux and intracellular K^+ depletion are critical steps in apoptotic cell shrinkage and downstream events. Physiological concentration of intracellular K^+ acts as a repressor of apoptotic effectors. A huge loss of cellular K^+, likely a common event in apoptosis of many cell types, may serve as a disaster signal allowing the execution of the suicide program by activating key events in the apoptotic cascade including caspase cleavage, cytochrome-c release, and endonuclease activation. The proapoptotic disruption of K^+ homeostasis can be mediated by overactivated K^+ channels and, most likely, accompanied by reduced K^+ uptake as a result of dysfunction of Na^+,K^+-adenosine triphosphatase. In addition to the K^+ channels in the plasma membrane, mitochondrial K^+ channels also play important roles in apoptosis. Investigations on the Ca^{2+} and K^+ regulation of apoptosis, together with the molecular mechanism, have provided a more comprehensive understanding of the apoptotic mechanism. Further studies are needed to address new questions and may afford novel therapeutic strategies for apoptosis-related diseases.

Key Words: Apoptosis; ionic homeostasis; intracellular K^+; endoplasmic reticulum; mitochondria; *Bcl-2* genes; channel; receptor; K^+ efflux.

1. Ca^{2+} HOMEOSTASIS IN APOPTOSIS

The interest in a potential role for Ca^{2+} in apoptosis has lasted for years because the discovery in the 1980s that accumulation of intracellular Ca^{2+} ($[Ca^{2+}]_i$)-mediated cell death *(1)*. Although it had been clear that the excessive Ca^{2+} influx and uncontrolled increases in $[Ca^{2+}]_i$ mediate the necrotic cell death induced by excitotoxicity *(2)*, evidence for the role of Ca^{2+} in apoptosis has been contradictory and inconclusive *(3)*. Both high and low $[Ca^{2+}]_i$ levels have been observed in apoptotic cells and an increase in $[Ca^{2+}]_i$ may either induce or block apoptosis. Progress in the past few years, nevertheless, shows that the Ca^{2+} concentration in cell organelles but not in the cytosol plays crucial roles in regulating and mediating apoptosis. Specifically, Ca^{2+} in the mitochondria

From: *The Cell Cycle in the Central Nervous System*
Edited by: D. Janigro © Humana Press Inc., Totowa, NJ

Table 1
Physiological and Pharmacological Profile of ER

Ligand or drug	Main target	Effect
Ins(1,4,5)P$_3$	IP$_3$ receptors	Channel opening, Ca^{2+} release
Adenophostin-A	IP$_3$ receptors	Channel opening, Ca^{2+} release
Acyclophostin	IP$_3$ receptors	Partial agonist (pH dependent)
Xestospongin	IP$_3$ receptors	Block of Ca^{2+} release
Heparin	IP$_3$ receptors	Block of Ca^{2+} release
Ry	Ry receptor	
Low concentrations	Ry receptor	Channel opening, Ca^{2+} release
High concentrations	Ry receptor	Inhibition of Ca^{2+} release
Caffeine	Ry receptor	Channel opening, Ca^{2+} release
Dantrolene	Ry receptor	Block of Ca^{2+} release
FK-506, repamycin	FKBP	Ry receptor stabilization
Thapsigargin	Ca^{2+}-ATPase	Block, Ca^{2+} efflux
Tunicamycin, brefeldin-A	Protein-folding apparatus	Inhibition of protein folding

and endoplasmic reticulum (ER) and the Ca^{2+} homeostasis among the organelles are important factors in apoptosis *(4,5)*.

1.1. ER and Mitochondria as Intracellular Ca^{2+} Storage Pools

ER is classically divided into two subtypes: "rough" ER, which contains ribosomes and is responsible for protein synthesis, and "smooth" ER, which can serve a particularly important role in Ca^{2+} signaling. ER is continuous with the outer nuclear membrane and is often associated intimately with plasma membrane and mitochondria, which suggests functional coupling between these structures *(6)*. Several proteins are housed in the ER that control movements of Ca^{2+} across its membrane under basal conditions and in response to environmental stimuli *(7)*. Ca^{2+} in the ER lumen is either free or bound to luminal proteins such as calreticulin and calnexin. Under resting conditions the concentration of Ca^{2+} in the ER lumen is considerably higher (10–100 µ*M*) than the Ca^{2+} concentration in the cytoplasm (approx 100 n*M*). This Ca^{2+} gradient is maintained by an adenosine triphosphate (ATP)-dependent pump called SERCA (smooth ER Ca^{2+} ATPase) in the ER membrane. ER also contains two types of Ca^{2+} channels associated with inositol (1,4,5)-trisphosphate receptors and ryanodine (Ry) receptors, which provide conduits for the rapid release of Ca^{2+}. Ca^{2+} can be released from inositol (1,4,5)-trisphosphate receptor- or Ry receptor-regulated pools via a positive-feedback mechanism known as Ca^{2+}-induced Ca^{2+} release. Knowledge of the pharmacology of ER Ca^{2+}-store regulation is rapidly growing and is providing an array of experimental tools to manipulate ER Ca^{2+} uptake and release (Table 1).

1.2. The ER-Mitochondrial Ca^{2+} Loop and Apoptosis

ER can respond to, and modify the function of, Ca^{2+}-regulatory systems in the plasma membrane and mitochondria (Fig. 1). Mitochondria in close physical proximity to ER rapidly take up Ca^{2+} released from ER (the ER-mitochondrial Ca^{2+} loop) *(6–8)*. Interestingly, mitochondrial Ca^{2+} uptake is stimulated by ER-mediated Ca^{2+} release only when the release is induced in a pulsatile manner, but not when levels of Ins(1,4,5)P$_3$ are gradually increased.

The ER can be an initiator of apoptosis when accumulation of unfolded proteins or inhibition of the ER–Golgi transport results in the so-called ER stress response *(9)*. Alterations of ER-mediated Ca^{2+} homeostasis are sufficient to induce apoptosis. For example, thapsigargin, a specific inhibitor of SERCA, can induce apoptosis in many cell types including neurons, and agents that suppress Ca^{2+} release from ER can protect neurons against apoptosis *(10)*. Overexpression of calreticulin, a major Ca^{2+}-binding chaperone in the ER lumen, sensitizes cells to mitochondrial cytochrome-*c* release and apoptosis induced by thapsigargin, staurosporine, and etoposide.

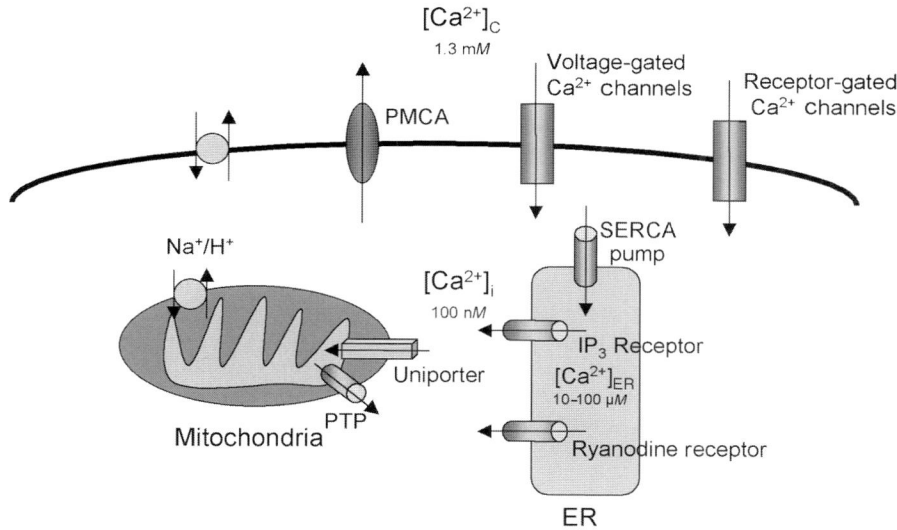

Fig. 1. The regulation of intracellular Ca^{2+} compartmentalization. Cellular Ca^{2+} imports through the plasma membrane occur largely by receptor-operated (e.g., glutamate receptors), voltage-sensitive, and store-operated channels. Once inside the cell, Ca^{2+} can either interact with Ca^{2+}-binding proteins or become sequestered into the ER or mitochondria. The largest Ca^{2+} store in cells is found in the ER or sarcoplasmic reticulum, with local Ca^{2+} concentrations reaching μM or even mM levels. Ca^{2+} levels in the ER are affected by the relative distribution of the Ca^{2+} pump SERCA and of $Ins(1,4,5)P_3$ receptors and Ry receptors, as well as by the relative abundance of Ca^{2+}-binding proteins (calreticulin, calsequestrin) in the ER or sarcoplasmic reticulum. The cytosolic free Ca^{2+} concentration in unstimulated cells is kept at 100 nM by both uptake into the ER and Ca^{2+} extrusion into the extracellular space by the plasma membrane Ca^{2+}-ATPase (PMCA). ER Ca^{2+} release is triggered by agonist stimulation through the generation of $Ins(1,4,5)P_3$ through hydrolysis of phosphatidylinositol-4,5-bisphosphate ($PtdIns[4,5]P_2$) operated by a phospholipase C. The mitochondria take up Ca^{2+} electrophoretically through a uniport transporter and can release it again through three different pathways: reversal of the uniporter, Na^+/H^+-dependent Ca^{2+} exchange, or as a consequence of permeability transition pore (PTP) opening. The PTP can also flicker to release small amounts of Ca^{2+}. Ca^{2+} efflux from cells is regulated primarily by the PMCA, which binds calmodulin and has a high affinity for Ca^{2+}. Ca^{2+} efflux might also be mediated by the Na^+/Ca^{2+} exchanger. (Please *see* companion CD for color version of this figure.)

Moreover, calreticulin-deficient mouse embryonic fibroblasts are resistant to cytochrome-*c* release and apoptosis-induced by etoposide and ultraviolet B *(11)*. Additional findings suggest that ER is potentially a regulator of the progression of apoptosis. This idea is supported at least for two reasons: ER is the main intracellular store of Ca^{2+}, and it is physically and physiologically interconnected with mitochondria. This spatial and functional organization impacts on the regulation of mitochondrial function and on complex cellular processes *(12,13)*. Through the ER-mitochondrial loop, mitochondrial Ca^{2+} uptake can shape the spatiotemporal distribution of cytosolic Ca^{2+} waves during signaling *(14,15)*. After the discharge of intracellular Ca^{2+} stores, clearance of cytosolic Ca^{2+} by strategically located mitochondria also modulates the opening of channels in the plasma membrane that are responsible for the capacitative entry of Ca^{2+} from the extracellular space *(16,17)*. On the other hand, Ca^{2+} modulates mitochondrial function. An adequate increase in mitochondrial matrix Ca^{2+} ($[Ca^{2+}]_m$) regulates metabolism, resulting in a net increase in ATP production *(13,18)*. Ca^{2+} is also a prominent modulator of the permeability transition (PT), a permeability change in the inner membrane to pass solutes up to 1.5 kDa *(19)*. PT has been implicated in both apoptotic and necrotic cell death following selected stimuli *(20–22)*. For example, PT may be one of the mechanisms for cytochrome-*c* release in apoptosis *(23)*.

1.3. Bcl-2 Regulation of ER Ca^{2+} in Apoptosis

The Bcl-2 protein, an antiapoptotic member in the Bcl-2 family, has been identified in the mitochondrial membranes and the ER membranes *(24,25)*. Bcl-2 may stabilize local Ca^{2+} homeostasis *(26)* and suppress oxidative stress *(27)*. Lam et al. *(24)* were the first to report that expression of Bcl-2 might act through regulation of Ca^{2+} release from ER. Bcl-2 overexpression in lymphoma cells prevented apoptosis induced by glucocorticoid hormone dexamethasone and by thapsigargin; the protective effect was correlated with an attenuation of a sustained increase of cytosolic Ca^{2+} in thapsigargin-treated cells *(24)*. Later, it was shown that sustained elevation of cytosolic Ca^{2+} owing to capacitative entry is not required for induction of apoptosis by thapsigargin. The ER Ca^{2+} pool depletion might trigger apoptosis and Bcl-2 inhibits apoptosis via maintaining Ca^{2+} homeostasis within the ER *(26,28)*. Bcl-2 decreases the free Ca^{2+} concentration within the ER lumen presumably by enhancing the permeability of the ER membrane to Ca^{2+} *(29,30)*. Overexpressing Bcl-2 not only reduces resting ER Ca^{2+} concentration but also the extent of capacitative Ca^{2+} entry and the levels of calreticulin and SERCA2b *(29–31)*.

Cells deficient in the proapoptotic genes Bax and Bak are resistant to death induced by thapsigargin and apoptotic stimuli *(32–34)*. Recent work from Korsmeyer's group *(35)* showed that mouse embryonic fibroblasts deficient for Bax and Bak have a reduced $[Ca^{2+}]_{ER}$ that resulted in decreased uptake of Ca^{2+} by mitochondria after Ca^{2+} release from the ER. Restoration of adequate ER Ca^{2+} levels by SERCA made the cells vulnerable to apoptosis *(35)*. When Bax and Bak are overexpressed, they induce ER Ca^{2+} release with subsequent increase in mitochondrial Ca^{2+} and augmented cytochrome-*c* release *(33,36)*. Following apoptotic signals, conformational activation of Bax and Bak might modulate the release of Ca^{2+}, conceivably reflecting their capacity to form ion-conductive pores *(37–39)* or regulate resident ER channels.

Thus, the regulation of steady state $[Ca^{2+}]_{ER}$ by Bcl-2 family members appears to be a crucial checkpoint for Ca^{2+}-dependent apoptotic stimuli, thereafter affects mitochondrial Ca^{2+} uptake and Ca^{2+}-controlled mitochondrial parameters *(29,30,35,40)*. Of note, reducing $[Ca^{2+}]_{ER}$ may not necessarily protective against apoptosis; for example, sustained depletion of Ca^{2+} in the ER by transfection with the Ry receptor and by treatment with Ry causes apoptosis, and this is inhibited by Bcl-XL without affecting the release of Ca^{2+} from the ER *(40a)*. SERCA1 truncated proteins unable to pump Ca^{2+} also decrease $[Ca^{2+}]_{ER}$ whereas inducing apoptosis *(41)*. On the other hand, Bcl-2 might enhance the retention of Ca^{2+} in the ER lumen *(26,28)*.

2. K^+ HOMEOSTASIS AND APOPTOSIS

2.1. Regulation of K^+ Homeostasis

K^+ is the predominant ion inside the cell (approx 140 m*M*); in striking contrast, intracellular Na^+, Ca^{2+}, and Cl^- concentrations are typically one to several orders of magnitude lower than K^+ (Fig. 2). Therefore, intracellular K^+ ($[K^+]_i$) and associated water movement are the major determinants of cytoplasm volume. The K^+ and Na^+ gradients across the membrane are mainly maintained by continuous activities of the Na^+,K^+-ATPase. Accordingly, an excessive K^+ efflux and/or dysfunction of Na^+,K^+-ATPase will lead to depletion of $[K^+]_i$ (Fig. 2). Meanwhile, Cl^- movement may also significantly affect cell volume regulation *(42,43)*.

2.2. K^+ Efflux and Cellular K^+ Depletion Induce Cell Shrinkage and Apoptosis

Evidence from our group and others revealed that excessive K^+ efflux triggered apoptosis in neurons and peripheral cells (reviewed in ref. *44*). For example, K^+ ionophores, such as valinomycin, trigger K^+ efflux and cell death of apoptotic features *(45–49)*. Hughes et al. *(50)* estimated a 95% loss in K^+ content associated with only a 33% loss of cell volume, arguing favorably for a decrease in $[K^+]_i$. The low K^+ concentration in apoptotic cells agrees with an early study that showed intracellular K^+ level was reduced to 50 m*M* during apoptosis in a fibroblast cell line *(51)*. In further studies, Hughes and Cidlowski *(52)* estimated that intracellular K^+ could

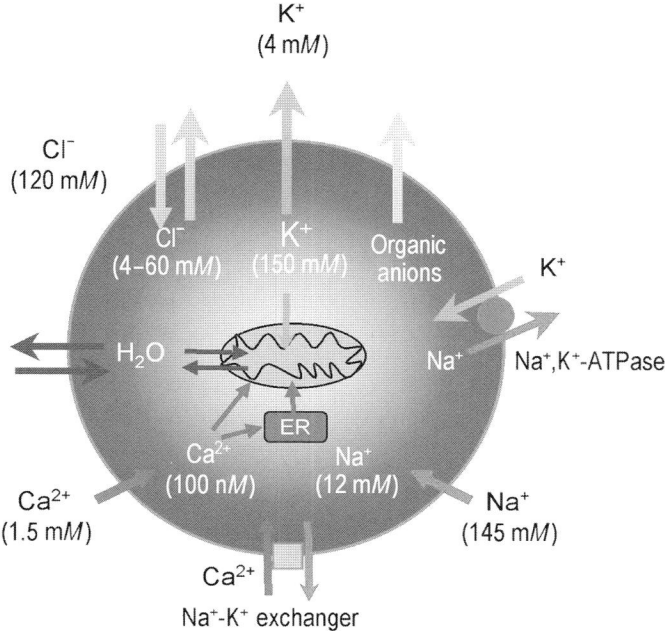

Fig. 2. Ionic distributions across the membranes and pathways for K$^+$ fluxes. Driven by their concentration gradients as well as electrical forces, K$^+$, Cl$^-$, and organic anions flow from the cytoplasm to extracellular space via ion channels and exchangers. Active ionic movements can be achieved by corresponding transporters/pumps. K$^+$ may also move into mitochondria on activation of mitochondrial K$^+$ channels such as the mK$_{ATP}$ channel. Water moves following the osmolarity changes in the cytoplasm and mitochondria. The K$^+$, Cl$^-$, and water efflux are most likely responsible for apoptotic cell volume decrease. Because of excessive K$^+$ and Cl$^-$ efflux over the water movement, there are significant decreases in intracellular concentrations of K$^+$ and Cl$^-$ in apoptotic cells. Moving in of K$^+$ and water into the mitochondrial matrix and intramitochondrial space may cause mitochondrial swelling, lost of mitochondrial potential, disruption of the outer membrane, and probably release of some apoptotic factors such as cytochrome-*c* into the cytoplasm (*see* Subheading 2.3.). Activation of Ca^{2+} and Na$^+$ channels in the plasma membrane results in influx of Ca^{2+} and Na$^+$, which is believed a trigger for necrosis. ER Ca^{2+} release and Ca^{2+} influx into the mitochondrial may also cause damage there related to apoptosis (*see* Fig. 1). The homeostasis of K$^+$, Ca^{2+}, and Na$^+$ is maintained by balance between efflux and influx across the membranes. The energy-dependent Na$^+$, K$^+$-ATPase may play a critical role in K$^+$ and Na$^+$ homeostasis and consequently, affect Ca^{2+} homeostasis. The cytoskeleton and other membrane channels/transporters may contribute to the cell volume control and various apoptotic events, which are not the topics of this review and are not illustrated in the figure. (Please *see* companion CD for color version of this figure.)

be as low as 35 m*M* in shrunken apoptotic thymocytes. Supporting a low K$^+$ concentration in apoptotic cells, Montague et al. *(53)* reported that shrunken cells lost K$^+$ and became hypotonic, consistent with an excess of K$^+$ efflux over water efflux and a reduction in [K$^+$]$_i$.

2.3. Intracellular K$^+$ Depletion and Activation of Key Apoptotic Enzymes

The clues for the link between a massive K$^+$ loss and induction of apoptosis were scattered in early studies which showed that in several cell types reagents promoting cellular K$^+$ loss stimulated maturation and release of the inflammatory factor interleukin (IL)-1β later named caspase-1) *(54,55)*.

K$^+$ inhibited the in vitro activation of procaspase-3 like enzymes in a dose-dependent manner ($K_i \cong 40$ m*M*) and that caspase and nuclease activations were totally inhibited at physiological intracellular K$^+$ concentrations (e.g., 140 m*M*) *(50)*. DNA fragmentation was detected after

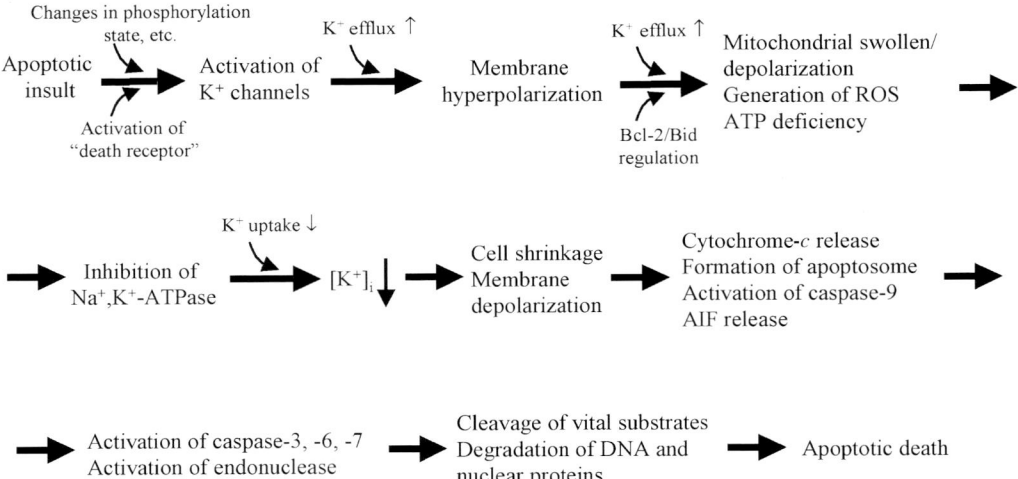

Fig. 3. Chronological events associated with the K$^+$ regulation of apoptosis. A number of apoptotic insults stimulate activation of a variety of K$^+$ channels and/or ionotropic glutamate receptors, likely mediated by signal transduction systems including those that change phosphorylation states of the targeted channels. Activation of "death receptors" such as tumor necrosis factor-α and Fas receptors may also initiate the K$^+$ mechanism for apoptosis. The higher K$^+$ conductance initially drags the membrane potential toward negative direction of the K$^+$ equilibrium potential. The mitochondrial PT including inflow of K$^+$ and water into the mitochondria cause mitochondrial swelling and depolarization. This apoptotic process may be counter-regulated by bcl-2 and Bid. The insult and damage to mitochondria lead to ROS generation and deficient energy production, which consequently block the Na$^+$,K$^+$-ATPase. Most likely, the dysfunction of Na$^+$,K$^+$-ATPase delivers a blow to the already deteriorating situation of excessive K$^+$ efflux, resulting in depletion of intracellular K$^+$, cell volume decrease, and cell membrane depolarization. In the presence of certain minimum level of dATP/ATP, mitochondrial released cytochrome-c and Apaf-1 form apoptosome and cleave procaspase-9. Release of apoptosis-inducing factor causes caspase-independent apoptosis that may also be regulated by K$^+$ homeostasis. In reduced intracellular K$^+$ concentration (approx 50 mM), activation of execution caspases such as caspase-3 and nucleases takes place. DNA damage and nuclear collapse eventually lead to apoptotic death. Information about potential contributions from other ions such as Ca^{2+}, Na$^+$, Mg^{2+}, and Cl$^-$ can be found in the text and listed reviews, and are not listed in the diagram.

[K$^+$]$_i$ was reduced from the normal level of approx 140–156 mM, which was suppressed by KCl with a K$_i$ of approx 70 mM *(50)*. Recent studies even suggest that intracellular K$^+$ may control cytochrome-c release and formation of the apoptosome *(56,57)* (Fig. 3).

The proapoptotic K$^+$ mechanism has been indicated in both the intrinsic (cytochrome-c release via mitochondrial disruption) and extrinsic ("death" receptor-mediated) apoptotic pathways (reviewed in ref. *44*).

2.4. Membrane K$^+$ Channels Provide a Pathway for Proapoptotic K$^+$ Efflux

The voltage-gated outward delayed rectifier or I_K channels have been identified as a major pathway for the proapoptosis K$^+$ efflux. Our group reported the first evidence revealing a critical role of I_K channels in the induction of neuronal apoptosis. We demonstrated that serum deprivation- or staurosporine-induced apoptosis of cortical neurons (death in 24–48 h) was associated with an early enhancement (during 3–12-h exposures) of the I_K current and loss of cellular K$^+$ *(48)*. Attenuating the outward K$^+$ current with tetraethylammonium (TEA) or elevated extracellular K$^+$ reduced apoptosis, even if associated increases in [Ca^{2+}]$_i$ were prevented *(48,58,59)*.

The enhancement of K$^+$ currents has been identified in many cells treated with a variety of apoptotic stimuli. For example, enhancement of outward K$^+$ currents may mediate ultraviolet light irradiation-induced apoptosis in myeloblastic leukemia cells *(60)* and corneal epithelial cells *(61)*. The Alzheimer's disease-associated β-amyloid fragments also enhanced K^+ conductance in neurons and astrocytes *(62–64,58)*. In heart-derived H9c2 cells, overexpression of apoptotic repressor with caspase recruitment domain arcuate nucleus, an antiapoptotic protein, inhibited apoptosis by reducing K^+ efflux mediated by voltage-gated K^+ channels *(65)*. Tumor necrosis factor (TNF)-α-induced apoptosis in rat liver cell line hepatoma tissue cells was accompanied by two and fivefold increases in K^+ and Cl^- currents, respectively *(66)*. Consistently, stimulating K^+ efflux by nigericin, valinomycin, or ouabain enhanced TNF-induced apoptosis in human and rodent tumor cell lines *(67)*.

Bcl-2 may inhibit apoptosis via regulating the K^+ flux. Overexpression of Bcl-2 attenuated staurosporine-induced K^+ current increase in pulmonary artery smooth muscle cells *(68)*. Bcl-2 decreased the amplitude and current density of delayed rectifier K^+ currents, accelerated their inactivation, and downregulated the mRNA expression of the pore-forming α-subunits of the I_K family members Kv1.1, Kv1.5, and Kv2.1 *(68)*.

Transfection and modification of K^+ channel genes in cells provide a way of selectively verifying the involvement of a particular channel in apoptosis. Expression of the Kv2.1 channel in HEK293 cells increased the vulnerability of these cells to the apoptotic insult C_2-ceramide *(69)*; Kv2.1 gene transfection in Chinese hamster ovary cells enhanced vulnerability to apoptosis induced by an nitrous oxide donor 3-morpholinosydnonimine (Sin-1) or oxidative stress *(70)*. Similarly, overexpression of another delayed-rectifier channel Shaker Kv1.5 channel caused basal apoptosis, increased the outward K^+ currents, sensitized staurosporine-induced caspase-3 activation and apoptosis in rat pulmonary artery smooth muscle cells and African green monkey kidney 7 cells *(71)*. Neurons deficient in functional Kv2.1-encoded K^+ channels were protected from oxidant- and staurosporine-induced apoptosis *(70)*. T-leukemic cells genetically deficient for the K^+ channel Kv1.3 were resistant to DNA fragmentation, cytochrome-*c* release, mitochondrial membrane potential change, and apoptotic death initiated by the cytostatic drug actinomycin-D, whereas transfection of Kv1.3 restored the sensitivity of these cells to actinomycin-D *(72)*. Overexpression of the inward rectifier K^+ channel ROMK1 (Kir 1.1) in rat hippocampal neurons-induced apoptosis in 48–72 h *(73)*. Ionomycin- or glucose-deprivation-induced erythrocyte apoptosis is accompanied with activation of inwardly rectifying K^+ channels; phosphatidylserine externalization and cell shrinkage, two landmarks of apoptosis, were attenuated by raising external K^+ concentration and adding a channel blocker such as charybdotoxin or clotrimazole *(74)*.

The K^+ channels capable of mediating the proapoptotic K^+ efflux are not limited to the delayed rectifier channels and, in some cases, the A-type I_A channels. The large-conductance, voltage- and Ca^{2+}-sensitive K^+ (maxi-K) channels, may mediate the proapoptotic K^+ efflux in vascular smooth muscle cells. Two-pore domain K^+ channels may underlie the K^+ efflux-related to apoptotic shrinkage in mouse embryos and other cells *(75,76)*. The human *ether-a-go-go*-related gene K^+ channel markedly promotes H_2O_2-induced apoptosis of various tumor cells *(77,78)*. FCCP, which dissipates the H^+ gradient across the inner membrane of mitochondria, increased the maxi-K currents in these cells and induced apoptosis; both of which were blocked by the K^+ channel blockers TEA and iberiotoxin *(79)*. This mechanism may also mediate the apoptosis induced by nitric oxide in these cells *(79)*. A recent study showed in T-lymphocytes that the Ca^{2+}-activated K^+ channel IKCa1-mediated the apoptotic shrinkage and cell death induced by the Ca^{2+} ionophore calcimycin. Blocking the IKCa1 channel completely prevented cell shrinkage, phosphatidylserine translocation and cell death *(80)*. Interestingly, the apoptosis tested in this study was caspase-independent, suggesting that the K^+ mechanism might regulate different forms of apoptosis.

We demonstrated that ligand-gated channels, that is, ionotropic glutamate receptors may also mediate the proapoptotic K^+ efflux. Stimulation of *N*-methyl-D-aspartate, α-amino-3-hydroxy-5-methyl-isoxazole-4-propionic acid, or kainite receptors may cause 50–80% of intracellular K^+ loss and apoptosis *(81,82)*.

2.5. Phosphorylation State and Proapoptotic Modulation of K⁺ Channels

As an important mechanism in cell survival and apoptosis, protein phosphorylation may play a principal role in the proapoptotic regulation of K⁺ channels and K⁺ efflux. In cortical neurons, we have shown that tyrosine phosphorylation-mediated the modulation of I_K channels during apoptosis, both the I_K current enhancement and apoptotic death were attenuated by the tyrosine kinase inhibitor herbimycin A or lavendustin A *(59)*. In neuronal apoptosis induced by 2,2′-dithiodipyridine, a compound inducing intracellular zinc release, an early and robust increase in TEA-sensitive K⁺ currents was likely mediated by p38 phosphorylation *(83)*. In addition to tyrosine kinases, protein kinase C (PKC) may mediate TNF-induced K⁺ current enhancement *(66)*. Fas-induced cell shrinkage of Jurkat cells and cellular K⁺ loss were blocked by PKC stimulation *(84)*. Conversely, inhibition of PKC enhanced the anti-Fas-induced cell shrinkage suggesting an underlying effect of PKC on cellular K⁺ loss.

2.6. Attenuating K⁺ Efflux and Antiapoptosis Effects

Concordant with a key role of K⁺ channels in apoptosis, K⁺ channel blockers are protective against apoptosis in many cell types. In addition to the classical K⁺ channel blocker TEA or 4-AP, other K⁺ channel blockers have been shown to be antiapoptotic, including the TEA analog tetrapentylammonium and tetrahexylammonium *(85,86)*, clofilium, quinine, bretylium tosylate, clotrimazole, and charybdotoxin (reviewed in ref. *44*). A recent study showed that K⁺ channel blockers might protect hippocampal neurons by reducing microglia-released diffusible messengers such as nitrous oxide, TNF-α, and reactive oxygen species *(87)*.

TEA and clofilium showed the protective effect of reducing infarct volume after global and focal brain ischemia in animal models *(88,89)*, which is consistent with the observation that hypoxia/anoxia or ischemia/reperfusion may activate various K⁺ currents in central neurons *(90–93)*. It is possible that altered K⁺ channels might be a pathological mechanism triggering neurodegenerations in scrapie-related diseases and that these channels might be targeted as a therapeutic approach *(94)*.

In addition to K⁺ channel blockers, raising extracellular K⁺ typically improves the survival of cultured neurons and attenuates neuronal apoptosis *(95)*. The antiapoptotic effect of high K⁺ medium was initially attributed to its ability to increase Ca^{2+} influx, primarily through voltage-gated Ca^{2+} channels *(96,97)*. This Ca^{2+} mechanism, however, was excluded from the neuronal protection of high K⁺ medium in cortical neurons *(58,59,98)*. A recent study discovered that the cerebellar granule cell death in physiological K⁺ parallels the developmental expression the novel K⁺ channel subunits TWIK-related, acid-sensitive K⁺ (TASK)-1 and TASK-3 that encode the pH-sensitive outward K⁺ current IKso *(99)*. Genetic transfer of the TASK subunits in hippocampal neurons lacking IKso-induced cell death, whereas their genetic inactivation protects these neurons. Neuronal death was also prevented by conditions that specifically reduced K⁺ efflux *(99)*.

Conflicting with the data showing the antiapoptotic effect of K⁺ channel blockers, a few reports have shown that K⁺ channel blockers 4-AP and clofilium induce apoptosis in some cells *(100–104)*. Whether these toxic effects are because of the blocking the K⁺ channels or a result from nonspecific actions of these compounds remains to be determined. Our recent studies show that the cell death induced by 4-AP and clofilium is likely a result from their potent inhibitory actions on the Na⁺,K⁺-ATPase at similar concentrations that inhibit K⁺ channels *(105–108)*.

2.7. K⁺ Channel Openers and Blockers: Mechanisms of Protection

In the past few years, a class of drugs with the property of opening the K_{ATP} channels in the plasma membrane or mitochondrial membrane has been shown to be protective against ischemic injury in the heart *(109–111)* as well as the brain *(112,113)*. Similarly, opening of the Ca^{2+}-sensitive maxi-K channel was proposed as a therapeutic treatment of ischemic stroke. K⁺ channel openers can hyperpolarize the membrane as long as K⁺ homeostasis remains relatively normal. With a hyperpolarized membrane potential, voltage-dependent Ca^{2+} and Na^+ influx will be eliminated or attenuated, thus preventing ischemia-induced necrotic cell death in the heart *(114,115)* or brain *(116–118)*.

The K^+ channel openers including diazoxide, levocromakalim, and pinacidil may also act as antioxidants; their protective effect remained even in the presence of a K^+ channel blocker *(119)*. Of note, a pretreatment of these K_{ATP} channel openers is usually required for the protective effect, suggesting that they likely trigger a preconditioning mechanism *(121)*. In other words, the K^+ efflux stimulated by K^+ channel openers may generate an insult. Promoting K^+ efflux via the plasmalemmal K_{ATP} channel and various consequences in the mitochondria following opening the mK_{ATP} channel may serve as mediators of the preconditioning effect. This opinion is supported by a recent report that diazoxide pretreatment induced a transient and mild depolarization of the mitochondrial membrane potential *(122)*. Other mechanisms not associated with a preconditioning effect and membrane hyperpolarization may be possible but remain to be specifically substantiated.

3. OTHER IONS IN APOPTOSIS

The ionic mechanism underlying apoptosis may not be specific to Ca^{2+} and K^+. Alternations of other ion channel activities may contribute to apoptotic process. Other possible players may include Na^+, Cl^-, and Zn^{2+} *(44,84,105,123)*. For example, apoptotic cells are likely to lose intracellular Cl^- that may contribute to apoptotic cell volume decrease and some other events *(42,43,105)*.

4. CONCLUSION AND REMARKS

The findings described herein suggest many exciting directions for future research into the ionic homeostasis in mitochondria and ER, and its novel roles in neuronal function and survival. Additional goals of future research are to understand the signaling mechanisms that regulate the ER-mitochondria Ca^{2+} loop movement of ER within and between cell compartments; to elucidate the mechanisms that govern interactions between ER, mitochondria, and plasma membrane; and to understand how dysfunction of ion channels/transporters and the disruption of ion homeostasis contributes to neurological disorders. It will be also necessary to delineate how the Ca^{2+} and K^+ mechanisms interact during apoptotic process.

REFERENCES

1. Olney JW. Excitotoxicity: an overview. Can Dis Wkly Rep 1990;16(Suppl 1E):47–57; discussion 57–58.
2. Choi DW. Excitotoxic cell death. J Neurobiol 1992;23:1261–1276.
3. Yu SP, Canzoniero LM, Choi DW. Ion homeostasis and apoptosis. Curr Opin Cell Biol 2001;13: 405–411.
4. Ferri KF, Kroemer G. Organelle-specific initiation of cell death pathways. Nat Cell Biol 2001;3: E255–E263.
5. Oakes SA, Opferman JT, Pozzan T, Korsmeyer SJ, Scorrano L. Regulation of endoplasmic reticulum Ca^{2+} dynamics by proapoptotic BCL-2 family members. Biochem Pharmacol 2003;66:1335–1340.
6. Rizzuto R, Pinton P, Carrington W, et al. Close contacts with the endoplasmic reticulum as determinants of mitochondrial Ca^{2+} responses. Science 1998;280:1763–1766.
7. Berridge MJ. Neuronal calcium signaling. Neuron 1998;21:13–26.
8. Csordas G, Thomas AP, Hajnoczky G. Quasi-synaptic calcium signal transmission between endoplasmic reticulum and mitochondria. EMBO J 1999;18:96–108.
9. Oyadomari S, Araki E, Mori M. Endoplasmic reticulum stress-mediated apoptosis in pancreatic beta-cells. Apoptosis 2002;7:335–345.
10. Guo Q, Sopher BL, Furukawa K, et al. Alzheimer's presenilin mutation sensitizes neural cells to apoptosis induced by trophic factor withdrawal and amyloid beta-peptide: involvement of calcium and oxyradicals. J Neurosci 1997;17:4212–4222.

11. Nakamura K, Bossy-Wetzel E, Burns K, et al. Changes in endoplasmic reticulum luminal environment affect cell sensitivity to apoptosis. J Cell Biol 2000;150:731–740.
12. Rutter GA, Theler JM, Murgia M, Wollheim CB, Pozzan T, Rizzuto R. Stimulated Ca^{2+} influx raises mitochondrial free Ca^{2+} to supramicromolar levels in a pancreatic beta-cell line. Possible role in glucose and agonist-induced insulin secretion. J Biol Chem 1993;268:22,385–22,390.
13. Hajnoczky G, Robb-Gaspers LD, Seitz MB, Thomas AP. Decoding of cytosolic calcium oscillations in the mitochondria. Cell 1995;82:415–424.
14. Jouaville LS, Ichas F, Holmuhamedov EL, Camacho P, Lechleiter JD. Synchronization of calcium waves by mitochondrial substrates in *Xenopus laevis* oocytes. Nature 1995;377:438–441.
15. Berridge MJ, Lipp P, Bootman MD. The versatility and universality of calcium signalling. Nat Rev Mol Cell Biol 2000;1:11–21.
16. Chavis P, Fagni L, Lansman JB, Bockaert J. Functional coupling between ryanodine receptors and L-type calcium channels in neurons. Nature 1996;382:719–722.
17. Hoth M, Fanger CM, Lewis RS. Mitochondrial regulation of store-operated calcium signaling in T lymphocytes. J Cell Biol 1997;137:633–648.
18. Jouaville LS, Pinton P, Bastianutto C, Rutter GA, Rizzuto R. Regulation of mitochondrial ATP synthesis by calcium: evidence for a long-term metabolic priming. Proc Natl Acad Sci USA 1999;96:13,807–13,812.
19. Bernardi P, Scorrano L, Colonna R, Petronilli V, Di Lisa F. Mitochondria and cell death. Mechanistic aspects and methodological issues. Eur J Biochem 1999;264:687–701.
20. Zamzami N, Marchetti P, Castedo M, et al. Inhibitors of permeability transition interfere with the disruption of the mitochondrial transmembrane potential during apoptosis. FEBS Lett 1996;384:53–57.
21. Scorrano L, Petronilli V, Di Lisa F, Bernardi P. Commitment to apoptosis by GD3 ganglioside depends on opening of the mitochondrial permeability transition pore. J Biol Chem 1999;274:22,581–22,585.
22. Scorrano L, Penzo D, Petronilli V, Pagano F, Bernardi P. Arachidonic acid causes cell death through the mitochondrial permeability transition. Implications for tumor necrosis factor-alpha aopototic signaling. J Biol Chem 2001;276:12,035–12,040.
23. Gogvadze V, Robertson JD, Zhivotovsky B, Orrenius S. Cytochrome-c release occurs via Ca^{2+}-dependent and Ca^{2+}-independent mechanisms that are regulated by Bax. J Biol Chem 2001;276; 19,066–19,071.
24. Lam M, Dubyak G, Chen L, Nunez G, Miesfeld RL, Distelhorst CW. Evidence that BCL-2 represses apoptosis by regulating endoplasmic reticulum-associated Ca^{2+} fluxes. Proc Natl Acad Sci USA 1994;91:6569–6573.
25. Tagami S, Eguchi Y, Kinoshita M, Takeda M, Tsujimoto Y. A novel protein, RTN-XS, interacts with both Bcl-XL and Bcl-2 on endoplasmic reticulum and reduces their anti-apoptotic activity. Oncogene 2000;19:5736–5746.
26. He H, Lam M, McCormick TS, Distelhorst CW. Maintenance of calcium homeostasis in the endoplasmic reticulum by Bcl-2. J Cell Biol 1997;138:1219–1228.
27. Bruce-Keller AJ, Begley JG, Fu W, et al. Bcl-2 protects isolated plasma and mitochondrial membranes against lipid peroxidation induced by hydrogen peroxide and amyloid beta-peptide. J Neurochem 1998;70:31–39.
28. Distelhorst CW, McCormick TS. Bcl-2 acts subsequent to and independent of Ca^{2+} fluxes to inhibit apoptosis in thapsigargin- and glucocorticoid-treated mouse lymphoma cells. Cell Calcium 1996;19:473–483.
29. Pinton P, Ferrari D, Magalhaes P, et al. Reduced loading of intracellular Ca^{2+} stores and downregulation of capacitative Ca^{2+} influx in Bcl-2-overexpressing cells. J Cell Biol 2000;148:857–862.
30. Foyouzi-Youssefi R, Arnaudeau S, Borner C, et al. Bcl-2 decreases the free Ca^{2+} concentration within the endoplasmic reticulum. Proc Natl Acad Sci USA 2000;97:5723–5728.
31. Vanden Abeele F, Skryma R, Shuba Y, et al. Bcl-2-dependent modulation of Ca^{2+} homeostasis and store-operated channels in prostate cancer cells. Cancer Cell 2002;1:169–179.
32. Wei MC, Zong WX, Cheng EH, et al. Proapoptotic BAX and BAK: a requisite gateway to mitochondrial dysfunction and death. Science 2001;292:727–730.
33. Nutt LK, Chandra J, Pataer A, et al. Bax-mediated Ca^{2+} mobilization promotes cytochrome-*c* release during apoptosis. J Biol Chem 2002;277:20,301–20,308.
34. Nutt LK, Pataer A, Pahler J, et al. Bax and Bak promote apoptosis by modulating endoplasmic reticular and mitochondrial Ca^{2+} stores. J Biol Chem 2002;277:9219–9225.

35. Scorrano L, Oakes SA, Opferman JT, et al. BAX and BAK regulation of endoplasmic reticulum Ca^{2+}: a control point for apoptosis. Science 2003;300:135–139.
36. Aiba-Masago S, Liu Xb XB, Masago R, et al. Bax gene expression alters Ca^{2+} signal transduction without affecting apoptosis in an epithelial cell line. Oncogene 2002;21:2762–2767.
37. Minn AJ, Velez P, Schendel SL, et al. Bcl-x(L) forms an ion channel in synthetic lipid membranes. Nature 1997;385:353–357.
38. Antonsson B, Conti F, Ciavatta A, et al. Inhibition of Bax channel-forming activity by Bcl-2. Science 1997;277:370–372.
39. Schlesinger PH, Gross A, Yin XM, et al. Comparison of the ion channel characteristics of proapoptotic BAX and antiapoptotic BCL-2. Proc Natl Acad Sci USA 1997;94:11,357–11,362.
40. Pinton P, Ferrari D, Rapizzi E, Di Virgilio F, Pozzan T, Rizzuto R. The Ca^{2+} concentration of the endoplasmic reticulum is a key determinant of ceramide-induced apoptosis: significance for the molecular mechanism of Bcl-2 action. EMBO J 2001;20:2690–2701.
40a. Pan Z, Damron D, Nieminen AL, Bhat MB, Ma J. Depletion of intracellular Ca^{2+} by caffeine and ryanodine induces apoptosis of Chinese hamster ovary cells transfected with ryanodine receptor. J Biol Chem 2000;275:19,978–19,984.
41. Chami M, Gozuacik D, Lagorce D, et al. SERCA1 truncated proteins unable to pump calcium reduce the endoplasmic reticulum calcium concentration and induce apoptosis. J Cell Biol 2001;153:1301–1314.
42. Yu SP, Choi DW. Ions, cell volume, and apoptosis. Proc Natl Acad Sci USA 2000;97:9360–9362.
43. Okada Y, Maeno E. Apoptosis, cell volume regulation and volume-regulatory chloride channels. Comp Biochem Physiol A Mol Integr Physiol 2001;130:377–383.
44. Yu SP. Regulation and critical role of potassium homeostasis in apoptosis. Prog Neurobiol 2003;70:363–386.
45. Ojcius DM, Zychlinsky A, Zheng LM, Young JD. Ionophore-induced apoptosis: role of DNA fragmentation and calcium fluxes. Exp Cell Res 1991;197:43–49.
46. Duke RC, Witter RZ, Nash PB, Young JD, Ojcius DM. Cytolysis mediated by ionophores and pore-forming agents: role of intracellular calcium in apoptosis. FASEB J 1994;8:237–246.
47. Que FG, Gores GJ, LaRusso NF. Development and initial application of an in vitro model of apoptosis in rodent cholangiocytes. Am J Physiol 1997;272:G106–G115.
48. Yu SP, Yeh CH, Sensi SL, et al. Mediation of neuronal apoptosis by enhancement of outward potassium current. Science 1997;278:114–117.
49. Marklund L, Behnam-Motlagh P, Henriksson R, Grankvist K. Bumetanide annihilation of amphotericin B-induced apoptosis and cytotoxicity is due to its effect on cellular K$^+$ flux. J Antimicrob Chemother 2001;48:781–786.
50. Hughes FM, Jr, Bortner CD, Purdy GD, Cidlowski JA. Intracellular K$^+$ suppresses the activation of apoptosis in lymphocytes. J Biol Chem 1997;272:30,567–30,576.
51. Barbiero G, Duranti F, Bonelli G, Amenta JS, Baccino FM. Intracellular ionic variations in the apoptotic death of L cells by inhibitors of cell cycle progression. Exp Cell Res 1995;217:410–418.
52. Hughes FM, Jr, Cidlowski JA. Potassium is a critical regulator of apoptotic enzymes in vitro and in vivo. Adv Enzyme Regul 1999;39:157–171.
53. Montague JW, Bortner CD, Hughes FM, Jr, Cidlowski JA. A necessary role for reduced intracellular potassium during the DNA degradation phase of apoptosis. Steroids 1999;64:563–569.
54. Perregaux D, Gabel CA. Interleukin-1 beta maturation and release in response to ATP and nigericin. Evidence that potassium depletion mediated by these agents is a necessary and common feature of their activity. J Biol Chem 1994;269:15,195–15,203.
55. Walev I, Klein J, Husmann M, et al. Potassium regulates IL-1 beta processing via calcium-independent phospholipase A2. J Immunol 2000;164:5120–5124.
56. Thompson GJ, Langlais C, Cain K, Conley EC, Cohen GM. Elevated extracellular [K$^+$] inhibits death-receptor- and chemical-mediated apoptosis prior to caspase activation and cytochrome-*c* release. Biochem J 2001;357:137–145.
57. Adams JM, Cory S. Apoptosomes: engines for caspase activation. Curr Opin Cell Biol 2002;14:715–720.
58. Yu SP, Farhangrazi ZS, Ying HS, Yeh CH, Choi DW. Enhancement of outward potassium current may participate in beta-amyloid peptide-induced cortical neuronal death. Neurobiol Dis 1998;5:81–88.
59. Yu SP, Yeh CH, Gottron F, Wang X, Grabb MC, Choi DW. Role of the outward delayed rectifier K$^+$ current in ceramide-induced caspase activation and apoptosis in cultured cortical neurons. J Neurochem 1999;73:933–941.

60. Wang L, Xu D, Dai W, Lu L. An ultraviolet-activated K$^+$ channel mediates apoptosis of myeloblastic leukemia cells. J Biol Chem 1999;274:3678–3685.
61. Wang L, Li T, Lu L. UV-induced corneal epithelial cell death by activation of potassium channels. Invest Ophthalmol Vis Sci 2003;44:5095–5101.
62. Furukawa K, Barger SW, Blalock EM, Mattson MP. Activation of K$^+$ channels and suppression of neuronal activity by secreted beta-amyloid-precursor protein. Nature 1996;379:74–78.
63. Jalonen TO, Charniga CJ, Wielt DB. beta-Amyloid peptide-induced morphological changes coincide with increased K$^+$ and Cl$^-$ channel activity in rat cortical astrocytes. Brain Res 1997;746:85–97.
64. Colom LV, Diaz ME, Beers DR, Neely A, Xie WJ, Appel SH. Role of potassium channels in amyloid-induced cell death. J Neurochem 1998;70:1925–1934.
65. Ekhterae D, Platoshyn O, Zhang S, Remillard CV, Yuan JX. Apoptosis repressor with caspase domain inhibits cardiomyocyte apoptosis by reducing K$^+$ currents. Am J Physiol Cell Physiol 2003;284:C1405–C1410.
66. Nietsch HH, Roe MW, Fiekers JF, Moore AL, Lidofsky SD. Activation of potassium and chloride channels by tumor necrosis factor alpha. Role in liver cell death. J Biol Chem 2000;275:20,556–20,561.
67. Penning LC, Denecker G, Vercammen D, Declercq W, Schipper RG, Vandenabeele P. A role for potassium in TNF-induced apoptosis and gene-induction in human and rodent tumour cell lines. Cytokine 2000;12:747–750.
68. Ekhterae D, Platoshyn O, Krick S, Yu Y, McDaniel SS, Yuan JX. Bcl-2 decreases voltage-gated K$^+$ channel activity and enhances survival in vascular smooth muscle cells. Am J Physiol Cell Physiol 2001;281:C157–C165.
69. Yu SP, Choi DW. K$^+$ efflux mediated by delayed rectifier K$^+$ channels contributes of neuronal death, Amsterdam, New York: John Wiley & Sons Ltd., 1999.
70. Pal S, Hartnett KA, Nerbonne JM, Levitan ES, Aizenman E. Mediation of neuronal apoptosis by Kv2.1-encoded potassium channels. J Neurosci 2003;23:4798–4802.
71. Brevnova EE, Platoshyn O, Zhang S, Yuan JX. Overexpression of human KCNA5 increases IK V and enhances apoptosis. Am J Physiol Cell Physiol 2004;287:C715–C722.
72. Bock J, Szabo I, Jekle A, Gulbins E. Actinomycin D-induced apoptosis involves the potassium channel Kv1.3. Biochem Biophys Res Commun 2002;295:526–531.
73. Nadeau H, McKinney S, Anderson DJ, Lester HA. ROMK1 (Kir1.1) causes apoptosis and chronic silencing of hippocampal neurons. J Neurophysiol 2000;84:1062–1075.
74. Lang PA, Kaiser S, Myssina S, Wieder T, Lang F, Huber SM. Role of Ca^{2+}-activated K$^+$ channels in human erythrocyte apoptosis. Am J Physiol Cell Physiol 2003;285:C1553–C1560.
75. Trimarchi JR, Liu L, Smith PJ, Keefe DL. Apoptosis recruits two-pore domain potassium channels used for homeostatic volume regulation. Am J Physiol Cell Physiol 2002;282:C588–C594.
76. Patel AJ, Lazdunski M. The 2P-domain K$^+$ channels: role in apoptosis and tumorigenesis. Pflugers Arch 2004;448:261–273.
77. Wang H, Zhang Y, Cao L, et al. HERG K$^+$ channel, a regulator of tumor cell apoptosis and proliferation. Cancer Res 2002;62:4843–4848.
78. Han H, Wang J, Zhang Y, et al. HERG K channel conductance promotes H$_2$O$_2$-induced apoptosis in HEK293 cells: cellular mechanisms. Cell Physiol Biochem 2004;14:121–134.
79. Krick S, Platoshyn O, Sweeney M, et al. Nitric oxide induces apoptosis by activating K$^+$ channels in pulmonary vascular smooth muscle cells. Am J Physiol Heart Circ Physiol 2002;282:H184–H193.
80. Elliott JI, Higgins CF. IKCa1 activity is required for cell shrinkage, phosphatidylserine translocation and death in T lymphocyte apoptosis. EMBO Rep 2003;4:189–194.
81. Yu SP, Yeh C, Strasser U, Tian M, Choi DW. NMDA receptor-mediated K$^+$ efflux and neuronal apoptosis. Science 1999;284:336–339.
82. Xiao AY, Homma M, Wang XQ, Wang X, Yu SP. Role of K$^+$ efflux in apoptosis induced by AMPA and kainate in mouse cortical neurons. Neuroscience 2001;108:61–67.
83. McLaughlin B, Pal S, Tran MP, et al. p38 activation is required upstream of potassium current enhancement and caspase cleavage in thiol oxidant-induced neuronal apoptosis. J Neurosci 2001;21:3303–3311.
84. Gomez-Angelats M, Bortner CD, Cidlowski JA. Protein kinase C (PKC) inhibits fas receptor-induced apoptosis through modulation of the loss of K$^+$ and cell shrinkage. A role for PKC upstream of caspases. J Biol Chem 2000;275:19,609–19,619.
85. Dallaporta B, Marchetti P, de Pablo MA, et al. Plasma membrane potential in thymocyte apoptosis. J Immunol 1999;162:6534–6542.

86. Wang X, Xiao AY, Ichinose T, Yu SP. Effects of tetraethylammonium analogs on apoptosis and membrane currents in cultured cortical neurons. J Pharmacol Exp Ther 2000;295:524–530.
87. Fordyce CB, Jagasia R, Schlichter JC. Modulation of microglia- induced neurotoxicity. Soc Neurosci Abstr 2002;892:17.
88. Huang H, Gao TM, Gong L, Zhuang Z, Li X. Potassium channel blocker TEA prevents CA1 hippocampal injury following transient forebrain ischemia in adult rats. Neurosci Lett 2001;305:83–86.
89. Wei L, Yu SP, Gottron F, Snider BJ, Zipfel GJ, Choi DW. Potassium channel blockers attenuate hypoxia- and ischemia-induced neuronal death in vitro and in vivo. Stroke 2003;34:1281–1286.
90. Cowan AI, Martin RL. Ionic basis of membrane potential changes induced by anoxia in rat dorsal vagal motoneurones. J Physiol 1992;455:89–109.
91. Haddad GE, Petrich ER, Zumino AP, Schanne OF. Background K^+ currents and response to metabolic inhibition during early development in rat cardiocytes. Mol Cell Biochem 1997;177:159–168.
92. Guatteo E, Federici M, Siniscalchi A, Knopfel T, Mercuri NB, Bernardi G. Whole cell patch-clamp recordings of rat midbrain dopaminergic neurons isolate a sulphonylurea- and ATP-sensitive component of potassium currents activated by hypoxia. J Neurophysiol 1998;79:1239–1245.
93. Xuan Chi X, Xu ZC. Potassium currents in CA1 neurons of rat hippocampus increase shortly after transient cerebral ischemia. Neurosci Lett 2000;281:5–8.
94. Johnston AR, Fraser JR, Jeffrey M, MacLeod N. Alterations in potassium currents may trigger neurodegeneration in murine scrapie. Exp Neurol 1998;151:326–333.
95. Franklin JL, Fickbohm DJ, Willard AL. Long-term regulation of neuronal calcium currents by prolonged changes of membrane potential. J Neurosci 1992;12:1726–1735.
96. Koike T, Martin DP, Johnson EM, Jr. Role of Ca^{2+} channels in the ability of membrane depolarization to prevent neuronal death induced by trophic-factor deprivation: evidence that levels of internal Ca^{2+} determine nerve growth factor dependence of sympathetic ganglion cells. Proc Natl Acad Sci USA 1989;86:6421–6425.
97. Johnson EM, Jr, Koike T, Franklin J. A "calcium set-point hypothesis" of neuronal dependence on neurotrophic factor. Exp Neurol 1992;115:163–166.
98. Yu SP, Sensi SL, Canzoniero LM, Buisson A, Choi DW. Membrane-delimited modulation of NMDA currents by metabotropic glutamate receptor subtypes 1/5 in cultured mouse cortical neurons. J Physiol 1997;499(Pt 3):721–732.
99. Lauritzen I, Zanzouri M, Honore E, et al. K^+-dependent cerebellar granule neuron apoptosis. Role of task leak K^+ channels. J Biol Chem 2003;278:32,068–32,076.
100. Chin LS, Park CC, Zitnay KM, et al. 4-Aminopyridine causes apoptosis and blocks an outward rectifier K^+ channel in malignant astrocytoma cell lines. J Neurosci Res 1997;48:122–127.
101. Choi BY, Kim HY, Lee KH, Cho YH, Kong G. Clofilium, a potassium channel blocker, induces apoptosis of human promyelocytic leukemia (HL-60) cells via Bcl-2-insensitive activation of caspase-3. Cancer Lett 1999;147:85–93.
102. Kim JA, Kang YS, Jung MW, Kang GH, Lee SH, Lee YS. Ca^{2+} influx mediates apoptosis induced by 4-aminopyridine, a K^+ channel blocker, in HepG2 human hepatoblastoma cells. Pharmacology 2000;60:74–81.
103. Abdul M, Hoosein N. Expression and activity of potassium ion channels in human prostate cancer. Cancer Lett 2002;186:99–105.
104. Manikkam M, Li Y, Mitchell BM, Mason DE, Freeman LC. Potassium channel antagonists influence porcine granulosa cell proliferation, differentiation, and apoptosis. Biol Reprod 2002;67:88–98.
105. Wei L, Xiao AY, Jin C, Yang A, Lu ZY, Yu SP. Effects of chloride and potassium channel blockers on apoptotic cell shrinkage and apoptosis in cortical neurons. Pflugers Arch 2004;448:325–334.
106. Yang A, Wang XQ, Sun CS, Wei L, Yu SP. Inhibitory effects of clofilium on membrane currents associated with Ca channels, NMDA receptor channels and Na^+, K^+-ATPase in cortical neurons. Pharmacology 2005;73:162–168.
107. Wang XQ, Xiao AY, Sheline C, et al. Apoptotic insults impair Na^+, K^+-ATPase activity as a mechanism of neuronal death mediated by concurrent ATP deficiency and oxidant stress. J Cell Sci 2003;116:2099–2110.
108. Wang XQ, Xiao AY, Yang A, LaRose L, Wei L, Yu SP. Block of Na^+,K^+-ATPase and induction of hybrid death by 4-aminopyridine in cultured cortical neurons. J Pharmacol Exp Ther 2003;305:502–506.
109. Grover GJ, Garlid KD. ATP-Sensitive potassium channels: a review of their cardioprotective pharmacology. J Mol Cell Cardiol 2000;32:677–695.

110. Akao M, Ohler A, O'Rourke B, Marban E. Mitochondrial ATP-sensitive potassium channels inhibit apoptosis induced by oxidative stress in cardiac cells. Circ Res 2001;88:1267–1275.
111. Dos Santos P, Kowaltowski AJ, Laclau MN, et al. Mechanisms by which opening the mitochondrial ATP- sensitive K$^+$ channel protects the ischemic heart. Am J Physiol Heart Circ Physiol 2002;283:H284–H295.
112. Murphy KP, Greenfield SA. ATP-sensitive potassium channels counteract anoxia in neurones of the substantia nigra. Exp Brain Res 1991;84:355–358.
113. Shimizu K, Lacza Z, Rajapakse N, Horiguchi T, Snipes J, Busija DW. MitoK(ATP) opener, diazoxide, reduces neuronal damage after middle cerebral artery occlusion in the rat. Am J Physiol Heart Circ Physiol 2002;283:H1005–H1011.
114. Grover GJ. Pharmacology of ATP-sensitive potassium channel (KATP) openers in models of myocardial ischemia and reperfusion. Can J Physiol Pharmacol 1997;75:309–315.
115. Garlid KD, Paucek P. The mitochondrial potassium cycle. IUBMB Life 2001;52:153–158.
116. Fujita H, Takizawa S, Nanri K, Matsushima K, Ogawa S, Shinohara Y. Potassium channel opener reduces extracellular glutamate concentration in rat focal cerebral ischemia. Brain Res Bull 1997;43:365–368.
117. Takaba H, Nagao T, Yao H, Kitazono T, Ibayashi S, Fujishima M. An ATP-sensitive potassium channel activator reduces infarct volume in focal cerebral ischemia in rats. Am J Physiol 1997;273:R583–R586.
118. Lauritzen I, De Weille JR, Lazdunski M. The potassium channel opener (-)-cromakalim prevents glutamate-induced cell death in hippocampal neurons. J Neurochem 1997;69:1570–1579.
119. Goodman Y, Mattson MP. K$^+$ channel openers protect hippocampal neurons against oxidative injury and amyloid beta-peptide toxicity. Brain Res 1996;706:328–332.
120. Mizumura T, Nithipatikom K, Gross GJ. Bimakalim, an ATP-sensitive potassium channel opener, mimics the effects of ischemic preconditioning to reduce infarct size, adenosine release, and neutrophil function in dogs. Circulation 1995;92:1236–1245.
121. Menasche P, Kevelaitis E, Mouas C, Grousset C, Piwnica A, Bloch G. Preconditioning with potassium channel openers. A new concept for enhancing cardioplegic protection? J Thorac Cardiovasc Surg 1995;110:1606–1613.
122. Ichinose M, Yonemochi H, Sato T, Saikawa T. Diazoxide triggers cardioprotection against apoptosis induced by oxidative stress. Am J Physiol Heart Circ Physiol 2003;284:H2235–H2241.
123. Bortner CD, Hughes FM, Jr, Cidlowski JA. A primary role for K$^+$ and Na$^+$ efflux in the activation of apoptosis. J Biol Chem 1997;272:32,436–32,442.

11
Mammalian Neural Stem Cell Renewal

Yvan Arsenijevic, PhD

SUMMARY

Neural stem cells (NSCs) responsible for the growth of the brain modify their competence throughout embryogenesis, but recent data show that these changes also occur after birth. Both environmental and intrinsic factors tightly control NSC proliferation with specifications intervening at specific times of brain development. The change of NSC competence also parallels different stages of neurogenesis, demonstrating that the control of NSC may have important consequential effects on the number of cells generated, as well as on their fate. This link is poorly understood, except for the retina, but several actors have already been identified bringing new hope toward understanding how the NSC controls neurogenesis and maintains its ability to self-renew.

Key Words: Stem cell; neural precursors; neurogenesis; retina; spinal cord; telencephalon; neurons; glia.

1. INTRODUCTION

The past years have been extremely prolific in generating new data concerning factors controlling (intrinsic and extrinsic) neural stem cell (NSC) renewal. A general picture now starts to emerge showing that NSCs are regulated in a region-specific manner. NSCs are also differently regulated during adulthood as compared with fetal life. Lineage analysis using nonreplicative retroviruses have revealed, many years ago, that neural and retinal stem cells (RSCs) change their competence during development to successively generate neurons, astrocytes, and oligodendrocytes or different neurons and glia of the retina *(1–3)*. Only recently have the molecular mechanisms underlining these changes during development and adulthood began to be identified to reveal the complexity of NSC quiescence and proliferation.

Before discussing the different mechanisms leading to the conservation of stemness and to the proliferation of NSCs, I would like to recall the main characteristics that allow a NSC to be a stem cell. NSCs and RSCs participate in brain, spinal cord, and eye organogenesis by generating neurons and glia in a precise spatial and temporal manner. This implicates that the NSCs have to proliferate extensively without losing their identity. Such action can be achieved through either asymmetric cell division or symmetric division to increase the population of NSCs. It appears that the NSCs must renew throughout development and even throughout life, as shown for NSCs that continuously add new neurons in the olfactory bulb *(4,5)* and in the hippocampus *(6,7)*. The second main characteristic of NSCs is multipotentiality; that is, the potential to generate neurons, astrocytes, and oligodendrocytes. This chapter is dedicated to the molecular control of the NSC renewal.

Our capacity to dissect the function of environmental and intrinsic factors that control NSC renewal is mainly because of the easy model and methods developed by Reynolds and Weiss to isolate and characterize NSCs *(8–10)*. This method is summarized in Fig. 1. Thanks to the facility to

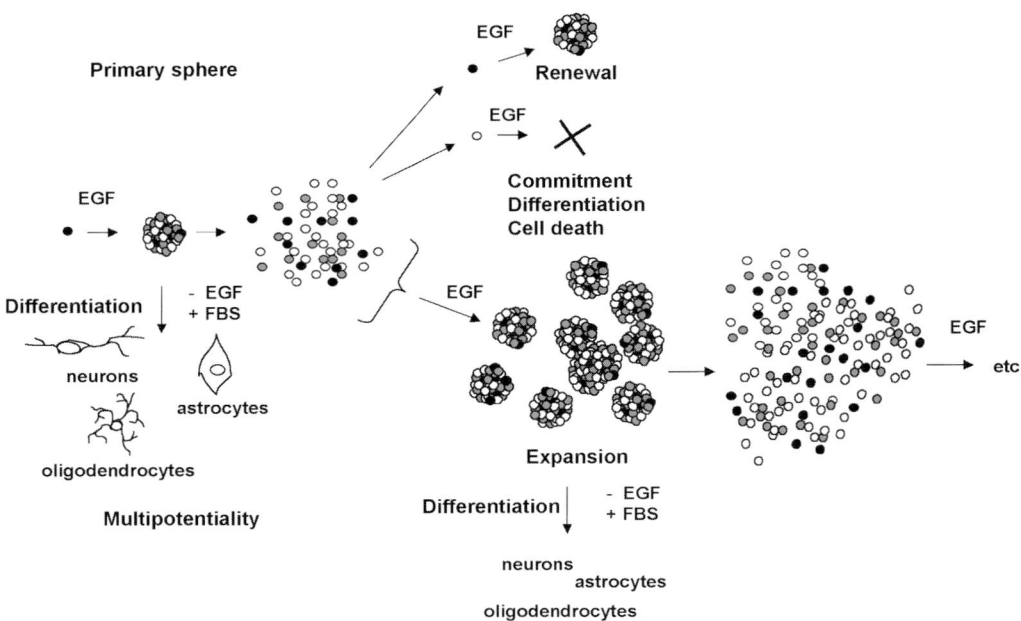

Fig. 1. In the presence of EGF, a single NSC generates numerous cells able to produce neurons, astrocytes, and oligodendrocytes revealing the multipotentiality of the stem cell. The floating colony (named sphere) formed by the NSC also contains one or more stem cells able to generate new spheres with the capacity to form neurons, astrocytes, and oligodendrocytes, demonstrating the renewal capacity of the NSC. Other cells are either committed to form a specific cell type or to die. The newly generated spheres can again be dissociated and induced by EGF to form new spheres to expand the population of NSCs. Comment: working at clonal density allows to define whether a factor acts directly on the NSC. Then, it is easy to manually catch the colony with a pipet and to challenge its capacity to generate a new sphere or to determine which cell phenotype can be derived from this colony.

isolate NSCs in serum-free medium and to study them, NSCs provided numerous new insights on neurogenesis as reviewed in refs. *11–15*. NSCs of different regions of the developing mouse brain, including striatum, cortex, mesencephalon, and spinal cord can be studied by such an approach.

This chapter discusses the role of the various ligands that control NSC renewal, as well as the receptors that mediate their stimulation, and presents the specific role of components of the cell cycle machinery, which control NSC proliferation in different brain regions. This includes tumor suppressors and the transcription factors controlling their expression. Figure 2 may help the reader all along the description of the different actors controlling NSC renewal.

2. BIRTH, EXPANSION, AND POPULATION RESTRICTION OF NSCs DURING MAMMALIAN LIFE

To understand and identify which factors control the development of NSCs, it is important to delineate the chronological appearance and life of NSCs. From the different NSC isolation and characterization studies, it seems that the NSCs change their state at least four times during development. Thus,

1. Pluripotent NSCs, capable of producing neural and nonneural phenotypes, can be generated from embryonic stem cells (ES cells) with leukemia-inducing factor (LIF) and fibroblast growth factor-2 (FGF-2) *(16)*; then after the neural tube formation.

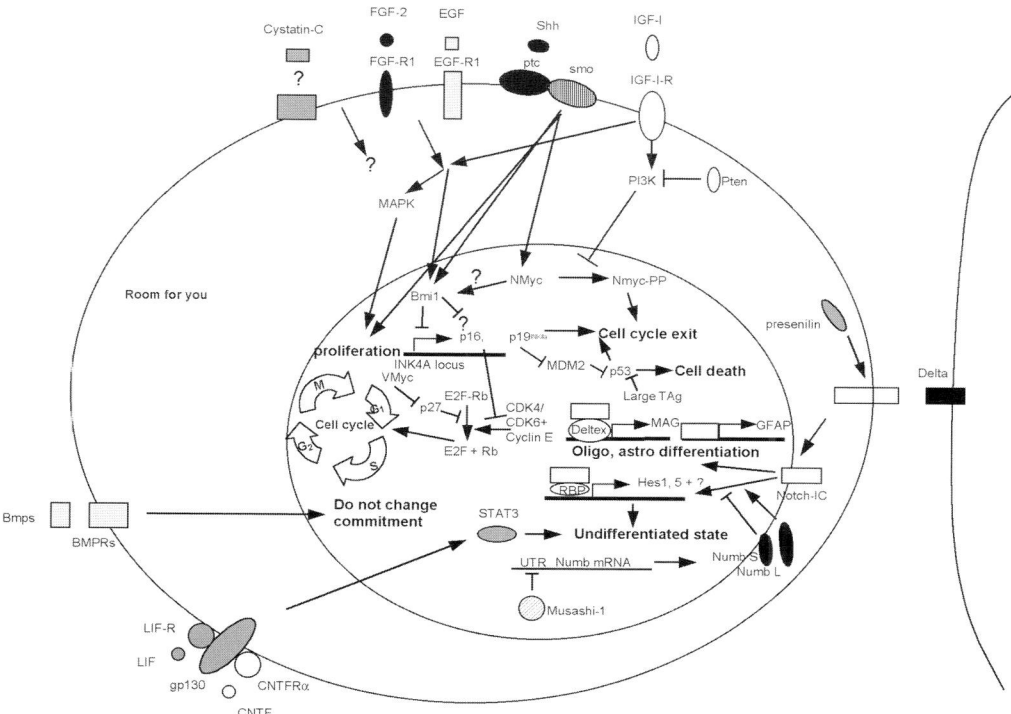

Fig. 2. This scheme summarizes the action of several pathways that are necessary to maintain NSC characteristics and stimulate NSC proliferation. The results are taken from different NSCs at various stages and are integrated into a single "ideal" NSC. The balance between cell differentiation and NSC characteristic maintenance is illustrated with the *Notch* pathway (including the roles of *Numb* and Musashi) and the activation of the gp130 pathway preserving NSC characteristics. The stimulation of NSC proliferation occurs by the MAPK pathway through the activation of EGF and IGF-I, or FGF-2 and IGF-I. *Shh* on its own can also stimulate NSC proliferation. IGF-I and *Shh* play an important role in the maintenance of the cell cycle by regulating the stability of Nmyc. The action of these different mitogens converges on Bmi1, which regulates the cell cycle potential by inhibiting the action of p16, of an unknown tumor suppressor, and probably of p19^{INK4a}. The action of the *Bmp* family is unclear as a general function, but the different data suggest that *Bmp*s may preserve the change of NSC state (FGF-2-responsive vs EGF-responsive NSC and generate preferentially neurons vs generate preferentially glia); further data are necessary to identify the precise function of this pathway.

2. At E8.5, NSCs can be isolated with FGF-2 alone *(17–19)*.
3. Later with epidermal growth factor (EGF) *(9,17–19)*.
4. Finally, after birth the dissection of the function of specific genes (*see* Heading 5) reveals that postnatal/adult NSCs are differently regulated, showing that during development and life the NSCs change their responsiveness to environmental and intrinsic factors.

It remains to be determined whether the generated NSCs also behave differently during the neuroectoderm induction compared with other developmental stages. During development and adulthood, the number of NSCs decreases drastically to reach only a few hundred/thousand stem cells in specific regions, out of 10- to 100-fold more mature cells. These modifications of NSC responsiveness and regulation parallel certain changes in competences of NSCs during neurogenesis. Indeed, NSCs isolated at different times of development preferentially generate one phenotype: neurons during early neurogenesis, then astrocytes, and after birth oligodendrocytes *(20)* (*see also* Chapter 3). It appears that throughout life, NSCs change their function and its cell cycle regulation.

3. EPIGENETIC FACTORS CONTROLLING STEM CELL RENEWAL AND EXPANSION

3.1. Activation of the Tyrosine Kinase Pathway is Essential for NSC Proliferation: The Roles of EGF, FGF-2, and IGF-I

Since the isolation and discovery of NSCs in 1992 by Reynolds et al. *(9)*, with the EGF mitogen, several other factors were shown to either be essential for the control of NSC renewal or to modulate proliferation. The role of these factors depends on the developmental state of the brain and its environment. FGF-2 was shown to activate NSCs early during brain organogenesis, at E8.5, and then NSCs become progressively responsive to EGF *(18–22)*. Several efforts attempted to determine whether the developing brain contains different stem cells or whether the same stem cells acquire the competence to respond first to FGF-2 and then to EGF. The work performed by Tropepe et al. *(19)* showed that at low cell culture density, EGF and FGF-2 have additive actions on E14.5 NSCs, but such phenomenon did not occur at high cell culture density, suggesting that two populations of NSCs are located in the same brain region. Isolation of NSCs from FGF-2 and EGF knockout mice confirmed the successive appearances of FGF-2 and EGF-responsive NSCs. The study of cAMP-responsive element binding protein activation by the FGF-2 and EGF mitogens also reveals the presence of two distinct NSC populations appearing in chronological order during development *(18)*. During brain organogenesis, the generation of EGF-responsive stem cells parallels the formation of glial cells that follows the generation of neurons. Moreover, the manipulation of EGF receptors (EGFR) in organotypic cultures has shown that glia formation is tightly linked to EGFR expression *(21)*. It appears that during development, EGF and FGF-2 are essential for stem cell formation and that EGFRs participate in the cell fate decision.

During mouse adulthood, as well as during development, striatal NSCs can also proliferate under the stimulation of EGF or FGF-2 *(8,23)*. NSCs in other regions, such as the spinal cord and around the fourth ventricle, necessitate the presence of the two factors during adulthood to undergo proliferation *(24)*. Such is the case for several multipotent progenitor/stem cells located in the human brain *(25–28)*. The adult rat hippocampus also requires at least two factors to proliferate. Indeed, a new trophic factor, the cystatin-C protein, was identified to be secreted by NSC culture, which sustains NSC proliferation. Cystatin-C is known as a protease but it is not this characteristic that allows NSC proliferation, yet another specific domain of the protein *(29)*. Concerning the human hippocampal progenitor, a marked cell expansion was only observed in the presence of cystatin-C *(30)*. Intriguingly, NSCs from other regions of the central nervous system (CNS) do not require the exogenous presence of FGF-2 or EGF to proliferate: this is the case of the RSC located in the adult mouse pigmented epithelium of the ciliary margin *(31)*. The sole proteins present in the medium, insulin and transferring, seem to be sufficient in stimulating RSC proliferation. In fact, RSCs release FGF-2, which acts as an autocrine factor *(31)*. It appears that during development and adulthood, NSCs in different regions of the CNS necessitate different combinations of factors to proliferate. As we proceed we will observe that NSC proliferation and generation are also controlled by numerous other factors.

Several factors and their respective receptors are expressed in developing neuroepithelium and the ventricular zones (VZ) and subventricular zones (SVZ), suggesting that these factors may interact to control NSC activity. Among them, the insulin-like growth factor (IGF) family was shown to have a fundamental role in the control of NSC proliferation. For E14 mouse striatum, in the absence of IGF-I or insulin (high doses of insulin stimulate the *IGF-I* receptor), NSCs do not proliferate, even in the presence of EGF or FGF-2 *(32)*. Recent studies have shown that insulin and proinsulin can also have a redundant action on NSCs *(33)* and personal data. The compensatory actions of these molecules would explain why IGF-I knockout mice are smaller (including the brain), but nonetheless present normal organs *(34,35)*. Members of the IGF family seem to act on numerous different stem cells and play a crucial role to modulate organ size (for review *see* Arsenijevic, in press). Interestingly, the use of the colony-forming assay (sphere formation) at low cell density allows dissection of the respective roles of IGF-I, FGF-2, and EGF on mouse NSCs. Neither IGF-I nor EGF has any effect on NSC survival, whereas FGF-2 promotes NSC survival or maintenance. Only the presence of IGF-I and FGF-2 or IGF-I and EGF induces NSC proliferation *(32)*. In the adult rat hippocampal NSCs, FGF-2 increases the expression of the IGF-I receptors, and IGF-I, in the presence of insulin and FGF-2, also increases the rate of cell proliferation *(36)*. Such a mechanism occurs through the activation of the mitogen-activated protein kinase kinase (MAPKK) and PI3-K pathways, the MAPK pathway being essential. It appears that NSC division is controlled by several essential factors.

The EGFR can also be stimulated by transforming growth factor-α, which is present in the SVZ *(37)*. In the *transforming growth factor-α* knockout mice, studies with the thymidin analog, bromodeoxyuridine (BrdU), revealed a lengthening of the NSC mitotic cycle, whereas the number of NSCs in the lateral corner of the SVZ is not affected in adult mice *(37)*. It is interesting to note that amphiregulin-I, which belongs to the EGF family, is expressed in the choroid plexus and the hippocampus, and can also reach adult NSCs. Amphiregulin shows the same potency as EGF to induce proliferation of adult NSCs *(38)*.

The complexity of NSC division is further demonstrated by the fact that several other trophic factors modulate this phenomenon. Indeed, bone morphogenic protein (Bmp) 2, 4, and 7, Noggin, ciliary neurotrophic factor (CNTF), LIF, Delta/Jagged, and sonic hedgehog (Shh) also regulate the stemness of the NSCs and are discussed in Subheadings 3.2–3.4.

3.2. EGFR Expression Regulation

Several factors may stimulate or enhance NSC proliferation. During cortical development, the number of EGFRs increase progressively, paralleling the appearance of EGF-responsive stem cells *(21)*. It was shown that young cortical cells secrete Bmp4, a factor inhibiting EGFR expression *(22)*. Bmp4 is a bone morphogenic protein, which is required for the dorsalization of the CNS *(39,40)*. We previously learned that FGF-2 stimulates the appearance of EGF-responsive NSCs, and other data have shown that Bmp4 has the opposite action on the NSC population in comparison with FGF-2 *(22)*. It appears that during early developmental stages, Bmp4 prevents the generation of EGF-responsive stem cells, then FGF-2, which is secreted during neurogenesis, counteracts the inhibition imposed by Bmp4 and participates in inducing gliogenesis. During adulthood, the receptors for Bmp2 and 4 are still present in the SVZ and may participate to maintain cells in a nondifferentiated state by preventing neurogenesis *(41)*. Indeed, Bmps inhibit neuronal differentiation in vitro and reduce SVZ cell proliferation in vivo by 50%. Noggin, by inhibiting Bmp action, promotes in vivo neuronal differentiation. It would be very interesting to investigate in vitro the target of Bmps to understand whether Bmp acts directly on NSCs or on other early progenitors. All these experiments suggest that Bmp signaling regulates neurogenesis throughout development and adulthood.

3.3. gp130 and Stemness Maintenance

The loss of a component of another neurotrophic pathway can also affect the pool of NSCs in the brain. This is the case for the LIF receptor (LIFR), which transmits the stimulation of LIF by the dimerization of the LIF/LIFR complex with gp130. CNTF can also act on the LIFR: at low-concentration CNTF binds the CNTF-R-α, glycoprotein 130 (gp130), and LIFR complex, and at high concentration to the gp130/LIFR dimer *(42)*. On adult *LIFR$^{-/-}$* mice, the number of spheres generated from the subventricular zone is lower in comparison with wild-type animals. This is accompanied by a reduced number of interneurons in the olfactory bulb *(43)*. Interestingly, SVZ cell stimulation by CNTF increases in vivo and in vitro the number of proliferating NSCs by stimulating stem cell renewal; thus, CNTF also seems to prevent progression to glial cell fate induced by EGF *(43)*. It appears that CNTF plays a role in maintaining stem cell characteristics. Recently, it was shown that CNTF maintains NSC renewal by stimulating *Notch* expression *(44)*. *Notch* is an important receptor to control the undifferentiated or the differentiated state of a cell (*see* Heading 4). It is also known that LIF has an action on ES cell maintenance (for review *see* ref. *45*), and certain groups use LIF to induce sustained proliferation of human fetal NSCs *(46)*. Interestingly, the generation in vitro of NSCs derived from ES cells necessitates the presence of FGF-2 and LIF *(16)*. The proliferation of such NSCs leads to the formation of spheres containing cells able to produce specialized cells of the three germ layers. The derivation of FGF-2 spheres from the FGF-2/LIF-generated spheres leads to the formation of restricted NSCs dedicated to generating neural cells only *(16)*. The gp130 pathway appears to be an important regulator of the NSC pool as well as of stemness.

3.4. Shh Balance Between Activation and Modulation

Shh was shown to control the proliferation of various neural progenitors and certain mutations in the Shh intracellular pathway lead to tumor formation *(47–52)*. The smoothened (smo) receptor, on its own, constitutively activates the *Shh* intracellular pathway. However, the patched (ptc) receptor inhibits smo in absence of Shh, and as a consequence, the intracellular pathway is shut down if enough ptc receptors are present. To activate the system, *Shh* binds specifically to *ptc* and prevents ptc-smo inhibition (for review *see* ref. *53*). In vitro, Shh markedly induces the proliferation of cerebellar granule cell (CGC) precursors and the blocking of Shh in vivo by anti-Shh hybridomas leads to a dramatic impairment of the external germinal layer growth *(54)*, showing that Shh is required for the generation of cerebellar cells during development. In vitro, Shh also increases the proliferation of fetal telencephalic progenitors stimulated by FGF-2 *(52)* and adult hippocampal progenitor in absence of FGF-2 *(55)*, but not of cortical NSC *(56)*. However, Shh synergizes at low doses with EGF to induce cortical NSC proliferation. Moreover, EGF generated neurospheres express Shh and neural cells derived from mice mutant for Shh, as well as from mice lacking the downstream transcription factors *Gli*1 and *Gli*3 (glioma-associated oncogene homologs), have an impaired ability of forming neurosphere with stable renewal ability *(56)* showing a crosstalk between the EGF and Shh pathways. Overexpression of Shh in transgenic mice produces a hyperplasia of the spinal cord and an expansion of the SVZ with an increased number of proliferating cells *(50)*. In cerebellar cells, Shh stimulates the transcription of cyclin D_1, cyclin D_2, and cyclin E, but cyclin D_1 and cyclin D_2 are not required individually to allow Shh function *(57)* suggesting a redundancy effect between members of the cyclin D family. On the other hand, the activation of proliferation by Shh does not happen via the activation of the MAPK pathway *(57)*. Intrastriatal injection of a Shh analog markedly increases *ptc* expression in SVZ cells, but not progenitor proliferation *(58)*. These data show that the Shh pathway can be activated in the adult SVZ. In contrast, in the hippocampus, the delivery of an adeno-associated virus carrying the *Shh* complementary (c)DNA induces an increase in cell proliferation and in the generation of new neurons in the dentate gyrus *(55)*. In conditional null mutants for *smo*, in which the deletion is driven by the expression of nestin, the medial ganglionic eminence is reduced by 50% at E12.5, but no changes of cell death or proliferation were observed in mice *(59)*. At P15, the regions known to undergo neurogenesis postnatally, the SVZ, the rostral migratory stream, and the hippocampus have a reduced number of cells as well as proliferating cells, and an increased number of dying cells (around 50–80%). At P15, the number of NSCs is greatly decreased in conditional *smo* null mice. Also the injection of cyclopamine, an inhibitor of the Shh pathway, leads to a decrease of about 50% of endogenous proliferating cells in the hippocampus *(55)*. A major study established the link between the Shh and the IGF-I pathways for the control of cerebellar granule cell proliferation *(60)*. Shh stimulates the proliferation of postnatal mouse CGCs by upregulating *N*-Myc expression *(61)*. *N*-Myc stimulates neural cell proliferation, whereas the phosphorylation of the oncogen in two sites inhibits *N*-Myc action *(60)*. In a separate pathway, PI3-K is known to also activate CGC division *(62)* and plays an important role in the maintenance of the unphosphorylated form of *N*-Myc *(60)*. Because IGF-I (or high insulin concentrations) activates PI3K *(62)*, this growth factor also prevents the phosphorylation and the destabilization of *N*-Myc *(60)* showing how IGF-I and Shh interact for the control of CGC proliferation.

In an eye preparation, in vitro, Shh also increases the proliferation of E18 and P1 retinal progenitors by more than 60 and 80%, respectively *(48)*. At the periphery of the retina, the ciliary margin zone (CMZ) contains RSCs *(31)*. In wild-type mice, this region is poorly proliferative after birth in contrast to birds *(63)*. However in *ptc±* mice, a marked conservation of proliferating cells in the CMZ is observed at P16, whereas only a slight change in the number of dividing cells is observed after birth *(64)*. At P10, the CMZ of *ptc±* mice contains more than 2.5-fold more progenitors responding to the mitogenic action of EGF. More interestingly, when these mice are crossed with mice undergoing a retinal degeneration owing to a mutation on the rhodopsin gene (P23H mice), these mice have more proliferating cells in the CMZ and generate new photoreceptors (not quantified) at P13–15. The ability to generate new retinal cells during adulthood remains to be determined.

These findings shows that the Shh pathway controls the proliferation of NSCs throughout the CNS and has a major role in controlling the pool of NSCs and, during adulthood, in regulating the formation of new neurons by modulating both cell proliferation and differentiation. These studies also show that the MAPK and Shh pathways are required for NSC renewal, that PI3K modulates the proliferation, and that the activation of gp130-Stat3, as well as the Smad pathways, seem to favor cell commitment maintenance. Thus, NSCs are differently regulated from the rostral to the caudal apex of the CNS, as well as between rodents and human, and understanding of the intrinsic mechanisms of the different NSCs should allow the identification of crucial genes necessary for NSC renewal.

3.5. Other Environmental Factors

The ephrin family consist of eight ligands as well as 14 receptors and are involved in brain organogenesis by controlling axon targeting *(65)*, cell migration *(66)*, and brain boundary *(67)*. Their localization during adulthood suggests that they might regulate NSC functions. Indeed, receptors for ephrin as well as the ephrins-B2/3 are present in the adult SVZ. Intracerebral infusion of the clustered ectodomain of ephrin-B2 (an active form of ephrin) induces cell proliferation in the SVZ and increases the number of astrocytes when migratory neuroblasts decrease *(68)*. The generated astrocytes have the characteristics of the phenotype named type-B cell, identified as NSCs. These results show that ephrin can induce NSC division in vivo; nonetheless, in vitro studies are necessary to determine whether the ephrin acts directly on NSCs.

Another component of the extracellular matrix, the heparin-binding growth-associated molecule (HBGAM), is expressed during early neurogenesis in the neuroepithelium *(69)*. In *HBGAM* knockout mice, the cell density is increased in the cortex at birth and the number of mitotic cells also at E12 *(69)*. Clonal cultures of E12.5 cortical cells revealed that HBGAM inhibits in a dose-dependent manner the number of proliferating NSCs stimulated by FGF-2, but not by EGF, and reduces the phosphorylation of the FGF-R1. The data shows that components of the extracellular matrix can importantly modulate the action of FGF-2 on NSC proliferation.

The hepatocyte growth factor (HGF) and its receptor c-Met are present in the developing CNS and during adulthood. HGF potency to induce NSC proliferation was challenged in the absence and presence of EGF and FGF-2 *(70)*. In the absence of growth factors, HGF stimulates the appearance of neurospheres at a density of 75,000 cells per well (24-well plate) and increased the number of neurospheres generated when coincubated with EGF, FGF-2, or EGF and FGF-2. Subcultures of the primary neurospheres revealed that the number of secondary spheres decreased when continuously exposed to HGF, suggesting that HGF acts predominantly on progenitor cells rather than NSCs. These results also show that the subculture strategy is necessary to identify which population a factor is acting upon.

The pituitary adenylate cyclase-activating polypeptide (PACAP) binds with high affinity to the PACAP receptor-1 and to the vasoactive intestinal peptide receptor-1 and -2 *(71)*. In the adult mouse brain, PACAP receptor-1 is expressed in the adult SVZ and the dentate gyrus *(72)*. In vitro, PACAP stimulates the formation of neurospheres, but it is four times less effective than EGF and the spheres are much smaller and generate neurons, astrocytes, or oligodendrocytes *(73)*. The potential to renew was not challenged and it is difficult to assess whether PACAP acts on NSCs. In vivo PACAP infusion leads to an increased proliferation in the dentate gyrus and the SVZ, but the target of PACAP, as in vitro, remains to be clarified.

Unidentified factors from endothelial cells are also involved in the control of NSC renewal and cell fate induction. During development and adulthood, blood vessels are thought to participate in brain neurogenesis *(74)*, as observed in the singing birds *(75)*. Coculture of bovine endothelial cells or a mouse brain endothelial cell line with E10 and EC11 cortical cells cultured at clonal density, increases the clone size, and the removal of endothelial cells allows cell differentiation with a markedly enhanced percentage of neurons generated *(76)*. Moreover, endothelial factors induce the induction of specific neuronal subtypes. Remarkably, time-lapse video recordings reveal that the division of progenitor/stem cells induced by endothelial cells occurs by symmetric division (generation of nondifferentiated cells only) whereas cell division in the absence of endothelial cells produces both neurons and nondifferentiated cells. These results show that endothelial cells control progenitor/NSC division by enhancing symmetric division and favoring the appearance of a specific neuronal phenotype. Such induction does not seem to be dependant on the developmental stage of endothelial cells, because cells derived either from the bovine pulmonary artery or from the mouse brain have similar effects.

The interactions of these newly identified factors, with the others listed previously, remain to be clarified and, hopefully, will help to dissect precise functions during development.

4. THE NOTCH PATHWAY IS ESSENTIAL FOR NSC RENEWAL

To be a NSC, it is necessary to maintain a nondifferentiated state. Studies in flies, and from invertebrates to mammals, revealed that the *Notch* receptor is a key factor to remain a primitive cell, for review *see* ref. *(77)*. Indeed, the activation of Notch by Delta or Jagged prevents the induction of differentiation, via the cleavage of the intracellular domain of Notch (NotchIC). Often the cleavage of Notch necessitates the presenilin protease. *(78)*. The NotchIC then binds RBP-jk to activate a gene expression, such as the *Hes-1* gene, which is a basic Helix–Loop–Helix transcription factor. Overexpression of Notch by a nonreplicative retrovirus in the developing rat eye results in an abnormal growth of the infected retina suggesting a longer maintenance of the nondifferentiated state *(79)*, which leads to an abnormal cell differentiation. Notch is also expressed in neurospheres *(80,81)* and the forebrain tissue of *Notch1*$^{-/-}$, or *RBP-jk*$^{-/-}$ knockout mice is almost depleted of FGF-2-responsive NSCs, and their number is greatly reduced in *presenilin*$^{-/-}$ mice. The few NSCs present at E14.5 were not able to renew. These results show that the Notch pathway is of prime importance for stem cell generation, maintenance, and renewal *(81)*. Profound defects of NSC number and renewal were also recorded when two target genes of *Notch*IC/*RBP-jk*, *Hes-1* and *Hes-5*, are absent *(82)*. Whereas the loss of *Hes-1* alone has only a partial effect on the control of NSC proliferation *(83)*. These experiments show that *Hes-1* and *Hes-5* are components of the Notch pathway to maintain a NSC state.

Interactions between environmental stimuli and the Notch pathway were reported and reveal that the majority of the factors inducing NSC proliferation or enhancing symmetric divisions regulate *Notch* expression. Indeed, FGF-2 stimulates the expression of *Notch*1 and *Notch*3, and decreases Delta-1 in neuroepithelial cells *(84)*. Even if these studies were not performed directly on NSCs, they uncover a direct link between FGF-2 and Notch. In the brain, Shh is thought to regulate the proliferation of cerebellar granule neuron precursors by maintaining the expression of Notch *(85)*. Other pathways are involved in the control of Notch expression in NSCs. Indeed, the stimulation of the gp130 by CNTF rapidly activates the expression of Notch1 *(44)* and maintains the striatal NSC in an undifferentiated state *(43)*. Intriguingly, the action of CNTF does not stimulate the expression of *Hes-1* or *Hes-5*, thus suggesting that another member of the Hes family may mediate the effect of gp130 activation. All of these results show that different intracellular pathway activations can lead to the stimulation of Notch expression, and that Notch is required to maintain NSC characteristics in different brain locations.

Other studies have revealed that Notch1 is involved in the process of glia differentiation in vitro *(86)* and in vivo for the generation of cortical astrocytes *(87,88)*, Müller cells in the retina *(89)*, or Schwann cells in the peripheral nervous system *(90)*, as well as oligodendrocytes differentiation *(80,91)*, but not for olfactory bulb glia *(87)*. The studies of Ge et al. *(92)* demonstrated that the NotchIC-RBPJ complex can activate genes without the transactivation of Hes-1. The complex can bind directly and activate the glial fibrillary acidic protein promoter *(92)*. Concerning oligodendrocyte formation, new ligands for Notch have been identified, F3/contactin and NB-3, which activate Notch and Deltex as downstream effectors *(91)*. These results show that the Notch receptor may mediate different responses depending on the extra- and intracellular partners. The understanding of such pathways may help to conciliate the apparent contradictory results showing that Notch activation can both prevent cell differentiation and induce glia formation. To be sure that Notch activation can induce astrocyte formation and not stem cells expressing glial fibrillary acidic protein *(4)* it would be interesting to better characterize the generated astrocytes to reveal whether they are metabolically competent (glucose uptake) and express specific transporters such as glutamate and aspartate transporter *(93)*, or if they have NSC characteristics.

Numb is considered a major regulator of Notch action from *Drosophila (94)* to mammals. During cortical and retinal neurogenesis the NSC undergoes symmetric division to increase its population, or asymmetric division to generate a cell committed to a specific lineage. During these processes, it was observed that the stem cell divides in different planes depending on the fate of the daughter cell: symmetric division implies that the division plane is perpendicular to the ventricle, and parallel during asymmetric division. This cell orientation is a tendency but not an absolute requirement (*see* ref. *95*). During asymmetric division, Numb is also distributed asymmetrically in the cell, and remains in the future daughter NSC, whereas during symmetric division Numb is distributed in both daughter cells *(96–99)*. In *Numb* knockout mice, animals die at E11.5, and have an open neural tube with impaired neuronal differentiation showing that *Numb* loss interferes with mammalian neurogenesis *(100)*. In *Numb* conditional knockout mice driven by the expression of nestin, neurons appear prematurely and the mice present a smaller brain due probably to a depletion of neural progenitors and cell growth arrest *(101)*. A much more pronounced defect is observed when the *Numb* homolog, *Numblike*, is also deleted with *Numb* (double conditional knockout): embryos die at around E11.5, are smaller, and the early neurons are generated normally. An almost complete depletion of neural progenitor cells occurs after this event, leading to the formation of a very small brain containing mainly early-born neurons *(102)*. In this study, the loss of *Numb* and *Numblike* functions parallels the loss of NSCs in the brain thus they are believed to maintain the NSC population. These results contrast with those obtained with mice, which downregulate *Numb* and *Numblike* expressions under the activity of the *Emx1* gene *(103)*. In these mice, the cortex is poorly differentiated and presents an increased number of progenitor cells. These cells proliferate more at E12.5 and E14.5 meaning that *Numb* inhibits progenitor cell proliferation at this age. It is important to point out that different isoforms of the human Numb regulate differently proliferation and differentiation *(104)* and *Numb* and *Numblike* were also shown to promote neuronal differentiation in cells that are already committed to a neuronal lineage *(98)* allowing broad actions of *Numb*. Indeed, the overexpression in vitro of the short Numb isoforms in P19 embryonic carcinoma cells or in primary neural crest stem cells markedly enhances the percentage of clones expressing neurons only, whereas the long isoforms did not, but increased clone size *(104)*. Supporting a dual role for Numb, in vitro clonal analysis revealed that Numb is localized both in progenitor and neuronal cells at E10 in mouse cortical cells and only in progenitors at E13 *(99)*. From these different studies, it appears that Numb can control both neural progenitor and neuron formation. Numb and Numblike or their RNAs are thought to interact with numerous intracellular partners such as Notch, β-amyloid precursor, LNX, GRIP-1, Musashi, NAK, α-adaptin, Lux2, Siah-1, Esp15, and E3 ligase Itch (*see* discussion in ref. *103*). The balance of expression of these molecules can be implicated in the regulation of the different functions of Numb. As *Emx2 Numb* and *Numblike* seem to have opposite actions depending on stage of development, numerous intracellular candidates may be involved in such change of function for *Numb* and *Numblike*. A direct analysis of Numb action on NSCs remains to be determined and an in vitro model of neurogenesis, from the NSC to the neuron, should help to clarify the Numb function.

The Notch pathway is also regulated by *Musashi-1*, which is involved in the fate decision of the sensory neurons in *Drosophila (105)*. In mammals, Musashi binds RNA and modulates protein expression by regulating the process of translation *(105,106)*. Cell isolation based on Musashi expression revealed that progenitors and NSCs express Musashi-1 *(105,107)*. Fetal human neural cells expressing green fluorescent protein driven by the *Musashi* promoter activity are able to renew (36–42 cell doubling) and are multipotent. The loss of Musashi expression in $Msi1^{-/-}$ mice does not disturb NSC renewal, and moderately decreased the frequency of multipotent neurospheres *(105,108)*. However, the down regulation of *Musashi-2* in $Msi1^{-/-}$ neural cells provokes a reduction of about 60% of the neurosphere number generated with EGF and FGF-2. These series of data show that the Musashi family participates in the regulation of neural stemness. In vitro, it was demonstrated that Musashi binds the 3′-untranslated sequence of the *Numb* mRNA and decreases the synthesis of Numb protein *(106)*, whereas Musashi-1 enhances *Notch1* transactivation *(106)*. These data are in agreement with the functions of *Notch* and *Musashi* on stem cell renewal, but seem to be contradictory in view of data showing that ablation of *Numb* and *Numblike* actions leads to a loss of progenitor cells during early neurogenesis. Nonetheless, this mechanism supports the postulated role of *Numb* during late neurogenesis, when cells are committed to a neuronal lineage *(98)*. Conditional double knockout mice for both *Musashi*-1 and *Musashi*-2 should clearly help to understand the interactions between *Notch*, *Musashi*, and *Numb* during different stages of neurogenesis.

The data cited above give a general picture of the NSC control as well as the actions of NSCs during development. It appears that NSCs have multiple functions throughout embryogenesis, and life, and that these changes are accompanied by specific molecular mechanisms. The links and transitions between these modifications of NSC competence are still obscure, but the main actors are identified and can serve to dissect these mechanisms.

5. INTRINSIC CONTROL OF STEM CELL RENEWAL

The role of environmental factors and their intracellular pathways were described in the previous sections. The intrinsic factors that control different states of the NSC, as well as the generation of specific cell fates, are presented here.

5.1. The Homeodomain Transcription Factors

During development, the homeodomain transcription factors control body segmentation, organogenesis, cell proliferation, specification, and differentiation. Some of them are expressed in the developing neuroepithelium, as well as in the adult SVZ, and regulate NSC proliferation, such is the case for *Emx2 (109–112)*. From different studies, it appeared that *Emx2* has different functions as a result of its temporal and spatial expression and can have opposite functions on NSC renewal. Adult mouse SVZ neurospheres contain Emx2, but its expression is lost during cell differentiation suggesting that *Emx2* may have a role on progenitors or stem cells present in the neurospheres *(109)*. Clonal and subcloning analyses of *Emx2* knockout mice neural cells and Emx2 expressing NSCs reveal that Emx2 decreases the ability of primary stem cells to generate secondary stem cells, this apparently by favoring asymmetric division rather than cell survival *(109)*. Supporting this hypothesis, an increase of SVZ cell proliferation is observed in vivo in *Emx2* knockout mice at the expense of neuroblast generation, dedicated to the olfactory bulb. In contrast, another picture of *Emx2* action is observed during development. Indeed, overexpression of Emx2 in cortical E14 progenitors in vitro increases clone size. BrdU incorporation analyses revealed that Emx2 instructs the cells to divide symmetrically *(95)*. Such action seems to be specific to cortical cells, in view of the fact that Emx2 fails to stimulate cells derived from the ganglionic eminences. In knockout mice in vivo, E14 cortical cells show a reduced ability by 30% to divide symmetrically. Similar cell behavior was observed in vitro with a decreased capacity in multipotentiality: in absence of *Emx2* only 25% of the progenitors are multipotent, the others generating neurons, whereas 60% of wild-type progenitor cells can give rise to neurons, astrocytes, and oligodendrocytes *(15)*.

The search for Emx2 binding sites in various promoters, by using the chip-to-chip or the DNA shift assays at different time of the development (to reveal the available Emx2 binding sites in specific brain regions), should help to determine why Emx2 has opposite actions on NSCs. The use of the yeast two-hybrid system may also allow to reveal whether this difference in Emx2 action depends on protein interactions.

Sox2 is another homeodomain transcription factor that regulates NSC renewal. During development, Sox2 is expressed when the neural tube is formed *(113)* and then, around the lateral ventricles during adulthood *(114)*. E14.4 dpc dorsal telencephalic cells of transgenic mice expressing the β-galactosidase and the neomycine transgenes under the control of *Sox2* expression, selected with G418, show NSC characteristics *(114)*. No NSCs of the spinal cord survived to G418 selections. These experiments demonstrate that Sox2 is expressed in NSCs and specifically in telencephalic NSCs. This transcription factor should play an important role in the stem cell state or cell fate specification, but its specific function remains to be determined.

The proliferation of progenitor cells of the eye is controlled by different homeobox genes such as *Six3*, *Rx*, *Pax6*, *Prox1*, and CEH 10 homeodomain-containing homolog (CHX10). In vertebrates *Pax6*, *Rx*, or *Six3* can generate an ectopic retina or eye when overexpressed during development *(115–117)*. Concerning *Rx*, its overexpression in the retina leads to hyperproliferation *(117)*. In the mouse, *Rx* was shown to be necessary for the induction of retinal progenitors and for *Pax6* expression *(118)*. Because *Rx* is expressed the whole proliferating retina at E15.5 *(117)*, it would be very interesting to test whether *Rx* also controls mammalian RSC proliferation. This reasoning can also be applied to the *Six3* gene *(116)*. *Pax6* is considered as a master gene for eye development, but intriguingly its role was never directly tested in retinal progenitors. Nonetheless, it was shown that *Pax6* is necessary for cortical progenitor proliferation *(119)*, as well as for the generation of radial glia *(120)*, which share characteristics of NSCs *(121)*. *Pax6* was shown to be required for neurogenesis derived from radial glia *(122)*. It also would be interesting to uncover the role of *Pax6* in stem cells. Likewise, the role of *ChX10* was clearly identified as a transcription factor necessary for retinal progenitor proliferation. Indeed, null mutation in the *ChX10* allele provokes a reduction in cell proliferation and an absence of bipolar cells only, one of the seven cell phenotypes composing the retina, showing that *ChX10* controls a specific stage of retinal progenitor proliferation during retinogenesis *(123)*.

Homeodomain transcription factors are not all dedicated to enhancing cell proliferation. It was recently shown that *Prox1* is necessary for the cell cycle exit of retinal progenitor cells and regulates the generation of horizontal cells, a group of neurons born during early retinogenesis *(124)*. Indeed, the loss of function of *Prox1* leads to an increased cell proliferation of progenitor cells which generate less early neurons (horizontal cells) and more late cells (rod photoreceptors). Overexpression of *Prox1* in the newborn developing rat retina using nonreplicative retroviruses produced more early neurons at the expense of late neurons.

The polycomb transcriptional repressor *Bmi1* promotes cell cycle progression, and controls cell senescence *(125)*. Loss of *Bmi1* leads to a decreased brain size, particularly the cerebellum, and causes progressive ataxia and epilepsy *(126)*. Growth retardation occurs mainly postnatally, and recent works demonstrated that Bmi1 is essential for adult hematopoietic stem cell renewal *(127)*. In *Bmi1* knockout mice, the frequency of NSCs in the neural cell population is slightly decreased at E14 and P0, but markedly reduced at P30 *(128)*. When the NSC are grown in EGF only, a deficit on NSC proliferation is already observed at E14 (personal communication). The self-renewal ability of the generated neurospheres is greatly impaired at all developmental ages and cell division in vivo in the SVZ is more affected at P30 than P0 *(128)*. All these data show that the greatly decreased capacity to renew leads to a depletion of NSCs during early adulthood. Interestingly, the loss of *Bmi1* does not affect the proliferation capacity of restricted precursors in vitro *(128)*. The CGCs depend also on the presence of Bmi1 to proliferate adequately. Indeed, in *Bmi1$^{-/-}$* mice the cerebellum is smaller owing to reduced ability of CGCs to proliferate as well as regarding an increased cell death. Shh, known to induce the proliferation of CGCs *(129)*, stimulates Bmi1 expression in 4 h and is less effective in CGCs of *Bmi1$^{-/-}$* mice.

Overexpression of Bmi1 in CGCs-*Bmi1*^(−/−) rescues the proliferation capacity partially. It has been confirmed that certain medulloblastoma are the consequence of a deregulation of the Shh pathway. Interestingly, in three out of three medulloblastoma cell lines and in 8 out of 12 primary tumors, Bmi1 is strongly expressed, whereas in 10 glioblastoma Bmi1 is not or only slightly expressed. All these findings reveal that Bmi1 is necessary for NSC renewal, but usually not for restricted precursors, and is tightly related to mechanisms leading to the pathogenesis of medulloblastoma.

5.2. Other Intrinsic Factors

By using a complementary DNA subtractive screen, Tsai and McKay have identified another gene expressed in NSCs: the *nucleostemin (130)*. Nucleostemin is expressed in ES cells, in the neuroepithelium at E8.5 before the expression of nestin, and rapidly decreases during development. Nucleostemin is located in the nucleoli of NSCs and neural progenitors and the down regulation of its expression by siRNA results in an increase of noncycling cells (from 7% controls to 16.5% small interference RNA-treated group). The effect is much more pronounced in the U2OS tumor cell line, reducing by sixfold the number of cells in S phase. Overexpression of Nucleostemin also reduces the number of noncycling NSCs (from 5% to 13.6%). These actions are thought to be mediated by guanosine triphosphate. Interestingly, glutathione pull-down experiments show that nucleostemin forms a complex with p53 and suggest that nucleostemin may regulate NSC proliferation and survival through its interactions with guanosine triphosphate and p53.

T-cell leukemia homeobox (TLX) is an orphan nuclear receptor, which is expressed in neurogenic regions of the adult rodent brain *(131)*. *TLX*-null mutation leads to reduced cerebral hemispheres, hippocampus and olfactory bulb. Cells sorted under the TLX expression and analyzed clonally show proliferation ability and multipotency, whereas cells of $TLX^{-/-}$ mice do not proliferate. In vitro, stem cell characteristics of $TLX^{-/-}$ cells isolated by LacZ expression, the β-*galactosidase* gene being knocked into the *TLX* locus, can be restored by lentiviral vectors expressing TLX, showing that TLX intrinsically controls NSC renewal.

These new actors on NSC proliferation have to be linked to other systems that control NSCs. They promise to reveal new interesting interactions between components of the intracellular pathways and may help to understand how different stages of neurogenesis take place.

5.3. Telomerase and NSC Life-Span

The extent of the stem cell proliferation potential is one of the main characteristics for a stem cell to be a stem cell. Indeed, it is expected that a stem cell is active all along development and during adulthood for tissue repair or growth. This is also the case for NSCs. The maintenance of cell division depends on a critical control of the chromosomal extremities that, at cell birth, are formed by telomeres. The telomeres are comprised of a repetitive CCTTAG sequence, which are partially removed at each cell division. During this phenomenon, the cells lose 8–12 bp on each chromosome *(132)* and telomeres shorten *(133,134)*. The shortening of the telomeres usually leads to senescence to prevent cell transformation as a result of the possible fusion of chromosomes in absence of telomeres *(135–137)*. The induction of cell senescence occurs through the activation of p19 and p16, and then p53 (reviewed in ref. *138*). To prevent degradation of the genes, the telomerase enzyme, through its catalytic subunit Tert, adds new CCTTAG sequences after each cell division *(139–141)*. *Tert* gene transfer in several cell types can lead to immortalization *(142,143)*. Telomerase activity is present during brain development and decreases during brain maturation *(144,145)*, but remains in discrete areas such as the SVZ *(146)*. In vitro, telomerase activity is detected in rodent neurospheres throughout cell passages *(147)*, whereas human NSC cultures show a decreased telomerase activity during culture expansion *(147)*. In adult mice, in vitro studies demonstrated that telomerase is expressed in migratory neuroblasts, and in vivo, treatment to temporarily eliminate the neuroblasts population revealed that telomerase is expressed in the B (NSC) and C (progenitors) mixed cell cultures *(146)*. It appears that the telomerase activity reflects the potential of NSCs to proliferate in the culture conditions used and might serve to characterize NSC. Interestingly, the environment can control telomerase expression, as revealed by experiments attempting to stimulate cortical neural progenitors by FGF-2: such stimulation leads to an increased expression of telomerase *(148)*.

6. CELL CYCLE MACHINERY AND STEM CELL RENEWAL

6.1. Cell Cycle Arrest and Replicative Senescence

The control of NSC proliferation implies a tight control of the cell cycle entry and progression. The major activator of the cell cycle is the E2F family. When released from the retinoblastoma (Rb) tumor suppressor by phosphorylation of Rb, the E2Fs transcription factors induces gene expression of factors participating in the cell division process. Rb and E2F are the key regulators of the cell cycle. The cyclin D/cdk4/6 and the cyclin E/cdk2 complexes are both necessary to phosphorylate Rb on two different specific sites to allow the dissociation of the Rb–E2F complex. The p21/p27 tumor suppressors prevent the activation of both cyclin/cdk complexes, whereas p16 blocks the action of cyclin D/cdk4/6 only (for review *see* ref. *149*). It is important to note that cell division can be transiently blocked by the p21 and p27 tumor suppressors, and definitively by p16 and p14 (p19 in mouse), and that p53 induces cell death *(150–152)*. Irreversible growth arrest is termed replicative senescence and can be induced by different environmental stresses such as oxidative damage *(134,153,154)*, and oncogene activation *(151,155,156)*, and also by trophic factor removal and telomere shortening *(133,136,155–158)*. Replicative senescence is considered to prevent cell transformation and oncogenicity. Besides p21 and p27, which are commonly involved in the cell proliferation control of an organ, several other tumor suppressors regulate the cell cycle exit of specific cell populations. Studies revealing that different functions of NSCs in the brain and the retina are controlled by different tumor suppressors are presented here.

6.2. The E2F1 Transcription Factor

Surprisingly, the loss of the E2F1 transcription factor has no obvious deleterious effect on the brain development. Only adult brains present a reduced weight *(159)*. Cell proliferation studies using BrdU incorporation show reduced cell division in the subventricular zone and the hippocampus with a decreased number of newborn neurons in the olfactory bulb and the dentate gyrus *(159)*. Nonetheless, these studies did not allow the determination regarding whether the NSCs are directly controlled by E2F1 or whether these phenomena are owing to an impairment of progenitor activity. It is probable that the other members of the E2F family play a crucial role during brain development.

6.3. The Tumor Suppressors p27, p53, p16, p19, p57, RB, and Pten

p53 and p27 have been widely studied for a long time because inhibition of their expression can lead to cell immortalization of neural cells *(160,161)*. The fact that multipotent neural progenitors can be immortalized by the large T-antigen, which inhibits p53, or by v-myc, which inhibits p27, suggests that the natural NSC possesses mechanisms that prevent the expression of these tumor suppressors to allow long-term renewal, one of them could be in part mediated by Bmi1 *(128,129)* (*see* following paragraph). The regulation of multipotent NSCs by tumor suppressors seems to be different between mammals. Indeed, the Sv40 large T-antigen does not induce immortalization of fetal human neurospheres *(162)*, whereas the inhibition of p27 by v-myc leads to the immortalization of rodent and human multipotent progenitors *(162–164)* suggesting that p27 is apparently rapidly expressed when the NSC exits its state. Supporting this hypothesis, p27 was observed to be expressed in the adult transit-amplifying progenitors of the SVZ but not of stem cells *(165)*.

The absence of p16 does not change the renewal characteristics of NSCs. Nonetheless, *Bmi1*$^{-/-}$ mice crossed with *p16*$^{-/-}$ mice only partially restored the renewal ability of *Bmi1*$^{-/-}$ NSCs. These results show that Bmi1 controls, in NSCs, not only p16 but also another unidentified tumor suppressor, and that p16 can participate in NSC cycle arrest *(128)*.

Interestingly, other tumor suppressors control the proliferation of populations of cells already committed to a specific fate such as p19^{INK4d} for neuroblasts *(166)*, or participate in controlling both the proliferation of retinal progenitor cells and the cell fate determination such as p57 for retinal amacrine cells *(167,168)*, p27 in proliferative zones of the retina and in Müller cells *(168,169)*, and Rb in photoreceptors *(170)*. Interestingly, Rb was shown to also be essential for the differentiation of the photoreceptors.

The Pten tumor suppressor is a phosphatase of phosphatidylinositol-triphosphate that can antagonize the activated PI3-K signaling pathway (for review *see* ref. *171*). Mutation of the *Pten* gene can lead to tumor formation, such as glioblastoma *(172)*. Conditional *Pten* knockout mice driven by the expression of nestin provokes an increase of brain growth by stimulating the proliferation of progenitor cells in the SVZ, as well as stem cells by around 70% *(173)*. Interestingly, in vitro neurospheres that have lost Pten function show a greater ability, of around 6.5-fold, to generate more new spheres after dissociation. It appears that Pten reduces the capacity of NSCs to renew and probably favor an asymmetric division of the stem cell.

It appears that different tumor suppressors control the cell arrest of various cell types, including NSCs, progenitor and precursor cells, and that such tumor suppressors may also be involved in the control of cell differentiation. Concerning the NSCs, it seems that at each development time various tumor suppressors control NSC division and a number of them still have to be identified.

7. CONCLUSIONS AND PERSPECTIVES

The study of the regulation of NSC activity reveals that NSCs change their competence several times during development and that they acquire a specific control of their renewal depending on their location in the brain. The numerous factors regulating NSC renewal seem to be comprised of labyrinth of events, but different categories of this control can be identified and summarized as follows in an ideal theoretical NSC, which could integrate all the signals identified in all the NSCs in the brain (*see* Fig. 2). Notch activation maintains the NSC in a nondifferentiated state, but the cell fate decision can be modulated by intracellular factors such as Deltex and Numb. From the different studies on NSC and ES cells, it appears that the gp130-Stat3 activation also leads to the maintenance of the stem cell state. Activation of the MAPK pathway by EGF, FGF-2, and IGF-I is required to induce NSC proliferation, probably through the maintenance of Bmi1 expression. Such action can also be reached by the stimulation of the Shh pathway through the activation of *N*-Myc and Bmi1. The stability of *N*-Myc can be controlled by IGF-I through PI3-K to assure the proliferation potential. The Bmp Smad pathway is the most difficult to categorize, but my personal hypothesis is that Bmp prevents the activation of signals driving the cell to change its competence. Obviously this summary is schematic and does not reflect what really happens in a given NSC, but may serve as a lead. All these factors fluctuate during different stages of neurogenesis and one pathway can be preferentially activated to favor the commitment of a specific cell phenotype. At different stages of retinogenesis several homeodomain transcription factors and tumor suppressors control cell proliferation and cell cycle exit coupling with the generation of specific cell phenotypes. Such mechanisms probably also exist in the developing brain, but still remain poorly understood. The identification of the different actors of NSC control will surely help to dissect and understand how a NSC successively generates different cell phenotypes.

The understanding of NSC renewal not only helps to understand neurogenesis, but also serves to reveal the mechanisms leading to tumorigenesis. Such is the case for the Shh pathway, as well as *N*-Myc and Bmi1. All control NSC renewal and all play a role or are detected in medulloblastoma. The balance between a NSC and a cancer stem cell *(174)* is delicate and the study of the subtle actions occurring during stem cell proliferation may reveal the critical nodes that can push a stem cell to be tumorigenic. The recent advances in the understanding of NSC renewal will surely have favorable consequences on regeneration perspectives, but also for the fight against brain cancers.

ACKNOWLEDGMENTS

I would like to thank my colleagues for helpful discussions and their critics, and Dana Hornfeld for her assistance in the writing of this manuscript. This work was supported by the Swiss National Science Foundation.

REFERENCES

1. Reid CB, Tavazoie SF, Walsh CA. Clonal dispersion and evidence for asymmetric cell division in ferret cortex. Development 1997;124:2441–2450.
2. Turner DL, Cepko CL. A common progenitor for neurons and glia persists in rat retina late in development. Nature 1987;328:131–136.
3. Livesey FJ, Cepko CL. Vertebrate neural cell-fate determination: lessons from the retina. Nat Rev Neurosci 2001;2:109–118.
4. Doetsch F, Caille I, Lim DA, Garcia-Verdugo JM, Alvarez-Buylla A. Subventricular zone astrocytes are neural stem cells in the adult mammalian brain. Cell 1999;97:703–716.
5. Lois C, Alvarez-Buylla A. Long-distance neuronal migration in the adult mammalian brain. Science 1994;264:1145–1148.
6. Altman J, Das GD. Autoradiographic and histological evidence of postnatal hippocampal neurogenesis in rats. J Comp Neurol 1965;124:319–335.
7. van Praag H, Schinder AF, Christie BR, Toni N, Palmer TD, Gage FH. Functional neurogenesis in the adult hippocampus. Nature 2002;415:1030–1034.
8. Reynolds BA, Weiss S. Generation of neurons and astrocytes from isolated cells of the adult mammalian central nervous system. Science 1992;255:1707–1710.
9. Reynolds BA, Tetzlaff W, Weiss S. A multipotent EGF-responsive striatal embryonic progenitor cell produces neurons and astrocytes. J Neurosci 1992;12:4565–4574.
10. Reynolds BA, Weiss S. Clonal and population analyses demonstrate that an EGF-responsive mammalian embryonic CNS precursor is a stem cell. Dev Biol 1996;175:1–13.
11. Mehler MF, Kessler JA. Hematolymphopoietic and inflammatory cytokines in neural development. Trends Neurosci 1997;20:357–365.
12. Gage FH. Mammalian neural stem cells. Science 2000;287:1433–1438.
13. Reh TA, LevineEM. Multipotential stem cells and progenitors in the vertebrate retina. J Neurobiol 1998;36:206–220.
14. McKay R. Stem cells and the cellular organization of the brain. J Neurosci Res 2000;59:298–300.
15. Sommer L, Rao M. Neural stem cells and regulation of cell number. Prog Neurobiol 2002;66:1–18.
16. Tropepe V, Hitoshi S, Sirard C, Mak TW, Rossant J, van der Kooy D. Direct neural fate specification from embryonic stem cells: a primitive mammalian neural stem cell stage acquired through a default mechanism. Neuron 2001;30:65–78.
17. Ciccolini F, Svendsen CN. Fibroblast growth factor 2 (FGF-2) promotes acquisition of epidermal growth factor (EGF) responsiveness in mouse striatal precursor cells: identification of neural precursors responding to both EGF and FGF-2. J Neurosci 1998;18:7869–7880.
18. Ciccolini F. Identification of two distinct types of multipotent neural precursors that appear sequentially during CNS development. Mol Cell Neurosci 2001;17:895–907.
19. Tropepe V, Sibilia M, Ciruna BG, Rossant J, Wagner EF, van der Kooy D. Distinct neural stem cells proliferate in response to EGF and FGF in the developing mouse telencephalon. Dev Biol 1999;208:166–188.
20. Zhu G, Mehler MF, Mabie PC, Kessler JA. Developmental changes in progenitor cell responsiveness to cytokines. J Neurosci Res 1999;56:131–145.
21. Burrows RC, Wancio D, Levitt P, Lillien L. Response diversity and the timing of progenitor cell maturation are regulated by developmental changes in EGFR expression in the cortex. Neuron 1997;19:251–267.
22. Lillien L, Raphael H. BMP and FGF regulate the development of EGF-responsive neural progenitor cells. Development 2000;127:4993–5005.
23. Gritti A, Cova L, Parati EA, Galli R, Vescovi AL. Basic fibroblast growth factor supports the proliferation of epidermal growth factor-generated neuronal precursor cells of the adult mouse CNS. Neurosci Lett 1995;185:151–154.
24. Weiss S, Dunne C, Hewson J, et al. Multipotent CNS stem cells are present in the adult mammalian spinal cord and ventricular neuroaxis. J Neurosci 1996;16:7599–7609.

25. Arsenijevic Y, Villemure JG, Brunet JF, et al. Isolation of multipotent neural precursors residing in the cortex of the adult human brain. Exp Neurol 2001;170:48–62.
26. Johansson CB, Svensson M, Wallstedt L, Janson AM, Frisen J. Neural stem cells in the adult human brain. Exp Cell Res 1999;253:733–736.
27. Kukekov VG, Laywell ED, Suslov O, et al. Multipotent stem/progenitor cells with similar properties arise from two neurogenic regions of adult human brain. Exp Neurol 1999;156:333–344.
28. Pagano SF, Impagnatiello F, Girelli M, et al. Isolation and characterization of neural stem cells from the adult human olfactory bulb. Stem Cells 2000;18:295–300.
29. Taupin P, Ray J, Fischer WH, et al. FGF-2-responsive neural stem cell proliferation requires CCg, a novel autocrine/paracrine cofactor. Neuron 2000;28:385–397.
30. Palmer TD, Schwartz PH, Taupin P, Kaspar B, Stein SA, Gage FH. Cell culture. Progenitor cells from human brain after death. Nature 2001;411:42–43.
31. Tropepe V, Coles BL, Chiasson BJ, et al. Retinal stem cells in the adult mammalian eye. Science 2000;287:2032–2036.
32. Arsenijevic Y, Weiss S, Schneider B, Aebischer P. Insulin-like growth factor-I is necessary for neural stem cell proliferation and demonstrates distinct actions of epidermal growth factor and fibroblast growth factor-2. J Neurosci 2001;21:7194–7202.
33. Vicario-Abejon C, Yusta-Boyo MJ, Fernandez-Moreno C, de Pablo F. Locally born olfactory bulb stem cells proliferate in response to insulin-related factors and require endogenous insulin-like growth factor-I for differentiation into neurons and glia. J Neurosci 2003;23:895–906.
34. Baker J, Liu JP, Robertson EJ, Efstratiadis A. Role of insulin-like growth factors in embryonic and postnatal growth. Cell 1993;75:73–82.
35. Liu JP, Baker J, Perkins AS, Robertson EJ, Efstratiadis A. Mice carrying null mutations of the genes encoding insulin-like growth factor I (Igf-1) and type 1 IGF receptor (Igf1r). Cell 1993;75:59–72.
36. Aberg MA, Aberg ND, Palmer TD, et al. IGF-I has a direct proliferative effect in adult hippocampal progenitor cells. Mol Cell Neurosci 2003;24:23–40.
37. Tropepe V, Craig CG, Morshead CM, van der Kooy D. Transforming growth factor-alpha null and senescent mice show decreased neural progenitor cell proliferation in the forebrain subependyma. J Neurosci 1997;17:7850–7859.
38. Falk A, Frisen J. Amphiregulin is a mitogen for adult neural stem cells. J Neurosci Res 2002;69:757–762.
39. Liem KF, Jr, Tremml G, Roelink H, Jessell TM. Dorsal differentiation of neural plate cells induced by BMP-mediated signals from epidermal ectoderm. Cell 1995;82:969–979.
40. Liem KF, Jr, Tremml G, Jessell TM. A role for the roof plate and its resident TGF-beta-related proteins in neuronal patterning in the dorsal spinal cord. Cell 1997;91:127–138.
41. Lim DA, Tramontin AD, Trevejo JM, Herrera DG, Garcia-Verdugo JM, Alvarez-Buylla A. *Noggin* antagonizes BMP signaling to create a niche for adult neurogenesis. Neuron 2000;28:713–726.
42. Monville C, Coulpier M, Conti L, et al. Ciliary neurotrophic factor may activate mature astrocytes via binding with the leukemia inhibitory factor receptor. Mol Cell Neurosci 2001;17:373–384.
43. Shimazaki T, ShingoT, Weiss S. The ciliary neurotrophic factor/leukemia inhibitory factor/gp130 receptor complex operates in the maintenance of mammalian forebrain neural stem cells. J Neurosci 2001;21:7642–7653.
44. Chojnacki A, Shimazaki T, Gregg C, Weinmaster G, Weiss S. Glycoprotein 130 signaling regulates *Notch*1 expression and activation in the self-renewal of mammalian forebrain neural stem cells. J Neurosci 2003;23:1730–1741.
45. Burdon T, Smith A, Savatier P. Signalling, cell cycle and pluripotency in embryonic stem cells. Trends Cell Biol 2002;12:432–438.
46. Carpenter MK, Cui X, Hu ZY, et al. In vitro expansion of a multipotent population of human neural progenitor cells. Exp Neurol 1999;158:265–278.
47. Goodrich LV, Milenkovic L, Higgins KM, Scott MP. Altered neural cell fates and medulloblastoma in mouse patched mutants. Science 1997;277:1109–1113.
48. Jensen AM, WallaceVA. Expression of Sonic hedgehog and its putative role as a precursor cell mitogen in the developing mouse retina. Development 1997;124:363–371.
49. Raffel C, Jenkins RB, Frederick L, et al. Sporadic medulloblastomas contain PTCH mutations. Cancer Res 1997;57:842–845.
50. Rowitch DH, Jacques B, Lee SM, Flax JD, Snyder EY, McMahon AP. Sonic hedgehog regulates proliferation and inhibits differentiation of CNS precursor cells. J Neurosci 1999;19:8954–8965.

51. Vorechovsky I, Tingby O, Hartman M, et al. Somatic mutations in the human homologue of Drosophila patched in primitive neuroectodermal tumours. Oncogene 1997;15:361–366.
52. Zhu G, Mehler MF, Zhao J, Yu YS, Kessler JA. Sonic hedgehog and BMP2 exert opposing actions on proliferation and differentiation of embryonic neural progenitor cells. Dev Biol 1999;215:118–129.
53. Ingham PW. The patched gene in development and cancer. Curr Opin Genet Dev 1998;8:88–94.
54. Wechsler-Reya RJ, Scott MP. Control of neuronal precursor proliferation in the cerebellum by sonic hedgehog. Neuron 1999;22:103–114.
55. Lai K, Kaspar BK, Gage FH, Schaffer DV. Sonic hedgehog regulates adult neural progenitor proliferation in vitro and in vivo. Nat Neurosci 2003;6:21–27.
56. Palma V, Ruiz i Altaba A. Hedgehog-GLI signaling regulates the behavior of cells with stem cell properties in the developing neocortex. Development 2004;131:337–345.
57. Kenney AM, Rowitch DH. Sonic hedgehog promotes G(1) cyclin expression and sustained cell cycle progression in mammalian neuronal precursors. Mol Cell Biol 2000;20:9055–9067.
58. Charytoniuk D, Traiffort E, Hantraye P, Hermel JM, Galdes A, Ruat M. Intrastriatal sonic hedgehog injection increases Patched transcript levels in the adult rat subventricular zone. Eur J Neurosci 2002;16:2351–2357.
59. Machold R, Hayashi S, Rutlin M, et al. Sonic hedgehog is required for progenitor cell maintenance in telencephalic stem cell niches. Neuron 2003;39:937–950.
60. Kenney AM, Widlund HR, Rowitch DH. Hedgehog and PI-3 kinase signaling converge on Nmyc1 to promote cell cycle progression in cerebellar neuronal precursors. Development 2004;131:217–228.
61. Kenney AM, Cole MD, Rowitch DH. Nmyc upregulation by sonic hedgehog signaling promotes proliferation in developing cerebellar granule neuron precursors. Development 2003;130:15–28.
62. Dudek H, Datta SR, Franke TF, et al. Regulation of neuronal survival by the serine-threonine protein kinase Akt. Science 1997;275:661–665.
63. Kubota R, Hokoc JN, Moshiri A, McGuire C, Reh TA. A comparative study of neurogenesis in the retinal ciliary marginal zone of homeothermic vertebrates. Brain Res Dev Brain Res 2002;134:31–41.
64. Moshiri A, Reh TA. Persistent progenitors at the retinal margin of $ptc^{\pm}$ mice. J Neurosci 2004;24:229–237.
65. Tessier-Lavigne M, Goodman CS. The molecular biology of axon guidance. Science 1996;274:1123–1133.
66. Drescher U. The Eph family in the patterning of neural development. Curr Biol 1997;7:R799–R807.
67. O'Leary DD, Wilkinson DG. Eph receptors and ephrins in neural development. Curr Opin Neurobiol 1999;9:65–73.
68. Conover JC, Doetsch F, Garcia-Verdugo JM, Gale NW, Yancopoulos GD, Alvarez-Buylla A. Disruption of Eph/ephrin signaling affects migration and proliferation in the adult subventricular zone. Nat Neurosci 2000;3:1091–1097.
69. Hienola A, Pekkanen M, Raulo E, Vanttola P, Rauvala H. HB-GAM inhibits proliferation and enhances differentiation of neural stem cells. Mol Cell Neurosci 2004;26:75–88.
70. Kokuzawa J, Yoshimura S, Kitajima H, et al. Hepatocyte growth factor promotes proliferation and neuronal differentiation of neural stem cells from mouse embryos. Mol Cell Neurosci 2003;24:190–197.
71. Sherwood NM, Krueckl SL, McRory JE. The origin and function of the pituitary adenylate cyclase-activating polypeptide (PACAP)/ glucagon superfamily. Endocr Rev 2000;21:619–670.
72. Jaworski DM, Proctor MD. Developmental regulation of pituitary adenylate cyclase-activating polypeptide and PAC(1) receptor mRNA expression in the rat central nervous system. Brain Res Dev Brain Res 2000;120:27–39.
73. Mercer A, Ronnholm H, Holmberg J, et al. PACAP promotes neural stem cell proliferation in adult mouse brain. J Neurosci Res 2004;76:205–215.
74. Palmer TD, Willhoite AR, Gage FH. Vascular niche for adult hippocampal neurogenesis. J Comp Neurol 2000;425:479–494.
75. Louissaint A, Jr, Rao S, Leventhal C, Goldman SA. Coordinated interaction of neurogenesis and angiogenesis in the adult songbird brain. Neuron 2002;34:945–960.
76. Shen Q, Goderie S, Jin L, et al. Endothelial cells stimulate self-renewal and expand neurogenesis of neural stem cells. Science 2004;304(5675):1338–1340.
77. Artavanis-Tsakonas S, Rand MD, Lake RJ. *Notch* signaling: cell fate control and signal integration in development. Science 1999;284:770–776.
78. De Strooper B, Annaert W, Cupers P, et al. A presenilin-1-dependent gamma-secretase-like protease mediates release of *Notch* intracellular domain. Nature 1999;398:518–522.

79. Bao ZZ, Cepko CL. The expression and function of *Notch* pathway genes in the developing rat eye. J Neurosci 1997;17:1425–1434.
80. Cui XY, Hu QD, Tekaya M, et al. NB-3/*Notch*1 pathway via Deltex1 promotes neural progenitor cell differentiation into oligodendrocytes. J Biol Chem 2004;279(24):25,858–25,865.
81. Hitoshi S, Alexson T, Tropepe V, et al. *Notch* pathway molecules are essential for the maintenance, but not the generation, of mammalian neural stem cells. Genes Dev 2002;16:846–858.
82. Ohtsuka T, Sakamoto M, Guillemot F, Kageyama R. Roles of the basic helix-loop-helix genes Hes1 and Hes5 in expansion of neural stem cells of the developing brain. J Biol Chem 2001;276:30,467–30,474.
83. Nakamura Y, Sakakibara S, Miyata T, et al. The bHLH gene hes1 as a repressor of the neuronal commitment of CNS stem cells. J Neurosci 2000;20:283–293.
84. Faux CH, Turnley AM, Epa R, Cappai R, Bartlett PF. Interactions between fibroblast growth factors and *Notch* regulate neuronal differentiation. J Neurosci 2001;21:5587–5596.
85. Solecki DJ, Liu XL, Tomoda T, Fang Y, Hatten ME. Activated *Notch*2 signaling inhibits differentiation of cerebellar granule neuron precursors by maintaining proliferation. Neuron 2001;31:557–568.
86. Tanigaki K, Nogaki F, Takahashi J, Tashiro K, Kurooka H, Honjo T. *Notch*1 and *Notch*3 instructively restrict bFGF-responsive multipotent neural progenitor cells to an astroglial fate. Neuron 2001;29:45–55.
87. Chambers CB, Peng Y, Nguyen H, Gaiano N, Fishell G, Nye JS. Spatiotemporal selectivity of response to *Notch*1 signals in mammalian forebrain precursors. Development 2001;128:689–702.
88. Gaiano N, Nye JS, Fishell G. Radial glial identity is promoted by *Notch*1 signaling in the murine forebrain. Neuron 2000;26:395–404.
89. Furukawa T, Mukherjee S, Bao ZZ, Morrow EM, Cepko CL. rax, Hes1, and notch1 promote the formation of Muller glia by postnatal retinal progenitor cells. Neuron 2000;26:383–394.
90. Morrison SJ, Perez SE, Qiao Z, et al. Transient *Notch* activation initiates an irreversible switch from neurogenesis to gliogenesis by neural crest stem cells. Cell 2000;101:499–510.
91. Hu QD, Ang BT, Karsak M, et al. F3/contactin acts as a functional ligand for *Notch* during oligodendrocyte maturation. Cell 2003;115:163–175.
92. Ge W, Martinowich K, Wu X, et al. *Notch* signaling promotes astrogliogenesis via direct CSL-mediated glial gene activation. J Neurosci Res 2002;69:848–860.
93. Anderson CM, Swanson RA. Astrocyte glutamate transport: review of properties, regulation, and physiological functions. Glia 2000;32:1–14.
94. Frise E, Knoblich JA, Younger-Shepherd S, Jan LY, Jan YN. The Drosophila *Numb* protein inhibits signaling of the *Notch* receptor during cell–cell interaction in sensory organ lineage. Proc Natl Acad Sci USA 1996;93:11,925–11,932.
95. Heins N, Cremisi F, Malatesta P, et al. *Emx2* promotes symmetric cell divisions and a multipotential fate in precursors from the cerebral cortex. Mol Cell Neurosci 2001;18:485–502.
96. Cayouette M, Whitmore AV, Jeffery G, Raff M. Asymmetric segregation of *Numb* in retinal development and the influence of the pigmented epithelium. J Neurosci 2001;21:5643–5651.
97. Zhong W, Feder JN, Jiang MM, Jan LY, JanYN. Asymmetric localization of a mammalian numb homolog during mouse cortical neurogenesis. Neuron 1996;17:43–53.
98. Zhong W, Jiang MM, Weinmaster G, Jan LY, Jan YN. Differential expression of mammalian *Numb*, *Numblike* and *Notch*1 suggests distinct roles during mouse cortical neurogenesis. Development 1997;124:1887–1897.
99. Shen Q, Zhong W, JanYN, Temple S. Asymmetric *Numb* distribution is critical for asymmetric cell division of mouse cerebral cortical stem cells and neuroblasts. Development 2002;129:4843–4853.
100. Zilian O, Saner C, Hagedorn L, et al. Multiple roles of mouse *Numb* in tuning developmental cell fates. Curr Biol 2001;11:494–501.
101. Zhong W, Jiang MM, Schonemann MD, et al. Mouse numb is an essential gene involved in cortical neurogenesis. Proc Natl Acad Sci USA 2000;97:6844–6849.
102. Petersen PH, Zou K, Hwang JK, Jan YN, Zhong W. Progenitor cell maintenance requires numb and numblike during mouse neurogenesis. Nature 2002;419:929–934.
103. Li HS, Wang D, Shen Q, et al. Inactivation of *Numb* and *Numblike* in embryonic dorsal forebrain impairs neurogenesis and disrupts cortical morphogenesis. Neuron 2003;40:1105–1118.
104. Verdi JM, Bashirullah A, Goldhawk DE, et al. Distinct human NUMB isoforms regulate differentiation vs proliferation in the neuronal lineage. Proc Natl Acad Sci USA 1999;96:10,472–10,476.
105. Okano H, Imai T, Okabe M. Musashi: a translational regulator of cell fate. J Cell Sci 2002;115:1355–1359.

106. Imai T, Tokunaga A, Yoshida T, et al. The neural RNA-binding protein Musashi1 translationally regulates mammalian numb gene expression by interacting with its mRNA. Mol Cell Biol 2001;21: 3888–3900.
107. Keyoung HM, Roy NS, Benraiss A, et al. High-yield selection and extraction of two promoter-defined phenotypes of neural stem cells from the fetal human brain. Nat Biotechnol 2001;19: 843–850.
108. Sakakibara S, Nakamura Y, Yoshida T, et al. RNA-binding protein Musashi family: roles for CNS stem cells and a subpopulation of ependymal cells revealed by targeted disruption and antisense ablation. Proc Natl Acad Sci USA 2002;99:15,194–15,199.
109. Galli R, Fiocco R, De Filippis L, et al. *Emx2* regulates the proliferation of stem cells of the adult mammalian central nervous system. Development 2002;129:1633–1644.
110. Gulisano M, Broccoli V, Pardini C, Boncinelli E. *Emx1* and *Emx2* show different patterns of expression during proliferation and differentiation of the developing cerebral cortex in the mouse. Eur J Neurosci 1996;8:1037–1050.
111. Mallamaci A, Iannone R, Briata P, et al. EMX2 protein in the developing mouse brain and olfactory area. Mech Dev 1998;77:165–172.
112. Simeone A, Gulisano M, Acampora D, Stornaiuolo A, Rambaldi M, Boncinelli E. Two vertebrate homeobox genes related to the Drosophila empty spiracles gene are expressed in the embryonic cerebral cortex. EMBO J 1992;11:2541–2550.
113. Wood HB, Episkopou V. Comparative expression of the mouse *Sox1*, *Sox2* and *Sox3* genes from pregastrulation to early somite stages. Mech Dev 1999;86:197–201.
114. Zappone MV, Galli R, Catena R, et al. *Sox2* regulatory sequences direct expression of a (beta)-geo transgene to telencephalic neural stem cells and precursors of the mouse embryo, revealing regionalization of gene expression in CNS stem cells. Development 2000;127:2367–2382.
115. Chow RL, Altmann CR, Lang RA, Hemmati-Brivanlou A. *Pax6* induces ectopic eyes in a vertebrate. Development 1999;126:4213–4222.
116. Loosli F, Winkler S, Wittbrodt J. *Six3* overexpression initiates the formation of ectopic retina. Genes Dev 1999;13:649–654.
117. Mathers PH, Grinberg A, Mahon KA, Jamrich M. The *Rx* homeobox gene is essential for vertebrate eye development. Nature 1997;387:603–607.
118. Zhang L, Mathers PH, Jamrich M. Function of *Rx*, but not *Pax6*, is essential for the formation of retinal progenitor cells in mice. Genesis 2000;28:135–142.
119. Warren N, Caric D, Pratt T, et al. The transcription factor, *Pax6*, is required for cell proliferation and differentiation in the developing cerebral cortex. Cereb Cortex 1999;9:627–635.
120. Gotz M, Stoykova A, Gruss P. *Pax6* controls radial glia differentiation in the cerebral cortex. Neuron 1998;21:1031–1044.
121. Gotz M. Glial cells generate neurons—master control within CNS regions: developmental perspectives on neural stem cells. Neuroscientist 2003;9:379–397.
122. Heins N, Malatesta P, Cecconi F, et al. Glial cells generate neurons: the role of the transcription factor *Pax6*. Nat Neurosci 2002;5:308–315.
123. Burmeister M, Novak J, Liang MY, et al. Ocular retardation mouse caused by Chx10 homeobox null allele: impaired retinal progenitor proliferation and bipolar cell differentiation. Nat Genet 1996;12:376–384.
124. Dyer MA, Livesey FJ, Cepko CL, Oliver G. *Prox1* function controls progenitor cell proliferation and horizontal cell genesis in the mammalian retina. Nat Genet 2003;34:53–58.
125. Jacobs JJ, Kieboom K, Marino S, DePinho RA, van Lohuizen M. The oncogene and Polycomb-group gene bmi-1 regulates cell proliferation and senescence through the ink4a locus. Nature 1999;397:164–168.
126. van der Lugt NM, Domen J, Linders K, et al. Posterior transformation, neurological abnormalities, and severe hematopoietic defects in mice with a targeted deletion of the bmi-1 proto-oncogene. Genes Dev 1994;8:757–769.
127. Park IK, Qian D, Kiel M, et al. Bmi-1 is required for maintenance of adult self-renewing haematopoietic stem cells. Nature 2003;423:302–305.
128. Molofsky AV, Pardal R, Iwashita T, Park IK, Clarke MF, Morrison SJ. Bmi-1 dependence distinguishes neural stem cell self-renewal from progenitor proliferation. Nature 2003;425:962–967.
129. Leung C, Lingbeek M, Shakhova O, et al. Bmi1 is essential for cerebellar development and is overexpressed in human medulloblastomas. Nature 2004;428:337–341.

130. Tsai RY, McKay RD. A nucleolar mechanism controlling cell proliferation in stem cells and cancer cells. Genes Dev 2002;16:2991–3003.
131. Shi Y, Chichung LD, Taupin P, et al. Expression and function of orphan nuclear receptor TLX in adult neural stem cells. Nature 2004;427:78–83.
132. Makarov VL, Hirose Y, Langmore JP. Long G tails at both ends of human chromosomes suggest a C strand degradation mechanism for telomere shortening. Cell 1997;88:657–666.
133. Harley CB, Futcher AB, Greider CW. Telomeres shorten during ageing of human fibroblasts. Nature 1990;345:458–460.
134. von Zglinicki T, Saretzki G, Docke W, Lotze C. Mild hyperoxia shortens telomeres and inhibits proliferation of fibroblasts: a model for senescence? Exp Cell Res 1995;220:186–193.
135. Blasco MA, Lee HW, Hande MP, et al. Telomere shortening and tumor formation by mouse cells lacking telomerase RNA. Cell 1997;91:25–34.
136. Greider CW. Telomeres, telomerase and senescence. Bioessays 1990;12:363–369.
137. Hemann MT, Strong MA, Hao LY, Greider CW. The shortest telomere, not average telomere length, is critical for cell viability and chromosome stability. Cell 2001;107:67–77.
138. Lundberg AS, Hahn WC, Gupta P, Weinberg RA. Genes involved in senescence and immortalization. Curr Opin Cell Biol 2000;12:705–709.
139. Greider CW, Blackburn EH. Identification of a specific telomere terminal transferase activity in Tetrahymena extracts. Cell 1985;43:405–413.
140. Greider CW, Blackburn EH. The telomere terminal transferase of Tetrahymena is a ribonucleoprotein enzyme with two kinds of primer specificity. Cell 1987;51:887–898.
141. Greider CW, Blackburn EH. A telomeric sequence in the RNA of Tetrahymena telomerase required for telomere repeat synthesis. Nature 1989;337:331–337.
142. Bodnar AG, Ouellette M, Frolkis M, et al. Extension of life-span by introduction of telomerase into normal human cells. Science 1998;279:349–352.
143. Wang J, Hannon GJ, Beach DH. Risky immortalization by telomerase. Nature 2000;405:755–756.
144. Fu W, Killen M, Culmsee C, Dhar S, Pandita TK, Mattson MP. The catalytic subunit of telomerase is expressed in developing brain neurons and serves a cell survival-promoting function. J Mol Neurosci 2000;14:3–15.
145. Klapper W, Shin T, Mattson MP. Differential regulation of telomerase activity and TERT expression during brain development in mice. J Neurosci Res 2001;64:252–260.
146. Caporaso GL, Lim DA, Alvarez-Buylla A, Chao MV. Telomerase activity in the subventricular zone of adult mice. Mol Cell Neurosci 2003;23:693–702.
147. Ostenfeld T, Caldwell MA, Prowse KR, Linskens MH, Jauniaux E, Svendsen CN. Human neural precursor cells express low levels of telomerase in vitro and show diminishing cell proliferation with extensive axonal outgrowth following transplantation. Exp Neurol 2000;164:215–226.
148. Haik S, Gauthier LR, Granotier C, et al. Fibroblast growth factor 2 up regulates telomerase activity in neural precursor cells. Oncogene 2000;19:2957–2966.
149. Boulaire J, Fotedar A, Fotedar R. The functions of the cdk-cyclin kinase inhibitor p21WAF1. Pathol Biol (Paris) 2000;48:190–202.
150. Aloyz RS, Bamji SX, Pozniak CD, et al. p53 is essential for developmental neuron death as regulated by the TrkA and p75 neurotrophin receptors. J Cell Biol 1998;143:1691–1703.
151. Serrano M, Lin AW, McCurrach ME, Beach D, Lowe SW. Oncogenic ras provokes premature cell senescence associated with accumulation of p53 and p16INK4a. Cell 1997;88:593–602.
152. Sherr CJ, Weber JD. The ARF/p53 pathway. Curr Opin Genet Dev 2000;10:94–99.
153. Chen Q, Ames BN. Senescence-like growth arrest induced by hydrogen peroxide in human diploid fibroblast F65 cells. Proc Natl Acad Sci USA 1994;91:4130–4134.
154. Saito H, Hammond AT, Moses RE. The effect of low oxygen tension on the in vitro-replicative life span of human diploid fibroblast cells and their transformed derivatives. Exp Cell Res 1995;217:272–279.
155. Lin AW, Barradas M, Stone JC, van Aelst L, Serrano M, Lowe SW. Premature senescence involving p53 and p16 is activated in response to constitutive MEK/MAPK mitogenic signaling. Genes Dev 1998;12:3008–3019.
156. Zhu J, Woods D, McMahon M, Bishop JM. Senescence of human fibroblasts induced by oncogenic Raf. Genes Dev 1998;12:2997–3007.
157. Allsopp RC, Vaziri H, Patterson C, et al. Telomere length predicts replicative capacity of human fibroblasts. Proc Natl Acad Sci USA 1992;89:10,114–10,118.

158. Vaziri H, Schachter F, Uchida I, et al. Loss of telomeric DNA during aging of normal and trisomy 21 human lymphocytes. Am J Hum Genet 1993;52:661–667.
159. Cooper-Kuhn CM, Vroemen M, Brown J, et al. Impaired adult neurogenesis in mice lacking the transcription factor E2F1. Mol Cell Neurosci 2002;21:312–323.
160. Cattaneo E, McKay R. Identifying and manipulating neuronal stem cells. Trends Neurosci 1991;14:338–340.
161. Martinez-Serrano A, Bjorklund A. Immortalized neural progenitor cells for CNS gene transfer and repair. Trends Neurosci 1997;20:530–538.
162. Villa A, Snyder EY, Vescovi A, Martinez-Serrano A. Establishment and properties of a growth factor-dependent, perpetual neural stem cell line from the human CNS. Exp Neurol 2000;161:67–84.
163. Hoshimaru M, Ray J, Sah DW, Gage FH. Differentiation of the immortalized adult neuronal progenitor cell line HC2S2 into neurons by regulatable suppression of the v-myc oncogene. Proc Natl Acad Sci USA 1996;93:1518–1523.
164. Snyder EY, Deitcher DL, Walsh C, Arnold-Aldea S, Hartwieg EA, Cepko CL. Multipotent neural cell lines can engraft and participate in development of mouse cerebellum. Cell 1992;68:33–51.
165. Doetsch F, Verdugo JM, Caille I, Alvarez-Buylla A, Chao MV, Casaccia-Bonnefil P. Lack of the cell cycle inhibitor p27Kip1 results in selective increase of transit-amplifying cells for adult neurogenesis. J Neurosci 2002;22:2255–2264.
166. Coskun V, Luskin MB. The expression pattern of the cell cycle inhibitor p19(INK4d) by progenitor cells of the rat embryonic telencephalon and neonatal anterior subventricular zone. J Neurosci 2001;21:3092–3103.
167. Dyer MA, Cepko CL. p57(Kip2) regulates progenitor cell proliferation and amacrine interneuron development in the mouse retina. Development 2000;127:3593–3605.
168. Dyer MA, Cepko CL. p27Kip1 and p57Kip2 regulate proliferation in distinct retinal progenitor cell populations. J Neurosci 2001;21:4259–4271.
169. Levine EM, Close J, Fero M, Ostrovsky A, Reh TA. p27(Kip1) regulates cell cycle withdrawal of late multipotent progenitor cells in the mammalian retina. Dev Biol 2000;219:299–314.
170. Zhang J, Gray J, Wu L, et al. Rb regulates proliferation and rod photoreceptor development in the mouse retina. Nat Genet 2004;36:351–360.
171. Leslie NR, Downes CP. PTEN: The down side of PI 3-kinase signalling. Cell Signal 2002;14:285–295.
172. Li J, Yen C, Liaw D, et al. PTEN, a putative protein tyrosine phosphatase gene mutated in human brain, breast, and prostate cancer. Science 1997;275:1943–1947.
173. Groszer M, Erickson R, Scripture-Adams DD, et al. Negative regulation of neural stem/progenitor cell proliferation by the *Pten* tumor suppressor gene in vivo. Science 2001;294:2186–2189.
174. Reya T, Morrison SJ, Clarke MF, Weissman IL. Stem cells, cancer, and cancer stem cells. Nature 2001;414:105–111.

III
CONTROL OF THE CELL CYCLE AND APOPTOSIS IN GLIA

12
Methods of Determining Apoptosis in Neuro-Oncology
Review of the Literature

Brian T. Ragel, MD, Bardia Amirlak, MD, Ganesh Rao, MD, and William T. Couldwell, MD, PhD

SUMMARY

Tumor growth depends not only on the rate of cell proliferation but also on the rate of programmed cell death (apoptosis). Thus, treatments focusing either on the restoration of apoptosis or on triggering the apoptotic pathways may provide new treatment targets. Therefore, the ability to detect apoptosis in experimental studies is vital in our assessment of treatment success. Apoptosis detection is important in assessing treatment outcomes in both the experimental and clinical settings. Furthermore, the analysis of surgically removed specimens for apoptosis provides patient-specific data, allowing for tailored treatment plans. Many techniques are currently used to detect apoptosis, including identifying those cells exhibiting the pathognomonic morphological characteristics of apoptosis, identifying DNA breaks by the terminal deoxynucleotidyl transferase-mediated dUTP nick-end labeling assay, labeling externalized phosphatidylserine, flow cytometry techniques, laser scanning cytometry, poly(ADP-ribose) polymerase cleavage, detection of caspase activation, membrane permeability changes, mitochondrial failure, identifying denatured DNA, as well as a recent noninvasive technique for measuring apoptosis with a novel fluorescence reporter. This chapter provides an overview of these various apoptosis detection methods used in experimental studies with an emphasis on glioma research.

Key Words: Apoptosis; glioma; programmed cell death.

1. INTRODUCTION

Gliomas are among the most difficult tumors to treat. Despite advances in surgery, radiation, and chemotherapy, the prognosis for patients with malignant gliomas remains poor *(1,2)*. Patients with the most aggressive grade of glioma, glioblastoma multiforme (GBM), have a median survival of 1 yr *(1,3,4)*. Tumor growth depends not only on the rate of cell proliferation but also on the rate of programmed cell death (apoptosis) *(5,6)*. Furthermore, genomic alterations affecting apoptotic pathways have been implicated in tumorigenesis *(7–11)*. Thus, glioma treatments focusing either on the restoration of apoptosis or on triggering the apoptotic pathways may provide new treatment targets. Therefore, the ability to detect apoptosis in experimental studies is vital in our assessment of treatment success. This chapter provides an overview of the various apoptosis detection methods used in experimental studies with an emphasis on glioma research. Although we have focused on these methods in glioma research, the methods described are more broadly applicable to other nervous system tumors as well.

From: *The Cell Cycle in the Central Nervous System*
Edited by: D. Janigro © Humana Press Inc., Totowa, NJ

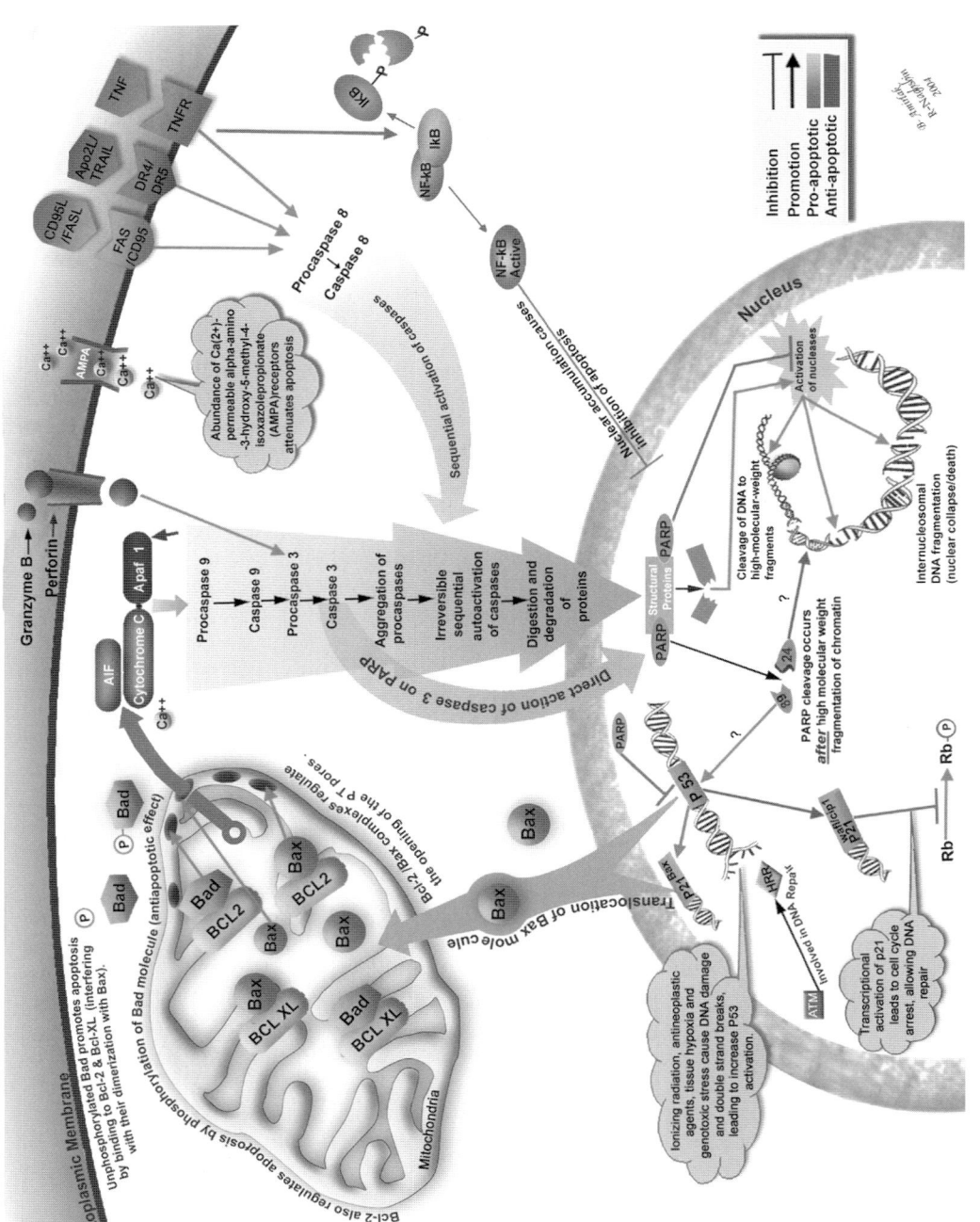

Fig. 1.

In 1972, Kerr et al. described the physiological process of apoptosis (a term derived from the Greek word for "falling off") *(12)*. Apoptosis, or programmed cell death, is critical for organ development, tissue homeostasis, and elimination of defective or potentially dangerous cells *(6,12,13)*. Cells undergoing apoptotic "cellular suicide" rapidly shrink and lose their normal intercellular contacts. They subsequently exhibit phosphatidyl serine (PS) externalization, dense chromatin condensation, nuclear fragmentation, cytoplasmic blebbing, and cellular fragmentation into small apoptotic bodies. These apoptotic bodies are phagocytosed and digested by neighboring cells or macrophages *(6,12)*. Apoptosis permits cell death without an inflammatory response, because no cytosolic components are released into the intercellular space *(6)*.

Many triggers for apoptosis have been identified, including developmental signals, cellular stress, and disruption of the cell cycle. Three classical molecular pathways for the initiation of programmed cell death have been identified, with all of these pathways culminating in the activation of a proteolytic cascade consisting of cysteine aspartic acid-specific proteases, called caspases (Fig. 1) *(14)*. The first molecular pathway involves growth factor withdrawal and is regulated by the Bcl-2 family proteins (e.g., Bax, Bcl-2). Decreased growth factor results in cytochrome-*c* release from the mitochondria, with activation of Apaf-1 and subsequent activation of the caspase cascade. The second pathway relies on cell "death" receptors, such as tumor necrosis factor (TNF) or Fas. Ligand binding of these receptors results in the downstream activation of caspases. The final pathway is initiated by DNA damage and involves p53 and ataxia telangiectasia mutated protein. Ataxia telangiectasia mutated facilitates the DNA repair process after damage by regulating homologous recombination repair. On the other hand, the process of apoptosis may occur based on the severity of DNA damage *(15,16)*. The precise mechanism by which DNA damage activates caspases, however, is unknown. In all of these pathways, the family of Bcl-2 proteins and the caspase cascade play a key role in the regulation and execution of apoptosis *(14,17–21)*. Apoptosis plays an important role in multiple physiological processes such as organ development, tissue homeostasis, and regulation of the immune system. Defects in the apoptotic mechanisms can extend cell life, contributing to neoplastic cell expansion independent of cell division *(22)*. Deficiencies in apoptosis also contribute to carcinogenesis by creating an environment for genetic instability and accumulation of gene mutations, thus permitting disregard for cell cycle checkpoints that would normally induce apoptosis *(22)*.

The ability to detect apoptosis is vital in assessing treatment outcomes in both the experimental and clinical settings. Furthermore, the analysis of surgically removed specimens for apoptosis provides patient-specific data. This patient-specific information allows for tailored treatment plans. Many techniques are currently used to detect apoptosis (*see* Table 1), including

---

Fig. 1. (Left) Apoptosis triggered by internal signals. **(A)** Tumor-suppressor-gene-mediated signals arising within the cell because of DNA stress and damage cause the translocation of the Bax molecule from the cytosol to the inner mitochondrial membrane. **(B)** Bax forms heterodimers with members of the Bcl-2 family, causing a shift in equilibrium toward apoptosis (proapoptotic Bax, Bad, Bcl-Xs vs antiapoptotic Bcl-2, Bcl-XL); these events lead to a drop in mitochondrial transmembrane potential and open the pores, causing the mitochondrial membrane to become leaky to certain molecules. **(C)** Cytochrome-*c*, apoptosis-inducing factor (AIF), and Ca^{2+} are released from the mitochondria to the cytoplasm. **(D)** Cytochrome-*c* (Apaf-2) associates with Apaf-1 and forms a complex that activates procaspase-9. **(E)** Procaspase-9 turns into caspase-9 (Apaf-3), which in turn activates downstream caspases, starting with caspase-3. **(F)** Over 15 identified caspases are activated in a particular sequence, causing the cleavage and digestion of a variety of structural proteins (DNA-dependent protein kinase, PARP, nuclear lamins, etc.) in the cytoplasm and nucleus, which lead to the death of the cell. (Right) Apoptosis triggered by external signals. **(G)** Binding of the apoptotic activators, either free soluble or cell-surface expressed, such as CD95L (Fas/ApoIL), TRAIL/Apo2L, and TNF, to receptors on the glioma cell surface. **(H)** Receptors transmit a signal to their cytoplasmic domain, forming complexes that lead to the activation of caspase-8; caspase-8 initiates a cascade of caspase activation leading to death of the cell. **(I)** Immune surveillance cells cause apoptosis of glioma cells via granule-mediated cytotoxicity (perforines and granzyme B). (Adapted with permission from ref. *11*.) (Please *see* companion CD for color version of this figure.)

Table 1
Methods of Detecting Apoptosis

Apoptosis marker	Characteristics	Method for detection
Cell morphology ("gold standard")	Early • Shrunken, dehydrated appearance • Chromatin condensation and fragmentation • Dilated endoplasmic reticulum	LM, EM
	Late • Membrane-bound apoptotic bodies (phagocytized by macrophages)	LM, EM
DNA fragments	DNA laddering • Electrophoresis of nuclear extract	GE
	TUNEL assay • Labeling of 3′ termini of DNA breaks	LM, FC, LSC
	Flow cytometry • Sub-G_1 peak	FC
Plasma membrane changes	Loss of phospholipid asymmetry, externalization of phosphatidylserine • Annexin V binding to externalized phosphatidylserine	FC, LSC
	Increased fluorescence staining noted with the fluorochromes AAD, PI, Hoechst 33342, and Hoechst 332258 indicates increased cell membrane permeability	FC, LSC
Changes in FC light scatter	Decreased forward scatter indicating rehydrated cells	FC
	Increased side scatter indicates an increase in chromatin condensation	FC
Changes in expression and location of apoptotic molecular markers	Increased Bax to Bcl-2 ratio	GE, FC, LSC
	Translocation of cytochrome-c from mitochondria to cytoplasm	LSC
	Translocation of Bax from cytoplasm to mitochondria	LSC
	Translocation of NF-κB from cytoplasm to nucleus	LSC
	Translocation of p53 from cytoplasm to nucleus	LSC
Detecting PARP cleavage products	Western blot analysis for the 85-k and 24-kD PARP, or the 85-kDa cleavage fragments	GE
	Anti-PARP cleavage fragment antibodies	LM, FC, LSC
Caspase activation detection	Fluorescence-labeled caspase substrates—activation causes fluorescences	FC, LSC
	Western blot for cleavage products of pro-enzyme	GE
	Labeled caspase inhibitors	FC, LSC
	Anticleaved caspase antibodies	FC
Detecting mitochondrial membrane permeability changes	Fluorochrome uptake indicates a functioning mitochondria	FC
	Translocation of cytochrome-c from mitochondria to cytoplasm	LSC
Increased DNA sensitivity to denaturation	Anti-single-stranded DNA antibody indicates denatured DNA	FC
	Acridine orange fluorochrome • Green fluorescence seen with double-stranded DNA staining • Red fluorescence seen with single-stranded DNA staining	FC

LM, light microscopy; EM, electron microscopy; GE, gel electrophoresis; FC, flow cytometry; LSC, laser scanning cytometry.

identifying those cells exhibiting the pathognomonic morphological characteristics of apoptosis, identifying DNA breaks by the terminal deoxynucleotidyl transferase-mediated dUTP nick-end labeling (TUNEL) assay, labeling externalized PS, flow cytometry (FC) techniques, laser scanning cytometry (LSC), poly(ADP-ribose) polymerase (PARP) cleavage, detection of caspase activation, membrane permeability changes, mitochondrial failure, identifying denatured DNA, as well as a recent noninvasive technique for measuring apoptosis with a novel fluorescence reporter.

2. CHANGES IN CELL MORPHOLOGY: THE "GOLD STANDARD" FOR IDENTIFICATION OF APOPTOSIS BY LIGHT MICROSCOPY

Apoptosis was originally defined as a specific mode of cell death based on very characteristic morphological changes *(12)*. The presence of these changes is still considered the "gold standard" for identification of apoptotic cells by light microscopy. The most specific of these changes is chromatin condensation; the chromatin of apoptotic cells is very "smooth" in appearance, and the structural framework that otherwise characterizes the cell nucleus is lost. Other changes that occur include reduced cell size, plasma membrane "blebbing," nuclear fragmentation, apoptotic bodies, dilatation of endoplasmic reticulum, and cell detachment from tissue culture flasks *(23)*.

3. DNA FRAGMENTATION: DETECTION OF DNA STRAND BREAKS (TUNEL ASSAY)

DNA fragmentation is considered a hallmark of apoptosis, and detection of endonuclease-mediated DNA breaks is perhaps the most widely used technique for confirming apoptosis. It is commonly used in both clinical and experimental settings *(24,25)*. Cells undergoing apoptosis activate endonucleases that cleave DNA at internucleosomal sections into fragments ranging from 50 to 300 kb *(26,27)*. Detection of DNA fragments can be achieved by either gel electrophoresis (i.e., DNA laddering) or by labeling the 3′-OH-free ends of cleaved DNA (TUNEL assay).

Gel electrophoresis of the nuclear extract of apoptotic cells shows a characteristic DNA laddering of the fragmented DNA bands. Disadvantages to this technique include lack of quantification and difficulty in visualizing DNA banding because of the development of smears. The TUNEL assay offers better resolution in identifying apoptosis.

The DNA breaks of apoptosis expose free 3′ termini, which can be detected by attaching fluorochrome-labeled nucleotides in a reaction catalyzed by exogenous terminal deoxynucleotidyl transferase. The reaction is commonly known as TUNEL *(23)*. Of all the markers used to label DNA breaks, 5-bromo-2′-deoxyuridine 5′-triphosphate (BrdUTP) appears to be the most advantageous with respect to sensitivity, low cost, and simplicity of reaction *(23)*. Generally, the DNA strand breaks are labeled with BrdUTP and then conjugated with fluorescein isothiocyanate (FITC) anti-BrdUTP antibody. Visualization of the labeled 3′-OH free ends of DNA can be accomplished with light microscopy, fluorescent microscopes, FC, or the laser scanning cytometer *(23,26,28–31)*.

The TUNEL assay for apoptosis is specific and relatively easy to use *(23)*. A main disadvantage of the assay is the loss of the fragmented DNA during the staining procedure, which could potentially give false-negative results *(32,33)*. Cell fixation with a crosslinking agent (e.g., formaldehyde) is critical in the detection of DNA strand breaks, especially when cells are in advanced stages of apoptosis. When formaldehyde is used, the subsequent cell washings during the procedure do not markedly diminish the DNA content of the apoptotic cells *(32,33)*. Another disadvantage of the TUNEL assay is that it shows different staining intensities, corresponding to different grades of DNA fragmentation of any kind, which makes reproducible quantification difficult to achieve *(34,35)*. Even with these drawbacks, the TUNEL method has proven its efficacy in malignant glioma cells exposed to various radiochemotherapeutic regimens, and commercial kits are readily available *(36–41)*.

4. EXTERNALIZATION OF PS IS AN EARLY MARKER OF APOPTOSIS

Eukaryotic plasma membranes exhibit significant phospholipid asymmetry, with choline-containing phospholipids (e.g., phosphatidylcholine and sphingomyelin) being expressed on the extracellular side of the cell and aminophospholipids (e.g., PS and phosphatidyl ethanolamine) confined to the inner side of the phospholipid bilayer *(42,43)*. An early hallmark of apoptosis is the externalization of PS from the cytoplasmic to the extracellular side of the lipid bilayer. In fact, this inner-to-outer switch of PS precedes cytoplasmic shrinkage, chromatin condensation, DNA fragmentation, and apoptotic bodies *(44)*. These externalized PS markers act as signals for phagocytic cells for subsequent removal *(44)*. Several ligands for PS have been developed to take advantage of this phenomenon, the most noted being annexin V.

Annexin V is a member of the calcium and phospholipid binding superfamily of annexin proteins *(44)*. This protein is mainly found intracellularly on the cytosolic side of plasma membranes and is expressed in a wide variety of cell types. The biological activity of annexin as an anticoagulant is attributed to its high affinity for negatively charged phospholipids, specifically PS *(44–47)*. This observation has led to the development of fluorescent and radioligand tags (e.g., FITC), which are now routinely used to identify apoptotic cells. Tagging apoptotic cells with annexin-conjugated molecules has allowed for identification of these cells with microscopy, immunohistochemistry, and FC. In glioma research, FC has been used to identify a fluorochrome-conjugated annexin V bound to apoptotic cells *(48)*. However, necrosis also results in externalization of PS, making it impossible to differentiate between apoptosis and necrosis by PS tagging alone.

5. FLOW CYTOMETRY

FC is commonly used to identify apoptotic cells. The flow cytometer makes rapid measurements on cells as they move in single file in a fluid stream past a sensor. A laser source is applied to the samples, and the resultant light scatter of the cells is measured. FC analysis provides separate measurements on each particle rather than average values for a whole population *(49)*. FC has the ability to measure multiple cellular parameters and to purify subpopulations. In the application of identifying cells undergoing apoptosis, FC is used primarily to measure immunocytochemically detected proteins (e.g., members of the Bcl-2/Bax family, products of tumor suppressor genes, and proto-oncogenes) that play a role in programmed cell death. It is also used to study cell functions such as mitochondrial changes, which are closely linked to mechanisms regulating apoptosis. FC offers the advantage of simultaneous analysis of several cell attributes that are studied during cell death on an individual cell. There are several disadvantages to FC measurement of apoptosis *(50)*. These include an inability to perform multiple measurements on the same cell, an inability to provide information on spatial resolution of fluorochromes within the cell, the inability to restain and remeasure cells previously stained, and a significant loss of cells from the repeated centrifugation processes that must be performed *(51)*. As mentioned previously, cells undergoing apoptosis demonstrate characteristic morphological changes. FC relies on the identification of a single biomechanical or molecular parameter to identify an apoptotic cell rather than cellular morphology. Other techniques, such as LSC (discussed in Section 6), offer the advantage of visualizing cell morphology.

5.1. Apoptotic Cells at Different Stages Have Different FC Light-Scatter Characteristics

Apoptotic cells undergo dehydration that causes them to shrink and elongate *(6,12)*. It is this cell shrinkage that causes the light scatter of the FC laser. Forward light scatter is a marker of cell size and is decreased in apoptotic cells. Because the decrease is proportional to the duration of apoptosis, minimal light scatter is seen in late apoptotic cells *(52,53)*.

Condensation of chromatin is followed by nuclear breakdown into several fragments of heterogenous size. As a result of this nuclear breakdown, the cell is able to scatter laser light at 90°, a phenomenon known as side scatter. Side scatter transiently increases during apoptosis and likely represents the change in reflectivity and light refraction of the condensed chromatin *(52)*. As apoptosis progresses, cells discard apoptotic bodies and rid themselves of chromatin fragments, thus

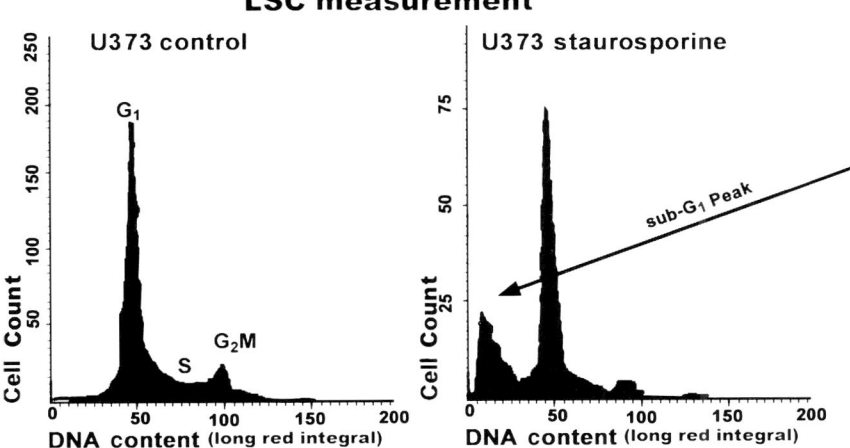

Fig. 2. Detection of apoptotic cells by measuring DNA content. Apoptosis of U373 cells was induced by their exposure to 100 μM staurosporine for 12 h. Sub-G_1 peak on the LSC represents cells with fractional DNA content (apoptotic). These cells are then located by the machine and their apoptosis is confirmed by morphological examination under the LSC microscope to separate the genuine from false-positive apoptotic cells. (Reprinted with permission from ref. *11*.) (Please *see* companion CD for color version of this figure.)

decreasing the intensity of side scatter. Although FC can analyze apoptotic cells based on forward and side light scatter without analyzing cell fluorescence, it cannot differentiate between the light scatter of late apoptotic cells and necrotic cells. LSC, however, can analyze the differences morphologically, and thus distinguish between apoptotic and necrotic cells.

5.2. The FC Sub-G_1 Peak, an Indication of Apoptosis

Fragmented DNA resulting from activation of endonucleases during apoptosis can be extracted from the cell after fixation and permeabilization. Staining the cell with DNA fluorochromes can detect the decreasing amount of DNA during apoptosis. Because the amount of DNA fragmentation varies according to cell type, the degree of separation of apoptotic from live cells identified by this method also varies. In glioma cells, it is possible to enhance the extraction of fragmented DNA by rinsing the cells in high-molarity phosphate–citrate buffer *(54)*. Cell permeabilization is also required with the use of detergents prior to staining. The sub-G_1 peak on the DNA frequency histogram located left of the peak representing G_1 cells represents the fractional DNA content of apoptotic cells (Fig. 2). This is a frequently used method to identify and quantify apoptotic cells by FC *(55–59)*. The sub-G_1 peak may also include cells which are damaged that have a lower DNA content, or that have a different chromatin structure, in which the accessibility of DNA fluorochromes is diminished. This lowers the specificity of FC in identifying apoptotic cells *(59)*. LSC has an advantage over FC in this regard because it can verify apoptotic cells based on their morphology. This feature is of special importance when markers for apoptotic cells are not very specific, as in the case of the sub-G_1 cell population.

5.3. Combined FC Staining for Early vs Late Apoptosis: Propidium Iodide and Annexin V

Propidium iodide (PI) is a membrane-impermeable dye that stains by intercalating into nucleic acid molecules, thus staining the DNA of cells with compromised cellular membranes (e.g., late apoptosis). By combining PI staining with a FITC-conjugated annexin V, it is possible to detect nonapoptotic live cells (PI-negative/FITC-negative), early apoptotic cells (PI-negative/FITC-positive), and late apoptotic cells (PI-positive/FITC-positive). The same staining method can be used in analysis by LSC. However, in FC, necrotic cells behave like late apoptotic cells (e.g., PI-positive/FITC-positive) *(45,60)*.

6. LASER SCANNING CYTOMETRY

LSC combines the advantages of FC with the ability to visualize cell morphology. The LSC uses a microscope-based cytofluorometer. Unlike FC, which requires cells be suspended in a fluid stream, LSC analysis is performed on slides. Therefore, cells may be reidentified and reexamined. Because the cells are mounted on slides, cellular morphology may be observed *(33)*. Major applications of LSC include analysis of enzyme kinetics, cellular drug uptake, ligand binding of molecules *(33)*, activation/deactivation/translocation of macromolecules during apoptosis (e.g., cytoplasmic–nuclear translocation of p53 or cytoplasmic–mitochondrial translocation of Bax) *(61–63)*, and analysis of tissue sections in pathology *(64,65)*. LSC has become the primary instrument used to study apoptosis, and as a result, many cytometric methods have been modified and adapted so that they can be used with the machine *(29,62,66,67)*.

6.1. LSC and Gliomas

Glioma cells are fixed in formaldehyde and ethanol. After mounting to microscope slides, they are incubated with certain fluorochromes. LSC can then identify nuclear or cellular fluorescence by measuring light intensity. Cells undergoing apoptosis demonstrate more intense nuclear fluorescence because of the higher degree of chromatin condensation *(29,33)*. Identification of apoptotic cells by LSC, based on the highest pixel value of DNA-associated fluorescence, is achieved in conjunction with DNA-content analysis. Thus, DNA ploidy and the cell cycle position of apoptotic and nonapoptotic cells can be determined at the same time as the estimate of apoptotic index. Furthermore, given that the staining procedure is simple, probing the cells with fluorochromes of different color to analyze other constituents of the cell can be combined with measurements of DNA-associated fluorescence *(67)*.

Molecular evaluation of metabolic events can be performed in cells that have not been fixed before analysis. These events include mitochondrial metabolism or plasma membrane transport functions *(68–72)*. The cells are again affixed to a microscope slide (unlike FC, which requires the cells be suspended in reaction media). Spatial relationships are therefore maintained and the cell can be examined for morphological information, probed by another fluorochrome, or analyzed by light microscopy. Live cells are adhered to slides by growing them up either on microscope slides or on cover slips. The materials used to culture cells on slides are commercially available and essentially involve a cell culture chamber that has a microscope slide as its base. The chamber can be removed from the bottom, leaving the cells adherent to the slide, which is then covered with a cover slip. Nevertheless, unfixed, live cells undergoing the late stages of apoptosis may detach from the slide and be lost for analysis.

LSC offers the ability to measure up to 100 cells per second with an accuracy and sensitivity comparable with FC *(72,73)*. Although both systems are able to measure fluorescence intensity, time of measurement, and light scatter, only FC is able to measure right-angle (side) scatter. LSC, however, is able to measure individual pixel values *(29)*. Individual pixel values provide information on fluorochrome distribution and the maximum concentration per area imaged on a single pixel. The ability to distinguish fluorescence between the nucleus and cytoplasm is also unique to LSC *(74)*.

As mentioned previously, the most striking characteristic of LSC that distinguishes it from FC is that cell analysis is done on a slide, offering the possibility of visual cell examination to assess morphology and correlate it with measured parameters. LSC also allows cell-image capture and comparison analysis. Furthermore, additional cytofluorimetric analysis of the same sample of cells is possible using new sets of markers and changing the contouring thresholds *(29)*. The results of the sequential measurements can then be integrated in list-mode fashion by means of the merge capability of the instrument.

7. PARP CLEAVAGE INDICATES EARLY APOPTOSIS

PARP is a nuclear enzyme involved with DNA repair and endonuclease inhibition that is activated in response to DNA damage *(23,75)*. In the early stages of apoptosis, PARP is cleaved by caspases, primarily caspase-3 *(23,76)*, resulting in distinct fragments (*see* details in Table 1) *(77,78)*. These PARP fragments can be detected both immunohistochemically and electrophoretically and are considered a hallmark for programmed cell death *(23,46,47,79,80)*. Furthermore, PARP overactivation consumes adenosine triphosphate, leading to energy depletion and necrosis, whereas PARP cleavage conserves energy and is seen in apoptosis *(46,81)*. Glioma studies have shown that PARP is used in repairing the DNA damage of radiation and chemotherapeutic agents *(40,82)*.

Quantification of apoptosis can be achieved by immunohistochemically staining for cleaved PARP fragments and measuring marked cells with FC or LSC *(83)*. FC and LSC not only allow for the detection of cells with PARP fragments but also allow for the direct comparison with phase of cell cycle, temporal sequence of apoptotic events, and time elapsed between the administration of the inducer and the observed effect. Western blot analysis of PARP can be accomplished with antibodies directed toward both the intact protein and its cleaved fragments *(83–87)*. Autoradiographs are digitized and analyzed with commercially available software. Progressive cleavage of PARP can be observed and quantified by comparing the density difference of bands corresponding to intact and cleaved fragments.

Pathways resulting in PARP cleavage, caspase activation, and DNA cleavage vary in different cell types. In some apoptotic cell types, there is a lack of association between the collapse of mitochondrial membrane potential and PARP cleavage or DNA fragmentation. In these cells, caspases can be activated in the absence of mitochondrial membrane potential *(88)*. Such studies have become possible because of the "file merge" capability of LSC, which allows for correlation of the decrease in the mitochondrial transmembrane potential (in the same cell) with the presence or absence of PARP p89. It has not yet clearly been recognized whether in glioma cells, change in mitochondrial transmembrane potential is a prerequisite for caspase activation and cleavage of PARP. Nevertheless, it has been documented that all of these biological systems are activated when glioma cells undergo apoptosis *(82,89–91)*.

8. DETECTING THE ACTIVATED CASPASES OF APOPTOSIS

Caspases are a family of cysteine-containing aspartate-specific proteases that play a central role in apoptosis. The caspases are synthesized intracellularly as inactive proenzymes and are activated by cleavage at specific aspartate-cleaving sites. The active protease can in turn sequentially activate the next caspase. Caspase activation occurs early in the apoptotic process, making detection methods highly specific. Furthermore, caspase activation does not occur in necrosis, making it more reliable than detection of apoptosis by PS externalization. Methods for detecting activated caspases include fluorescent substrates, Western blotting, radiolabeled inhibitors, and anticaspase-specific antibodies *(44,76,92–96)*.

Fluorescence detection requires a fluorochrome-labeled substrate whose fluorescence is a result of caspase-induced cleavage *(97–100)*. Western blotting can also be used to identify the transcatalytic cleavage products of the pro-enzyme *(100,101)*. These two methods provide no information, however, on individual cells, the heterogeneity of cell populations, or the correlation with other cell attributes, especially when a cell-by-cell analysis of caspase activity would be required. Fluorescence-labeled and radiolabeled caspase inhibitors have been able to overcome some of these problems.

Labeled caspase inhibitors have been constructed for different members of the caspase family, each with different specificity. However, the group of fluoromethylketone (fmk) caspase inhibitors appears to be the most promising. Benzyloxycarbonyl-Val-Ala-DL-Asp (*O*-methyl)-fmk (Z-VAD-fmk) is one of the more commonly used caspase inhibitors *(44)*. The two fluorochrome-labeled inhibitors of caspases (FLICA), FAM-VAD-fmk (generic caspase inhibitor) and FAM-VEID-fmk (caspase-6 inhibitor), have been used to measure caspase activation *(94,100,102)*.

FLICA can be quantified by LSC by using an excitation laser to measure maximal pixel intensities *(72,74,94)*. LSC can also compare FLICA side-by-side with DNA fragmentation, detected by the TUNEL assay. Such comparisons might permit discovery of where and how a specific apoptosis-inducing drug can affect the stimulation of the caspase system in glioma cells. Furthermore, the strong correlation between FLICA and TUNEL apoptotic indices *(94)* implies that FLICA may not only be a rapid and convenient method of assessing caspase activation in individual cells but can also be used to estimate the frequency of apoptosis, providing that the LSC is available for comparative studies.

Anticleaved caspase-3 antibodies conjugated to phycoerythrin have been used to identify apoptotic glioma cell lines in vitro with FC *(95)*. The advantages of using activated caspase-3 as a marker for apoptosis in gliomas are several. First, in glioma cell lines, annexin V binding is seen late in the apoptotic process, whereas activated caspase-3 occurs early, allowing for detection of cells in the initial stages of apoptosis *(95)*. Second, nuclear morphology, DNA laddering, TUNEL assay, and Western blot methods can be subjective, whereas FC of glioma cells tagged for activated caspase-3 is not only highly specific for apoptosis but also allows for easy quantification at different time points *(95,96)*.

9. PLASMA MEMBRANE PERMEABILITY INCREASES WITH APOPTOSIS

During apoptosis, the plasma membrane permeability of some cells, including glioma cells, is increased for certain fluorochromes (e.g., 7-amino actinomycin D, Hoechst 33342 and 33258, and PI) *(71,103–106)*. This increased permeability in apoptotic cells is probably a result of changes in the structural and transport functions of the plasma membrane. This characteristic can allow for the differentiation between live cells, early apoptotic cells, and late apoptotic cells by FC. The fluorescence staining of cells progressively strengthens as the cells approach the end of the apoptotic process. Therefore, a late apoptotic cell has a stronger fluorescence than an early one. All of these fluorochrome-using assays are excited by lasers emitting either a blue or a red light and could be adapted to LSC.

10. FAILURE OF THE MITOCHONDRIAL MEMBRANE: FLUOROCHROME UPTAKE AND THE TRANSLOCATION OF Bax AND CYTOCHROME-*c*

Mitochondrial membrane failure is thought to be the point of no return for programmed cell death *(11)*. This process results in increased permeability for fluorochromes and the translocation of Bax to the mitochondria and the escape of cytochrome-*c* to the cytosol. During the early stages of apoptosis, the pro-apoptotic regulatory protein Bax undergoes translocation into the mitochondria *(61,107)*, where it is involved in the decrease of mitochondrial transmembrane potential and opening of mitochondrial pores. This event results in increased mitochondrial permeability, releasing cytochrome-*c* from the mitochondria to the cytoplasm. Cytochrome-*c* is associated with activation of the caspase cascade *(91,108–110)*.

Increased mitochondrial permeability can be observed with several types of mitochondrial membrane-permeable fluorochromes. These cationic probes accumulate in the mitochondria of live cells. Therefore, the intensity of cellular fluorescence reflects the change in the mitochondrial transmembrane potential *(111,112)*. FITC-conjugated anti-Bax antibodies have been used in glioma cells and can be correlated with cell cycle by counterstaining with PI *(11)*. Thus, cellular DNA content analysis, followed by bivariate analysis of the scattergrams, makes it possible to correlate Bax translocation induced by the desired toxic agent with cell cycle position. The level of Bax and the relative levels of expression of its mRNA in glioma cells have been measured by means of Western and Northern blot analysis *(106,113)*. The maximal pixel of Bax immunofluorescence in the cytoplasm of the cells undergoing apoptosis can be measured by LSC *(61)*. Finally, LSC has been used to detect the translocation of cytochrome-*c* from the mitochondria to the cytoplasm *(114)*.

11. DNA DENATURATION INCREASES IN APOPTOTIC CELLS

The condensed DNA of apoptotic cells is more sensitive to denaturation than that of nonmitotic live cells *(91,115–117)*. Two methods of exploiting this phenomenon are available: immunohistochemistry and acridine orange (AO). Antibodies directed against single-stranded DNA can identify the significant amount of denatured DNA in apoptotic cells *(118)*. The DNA fluorochrome AO differentially stains double- and single-stranded nucleic acids *(117)*. A green fluorescence indicates double-stranded DNA staining, whereas red fluorescence indicates AO staining of single-stranded DNA. This process requires that cells be stained at a low pH to prevent DNA renaturation *(117)*. Because apoptotic cells have a larger fraction of DNA in their denatured form, they emit a stronger red and a weaker green fluorescence than nonapoptotic live cells *(115,117)*. Therefore, analyzing total cell fluorescence (i.e., red plus green, which represents total DNA content) vs the ratio of red fluorescence to the total fluorescence allows apoptosis to be related to cell cycle position.

12. NF-κB IS ACTIVATED AND TRANSLOCATED TO THE NUCLEUS TO PROTECT AGAINST APOPTOSIS

The transcriptional activator nuclear factor κB (NF-κB) is involved in the cellular response to a variety of stressful stimuli such as cytokines, cytotoxic agents, and some physiological agents *(119–122)*. NF-κB is sequestered in the cytoplasm and, in response to extracellular apoptotic-inducing signals (e.g., TNF), is released from its inhibitor (IκB) and translocated from the cytoplasm to the nucleus. This activity is believed to result in the protection of cells against apoptosis *(123–126)*. Activation of NF-κB can be detected immunocytochemically with FITC-tagged antibody, and the ratio of the nuclear/cytoplasmic fluorescence can be monitored by LSC *(62)*. In malignant glioma cells, activation of this anti-apoptotic gene protects the cell against the apoptotic effects of TNF *(120)*. The use of LSC is a novel approach for measuring activation and translocation of such factors. Unlike LSC and FC methods, other methods of measuring NF-κB activation suffer shortcomings such as the inability to provide information on cellular heterogeneity, on activation of different cell subpopulations, and on the relationship of factor activation to cell cycle position *(127)*. LSC also offers insight into the mechanism of activation of NF-κB by revealing the intracellular pathways involved in the response to TNF. Furthermore, this measurement can be combined with morphological identification and correlated to cell cycle position. Because activation of NF-κB complex plays a crucial role in the progression of cell cycle in malignant glioma cells *(120,125)*, identification and analysis of this event is important to the study of apoptosis.

13. p53 AS A MARKER FOR CELLULAR STRESS

In wild-type cells, the p53 tumor suppressor protein is expressed at low levels in the nucleus *(128–130)*. Because of increased half-life, glioma cells with p53 mutations have increased levels of this protein, which can be present in the cytoplasm *(131)*. Levels of p53 increase in response to cellular stress and DNA damage, and this increase could be used as a marker of the apoptotic and genotoxic effects of different agents *(132)*. By using the LSC to monitor the expression of both the nuclear and cytoplasmic expression of p53, the stress effects that treatments have on cell subtypes can be monitored. Consequently, in addition to DNA-associated fluorescence, the integrated p53 immunofluorescence emitted from the nucleus and cytoplasm should be measured separately.

A concern about the use of adenoviral-mediated p53 gene therapy is whether therapeutically effective levels of adenoviral p53 vector are delivered to the tumor. Using intracellular stains, LSC has proven a reliable and useful tool for measuring adenoviral protein expression when determining infectivity and for measuring p53 protein expression when determining the expression p53 transgene *(133)*.

14. A NOVEL FLUORESCENCE CONSTRUCT FOR NON-INVASIVE REAL-TIME IMAGING OF APOPTOSIS

Laxman et al. *(134)* developed a recombinant luciferase (Luc) reporter molecule whose luminescence is suppressed by two estrogen receptor regulatory domain sequences located on either side of the Luc sequence. A protease cleavage site for caspase-3 is placed on either side of the luciferase sequence. Caspase-3-specific cleavage of the estrogen sequences results in the release of the Luc molecule with activation of fluorescence. The fluorescence is detected using bioluminescence imaging. The cleavage site of this construct effectively reports on the activity of caspase-3-induced apoptosis. These investigators went on to create D54 human glioma cells expressing Luc. In vitro modeling using this cell line showed that Luc was activated on caspase-3 induction. In vivo studies showed that induction of apoptosis could be detected noninvasively by using this novel bioluminescence imaging technique *(134)*. The ability to detect apoptosis noninvasively and dynamically provides an opportunity to test proapoptotic and antiapoptotic compounds in both intact cells and animal models in an ongoing experimental manner. This is in contrast to other tools for detecting apoptosis that require fixed tissue.

15. CLINICAL APPLICATIONS OF DETECTING APOPTOSIS

15.1. Clinical Relevance of Apoptotic Studies in Gliomas

Given the existence of an association between clinical outcome in patients with malignant gliomas and the apoptotic rate of their tumors *(135)*, studies of apoptotic index are likely to yield important information for enhancing the design of more effective glioma therapies. Diagnosis in anatomical pathology relies heavily on the examiner's experience in analyzing histological sections. Therefore, great potential exists for the development of computer-assisted histological section analysis *(136,137)*.

15.2. DNA Fragmentation and Apoptosis

Identifying apoptosis by means of DNA strand breaks is currently the preferred method for many clinical applications *(24,138,139)*, including analysis of glioma tumor cells *(37,38,140)*. LSC allows samples to be collected at different times and then fixed, stored, transported, and examined at another time. The possibility of confirming apoptosis by image analysis and morphological criteria, as provided by LSC, is even more important for clinical material than for experimental model systems. The use of LSC in tissue-section analysis is particularly helpful to view specific areas of interest, which may be only a minor component of the total tumor section.

15.3. Bax, PARP, Caspases, and Other Molecules Involved in Apoptosis of Gliomas

In glioma cells, as in other cell types, the translocation of the proapoptotic member of the Bcl-2 protein family (Bax) to the mitochondria appears to be a critical step in the process of apoptosis *(91,116)*. This event triggers the irreversible step of caspase activation and ultimate programmed cell death. Bax translocation can be measured and analyzed using immunofluorescence by FC and LSC *(61)*. The different patterns of Bax and Bcl-2 expression may reflect histiogenic differences in different types of brain tumors *(141)*. The level of Bax expression in glioma cell lines may correlate with the sensitivity of these cells to different forms of treatment-induced apoptosis *(142)*. Modulation of Bax expression may potentially be a useful therapeutic modality for gliomas; human glioma cell lines that were initially resistant to apoptosis then transfected with proapoptotic gene Bax showed an increased sensitivity to apoptosis *(107,113)*.

PARP cleavage is almost always associated with the collapse of the mitochondrial membrane, Bax translocation, subsequent DNA cleavage, and apoptotic death *(85,87,88)*. Caspase-3 activation and PARP cleavage are detectable in some brain tumors much earlier than any other morphological and biochemical indices of apoptosis *(79,95)*. In glioma cells, PARP is also required for rapid accumulation of p53, activation of p53 sequence-specific DNA binding, and p53 transcriptional activity after DNA damage *(40)*. PARP cleavage could therefore be used as a sensitive parameter for identifying different types of cell death and as a marker for the activation of different death proteases.

Most malignant gliomas express a high level of the death receptor Fas, whereas the surrounding normal neurons and astrocytes express a very low level of Fas receptors *(143)*. Different subpopulations of gliomas may have different Fas/Fas-L expression patterns *(144,145)*, which could be used for assessment of vulnerability of these cells to apoptosis *(107,146–148)*.

15.4. Combination of the LSC and FC in Apoptotic Studies

Because the use of LSC can be extended to include monitoring the spatial localization of intracellular fluorochromes and the highest local concentrations by maximal pixel analysis, changes in the abundance and localization of particular proteins can be detected in glioma cells. LSC allows the integration and merging of the results of two or more measurements, and thus the nuclear expression and the cytoplasmic or total cellular expression of the measured factor can be compared.

LSC subjects cells to morphological examination—still the uncontested criterion for identifying apoptosis. On this basis, LSC is favored above other types of instrumentation for apoptotic analysis. However, because LSC is slower, FC is preferable when large numbers of samples must be analyzed in a limited time. In this context, LSC and FC complement each other. Used together, they expand the ability to study diverse biological materials.

16. CONCLUSION

Many new treatment strategies for glioma tumors stem from the use of techniques aimed at manipulating apoptosis. Used alone or in combination with other treatments, they may ultimately succeed in prolonging survival of patients with gliomas. Being able to assess the efficacy of experimental treatments with refined techniques and being able to use instruments that can provide accurate measurements of the apoptotic markers will open the door for discovering novel strategies with the potential to induce effective and selective cytotoxicity. The use of assays that directly probe apoptotic mechanisms should enable further understanding of the molecular determinants of cell death and provide avenues to manipulate these determinants.

ACKNOWLEDGMENTS

Portions of this chapter have been adapted from ref. *11*. The authors thank Kristin Kraus for her excellent editorial guidance and Rozita Naghshin for assistance in preparing figures.

REFERENCES

1. Davis FG, McCarthy BJ, Freels S, Kupelian V, Bondy ML. The conditional probability of survival of patients with primary malignant brain tumors: surveillance, epidemiology, and end results (SEER) data. Cancer 1999;85:485–491.
2. Nieder C, Grosu AL, Molls M. A comparison of treatment results for recurrent malignant gliomas. Cancer Treat Rev 2000;26:397–409.
3. Walker MD, Alexander E, Hunt WE, et al. Evaluation of BCNU and/or radiotherapy in the treatment of anaplastic gliomas. A cooperative clinical trial. J Neurosurg 1978;49:333–343.
4. Walker MD, Green SB, Byar DP, et al. Randomized comparisons of radiotherapy and nitrosoureas for the treatment of malignant glioma after surgery. N Engl J Med 1980;303:1323–1329.
5. Kaufmann SH, Gores GJ. Apoptosis in cancer: cause and cure. Bioessays 2000;22;1007–1017.
6. Majno G, Joris I. Apoptosis, oncosis, and necrosis. An overview of cell death. Am J Pathol 1995;146:3–15.
7. Ekert PG, Vaux DL. Apoptosis, haemopoiesis and leukaemogenesis. Baillieres Clin Haematol 1997;10:561–576.
8. Smith JS, Jenkins RB. Genetic alterations in adult diffuse glioma: occurrence, significance, and prognostic implications. Front Biosci 2000;5:D213–D231.

9. Vaux DL. Immunopathology of apoptosis—introduction and overview. Springer Semin Immunopathol 1997;19:271–278.
10. Shu HK, Kim MM, Chen P, Furman F, Julin CM, Israel MA. The intrinsic radioresistance of glioblastoma-derived cell lines is associated with a failure of p53 to induce p21(BAX) expression. Proc Natl Acad Sci USA 1998;95:14,453–14,458.
11. Amirlak B, Couldwell WT. Apoptosis in glioma cells: review and analysis of techniques used for study with focus on the laser scanning cytometer. J Neurooncol 2003;63:129–145.
12. Kerr JF, Wyllie AH, Currie AR. Apoptosis: a basic biological phenomenon with wide-ranging implications in tissue kinetics. Br J Cancer 1972;26:239–257.
13. Saraste A, Pulkki K. Morphologic and biochemical hallmarks of apoptosis. Cardiovasc Res 2000;45:528–537.
14. Dragovich T, Rudin CM, Thompson CB. Signal transduction pathways that regulate cell survival and cell death. Oncogene 1998;17:3207–3213.
15. Golding SE, Rosenberg E, Khalil A, et al. Double strand break repair by homologous recombination is regulated by cell cycle-independent signaling via ATM in human glioma cells. J Biol Chem 2004;279:15,402–15,410.
16. Tribius S, Pidel A, Casper D. ATM protein expression correlates with radioresistance in primary glioblastoma cells in culture. Int J Radiat Oncol Biol Phys 2001;50:511–523.
17. Takeuchi H, Kanzawa T, Kondo Y, et al. Combination of caspase transfer using the human telomerase reverse transcriptase promoter and conventional therapies for malignant glioma cells. Int J Oncol 2004;25:57–63.
18. Collins VP. Brain tumours: classification and genes. J Neurol Neurosurg Psychiat 2004;75(Suppl 2):ii2–ii11.
19. Strik H, Deininger M, Streffer J, et al. BCL-2 family protein expression in initial and recurrent glioblastomas: modulation by radiochemotherapy. J Neurol Neurosurg Psychiat 1999;67:763–768.
20. Roth W, Grimmel C, Rieger L, et al. Bag-1 and Bcl-2 gene transfer in malignant glioma: modulation of cell cycle regulation and apoptosis. Brain Pathol 2000;10:223–234.
21. Glaser T, Weller M. Caspase-dependent chemotherapy-induced death of glioma cells requires mitochondrial cytochrome c release. Biochem Biophys Res Commun 2001;281:322–327.
22. Reed JC. Mechanisms of apoptosis avoidance in cancer. Curr Opin Oncol 1999;11:68–75.
23. Pozarowski P, Grabarek J, Darzynkiewicz Z. Flow cytometry of apoptosis. In: Robinson P, Darzynkiewicz Z, Hyun W, Orfao A, Rabinovitch P (eds). Current Protocols in Cytometry. New York, NY: John Wiley & Sons, 2003, pp 7.19.11–7.19.33.
24. Li X, Gong J, Feldman E, Seiter K, Traganos F, Darzynkiewicz Z. Apoptotic cell death during treatment of leukemias. Leuk Lymphoma 1994;13(Suppl 1):65–70.
25. Seiter K, Feldman EJ, Halicka HD, et al. Phase I clinical and laboratory evaluation of topotecan and cytarabine in patients with acute leukemia. J Clin Oncol 1997;15:44–51.
26. Oberhammer F, Wilson JW, Dive C, et al. Apoptotic death in epithelial cells: cleavage of DNA to 300 and/or 50 kb fragments prior to or in the absence of internucleosomal fragmentation. EMBO J 1993;12:3679–3684.
27. Kuribayashi N, Sakagami H, Iida M, Takeda M. Chromatin structure and endonuclease sensitivity in human leukemic cell lines. Anticancer Res 1996;16:1225–1230.
28. Darzynkiewicz Z, Li X. Measurements of cell death by flow cytometry. In: Cotter TG (ed). Techniques in Apoptosis. A user's guide. London: Portland Press, 1996, pp 71–106.
29. Darzynkiewicz Z, Bedner E. Analysis of apoptotic cells by flow and laser scanning cytometry. Methods Enzymol 2000;322:18–39.
30. Darzynkiewicz Z, Bruno S, Del-Bino G, et al. Features of apoptotic cells measured by flow cytometry. Cytometry 1992;13:795–808.
31. Darzynkiewicz Z, Li X, Gong J. Assays of cell viability: discrimination of cells dying by apoptosis. Methods Cell Biol 1994;41:15–38.
32. Gorczyca W, Deptala A, Bedner E, Li X, Melamed MR, Darzynkiewicz Z. Analysis of human tumors by laser scanning cytometry. Methods Cell Biol 2001;64:421–443.
33. Bedner E, Melamed MR, Darzynkiewicz Z. Enzyme kinetic reactions and fluorochrome uptake rates measured in individual cells by laser scanning cytometry. Cytometry 1998;33:1–9.
34. Gold R, Schmied M, Giegerich G, et al. Differentiation between cellular apoptosis and necrosis by the combined use of in situ tailing and nick translation techniques. Lab Invest 1994;71:219–225.

35. Mundle SD, Raza A. The two in situ techniques do not differentiate between apoptosis and necrosis but rather reveal distinct patterns of DNA fragmentation in apoptosis. Lab Invest 1995;72:611–613.
36. Iida M, Doi H, Asamoto S, et al. Endonuclease activity and hydrogen peroxide-induced cytotoxicity in human glioblastoma and glioma cell lines. Anticancer Res 1999;19:1235–1240.
37. Casper D, Lekhraj R, Yaparpalvi US, et al. Acetaminophen selectively reduces glioma cell growth and increases radiosensitivity in culture. J Neuro-Oncol 2000;46:215–229.
38. Ehrmann J, Rihakova P, Hlobilkova A, Kala M, Kolar Z. The expression of apoptosis-related proteins and the apoptotic rate in glial tumors of the brain. Neoplasma 2000;47:151–155.
39. Thomas LB, Gates DJ, Richfield EK, O'Brien TF, Schweitzer JB, Steindler DA. DNA end labeling (TUNEL) in Huntington's disease and other neuropathological conditions. Exp Neurol 1995; 133:265–272.
40. Wang J, Hu L, Gupta N, et al. Induction and characterization of human glioma clones with different radiosensitivities. Neoplasia 1999;1:138–144.
41. Ge S, Rempel SA, Divine G, Mikkelsen T. Carboxyamido-triazole induces apoptosis in bovine aortic endothelial and human glioma cells. Clin Cancer Res 2000;6:1248–1254.
42. Lockshin RA, Zakeri Z, Tilly JL. When Cells Die: A Comprehensive Evaluation of Apoptosis and Programmed Cell Death. New York, NY: Wiley-Liss, 1998.
43. Fadok VA, Voelker DR, Campbell PA, Cohen JJ, Bratton DL, Henson PM. Exposure of phosphatidylserine on the surface of apoptotic lymphocytes triggers specific recognition and removal by macrophages. J Immunol 1992;148:2207–2216.
44. Lahorte CM, Vanderheyden JL, Steinmetz N, Van De Wiele C, Dierckx RA, Slegers G. Apoptosis-detecting radioligands: current state of the art and future perspectives. Eur J Nucl Med Mol Imaging 2004;31:887–919.
45. van Engeland M, Nieland LJ, Ramaekers FC, Schutte B, Reutelingsperger CP. Annexin V-affinity assay: a review on an apoptosis detection system based on phosphatidylserine exposure. Cytometry 1998;31:1–9.
46. Oliver FJ, Menissier de Murcia J, de Murcia G. Poly(ADP-ribose) polymerase in the cellular response to DNA damage, apoptosis, and disease. Am J Hum Genet 1999;64:1282–1288.
47. Lautier D, Lagueux J, Thibodeau J, Menard L, Poirier GG. Molecular and biochemical features of poly (ADP-ribose) metabolism. Mol Cell Biochem 1993;122:171–193.
48. Kawaguchi S, Mineta T, Ichinose M, Masuoka J, Shiraishi T, Tabuchi K. Induction of apoptosis in glioma cells by recombinant human Fas ligand. Neurosurgery 2000;46:431–438; discussion 438–439.
49. Ormerod MG. Flow Cytometry: A Practical Approach. Oxford, UK: Oxford University Press, 2000.
50. Darzynkiewicz Z, Tragos F. Measurement of apoptosis. In: Scheper T (ed). Advances in Biochemical Engineering/Biotechnology. Heidelberg, Germany: Springer-Verlag, 1998, pp 35–73.
51. Matsumura K, Kawamoto K. Long-term passage results of glioma cells and their cell kinetics. Hum Cell 1994;7:158–166.
52. Salzman GC, Singham SB, Johnston RG, Bohren CF. Light scattering and cytometry. In: Flow Cytometry and Cell Sorting (Melamed MR, Lindmo T, Mendelsohn ML, eds.), New York: Wiley-Liss, 1990; pp 81–107.
53. Ormerod MG, Paul F, Cheetham M, Sun XM. Discrimination of apoptotic thymocytes by forward light scatter. Cytometry 1995;21:300–304.
54. Gong J, Traganos F, Darzynkiewicz Z. A selective procedure for DNA extraction from apoptotic cells applicable for gel electrophoresis and flow cytometry. Anal Biochem 1994;218:314–319.
55. Nicoletti I, Migliorati G, Pagliacci MC, Grignani F, Riccardi C. A rapid and simple method for measuring thymocyte apoptosis by propidium iodide staining and flow cytometry. J Immunol Methods 1991;139:271–279.
56. Honda C, Tabuchi K. Effects of cisplatin on cultured glioma cells. Gan To Kagaku Ryoho 1986;13:1921–1926.
57. Kawamoto K. Flow cytometric analysis of cell cycle for the action mechanism of antineoplastic agents. Hum Cell 1995;8:85–88.
58. Kin Y, Chintala SK, Go Y, et al. A novel role for the urokinase-type plasminogen activator receptor in apoptosis of malignant gliomas. Int J Oncol 2000;17:61–65.
59. Bedner E, Burfeind P, Gorczyca W, Melamed MR, Darzynkiewicz Z. Laser scanning cytometry distinguishes lymphocytes, monocytes, and granulocytes by differences in their chromatin structure. Cytometry 1997;29:191–196.

60. Koopman G, Reutelingsperger CP, Kuijten GA, Keehnen RM, Pals ST, van Oers MH. Annexin V for flow cytometric detection of phosphatidylserine expression on B cells undergoing apoptosis. Blood 1994;84:1415–1420.
61. Bedner E, Li X, Kunicki J, Darzynkiewicz Z. Translocation of Bax to mitochondria during apoptosis measured by laser scanning cytometry. Cytometry 2000;41:83–88.
62. Deptala A, Bedner E, Gorczyca W, Darzynkiewicz Z. Activation of nuclear factor kappa B (NF-kappaB) assayed by laser scanning cytometry (LSC). Cytometry 1998;33:376–382.
63. Deptala A, Li X, Bedner E, Cheng W, Traganos F, Darzynkiewicz Z. Differences in induction of p53, p21WAF1 and apoptosis in relation to cell cycle phase of MCF-7 cells treated with camptothecin. Int J Oncol 1999;15:861–871.
64. Abdel-Moneim I, Melamed MR, Darzynkiewicz Z, Gorczyca W. Proliferation and apoptosis in solid tumors. Analysis by laser scanning cytometry. Anal Quant Cytol Histol 2000;22:393–397.
65. Clatch RJ, Foreman JR, Walloch JL. Simplified immunophenotypic analysis by laser scanning cytometry. Cytometry 1998;34:3–16.
66. Furuya T, Kamada T, Murakami T, Kurose A, Sasaki K. Laser scanning cytometry allows detection of cell death with morphological features of apoptosis in cells stained with PI. Cytometry 1997;29:173–177.
67. Li X, Melamed MR, Darzynkiewicz Z. Detection of apoptosis and DNA replication by differential labeling of DNA strand breaks with fluorochromes of different color. Exp Cell Res 1996;222:28–37.
68. Liang BC, Ullyatt E. Chemosensitization of glioblastoma cells to bis-dichloroethyl-nitrosourea with tyrphostin AG17. Clin Cancer Res 1998;4:773–781.
69. Miccoli L, Poirson Bichat F, Sureau F, et al. Potentiation of lonidamine and diazepam, two agents acting on mitochondria, in human glioblastoma treatment. J Natl Cancer Inst 1998;90:1400–1406.
70. Wolter KG, Hsu YT, Smith CL, Nechushtan A, Xi XG, Youle RJ. Movement of Bax from the cytosol to mitochondria during apoptosis. J Cell Biol 1997;139:1281–1292.
71. Ormerod MG, Sun XM, Snowden RT, Davies R, Fearnhead H, Cohen GM. Increased membrane permeability of apoptotic thymocytes: a flow cytometric study. Cytometry 1993;14:595–602.
72. Kamentsky LA, Burger DE, Gershman RJ, Kamentsky LD, Luther E. Slide-based laser scanning cytometry. Acta Cytol 1997;41:123–143.
73. Kamentsky LA, Kamentsky LD. Microscope-based multiparameter laser scanning cytometer yielding data comparable to flow cytometry data. Cytometry 1991;12:381–387.
74. Darzynkiewicz Z, Bedner E, Li X, Gorczyca W, Melamed MR. Laser-scanning cytometry: A new instrumentation with many applications. Exp Cell Res 1999;249:1–12.
75. Rice WG, Hillyer CD, Harten B, et al. Induction of endonuclease-mediated apoptosis in tumor cells by C-nitroso-substituted ligands of poly(ADP-ribose) polymerase. Proc Natl Acad Sci USA 1992;89:7703–7707.
76. Lazebnik YA, Kaufmann SH, Desnoyers S, Poirier GG, Earnshaw WC. Cleavage of poly(ADP-ribose) polymerase by a proteinase with properties like ICE. Nature 1994;371:346–347.
77. Rahaman SO, Harbor PC, Chernova O, Barnett GH, Vogelbaum MA, Haque SJ. Inhibition of constitutively active Stat3 suppresses proliferation and induces apoptosis in glioblastoma multiforme cells. 2002;21:8404–8413.
78. Ray SK, Patel SJ, Welsh CT, Wilford GG, Hogan EL, Banik NL. Molecular evidence of apoptotic death in malignant brain tumors including glioblastoma multiforme: upregulation of calpain and caspase-3. Neurosci Res 2002;69:197–206.
79. Bursztajn S, Feng JJ, Berman SA, Nanda A. Poly (ADP-ribose) polymerase induction is an early signal of apoptosis in human neuroblastoma. Brain Res Mol Brain Res 2000;76:363–376.
80. Jeggo PA. DNA repair: PARP—another guardian angel? Curr Biol 1998;8:R49–R51.
81. Ha HC, Snyder SH. Poly(ADP-ribose) polymerase is a mediator of necrotic cell death by ATP depletion. Proc Natl Acad Sci USA 1999;96:13,978–13,982.
82. Malapetsa A, Noe AJ, Poirier GG, Desnoyers S, Berger NA, Panasci LC. Identification of a 116 kDa protein able to bind 1,3-bis(2-chloroethyl)-1-nitrosourea-damaged DNA as poly(ADP-ribose) polymerase. Mutat Res 1996;362:41–50.
83. Sallmann FR, Bourassa S, Saint-Cyr J, Poirier GG. Characterization of antibodies specific for the caspase cleavage site on poly(ADP-ribose) polymerase: specific detection of apoptotic fragments and mapping of the necrotic fragments of poly(ADP-ribose) polymerase. Biochem Cell Biol 1997;75:451–456.
84. Affar EB, Duriez PJ, Shah RG, et al. Immunological determination and size characterization of poly(ADP-ribose) synthesized in vitro and in vivo. Biochim Biophys Acta 1999;1428:137–146.

85. Li X, Darzynkiewicz Z. Cleavage of poly(ADP-ribose) polymerase measured in situ in individual cells: relationship to DNA fragmentation and cell cycle position during apoptosis. Exp Cell Res 2000;255:125–132.
86. Duriez PJ, Desnoyers S, Hoflack JC, et al. Characterization of anti-peptide antibodies directed towards the automodification domain and apoptotic fragment of poly (ADP-ribose) polymerase. Biochim Biophys Acta 1997;1334:65–72.
87. Duriez PJ, Shah GM. Cleavage of poly(ADP-ribose) polymerase: a sensitive parameter to study cell death. Biochem Cell Biol 1997;75:337–349.
88. Li X, Du L, Darzynkiewicz Z. During apoptosis of HL-60 and U-937 cells caspases are activated independently of dissipation of mitochondrial electrochemical potential. Exp Cell Res 2000;257:290–297.
89. Hirsch T, Marchetti P, Susin SA, et al. The apoptosis-necrosis paradox. Apoptogenic proteases activated after mitochondrial permeability transition determine the mode of cell death. Oncogene 1997;15:1573–1581.
90. Hermisson M, Wagenknecht B, Wolburg H, Glaser T, Dichgans J, Weller M. Sensitization to CD95 ligand-induced apoptosis in human glioma cells by hyperthermia involves enhanced cytochrome c release. Oncogene 2000;19:2338–2345.
91. Ikemoto H, Tani E, Ozaki I, Kitagawa H, Arita N. Calphostin C-mediated translocation and integration of Bax into mitochondria induces cytochrome c release before mitochondrial dysfunction. Cell Death Differ 2000;7:511–520.
92. Kaufmann SH, Desnoyers S, Ottaviano Y, Davidson NE, Poirier GG. Specific proteolytic cleavage of poly(ADP-ribose) polymerase: an early marker of chemotherapy-induced apoptosis. Cancer Res 1993;53:3976–3985.
93. Alnemri ES, Livingston DJ, Nicholson DW, et al. Human ICE/CED-3 protease nomenclature. Cell 1996;87:171.
94. Zhivotovsky B, Samali A, Gahm A, Orrenius S. Caspases: their intracellular localization and translocation during apoptosis. Cell Death Differ 1999;6:644–651.
95. Zartman JK, Foreman NK, Donson AM, Fleitz JM. Measurement of tamoxifen-induced apoptosis in glioblastoma by cytometric bead analysis of active caspase-3. J Neurooncol 2004;67:3–7.
96. Belloc F, Belaud Rotureau MA, Lavignolle V, et al. Flow cytometry detection of caspase 3 activation in preapoptotic leukemic cells. Cytometry 2000;40:151–160.
97. Hug H, Los M, Hirt W, Debatin KM. Rhodamine 110-linked amino acids and peptides as substrates to measure caspase activity upon apoptosis induction in intact cells. Biochemistry 1999;38:13,906–13,911.
98. Liu J, Bhalgat M, Zhang C, Diwu Z, Hoyland B, Klaubert DH. Fluorescent molecular probes V: a sensitive caspase-3 substrate for fluorometric assays. Bioorg Med Chem Lett 1999;9:3231–3236.
99. Gorman AM, Hirt UA, Zhivotovsky B, Orrenius S, Ceccatelli S. Application of a fluorometric assay to detect caspase activity in thymus tissue undergoing apoptosis in vivo. J Immunol Methods 1999;226:43–48.
100. Earnshaw WC, Martins LM, Kaufmann SH. Mammalian caspases: structure, activation, substrates, and functions during apoptosis. Annu Rev Biochem 1999;68:383–424.
101. Bedner E, Smolewski P, Amstad P, Darzynkiewicz Z. Activation of caspases measured in situ by binding of fluorochrome-labeled inhibitors of caspases (FLICA): correlation with DNA fragmentation. Exp Cell Res 2000;259:308–313.
102. Budihardjo I, Oliver H, Lutter M, Luo X, Wang X. Biochemical pathways of caspase activation during apoptosis. Annu Rev Cell Dev Biol 1999;15:269–290.
103. Bowles AP, Pantazis CG, Wansley W, Allen MB. Chemosensitivity testing of human gliomas using a fluorescent microcarrier technique. J Neuro-Oncol 1990;8:103–112.
104. Fuse T, Yoon KW, Kato T, Yamada K. Heat-induced apoptosis in human glioblastoma cell line A172. Neurosurgery 1998;42:843–849.
105. Frey T. Nucleic acid dyes for detection of apoptosis in live cells. Cytometry 1995;21:265–274.
106. Yin D, Tamaki N, Kokunai T. Wild-type p53-dependent etoposide-induced apoptosis mediated by caspase-3 activation in human glioma cells. J Neurosurg 2000;93:289–297.
107. Shinoura N, Saito K, Yoshida Y, et al. Adenovirus-mediated transfer of bax with caspase-8 controlled by myelin basic protein promoter exerts an enhanced cytotoxic effect in gliomas. Cancer Gene Ther 2000;7:739–748.
108. Marchetti P, Castedo M, Susin SA, et al. Mitochondrial permeability transition is a central coordinating event of apoptosis. J Exp Med 1996;184:1155–1160.

109. Slee EA, Harte MT, Kluck RM, et al. Ordering the cytochrome c-initiated caspase cascade: hierarchical activation of caspases-2, -3, -6, -7, -8, and -10 in a caspase-9-dependent manner. J Cell Biol 1999; 144:281–292.
110. Castedo M, Hirsch T, Susin SA, et al. Sequential acquisition of mitochondrial and plasma membrane alterations during early lymphocyte apoptosis. J Immunol 1996;157:512–521.
111. Darzynkiewicz Z, Traganos F, Staiano-Coico L, Kapuscinski J, Melamed MR. Interaction of rhodamine 123 with living cells studied by flow cytometry. Cancer Res 1982;42:799–806.
112. Lizard G, Fournel S, Genestier L, et al. Kinetics of plasma membrane and mitochondrial alterations in cells undergoing apoptosis. Cytometry 1995;21:275–283.
113. Vogelbaum MA, Tong JX, Perugu R, Gutmann DH, Rich KM. Overexpression of bax in human glioma cell lines. J Neurosurg 1999;91:483–489.
114. Bedner E, Li X, Gorczyca W, Melamed MR, Darzynkiewicz Z. Analysis of apoptosis by laser scanning cytometry. Cytometry 1999;35:181–195.
115. Hotz MA, Gong J, Traganos F, Darzynkiewicz Z. Flow cytometric detection of apoptosis: comparison of the assays of in situ DNA degradation and chromatin changes. Cytometry 1994;15:237–244.
116. Jurgensmeier JM, Xie Z, Deveraux Q, Ellerby L, Bredesen D, Reed JC. Bax directly induces release of cytochrome c from isolated mitochondria. Proc Natl Acad Sci USA 1998;95:4997–5002.
117. Darzynkiewicz Z. Acid-induced denaturation of DNA in situ as a probe of chromatin structure. Methods Cell Biol 1994;41:527–541.
118. Frankfurt OS, Byrnes JJ, Seckinger D, Sugarbaker EV. Apoptosis (programmed cell death) and the evaluation of chemosensitivity in chronic lymphocytic leukemia and lymphoma. Oncol Res 1993;5:37–42.
119. Beg AA. Baltimore D. An essential role for NF-kappaB in preventing TNF-alpha-induced cell death. Science 1996;274:782–784.
120. Otsuka G, Nagaya T, Saito K, Mizuno M, Yoshida J, Seo H. Inhibition of nuclear factor-kappaB activation confers sensitivity to tumor necrosis factor-alpha by impairment of cell cycle progression in human glioma cells. Cancer Res 1999;59:4446–4452.
121. Nagata S. Apoptosis by death factor. Cell 1997;88:355–365.
122. Boland MP, Foster SJ, O Neill LA. Daunorubicin activates NF-kappaB and induces kappaB-dependent gene expression in HL-60 promyelocytic and Jurkat T lymphoma cells. J Biol Chem 1997;272:12,952–12,960.
123. Baldwin AS. The NF-kappa B and I kappa B proteins: new discoveries and insights. Annu Rev Immunol 1996;14:649–683.
124. Baeuerle PA, Henkel T. Function and activation of NF-kappa B in the immune system. Annu Rev Immunol 1994;12:141–179.
125. Miyakoshi J, Yagi K. Inhibition of I kappaB-alpha phosphorylation at serine and tyrosine acts independently on sensitization to DNA damaging agents in human glioma cells. Br J Cancer 2000;82:28–33.
126. Siebenlist U, Franzoso G, Brown K. Structure, regulation and function of NF-kappa B. Annu Rev Cell Biol 1994;10:405–455.
127. Lahiri DK, Ge Y. Electrophoretic mobility shift assay for the detection of specific DNA-protein complex in nuclear extracts from the cultured cells and frozen autopsy human brain tissue. Brain Res Brain Res Protoc 2000;5:257–265.
128. Olson DC, Levine AJ. The properties of p53 proteins selected for the loss of suppression of transformation. Cell Growth Differ 1994;5:61–71.
129. Lopes UG, Erhardt P, Yao R, Cooper GM. p53-dependent induction of apoptosis by proteasome inhibitors. J Biol Chem 1997;272:12,893–12,896.
130. Kubbutat MH, Vousden KH. Proteolytic cleavage of human p53 by calpain: a potential regulator of protein stability. Mol Cell Biol 1997;17:460–468.
131. Anker L, Ohgaki H, Ludeke BI, Herrmann HD, Kleihues P, Westphal M. p53 protein accumulation and gene mutations in human glioma cell lines. Int J Cancer 1993;55:982–987.
132. Zhan Q, Fan S, Bae I, et al. Induction of bax by genotoxic stress in human cells correlates with normal p53 status and apoptosis. Oncogene 1994;9:3743–3751.
133. Musco ML, Cui S, Small D, Nodelman M, Sugarman B, Grace M. Comparison of flow cytometry and laser scanning cytometry for the intracellular evaluation of adenoviral infectivity and p53 protein expression in gene therapy. Cytometry 1998;33:290–296.

134. Laxman B, Hall DE, Bhojani MS, et al. Noninvasive real-time imaging of apoptosis. Proc Natl Acad Sci USA 2002;99:16,551–16,555.
135. Korshunov A, Golanov A, Sycheva R, Pronin I. Prognostic value of tumour associated antigen immunoreactivity and apoptosis in cerebral glioblastomas: an analysis of 168 cases. J Clin Pathol 1999;52:574–580.
136. Clatch RJ, Walloch JL. Multiparameter immunophenotypic analysis of fine needle aspiration biopsies and other hematologic specimens by laser scanning cytometry. Acta Cytol 1997;41:109–122.
137. Gorczyca W, Darzynkiewicz Z, Melamed MR. Laser scanning cytometry in pathology of solid tumors. A review. Acta Cytol 1997;41:98–108.
138. Maciorowski Z, Delic J, Padoy E, et al. Comparative analysis of apoptosis measured by Hoechst and flow cytometry in non-Hodgkin's lymphomas. Cytometry 1998;32:44–50.
139. Seiter K, Feldman EJ, Traganos F, et al. Evaluation of in vivo induction of apoptosis in patients with acute leukemia treated on a phase I study of paclitaxel. Leukemia 1995;9:1961–1966.
140. Heesters MA, Koudstaal J, Go KG, Molenaar WM. Analysis of proliferation and apoptosis in brain gliomas: prognostic and clinical value. J Neuro-Oncol 1999;44:255–266.
141. Krajewski S, Krajewska M, Ehrmann J, et al. Immunohistochemical analysis of Bcl-2, Bcl-X, Mcl-1, and Bax in tumors of central and peripheral nervous system origin. Am J Pathol 1997;150:805–814.
142. McPake CR, Tillman DM, Poquette CA, George EO, Houghton JA, Harris LC. Bax is an important determinant of chemosensitivity in pediatric tumor cell lines independent of Bcl-2 expression and p53 status. Oncol Res 1998;10:235–244.
143. Weller M, Frei K, Groscurth P, Krammer PH, Yonekawa Y, Fontana A. Anti-Fas/APO-1 antibody-mediated apoptosis of cultured human glioma cells. Induction and modulation of sensitivity by cytokines. J Clin Invest 1994;94:954–964.
144. Roth W, Fontana A, Trepel M, Reed JC, Dichgans J, Weller M. Immunochemotherapy of malignant glioma: synergistic activity of CD95 ligand and chemotherapeutics. Cancer Immunol Immunother 1997;44:55–63.
145. Gratas C, Tohma Y, Van Meir EG, et al. Fas ligand expression in glioblastoma cell lines and primary astrocytic brain tumors. Brain Pathol 1997;7:863–869.
146. Glaser T, Castro MG, Lowenstein PR, Weller M. Death receptor-independent cytochrome c release and caspase activation mediate thymidine kinase plus ganciclovir-mediated cytotoxicity in LN-18 and LN-229 human malignant glioma cells. Gene Ther 2001;8:469–476.
147. Shinoura N, Yamamoto N, Asai A, Kirino T, Hamada H. Adenovirus-mediated transfer of Fas ligand gene augments radiation-induced apoptosis in U-373MG glioma cells. Jpn J Cancer Res 2000;91:1044–1050.
148. Shinoura N, Ohashi M, Yoshida Y, et al. Adenovirus-mediated overexpression of Fas induces apoptosis of gliomas. Cancer Gene Ther 2000;7:224–232.

13
Cell Cycle, Neurological Disorders, and Reactive Gliosis

Kerri L. Hallene, BS, and Damir Janigro, PhD

SUMMARY

Neurons and glia play a critical role in the normal development of the central nervous system and contribute to injury and disease via numerous pathological pathways. Internal and external stimuli induce signaling cascades that are incorporated into apoptotic mechanisms, ultimately culminating in cell death or cell survival. Growing bodies of evidence demonstrate that cell cycle proteins are expressed in dying neurons in both the developing and adult brain. Although the general mechanisms that control cell cycle have been studied at length, the regulation of progression of the cell cycle, as well as the reactivation of cell cycle pathways in neurons, remains to be fully identified. In this chapter, we will review the cell cycle in brief with a focus on the central nervous system, examine key contributors in the cell cycle that are involved in neurological disorders and gliosis, and look at the mechanism of cell cycle re-entry as a novel mechanism in neurological disorders.

Key Words: Cell cycle; neurodegenerative disorders; gliosis; epilepsy; p53; brain tumor.

1. INTRODUCTION

In the past few years the neurobiological correctness governing the concept of neurogenesis has undergone a revolution. Ideas once considered impossible or dubious at best, are now well accepted. It is, for example, recognized that a limited amount of neurogenesis confined to groups of neural stem cells occurs in the adult brain *(1,2)*. Nevertheless, controversy still exists on where these new neurons are formed, the impact that adult neurogenesis in the adult brain has on central nervous system (CNS) function, and the topographic and quantitative extent of mammalian neurogenesis.

Neurogenesis and gliogenesis are consequences of cell cycle in an organ in which mitosis (the process by which a cell divides and produces two daughter cells from a single parent cell) is rare at best. Historically, detection of mitotic cells in the adult brain was a prerogative of the neuropathologist, and increased cell cycle was often interpreted as a hallmark of hyperplastic or neoplastic disease. The discovery of neurogenesis has injected a new perspective into the field, and today mitosis in the adult brain is overwhelmingly perceived as a positive mechanism of plasticity or repair. Thus, the fact that following cerebral ischemia new cells are formed in or around the lesioned cortex is interpreted as a benign response of surviving cells, rather than a consequence of the ischemic pathophysiology (*see* refs. *3,4,* and Section 4). Similarly, seizure-induced neurogenesis *(5)* has not been interpreted as a pathogenetic mechanism, but rather as a response to widespread excitotoxic hippocampal death.

Although the old dogma predicted that newly created *neurons* exist only during brain development, the fact that "support" glial tissue may undergo cell division was typically accepted.

From: *The Cell Cycle in the Central Nervous System*
Edited by: D. Janigro © Humana Press Inc., Totowa, NJ

It is also now accepted that vascular endothelial cells may also proliferate in response to robust epileptic forces activity *(6)*. Interestingly, proliferation of glia and endothelial cells is controlled by electrical fields that mimic neuronal field potentials *(7)*. A picture thus emerges in which regulation of cell cycle in mammalian brain shares common cellular mechanisms with non-CNS cells, whereas mitosis is kept low by environmental cues such as neuronal firing.

As in other parts of the body, after brain cells undergo division, the offspring remain in either a mitotically quiescent state, enter the G_1 phase, or die. Based on current knowledge of the cell cycle, no other possibilities exist. The brain is characterized by a low mitotic index owing to a prolonged G_0 phase, seldom an S phase or an arrested G_2 phase. Therefore, the majority of cells in the adult human brain are in a nonproliferative state. This is dissimilar to other organs in which cells in G_1 phase have positioned themselves to divide again. Again, this low propensity to cell division may be regulated by the electrical activity of the brain itself *(7)*.

Nonproliferative cells may be characterized into three categories. A number of highly differentiated cells are thought to be permanently incapable of mitosis; consequently, the life span of a cell approximates the life of the organism, except if disease or injury occurs *(1)*. It is believed that most adult neurons belong to this category, although this issue is controversial. Polymorphonuclear leukocytes and other terminally differentiated cells, however, have a finite life span *(8)*. A third type of nonproliferative cells remains in an unstable G_0 phase: here, cells may potentially be recruited into the G_1 phase with an appropriate extracellular signal *(8)*. Noteworthy, stem cells share a commonality with neurons, in that they live as long as the organism. Although similar to unstable G_0 cells, they may occasionally, or on command, produce viable progeny. Stem cells are also termed *clonogenic cells* owing to their capacity for unlimited proliferation.

Stem cell recruitment signaling is often owing to physiological changes in the environment, such as extracellular stimuli, cell injury, cell death, or by drugs and/or hormones *(9–11)*. Neurogenesis itself is highly restricted to discrete locations within the adult brain and the production of neurons from multipotent stem cells or their committed progeny is entirely dependent on signaling within these neurogenic zones *(12–15)*. Similar to stem or totipotent cells, differentiated, adult astrocytes may undergo cell cycle re-entry when exposed to appropriate signals. These changes appear often under pathological conditions ("reactive gliosis" *[16]*). This chapter primarily describes the mechanisms of cell cycle, neurological disorders, and reactive gliosis as they relate to cell proliferation, turnover, and pathophysiology.

2. CELL CYCLE REGULATION: EMPHASIS ON THE CNS

In eukaryotic cells, cell cycle is a well-conserved mechanism of proliferation. In recent years, the most relevant mechanisms regulating cell cycle and the several proteins involved have been studied at length in mammalian cells *(17)*. The somatic cell cycle is identified as the period between two mitotic divisions, and is typically divided into four phases: G_1 (first gap-phase period characterized by cell growth), S (DNA replication), G_2 (second gap-phase cells are prepped for division), and M (mitosis; point in which two identical daughter cells are produced) *(18)*. An additional phase known as G_0 also exists and may be defined as the phase of the cell cycle in which cells remain in a quiescent state. Here, cells have degraded RNA and protein, unduplicated DNA, and low enzyme activity. Postembryonic cell proliferation rate is determined by the ability to switch between G_0 and G_1 and vice versa (Fig. 1A).

Cell cycle progression is regulated by several factors including phosphorylation and dephosphorylation by kinases and phosphatases, respectively, and chronological expression, activation and inhibition of nuclear cell cycle proteins (cyclin-dependent kinases [CDKs]; catalytic subunits with constitutively expressed levels throughout the cell cycle. Cyclins accumulate continuously through the cell cycle and are abruptly degraded at mitosis, unlike CDKs, and CDK inhibitors [CDKIs] [Table 1] *[1,19]*). Various other factors also dictate cell cycle progression, such as external and internal stimuli, e.g., nutrients, and DNA damage, all acting on cell cycle mechanisms *(17)*. If for some reason a cell aborts the cycle, it remains quiescent in G_0-nonproliferative phase for an indefinite period of time (Fig. 1C).

In the CNS, CDKs play crucial role in brain development and may be a part of the pathogenic process. They are involved in early differentiation processes (e.g., notochord formation) as well as in determining the quiescent nature of mature CNS cells. In particular, cyclins D_2 and D_3 are both important in early embryogenesis. CDKs are preferentially expressed in three major CNS regions, retina, cerebral cortex, and cerebellum *(20)*.

Checkpoints control how proliferating cells advance through the cell cycle. Checkpoints are generally tumor suppressor proteins, that is, p53 and retinoblastoma protein (Rb). These function as molecular "gatekeepers" to repress unrestrained growth and guarantee that crucial events in one phase of the cycle are fulfilled before progression into the next phase. However, the major inhibitory pathway in phase G_1 of the cell cycle is regulated via p21 activation (a CDK inhibitor) *(18)*. In response to damaged DNA, checkpoints aid by blocking mitosis and by controlling progression through mitosis itself. Hence, checkpoints are able to synchronize cell growth with cell proliferation *(1,19)*. In the event that cell arrest ensues at a checkpoint, one of two things occur: (1) cells will return to G_0 phase and redifferentiate (if arrest is at G_1 checkpoint), or (2) apoptotic death of cells is initiated *(1,21,22)*. The second method warrants that nonrepairable DNA modifications are unable to be passed on to progeny of damaged cells. In the CNS, p21 is not found in the cortex, but its expression is predominant in ependyma and olfactory bulb cells. p27 is present in the cerebellar granule cells external layer where precursors existing in the cell cycle are typically found. Postnatal Purkinje cells also express p21, suggesting that p27–p21 may cooperate to keep major cell types in a state of quiescence *(20)*.

Stimulation by mitogens, environment, and developmental cues provoke cell cycle entry and progression by triggering numerous signaling pathways and cyclin/CDK complexes. To date, there have been eight cyclins and seven CDKs described in mammals *(1)*. Expression of appropriate cyclin proteins assists in progression through the cell cycle, namely, cyclins D, E, B, and A. Subsequent to a mitogenic stimulus, synthesis, binding, and activation occurs between D-type cyclins and CDKs, specifically, CDKs 4 and 6. This activity permits cells to leave the quiescent state (G_0), for G_1 *(1,23)*. In late G_1 phase, the induction of cyclin E takes place and forms a complex with CDK2; this complex is essential for an S-phase transition *(1,23)*. Cyclin A is another crucial component of the cell cycle and is associated with DNA synthesis (S phase). Cell division control protein (CDC)2, (also known as CDK1) and cyclin A forms a complex and forces cells through G_2 phase *(1,23)*. At G_2/M phase transition, degradation of cyclin A occurs and CDC2 forms a complex with a recently created cyclin B. This newly synthesized complex is necessary for progression through mitosis, however, shortly after mitosis cyclin B is degraded.

Other proteins are crucial in the cell cycle as well. There are two specific families of proteins in mammals which inhibit cyclin-CDK complex activity: members of the KIP/CIP family, p21 and p27, and members of the INK4 family, p15 and p16 *(1,24,25)*. Members of the KIP/CIP family govern G_1 cyclin-CDK complex actions, and to a minor extent, cyclin B/CDC2 activity. INK4 members, however, explicitly hamper activity of CDKs 4 and 6. CDK1s on the other hand are highly controlled by translation, transcription and ubiquitin-mediated proteolysis. The CDK1s not only act as negative regulators (of cyclin E-CDK2, cyclin A-CDK, and cyclin B-CDK1) of the cell cycle, but as positive regulators too, where they control cyclin D-CDKp 4 and 6 by governing their timely G_1 assembly.

3. WHO IS WHO IN CELL CYCLE AND NEUROLOGICAL DISORDERS: MOLECULAR PLAYERS

In *human* neurological disorders, cell cycle re-entry may be an actual contributor in the pathological cascade of disease progression *(26)*. Many of the components of intracellular signaling pathways involved in the regulation of cell division are encoded by genes that were originally identified as cancer-promoting genes, or *oncogenes*, because mutations in them contribute

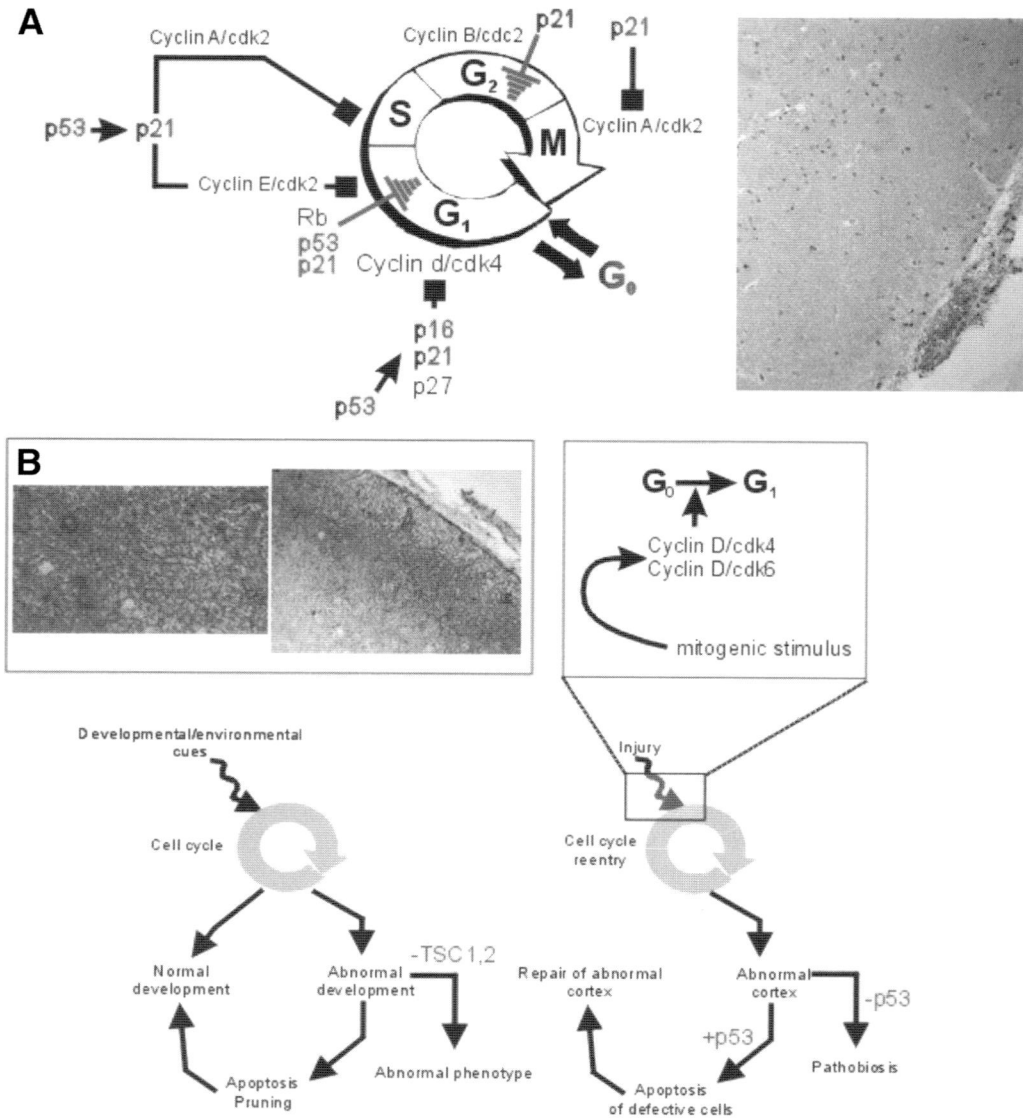

Fig. 1. Simplified view of cell cycle events highlighting the main steps and checkpoints discussed in this minireview. A number of CNS-specific examples are also shown. **(A)** Dual role for p21/p53 in the regulation of cell cycle. The micrograph shows a normal brain section stained with Cresyl violet and antibodies against MIB-1. Note the virtual absence of dividing cells in this normal human cortex. The small cells lining the pial surface are red cells that show artifactual positivity for MIB-1. **(B)** Reactive gliosis in human epileptogenic cortex (temporal lobe of a child following resection to relieve drug-resistant seizures). Note the abundant GFAP positivity and the abnormal shape of two adjacent astrocytes (arrows). Also, note the thick subpial layer of reactive, GFAP-positive cells. **(C)** Possible mechanisms linking external stimuli, molecular events, cell cycle, and neurological disorders. *See* Heading 5. (Please *see* companion CD for color version of this figure.)

to the development of cancer. A more realistic view suggests that in many instances these oncogenes may actually be part of a pathological process distinct from tumorigenesis. For example, the mutation of a single amino acid in Ras leads to constant stimulation of Ras-dependent signaling pathways, even in the absence of mitogenic stimulation. Similarly, mutations that cause

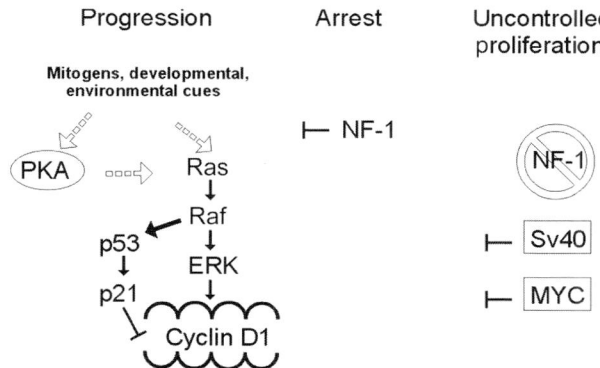

Fig. 2. Molecular players in CNS control of cell proliferation. *See* Heading 5.

an overexpression of Myc promote excessive cell growth and proliferation, and thereby promote the development of cancer. However, when Ras or Myc are experimentally hyperactivated in a normal tissue, the result is the activation of checkpoint mechanisms causing the cells to undergo either mitotic arrest or programmed cell death (PCD) or apoptosis (Fig. 2).

The *TP53* gene encodes a 53 kDa nuclear phosphoprotein, p53. p53 is the most frequently mutated gene in human cancers. The protein exists as a tetramer, and when associated with a mutant form, the tetramer takes on the mutant confirmation. Wild-type p53 is present at constitutive levels in all normal cells to restrain cell growth; however, it was originally classified as an oncogene because its mutations took an oncogenic phenotype. Several regions in the gene product mediate these processes, allowing p53 protein to negatively or positively regulate transcription of other genes *(27)*.

Today, it is well-known that the *p53* gene plays a role in a variety of biological functions, including cell cycle regulation, via transactivating important genes for both cell cycle (i.e., G_1 arrest via p21) and apoptosis (i.e., BAX) *(17)*, DNA damage response, cell death and differentiation, and neovascularization *(27)*. These are the essential features to ensure both normal brain development and to avoid cellular transformation.

Wild-type p53 is now recognized as a tumor suppressor protein because the missense mutations serve an inhibitory function *(28)*. In human tumors, an absence of p53 or nonfunctional p53 is fairly common and may possibly contribute to the progression of other tumor types with a lack of cell specificity. Primarily owing to deletions of both wild-type alleles or by a dominant mutation in one allele, p53 might also be silent *(28)*. Silent wild-type p53 is a possibility in many malignant and premalignant tissues as a consequence of abnormal sequestration in the cytoplasm where it is functionally muted.

p53 activity levels are increased at sites of DNA damage, although present in low levels in all cells; thus, activating one of two known pathways. Early in the cell cycle, p53 activates a checkpoint inhibiting cell cycle progression until damage has been repaired. DNA repair is activated by p53, which induces transcription of p21 and DNA repair enzymes *(29)*. However, if the cell has already entered S phase and is prepared to divide, PCD occurs, preventing proliferation of damaged cells. In the brain, loss of p53 influences the cell cycle, therefore allowing the survival of potentially damaged glial cells to proliferate *(30–34)*. Deletion of p53, however, may play a neuroprotective role in diverse paradigms of neuronal cell death, and neurodegenerative diseases including focal ischemia and spinal ataxia *(35,36)*.

The discovery of p21 arose as a senescent cell-derived inhibitor of DNA replication in diploid human fibroblasts *(37)*. p21 (WAF1/CIP1) is a 21 kDa protein found downstream of p53. Recognition sequences for the TP53 protein are found on the promoter region and thus, p21 is regulated by the levels of WT p53 present in the cell. Mutant p53, however, is unable to do so. As a universal inhibitor of CDK, the p21 protein binds to an extensive array of cyclin/CDK complexes, including but not limited to, CDK4,6-cyclin D and CDK2-cyclin E and G_1 cyclins (complexed to CDK2), which are most significant to TP53 mediated G_1/S phase arrest *(28,38)*. In cells perpetuated successfully in vitro, CDK-cyclin complexes tend to lack p21, suggesting that p21 may be implicated in the control of G_1/S phase *(28)*. Over expression of p21 causes growth arrest, also suggestive of the role of p21 in cell cycle progression control *(27)*.

The most recognized inhibitor of CDK is p27. It is responsible for inhibiting a diverse number of cyclin-CDK complexes, including cyclin E-CDK2 and cyclin A-CDK2, but predominantly binding to the cyclin D-CDK4 complex, ultimately leading to G_1 cell cycle arrest *(23,39)*. In the vertebrate CNS, relatively high levels of p27 have been shown *(40)*. Furthermore, p27 has also been demonstrated to inhibit CDK-activating kinase activity, resulting in arrest of cell cycle *(41)*.

Rb and its related family members are thought to play a role in cell cycle control as active checkpoints in the G_1/S transition point for proliferating cells *(1,17)*. Rb is nonphosphorylated in the active state and inhibits cell cycle progression by differential binding to, and repression of, the transcription factor E2F. Consequently, genes needed for cell cycle progression, DNA synthesis, etc., are not transcribed. Conversely, on mitogenic stimulation via cyclin D-CDK4/6 complexes, Rb becomes phosphorylated and release of E2F occurs. Hence, genes required for S phase entry may be transcribed *(1,17)*.

Recently, a role for Rb in abortive cell cycle re-entry and apoptosis has been demonstrated in numerous studies *(21)*. In several neuronal cultures, phosphorylated Rb was discovered after exposure to different death stimuli, e.g., excitotoxicity *(42)*. Rb null mice are embryonic lethal, exhibit massive cell death in both the CNS and peripheral nervous system, and display DNA synthesis. Together, these results suggest that Rb is an essential gate-keeper of cell cycle re-entry, as well as a component in apoptosis in the CNS *(43)*.

The family of E2F transcription factors is comprised of six members (E2F1–E2F6). In proliferating cells, the activity of E2F is essential in transcription of numerous genes required for cell cycle progression including cyclins, DNA replication enzymes and CDK. E2F proteins form heterodimers with DP transcription factors and bind to specific DNA sequences in promoters of many genes involved in cell cycle progression *(17,19,44)*. These complexes are, in turn, regulated by members of the Rb family; however, the interactions are limited, as Rb preferentially binds to E2F1–E2F3.

4. GLIA AND REACTIVE GLIOSIS

Three cell types exist in the vertebrate CNS: neurons, oligodendrocytes, and astrocytes (glia), which arise from multipotent neural stem cells. Perhaps the most abundant and essential, but least understood, are astrocytes. Unlike the majority of other cell types, many astrocytes never fully exit the cell cycle in the adult CNS; consequently, cell contact may negatively regulate their proliferation. Historically, glial cells were considered a homogeneous population of cells that played a passive role in brain activity, serving as metabolic and support cells for neurons. In recent years, it has been established that they play a more dynamic role in CNS differentiation than previously believed: for example, astroglia take part in a significant role in blood–brain barrier (BBB) formation by releasing soluble factors to induce the BBB phenotype along with other cellular properties in endothelial cells *(45)*. It has also become apparent that astroglial response to brain injury, aging, and plasticity is much more multifaceted than previously believed *(46)*.

The most commonly studied proliferative change in adult glia is the transformation into a neoplastic phenotype. It has been recently suggested that preneoplastic or nonmalignant changes in astrocytes may be the underlying mechanism of many neurological disorders including epilepsy *(26)*. The following will describe examples of nonneoplastic changes; crucial molecular events that cause the transformation of normal glia into a neoplastic phenotype will be later discussed in Aeder and Hussaini's chapter.

Glial cells are involved in regulating the brain microenvironment, in particular, they provide extensive structural, physiological, and trophic support to neurons by phagocytizing neuronal debris, acting as free radical and excess glutamate scavengers, maintaining both homeostasis and extracellular levels of ions, neurotransmitter uptake, controlling synaptogenesis, synapse number, function, and plasticity *(47–52)*. Glia are also involved in regulation of energy metabolism, metal sequestration, and developing and maintaining the BBB *(53)*. Recently, glia have also been found to promote neovascularization and remyelination in the damaged CNS, and also to kindle neurogenesis from neural stem cells *(54)*. It is clear that these neuronal–glial interactions are critical to optimize neuronal function. Additionally, it has been found that glial cells orchestrate responses to cerebral inflammatory diseases and other CNS injuries *(55)*, as well as involvement in antioxidant functions *(56)*. Astrocytes are involved in almost all aspects of brain function and development *(12)*. Thus, they are developmentally complex and have a full range of physiological capacities *(56)*, entertaining the idea of now being recognized as a heterogeneous population of cells. In addition to the role of glia in normal brain, an increasingly recognized role is now being accepted in the pathogenesis of common neurological disorders *(57–59)*.

A function attributed to astrocytes is extracellular (K^+) buffering which appears to be of critical importance to assure normal neuronal excitability *(57)*. Previously, it has been shown that astrocytes are liberally endowed with voltage-activated K^+ channels *(60)*, which mediate diffusional uptake of K^+. In particular, inwardly rectifying K (K_{IR}) channels play an important role in K^+ buffering by astrocytes *(57)*. Although both the biochemical and antigenic changes that occur in conjunction with reactive gliosis have been well studied *(61)*, the changes in biophysical properties of reactive astrocytes are currently being discovered. Decreases in K_{IR} current amplitudes have been reported after injury of cultured astrocytes *(62)* and after a posttraumatic injury in vivo *(57)*, leading to impaired K^+ homeostasis and seizures *(63)*. Intercellular coupling by gap junctions, another important aspect of potassium buffering *(64)*, is also affected after injury *(65,66)*.

Astrocytes frequently undergo remarkable transformations termed astrogliosis, or are referred to as reactive astrocytes. This is one of the most outstanding cellular responses to an array of CNS injury. Reactive gliosis is found in virtually all CNS lesions, including ischemia, trauma, demyelinative disorders (i.e., multiple sclerosis), infections, and degenerative conditions (i.e., Alzheimer's disease [AD]) *(53)*. Gliosis is also found in response to toxin exposure, such as quinolinic and kainic acids, as well as exposure to heavy metals (i.e., iron, lead, and cobalt).

Reactive gliosis may also transpire in a physiological state; for example, many groups have demonstrated increased glial fibrillary acidic protein (GFAP) immunoreactivity (an astrocyte marker) following excessive neuronal activity produced by electrically induced seizures. However, depressed neuronal activity has also been correlated with astrogliosis *(67)*.

Several studies have shown that reactive gliotic changes include attempted re-entry into the cell cycle *(62,68,69)*. In a reactive gliosis model, Sontheimer and co-workers found that proliferative astrocytes were concentrated in specific regions surrounding the lesion, where these glia lacked expression of K_{IR} channels and were not coupled by gap junctions as judged from dye injections *(62)*. In contrast, astrocytes in the hyperexcitable zone were not proliferating and displayed increased intercellular coupling associated with expression of K_{IR}. These findings emphasize the need for accurate localization of electrical activity in tissue resections from human brain, and underscore the fact that astrocytes form a pathologically heterogeneous population of cells.

5. CELL CYCLE RE-ENTRY AS A MECHANISM IN NEUROLOGICAL DISEASE

Although in normal brain cell cycle is low, as is apoptotic cell death, following a number of injurious triggers, both apoptosis and cell cycle markers become evident. Whether increased signals for cell cycle truly depict cell division or cell damage, remains a matter of considerable dispute *(13–15,70–72)*. In fact, cells may attempt to re-enter the cell cycle without actually proliferating. For example, whereas neurons retain the capability to reactivate the cell cycle, engagement of cell cycle apparatus infrequently allows proliferation of neurons, and usually induces apoptosis *(1,73,74)*. Recent evidence suggests that, although re-entry into the cell cycle is not specific to a particular neurodegenerative disease or distinct neuronal populations, it characterizes a common pathway involved in neuronal apoptosis. Reactivation of cell cycle components is commonly observed in dying neurons and possibly contributes to neuronal apoptosis in both neuronal degeneration and development in adult brain *(1)*.

Furthermore, substantial studies have pointed to an association of cell cycle abnormalities, such as cell cycle re-entry, G_1/S progression, DNA synthesis, and recurrent appearances of G_2/M markers within mouse and human neurodegenerative diseases including Parkinson's disease (PD), Pick's disease, intractable temporal lobe epilepsy (TLE), frontotemporal dementia linked to chromosome 17, stroke, AD, and Parkinson-amyotrophic lateral sclerosis of Guam *(22,75,76)*. Down's syndrome is another disease in which neuronal cell death has been linked to aberrant re-entry of neurons into the cell cycle. In these patients, hippocampal pyramidal neurons that contain neurofibrillary tangles show increased levels of CDK4 and the cell proliferation marker Ki-67 *(77)*.

Epilepsy is a brain disorder in which abnormal firing of neurons occurs. There is a great diversity in the disorder, e.g., mild-to-severe convulsions, muscle spasms, loss of consciousness, memory, learning, and behavioral deficits. Epilepsy is one of the most prevalent neurological disorders affecting approx 40×10^6 people worldwide, with 1% being Americans. Several drug treatment options are available; however, one-third of epileptics are not drug-respondent (refractory). Surgery is often a last resort for these cases.

One of the most common forms of drug refractory epilepsy is TLE. The hippocampus tends to be the most affected area. The onset of TLE generally occurs during early childhood or is associated with prolonged episodes of seizures in early life, followed by a latency period, then epileptic development.

Recent studies suggest that delays in neuronal death in epilepsy and other neurological disorders include apoptotic components which involve PCD, and may thus be a characteristic of CNS degenerative disorders. PCD results at least in part, from aberrant cell cycle control *(78)*. The tumor suppressor p53 is involved in PCD, suggesting that PCD may be implicated with both cell cycle and activation abnormalities. D_1, D_2, and D_3 cyclins also play a crucial role in the cell cycle, specifically in G_1 progression, and have therefore been examined at length. The apparent link between cell cycle and apoptosis has prompted a number of investigations.

In human studies by Nagy and co-workers, cyclin expression was examined in AD, Down's syndrome, and epileptic patients. The group detected cyclin expression in the dentate gyrus of elderly, drug-respondent epileptic patients. A different patient population suffering from intractable TLE examined by the same group, highly expressed cyclin B in neurons, the dentate gyrus, and CA1 and CA4 hippocampal regions *(79,80)*. Small numbers of neurons in both disease populations and controls had cyclin E positive nuclei. The presence of cyclins B and E (specific markers for G_1 and G_2 phases of the cell cycle) implies that some neurons have reentered the cell cycle *(79)*. Differences between the two patient populations were astounding; although cyclin E expression was low and not dissimilar between the two, the former demonstrated some cyclin expression associated with an AD-related pathology, whereas the latter revealed a great number of neurons with cell cycle protein expression, cell death, and immense synaptic loss *(79,80)*.

In an animal model of kainite-induced seizures, Timsit et al. detected cyclin D_1 mRNA upregulation in the CA3 hippocampal region, the amygdala, around the site of kainate injection, and in the thalamus *(78)*. These findings suggest that upregulation of cyclin D_1 is important in signaling the molecular cascade of excitotoxic cell death. Liu et al. demonstrated increased cyclin D_1 immunoreactivity in the entire hippocampus in another model of kainic acid *(81)*.

Cell cycle re-entry may represent nonspecific response mechanisms to insults that may result in cell death as in TLE, or in cases of the absence of killer proteins, generation of an AD-related pathology *(79)*. Collectively, these results suggest that cell cycle proteins may promote "unscheduled" or improper progression/aberrant re-entry, etc., and could participate in combination with other apoptotic modulators in cell death commitment.

Cell death of neurons following ischemic stroke is traditionally defined as necrosis *(73,82)*, however, biochemical and morphological verification indicate that apoptosis is also occurring in this brain injury *(82,83)*. It is believed that apoptosis occurs in the penumbra where there is more blood supply than in the lesioned core, whereas in the core of the lesion, necrosis occurs *(84)*. Existing knowledge of cell cycle events in ischemia are derived from rodent animal models. Cyclin D and G, CDKs 4 and 2, p21, etc., occur prior to apoptosis *(85–88)*. These findings imply that peculiar cell cycle re-entry triggers neuronal cell death following ischemic stroke. The postulated role of glia in stroke has been to protect neurons by antioxidant mechanisms *(56)*.

The most common human motor neuron disease is amyotrophic lateral sclerosis (ALS) *(1)*. The ALS etiology is thought to be multifactorial and possible causes include oxidative stress, peripheral deficiencies leading to retrograde degeneration, and glutamate excitotoxicity, all of which lead to the eventual paralysis of the body's general musculature *(1)*. Numerous studies have implicated the contribution of cell cycle regulators in the pathogenesis of ALS. In particular, are the G_1/S phase proteins cyclin D, CDK4, hyperphosphorylated Rb and E2F-1 which have been identified in spinal cord and motor cortex of patients with ALS *(1,85,89)*.

In a fraction of these patients, ALS is a result of a missense mutation in the gene that encodes superoxide dismutase 1 (SOD1). SOD1 is an enzyme that scavenges free radicals and aides in cell protection owing to oxidative stress. Thus, such a mutation leads to enhanced oxidative stress, and in turn leads to cell cycle re-entry and consequently causes affected neurons to undergo apoptosis *(1)*. Although neurotoxic mechanisms still remain controversial in ALS cell cycle re-entry may prove to be an essential pathway involved.

The age-dependent neurodegenerative disorder PD, is characterized by degeneration of dopaminergic neurons. The specific etiology of PD is unknown, however, it has been hypothesized that several factors may contribute to this disorder, such as environment and genetics. Parkin, an ubiquitin ligase associated with early-onset PD, is primarily found to be expressed in astrocytes and neurons. Remarkably, cyclin E has recently been demonstrated to be a substrate of Parkin *(90)*. Consequently, an accumulation of cyclin E has been found in the extracts of substantia nigra from patients with PD who express malfunctioning Parkin *(90)*. These results help provide a direct correlation between cell cycle re-entry, neurodegeneration, and apoptosis *(1)*.

In vitro, Parkin is upregulated only in astrocytes after induced cytotoxic stress. Such a dysfunction may lead to nigrostriatal degeneration via loss of trophic support, free radical damage, and/or release of cytokines by astrocytes *(56)*. Similar mechanisms involving astrocytes may perhaps explain the lack of ALS-like pathology when the *SOD1* gene was targeted to motor neurons *(56,91)*. Additionally, cell cycle events happening in neurodegenerative diseases which affect adult brain have been associated to spinal cord injuries *(92)*.

Among the neurodegenerative disorders, AD is the most intensely studied disease in which cell cycle events are involved *(1)*. Degeneration of neurons, cognitive impairment, deficits in memory, attention, and planning, followed by a failure in language, general intelligence, and spatial–visual functions are a few of the characteristics of AD, the leading cause of senile dementia. PCD is thought to be a contributing factor to AD pathogenesis, as the presence of apoptotic neurons has been discovered in AD brain *(1,93)*. What causes neurodegeneration in AD is under constant debate; however, several factors have been implicated in AD pathogenesis, for example, τ protein, apolipoprotein E, and presenilins.

Initial studies by both Busser and Vincent et al. demonstrated induction and activation of both cyclin B and its complex partner CDC2 in terminally differentiated neurons in the AD brain, implicating the occurrence of cell cycle events *(1,94,95)*. τ Phosphorylation via CDC2 also suggests a direct link between AD pathogenesis and cell cycle, in that, τ hyperphosphorylation coupled with the formation of neurofibrillary tangles are thought to result in neurodegeneration in many dementias. Aberrant CDC2/cyclin B activation has also been reported in several other neurodegenerative disorders with τ pathology including Down's syndrome and Pick's disease, as previously mentioned *(77)*. Several studies have also demonstrated that AD neurons undergo cell cycle re-entry before apoptosis. This is evident by the expression of numerous cell cycle components including CDKs 4/6, cyclin D, and Ki-67. Also, the presence of late G_1 markers, CDK2 and cyclin E, are consistent with AD neuronal progression through G_1 *(76,77)*.

6. CONCLUSIONS AND UNANSWERED QUESTIONS

Taken together, several studies suggest that reactivation of the cell cycle or abortive re-entry is an essential step and common feature in the induction of the death program in postmitotic *neurons*. However, these processes are not specific to any particular neurodegenerative disorder, and the degree of cell cycle re-entry or progression may be dependent on several factors, including the state of the neuron/glia itself. Reactivation of the cell cycle seems to play a crucial role in arbitrating a vast array of stimuli to induce neuronal apoptosis and neurodegeneration in the human brain *(1)*.

Numerous neuropathic studies have used human brain samples which have revealed impressive correlations between cell cycle protein expression, neuronal cell death, and neurodegenerative diseases *(1)*. However, postmortem neuropathological studies, using samples of human brain, reveal that *glial* cell death and degeneration may also play a role in several neurological diseases.

In summary, several interactions, both extrinsic and intrinsic, determine whether cells will commit to the cycle and differentiate or die. Cell fate is decided from unique and particular combinations of such signals. Nevertheless, it is clear that aberrant cell cycle re-entry and reactivation of cell cycle machinery appear to be crucial components in neuronal *and perhaps* glial apoptosis, as well as in development and disease. The following questions remain fully or partly unanswered:

1. What is the origin of extracellular mitogenic signals in the CNS? Are growth factors released locally, or even in an autocrine fashion? Or is extravasation of mitogens across a leaky BBB necessary?
2. Are all mitotic figures in the diseased or normal CNS attributable to cell cycle occurring in CNS cells? Are any of these cells of extra-CNS (lymphatic) origin?
3. Are reactive or dividing cells an attempt to overcome injury-induced deficits or are these glia a part of the etiological process?
4. In general, under what conditions should one wish to enhance or decrease cell cycle propensity and/or proliferation of precursors in the CNS?

ACKNOWLEDGMENTS

This work was supported by grant nos. NIH-2RO1 HL51614, NIH-RO1 NS43284, and NIH-RO1 NS38195 to D.J.

REFERENCES

1. Becker EB, Bonni A. Cell cycle regulation of neuronal apoptosis in development and disease. Prog Neurobiol 2004;72:1–25.
2. Pevny L, Rao MS. The stem-cell menagerie. Trends Neurosci 2003;26:351–359.
3. Kondziolka D, Wechsler L, Achim C. Neural transplantation for stroke. J Clin Neurosci 2002;9:225–230.
4. Fisher M. Stem cell transplantation for stroke: does it work, and if so, how? Stroke 2003;34:2083.
5. Scharfman HE. Functional implications of seizure-induced neurogenesis. Adv Exp Med Biol 2004;548:192–212.
6. Hellsten J, Wennstrom M, Bengzon J, Mohapel P, Tingstrom A. Electroconvulsive seizures induce endothelial cell proliferation in adult rat hippocampus. Biol Psychiat 2004;55:420–427.
7. Cucullo L, Dini G, Hallene K, et al. Very low intensity alternating current decreases cell proliferation. Glia 1995;51:65–72.
8. Blow JJ, Hodgson B. Replication licensing—defining the proliferative state? Trends Cell Biol 2002;12:72–78.
9. Hernandez-Lopez C, Varas A, Sacedon R, et al. Stromal cell-derived factor 1/CXCR4 signaling is critical for early human T-cell development. Blood 2002;99:546–554.
10. Li K, Miller C, Hegde S, Wojchowski D. Roles for an Epo receptor Tyr-343 Stat5 pathway in proliferative co-signaling with kit. J Biol Chem 2003;278:40,702–40,709.
11. Shakoory B, Fitzgerald SM, Lee SA, Chi DS, Krishnaswamy G. The role of human mast cell-derived cytokines in eosinophil biology. J Interferon Cytokine Res 2004;24:271–281.

12. Rakic P. Adult neurogenesis in mammals: an identity crisis. J Neurosci 2002;22:614–618.
13. Rakic P. Neuroscience: immigration denied. Nature 2004;427: 685–686.
14. Rakic P. Neurogenesis in adult primates. Prog Brain Res 2002;138:3–14.
15. Rakic P. Neurogenesis in adult primate neocortex: an evaluation of the evidence. Nat Rev Neurosci 2002;3:65–71.
16. Ridet JL, Malhotra SK, Privat A, Gage FH. Reactive astrocytes: cellular and molecular cues to biological function. Trends Neurosci 1997;20:570–577.
17. Galderisi U, Jori FP, Giordano A. Cell cycle regulation and neural differentiation. Oncogene 2003;22:5208–5219.
18. Cell cycle and growth regulation, In: Lewin B (ed). Genes VII. New York, NY: Oxford University Press, 2004 pp 835–874.
19. Lees E. Cyclin dependent kinase regulation. Curr Opin Cell Biol 1995;7:773–780.
20. Cunningham JJ, Roussel MF. Cyclin-dependent kinase inhibitors in the development of the central nervous system. Cell Growth Differ 2001;12:387–396.
21. Liu DX, Greene LA. Neuronal apoptosis at the G1/S cell cycle checkpoint. Cell Tissue Res 2001;305:217–228.
22. Nagy Z. Cell cycle regulatory failure in neurones: causes and consequences. Neurobiol Aging 2000;21:761–769.
23. Sherr CJ. G1 phase progression: cycling on cue. Cell 1994;79:551–555.
24. Cheng M, Olivier P, Diehl JA, et al. The p21(Cip1) and p27(Kip1) CDK 'inhibitors' are essential activators of cyclin D-dependent kinases in murine fibroblasts. EMBO J 1999;18:1571–1583.
25. Lee MH, Reynisdottir I, Massague J. Cloning of p57KIP2, a cyclin-dependent kinase inhibitor with unique domain structure and tissue distribution. Genes Dev 1995;9:639–649.
26. Marroni M, Agarwal M, Kight K, et al. Relationship between expression of multiple drug resistance proteins and p53 tumor suppressor gene proteins in human brain astrocytes. Neuroscience 2003;121:605–617.
27. Kleihues P, Cavenee WK (eds). Pathology and Genetics of the Nervous System. Lyon: IARC Press, 2000.
28. Oncogenes and Cancer. In: Lewin B, ed. Genes VII. New York: Oxford University Press, 2004 pp. 875–912.
29. Wales MM, Biel MA, el Deiry W, et al. p53 activates expression of HIC-1, a new candidate tumour suppressor gene on 17p13.3. Nat Med 1995;1:570–577.
30. Chin KV, Ueda K, Pastan I, Gottesman MM. Modulation of activity of the promoter of the human MDR1 gene by Ras and p53. Science 1992;255:459–462.
31. de Kant E, Heide I, Thiede C, Herrmann R, Rochlitz CF. MDR1 expression correlates with mutant p53 expression in colorectal cancer metastases. J Cancer Res Clin Oncol 1996;122:671–675.
32. Nguyen KT, Liu B, Ueda K, Gottesman MM, Pastan I, Chin KV. Transactivation of the human multidrug resistance (MDR1) gene promoter by p53 mutants. Oncol Res 1994;6:71–77.
33. Thottassery JV, Zambetti GP, Arimori K, Schuetz EG, Schuetz JD. p53-dependent regulation of MDR1 gene expression causes selective resistance to chemotherapeutic agents. Proc Natl Acad Sci USA 1997;94:11,037–11,042.
34. Chevillard S, Lebeau J, Pouillart P, et al. Biological and clinical significance of concurrent p53 gene alterations, MDR1 gene expression, and S-phase fraction analyses in breast cancer patients treated with primary chemotherapy or radiotherapy. Clin Cancer Res 1997;3:2471–2478.
35. Crumrine RC, Thomas AL, Morgan PF. Attenuation of p53 expression protects against focal ischemic damage in transgenic mice. J Cereb Blood Flow Metab 1994;14:887–891.
36. Shahbazian MD, Orr HT, Zoghbi HY. Reduction of Purkinje cell pathology in SCA1 transgenic mice by p53 deletion. Neurobiol Dis 2001;8:974–981.
37. el Deiry WS, Tokino T, Velculescu VE, et al. WAF1, a potential mediator of p53 tumor suppression. Cell 1993;75:817–825.
38. Kleihues P, Louis DN, Scheithauer BW, et al. The WHO classification of tumors of the nervous system J Neuropathol Exp Neurol 2002;61:215–225
39. Nakatsuji Y, Miller RH. Density dependent modulation of cell cycle protein expression in astrocytes. J Neurosci Res 2001;66:487–496.
40. Nakayama K, Ishida N, Shirane M, et al. Mice lacking p27(Kip1) display increased body size, multiple organ hyperplasia, retinal dysplasia, and pituitary tumors. Cell 1996;85:707–720.
41. Kato JY, Matsuoka M, Polyak K, Massague J, Sherr CJ. Cyclic AMP-induced G1 phase arrest mediated by an inhibitor (p27Kip1) of cyclin-dependent kinase 4 activation. Cell 1994;79:487–496.

42. Park DS, Morris EJ, Bremner R, et al. Involvement of retinoblastoma family members and E2F/DP complexes in the death of neurons evoked by DNA damage. J Neurosci 2000;20:3104–3114.
43. Lee EY, Chang CY, Hu N, et al. Mice deficient for Rb are nonviable and show defects in neurogenesis and haematopoiesis. Nature 1992;359:288–294.
44. Dick FA, Sailhamer E, Dyson NJ. Mutagenesis of the pRB pocket reveals that cell cycle arrest functions are separable from binding to viral oncoproteins. Mol Cell Biol 2000;20:3715–3727.
45. Grant GA, Janigro D. The blood-brain barrier, In: Winn HR (ed). Youmans Neurological Surgery. Saunders, Philadelphia: 2004 153–174.
46. Rakic P. Glial cells in development. In vivo and in vitro approaches. Ann NY Acad Sci 1991;633:96–99.
47. Araque A, Parpura V, Sanzgiri RP, et al. Tripartite synapses: glia, the unacknowledged partner. Trends Neurosci 1999;22:208–215.
48. Bezzi P, Carmignoto G, Pasti L, et al. Prostaglandins stimulate calcium-dependent glutamate release in astrocytes. Nature 1998;391:281–285.
49. Parpura V, Basarsky TA, Liu F, Jeftinija K, Jeftinija S, Haydon PG. Glutamate-mediated astrocyte-neuron signaling. Nature 1994;369:744–747.
50. Chen ZL, Strickland S. Neuronal death in the hippocampus is promoted by plasmin-catalyzed degradation of laminin. Cell 1997; 91:917–925.
51. del Zoppo GJ, Mabuchi T. Cerebral microvessel responses to focal ischemia. J Cereb Blood Flow Metab 2003;23:879–894.
52. Privat A. Astrocytes as support for axonal regeneration in the central nervous system of mammals. Glia 2003;43:91–93.
53. Norenberg MD. The reactive astrocyte. In: Aschner M, Costa LG (eds). The Role of Glia in Neurotoxicity. Boca Raton: CRC Press, 2005 pp 73–92.
54. Liberto CM, Albrecht PJ, Herx LM, Yong VW, Levison SW. Pro-regenerative properties of cytokine-activated astrocytes. J Neurochem 2004;89:1092–1100.
55. Descamps L, Coisne C, Dehouck B, Cecchelli R, Torpier G. Protective effect of glial cells against lipopolysaccharide-mediated blood-brain barrier injury. Glia 2003;42:46–58.
56. Bachoo RM, Kim RS, Ligon KL, et al. Molecular diversity of astrocytes with implications for neurological disorders. Proc Natl Acad Sci USA 2004;101:8384–8389.
57. D'Ambrosio R, Maris DO, Grady MS, Winn HR, Janigro D. Impaired K homeostasis and altered electrophysiological properties of post-traumatic hippocampal glia. J Neurosci 1999;19:8152–8162.
58. D'Ambrosio R, Maris DO, Grady MS, Winn HR, Janigro D. Selective loss of hippocampal long-term potentiation, but not depression, following fluid percussion injury. Brain Res 1998;786: 64–79.
59. Dombrowski S, Najm I, Janigro D. CNS microenvironment and neuronal excitability. In: Walz W (ed). The Neuronal Environment: Brain Homeostasis in Health and Disease. Totowa, NJ: Humana Press Inc. 2002; pp 3–24.
60. Barres BA. Glial ion channels. Curr Opin Neurobiol 1991;1: 354–359.
61. Roitbak T, Syková E. Diffusion barriers evoked in the rat cortex by reactive astrogliosis. Glia 1999;28:40–48.
62. MacFarlane SN, Sontheimer H. Electrophysiological changes that accompany reactive gliosis in vitro. J Neurosci 1997;17:7316–7329.
63. D'Ambrosio R, Fairbanks JP, Fender JS, Born DE, Doyle DL, Miller JW. Post-traumatic epilepsy following fluid percussion injury in the rat. Brain 2004;127:304–314.
64. McKhann GM, D'Ambrosio R, Janigro D. Heterogeneity of astrocyte resting membrane potentials and intercellular coupling revealed by whole-cell and gramicidin-perforated patch recordings from cultured neocortical and hippocampal slice astrocytes. J Neurosci 1997;17:6850–6863.
65. Lee SH, Kim WT, Cornell-Bell AH, Sontheimer H. Astrocytes exhibit regional specificity in gap-junction coupling. Glia 1994;11:315–325.
66. Lee SH, Magge S, Spencer DD, Sontheimer H, Cornell-Bell AH. Human epileptic astrocytes exhibit increased gap junction coupling. Glia 1995;15:195–202.
67. Canady KS, Rubel EW. Rapid and reversible astrocytic reaction to afferent activity blockade in chick cochlear nucleus. J Neurosci 1992;12:1001–1009.
68. Bordey A, Sontheimer H. Postnatal development of ionic currents in rat hippocampal astrocytes in situ. J Neurophys 1997;78:461–477.
69. MacFarlane SN, Sontheimer H. Changes in ion channel expression accompany cell cycle progression of spinal cord astrocytes. Glia 2000;30:39–48.

70. Gould E, Reeves AJ, Graziano MS, Gross CG. Neurogenesis in the neocortex of adult primates. Science 1999;286:548–552.
71. Shors TJ, Miesegaes G, Beylin A, Zhao M, Rydel T, Gould E. Neurogenesis in the adult is involved in the formation of trace memories. Nature 2001;410:372–376.
72. Gould E, Gross CG. Neurogenesis in adult mammals: some progress and problems. J Neurosci 2002;22:619–623.
73. Endres M, Dirnagl U. Ischemia and stroke. Adv Exp Med Biol 2002;513:455–473.
74. Wartiovaara A, Syvanen AC. Analysis of nucleotide sequence variations by solid-phase minisequencing. Methods Mol Biol 2002;187: 57–63.
75. Nagy ZS, Esiri MM. Apoptosis-related protein expression in the hippocampus in Alzheimer's disease. Neurobiol Aging 1997;18:565–571.
76. Smith MZ, Nagy Z, Esiri MM. Cell cycle-related protein expression in vascular dementia and Alzheimer's disease. Neurosci Lett 1999;271:45–48.
77. Nagy Z. Mechanisms of neuronal death in Down's syndrome. J Neural Trans Suppl 1999;57:233–245.
78. Timsit S, Rivera S, Ouaghi P, et al. Increased cyclin D1 in vulnerable neurons in the hippocampus after ischaemia and epilepsy: a modulator of in vivo programmed cell death? Eur J Neurosci 1999;11:263–278.
79. Nagy Z, Esiri MM. Neuronal cyclin expression in the hippocampus in temporal lobe epilepsy. Exp Neurol 1998;150:240–247.
80. Nagy Z, Esiri MM, Smith AD. The cell division cycle and the pathophysiology of Alzheimer's disease. Neuroscience 1998;87:731–739.
81. Liu W, Bi X, Tocco G, Baudry M, Schreiber SS. Increased expression of cyclin D1 in the adult rat brain following kainic acid treatment. Neuroreport 1996;7:2785–2789.
82. Dirnagl U, Iadecola C, Moskowitz MA. Pathobiology of ischaemic stroke: an integrated view. Trends Neurosci 1999;22:391–397.
83. Li Y, Chopp M, Jiang N, Zhang ZG, Zaloga C. Induction of DNA fragmentation after 10 to 120 minutes of focal cerebral ischemia in rats. Stroke 1995;26:1252–1257.
84. O'Hare M, Wang F, Park DS. Cyclin-dependent kinases as potential targets to improve stroke outcome. Pharmacol Therapeut 2002;93:135–143.
85. Becker EB, Bonni A. Cell cycle regulation of neuronal apoptosis in development and disease. Prog Neurobiol 2004;72:1–25.
86. Guegan C, Levy V, David JP, Ajchenbaum-Cymbalista F, Sola B. c-Jun and cyclin D1 proteins as mediators of neuronal death after a focal ischaemic insult. Neuroreport 1997;8:1003–1007.
87. Osuga H, Osuga S, Wang F, et al. Cyclin-dependent kinases as a therapeutic target for stroke. Proc Natl Acad Sci USA 2000;97:10,254–10,259.
88. Becker F, Murthi K, Smith C, et al. A three-hybrid approach to scanning the proteome for targets of small molecule kinase inhibitors. Chem Biol 2004;11:211–223.
89. Ranganathan S, Bowser R. Alterations in G(1) to S phase cell cycle regulators during amyotrophic lateral sclerosis. Am J Pathol 2003;162:823–835.
90. Staropoli JF, McDermott C, Martinat C, Schulman B, Demireva E, Abeliovich A. Parkin is a component of an SCF-like ubiquitin ligase complex and protects postmitotic neurons from kainate excitotoxicity. Neuron 2003;37:735–749.
91. Clement AM, Nguyen MD, Roberts EA, et al. Wild-type nonneuronal cells extend survival of SOD1 mutant motor neurons in ALS mice. Science 2003;302:113–117.
92. Sakurai M, Hayashi T, Abe K, Itoyama Y, Tabayashi K, Rosenblum WI. Cyclin D1 and Cdk4 protein induction in motor neurons after transient spinal cord ischemia in rabbits. Stroke 2000;31:200–207.
93. Su JH, Anderson AJ, Cummings BJ, Cotman CW. Immunohistochemical evidence for apoptosis in Alzheimer's disease. Neuroreport 1994;5:2529–2533.
94. Busser J, Geldmacher DS, Herrup K. Ectopic cell cycle proteins predict the sites of neuronal cell death in Alzheimer's disease brain. J Neurosci 1998;18:2801–2807.
95. Vincent I, Jicha G, Rosado M, Dickson DW. Aberrant expression of mitotic cdc2/cyclin B1 kinase in degenerating neurons of Alzheimer's disease brain. J Neurosci 1997;17:3588–3598.

14
Potassium Channels, Cell Cycle, and Tumorigenesis in the Central Nervous System

Gabriele Dini, PhD, Erin V. Ilkanich, BS, and Damir Janigro, PhD

SUMMARY

In addition to the well-understood role in action potential repolarization and control of brain homeostasis, potassium channels also play a fundamental role in controlling cellular proliferation. Here we wish to summarize evidence supporting potassium channels' involvement in tumorigenesis. We will first introduce basic concepts in cell cycle in the central nervous system, and then describe the multiple roles of potassium channels and their modulators in the central nervous system. Finally, we will focus on how brain tumors are controlled by potassium channel regulation, linking tumorigenesis process to cell cycle re-entry. We also introduce a novel possible modulator of potassium channels, and thus of tumorigenesis.

Key Words: Ion channels; potassium channels; cell cycle; tumors; CNS.

1. CELL CYCLE IN THE CENTRAL NERVOUS SYSTEM

In order for an organism to grow and survive, a proper balance between cell division, differentiation, and controlled cell death (apoptosis) is crucial. Cell cycle is a highly conserved mechanism by which eukaryotic cells proliferate. It is typically divided into four distinct phases: G_1 (first gap), S (DNA synthesis), G_2 (second gap), and M (mitosis); cells that exit the cell cycle are said to be in a quiescent (G_0) state. Cell cycle progression is tightly regulated by control checkpoints, which ensure the correct completion of each phase before the progression to the next one.

A key role is played by cyclins (D, E, A, and B), cyclin-dependent kinases (CDKs), and their inhibitors (CDKIs), which are sequentially expressed to coordinate the orderly progression through cell cycle *(1,2)*. The sequential activation/inhibition of different CDKs and their activity are regulated by several mechanisms, including growth factors and various signaling molecules (Rb, E2F, p21, p27, and p53). Passage beyond the restriction point in G_1 is controlled by a number of complex transcription factors as well as the expression of cell cycle-related proteins *(2,3)*. The E2F family of transcription factors plays a central role in regulating cellular proliferation by controlling gene expression *(4)*. E2F activity is regulated in a cell cycle-dependent manner, principally through its association with the retinoblastoma tumor suppressor protein (Rb), which is in turn regulated by other CDKs and CDKIs. p53 is a tumor suppressor protein which is activated in response to cellular stress and prevents further progression through the cell cycle until cellular damage has been repaired.

If the cell has entered the S phase and is prepared to divide, programmed cell death occurs, preventing proliferation of damaged cells. In the brain, loss of p53 influences the cell cycle and therefore allows the survival of potentially damaged glial cells to proliferate *(5)*. p21 is found downstream of p53 and is a universal inhibitor of CDKs; it binds to an extensive array of

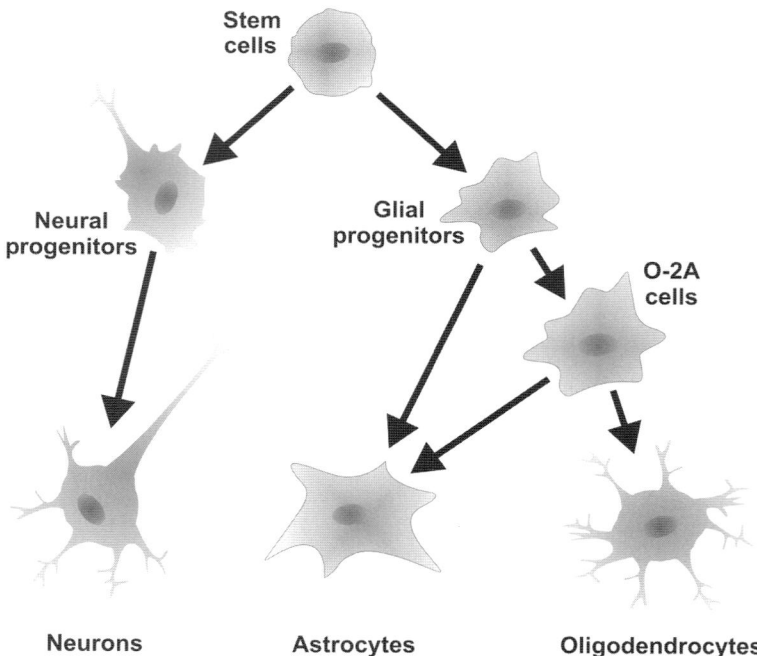

Fig. 1. Simplified model of differentiation of neural stem cells. Their fate is dependent on a number of factors involved in cell cycle regulation.

cyclin/CDK complexes which are most pertinent to G_1/S phase arrest. p27 is another member of the CDKIs; it interacts with all subtypes of cyclin/CDK, causing cell cycle arrest in the G_1 phase.

Developmentally, neural progenitor stem cells differentiate into neurons or glial cells over a protracted period (Fig. 1), and the mature phenotype of a cell is often determined postmitotically, or in adult stages. Cell cycle and cell fate determination are closely related events, and the same machinery and regulatory factors that govern the cell cycle are also fundamental to trigger cell differentiation (6,7). Specific cell cycle components differentially affect neuronal determination and differentiation; p27, for example, not only modulates various cyclins/CDKs complexes, but is also fundamental for the differentiation of glial retinal cells (6).

Apoptosis (or programmed cell death) and necrosis are fundamental for the normal development and function of multicellular organisms, balancing cell proliferation. Contrary to the necrotic process, apoptotic cells die in a controlled, regulated fashion. Several steps are characteristic of the apoptotic process: cell shrinkage, nuclear condensation, nuclear fragmentation, cell blebbing, and phagocytosis by macrophages. In the central nervous system (CNS), neuronal apoptosis is indispensable for a normal development, and the main function of this postnatal programmed cell death appears to be the adjustment of the number of neurons, possibly eliminating redundancy (8).

Interestingly, the cell cycle machinery shares some common participants with the apoptotic pathway, suggesting that both processes might share similar regulatory mechanisms. The decision to remain quiescent, resume active proliferation, or enter apoptosis is dictated by common extracellular and intracellular factors (7). A diagrammatic representation of the major players involved in cell cycle re-entry and cellular proliferation is shown in Fig. 2.

In most brain regions, neurogenesis is generally confined to a discrete developmental period, during which neuronal precursor cells proliferate and differentiate into neurons and glial cells within designated germinal zones. Once the neuronal precursors exit the cell cycle and enter a postmitotic state, these cells migrate out of their proliferative zones and remain in a terminally

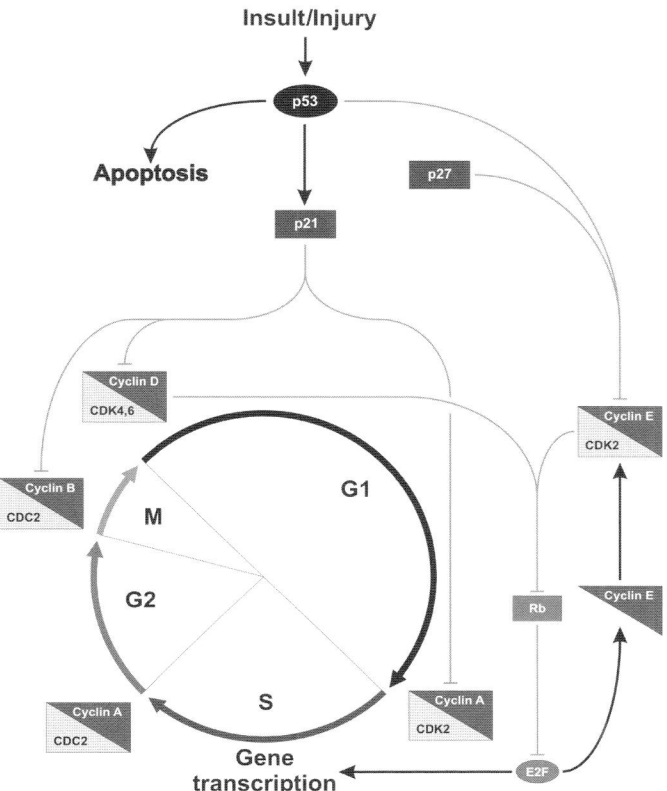

Fig. 2. Principal regulators of the cell cycle. The proper activation/inactivation of various cyclins, CDKs, and CDKIs in necessary for cell cycle progression or arrest.

differentiated state. However, a certain amount of neurogenesis takes place during adulthood. It has been shown that neurons and glial cells are generated in well-defined zones of several species, well into the postnatal and adult period. The presence of proliferating cells in the adult human brain has been demonstrated in both the dentate gyrus (neurons, astrocytes, and asymmetrically dividing progenitor cells) and in the subventricular zone (progenitor cells only) *(9–11)*. The newly generated cells in the adult brain have been associated with a functional role in a neuronal remodeling process, necessary for adaptation to novel environments *(12)*, memory formation *(13)*, and learning *(14,15)*.

Abnormal attempts at cell cycle re-entry by terminally differentiated cells may occur in response to perturbations in the extracellular environment, and usually results in apoptosis rather than proliferation *(16)*. Common triggering factors for cell cycle re-entry and apoptosis are the absence of factors that promote cell survival (such as neurotrophins or electrical activity), the presence of extrinsic cues that actively promote cell death (oxidative stress, glutamate toxicity, and extracellular proapoptotic signals), ionic imbalances, aggregation of misfolded proteins that interfere with normal cell life, injury, and disease *(17,18)*. Apoptosis and neural proliferation are a prominent feature in a number pathological states, such as Alzheimer's disease *(19,20)*, adult Huntington's disease *(21)*, stroke *(8)*, transient cerebral ischemia *(22–24)*, and epilepsy *(25)*.

2. POTASSIUM CHANNELS IN THE CNS

Potassium (K) channels are a family of membrane proteins that, by allowing selective permeation of K^+ across the cell membrane, play critical roles in a wide variety of physiological processes. More than 200 genes encoding a variety of K^+ channels have been identified, all sharing a set of common features: (1) a pore that allows K^+ to flow across the cell membrane; (2) a selectivity filter for K^+; and (3) a gating mechanism to switch between the open and closed channel conformations (26). The latter can be represented either by changes in transmembrane potential, blocking of the pore by intracellular factors (ions, polyamines), or modulation by phosphorylating proteins (see Heading 4.).

Three major groups of channels are recognized: (1) six-transmembrane domains, one-pore channels, in which depolarization causes conformational changes leading to channel opening, allowing ions to flow; (2) two-transmembrane domains, one-pore channels: they allow K^+ currents to flow more in the inward direction than outwardly, owing to a rectification mechanism attributed to internal Mg^{2+} and polyamines that occlude the conductive pore (26,27); and (3) four-transmembrane domains, two-pore channels: perhaps the most abundant class of potassium channels (26), they include weak inward rectifier K^+ channels, which may be responsible for background leak currents. Each group further comprises a variety of channel subtypes, divided accordingly to their current characteristics and gating properties. K_{IR} channels have been classified into four subfamilies: (1) classic inward rectifying K^+ channels ($K_{IR}2$), (2) G-protein-coupled inward rectifying channels ($K_{IR}3$), (3) adenosine triphosphate (ATP)-dependent channels ($K_{IR}1$, $K_{IR}4$), and (4) ATP-sensitive channels ($K_{IR}6$). Three categories of voltage-gated channels have been identified: (1) calcium (Ca^{2+})-activated channels; (2) ether-a-go-go channels, and (3) voltage-gated K^+ channels (K_V) (26).

Potassium channels play a fundamental role in the function of numerous cell types, and are involved in modulation of neuronal excitability, cellular signaling, regulation of membrane potential, and cell volume. K^+ channel gene mutations are linked to various pathogenic processes, which usually involve altered channel regulation leading to perturbations in membrane excitability and neuronal function (26,28). For example, in the weaver mouse, a single missense mutation in the gene encoding the G protein-coupled inward rectifier potassium channel $K_{IR}3.2$ (or GIRK2) results in a constitutive activation of the channel, leading to seizures, chronic depolarization of affected neurons, and massive cell death (8,26).

Neurons and glia express different types of potassium channels, reflecting the variety of functions of these cell populations in the brain. Given their role in the CNS, neurons are richly endowed with voltage-gated channels, which are fundamental components in the regulation of their excitability. Glial cells express a variety of both voltage-gated and inwardly rectifying channels (29). Table 1 shows the most relevant types of potassium channels expressed in the brain and their distribution across neurons, astrocytes, and oligodendrocytes.

K_{IR} channels, which are the only channels with a high, open probability at the hyperpolarized resting membrane potential (RMP), are essential to preserve neuronal firing by buffering excess $[K^+]_{out}$, which would depolarize neurons (62). A second important function is their ability to maintain a very negative RMP in glial cells, which establishes the Na^+ gradient across the membrane used as driving force for several transporters (27,62) and prevents generation of action potentials (63).

Owing to their role in determination of intracellular osmolarity, K^+ ions have a major role in regulation cell volume. Increases in outward K^+ currents would then cause a decrease in intracellular potassium and lead to cell shrinkage; a process which is involved in the signal transduction cascade that leads to apoptosis (64–66). Induction of the apoptotic pathway is a multistep process in which a reduction of intracellular potassium leads alterations in the cell homeostasis, eventually resulting in membrane depolarization, cell shrinkage, and activation of the apoptotic effectors (67).

3. POTASSIUM CHANNELS AND CELL CYCLE RE-ENTRY

There is increasing evidence that potassium channel activity can influence cell cycle progression. The initial phases of cellular proliferation and differentiation are strongly dependent on

Table 1
Predominant Potassium Channels in the Adult CNS, and Their Relative Expression Across Cell Populations of the Brain

Channel type	Channel name	Gene	WB	N	A	O	References
Inward rectifier	$K_{IR}2.1$	KCNJ2	+	+	(+)	(+)	(26,27,30–34)
	$K_{IR}2.2$	KCNJ3	+	+	(+)	(+)	
	$K_{IR}2.3$	KCNJ4	+	+	(+)		
	$K_{IR}2.4$	KCNJ14	+				
G protein-coupled IR	$K_{IR}3.1$	KCNJ3	(+)	+	+		(26,27,30,33–36)
	$K_{IR}3.2$	KCNJ6	+	+			
	$K_{IR}3.3$	KCNJ9	+	+			
	$K_{IR}3.4$	KCNJ5	(+)	(+)			
ATP-dependent IR	$K_{IR}4.1$	KCNJ10	+	(+)	+	+	(26,27,30–32,34, 35,37,38)
	$K_{IR}4.2$	KCNJ15	+				
	$K_{IR}5.1$	KCNJ16	+	(+)	+	(+)	
ATP-sensitive IR	$K_{IR}6.1$	KCNJ8	+	(+)	+		(26,27,34,35,39,40)
	$K_{IR}6.2$	KCNJ11	+	+			
	$K_{IR}7.1$	KCNJ13	+				
Voltage-gated (Shaker)	$K_V1.1$	KCNA1	++	+			(26,29,31,33,35, 41–45)
	$K_V1.2$	KCNA2	++	+	+	+	
	$K_V1.3$	KCNA3	+		+		
	$K_V1.4$	KCNA4	++	+		+	
	$K_V1.5$	KCNA5	+		+	+	
	$K_V1.6$	KCNA6	+	+	+	+	
Voltage/cGMP-gated	$K_V1.10$	KCNA10	+				(26,46)
Shab	$K_V2.x$	KCNB1	++	+			(26,42,47)
Shaw	$K_V3.x$	KCNC1-4	+	+			(26,42)
Shal	$K_V4.x$	KCND1-3	+	+			(26,42)
Silent	$K_V5.1$	KCNF1	+	*	*	*	(26,47–49)
	$K_V6.1$	KCNG1	+	*	*	*	
	$K_V8.1$	KCNV1	+	*	*	*	
	$K_V9.x$	KCNS1-3	+	*	*	*	
Ether-a-go-go (EAG)	EAG	KCNH1	+	+	(+)	(+)	(26,50,51)
EAG-related	ERG	KCNH2	+	+	(+)		(26,45,50–52)
	BEC1	KCNH3	+	+			
	BEC2	KCNH4	+	+			
K_VLQT-related	K_VLQT1	KCNQ1	+	+			(26,35,51,53)
	K_VLQT2	KCNQ2	+	+			
	K_VLQT3	KCNQ3	+	+			
Ca^{2+}-activated	Slo	KCNMA1	+	+			(26,54,55)
	SKCa	KCNN1-3	+	+			
	IKCa	KCNN4	+	+			
Two-pore channels	TWIK-1	KCNK1	+	+	(+)	(+)	(26,56–61)
	TREK	KCNK2	+	+	(+)	(+)	
	TASK	KCNK3	+	+	(+)	(+)	
	TRAAK	KCNK7	+	+	(+)	(+)	

WB, whole brain; N, neurons; A, astrocytes; O, oligodendrocytes; +, generalized expression; (+), low expression levels or limited expression; ++, very strong expression; *, K_V5.1–9.x are electrically silent, and act as modulators for other K_V subunits.

transient changes in membrane ionic permeability and intracellular cations. These play important roles in signaling pathways leading to proliferation, differentiation and apoptosis, and have been demonstrated in a variety of CNS and non-CNS cells *(68)* over the course of development and differentiation. K^+ channel activity has been linked to active proliferation in retinal glial cells *(69)*, spinal cord astrocytes *(70)*, neural crest precursors *(71)*, reactive brain astrocytes *(72)*, pituitary cell lines *(73)*, pituitary cells *(3)*, neuroblastoma cells *(74)*, Schwann cells, and oligodendrocyte progenitors *(1,75)*.

Cell cycle-specific changes in K^+ currents have been described in both developing cells and tumors; alterations of inwardly and outwardly rectifying K^+ and Na^+ currents are concurrent with changes in RMP and necessary for progression through cell cycle checkpoints *(1,72,76)*. In many cell types, these changes in RMP correlate with cell cycle state and cell activation, but the precise cause–effect relationships are still unclear *(77)*. It has been hypothesized that changes in K^+ channels activity maintain permissive RMP at critical cell cycle checkpoints *(26,76)*. Alternatively, potassium channel activity might control proliferation by regulating intracellular K^+ concentration, which plays an important role in cell cycle re-entry. A decrease in potassium is a permissive event in initiating proliferation, whereas physiological concentrations are inhibitory for cell cycle re-entry *(78,79)*. Finally, it has also been hypothesized that K^+ channels regulate cellular volume as well as the concentration of other intracellular solutes that may be critical for cell metabolism *(3,70)*.

A common feature of cell cycle regulation is that cells arrested at specific stages of the cell cycle exhibit distinctly different current and RMP profiles. Cells in the G_1 phase demonstrate a marked inward rectification, attributed to prominent K_{IR} currents. Progression through the G_1/S border shows a slight increase in K_V currents (without changes in RMP). Outward potassium currents are further increased in cells in the S phase, whereas inward currents decrease. Pharmacological blockade of K_V prevents S phase entry, and K_{IR} inhibition by Cs^+ or Ba^{2+} promotes cell cycle progression through the G_1/S checkpoint. Chemical arrest of actively proliferating astrocytes in G_1/G_0 induces premature expression of inwardly rectifying K^+ currents *(62)*.

Membrane hyperpolarization seems to be an essential requirement for cell cycle progression *(76)*, but not for cell cycle re-entry. Membrane depolarization, either by blockade of K^+ channels or high $[K^+]_{out}$, arrests cells in the G_1 phase without preventing cell cycle re-entry and cyclin D accumulation *(1,73,75)*. Conversely, conditions that increase $[Na^+]_{in}$, such as inhibition of the Na^+/K^+-ATPase or low $[K^+]_{out}$, facilitate cell proliferation by playing an important role during S phase *(72)*.

Changes in RMP and K^+ channel activity are necessary to modify the levels of CDKIs in mitotically active cells of the mammalian CNS, probably through a mechanism involving p27 and p21 *(1,80)*. Most cells display outward potassium currents during their proliferative phase, whereas postnatal differentiation of precursor cells, neurons, and glia is accompanied by important changes in K^+ channel activity, and in particular by an enhanced amplitude of K_{IR} currents *(69,75,81,82)*.

In glial cells, expression of K_{IR} currents is a typical feature of differentiated, mature cells: they are absent in immature cells and begin to appear gradually, correlating with cell differentiation. Cell injury or gliosis can cause downregulation of channel activity, resulting in impaired K^+ buffering, large RMP oscillations, and enhanced glial proliferation. Conversely, large K_V currents are typical of proliferating cells, and their activity is increased following cell injury and gliosis *(72,83)*. Analogous observations correlate functional state of different types of cells with potassium channel expression: quiescent cells primarily express K_{IR} channels, which are downregulated in proliferating, active cells, although the opposite is observed for K_V currents *(69,71,72,77,84,85)*. The most significant changes in glial RMP, K_{IR}, and K_V currents are summarized in Table 2.

4. MODULATION OF POTASSIUM CHANNELS

Potassium channels' function is modulated by a diversity of factors, depending on their type and characteristics. Fast channel regulation can be achieved through changes in channel activity, whereas up-downregulation in channel expression modulates their long-term response. Major known regulators of activity are intracellular and extracellular ionic concentrations (i.e., Mg^{2+}), ATP/adenosine diphosphate, pH, cellular volume, growth factors, polyamines, and a variety of

Table 2
Changes in Glial Resting Membrane Potential and Inward/Outward Potassium Currents in the Early Phases of the Cell Cycle

	Quiescence	Cell cycle re-entry		Proliferation	
	G_0	G_0/G_1	G_1	G_1/S	S
RMP	Polarized	Hyperpolarized	Polarized	Depolarized	Depolarized
K_{IR}	High	–	–	– –	– –
K_V	Low	+	+	++	++

After a transient hyperpolarization, the membrane is significantly depolarized once the cell starts proliferating. Inward currents are progressively reduced, whereas outward currents increase (*[1,62,69,71–77,80–85], see text for details*).

phosphorylating proteins *(80,86,87)*. Expression of the voltage-dependent potassium channel $K_V 1.3$, for example, is modulated by the epidermal growth factor receptor and the insulin receptor tyrosine kinase, through a mechanism associated with tyrosine phosphorylation of the channel *(88)*. Similarly, platelet-derived growth factor upregulates $K_V 1.5$–$K_V 1.6$ mRNA transcription in oligodendrocyte progenitors and increases outward currents *(87)*. Protein kinase C (PKC) isoforms modulate both inwardly rectifying and voltage-gated channels *(89–92)*, and has been demonstrated to be required for changes in K_V currents in neuroblastoma PC12 cells exposed to neural growth factor *(93)*.

The PKC family comprises a variety of serin/threonine kinase enzymes responsible for transducing many cellular signals during mitogenesis, cellular metabolism, differentiation, tumor promotion, and apoptosis. In addition to being a regulator of both inward and outward, PKC is known to regulate the activity of different plasma membrane-related cytoskeletal proteins *(82)*. Various PKC isoforms have been related to modulation of potassium channels through phosphorylation of their active sites. In general, K_{IR} channel activity is downregulated by PKC *(85,92,94–96)*, whereas K_V currents seem to be upregulated *(3,89,91,94)*. Finally, PKC is fundamental during the transduction of cell death signals by modulating K^+ channel activity and cell shrinkage, which are early requisites for the activation of the apoptotic cascade *(82)*.

Other cell cycle-related proteins may link potassium channels and proliferation. It has been demonstrated that p62 binds to subunits of both PKC and K_V channels, recruiting the regulator protein kinase to the target ion channel and modulating the channel itself *(93)*.

5. TUMORS OF THE CNS

The process of tumorigenesis is initiated when alterations in three types of genes (oncogenes, tumor-suppressor, and stability genes) endow cells with a selective growth advantage. Mutated oncogenes are either constitutively active or activated by conditions by which the wild-type gene would not. Conversely, mutations in tumor-suppressor genes reduce their activity, facilitating cell cycle progression. Mutations in stability genes, which are responsible for repairing DNA and keep genetic alterations to a minimum, let other mutations occur and propagate at a higher rate (for an extensive review of tumorigenesis pathways, *see* ref. 97).

In the brain, majority of primary tumors arise from glial cells, and are collectively named gliomas *(98)*. They are characterized by defects in genes involved in cell cycle regulation *(99,100)* and have lost the ability to differentiate spontaneously. Gliomas can arise from astrocytes, oligodendrocytes, or their precursors, and are classified according to their hypothesized line of differentiation *(98)*. Their prognosis is very poor, owing to their invasive nature, and they are considered among the deadliest of the human tumors *(98,101)*. The factors that induce glial transformation into glioma are not well understood. The presence of stem cells in the adult human brain, which have proliferative and migratory capabilities, represents a possible origin of gliomas. Another possibility is that mature glial cells may dedifferentiate in response to insults or genetic mutations *(101)*.

A common feature of cancer cells is the loss of cell cycle checkpoints as a result of altered expression of positive (cyclins, CDKs) or negative regulators; virtually all tumors present inactivation of Rb and/or p53 pathways *(102–104)*. The most benign variant of astrocytoma, the pilocytic astrocytoma (grade I), is not characterized by consistent cell cycle aberrations. Grade III tumors (anaplastic astrocytomas), however, present aberrations of genes involved in the control of entry into the S phase in approx 50% of cases, and about 80% of glioblastomas (grade IV) have genetic alterations involving progression from G_1 to the S phase of the cell cycle, along with constitutive amplification of growth factor receptors (mainly epidermal growth factor receptor and platelet-derived growth factor receptor) *(86,100,103,104)*. Finally, the Ras signaling pathway is commonly altered in gliomas *(2)*.

6. POTASSIUM CHANNELS AND TUMORIGENESIS

Among the ion channels activated by growth factor stimulation, K^+ channels have been implicated in the growth and proliferation of several CNS and non-CNS cancer cell lines. Alterations in ion channel expression and activity are associated with cell proliferation, and possibly essential for tumorigenesis to occur. Overexpression of K_V channels has been demonstrated in a variety of tumors *(41,66,105)*, and their blockade by pharmacological agents inhibits proliferation in gliomas *(106,107)*, human melanoma cells *(108)*, small lung cancer cells *(109)*, squamous cell carcinoma *(105)*, prostatic cells *(26)*, and other tumoral cell lines *(110)*.

The magnitude and expression of outward potassium currents seem to correlate negatively with tumor grade, whereas comparable currents have been found in normal astrocytes and astrocytomas *(41,50,63)*, the channels of higher-grade tumors have lower activity *(41,50)*, maybe as a result of a more extensive cellular degeneration and DNA damage. In comparison to normal glia, brain tumors lack the K_{IR} currents typically expressed in astrocytes and have significantly more depolarized membrane potentials *(62,63)*, matching cells that are re-entering the cell cycle. The lack of K_{IR} currents could be either owing to a modulation of channel activity, or to downregulation of its expression by cell cycle-related agents (protein kinases, chaperonines, etc.). It has also been suggested that the lack of K_{IR} currents might be owing to a mislocalization of the channel, rather than its downregulation *(62)*. This could possibly be a result of alterations in the potassium channel assembly and membrane anchoring (for a description of the potassium channel ontogeny *see* the review by Deutsch *[111]*).

The loss of functional K_{IR} currents in glioma cells may have important physiological consequences, as potassium buffering and cell volume regulation would be impaired in tumors *(62,63)*. Significantly, glioma cells had a more rounded-up morphology and appeared swollen *(63)*. The lack significant K^+ uptake and clearance can also possibly exacerbate symptoms associated with their malignancy, such as seizures *(62)*. In addition, ion channels may contribute to the invasive behavior of gliomas by influencing salt and water movements between intracellular and extracellular compartments.

7. CELL CYCLE RE-ENTRY AND TUMORIGENESIS

Expression of K_V channels, lack of K_{IR} currents, and depolarized RMP in glioma cells are of particular interest, because they closely resemble the phenotype assumed by glial cells whereas proliferating or responding to injury and/or disease (Fig. 3). Thus, it can be hypothesized that the process of tumorigenesis and cellular proliferation in response to injury are closely related. As an example, both processes are associated with alterations in K_{IR} and K_V channels activity. Figure 3 shows a possible working model linking injury, cell cycle re-entry, cellular proliferation, and tumorigenesis. Cell cycle re-entry and cellular dedifferentiation could be triggered by either insult/injury or alterations in endogenous factors, leading to cellular proliferation and redifferentiation or, more likely, to apoptosis. If the proliferating cells have mutations in either tumors suppressor factors or oncogenes, they may develop a tumoral

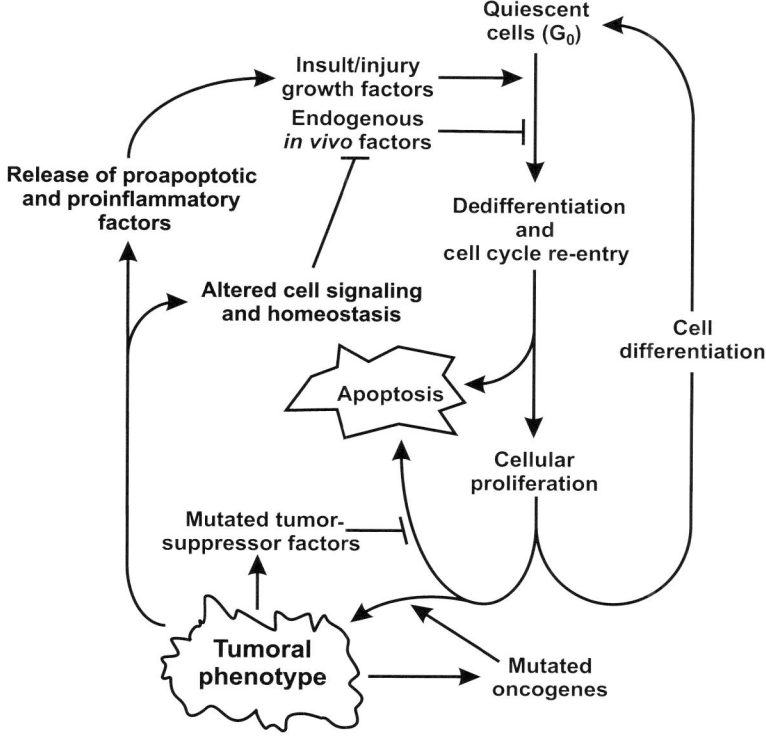

Fig. 3. Relationship between cell proliferation and tumorigenesis.

phenotype, which in turn may lead to further mutations and changes in the extracellular environment.

8. ELECTRICAL ACTIVITY: A NOVEL MODULATOR OF POTASSIUM CHANNELS AND TUMORIGENESIS

Recent work from our group *(112–114)* describes the antiproliferative effect of low-amplitude, low-frequency electrical stimulation on a variety of cells (Fig. 4). Similar results were independently reported by others *(115)*. A significant reduction in cell proliferation was observed in glial and prostate cancer cells, with a concomitant increase of the levels of expression of $K_{IR}3.2$. Proliferation was not altered in lung cancer cells, which did not express any significant level of $K_{IR}3.2$ protein either before or after stimulation. These antiproliferative effects of electrical stimulation were countered by blockade of inward rectifier channels with Cs^+, further demonstrating a functional involvement of K_{IR} channels in cellular proliferation (not shown).

The mechanism through which electrical stimulation affects cell cycle remains unknown. The paucity of tumors associated with excitable cells *(25,116)* hints that electrical excitability is associated with low mitotic activity. Interestingly, although brain tumors are most devastating, their frequency in the general population is low *(116)*. Clinical evidence also shows that seizures associated with long-term chronic epilepsy are positive prognostic factors in decreasing glioma mortality *(117)*. This suggests that electrical activity is not permissive for cell proliferation, and its effects can be mimicked by electrical stimulation.

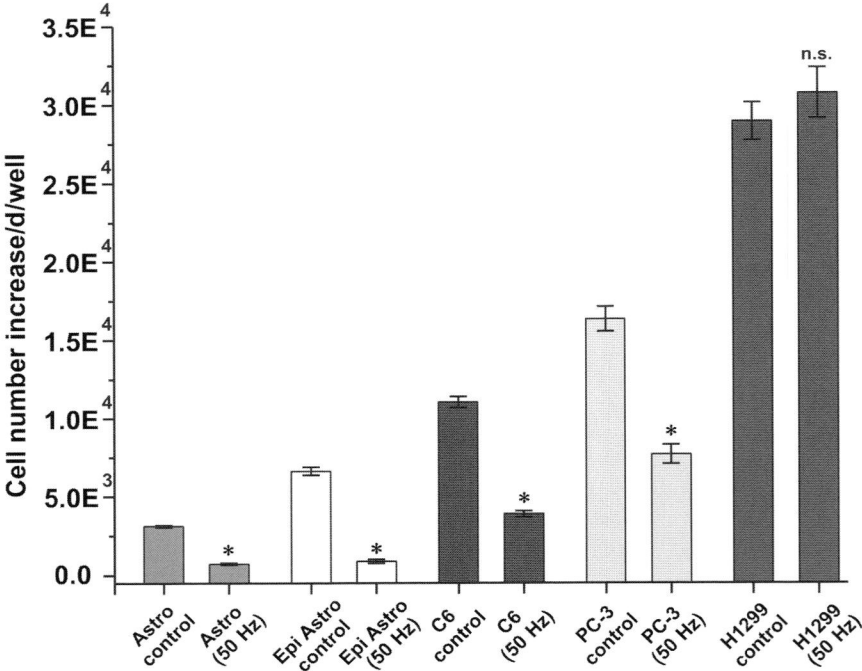

Fig. 4. Effect of electrical stimulation on cell proliferation on various cell populations. Astro, normal cortical astrocytes; Epi Astro, epileptic astrocytes; C6, rat glioma; PC3, prostate cancer; H1299, lung cancer; *, $p < 0.001$.

A possible explanation for the effects of electrical stimulation on potassium channels, and thus on proliferation, may be represented by PKC. Oscillating electric fields analogous to the ones we used have been shown to result in a significant decrease in PKC activity in cytosolic fraction *(118)*. It has also been demonstrated that electrical stimulation induces cellular differentiation, with a mechanism mediated by PKC *(4)*.

Interestingly, inhibition of a particular PKC isoform blocks proliferation in glioblastoma cell lines, further linking potassium channel activity to the kinase; conversely, PKC activators have been shown to decrease inward currents associated with $K_{IR}2.1$ channels *(109)*. PKC-mediated phosphorylation of ion channels seems therefore to be essential to modulate the channels activity, as shown in Fig. 5.

9. CONCLUSIONS, QUESTIONS, AND PERSPECTIVES

Changes in potassium channel activity regulate cellular proliferation, differentiation and activation. Channel expression can change in response to various stimuli, thus governing cell cycle re-entry. Normally, attempts at cell cycle re-entry lead to apoptosis, rather than proliferation; however, if underlying conditions involving cell cycle misregulation exist, the proliferating cells can assume a progressively worse tumoral phenotype. The exact cause–effect relationship linking tumorigenesis and potassium channels regulation needs to be further investigated. Although it is clear that K^+ currents are directly involved in the cell cycle, it is unclear whether mutations associated to constitutive activation/inactivation of the channels can lead to tumorigenesis *per se*, or simply are facilitating conditions for other mutations.

Potassium channel activity regulates cell proliferation, differentiation, and tumorigenesis. Therefore, altering the expression of K_V and K_{IR} channels may represent a novel approach to control cell proliferation in neoplastic tissues and, in particular, of brain tumors.

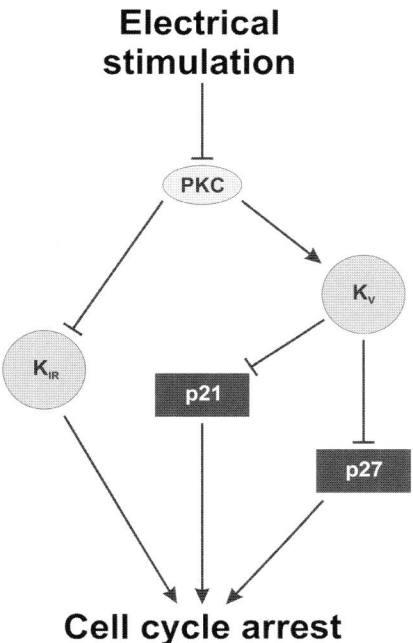

Fig. 5. Electrical stimulation may affect cell cycle by downregulation of PKC activity, which is needed for cell cycle progression.

REFERENCES

1. Ghiani CA, Yuan X, Eisen AM, et al. Voltage-activated K+ channels and membrane depolarization regulate accumulation of the cyclin-dependent kinase inhibitors p27(Kip1) and p21(CIP1) in glial progenitor cells. J Neurosci 1999;19:5380–5392.
2. Stevens B, Fields RD. Regulation of the cell cycle in normal and pathological glia. Neuroscientist 2002;8:93–97.
3. Czarnecki A, Vaur S, Dufy-Barbe L, Dufy B, Bresson-Bepoldin L. Cell cycle-related changes in transient K(+) current density in the GH3 pituitary cell line. Am J Physiol Cell Physiol 2000;279: C1819–C1828.
4. Stevens C, La Thangue NB. The emerging role of E2F-1 in the DNA damage response and checkpoint control. DNA Repair (Amst) 2004;3:1071–1079.
5. Marroni M, Agrawal ML, Kight K, et al. Relationship between expression of multiple drug resistance proteins and p53 tumor suppressor gene proteins in human brain astrocytes. Neuroscience 2003;121:605–617.
6. Ohnuma S, Philpott A, Harris WA. Cell cycle and cell fate in the nervous system. Curr Opin Neurobiol 2001;11:66–73.
7. Galderisi U, Jori FP, Giordano A. Cell cycle regulation and neural differentiation. Oncogene 2003;22:5208–5219.
8. Becker EB, Bonni A. Cell cycle regulation of neuronal apoptosis in development and disease. Prog Neurobiol 2004;72:1–25.
9. Bedard A, Parent A. Evidence of newly generated neurons in the human olfactory bulb. Brain Res Dev Brain Res 2004;151:159–168.
10. Pevny L, Rao MS. The stem-cell menagerie. Trends Neurosci 2003;26:351–359.
11. Eriksson PS, Perfilieva E, Bjork-Eriksson T, et al. Neurogenesis in the adult human hippocampus. Nat Med 1998;4:1313–1317.
12. Schaffer DV, Gage FH. Neurogenesis and neuroadaptation. Neuromolecular Med 2004;5:1–9.
13. Shors TJ, Miesegaes G, Beylin A, Zhao M, Rydel T, Gould E. Neurogenesis in the adult is involved in the formation of trace memories. Nature 2001;410:372–376.

14. Prickaerts J, Koopmans G, Blokland A, Scheepens A. Learning and adult neurogenesis: survival with or without proliferation? Neurobiol Learn Mem 2004;81:1–11.
15. van Praag H, Schinder AF, Christie BR, Toni N, Palmer TD, Gage FH. Functional neurogenesis in the adult hippocampus. Nature 2002;415:1030–1034.
16. Zhu X, Raina AK, Smith MA. Cell cycle events in neurons. Proliferation or death? Am J Pathol 1999;155:327–329.
17. Parent JM. Injury-induced neurogenesis in the adult mammalian brain. Neuroscientist 2003;9:261–272.
18. Chirumamilla S, Sun D, Bullock MR, Colello RJ. Traumatic brain injury induced cell proliferation in the adult mammalian central nervous system. J Neurotrauma 2002;19:693–703.
19. Zhu X, McShea A, Harris PL, et al. Elevated expression of a regulator of the G2/M phase of the cell cycle, neuronal CIP-1-associated regulator of cyclin B, in Alzheimer's disease. J Neurosci Res 2004;75:698–703.
20. Raina AK, Pardo P, Rottkamp CA, Zhu X, Pereira-Smith OM, Smith MA. Neurons in Alzheimer disease emerge from senescence. Mech Ageing Dev 2001;123:3–9.
21. Curtis MA, Penney EB, Pearson AG, et al. Increased cell proliferation and neurogenesis in the adult human Huntington's disease brain. Proc Natl Acad Sci USA 2003;100:9023–9027.
22. Wen Y, Yang S, Liu R, Brun-Zinkernagel AM, Koulen P, Simpkins JW. Transient cerebral ischemia induces aberrant neuronal cell cycle re-entry and Alzheimer's disease-like tauopathy in female rats. J Biol Chem 2004;279:22,684–22,692.
23. Love S. Neuronal expression of cell cycle-related proteins after brain ischaemia in man. Neurosci Lett 2003;353:29–32.
24. Sharp FR, Liu J, Bernabeu R. Neurogenesis following brain ischemia. Brain Res Dev Brain Res 2002;134:23–30.
25. Blumcke I, Wiestler OD. Gangliogliomas: an intriguing tumor entity associated with focal epilepsies. J Neuropathol Exp Neurol 2002;61:575–584.
26. Shieh CC, Coghlan M, Sullivan JP, Gopalakrishnan M. Potassium channels: molecular defects, diseases, and therapeutic opportunities. Pharmacol Rev 2000;52:557–594.
27. Nichols CG, Lopatin AN. Inward rectifier potassium channels. Annu Rev Physiol 1997;59:171–191.
28. Jan LY, Jan YN. Voltage-gated and inwardly rectifying potassium channels. J Physiol 1997;505(Pt 2): 267–282.
29. Verkhratsky A, Steinhauser C. Ion channels in glial cells. Brain Res Brain Res Rev 2000;32:380–412.
30. Doupnik CA, Davidson N, Lester HA. The inward rectifier potassium channel family. Curr Opin Neurobiol 1995;5:268–77.
31. Horio Y. Potassium channels of glial cells: distribution and function. Jpn J Pharmacol 2001;87:1–6.
32. Schroder W, Seifert G, Huttmann K, Hinterkeuser S, Steinhauser C. AMPA receptor-mediated modulation of inward rectifier K+ channels in astrocytes of mouse hippocampus. Mol Cell Neurosci 2002;19:447–458.
33. Karschin C, Dissmann E, Stuhmer W, Karschin A. IRK(1-3) and GIRK(1-4) inwardly rectifying K+ channel mRNAs are differentially expressed in the adult rat brain. J Neurosci 1996;16:3559–3570.
34. Isomoto S, Kondo C, Kurachi Y. Inwardly rectifying potassium channels: their molecular heterogeneity and function. Jpn J Physiol 1997;47:11–39.
35. Wickenden AD. Potassium channels as anti-epileptic drug targets. Neuropharmacology 2002;43:1055–1060.
36. Leaney JL. Contribution of Kir3.1, Kir3.2A and Kir3.2C subunits to native G protein-gated inwardly rectifying potassium currents in cultured hippocampal neurons. Eur J Neurosci 2003;18:2110–2118.
37. Kalsi AS, Greenwood K, Wilkin G, Butt AM. Kir4.1 expression by astrocytes and oligodendrocytes in CNS white matter: a developmental study in the rat optic nerve. J Anat 2004;204:475–485.
38. Hibino H, Fujita A, Iwai K, Yamada M, Kurachi Y. Differential assembly of inwardly rectifying K+ channel subunits, Kir4.1 and Kir5.1, in brain astrocytes. J Biol Chem 2004;279:44,065–44,073.
39. Thomzig A, Wenzel M, Karschin C, et al. Kir6.1 is the principal pore-forming subunit of astrocyte but not neuronal plasma membrane K-ATP channels. Mol Cell Neurosci 2001;18:671–690.
40. Thomzig A, Pruss H, Veh RW. The Kir6.1-protein, a pore-forming subunit of ATP-sensitive potassium channels, is prominently expressed by giant cholinergic interneurons in the striatum of the rat brain. Brain Res 2003;986:132–138.
41. Preussat K, Beetz C, Schrey M, et al. Expression of voltage-gated potassium channels Kv1.3 and Kv1.5 in human gliomas. Neurosci Lett 2003;346:33–36.

42. Trimmer JS, Rhodes KJ. Localization of voltage-gated ion channels in mammalian brain. Annu Rev Physiol 2004;66:477–519.
43. Attali B, Wang N, Kolot A, Sobko A, Cherepanov V, Soliven B. Characterization of delayed rectifier Kv channels in oligodendrocytes and progenitor cells. J Neurosci 1997;17:8234–8245.
44. Coleman SK, Newcombe J, Pryke J, Dolly JO. Subunit composition of Kv1 channels in human CNS. J Neurochem 1999;73:849–858.
45. Miyake A, Mochizuki S, Yokoi H, Kohda M, Furuichi K. New ether-a-go-go K(+) channel family members localized in human telencephalon. J Biol Chem 1999;274:25,018–25,025.
46. Yao X, Segal AS, Welling P, et al. Primary structure and functional expression of a cGMP-gated potassium channel. Proc Natl Acad Sci USA 1995;92:11,711–11,715.
47. Salinas M, Duprat F, Heurteaux C, Hugnot JP, Lazdunski M. New modulatory alpha subunits for mammalian Shab K+ channels. J Biol Chem 1997;272:24,371–24,379.
48. Sano A, Mikami M, Nakamura M, Ueno S, Tanabe H, Kaneko S. Positional candidate approach for the gene responsible for benign adult familial myoclonic epilepsy. Epilepsia 2002;43(Suppl 9):26–31.
49. Su K, Kyaw H, Fan P, et al. Isolation, characterization, and mapping of two human potassium channels. Biochem Biophys Res Commun 1997;241:675–681.
50. Patt S, Preussat K, Beetz C, et al. Expression of ether a go-go potassium channels in human gliomas. Neurosci Lett 2004;368:249–253.
51. Saganich MJ, Machado E, Rudy B. Differential expression of genes encoding subthreshold-operating voltage-gated K+ channels in brain. J Neurosci 2001;21:4609–4624.
52. Papa M, Boscia F, Canitano A, et al. Expression pattern of the ether-a-gogo-related (ERG) K+ channel-encoding genes ERG1, ERG2, and ERG3 in the adult rat central nervous system. J Comp Neurol 2003;466:119–135.
53. Robbins J. KCNQ potassium channels: physiology, pathophysiology, and pharmacology. Pharmacol Ther 2001;90:1–19.
54. Knaus HG, Schwarzer C, Koch RO, et al. Distribution of high-conductance Ca(2+)-activated K+ channels in rat brain: targeting to axons and nerve terminals. J Neurosci 1996;16:955–963.
55. Rimini R, Rimland JM, Terstappen GC. Quantitative expression analysis of the small conductance calcium-activated potassium channels, SK1, SK2 and SK3, in human brain. Brain Res Mol Brain Res 2000;85:218–20.
56. Rusznak Z, Pocsai K, Kovacs I, et al. Differential distribution of TASK-1, TASK-2 and TASK-3 immunoreactivities in the rat and human cerebellum. Cell Mol Life Sci 2004;61:1532–1542.
57. Hervieu GJ, Cluderay JE, Gray CW, et al. Distribution and expression of TREK-1, a two-pore-domain potassium channel, in the adult rat CNS. Neuroscience 2001;103:899–919.
58. Duprat F, Lesage F, Fink M, Reyes R, Heurteaux C, Lazdunski M. TASK, a human background K+ channel to sense external pH variations near physiological pH. EMBO J 1997;16:5464–5471.
59. Lesage F, Lauritzen I, Duprat F, et al. The structure, function and distribution of the mouse TWIK-1 K+ channel. FEBS Lett 1997;402:28–32.
60. Fink M, Lesage F, Duprat F, et al. A neuronal two P domain K+ channel stimulated by arachidonic acid and polyunsaturated fatty acids. EMBO J 1998;17:3297–3308.
61. Medhurst AD, Rennie G, Chapman CG, et al. Distribution analysis of human two pore domain potassium channels in tissues of the central nervous system and periphery. Brain Res Mol Brain Res 2001;86:101–114.
62. Olsen ML, Sontheimer H. Mislocalization of Kir channels in malignant glia. Glia 2004;46:63–73.
63. Bordey A, Sontheimer H. Electrophysiological properties of human astrocytic tumor cells In situ: enigma of spiking glial cells. J Neurophysiol 1998;79:2782–2793.
64. Remillard CV, Yuan JX. Activation of K+ channels: an essential pathway in programmed cell death. Am J Physiol Lung Cell Mol Physiol 2004;286:L49–L67.
65. Lang F, Ritter M, Gamper N, et al. Cell volume in the regulation of cell proliferation and apoptotic cell death. Cell Physiol Biochem 2000;10:417–428.
66. Wang H, Zhang Y, Cao L, et al. HERG K+ channel, a regulator of tumor cell apoptosis and proliferation. Cancer Res 2002;62: 4843–4848.
67. Yu SP. Regulation and critical role of potassium homeostasis in apoptosis. Prog Neurobiol 2003;70:363–386.
68. Gamper N, Fillon S, Huber SM, et al. IGF-1 up-regulates K+ channels via PI3-kinase, PDK1 and SGK1. Pflugers Arch 2002;443:625–634.

69. Bringmann A, Francke M, Pannicke T, et al. Role of glial K(+) channels in ontogeny and gliosis: a hypothesis based upon studies on Muller cells. Glia 2000;29:35–44.
70. Pappas CA, Ullrich N, Sontheimer H. Reduction of glial proliferation by K+ channel blockers is mediated by changes in pHi. Neuroreport 1994;6:193–196.
71. Arcangeli A, Rosati B, Cherubini A, et al. HERG- and IRK-like inward rectifier currents are sequentially expressed during neuronal development of neural crest cells and their derivative. Eur J Neurosci 1997;9:2596–2604.
72. MacFarlane SN, Sontheimer H. Electrophysiological changes that accompany reactive gliosis in vitro. J Neurosci 1997;17:7316–7329.
73. Vaur S, Bresson-Bepoldin L, Dufy B, Tuffet S, Dufy-Barbe L. Potassium channel inhibition reduces cell proliferation in the GH3 pituitary cell line. J Cell Physiol 1998;177:402–410.
74. Arcangeli A, Bianchi L, Becchetti A, et al. A novel inward-rectifying K+ current with a cell cycle dependence governs the resting potential of mammalian neuroblastoma cells. J Physiol 1995;489 (Pt 2):455–471.
75. Knutson P, Ghiani CA, Zhou JM, Gallo V, McBain CJ. K+ channel expression and cell proliferation are regulated by intracellular sodium and membrane depolarization in oligodendrocyte progenitor cells. J Neurosci 1997;17:2669–2682.
76. Wonderlin WF, Strobl JS. Potassium channels, proliferation and G1 progression. J Membr Biol 1996;154:91–107.
77. Walz W, Bekar LK. Ion channels in cultured microglia. Microsc Res Tech 2001;54:26–33.
78. McCabe RD, Young DB. Potassium inhibits cultured vascular smooth muscle cell proliferation. Am J Hypertens 1994;7:346–350.
79. Cone CD, Jr, Cone CM. Induction of mitosis in mature neurons in central nervous system by sustained depolarization. Science 1976;192:155–158.
80. Chew LJ, Gallo V. Regulation of ion channel expression in neural cells by hormones and growth factors. Mol Neurobiol 1998;18:175–225.
81. Sobko A, Peretz A, Shirihai O, et al. Heteromultimeric delayed-rectifier K+ channels in schwann cells: developmental expression and role in cell proliferation. J Neurosci 1998;18:10,398–10,408.
82. Gomez-Angelats M, Bortner CD, Cidlowski JA. Protein kinase C (PKC) inhibits fas receptor-induced apoptosis through modulation of the loss of K+ and cell shrinkage. A role for PKC upstream of caspases. J Biol Chem 2000;275:19,609–19,619.
83. D'Ambrosio R, Maris DO, Grady MS, Winn HR, Janigro D. Impaired K(+) homeostasis and altered electrophysiological properties of post-traumatic hippocampal glia. J Neurosci 1999;19: 8152–8162.
84. Vicente R, Escalada A, Coma M, et al. Differential voltage-dependent K+ channel responses during proliferation and activation in macrophages. J Biol Chem 2003;278:46,307–46,320.
85. Eder C. Ion channels in microglia (brain macrophages). Am J Physiol 1998;275:C327–C342.
86. Tang P, Steck PA, Yung WK. The autocrine loop of TGF-alpha/EGFR and brain tumors. J Neuro-Oncol 1997;35:303–314.
87. Soliven B, Ma L, Bae H, Attali B, Sobko A, Iwase T. PDGF upregulates delayed rectifier via Src family kinases and sphingosine kinase in oligodendroglial progenitors. Am J Physiol Cell Physiol 2003;284:C85–C93.
88. Bowlby MR, Fadool DA, Holmes TC, Levitan IB. Modulation of the Kv1.3 potassium channel by receptor tyrosine kinases. J Gen Physiol 1997;110:601–610.
89. Chung I, Schlichter LC. Native Kv1.3 channels are upregulated by protein kinase C. J Membr Biol 1997;156:73–85.
90. Fakler B, Brandle U, Glowatzki E, Zenner HP, Ruppersberg JP. Kir2.1 inward rectifier K+ channels are regulated independently by protein kinases and ATP hydrolysis. Neuron 1994;13:1413–1420.
91. Hoffman DA, Johnston D. Downregulation of transient K+ channels in dendrites of hippocampal CA1 pyramidal neurons by activation of PKA and PKC. J Neurosci 1998;18:3521–3528.
92. Light PE, Bladen C, Winkfein RJ, Walsh MP, French RJ. Molecular basis of protein kinase C-induced activation of ATP-sensitive potassium channels. Proc Natl Acad Sci USA 2000;97:9058–9063.
93. Kim Y, Uhm DY, Shin J, Chung S. Modulation of delayed rectifier potassium channel by protein kinase C zeta-containing signaling complex in pheochromocytoma cells. Neuroscience 2004;125:359–368.
94. Yoo AS, McLarnon JG, Xu RL, Lee YB, Krieger C, Kim SU. Effects of phorbol ester on intracellular Ca2+ and membrane currents in cultured human microglia. Neurosci Lett 1996;218:37–40.
95. Mao J, Wang X, Chen F, et al. Molecular basis for the inhibition of G protein-coupled inward rectifier K(+) channels by protein kinase C. Proc Natl Acad Sci USA 2004;101:1087–1092.

96. Zhu G, Qu Z, Cui N, Jiang C. Suppression of Kir2.3 activity by protein kinase C phosphorylation of the channel protein at threonine 53. J Biol Chem 1999;274:11,643–11,646.
97. Vogelstein B, Kinzler KW. Cancer genes and the pathways they control. Nat Med 2004;10:789–799.
98. Kleihues P, Louis DN, Scheithauer BW, et al. The WHO classification of tumors of the nervous system. J Neuropathol Exp Neurol 2002;61:215–225.
99. Tang XD, Santarelli LC, Heinemann SH, Hoshi T. Metabolic regulation of potassium channels. Annu Rev Physiol 2004;66:131–159.
100. von Deimling A, Louis DN, Wiestler OD. Molecular pathways in the formation of gliomas. Glia 1995;15:328–338.
101. Maher EA, Furnari FB, Bachoo RM, et al. Malignant glioma: genetics and biology of a grave matter. Genes Dev 2001;15:1311–1333.
102. Dirks PB, Rutka JT. Current concepts in neuro-oncology: the cell cycle—a review. Neurosurgery 1997;40:1000–1013.
103. Burton EC, Lamborn KR, Forsyth P, et al. Aberrant p53, mdm2, and proliferation differ in glioblastomas from long-term compared with typical survivors. Clin Cancer Res 2002;8:180–187.
104. Collins VP. Gliomas. Cancer Surv 1998;32:37–51.
105. Chang KW, Yuan TC, Fang KP, et al. The increase of voltage-gated potassium channel Kv3.4 mRNA expression in oral squamous cell carcinoma. J Oral Pathol Med 2003;32:606–611.
106. Chin LS, Park CC, Zitnay KM, et al. 4-Aminopyridine causes apoptosis and blocks an outward rectifier K+ channel in malignant astrocytoma cell lines. J Neurosci Res 1997;48:122–127.
107. Ransom CB, Sontheimer H. BK channels in human glioma cells. J Neurophysiol 2001;85:790–803.
108. Gavrilova-Ruch O, Schonherr K, Gessner G, et al. Effects of imipramine on ion channels and proliferation of IGR1 melanoma cells. J Membr Biol 2002;188:137–149.
109. Sakai H, Shimizu T, Hori K, Ikari A, Asano S, Takeguchi N. Molecular and pharmacological properties of inwardly rectifying K+ channels of human lung cancer cells. Eur J Pharmacol 2002;435:125–133.
110. Pancrazio JJ, Ma W, Grant GM, et al. A role for inwardly rectifying K+ channels in differentiation of NG108-15 neuroblastoma x glioma cells. J Neurobiol 1999;38:466–474.
111. Deutsch C. Potassium channel ontogeny. Annu Rev Physiol 2002;64:19–46.
112. Dini G, Cucullo L, Ilkanich EV, et al. Potential role of the inward rectifying potassium channel GIRK2 in the regulation of cell proliferation during AC electrical stimulation. 34th Annual Society for Neuroscience Meeting (abstract), 2004.
113. Cucullo L, Dini G, Hallene KL, et al. Very low intensity alternating current decreases cell proliferation. Glia 2005;51:65–72.
114. Cucullo L, Dini G, Fazio V, Hallene KL, Janigro D. Do seizures act as a deterrent to tumor cell proliferation? 58th Annual American Epilepsy Society Meeting (abstract), 2004.
115. Kirson ED, Gurvich Z, Schneiderman R, et al. Disruption of cancer cell replication by alternating electric fields. Cancer Res 2004;64:3288–3295.
116. Jemal A, Tiwari RC, Murray T, et al. Cancer statistics, 2004. CA Cancer J Clin 2004;54:8–29.
117. Luyken C, Blumcke I, Fimmers R, et al. The spectrum of long-term epilepsy-associated tumors: long-term seizure and tumor outcome and neurosurgical aspects. Epilepsia 2003;44:822–830.
118. Walter RJ, Shtil AA, Roninson IB, Holian O. 60-Hz electric fields inhibit protein kinase C activity and multidrug resistance gene (MDR1) up-regulation. Radiat Res 1997;147:369–375.

IV

ADULT NEUROGENESIS: A MECHANISM FOR BRAIN REPAIR?

15
Enhanced Neurogenesis Following Neurological Disease

Philippe Taupin, PhD

SUMMARY

Neurological diseases are disorders of the nervous system that have biological origins, and impair one's ability to live a normal life. Neurological diseases originate from neurochemical alterations or the loss of specific cell populations of the nervous system. The causes for these disorders remain poorly understood. Their symptoms span from psychiatric to neurological to physical impairments. For decades, finding a cure for neurological diseases was synonymous with compensating for the neurochemical loss, by stimulating endogenous cells, replacing the dysfunctional or degenerated cells, or by preventing the degeneration of cell populations in neurodegenerative diseases, to restore brain functions. There is still no cure for neurological diseases, albeit in some cases treatments are available to alleviate some of the symptoms. With recent evidences that neurogenesis occurs in the adult brain and that neural stem cells (NSCs) reside in the adult brain, new theories have emerged regarding the possible involvement of newly generated neuronal cells in neurological diseases, and the potential of NSCs for the treatment of neurological diseases. Studies have shown that adult neurogenesis is increased in neurological diseases. Hence, researchers are now integrating these new data to understand the contribution of adult neurogenesis in neurological diseases, its significance, and potential for functional recovery. In this chapter, we will review the literature on adult neurogenesis in neurological diseases. This chapter discusses the functions of adult neurogenesis in the diseased brain, and its potential for the treatment of neurodegenerative diseases. Although many questions regarding the involvement of this plasticity in neurological diseases remain unanswered, these developments point to new hypotheses and concepts regarding neurological diseases, their origins, developments, and treatments.

Key Words: Neural stem cell; progenitor; hippocampus; neurodegenerative disease; self-repair; transplantation.

1. ADULT NEUROGENESIS IN ALZHEIMER'S DISEASE AND IN NEURODEGENERATIVE DISEASES OF THE BRAIN

Alzheimer's disease (AD) is a progressive, neurodegenerative disease characterized by amyloid plaque deposits and neurofibrillary tangles *(1,2)*. AD is associated with a loss of nerve cells in areas of the brain that are vital to memory and other mental abilities, such as the hippocampus. It is a slowly progressing disease, starting with mild memory problems and ending with severe brain damage. AD is the most common form of dementia among older people. The disease usually begins after age 60, and risk increases with age. On an average, patients with AD live for 8–10 yr after they are diagnosed, although the disease can last for as many as 20 yr. Three genes, presenilin 1 (PS 1), presenilin 2 (PS 2), and amyloid precursor protein (APP), have been discovered that cause early onset of the disease, that is, familial form of AD. Other genetic mutations that cause excessive accumulation of amyloid protein are associated with age-related form of the disease, that is, sporadic form of AD. There are several animal models that have been devised to study the genes known to be involved in AD, such as gene-altered mice, knockout, or expressing mutant forms of PS 1 *(3,4)* or APP *(5–7)*.

Recent investigations of autopsies of AD brain patients have reported an increase in the expression of markers for immature neuronal cells, such as doublecortin, polysialylated nerve cell adhesion molecule (PSA-NCAM), neurogenic differentiation factor, and turned on after division (TOAD-64, also known as TUC-4), in the subgranular zone and the granular layer of the dentate gyrus (DG), as well as in the CA1 region of Ammon's horn, suggesting that neurogenesis is increased in the hippocampus of patients suffering from AD *(8)*. These data conflict with previous reports aiming at characterizing adult neurogenesis in animal models of AD mutant *(9,10)* and knockout *(4)* for PS 1 and mutant for APP *(6)*. These studies, where neurogenesis was assessed by bromodeoxyuridine (BrdU) staining, showed decrease in neurogenesis in the DG and subventricular zone (SVZ). They also showed decrease in the number of differentiated neurons in the hippocampus, as detected by immunoreactivity for β-tubulin, a neuronal marker. Such discrepancies between the studies could be explained by the limitation of the transgenic animal models as representative of AD. Both are single-gene-deficient transgenic mice models, thus they neither fully reproduce the features of familial AD, nor the sporadic form of AD. To this aim, studies from other models such as transgenic mice co-expressing a mutant τ protein, involved in neurofibrillary tangles formation and progressive motor disturbance, and a mutant form of APP *(5)*, may provide better systems for studying adult neurogenesis in AD. A more recent study reports an increase in neurogenesis in the DG and SVZ of transgenic mouse models *(11)*, which express the Swedish and Indiana APP mutations *(7)*. Although the observed increase in neurogenesis correlates the observation made from human tissues *(8)*, concerns have been raised regarding the validity of such models, as representative of AD *(12)*. These data show the difficulties encountered in using animal models for studying neurological diseases and disorders, and particularly AD *(13,14)*. Further analysis of the adult SVZ in the APP transgenic mice reports a thinning of the ventricular zone *(15)*, suggesting a decrease in the number of progenitor cells during early neurogenesis, resulting in a partial depletion of the neural progenitor cell population in the adult. This depletion of the neural progenitor cell population may underlie the decrease in neurogenesis observed in the adult transgenic mice for PS 1 and APP. The authors observed that the rate of proliferation of the newly generated neuronal cells in the adult mutants is similar to the wild type, suggesting that PS 1 and APP are not involved in the regulation of neural progenitor proliferation, but rather in the regulation of neuronal survival, migration, or differentiation. The involvement of PS 1 and APP in adult neurogenesis is further highlighted by the expression of PS 1 in newly generated neuronal cells in the adult DG *(16)*, and by the involvement of APP in the regulation of adult neurogenesis in vitro, and in vivo *(17,18)*. However, the function of these molecules in adult neurogenesis is at present unclear, and remains to be investigated.

1.1. Huntington's Disease

Huntington's disease (HD) results from genetically programmed degeneration of neuronal cells in certain areas of the brain *(19)*. The caudate nucleus is the part of the brain that is, most severely, and preferentially affected in HD. This degeneration causes uncontrolled movements, loss of intellectual faculties, and emotional disturbance. HD is a familial disease, passed from parent to child through a mutation; a polyglutamine repeat (poly Gln, polyQ, or p[CAG]n) expansion that lengthens a glutamine (CAG) segment in the novel huntingtin protein *(20)*. Actual treatments for HD consist of controlling emotional and movement problems associated with the disease. Models such as transgenic HD mice for the 5′-end of the human HD gene carrying CAG repeat expansions (R6/1) *(21)*, and quinolinic acid lesion-induced striatal cell loss rat model of HD have been devised.

In the SVZ close to the caudate nucleus of adult human postmortem samples, confocal immunofluorescence analysis for molecular markers, such as the cell cycle marker proliferating cell nuclear antigen (PCNA), the neuronal marker β-tubulin, and the glial cell marker glial fibrillary acidic protein, has shown that progenitor cell proliferation and neurogenesis are increased in the SVZ of HD brains *(22)*. Furthermore, the authors correlated the degree of cell proliferation with pathological severity and length of polyglutamine repeats in the HD gene. Studies from R6/1 transgenic mouse model of HD reported that the level of PCNA mRNA is increased approximately twofold in the striatum of 12-wk-old mice *(23)*, and a decrease in neurogenesis in the DG *(24)*. However, these data are difficult to extrapolate to adult neurogenesis in HD, as it has been reported that mutated forms of huntingtin affect brain development *(25)*. Thus, the consequence of the mutation of huntingtin during brain development could underlie the decrease in neurogenesis observed in the adult transgenic mice. More recently, Tattersfield et al. reported that quinolinic acid striatal lesioning of the adult brain increases SVZ neurogenesis, leading to the putative migration of neuroblasts to damaged areas of the striatum, and the formation of new neurons *(26)*, as previously observed in patients with HD *(22)*. Further confirming that adult neurogenesis is increased in HD.

1.2. Parkinson's Disease

Parkinson's disease (PD) is a chronic and progressive neurodegenerative disease of the brain primarily associated with the loss of a specific type of dopamine neurons in the substantia nigra (SN) *(27)*. The SN is in the ventral midbrain and contains neurons that use dopamine as a neurotransmitter and send their axons to the striatum. It is believed that a gradual decline in the number of nigral dopamine neurons occurs with normal aging in humans, and that PD is caused by an abnormally rapid rate of cell death. PD belongs to a group of conditions called motor system disorders. The four primary symptoms of PD are tremor or trembling in hands, arms, legs, jaw, and face; rigidity or stiffness of the limbs and trunk; bradykinesia, or slowness of movement; and postural instability, or impaired balance and coordination. The disease is considerably more common in the age group of over 50 yr. PD is not usually inherited. A variety of medications provide dramatic relief from the symptoms. However, no drug yet can stop the progression of the disease, and in many cases medications lose their benefit over time. In such cases, surgery may be considered. There are several animal models of PD. Among them, are the 1-methyl-4-phenyl-1,2,3,6-tetrahydropyridine and 6-hydroxydopamine-lesioned models, known to kill significant proportion of the nigral dopaminergic nerve cell population *(28)*.

In a recent study that reports the generation of new dopaminergic neuronal cells in the adult rat SN *(29)*, the authors have also investigated how the generation of new dopaminergic neuronal cells would be affected following lesion of the SN. They showed that the rate of neurogenesis, as measured by BrdU labeling, is increased twofold 3 wk following lesion induced by a systemic dose of 1-methyl-4-phenyl-1,2,3,6-tetrahydropyridine. However, a more recent study finds no evidence of new dopaminergic neurons in the SN of 6-hydroxydopamine-lesioned hemi-Parkinsonian rodents *(30)*, contradicting these results. Thus, the generation of new dopaminergic neurons following lesion of the SN, as well as in the adult SN, remains a source of controversy *(30–32)*, and needs to be further confirmed.

2. ADULT NEUROGENESIS IN EPILEPSY

Epilepsy is a brain disorder in which populations of neurons signal abnormally. In the affected individual, this translates into a variety of seizures that range from mild behavioral changes to more severe symptoms, such as convulsions, muscles spasms, and loss of consciousness. Long-term abnormalities in learning, memory, and behavior have been reported *(33)*. This ailment is one of the most prevalent neurological disorders, affecting approx 1% of Americans. Treatments are available for patients with epilepsy, and seizures can be controlled in most cases by modern medicine and surgical techniques. It is hypothesized that abnormal brain wiring and/or neurotransmitter imbalance are underlying factors in epilepsy. The hippocampal formation is a critical area in the pathology of epilepsy; it has been suggested that the DG may function as a gate with respect to controlling the propagation of seizures *(33,34)*. Granule cells can regulate the throughput of epileptiform activity transiting through the hippocampal formation, by virtue of specific feed-forward inhibitory pathways *(35)*. One of the most common forms of epilepsy is the temporal lobe epilepsy (TLE). TLE often has its onset during childhood or is associated with a prolonged seizure episode early in life that is followed, after a variable latent period, by the development of epilepsy. In patients suffering from TLE, dispersed granular cell layer was reported *(36,37)*. Neuronal cell death has been reported in both the granular and pyramidal layers, with granule cell loss occurring at a lesser rate than in other hippocampal areas *(37)*. Ectopic granule-like neuronal cells, as defined by their expression of calcium-binding protein calbindin D28K, are found in the hilus and inner molecular layer *(36)*. The dentate granule cells give rise to abnormal axonal projections, a process described as mossy fiber (MF) sprouting, to the supragranular inner molecular layer of the DG, and the basal dendrites of CA3 pyramidal cells in *stratum oriens (38–40)*. There are several models that reproduce some of the traits of epileptic seizures observed in human, including the kainic model of epilepsy *(41–43)*, the electrical kindling *(44)*, and the pilocarpine seizure model *(45,46)*. Systemic kainic acid and pilocarpine injection produce seizures with many similarities to TLE, including an initial episode of prolonged status epilepticus (SE), followed by a latent period and spontaneously recurrent seizures, and temporal lobe pathology similar to that seen in humans. In both models, as in TLE, limbic seizures cause apoptosis of granule and pyramidal cells, with granule cell loss occurring at a lesser rate than in other hippocampal areas *(47–51)*. In these models, ectopic granule-like cells in the hilus and inner molecular layer, as well as aberrant growth (sprouting) of granule cell axons (MFs) in the supragranular dentate inner molecular layer, and the basal dendrites of CA3 pyramidal cells in stratum oriens have been reported *(49,52–55)*, as in patients with epilepsy *(38–40)*. MF sprouting begins during week 2 after SE, and peaks after 2 mo *(45,49)*.

It is postulated that a reduction in inhibition by loss of interneurons *(48)* and/or the development of recurrent excitatory circuitry by the sprouting of MF into ectopic positions after loss of their normal targets, and subsequent hippocampal hyperexcitability, are the determining events in the pathogenesis of limbic epilepsy *(40,52,53)*. Alternatively, it has also been proposed that aberrant granule cell axonal projections stabilize the network by preferentially innervating inhibitory neurons, and thereby restoring recurrent inhibition *(55)*. The implication of MF sprouting in seizures has been challenged by recent data showing that spontaneous recurrent seizures are still observed when MF sprouting is prevented by pretreatment with cycloheximide, a protein synthesis inhibitor, in pilocarpine- or kainate-treated animals *(56,57)*. Therefore, the origin, precise mechanisms, and relative contribution of cell death, ectopic granule-like cells, and MF sprouting in epileptic seizures remain to be defined.

Recently, a number of investigators used the BrdU paradigm to study the effect of experimentally induced seizures on adult neurogenesis. They have reported an increase in dentate granule cell neurogenesis following limbic-induced seizures in adult rodents. The first investigation relating neurogenesis and seizure activity in adult rodents was reported by Parent et al. in the DG of pilocarpine-treated rats *(58)*. The authors reported ectopic granule-like BrdU-immunolabeled cells in the hilus, as far as the CA3 cell layer, and a marked increase in dentate granule cell neurogenesis following seizure activity. The number of BrdU-immunolabeled cells in the hilus/CA3 layer was reported to increase with time after BrdU administration, showing that ectopic granule-like cells in the hilus derived from newly generated neuronal cells that are born after seizures, and can migrate into the hilus, as far as the CA3 cell layer. In the subgranular zone, cell proliferation was reported to increase within 3 d, and persists for at least 2 wk after pilocarpine treatment. The increase of neurogenesis observed in the DG, and the subsequent differentiation of the newly generated neuronal cells induced by the seizure activity, follows the same time-course than the MF remodeling *(45,49)*, taking into account the additional time required for the differentiation of these cells into mature neurons *(59)*. Thus, the authors hypothesized that MF remodeling derives from newly born granule neurons rather than from preexisting, mature dentate granule cells, as previously suggested *(38–40,49,52–55)*. The authors further supported their hypothesis by reporting that MF-like processes immunostained for TOAD-64, a marker for newly generated neuronal cells, were detected in the granule cell layer of the stratum oriens of CA3 area and the inner molecular layer of the DG. These data suggest that hippocampal network plasticity associated with chronic seizures originate from newly generated cells in the DG. It remains to be determined whether the newly generated granule cells account for all the "ectopic" granule cells observed in SE, and whether seizure-induced MF remodeling arises primarily from the developing axons of newly generated dentate granule cells.

Further studies have confirmed increased neurogenesis in dentate granule cell layer, and the presence of newly generated dentate granule-like cells in ectopic locations, such as the hilus, the inner molecular layer of the DG, and the hilar/CA3 border, following seizure activity in different models of epilepsy—the kainic model of epilepsy *(60–64)*, the electrical kindling *(60,65,66)*, pilocarpine seizure model *(62,64)*, and in other models *(67,68)*. It was further reported that (1) unilaterally induced seizures (such as intracerebroventricular kainic acid injection induce bilateral granule cell progenitor proliferation) *(61)*, (2) SE not only stimulates neurogenesis in the DG, but also in the SVZ *(69)*, (3) seizures preferentially stimulate proliferation of radial glia-like astrocytes in the adult DG *(70)*, (4) a population of cells in the developing CNS reported to have neural stem cell (NSC) capabilities *(71,72)*, and (5) the severity of SE affect the outcome of the newly generated cells in the DG, with the most severe SE leading to the death of most of the newly generated neuronal cells within 4 wk post-SE *(73)*.

In a follow-up study, Parent et al. aimed to examine whether seizure-induced MF remodeling arises primarily from the developing axons of newly generated dentate granule cells and/or from mature granule cells *(74)*. The authors applied low-dose, whole-brain X-ray irradiation, to inhibit dentate granule cell neurogenesis, in adult rats after pilocarpine-induced SE. The authors reported that low-dose radiation treatment reduces dentate granule cell neurogenesis, and had no effect on seizure-induced MF sprouting. Thus, MF reorganization after pilocarpine-induced SE occurs even in the absence of dentate granule cell neurogenesis, suggesting that sprouting arises also from the mature granule cells, and not primarily from newly generated neuronal cells as previously suggested *(58)*. The effect of low-dose radiation treatment on dentate granule cell neurogenesis *(75)* and seizure activity was further confirmed in recent studies *(76)*. Altogether, these data show that adult neurogenesis is increased following seizures in animal models and that the newly generated neuronal cells elicit the two main features of epilepsy: formation of aberrant axonal projections and migration to ectopic locations. These data also show that although increased hippocampal dentate granule cell neurogenesis, ectopic granule-like cell migration, and abnormal synaptogenesis are prominent features of animal models of TLE, they might represent independent events. Thus, it remains to be determined, what are the function(s) of the newly generated cells in SE. Furthermore, these investigations and analysis remain to be validated in human patients with epilepsy.

Altogether, these studies show that adult neurogenesis is increased in neurological diseases, albeit there is no functional recovery. Why if new neurons are generated is there no functional recovery? What are the function(s) of the newly generated cells in neurodegenerative diseases and how functional recovery could be promoted remains to be determined.

3. FUNCTION OF THE NEWLY GENERATED NEURONAL CELLS IN NEUROLOGICAL DISEASE

In the adult CNS, the function(s) of the newly generated neuronal cells remains the center of intense research. Recent evidence suggests that newly generated neuronal cells in the adult DG are involved in memory *(77,78)*, stress *(79)*, and depression *(80,81)*, whereas, newly generated neuronal cells in the olfactory bulb are involved in odor perception and memory *(82)*. However, the function of the newly generated neuronal cells in the diseased brain remains to be determined. There are few speculations and hypotheses that can be raised in the context of their attributes in neurological diseases.

The increase in neurogenesis, observed in neurological diseases and in models of neurological diseases, may represent a mechanism directed toward the replacement of dead or damaged neurons. Such an increase has also been reported in other pathological conditions, such as stroke *(83)* and traumatic brain injury *(84)*, and may represent attempts by the CNS to self-repair. However, notwithstanding that neurogenesis appears to be increased in the brains of patients and animal models of neurological diseases, progressive cell losses are still occurring and no functional recovery is achieved. Thus, the increase of neurogenesis in itself is insufficient to promote functional recovery in neurological diseases. Several hypotheses can be raised to explain the limited capacity of the CNS to self-repair. First, the number of new neurons generated is too low to compensate for the neuronal loss. Second, the neurons that are produced may be nonfunctional because they do not develop into fully mature neurons, they do not develop into the right type of neurons, or they are incapable of integrating into the surviving brain circuitry. Third, the microenvironment of the diseased or injured brain may be toxic for the newly generated neuronal cells. Thus, neurogenesis in the adult CNS elicits only a limited repair capacity in neurological diseases. But the data presented show that the diseased brain has the potential for self-repair even in humans. This has important implications for cellular therapy applied to the adult CNS, and for designing new strategies to treat neurological diseases.

Recent evidence suggests that self-repair mechanisms may operate in the adult rodents' SN *(29)*, the area of the CNS affected in PD, although these data remain the source of controversies *(30–32)*. If such turnover of dopaminergic neuronal cells was confirmed, progression of the disease would then be determined not only by the rate of degeneration of SN neurons, but also by the efficacy in the formation of new dopamine neurons. Thus, disturbances of the equilibrium between cell genesis and cell death could result in neurodegenerative disorders. Therefore, in PD, neurogenesis might not only be a process for functional recovery, but it may also play a key role in the pathology of the disease. In contrast, a recent study reported that the induction of recurrent seizures following irradiation treatment prevents the seizure-induced increase of neurogenesis in the DG *(74)*. These data provide a strong argument against a critical role of adult neurogenesis in epileptogenesis. They particularly suggest that newly generated granule-like cells at the hilar/CA3 border are not critical to limbic seizures in pilocarpine-treated rats, although the newly generated granule cells elicit the prominent features of the response to seizures, such as ectopic granule-like cell migration, and abnormal synaptogenesis. Nonetheless, they could be a contributing factor, when present. These data argue against a critical role for neurogenesis in SE. Thus, neurogenesis may have different levels of involvement in neurological diseases. This remains to be thoroughly investigated, as a prerequisite for designing strategy to treat and cure neurological diseases.

Stimulation of neurogenesis might not only serve neuronal regeneration, but might be an attempt by the CNS to compensate for other neuronal functions associated with the disease. For example, in conditions such as AD and epilepsy, which are associated with memory impairment *(85,86)*, the ability of the diseased brain to mobilize new neurons in the hippocampus could have especially important consequences regarding memory function. In both AD and epilepsy, the hippocampus is the most affected brain area, and it has recently been suggested that memory function may depend on hippocampal neurogenesis *(77,78)*. Thus, the increase of neurogenesis in neurological diseases might also serve cognitive function recovery. Also, patients with neurological diseases, such as AD, epilepsy, HD, and PD, but also those recovering from stroke and injury, are at a greater risk of depression *(87–89)*. Stress *(79)* and depression *(80,81)* reduce granule cell neurogenesis in adult rodents, whereas ischemia stimulates neurogenesis *(83)*. Thus, the increase in neurogenesis in neurological diseases may also be an attempt by the organism to compensate for these symptoms.

4. THERAPY

In neurodegenerative diseases, dysfunction or loss of specific cells causes patients to present with psychiatric or neurological symptoms. With recent evidence that neurogenesis occurs in the adult brain and that cells with NSC properties can be isolated from the adult brain and cultured in vitro, new opportunities to treat neurodegenerative diseases have emerged. Cell therapeutic interventions might involve both the stimulation of the endogenous neural progenitor cells and cell transplantation.

4.1. Stimulation of Endogenous Neural Progenitor Cells

Although the brain does not regenerate following injury and in neurodegenerative disease, the findings that neurogenesis in stimulated in neurological diseases show that the diseased brain has the potential to self-repair. The process of neurogenesis is regulated by a variety of intrinsic and extrinsic stimuli, including steroid hormones, trophic factors, aging, stresses, and environmental enrichment *(79,90–96)*. It may be possible to stimulate neurogenesis in the diseased brain in the aim of promoting functional recovery or slowing the disease course. Several studies have reported that environmental enrichment has a beneficial effect in the diseased *(97,98)* and injured brain *(99)*. Pharmacological agents and compounds, such as lithium *(100)*, would provide means to stimulate neurogenesis, alone or in combination with trophic factors, steroids, and environmental enrichment *(90–96)*. Although no data have been reported suggesting that adequate numbers of cells can be generated to repopulate a diseased/injured area and promote functional recovery, strategies designed to enhance neurogenesis could have therapeutic value in neurological diseases.

4.2. Transplantation

Cell therapy is a prominent area of investigation in the biomedical field, particularly for the treatment of otherwise incurable neurodegenerative diseases. In this view, both fetal derived cells and NSCs are being proposed as elective sources of brain cells for transplantation. Recent data suggest that the use of fetal cells presents some limitation for transplantation *(101)*, leaving the NSC a model of choice for future therapy. Neural progenitor and stem cells have been transplanted in animal models, and show potent engraftment, proliferation, migration, and neural differentiation *(102–106)*. Classical neural transplantation approach consists in grafting cells in the proximity of the site of the degeneration or lesion, or into its target area, such as in PD or in focal injuries. In other neurodegenerative diseases, such as AD, HD, and multiple sclerosis, where the degenerative area is widespread, such strategy is not applicable. Because of the property of neural progenitor and stem cells to migrate to tumor sites *(107,108)* and diseased areas *(109)*, NSC therapy offers a much broader potential to cure neurodegenerative diseases. A recent study has reported that the systemic injection of neural progenitors and stem cells may provide significant clinical benefit in an animal model of multiple sclerosis *(109)*. Thus, NSCs may provide a therapeutic tool for the treatment of a broad range of neurodegenerative diseases.

5. CONCLUSIONS

The promise of adult stem cell research is as important therapeutically, as for our understanding and knowledge of developmental biology. The recent development in adult neurogenesis and NSCs has directed researchers to investigate whether newly generated neuronal cells were involved in neurological diseases. Taken together, the data reviewed here show that neurogenesis is stimulated in neurological diseases, albeit there is no functional recovery, and progressive cell losses and damages are still occurring. These studies have forced us to rethink and redefine the origins, mechanisms, and treatments for neurological diseases, and it has also forced us to reevaluate CNS plasticity.

However, the main issues remain to be addressed:

1. Are the newly generated cells induced to proliferate in neurological diseases from the same pool as in adult neurogenesis?
2. What are the mechanisms underlying the increase of neurogenesis in neurological diseases?
3. What are the functions of these newly generated neuronal cells in the diseased and injured brain?
4. How can we promote functional recovery?

The field of adult neurogenesis and NSCs is a challenging one. Future studies will bring new developments and NSC research closer to therapy.

ACKNOWLEDGMENTS

PT was supported by grants from the NMRC, BMRC, and the Juvenile Diabetes Research Foundation.

REFERENCES

1. Fukutani Y, Kobayashi K, Nakamura I, Watanabe K, Isaki K, Cairns NJ. Neurons, intracellular and extracellular neurofibrillary tangles in subdivisions of the hippocampal cortex in normal ageing and Alzheimer's disease. Neurosci Lett 1995;200:57–60.
2. Hardy J, Selkoe DJ. The amyloid hypothesis of Alzheimer's disease: progress and problems on the road to therapeutics. Science 2002;297:353–356. Erratum in: Science 2002;297:2209.
3. Borchelt DR, Thinakaran G, Eckman CB, et al. Familial Alzheimer's disease-linked presenilin 1 variants elevate Abeta1-42/1-40 ratio in vitro and in vivo. Neuron 1996;17:1005–1013.
4. Feng R, Rampon C, Tang YP, et al. Deficient neurogenesis in forebrain-specific presenilin-1 knock-out mice is associated with reduced clearance of hippocampal memory traces. Neuron 2001;32:911–926. Erratum in: Neuron 2002;33:313.
5. Lewis J, Dickson DW, Lin WL, et al. Enhanced neurofibrillary degeneration in transgenic mice expressing mutant tau and APP. Science 2001;293:1487–1491.
6. Haughey NJ, Nath A, Chan SL, Borchard AC, Rao MS, Mattson MP. Disruption of neurogenesis by amyloid beta-peptide, and perturbed neural progenitor cell homeostasis, in models of Alzheimer's disease. J Neurochem 2002;6:1509–1524.
7. Hsia AY, Masliah E, McConlogue L, et al. Plaque-independent disruption of neural circuits in Alzheimer's disease mouse models. Proc Natl Acad Sci USA 1999;96:3228–3233.
8. Jin K, Peel AL, Mao XO, et al. Increased hippocampal neurogenesis in Alzheimer's disease. Proc Natl Acad Sci USA 2004;101:343–347.
9. Wen PH, Shao X, Shao Z, et al. Overexpression of wild type but not an FAD mutant presenilin-1 promotes neurogenesis in the hippocampus of adult mice. Neurobiol Dis 2002;10:8–19.
10. Wen PH, Hof PR, Chen X, et al. The presenilin-1 familial Alzheimer disease mutant P117L impairs neurogenesis in the hippocampus of adult mice. Exp Neurol 2004;188:224–237.
11. Jin K, Galvan V, Xie L, et al. Enhanced neurogenesis in Alzheimer's disease transgenic (PDGF-APPSw, Ind) mice. Proc Natl Acad Sci USA 2004;101:13,363–13,367.
12. Schwab C, Hosokawa M, McGeer PL. Transgenic mice overexpressing amyloid beta protein are an incomplete model of Alzheimer disease. Exp Neurol 2004;188:52–64.
13. Janus C, Westaway D. Transgenic mouse models of Alzheimer's disease. Physiol Behav 2001;73:873–86.
14. Dodart JC, Mathis C, Bales KR, Paul SM. Does my mouse have Alzheimer's disease? Genes Brain Behav 2002;1:142–155.
15. Haughey NJ, Liu D, Nath A, Borchard AC, Mattson MP. Disruption of neurogenesis in the subventricular zone of adult mice, and in human cortical neuronal precursor cells in culture, by amyloid beta-peptide: implications for the pathogenesis of Alzheimer's disease. Neuromol Med 2002;1:125–135.
16. Wen PH, Friedrich VL, Jr, Shioi J, Robakis NK, Elder GA. Presenilin-1 is expressed in neural progenitor cells in the hippocampus of adult mice. Neurosci Lett 2002;318:53–56.
17. Caille I, Allinquant B, Dupont E, et al. Soluble form of amyloid precursor protein regulates proliferation of progenitors in the adult subventricular zone. Development 2004;131:2173–2181.
18. Yasuoka K, Hirata K, Kuraoka A, He JW, Kawabuchi M. Expression of amyloid precursor protein-like molecule in astroglial cells of the subventricular zone and rostral migratory stream of the adult rat forebrain. J Anat 2004;205:135–146.
19. Sawa A, Tomoda T, Bae BI. Mechanisms of neuronal cell death in Huntington's disease. Cytogenet Genome Res 2003;100:287–295.
20. Li SH, Li XJ. Huntingtin-protein interactions and the pathogenesis of Huntington's disease. Trends Genet 2004;20:146–154.
21. Mangiarini L, Sathasivam K, Seller M, et al. Exon 1 of the HD gene with an expanded CAG repeat is sufficient to cause a progressive neurological phenotype in transgenic mice. Cell 1996;87:493–506.

22. Curtis MA, Penney EB, Pearson AG, et al. Increased cell proliferation and neurogenesis in the adult human Huntington's disease brain. Proc Natl Acad Sci USA 2003;100:9023–9027.
23. Luthi-Carter R, Strand A, Peters NL, et al. Decreased expression of striatal signaling genes in a mouse model of Huntington's disease. Hum Mol Genet 2000;9:1259–1271.
24. Lazic SE, Grote H, Armstrong RJ, et al. Decreased hippocampal cell proliferation in R6/1 Huntington's mice. Neuroreport 2004;15:811–813.
25. White JK, Auerbach W, Duyao MP, et al. Huntingtin is required for neurogenesis and is not impaired by the Huntington's disease CAG expansion. Nat Genet 1997;17:404–410.
26. Tattersfield AS, Croon RJ, Liu YW, Kells AP, Faull RL, Connor B. Neurogenesis in the striatum of the quinolinic acid lesion model of Huntington's disease. Neuroscience 2004;127:319–332.
27. Fernandez-Espejo E. Pathogenesis of Parkinson's disease: prospects of neuroprotective and restorative therapies. Mol Neurobiol 2004;29:15–30.
28. Beal MF. Experimental models of Parkinson's disease. Nat Rev Neurosci 2001;2:325–334.
29. Zhao M, Momma S, Delfani K, et al. Evidence for neurogenesis in the adult mammalian substantia nigra. Proc Natl Acad Sci USA 2003;100:7925–7930.
30. Frielingsdorf H, Schwarz K, Brundin P, Mohapel P. No evidence for new dopaminergic neurons in the adult mammalian substantia nigra. Proc Natl Acad Sci USA 2004;101:10,177–10,182.
31. Lie DC, Dziewczapolski G, Willhoite AR, Kaspar BK, Shults CW, Gage FH. The adult substantia nigra contains progenitor cells with neurogenic potential. J Neurosci 2002;22:6639–6649.
32. Lindvall O, McKay R. Brain repair by cell replacement and regeneration. Proc Natl Acad Sci USA 2003;100:7430–7431.
33. Majak K, Pitkanen A. Do seizures cause irreversible cognitive damage? Evidence from animal studies. Epilepsy Behav 2004;5:S35–S44.
34. Heinemann U, Beck H, Dreier JP, Ficker E, Stabel J, Zhang CL. The dentate gyrus as a regulated gate for the propagation of epileptiform activity. Epilepsy Res 1992;7:273–280.
35. Sloviter RS. Feedforward and feedback inhibition of hippocampal principal cell activity evoked by perforant path stimulation: GABA-mediated mechanisms that regulate excitability in vivo. Hippocampus 1991;1:31–40.
36. Houser CR. Granule cell dispersion in the dentate gyrus of humans with temporal lobe epilepsy. Brain Res 1990;535:195–204.
37. de Lanerolle NC, Kim JH, Robbins RJ, Spencer DD. Hippocampal interneuron loss and plasticity in human temporal lobe epilepsy. Brain Res 1989;495:387–95.
38. Sutula T, Cascino G, Cavazos J, Parada I, Ramirez L. Mossy fiber synaptic reorganization in the epileptic human temporal lobe. Ann Neurol 1989;26:321–330.
39. Represa A, Tremblay E, Ben-Ari Y. Sprouting of mossy fibers in the hippocampus of epileptic human and rat. Adv Exp Med Biol 1990;268:419–424.
40. Babb TL, Kupfer WR, Pretorius JK, Crandall PH, Levesque MF. Synaptic reorganization by mossy fibers in human epileptic fascia dentata. Neuroscience 1991;42:351–363.
41. Ben-Ari Y, Tremblay E, Riche D, Ghilini G, Naquet R. Electrographic, clinical and pathological alterations following systemic administration of kainic acid, bicuculline or pentetrazole: metabolic mapping using the deoxyglucose method with special reference to the pathology of epilepsy. Neuroscience 1981;6:1361–1391.
42. Nadler JV. Kainic acid as a tool for the study of temporal lobe epilepsy. Life Sci 1981;29:2031–2042.
43. Ben-Ari Y. Limbic seizure and brain damage produced by kainic acid: mechanisms and relevance to human temporal lobe epilepsy. Neuroscience 1985;14:375–403.
44. Represa A, Le Gall La Salle G, Ben-Ari Y. Hippocampal plasticity in the kindling model of epilepsy in rats. Neurosci Lett 1989;99:345–350.
45. Cavalheiro EA, Leite JP, Bortolotto ZA, Turski WA, Ikonomidou C, Turski L. Long-term effects of pilocarpine in rats: structural damage of the brain triggers kindling and spontaneous recurrent seizures. Epilepsia 1991;32:778–782.
46. Turski L, Ikonomidou C, Turski WA, Bortolotto ZA, Cavalheiro EA. Cholinergic mechanisms and epileptogenesis. The seizures induced by pilocarpine: a novel experimental model of intractable epilepsy. Synapse 1898;3:154–171.
47. Sloviter RS. "Epileptic" brain damage in rats induced by sustained electrical stimulation of the perforant path. I. Acute electrophysiological and light microscopic studies. Brain Res Bull 1983;10:675–697.
48. Sloviter RS. Decreased hippocampal inhibition and a selective loss of interneurons in experimental epilepsy. Science 1987;235:73–76.

49. Mello LE, Cavalheiro EA, Tan AM, et al. Circuit mechanisms of seizures in the pilocarpine model of chronic epilepsy: cell loss and mossy fiber sprouting. Epilepsia 1993;34:985–995.
50. Cavazos JE, Das I, Sutula TP. Neuronal loss induced in limbic pathways by kindling: evidence for induction of hippocampal sclerosis by repeated brief seizures. J Neurosci 1994;14:3106–3121.
51. Sloviter RS, Dean E, Sollas AL, Goodman JH. Apoptosis and necrosis induced in different hippocampal neuron populations by repetitive perforant path stimulation in the rat. J Comp Neurol 1996;366:516–533.
52. Tauck DL, Nadler JV. Evidence of functional mossy fiber sprouting in hippocampal formation of kainic acid-treated rats. J Neurosci 1985;5:1016–1022.
53. Cronin J, Dudek FE. Chronic seizures and collateral sprouting of dentate mossy fibers after kainic acid treatment in rats. Brain Res 1988;474:181–184.
54. Represa A, Ben-Ari Y. Kindling is associated with the formation of novel mossy fibre synapses in the CA3 region. Exp Brain Res 1992;92:69–78.
55. Sloviter RS. Possible functional consequences of synaptic reorganization in the dentate gyrus of kainate-treated rats. Neurosci Lett 1992;137:91–96.
56. Longo BM, Mello LE. Blockade of pilocarpine- or kainate-induced mossy fiber sprouting by cycloheximide does not prevent subsequent epileptogenesis in rats. Neurosci Lett 1997;226:163–166.
57. Longo BM, Mello LE. Supragranular mossy fiber sprouting is not necessary for spontaneous seizures in the intrahippocampal kainate model of epilepsy in the rat. Epilepsy Res 1998;32:172–182.
58. Parent JM, Yu TW, Leibowitz RT, Geschwind DH, Sloviter RS, Lowenstein DH. Dentate granule cell neurogenesis is increased by seizures and contributes to aberrant network reorganization in the adult rat hippocampus. J Neurosci 1997;17:3727–3738.
59. Cameron HA, Woolley CS, McEwen BS, Gould E. Differentiation of newly born neurons and glia in the dentate gyrus of the adult rat. Neuroscience 1993;56:337–344.
60. Bengzon J, Kokaia Z, Elmer E, Nanobashvili A, Kokaia M, Lindvall O. Apoptosis and proliferation of dentate gyrus neurons after single and intermittent limbic seizures. Proc Natl Acad Sci USA 1997;94:10,432–10,437.
61. Gray WP, Sundstrom LE. Kainic acid increases the proliferation of granule cell progenitors in the dentate gyrus of the adult rat. Brain Res 1998;790:52–59.
62. Scharfman HE, Goodman JH, Sollas AL. Granule-like neurons at the hilar/CA3 border after status epilepticus and their synchrony with area CA3 pyramidal cells: functional implications of seizure-induced neurogenesis. J Neurosci 2000;20:6144–6158.
63. Nakagawa E, Aimi Y, Yasuhara O, et al. Enhancement of progenitor cell division in the dentate gyrus triggered by initial limbic seizures in rat models of epilepsy. Epilepsia 2000;41:10–18.
64. Covolan L, Ribeiro LT, Longo BM, Mello LE. Cell damage and neurogenesis in the dentate granule cell layer of adult rats after pilocarpine- or kainate-induced status epilepticus. Hippocampus 2000;10:69–80.
65. Parent JM, Janumpalli S, McNamara JO, Lowenstein DH. Increased dentate granule cell neurogenesis following amygdala kindling in the adult rat. Neurosci Lett 1998;247:9–12.
66. Scott BW, Wang S, Burnham WM, De Boni U, Wojtowicz JM. Kindling-induced neurogenesis in the dentate gyrus of the rat. Neurosci Lett 1998;248:73–76.
67. Jiang W, Wan Q, Zhang ZJ, et al. Dentate granule cell neurogenesis after seizures induced by pentylenetrazol in rats. Brain Res 2003;977:141–148.
68. Ferland RJ, Gross RA, Applegate CD. Increased mitotic activity in the dentate gyrus of the hippocampus of adult C57BL/6J mice exposed to the flurothyl kindling model of epileptogenesis. Neuroscience 2002;115:669–683.
69. Parent JM, Valentin VV, Lowenstein DH. Prolonged seizures increase proliferating neuroblasts in the adult rat subventricular zone-olfactory bulb pathway. J Neurosci 2002;22:3174–3188.
70. Huttmann K, Sadgrove M, Wallraff A, et al. Seizures preferentially stimulate proliferation of radial glia-like astrocytes in the adult dentate gyrus: functional and immunocytochemical analysis. Eur J Neurosci 2003;18:2769–2778.
71. Hartfuss E, Galli R, Heins N, Gotz M. Characterization of CNS precursor subtypes and radial glia. Dev Biol 2001;229:15–30.
72. Anthony TE, Klein C, Fishell G, Heintz N. Radial glia serve as neuronal progenitors in all regions of the central nervous system. Neuron 2004;41:881–890.
73. Mohapel P, Ekdahl CT, Lindvall O. Status epilepticus severity influences the long-term outcome of neurogenesis in the adult dentate gyrus. Neurobiol Dis 2004;15:196–205.

74. Parent JM, Tada E, Fike JR, Lowenstein DH. Inhibition of dentate granule cell neurogenesis with brain irradiation does not prevent seizure-induced mossy fiber synaptic reorganization in the rat. J Neurosci 1999;19:4508–4519.
75. Tada E, Parent JM, Lowenstein DH, Fike JR. X-irradiation causes a prolonged reduction in cell proliferation in the dentate gyrus of adult rats. Neuroscience 2000;99:33–41.
76. Ferland RJ, Williams JP, Gross RA, Applegate CD. The effects of brain-irradiation-induced decreases in hippocampal mitotic activity on flurothyl-induced epileptogenesis in adult C57BL/6J mice. Exp Neurol 2003;179:71–82.
77. Gould E, Beylin A, Tanapat P, Reeves A, Shors TJ. Learning enhances adult neurogenesis in the hippocampal formation. Nat Neurosci 1999;2:260–265.
78. Shors TJ, Miesegaes G, Beylin A, Zhao M, Rydel T, Gould E. Neurogenesis in the adult is involved in the formation of trace memories. Nature 2001;410:372–376. Erratum in: Nature 2001;414:938.
79. Gould E, McEwen BS, Tanapat P, Galea LA, Fuchs E. Neurogenesis in the dentate gyrus of the adult tree shrew is regulated by psychosocial stress and NMDA receptor activation. J Neurosci 1997;17:2492–2498.
80. Jacobs BL, Praag H, Gage FH. Adult brain neurogenesis and psychiatry: a novel theory of depression. Mol Psychiat 2000;5:262–269.
81. Santarelli L, Saxe M, Gross C, et al. Requirement of hippocampal neurogenesis for the behavioral effects of antidepressants. Science 2003;301:805–809.
82. Rochefort C, Gheusi G, Vincent JD, Lledo PM. Enriched odor exposure increases the number of newborn neurons in the adult olfactory bulb and improves odor memory. J Neurosci 2002;22:2679–2689.
83. Liu J, Solway K, Messing RO, Sharp FR. Increased neurogenesis in the dentate gyrus after transient global ischemia in gerbils. J Neurosci 1998;18:7768–7778.
84. Dash PK, Mach SA, Moore AN. Enhanced neurogenesis in the rodent hippocampus following traumatic brain injury. J Neurosci Res 2001;63:313–319.
85. Wang R, Dineley KT, Sweatt JD, Zheng H. Presenilin 1 familial Alzheimer's disease mutation leads to defective associative learning and impaired adult neurogenesis. Neuroscience 2004;126:305–312.
86. Kotloski R, Lynch M, Lauersdorf S, Sutula T. Repeated brief seizures induce progressive hippocampal neuron loss and memory deficits. Prog Brain Res 2002;135:95–110.
87. Gilliam FG, Santos J, Vahle V, Carter J, Brown K, Hecimovic H. Depression in epilepsy: ignoring clinical expression of neuronal network dysfunction? Epilepsia 2004;45:28–33.
88. Sawabini KA, Watts RL. Treatment of depression in Parkinson's disease. Parkinsonism Relat Disord 2004;10:S37–S41.
89. Perna RB, Rouselle A, Brennan P. Traumatic brain injury: depression, neurogenesis, and medication management. J Head Trauma Rehabil 2003;18:201–203.
90. Cameron HA, Gould E. Adult neurogenesis is regulated by adrenal steroids in the dentate gyrus. Neuroscience 1994;61:203–209.
91. Craig CG, Tropepe V, Morshead CM, Reynolds BA, Weiss S, van der Kooy D. In vivo growth factor expansion of endogenous subependymal neural precursor cell populations in the adult mouse brain. J Neurosci 1996;16:2649–2658.
92. Kuhn HG, Winkler J, Kempermann G, Thal LJ, Gage FH. Epidermal growth factor and fibroblast growth factor-2 have different effects on neural progenitors in the adult rat brain. J Neurosci 1997;17:5820–5829.
93. Taupin P, Ray J, Fischer WH, et al. FGF-2-responsive neural stem cell proliferation requires CCg, a novel autocrine/paracrine cofactor. Neuron 2000;28:385–397.
94. Kuhn HG, Dickinson-Anson H, Gage FH. Neurogenesis in the dentate gyrus of the adult rat: age-related decrease of neuronal progenitor proliferation. J Neurosci 1996;16:2027–2033.
95. Kempermann G, Kuhn HG, Gage FH. More hippocampal neurons in adult mice living in an enriched environment. Nature 1997;386:493–495.
96. van Praag H, Kempermann G, Gage FH. Running increases cell proliferation and neurogenesis in the adult mouse dentate gyrus. Nat Neurosci 1999;2:266–270.
97. Auvergne R, Lere C, El Bahh B, et al. Delayed kindling epileptogenesis and increased neurogenesis in adult rats housed in an enriched environment. Brain Res 2002;954:277–85.
98. Faverjon S, Silveira DC, Fu DD, et al. Beneficial effects of enriched environment following status epilepticus in immature rats. Neurology 2002;59:1356–1364.
99. Will B, Galani R, Kelche C, Rosenzweig MR. Recovery from brain injury in animals: relative efficacy of environmental enrichment, physical exercise or formal training (1990–2002). Prog Neurobiol 2004;72:167–182.

100. Chen G, Rajkowska G, Du F, Seraji-Bozorgzad N, Manji HK. Enhancement of hippocampal neurogenesis by lithium. J Neurochem 2000;75:1729–1734.
101. Olanow CW, Goetz CG, Kordower JH, et al. A double-blind controlled trial of bilateral fetal nigral transplantation in Parkinson's disease. Ann Neurol 2003;54:403–414.
102. Suhonen JO, Peterson DA, Ray J, Gage FH. Differentiation of adult hippocampus-derived progenitors into olfactory neurons in vivo. Nature 1996;383:624–627.
103. Fricker RA, Carpenter MK, Winkler C, Greco C, Gates MA, Bjorklund A. Site-specific migration and neuronal differentiation of human neural progenitor cells after transplantation in the adult rat brain. J Neurosci 1999;19:5990–6005.
104. Armstrong RJ, Tyers P, Jain M, et al. Transplantation of expanded neural precursor cells from the developing pig ventral mesencephalon in a rat model of Parkinson's disease. Exp Brain Res 2003;151:204–217.
105. Uchida N, Buck DW, He D, et al. Direct isolation of human central nervous system stem cells. Proc Natl Acad Sci USA 2000;97:14,720–14,725.
106. Fricker-Gates RA, Winkler C, Kirik D, Rosenblad C, Carpenter MK, Bjorklund A. EGF infusion stimulates the proliferation and migration of embryonic progenitor cells transplanted in the adult rat striatum. Exp Neurol 2000;165:237–247.
107. Aboody KS, Brown A, Rainov NG, et al. Neural stem cells display extensive tropism for pathology in adult brain: evidence from intracranial gliomas. Proc Natl Acad Sci USA 2000;97:12,846–12,851. Erratum in: Proc Natl Acad Sci USA 2001;98:777.
108. Brown AB, Yang W, Schmidt NO, et al. Intravascular delivery of neural stem cell lines to target intracranial and extracranial tumors of neural and non-neural origin. Hum Gene Ther 2003;14:1777–1785.
109. Pluchino S, Quattrini A, Brambilla E, et al. Injection of adult neurospheres induces recovery in a chronic model of multiple sclerosis. Nature 2003;422:688–694.

16
Endothelial Injury and Cell Cycle Re-Entry

Ljiljana Krizanac-Bengez, MD, PhD

SUMMARY

Endothelial cells (ECs) affect the homeostasis of the vessel wall in terms of vasomotor tone, platelet and monocyte adhesion, growth of vascular smooth muscle cells, and extracellular matrix production. They, thereby, provide an antithrombotic and anti-inflammatory barrier for the normal vessel wall. Migration and proliferation of ECs is critical in the repair of injured vessels and in angiogenesis and vasculogenesis during development, tumor growth, and tissue repair. Regulation of the cell cycle is achieved through a complex and ordered sequence of events controlled by cyclin-dependent kinases (CDKs), the activation of which depends on regulatory phosphorylation, and their association with protein subunits, cyclins. The activation of CDKs is in turn negatively regulated by several CDK inhibitors. This chapter summarizes the most pertinent findings related to the control of cell growth and migration of vascular ECs, as relevant for maintaining the homeostasis of the vessel wall during angiogenesis and vasculogenesis, as well as for vascular remodeling during wound healing, tumor in-growth, atherogenesis, and restenosis.

Key Words: Endothelial cells; injury; cell cycle; neovascularization; wound healing; atherogenesis.

1. REGULATION OF THE CELL CYCLE IN ENDOTHELIAL CELLS

Regulation of the cell cycle is achieved through a complex and ordered sequence of events controlled by cyclin-dependent kinases (CDKs), the activation of which depends on their association with protein subunits, cyclins, and on regulatory phosphorylation *(1)*. Under normal conditions during adult life, the endothelial cells (ECs) of many organ systems remain in a quiescent state *(2)*. This is opposite to a rapid proliferation of ECs in microvascular morphogenesis during vascular development, or in adulthood when EC monolayer is disrupted, such as during abrasion by balloon angioplasty, or with abnormal in-growth of new vessels during tumor-induced angiogenesis *(3)*. Despite the importance of EC proliferation during pathogenesis and normal development, little is known about the mechanisms that tightly couple changes in cell cycle activity with changes in the endothelial microenvironment.

Cell cycle progression is controlled by the periodic activation of CDKs, which become activated by their association with activating subunits, cyclins *(4)*. The CDK4/cyclin D complex functions in the early G_1 phase, whereas CDK2 performs key regulatory roles in the G_1 and S phases of the cell cycle. CDK2/cyclin E is activated in late G_1 phase, when the retinoblastoma gene product becomes phosphorylated. Cyclin E plays a key role regulating the transition from G_1 to S phase of cell cycle. Cyclin A, another CDK2 partner, is upregulated when cyclin E levels are downregulated. Expression of the CDK2/cyclin A holoenzyme peaks in the S phase and is required for the elongation of initiated replication and continuation of the S phase *(5)*.

The activation of CDKs is negatively regulated by several CDK inhibitors (CKIs), which modulate CDK activity through association with them *(6)*. The major mammalian CKIs fall into two classes: (a) those that have a broader specificity, such as $p21^{waf}$, $p27^{kip1}$, and p57, which are proteins with a preference for CDK2- and CDK4-cyclin complexes *(7–9)*; and (b) $p15^{INK4B}$, $p16^{INK4}$, p18, and p19, which are closely related CKIs that are specific for CDK4- and CDK6-cyclin complexes *(1)*. These inhibitors play a key role in cell growth inhibition during cell differentiation or in response to γ-irradiation, growth factors, and cytokines *(1)*.

1.1. Regulation of CDK2 Activity in ECs: Role of p27 and p21

CDK2 and cyclin A- and cyclin E-associated kinase activities are markedly reduced in contact-inhibited ECs, although levels of CDK2 protein are not affected *(10)*. By contrast, cyclin A protein levels decline during cell cycle contact inhibition, although levels of cyclin E do not change. Thus, cyclin A is expressed at much lower levels in contact inhibited cells. A low level of cyclin A protein likely contributes to the reduced CDK2/cyclin A kinase activity in contact-inhibited ECs *(10)*. p27 expression is influenced by cell–cell contact, as confirmed by low levels of p27 in proliferating, subconfluent EC cultures, or in scrap-induced injury of EC monolayer *(10)*.

p27 is downregulated and cyclin A upregulated in EC monolayers after a scraping injury that induces CDK2 activity *(10)*. The levels of p27 are responsible for the levels of CDK2/cyclin E and partially responsible for CDK2/cyclin A activity; increased levels of p27 can inhibit CDK2/cyclin E activity and release p107 and the retinoblastoma tumor suppressor protein that can inhibit E2F activity in the cyclin A promoter *(11)*. This may not be the only pathway, because the cAMP response element-binding site in the cyclin A promoter has also been shown to be required for the repression of cyclin A transcription in confluent ECs *(12)*. Thus, alterations in p27 and cyclin A expression regulate EC proliferation during the transition from contact inhibition to the proliferative state. Furthermore, the coordinate regulation of p27 and cyclin A may have a role in controlling EC proliferation during wound healing. This pattern has been found in different types of ECs, including bovine aortic ECs, human umbilical vein ECs, or human microvascular ECs *(10)*. Thus, p27 plays a key role in maintaining the growth-inhibitory status of ECs under cell–cell contact inhibition, whereas its downregulation is involved in re-entering the cell cycle.

In contrast to findings in muscle and other cells types *(13,14)*, analyses by Western blot failed to detect significant levels of p21 and p57 expression in ECs *(10)*. Expression of the p21 CKI can be regulated by the extracellular matrix (ECM) through a p53-dependent pathway in ECs. Agonists of $\alpha_V\beta_3$ suppress p53 transcription, inhibit p21 expression, and increase the bcl-2/bax level, which promotes EC survival *(15)*. Conversely, $\alpha_V\beta_3$ antagonists activate p53 and increase p21 expression, leading to EC apoptosis *(16)*. Recent data also support the role of p21 in mediating inhibition of cell proliferation that is associated with laminar shear stress *(17)*. Therefore, p27 is involved in cell–cell contact mediated EC growth regulation, whereas p21 is involved in EC survival mediated by the ECM.

1.2. Cyclin A Transcriptional Repression as Mechanism of Cell Cycle Regulation in Growth Factors and Cytokine-Stimulated ECs

EC proliferation is regulated by soluble growth factors, such as vascular endothelial growth factor (VEGF) and basic fibroblast growth factor (bFGF) *(2)*, as well as insoluble components of the ECM *(18)*. In vitro stimulation of EC proliferation by VEGF and bFGF requires activation of protein kinase C *(19)*. Forced expression of protein kinase C δ in ECs resulted in delayed passage through the S phase of the cell cycle. These data suggest that a diverse array of cell cycle proteins is involved in regulating EC proliferation in response to different environmental conditions.

Cytokines, such as transforming growth factor (TGF)-β, interferon-γ, and tumor necrosis factor (TNF)-α modulate cyclin A expression in various cell types *(20,21)*. TNF-α regulates cell proliferation, differentiation, and apoptosis in various cell types, including ECs *(22)*. TNF-α is known to be secreted by macrophages, activated T-cells *(23)*, and vascular smooth muscle cells (VSMCs) after vascular injury *(24)*. The upregulation of TNF-α expression in animal models of arterial injury by balloon angioplasty *(24)* and in human coronary artery restenotic lesions *(25)* provides further evidence that regulation of the expression of this cytokine may be functionally important in vivo. TNF-α is expressed in arteries during atherosclerosis and restenosis; blockade of TNF-α improves re-endothelization, whereas in vitro TNF-α arrests ECs in late G_1 phase, accompanied by enhanced apoptosis *(26)*.

TNF-α-mediated effects on ECs have been shown to involve the repression of cyclin A *(26)*. The short promoter fragment of the cyclin A gene has been shown to harbor most regulatory elements required for cell cycle regulation and efficient transcription *(27)*. These elements comprise two contiguous repressor binding sites: cell cycle-dependent (CDE) and chimeric receptor (CHR) *(28)*. Modulation in the binding of transfactors, such as CDE-F, to these repressor elements regulate phase-specific expression of cyclin A *(29)*. Inhibition of cyclin A promoter activity by TGF-β1 requires an intact amino-terminal fragment of urokinase site, and this effect involves decreased phosphorylation of cyclic adenosine 3′,5′-monophosphate response element-binding protein and amino-terminal fragment-1 *(21)*. By contrast, CDE-CHR corepressor *cis*-elements in the cyclin A promoter are the target of TNF-α-mediated transcriptional repression, owing to induction of a novel 84-kDa protein that binds specifically to CHR corepressor element in the cyclin A promoter. This protein binds downstream from the major initiation sites and may interfere with RNA-polymerase progression *(30)*. Targeted disruption of this protein could potentially be a therapeutic strategy to rescue EC proliferation in vivo. It is not known, however, if other cell cycle-regulated genes such as cdc2, B-myb, or cdc25C are the target of TNF-α *(30)*.

1.3. Engineering the Response to Vascular Injury

Vascular injury stimulates proliferation and migration of VSMCs likely owing to a release of TNF-α, which is produced by activated macrophages and VSMCs *(31)*. At the same time, TNF-α induces apoptosis in ECs. Repression of E2F1 activity seems to play a major role in these pathologies. E2F1 is a member of E2F family of transcription factors, which are involved in the regulation of gene expression, proliferation, differentiation, and apoptosis *(32)*. Adenovirus-mediated restoration of E2F1 activity (Ad-E2F1) rescued ECs from TNF-α-induced cell cycle arrest and apoptosis, whereas induced apoptosis and inhibited cell cycle progression in VSMCs *(33)*.

Differential activation of nuclear factor-κB (NF-κB) may play a key role in mediating these opposing effects. E2F1-mediated inhibition of NF-κB causes increase in apoptosis in various cell types *(34)*. E2F1 competes with p50 for binding to p65 subunit of NF-κB, and the physical interaction of E2F1/p65 inhibits NF-κB transcriptional activates *(35)*. Nuclear translocation of NF-κB was markedly attenuated in Ad-E2F1-transduced VSMCs, whereas it remained active in similarly treated ECs exposed to TNF-α. Thus, there is a divergence in the TNF-α/E2F1 signaling pathways between ECs and VSMCs. By exploiting this and other signaling pathways in ECs and VSMCs, it may be possible to develop an approach to inhibit neointimal thickening, whereas encouraging recovery of a functional endothelium during vascular injury.

2. ENDOTHELIAL GROWTH FACTORS IN CELL CYCLE AND ENDOTHELIAL REMODELING

EC repair of mechanical damage is fundamental for the reconstitution of vessel integrity. Cell migration and proliferation are controlled by multiple factors, such as FGF and TGF-β *(36)*, and cell adhesive interactions, mediated by cell surface receptors such as integrins *(37)*. Fibrobast growth factor-2 (FGF-2) is one of the most potent stimulators of angiogenesis in wound healing and tumor growth *(38)*. FGF-2 induces all the important components of blood vessel growth, including degradation of the ECM, EC migration and proliferation, and differentiation into vascular tubes. It signals through the FGF receptors, that is, transmembrane receptor tyrosine kinases; there are four of them, and FGFR1 is the most prevalent in ECs *(39)*.

Majority of growth factors exert their biological effects mostly by activating the ERK and the phosphoinositide 3 kinase (PI3-K) signaling pathways. These pathways have been implicated in cell proliferation, migration, differentiation, and survival *(40)*. EC proliferation requires activation of the ras/ERK and PI3-K/Akt pathways. Exogenous FGF-2 induces a variety of EC responses through activation of the ERK pathway. FGF-2 alone regulates skin endothelial wound repair and this may be mediated by FGF-2 derived from ECM or released from damaged cells at the wound edge *(40)*. UO126, a synthetic inhibitor of mitogen-activated protein kinase (MAPK), (MEK)-1/2, downregulates cell migration and abolishes ERK1/2 activation in wild-type and FGF-2-negative cells *(40)*. FGF-2 controls EC migration through ERK activation without affecting proliferation, and FGF-2-mediated ERK-1/2 activation after wounding occurs exclusively at the wound edge *(40)*.

3. THE ROLE OF ECM IN THE CELL CYCLE AND ENDOTHELIAL REMODELING

The initiation of cell migration may involve a change in cell-matrix interaction, characterized by ECM remodeling directed in part by growth factors. Proteoglycans (PGs), fibronectin, laminin, and collagens play a major role in these interactions.

3.1. The Role of PGs in Endothelial Growth and Remodeling

P-glycoproteins are group of proteins that bear anionic glycosaminoglycan chains covalently bound to core proteins. Members of the small leucine-rich PG family, such as biglycan and decorin, which bear dermatan or chondroitin sulfate side chains, respectively, bind fibrillar proteins, such as collagens *(41)*. As ECs migrate after wounding, biglycan expression, synthesis and proteolytic processing are selectively and transiently increased *(42)*. As constituents of the ECM, the core proteins of decorin and biglycan are thought to interact with several matrix proteins and influence matrix assembly *(43)*.

The interaction of decorin and/or biglycan with fibronectin modulates cellular adhesion, perhaps by influencing the binding of RGD-dependent cell surface receptors to sites of fibronectin *(44)*. This would destabilize focal adhesions and modulate cell migration *(45)*. Both of these PGs bind to growth factors that are important in vascular cell migration, including TGF-β *(46)*. This may affect the availability or activity of these growth factors, which are important to the control of cell growth and migration *(36)*. TGF-β1, which is an antagonist of bFGF, activates gene transcription and synthesis and deposition of fibronectin, collagens, and the PGs (versican and biglycan *[36]*), and is activated by proteases induced by bFGF *(47)*.

Expression of decorin is not upregulated as macrovascular borine aortic EC migrate after wounding, in contrast to PG expression by these cells during in vitro angiogenesis *(48)*. Release of endogenous bFGF is required for the modulation of these processes during cell migration. In addition to increased sulfate incorporation into the glycosamino-glycans chains of biglycan, biglycan RNA expression and protein core synthesis are also elevated after bFGF treatment or wounding *(49)*. Both metalloproteinases and serine proteases, such as urokinase-plasminogen activators, are strongly induced in ECs by bFGF *(49)*, which promotes cell migration and angiogenesis *(50)*. Serine protease inhibitor, aprotinin, inhibits the cleavage of biglycan in FGF-stimulated EC cultures, raising the possibility that plasmin, which is activated by urokinase induced by bFGF, may be responsible for biglycan processing during cell migration.

3.2. Heparan Sulfate PGs as Regulators of FGF-2 Signaling in Brain EC

Stable binding of FGF to its receptor on ECs requires the presence of heparan sulfate glycosaminoglycans (HSPGs) *(51)*. Both stimulatory and inhibitory moieties are embedded within HS chains and their relative balance determines the net effect on FGF signaling *(52)*. HSPGs can be divided into cell surface forms (syndecan and glypicans) and secreted ECM forms (perlecan). Perlecan has been described as a proangiogenic molecule *(53)*, as well as an inhibitor of

FGF-2 signaling in blood vessels *(54)*. Syndecan-4 has an exclusive role in FGF-2-induced angiogenesis *(55)*.

Although all HSPGs on ECs have the potential to promote binding of FGF-2 to its receptor on ECs *(56)*, only glypican-1 is overexpressed in glioma vessels, which have specific ability to sensitize brain EC to mitogenic FGF-2 stimulation. Both VEGF and FGF-2 are similarly potent in stimulating tumor angiogenesis; whereas FGF-2 predominates in small tumors and at the tumor periphery, VEGF is observed in larger tumors and in the tumor center, consistent with transcriptional regulation of VEGF via hypoxia-induced factors (HIFs) *(57)*. Glypican-1 overexpression in EC results in enhanced growth and response to FGF-2 *(56)*. In addition, shed glypican-1 can protect VEGF from oxidative change in a chaperone-like function *(58)* and has been identified as an essential low-affinity coreceptor for potent angiogenesis inhibitor endostatin *(59)*. Because FGF-2 is a potent EC survival factor during irradiation *(60)*, potentiation of this survival signal by glypican-1 may be partially responsible for radiation resistance developing in gliomas. The central role of glypican-1 in signaling of both VEGF and FGF-2 may present an attractive opportunity for therapeutic intervention *(60)*.

3.3. Role of Thrombospondin in Microvascular EC Proliferation

The inhibitory role of thrombospondin-1 (TSP-1) in angiogenesis of cultured ECs and in vivo has been well documented *(61)*. Studies with cultured skin ECs from TSP-2-null mice indicate that TSP-2 also contains antiangiogenic activity *(62)*. TSP-2 preferentially inhibits vascularization of fibroblast-rich connective tissues in healing wounds, the foreign body response and tumors *(63)*. TSP-2 is largely absent from cultured ECs, but is expressed at high levels by fibroblasts *(64)*. Both TSP-1 and TSP-2 increase caspase activity and impair viability in human microvascular endothelial cells (HMVECs) in the absence of VEGF, but inhibit VEGF-stimulated cell cycle progression in a caspase-independent manner.

TSP-2 promotes EC death, because it decreases cell viability in the presence of bFGF, insulin-like growth factor-1 (IGF-1) and epithelial growth factor (EGF), as well as owing to the increased caspase activity *(64)*. In turn, VEGF protects HMVEC from TSP-2-mediated cell death; presence of VEGF blocks both TSP-2-mediated caspase activation and impairment of viability *(64)*. This may result from the greater ability of VEGF to activate the PI3-K/Akt pathway, which induces expression of cytoprotective proteins of the IAP and bcl-2 families *(65)*. However, neither VEGF nor caspase inhibitors protect ECs from TSP-2-mediated impairment of the cell cycle progression *(64)*.

TSP-1 is more prominently expressed in ECs during embryogenesis *(66)* and is synthesized in large amounts by cultured ECs *(67)*. Thus, unlike TSP-2, TSP-1 seems to be an autocrine inhibitor of EC function *(68)*. The observation that HMVEC exhibit reduced transition from G_0/G_1 to S phase, but maintain high viability in the presence of TSP-2 and VEGF, provides a potential mechanism for maintaining the capillary EC in a state of quiescence. It is conceivable that TSP-2 from fibroblasts, SMCs and pericytes (which produce TSP-2) inhibit proliferation and migration of ECs to prevent new vessel growth, whereas promote the stability of pre-existing vessels *(64)*.

3.4. The Aminoterminal Matrix Assembly Domain of Fibronectin Stabilizes Cell Shape and Prevents EC Cycle Progression

Adhesion to the ECM activates signal transduction pathways important in the regulation of cell growth *(69)*, death *(70)*, and differentiation *(71)*. Information from ECM is transduced largely through integrin receptors *(72)*. At such sites, integrin receptors are linked to the actin microfilament system in a functional unit which "senses" both mechanical and chemical changes in the ECM, transducing it to changes in gene expression and growth behavior *(73)*. Fibronectin matrix may be a potential modulator of actin organization inside the cell, functioning to influence signaling and growth in ECs *(74)*.

Cells which are anchorage-dependent for proliferation require both adhesion to the ECM and growth factor stimulation to progress through the cell cycle *(69)*. Integrin and growth factor signaling pathways both activate MAPK *(75)* and regulate the synthesis and activities of cyclin D-CDK 4/6, cyclin A, as well as cyclin E-CDK2 complexes *(76–78)*. The ability of 70 kDa amino-terminal domain of fibronectin stabilizes both fibrillar fibronectin and the internal actin skeleton. Binding of fibronectin's amino terminus may also modulate the integrin-dependent regulation of small-molecular-weight G proteins required for actin organization and focal adhesion assembly *(79)*. Increased levels of fibronectin matrix are associated with decreased rates of cell migration and cell cycle progression *(80)*.

4. THE HORMONAL INFLUENCE ON EC CYCLE AND VASCULAR REMODELING

4.1. The Protective Effect of Estrogen in Vascular Injury

The protective effect of estrogen on vasculature was first shown in women on estrogen replacement therapy and later confirmed in more details in animal models and in vitro *(81)*. In addition to cholesterol-lowering effect, estrogen promotes vasodilation, improves endothelial function, and is accompanied by inhibitory effect on VSMC proliferation *(82)*. Estrogen binds to estrogen receptors α and β (ERα and ERβ). VSMC proliferation is one of the major vascular remodeling processes after vascular injury, which may participate in the vaso-occlusive disorders associated with multiple vascular diseases *(83)*. After endothelial injury in experimental animal models, VSMCs start to proliferate by day 3 and also start to appear in the intima 5–7 d after injury *(84)*. It has been suggested that estradiol may help migrating cells to colonize in injured vessel and aid in re-endothelization by protection of ECs from apoptosis and enhancement of their adhesion to the ECs *(85)*.

A group of novel selective estrogen receptor modulators, such as idoxifene (pyrrolidino-4-iodofamaxifene), show antagonism or minimal agonism in reproductive tissue, while maintaining protective characteristics of estrogen on bone, lipid metabolism, and in vascular tissue *(86)*. In vitro, treatment with idoxifene resulted in S phase cell cycle arrest in serum-stimulated VSMC, and significantly protected ECs from TNF-α-induced apoptosis. In ovariectomized rats, idoxifene significantly enhanced re-endothelization in the injured carotid arteries. Furthermore, the production of nitric oxide (NO) from excised carotid arteries was significantly higher in idoxifene-treated animals *(84)*. Thus, idoxifene beneficially modulates vascular injury response through inhibition of VSMC proliferation and acceleration of endothelial proliferation and recovery *(84)*.

4.2. The Role of IGF-1 in the Adult Brain

Vascular dysfunction is a major suspect in the etiology of several important neurodegenerative diseases. IGF-1 is a wide spectrum growth factor with angiogenic actions: IGF-1 induces the growth of cultured brain ECs through HIF-1α and VEGF *(87)*. Hypoxia and proinflammatory cytokines also recruit the HIF-1α/ VEGF angiogenic pathway. Notably, in stroke patients and in Alzheimer's disease patients, serum IGF-1 levels are increased *(88,89)*, maybe to cope with the increased brain demand of circulating IGF-1. The systemic injection of IGF-1 in adult mice increases brain vessel density *(90)*. Physical exercise that stimulates widespread brain vessel growth in normal mice fails to do so in mice with low serum IGF-1 *(91)*, whereas blockade of IGF-1 input abrogates vascular growth at the injury site. In addition, low serum/brain IGF-1 levels are associated with old age and with several neurodegenerative diseases related to an increased risk of vascular dysfunction *(92)*. Collectively, these data imply that IGF-1 participates in vessel remodeling in the adult brain.

4.3. The Role of Diabetes in Endothelial Injury and Cell Cycle

Hypercholesterolemia and diabetes have been shown to be associated with a significant impairment in adaptive vascular growth of both capillary-like tube vessels and collateral vessels

(93). NF-κB has been linked to the onset of atherosclerosis *(94)*. It has been reported that both native and modified low-density lipoprotein (LDL) activate a series of NF-κB-dependent genes that are relevant to the pathophysiology of the vessel wall *(94)*. A similar set of genes are a target of the signal transducers and activators of transcription (STATs) *(95)*. STATs are a family of latent cytoplasmic proteins that, on activation, acquire DNA-binding activity, translocate into the nucleus, bind to specific promoter elements, and control the expression of target genes *(95)*.

4.3.1. Role of Diabetes-Modified LDL on Cell Cycle in ECs

Diabetes causes many metabolic changes, one of which is qualitative change in LDL, caused by glycation and/or oxidation (dm-LDL). dm-LDL treatment of cultured ECs leads to an accumulation of cells in G_1 by increasing the level of p21^{waf} *(1)*. Nuclear run-off experiments support a role for transcription in regulating p21^{waf} expression on dm-LDL stimulation *(96)*. In addition, dm-LDL promotes p21^{waf} transcription through STAT5 activation *(96)*. Thus, LDL from type 2 diabetes can maintain EC in quiescent state in G_1 through STAT5B-mediated p21^{waf} expression. The presence of a positive immunoreactivity for activated STAT5 and p21^{waf} in intraplaque neovessels supports the possibility that induction of STAT5-dependent genes may exert substantial atherogenic effects on the vessel wall *(96)*.

4.3.2. Role of High Glucose on EC Cycle in Diabetes

A reduction in antioxidant reserves has been attributed to EC dysfunction in diabetes, even in patients with well-controlled glucose levels *(97)*. Exposure of ECs to elevated glucose levels causes glucose oxidation and generation of excess reactive oxygen species (ROS). Hyperglycemia stimulates apoptosis in ECs by a mechanism that involves the generation of ROS and O_2^- formation (Table 1).

High glucose stimulates MAPK, which is associated with an enhancement in p27 protein and growth arrest *(98)*. High glucose induces p21 and p27 in control cells but not in cells overexpressing heme oxygenase-1 (HO-1). It was shown that overexpression of human gene for HO-1 in rabbit and rat coronary EC renders the cells resistant to oxidative stress *(99)* and enhances cell growth and angiogenesis *(100)*. HO-1 can be induced in ECs in response to oxidants, such as heme, H_2O_2, and TNF-α *(101)*. Expression of HO-1 gene participates in the regulation of cell cycle in high glucose exposed cells, presumably through heme metabolism and alteration in the levels of bilirubin, an antioxidant, and CO, a vasoactive molecule *(101)*. The glucose-mediated decrease in HO activity is not limited to ECs, but occurs also in the liver and kidneys of diabetic rats *(102)*. Glucose may result in deactivation of HO-1 proteins via ROS. It was recently shown that Bach 1, a heme-regulated transcriptional repressor, functions as a hypoxia-inducible repressor for the *HO-1* gene *(103)*. It is possible that Bach 1, upregulated by glucose, contributes to decreased HO-1 expression in diabetes.

Accordingly, HO-1 overexpression, attenuates cell death by oxidants such as H_2O_2 and TNF-α *(104)*, an effect attributed to CO, which has been shown to play an important role in controlling cell cycle progression *(104)*. By contrast, HO-2, which is constitutively expressed in the blood vessels, is unaffected by glucose or factors known to act as inducers of HO-1 *(105)*. In addition to increases in oxidative stress, high glucose enhances the levels of angiotensin II, which then results in an increase in HO-1, as protective mechanism against ROS in vitro and in vivo *(106)*. In conclusion, these data indicate that heme metabolism and HO-1 expression regulate signaling systems in ECs exposed to high glucose, and control cell cycle progression *(101)*.

5. CONCLUSIONS

The integrity of the endothelial monolayer as a key component in the development of atherosclerotic lesions *(83)* and the acceleration of endothelial healing after arterial injury can inhibit the formation of restenotic lesions *(107)*. Future studies on EC cycle regulation may have important implications for understanding the mechanisms of wound healing, tumor vasculogenesis, atherogenesis, and restenosis. By exploiting these and other signaling pathways, involved in the

Table 1
Summary of Biological Processes in Different Endothelial Cell Types

Biological process	Endothelial cell type	References
Downregulation of p27; upregulation of cyclin A; induction of CDK2 activity	BAEC, HUVEC, HMEC (dermal)	(10)
Regulation of p21 CDK expression by ECM through p53-dependent pathway	HUVEC	(15)
Role of p21 in cell cycle inhibition by laminar shear stress	BAEC, HUVEC	(17)
PKC activation by VEGF and bFGF in vitro stimulated EC	HUVEC	(19)
Cytokine-mediated modulation of cyclin A expression	VSMC from rat aorta	(20)
Regulation of EC proliferation, differentiation and apoptosis by TNF-α	BAEC, bovine brain capillary, HUVEC	(22)
Upregulation of TNF-α by balloon angioplasty	VSMC from rabbit aorta, human coronary arteries	(24,25)
Cell cycle arrest by TNF-α, accompanied by apoptosis	Rat carotid artery EC	(26)
Cell cycle regulation harbored by cyclin A gene	Human cell lines: HELA, SAOS-2, Huh7, HepG2; Simian: Vero; Mouse: NIH 3T3; Hamster: CCL39	(27,28)
Role of E2F expression in rescuing EC from TNF-α-induced apoptosis	HUVEC, BAEC	(33)
Role of E2F1 in inhibiting NF-κB transcriptional activities	Mouse or a rat (Sprague-Dawley) model of balloon carotid injury	(35)
Role of FGF and TGF-β in EC migration and proliferation	Pulmonary artery EC	(36)
FGFR1 prevalence in EC	Balb/c mouse aortic EC and brain microvascular EC	(39)
Role of FGF-2 in wound repair	Skin EC from FGF-2 deficient mouse	(40)
Increased biglycan expression on migrating EC after wounding	BAEC	(42)
Interaction of decorin and biglycan with fibronectin in cell adhesion	CHO, retinal EC	(44)
Role of TGF-β in synthesis and deposition of fibronectin, collagen, and proteoglycans	BAEC, pericytes from bovine retina, VSMC from bovine and rat aorta	(47)
Increase in biglycan RNA expression after bFGF treatment	Bovine adrenal gland capillary EC, BAEC, fibroblasts from bovine embryonic skin	(49)
Role of perlecan in EC-mediated inhibition of intimal hyperplasia	BAEC, bovine aorta VSMC	(54)
Role of heparan sulfate in FGF2 signaling in brain EC	Mouse or bovine brain EC, HUVEC, human skin MVEC, C57BL/6 mouse MVEC from gastrointestinal mucosa	(56)
Role of FGF2 in EC survival during irradiation	Human and mouse dermal MVEC	(60)
Inhibition of MVEC proliferation by TSP-2	HUVEC	(64)
Activation of PI-3K/Akt pathway by VEGF	BAEC; HUVEC	(65)
Inhibitory effect of oxidized LDL	Human dermal MVEC	(93,96)
Protective role of HO-1 in glucose-mediated cell growth arrest in human MVEC		(101)

BAEC, bovine aortic endothelial cells; HUVEC, human umbilical vein endothelial cells; VSMC, vascular smooth muscle cells; MVEC, microvascular endothelial cells.

regulation of EC cycle and growth, it may be possible to develop new approaches in molecular-based cell therapy of vascular wall.

REFERENCES

1. Morgan DO. Principles of CDK regulation. Nature 1995;374:131–134.
2. Folkman J, Shing Y, Angiogenesis. J Biol Chem 1992;267:10,931–10,934.
3. D'Amore PA. Mechanisms of endothelial growth control. Am J Respir Cell Mol Biol 1992;6:1–8.
4. Weinberg RA. The retinoblastoma protein and cell cycle control. Cell 1995;81:323–330.
5. Pagano M, Pepperkok R, Verde F, Ansorge W, Draetta G. Cyclin A is required at two points in the human cell cycle. EMBO J 1992;11:961–971.
6. Peter M, Herskowitz I. Joining the complex: cyclin-dependent kinase inhibitory proteins and the cell cycle. Cell 1994;79:181–184.
7. Chan FK, Zhang J, Cheng L, Shapiro DN, Winoto A. Identification of human and mouse p19, a novel CDK4 and CDK6 inhibitor with homology to p16ink4. Mol Cell Biol 1995;15:2682–2688.
8. Guan KL, Jenkins CW, Li Y, et al. Growth suppression by p18, a p16INK4/. Genes Dev 1994;8:2939–2952.
9. Hirai H, Roussel MF, Kato JY, Ashmun RA, Sherr CJ. Novel INK4 proteins, p19 and p18, are specific inhibitors of the cyclin D-dependent kinases CDK4 and CDK6. Mol Cell Biol 1995;15: 2672–2681.
10. Chen D, Walsh K, Wang J. Regulation of CDK2 activity in endothelial cells that are inhibited from growth by cell contact. Arterioscler Thromb Vasc Biol 2000;20:629–635.
11. Zerfass-Thome K, Schulze A, Zwerschke W, et al. p27KIP1 blocks cyclin E-dependent transactivation of cyclin A gene expression. Mol Cell Biol 1997;17:407–415.
12. Yoshizumi M, Hsieh CM, Zhou F, et al. The ATF site mediates downregulation of the cyclin A gene during contact inhibition in vascular endothelial cells. Mol Cell Biol 1995;15:3266–3272.
13. Guo K, Wang J, Andres V, Smith RC, Walsh K. MyoD-induced expression of p21 inhibits cyclin-dependent kinase activity upon myocyte terminal differentiation. Mol Cell Biol 1995;15:3823–3829.
14. Jiang H, Lin J, Su ZZ, Collart FR, Huberman E, Fisher PB. Induction of differentiation in human promyelocytic HL-60 leukemia cells activates p21, WAF1/CIP1, expression in the absence of p53. Oncogene 1994;9:3397–3406.
15. Stromblad S, Becker JC, Yebra M, Brooks PC, Cheresh DA. Suppression of p53 activity and p21WAF1/CIP1 expression by vascular cell integrin alphaVbeta3 during angiogenesis. J Clin Invest 1996;98:426–433.
16. Yang C, Chang J, Gorospe M, Passaniti A. Protein tyrosine phosphatase regulation of endothelial cell apoptosis and differentiation. Cell Growth Differ 1996;7:161–171.
17. Akimoto S, Mitsumata M, Sasaguri T, Yoshida Y. Laminar shear stress inhibits vascular endothelial cell proliferation by inducing cyclin-dependent kinase inhibitor p21$^{Sdi1/Cip1/Waf1}$. Circ Res 2000;86:185–190.
18. Ingber DE. Extracellular matrix as a solid-state regulator in angiogenesis: identification of new targets for anti-cancer therapy. Semin Cancer Biol 1992;3:57–63.
19. Kent KC, Mii S, Harrington EO, Chang JD, Mallette S, Ware JA. Requirement for protein kinase C activation in basic fibroblast growth factor-induced human endothelial cell proliferation. Circ Res 1995;77:231–238.
20. Sibinga NE, Wang H, Perrella MA, et al. Interferon-gamma-mediated inhibition of cyclin A gene transcription is independent of individual cis-acting elements in the cyclin A promoter. J Biol Chem 1999;274:12,139–12,146.
21. Yu C, Takeda M, Soliven B. Regulation of cell cycle proteins by TNF-alpha and TGF-beta in cells of oligodendroglial lineage. J Neuroimmunol 2000;108:2–10.
22. Sato N, Goto T, Haranaka K, et al. Actions of tumor necrosis factor on cultured vascular endothelial cells: morphologic modulation, growth inhibition, and cytotoxicity. J Natl Cancer Inst 1986;76: 1113–1121.
23. Steffen M, Ottmann OG, Moore MA. Simultaneous production of tumor necrosis factor-alpha and lymphotoxin by normal T cells after induction with IL-2 and anti-T3. J Immunol 1988;140:2621–2624.
24. Tanaka H, Sukhova G, Schwartz D, Libby P. Proliferating arterial smooth muscle cells after balloon injury express TNF-alpha but not interleukin-1 or basic fibroblast growth factor. Arterioscler Thromb Vasc Biol 1996;16:12–18.

25. Clausell N, de Lima VC, Molossi S, et al. Expression of tumour necrosis factor alpha and accumulation of fibronectin in coronary artery restenotic lesions retrieved by atherectomy. Br Heart J 1995;73:534–539.
26. Krasinski K, Spyridopoulos I, Kearney M, Losordo DW. In vivo blockade of tumor necrosis factor-alpha accelerates functional endothelial recovery after balloon angioplasty. Circulation 2001;104:1754–1756.
27. Henglein B, Chenivesse X, Wang J, Eick D, Brechot C. Structure and cell cycle-regulated transcription of the human cyclin A gene. Proc Natl Acad Sci USA 1994;91:5490–5494.
28. Liu N, Lucibello FC, Korner K, Wolfraim LA, Zwicker J, Muller R. CDF-1, a novel E2F-unrelated factor, interacts with cell cycle-regulated repressor elements in multiple promoters. Nucleic Acids Res 1997;25:4915–4920.
29. Zwicker J, Lucibello FC, Wolfraim LA, et al. Cell cycle regulation of the cyclin A, cdc25C and cdc2 genes is based on a common mechanism of transcriptional repression. EMBO J 1995;14:4514–4522.
30. Kishore R, Spyridopoulos I, Luedemann C, Losordo DW. Functionally novel tumor necrosis factor-alpha-modulated CHR-binding protein mediates cyclin A transcriptional repression in vascular endothelial cells. Circ Res 2002;91:307–314.
31. Warner SJ, Libby P. Human vascular smooth muscle cells. Target for and source of tumor necrosis factor. J Immunol 1989;142:100–109.
32. Field SJ, Tsai FY, Kuo F, et al. E2F-1 functions in mice to promote apoptosis and suppress proliferation. Cell 1996;85:549–561.
33. Spyridopoulos I, Principe N, Krasinski KL, et al. Restoration of E2F expression rescues vascular endothelial cells from tumor necrosis factor-alpha-induced apoptosis. Circulation 1998;98:2883–2890.
34. Tanaka H, Matsumura I, Ezoe S, et al. E2F1 and c-Myc potentiate apoptosis through inhibition of NF-kappaB activity that facilitates MnSOD-mediated ROS elimination. Mol Cell 2002;9:1017–1029.
35. Goukassian DA, Kishore R, Krasinski K, et al. Engineering the response to vascular injury: divergent effects of deregulated E2F1 expression on vascular smooth muscle cells and endothelial cells result in endothelial recovery and inhibition of neointimal growth. Circ Res 2003;93:162–169.
36. Roberts AB, Sporn MB. Regulation of endothelial cell growth, architecture, and matrix synthesis by TGF-beta. Am Rev Respir Dis 1989;140:1126–1128.
37. Hynes RO. Integrins: a family of cell surface receptors. Cell 1987;48:549–554.
38. Ornitz DM, Itoh N. Fibroblast growth factors. Genome Biol 2001;2:REVIEWS3005.
39. Bastaki M, Nelli EE, Dell'Era P, et al. Basic fibroblast growth factor-induced angiogenic phenotype in mouse endothelium. A study of aortic and microvascular endothelial cell lines. Arterioscler Thromb Vasc Biol 1997;17:454–464.
40. Pintucci G, Moscatelli D, Saponara F, et al. Lack of ERK activation and cell migration in FGF-2-deficient endothelial cells. FASEB J 2002;16:598–600.
41. Scott JE, Orford CR. Dermatan sulphate-rich proteoglycan associates with rat tail-tendon collagen at the d band in the gap region. Biochem J 1981;197:213–216.
42. Kinsella MG, Tsoi CK, Jarvelainen HT, Wight TN. Selective expression and processing of biglycan during migration of bovine aortic endothelial cells. The role of endogenous basic fibroblast growth factor. J Biol Chem 1997;272:318–325.
43. Vogel KG, Paulsson M, Heinegard D. Specific inhibition of type I and type II collagen fibrillogenesis by the small proteoglycan of tendon. Biochem J 1984;223:587–597.
44. Bidanset DJ, LeBaron R, Rosenberg L, Murphy-Ullrich JE, Hook M. Regulation of cell substrate adhesion: effects of small galactosaminoglycan-containing proteoglycans. J Cell Biol 1992;118:1523–1531.
45. Couchman JR, Austria MR, Woods A. Fibronectin-cell interactions. J Invest Dermatol 1990;94:7S–14S.
46. Ruoslahti E, Yamaguchi Y. Proteoglycans as modulators of growth factor activities. Cell 1991;64:867–869.
47. Sato Y, Rifkin DB. Inhibition of endothelial cell movement by pericytes and smooth muscle cells: activation of a latent transforming growth factor-beta 1-like molecule by plasmin during co-culture. J Cell Biol 1989;109:309–315.
48. Jarvelainen HT, Iruela-Arispe ML, Kinsella MG, Sandell LJ, Sage EH, Wight TN. Expression of decorin by sprouting bovine aortic endothelial cells exhibiting angiogenesis in vitro. Exp Cell Res 1992;203:395–401.
49. Gross JL, Moscatelli D, Jaffe EA, Rifkin DB. Plasminogen activator and collagenase production by cultured capillary endothelial cells. J Cell Biol 1982;95:974–981.

50. Folkman J, Klagsbrun M. Angiogenic factors. Science 1987;235:442–447.
51. Ornitz DM, Yayon A, Flanagan JG, Svahn CM, Levi E, Leder P. Heparin is required for cell-free binding of basic fibroblast growth factor to a soluble receptor and for mitogenesis in whole cells. Mol Cell Biol 1992;12:240–247.
52. Liu D, Shriver Z, Venkataraman G, El Shabrawi Y, Sasisekharan R. Tumor cell surface heparan sulfate as cryptic promoters or inhibitors of tumor growth and metastasis. Proc Natl Acad Sci USA 2002;99:568–573.
53. Sharma B, Handler M, Eichstetter I, Whitelock JM, Nugent MA, Iozzo RV. Antisense targeting of perlecan blocks tumor growth and angiogenesis in vivo. J Clin Invest 1998;102:1599–1608.
54. Nugent MA, Nugent HM, Iozzo RV, Sanchack K, Edelman ER. Perlecan is required to inhibit thrombosis after deep vascular injury and contributes to endothelial cell-mediated inhibition of intimal hyperplasia. Proc Natl Acad Sci USA 2000;97:6722–6727.
55. Volk R, Schwartz JJ, Li J, Rosenberg RD, Simons M. The role of syndecan cytoplasmic domain in basic fibroblast growth factor-dependent signal transduction. J Biol Chem 1999;274:24,417–24,424.
56. Qiao D, Meyer K, Mundhenke C, Drew SA, Friedl A. Heparan sulfate proteoglycans as regulators of fibroblast growth factor-2 signaling in brain endothelial cells. Specific role for glypican-1 in glioma angiogenesis. J Biol Chem 2003;278:16,045–16,053.
57. Kumar R, Kuniyasu H, Bucana CD, Wilson MR, Fidler IJ. Spatial and temporal expression of angiogenic molecules during tumor growth and progression. Oncol Res 1998;10:301–311.
58. Gengrinovitch S, Berman B, David G, Witte L, Neufeld G, Ron D. Glypican-1 is a VEGF165 binding proteoglycan that acts as an extracellular chaperone for VEGF165. J Biol Chem 1999;274:10,816–10,822.
59. Karumanchi SA, Jha V, Ramchandran R, et al. Cell surface glypicans are low-affinity endostatin receptors. Mol Cell 2001;7:811–822.
60. Paris F, Fuks Z, Kang A, et al. Endothelial apoptosis as the primary lesion initiating intestinal radiation damage in mice. Science 2001;293:293–297.
61. Chen H, Herndon ME, Lawler J. The cell biology of thrombospondin-1. Matrix Biol 2000;19: 597–614.
62. Bornstein P, Kyriakides TR, Yang Z, Armstrong LC, Birk DE. Thrombospondin 2 modulates collagen fibrillogenesis and angiogenesis. J. Investig. Dermatol Symp Proc 2000;5:61–66.
63. Hawighorst T, Velasco P, Streit M, et al. Thrombospondin-2 plays a protective role in multistep carcinogenesis: a novel host anti-tumor defense mechanism. EMBO J 2001;20:2631–2640.
64. Armstrong LC, Bjorkblom B, Hankenson KD, Siadak AW, Stiles CE, Bornstein P. Thrombospondin 2 inhibits microvascular endothelial cell proliferation by a caspase-independent mechanism. Mol Biol Cell 2002;13:1893–1905.
65. Mesri M, Morales-Ruiz M, Ackermann EJ, et al. Suppression of vascular endothelial growth factor-mediated endothelial cell protection by survivin targeting. Am J Pathol 2001;158:1757–1765.
66. Iruela-Arispe ML, Liska DJ, Sage EH, Bornstein P. Differential expression of thrombospondin 1, 2, and 3 during murine development. Dev Dyn 1993;197:40–56.
67. Mosher DF, Doyle MJ, Jaffe EA. Synthesis and secretion of thrombospondin by cultured human endothelial cells. J Cell Biol 1982;93:343–348.
68. Tolsma SS, Stack MS, Bouck N. Lumen formation and other angiogenic activities of cultured capillary endothelial cells are inhibited by thrombospondin-1. Microvasc Res 1997;54:13–26.
69. Howe A, Aplin AE, Alahari SK, Juliano RL. Integrin signaling and cell growth control. Curr Opin Cell Biol 1998;10:220–231.
70. Frisch SM, Ruoslahti E. Integrins and anoikis. Curr Opin Cell Biol 1997;9:701–706.
71. Roskelley CD, Bissell MJ. Dynamic reciprocity revisited: a continuous, bidirectional flow of information between cells and the extracellular matrix regulates mammary epithelial cell function. Biochem Cell Biol 1995;73:391–397.
72. Jockusch BM, Bubeck P, Giehl K, et al. The molecular architecture of focal adhesions. Annu Rev Cell Dev Biol 1995;11:379–416.
73. Chen CS, Mrksich M, Huang S, Whitesides GM, Ingber DE. Geometric control of cell life and death. Science 1997;276:1425–1428.
74. Roman J, LaChance RM, Broekelmann TJ, et al. The fibronectin receptor is organized by extracellular matrix fibronectin: implications for oncogenic transformation and for cell recognition of fibronectin matrices. J Cell Biol 1989;108:2529–2543.
75. Short SM, Talbott GA, Juliano RL. Integrin-mediated signaling events in human endothelial cells. Mol Biol Cell 1998;9:1969–1980.

76. Fang F, Orend G, Watanabe N, Hunter T, Ruoslahti E. Dependence of cyclin E-CDK2 kinase activity on cell anchorage. Science 1996;271:499–502.
77. Schulze A, Zerfass-Thome K, Berges J, Middendorp S, Jansen-Durr P, Henglein B. Anchorage-dependent transcription of the cyclin A gene. Mol Cell Biol 1996;16:4632–4638.
78. Zhu X, Ohtsubo M, Bohmer RM, Roberts JM, Assoian RK. Adhesion-dependent cell cycle progression linked to the expression of cyclin D1, activation of cyclin E-CDK2, and phosphorylation of the retinoblastoma protein. J Cell Biol 1996;133:391–403.
79. Barry ST, Flinn HM, Humphries MJ, Critchley DR, Ridley AJ. Requirement for Rho in integrin signalling. Cell Adhes Commun 1997;4:387–398.
80. Hansen LK, Mooney DJ, Vacanti JP, Ingber DE. Integrin binding and cell spreading on extracellular matrix act at different points in the cell cycle to promote hepatocyte growth. Mol Biol Cell 1994;5:967–975.
81. Grady D, Rubin SM, Petitti DB, et al. Hormone therapy to prevent disease and prolong life in postmenopausal women. Ann Intern Med 1992;117:1016–1037.
82. Farhat MY, Lavigne MC, Ramwell PW. The vascular protective effects of estrogen. FASEB J 1996;10:615–624.
83. Ross R. The pathogenesis of atherosclerosis: a perspective for the 1990s. Nature 1993;362:801–809.
84. Yue TL, Vickery-Clark L, Louden CS, et al. Selective estrogen receptor modulator idoxifene inhibits smooth muscle cell proliferation, enhances reendothelialization, and inhibits neointimal formation in vivo after vascular injury. Circulation 2000;102:III281–III288.
85. Alvarez RJ, Gips SJ, Moldovan N, et al. 17beta-estradiol inhibits apoptosis of endothelial cells. Biochem Biophys Res Commun 1997;237:372–381.
86. Chander SK, McCague R, Luqmani Y, et al. Pyrrolidino-4-iodotamoxifen and 4-iodotamoxifen, new analogues of the antiestrogen tamoxifen for the treatment of breast cancer. Cancer Res 1991;51:5851–5858.
87. Maxwell PH, Ratcliffe PJ. Oxygen sensors and angiogenesis. Semin Cell Dev Biol 2002;13:29–37.
88. Schwab S, Spranger M, Krempien S, Hacke W, Bettendorf M. Plasma insulin-like growth factor I and IGF binding protein 3 levels in patients with acute cerebral ischemic injury. Stroke 1997;28:1744–1748.
89. Tham A, Nordberg A, Grissom FE, Carlsson-Skwirut C, Viitanen M, Sara VR. Insulin-like growth factors and insulin-like growth factor binding proteins in cerebrospinal fluid and serum of patients with dementia of the Alzheimer type. J Neural Transm Park Dis Dement Sect 1993;5:165–176.
90. Lopez-Lopez C, LeRoith D, Torres-Aleman I. Insulin-like growth factor I is required for vessel remodeling in the adult brain. Proc Natl Acad Sci USA 2004;101:9833–9838.
91. Black JE, Isaacs KR, Anderson BJ, Alcantara AA, Greenough WT. Learning causes synaptogenesis, whereas motor activity causes angiogenesis, in cerebellar cortex of adult rats. Proc Natl Acad Sci USA 1990;87:5568–5572.
92. Sonntag WE, Lynch CD, Cooney PT, Hutchins PM. Decreases in cerebral microvasculature with age are associated with the decline in growth hormone and insulin-like growth factor 1. Endocrinology 1997;138:3515–3520.
93. Chen CH, Jiang W, Via DP, et al. Oxidized low-density lipoproteins inhibit endothelial cell proliferation by suppressing basic fibroblast growth factor expression. Circulation 2000;101:171–177.
94. Collins T, Cybulsky MI. NF-kappaB: pivotal mediator or innocent bystander in atherogenesis? J Clin Invest 2001;107:255–264.
95. Ihle JN, Kerr IM. Jaks and Stats in signaling by the cytokine receptor superfamily. Trends Genet 1995;11:69–74.
96. Brizzi MF, Dentelli P, Pavan M, et al. Diabetic LDL inhibits cell cycle progression via STAT5B and p21(waf). J Clin Invest 2002;109:111–119.
97. Baynes JW. Role of oxidative stress in development of complications in diabetes. Diabetes 1991;40:405–412.
98. Wolf G, Schroeder R, Zahner G, Stahl RA, Shankland SJ. High glucose-induced hypertrophy of mesangial cells requires p27(Kip1), an inhibitor of cyclin-dependent kinases. Am J Pathol 2001;158:1091–1100.
99. Wagener FA, da Silva JL, Farley T, de Witte T, Kappas A, Abraham NG. Differential effects of heme oxygenase isoforms on heme mediation of endothelial intracellular adhesion molecule 1 expression. J Pharmacol Exp Ther 1999;291:416–423.
100. Deramaudt BM, Braunstein S, Remy P, Abraham NG. Gene transfer of human heme oxygenase into coronary endothelial cells potentially promotes angiogenesis. J Cell Biochem 1998;68:121–127.

101. Abraham NG, Kushida T, McClung J, et al. Heme oxygenase-1 attenuates glucose-mediated cell growth arrest and apoptosis in human microvessel endothelial cells. Circ Res 2003;93:507–514.
102. Abraham NG, Lin JH, Schwartzman ML, Levere RD, Shibahara S. The physiological significance of heme oxygenase. Int. J Biochem 1988;20:543–558.
103. Kitamuro T, Takahashi K, Ogawa K, et al; Bach1 functions as a hypoxia-inducible repressor for the heme oxygenase-1 gene in human cells. J Biol Chem 2003;278:9125–9133.
104. Kushida T, Quan S, Yang L, Ikehara S, Kappas A, Abraham NG. A significant role for the heme oxygenase-1 gene in endothelial cell cycle progression. Biochem Biophys Res Commun 2002;291: 68–75.
105. Quan S, Yang L, Shenouda S, et al. Functional expression of human heme oxygenase-1 (HO-1) driven by HO-1 promoter in vitro and in vivo. J Cell Biochem 2002;85:410–421.
106. Ishizaka N, de Leon H, Laursen JB, et al. Angiotensin II-induced hypertension increases heme oxygenase-1 expression in rat aorta. Circulation 1997;96:1923–1929.
107. Asahara T, Bauters C, Pastore C, et al. Local delivery of vascular endothelial growth factor accelerates reendothelialization and attenuates intimal hyperplasia in balloon-injured rat carotid artery. Circulation 1995;91:2793–2801.

17
The Contribution of Bone Marrow-Derived Cells to Cerebrovascular Formation and Integrity

David Kobiler, PhD and John Glod, MD, PhD

SUMMARY

The contribution of bone marrow-derived circulating cells to the formation and maintenance of the vasculature, and the cerebrovasculature in particular, has been established. It is becoming evident that several different populations of cells including early progenitor-like cells, monocytic cells, and perhaps mesenchymal stem cells are responsible for the reported actions of "endothelial progenitor cells." Large variation in the relative contribution of bone marrow-derived cells to formation and repair of the vasculature in different experimental systems illustrates some of the factors that influence the behavior of bone marrow-derived cells.

Key Words: Endothelial progenitor cells; monocytic cells; BBB breakdown; cerebrovasculature repair.

1. INTRODUCTION

The vasculature of the central nervous system (CNS) is characterized by the existence of the blood–brain barrier (BBB), which is a complex system that regulates the movement of substances from the circulation to the brain parenchyma and vice versa. A complex cellular system of endothelial cells (ECs), astroglia, pericytes, perivascular macrophages, and the basement membrane *(1)* play a role in the formation and maintenance of the BBB; however, the anatomic substrate of the BBB is the interendothelial tight and adherens junctions that form a continuous sealing *(2,3)*.

The integrity of the BBB and the maintenance of the blood supply to the brain are crucial to the survival of higher vertebrates. Disruption of the cerebrovasculature leads to neurological dysfunction both in animal models and in human disease states. Gross disruption of vessel integrity is seen in pathological conditions, such as stroke or traumatic brain injury with devastating consequences *(4–7)*. Less dramatic impairment of BBB function has been implicated in other diseases including Alzheimer's disease, multiple sclerosis, and CNS infection *(8–12)*. It has been demonstrated that an inverse relationship exists between the degree of BBB breakdown and cognitive outcome in both animal models *(13–15)* and patients with brain injury *(16–18)*. An important facet of brain vascular repair is the replacement of damaged ECs. In this review we will address the role of the bone marrow-derived cells in this process.

2. EMBRYOGENESIS OF THE CNS VASCULATURE

The biology of blood vessel formation in the embryo has been extensively studied. Endothelial progenitors, or angioblasts, are formed at the periphery of the blood islands and migrate into the head region at embryonic day 9. There they form the perineural vascular plexus, covering the entire surface of the neural tube. Vascular sprouts from the perineural plexus then invade the

From: *The Cell Cycle in the Central Nervous System*
Edited by: D. Janigro © Humana Press Inc., Totowa, NJ

neuroectoderm forming undifferentiated capillaries in the developing brain *(19–21)*. The formation of the vascular system of the CNS illustrates two distinct mechanisms used to supply ECs to growing vasculature. In early embryogenesis, the vascular system develops by vasculogenesis. Vasculogenesis relies on endothelial progenitors that migrate to an area of vessel formation and then incorporate into the endothelial vascular lining, as illustrated in the formation of the vascular plexus. This is contrasted by angiogenesis. Angiogenesis relies on the division of pre-existing ECs present at the site of vessel growth as a supply of new ECs, as seen when vascular sprouts from the perineural plexus invade the proliferating neuroectoderm *(22–24)*.

Mesodermal cells have been isolated from embryonic day 7.5 mice and shown to have properties consistent with an "endothelial progenitor cell" or angioblast *(25–26)*. The cells have unlimited growth potential and a stable phenotype in culture. They express the early endothelial markers Tie-2 and thrombomodulin, differentiate to mature ECs, form vascular tubes in vitro, and build blood vessels after transplantation during embryogenesis *(25)*. Their gene expression profile (*Tie-2, thrombomodulin, GATA-4, GATA-6*, etc.) matches the pattern observed in the proximal lateral mesoderm that will give rise to the embryonic endocardium and myocardium. These cells were shown under a variety of conditions in vitro to be committed to the endothelial lineage. Transplantation studies in embryos showed differentiation of these cells only toward the endothelial lineage *(25)*.

Vasculogenesis and angiogenesis involve the highly regulated and coordinated interaction of multiple vascular growth factors that are critical for embryonal development *(26)*. Two families of growth factors, vascular endothelial growth factor (VEGFs) and angiopoietins, play a prominent role in vascular formation *(28–31)*. The VEGF family currently includes six known members and they exert their biological function via three related tyrosine kinase receptors, VEGFR-1 (flt-1), VEGFR-2 (flk-1), and VEGFR-3 *(29)*. VEGF (VEGF-A) was demonstrated as a major inducer of EC proliferation, migration, sprouting, tube formation, and permeability during vasculogenesis and angiogenesis. The angiopoietin family consists of four members that function as ligands for the Tie-2/tek receptors. Angiopoietin-1 (Ang-1) induces autophosphorylation of Tie-2 and has a strong leakage-resistant effect and a remarkable chemotactic effect on ECs, whereas Ang-2 competitively inhibits this effect *(32–34)*. In addition to these major vascularization factors, there are several cytokines which participate in vasculogenesis and angiogenesis *(35–37)*.

In the adult brain, proliferation of the cerebral ECs ceases, and the turnover rate of ECs is approx 3 yr *(38)*. However, during pathological conditions, such as brain injury and tumor growth, new blood vessel formation occurs. Although both angiogenesis and vasculogenesis have been recognized as important contributors to blood vessel formation during embryogenesis, until recently, only angiogenesis was thought to play a role in vessel formation in the adult.

3. VASCULOGENESIS IN THE ADULT ORGANISM

The initial suggestion that endothelial-like cells are present in the adult circulation came in the 1960s when several laboratories showed that Dacron grafts implanted in experimental animals became "vascularized" or lined with ECs *(39–40)*. Later experiments indicated that the ECs present on the Dacron grafts were not formed by the division and migration of ECs from the normal vessel adjacent to the graft *(41–42)*. Other works showed that the endothelial lining of transplanted organs can be replaced by the host-derived cells *(43–45)*. These observations suggest the presence of circulating ECs in the vascular system. Several possible sources for circulating ECs have been investigated.

4. MATURE ECs IN THE CIRCULATION

One possibility is that mature ECs are present in the circulation and can participate in neovascularization. Circulating mature ECs have been found in increased numbers in conditions that cause vascular injury such as sickle cell anemia, infection, and myocardial infarction *(46–48)*. Although they are present in the circulation (presumably after being sloughed from mature blood vessels because of vascular injury), it is unlikely that these cells contribute significantly to vasculogenesis because of their extremely low prevalence in the blood (1–3/ mL^{-1} of blood) and their inability to divide to any great extent in culture. The prevailing evidence indicates that these circulating mature ECs are not the supply of endothelium for vasculogenesis in the adult.

5. ENDOTHELIAL PROGENITOR CELLS FROM THE BONE MARROW

Another possible source of circulating endothelial progenitors is the bone marrow. It has long been recognized that bone marrow-derived stem cells have the capacity to reconstitute all hematopoietic lineages, and recently have also been shown to differentiate into multiple cell types including smooth muscle, neurons, astrocytes, skeletal muscle, hepatocytes, and vascular endothelium *(49–53)*. Asahara and colleagues hypothesized that the bone marrow could serve as a source of endothelial progenitors in addition to hematopoietic stem cells. Using antigenic markers that are common to both endothelial and hematopoietic progenitors (CD34 and Flk-1), they isolated mononuclear cells from human peripheral blood that were able to express EC surface markers when cultured under angiogenic conditions and could incorporate into growing vasculature in animal models *(54,55)*. These circulating endothelial progenitors are highly proliferative cells derived from the bone marrow with the capacity to adopt endothelial characteristics in vitro. They can also incorporate into neovasculature and differentiate into mature ECs in vivo. The cells expressed endothelial characteristics such as the markers CD31, Tie-2, Flk-1, von Willebrand factor (vWF), and endothelial nitric oxide synthetase (eNOS). The cells also took up DiI-labeled acetylated low density lipoprotein (DiI acLDL) and lost the hematopoietic cell surface marker CD45. This initial work was a prelude to a number of studies in the area that have used a variety of culture conditions to induce the expression of endothelial traits in bone marrow-derived cells *(56–67)*. These studies led to the identification of at least two broad categories of cells with the capacity to express endothelial traits; cells expressing hematopoietic stem cell markers and peripheral blood monocytes.

6. ISOLATION AND CHARACTERIZATION OF CIRCULATING ENDOTHELIAL PROGENITORS FROM ADULT PERIPHERAL BLOOD

Several groups have reported that expression of CD133, CD34, and the VEGF receptor flk-1 (VEGFR2) identifies a subpopulation of circulating cells with endothelial potential *(56–60,66)*. These cells have been isolated both from the bone marrow and peripheral blood, and have been found to adopt endothelial characteristics in culture. On culture with angiogenic cytokines such as VEGF, basic fibroblast growth factor (bFGF), and others, the precursor ECs loose the expression of hematopoietic stem cell markers CD133 and CD34 and begin expressing the endothelial markers VE-cadherin (CD144), vWF, and *Ulex europaeus* agglutinin-1 (UEA-1) binding. Additionally, these cells are able to show some functional endothelial characteristics in vitro including uptake of acLDL and tube formation in Matrigel (*see* Table 1).

7. A SUBPOPULATION OF BLOOD MONOCYTES ADOPTS EC CHARACTERISTICS

Another hematopoietic cell population that has been shown to adopt endothelial characteristics in vitro and participate in neovasculature formation in vivo is peripheral blood monocytes. The ability of a monocyte fraction of cells to assume endothelial characteristics in vitro was first demonstrated in 2001 by Harraz and colleagues, and has been subsequently repeated in other studies *(61–65,68)*. As with the CD34+/CD133+/FLK1+ cells, monocytes express endothelial markers when cultured with proangiogenic cytokines. These markers include vWF, VE-cadherin (CD144), endoglin (CD105), thrombospondin receptor (CD36), VEGFR1, and VEGFR2. However, unlike the (CD34+/CD133+/FLK1+)-derived cells, the monocyte-derived cells continue to express myeloid markers such as CD45 and CD14 even as they begin showing endothelial characteristics (*see* Table 1).

Table 1
Characteristics of Stem Cell-Derived and Monocyte-Derived Endothelial-Like Cells

Starting cell source	Cytokines	CD144	ecNOS	UEA-1	vWF	CD31	acLDL	CD45	CD68	CD14	References
Stem cell-derived endothelial-like cells											
Human peripheral blood mononuclear cells	VEGF, bFGF, IGF-1	+	+	+	+	+	+	–		–	(56)
CD34+ human peripheral blood	Bovine brain extract		+	+	+	+	+	–	–		(54)
CD34+ human bone marrow, fetal liver, cord blood, and peripheral blood	VEGF, bFGF, IGF-1				+		+				(57)
CD133+ human peripheral blood	VEGF, SCGF	+		+	+	+	+	–		–	(58)
CD133+ human bone marrow	VEGF, bFGF, IGF-1	+		+	+	+	+	–		–	(59)
CD34+/VEGFR2+ human peripheral blood, fetal liver, and cord blood	bFGF	+			+	+	+	–		–	(60)
Monocyte-derived endothelial-like cells											
Human peripheral blood	VEGF, IGF-1, bFGF, EGF	Low		+	+	+	+	+		+	(61)
CD14+ human peripheral blood monocytes	VEGF, IGF-1, bFGF	+	+		+	+	+	+	+	Low	(62)
CD34+ human peripheral blood	VEGF, bFGF	+	+	+			+			+	(63)
CD14+ human peripheral blood	Bovine brain extract	+	+	+		+	+	+			(64)
CD14+ human peripheral blood	VEGF, IGF-1, bFGF, EGF	+			+		+		+		(65)

Bone marrow-derived cells have been shown to express endothelial characteristics when cultured with proangiogenic cytokines. The two main starting cell populations used in these experiments are monocytes and hematopoietic stem cells. The table shows starting cell populations in addition to culture additives; vascular endothelial growth factor (VEGF), basic fibroblast growth factor (bFGF), insulin-like growth factor I (IGF-1), and epidermal growth factor (EGF). The expression pattern of endothelial and monocyte markers is indicated for each study.

In vivo, monocytic cells adhere to injured endothelium using a monocytic chemoattractant protein-1-dependent mechanism and facilitate re-endothelialization as EC progenitors *(69)*. As the monocytic cells comprise approx 10% of the bone marrow-derived cells, whereas the endothelial progenitor cells are only 0.01%, the contribution of monocytes to vascular regeneration may be more significant.

8. CONTRIBUTION OF CIRCULATING CELLS TO NEOVASCULATURE

Asahara and colleagues initially demonstrated that after systemic injection, endothelial progenitors are incorporated into sites of vessel growth and repair in a rodent hindlimb ischemia model. This finding has because been repeated in other systems *(70,71)*. Both undifferentiated early progenitor cells and cells differentiated in vitro to a more endothelial-like phenotype have been delivered systemically and directly into areas of vessel growth. Bone marrow transplant studies have also been performed to document the incorporation of bone marrow-derived cells into a growing endothelium. Cell tracking methods have also varied and include dye labeling, green fluorescent protein (GFP) expression, use of transgenic donor animals expressing GFP or β-galactosidase, and identification of the Y chromosome in sex-mismatched donor/recipient pairs. Some criticisms of these studies have been the possibility of dye uptake by other native cells and the inherent difficulty in demonstrating that a cell is functionally incorporated into the vasculature as opposed to spatially located in the vasculature. Despite these criticisms, the ability of several groups to derive essentially the same conclusions using varied experimental methodologies make the data compelling.

Although the potential for circulating cells to incorporate into new vasculature appears to be established, there continues to be a debate on the relative importance of angiogenesis vs vasculogenesis in the adult. An effort has been made to quantitate the relative amount of endothelium derived from circulating cells vs mature local ECs using rodent bone marrow transplant models. Several groups have used this strategy and found wide range of donor vs recipient contribution to new vascular endothelium *(66,72–76)*. In studies in human transplant patients, host contributions to endothelium in donor hearts ranging from 0 to 25% have been demonstrated and recipient-derived ECs have been reported in transplanted kidneys and livers *(43–45)*. Even among different patients in the same study, there are wide variations in the reported relative contribution of circulating cells to new vascular endothelium. This variation in transplant patients as well as rodent bone marrow transplant models suggests that there are many factors that influence the interplay between angiogenesis and vasculogenesis, and also illustrates the complexity and variability of these experiments. One prevailing theme in the literature is the absence or low level of bone marrow-derived cells incorporated into quiescent endothelium whereas higher percentages of marrow-derived cells are reported in angiogenic vasculature. Other than this very broad generalization, factors that may influence the relative contributions of angiogenesis and vasculogenesis to neovessel formation in different circumstances remain unclear. One variable that may play a role in this is the number of circulating endothelial progenitors.

In a broad sense, the number of circulating endothelial progenitors is increased by vascular injury. More specifically, the levels of several cytokines have been found to influence the numbers of circulating EPCs. Not surprisingly, VEGF was one of the first substances identified, which increases the EPC number. Increased plasma levels of VEGF in adult mice and humans have been shown to increase EPCs concentration *(77–79)*. VEGF-mediated re-endothelialization may be attributable to its ability to mobilize EPCs or augment NO release from the endothelium. EC migration, sprouting from locally residing endothelium, and recruitment of circulating EPCs play an important role in re-endothelization in vascular repair. Granulocyte colony stimulating factor and erythropoietin, cytokines already in use clinically to promote mobilization of hematopoietic cells can also increase EPCs *(80,81)*. EPC mobilization can also be increased by hydroxy-3-methylglutaryl-coenzyme A reductase inhibitors (statins) *(82)*. Further dissection of the mobilization of EPCs by Heissig and colleagues shows that mobilization of both hematopoietic and endothelial precursors is dependent on MMP-9-mediated release of soluble kit ligand in the bone marrow *(83)*.

9. ARE CIRCULATING CELLS CRITICAL TO THE FORMATION OF NEW VASCULATURE?

Although there have been numerous studies, which demonstrate that circulating endothelial progenitors may be involved to a greater or lesser extent in vessel formation in the adult, few have unequivocally demonstrated their necessity in this process. Mice with reduced gene dosages of the Id transcription factors are unable to support neoangiogenesis in tumors. In an important study by Rafii and colleagues it was shown that bone marrow transplantation with wild-type marrow could reconstitute tumor angiogenesis in $Id1^{+/-}Id3^{+/-}$ mice *(72)*. These studies were the first to imply that not only were marrow-derived cells involved in new vessel formation in the adult, but marrow-derived cells have a critical function in angiogenesis. In this work approx 90% of the tumor vessels were found to contain cells derived from the donor bone marrow *(72)*.

10. CONTRIBUTION OF CIRCULATING CELLS TO CEREBROVASCULATURE

Specific attention has been focused on the role of the bone marrow in providing ECs for blood vessel growth in the CNS. This facet was studied in various in vivo models that result in damage to the BBB. BBB disruption and exudation of intravascular fluids into the interstitial space result in cerebral edema, which is a primary factor in determining the outcome of acute cerebral disorders such as head injury or cerebral infarction.

Several groups have investigated the contribution of bone marrow cells to neovascularization after brain injury. Zhang and colleagues used a murine bone marrow transplant model to show that bone marrow-derived cells expressing β-galactosidase under the transcriptional regulation of the Tie-2 promoter could incorporate into sites of neovascularization at an infarct border zone *(71)*. Transplanted cells were not detected in the nonischemic parenchyma. In similar experiments using GFP-labeled donor bone marrow Hess et al. demonstrated the presence of GFP expressing cells that stained with CD31 or vWF lining the neovasculature between 3 and 14 d after middle cerebral artery (MCA) occlusion *(84)*. In these experiments the percentage of bone marrow-derived cells in the vasculature decreased from 42% at 3 d to 26% at 14 d. In studies looking at longer periods of time after stroke few bone marrow-derived ECs (6 wk) or no bone marrow-derived ECs (6 mo) can be seen *(85)*. These studies suggest that the presence of bone marrow-derived ECs after injury may be transient. In addition to their contribution to vascular repair after stroke, the contribution of bone marrow-derived cells to tumor vasculature in the CNS has also been investigated. Endothelial progenitors derived from human cord blood have been shown to specifically "home" to the vasculature of an intracranial glioma when administered intravenously *(86)*. Limited numbers of bone marrow-derived cells expressing the endothelial markers have also been found in the vasculature of intracranial glioma xenograft models *(73,74)* and several studies have demonstrated the incorporation of bone marrow-derived cells into the endothelium during choroidal neovascularization *(87–89)*.

Although there is a body of work demonstrating the presence of bone marrow-derived ECs in neovasculature in the CNS, it is notable that other investigators have obtained conflicting results. In murine bone marrow transplant experiments numerous marrow-derived cells were seen in the brains of the animals from one to twelve months after transplant. However, the majority of these cells were identified as perivascular macrophages and no ECs were seen *(90)*. In another thorough study of bone marrow-derived ECs in a murine transplant model no bone marrow-derived cells were found in the brain vasculature despite the presence of bone marrow-derived ECs in other organs *(66)*. An important difference between investigations that show incorporation of bone marrow-derived cells into the CNS vasculature and those that do not is the presence of an injury. Given the slow turnover of brain vascular ECs it may be that vascular injury is necessary to induce the mobilization and recruitment of endothelial progenitors to the CNS.

11. AUGMENTATION OF CNS NEOVASCULARIZATION THROUGH THE DELIVERY AND MOBILIZATION OF BONE MARROW-DERIVED CELLS

The ability of bone marrow-derived cells to participate in neovascularization has led to a number of studies investigating the effect of bone marrow-derived cells in augmenting neovascularization. The most successful of these to date have been in the treatment of ischemic cardiac disease. Promising results have been obtained in animal models and the work has progressed to clinical trials *(49,70,91)*. Similar strategies are being tested in stroke and brain injury models. Whole bone marrow cells administered intravenously *(92,93)* and intracranially *(94,95)* have been shown to augment neovascularization after stroke. More importantly, recent studies have shown an improved neurocognitive outcome using these techniques *(92,96)*. Several groups have used a more defined population of bone marrow-derived cells, mesenchymal stem cells (MSCs), in these experiments *(71,93,94,97–99)*. MSCs are an uncharacterized mixed population of plastic-adherent cells. They selectively target injured tissue (using the signals that direct inflammatory cells to injury sites), and activate endogenous restorative responses in injured brain, which include angiogenesis, neurogenesis, and synaptogenesis. Injection of these cells has been shown to facilitate endogenous repair and plasticity of the brain. These cells have progenitor-like characteristics and have a distinct cytokine expression profile. However, it is not clear if the ability of these cells to incorporate into the vascular lining is important in their mechanism of action. Although MSCs are known to adopt endothelial characteristics under certain conditions, no MSC-derived ECs are reported in studies using MSCs to facilitate repair after brain injury. Other functions of these cells including the secretion of cytokines may be the main mechanism of their action in augmenting cerebral vessel repair *(97,100)*.

After injury the integrity of the BBB is disrupted leading to the leakage of large molecules such as albumin into the brain parenchyma. The reformation of the BBB after injury is dependent on re-establishing tight junctions between the endothelial and repairing cells. Human CD14+ peripheral blood monocytes that have been cultured with proangiogenic cytokines will home to sites of vascular repair in an area of brain injury and can be identified in the wall of the repairing vasculature. These cells form a low permeability barrier in vitro, developing a transcellular electrical resistance (TCER) and expressing the tight junction proteins occludin and ZO-1, which suggests that they may contribute to reconstitution of the BBB *(68)*.

12. CONCLUSIONS

Huge advances in understanding of the role of bone marrow-derived cells in the formation and maintenance of the vasculature, and the cerebrovasculature in particular, have occurred in the last decade. Despite the rapid advancement of the field there are large gaps in our understanding of these cells. It is becoming evident that several different populations of cells including early progenitor-like cells, monocytic cells, and perhaps MSCs are responsible for the reported actions of endothelial progenitor cells. Sorting out the roles and relative importance of these different cell types remains an important area of investigation. The large variation in the relative contribution of bone marrow-derived cells to formation and repair of the vasculature in different experimental systems serves to illustrate some of the factors that may influence the behavior of bone marrow-derived cells in this capacity. These factors include the presence or absence of an injury and perhaps even the type of injury, the time frame after injury that is being observed, and the organ being studied. For example, a specific requirement for an efficient repair in case of brain injury is the ability of the circulating cells to re-establish the BBB. Other important questions such as a detailed description of the mechanisms of homing to areas of vascular growth and the rate of expression of endothelial characteristics in vivo remain to be uncovered. It is becoming clear that understanding the biology of circulating endothelial progenitor cells is critical to understanding the mechanisms of vessel formation, a process important in many pathological and physiological processes. Continued advances in this area are likely to pay significant clinical dividends in the future.

REFERENCES

1. Rubin LL, Staddon JM. The cell biology of the blood-brain barrier. Annu Rev Neurosci 1999;22:11–28.
2. Kniesel U, Wolburg H. Tight junctions of the blood-brain barrier. Cell Mol Biol 2000;20:57–76.
3. Petty MA, Lo EH. Junctional complexes of the BBB: permeability changes in neuroinflammation. Prog Neurobiol 2002;68:311–323.
4. Petty MA, Wettstein JG. Elements of cerebral microvascular ischaemia. Brain Res 2001;36:23–34.
5. Rosenberg GA. Ischemic brain edema. Prog Cardiovasc Dis 1999; 42:209–216.
6. Cervos-Nararro J, Lafuente JV. Traumatic Brain injuries: structural changes. J Neurol Sci 1991;103(Suppl):S3–S14
7. Atwood CS, Bowen RL, Smith MA, Perry G. Cerebrovascular requirement for sealant, anti-coagulant and remodeling molecules that allow for the maintainance of vascular integrity and blood supply. Brain Res Rev 2003;43:164–178.
8. Ujiie M, Dickstein DL, Carlow DA, Jefferies WA. Blood-brain barrier permeability precedes senile plaque formatin in an Alzheimer disease model. Microcirculation 2003;10:463–470.
9. Jellinger KA. Alzheimer disease and cerebrovascular pathology: an update. J Neural Transm 2002;109:813–836.
10. Floris S, Bleezer EL, Schreibelt G, et al. Blood-brain barrier permeability and monocyte infiltration in experimental allergic encephalomyelitis: a quantitative MRI study. Brain 2004;127: 616–627.
11. Annunziata P. Blood-brain barrier changes during invasion of the central nervous system by HIV-1. Old and new insights into the mechanism. J Neurol 2003;250:901–906
12. Adams S, Brown H, Turner G. Breaking down the blood-brain barrier: signaling a path to cerebral malaria? Trends Parasitol 2002;18: 360–366.
13. Calapai G, Marciano MC, Corica F, et al. Erythropoietin protects against brain ischemic injury by inhibition of nitric oxide formation. Eur J Pharmacol 2000;401:349–356.
14. Ding-Zhou L, Margaill I, Palmier B, Pruneau D, Plotkine M, Marchand-Verrecchia C. LF 16-0687-Ms, a bradykinin B2 receptor antagonist, reduces ischemic brain injury in a murine model of transient focal cerebral ischemia. Br J Pharmacol 2003;139:1539–1547.
15. Jacobson JR, Dudek SM, Birukov KG, et al. Cytoskeletal activation and altered gene expression in endothelial barrier regulation by simvistatin. Am J Respir Cell Mol Biol 2004;30:662–670.
16. Zhang ZG, Zhang L, Tsang W, et al. Correlation of VEGF and angiopoietin expression with disruption of blood brain barrier and angiogenesis after focal cerebral ischemia. J Cereb Blood Flow Metab 2002;22:379–392.
17. Marchi N, Rasmussen P, Kapural M, et al. Peripheral markers of brain damage and blood-brain barrier dysfunction. Restor Neurol Neurosci 2003;21(3,4):109–121.
18. Pfefferkorn T, Rosenberg GA. Closure of the blood-brain barrier by matrix metalloproteinase inhibition reduces rtPA-mediated mortality in cerebral ischemia with delayed reperfusion. Stroke 2003;34:2025–2030.
19. Engelhardt B. Development of the blood-brain barrier. Cell Tissue Res 2003;314:119–129.
20. Dambska M. The vascularization of the developing human brain. Folia Neuropathol 1995;33:189–193.
21. Bauer HC, Steiner M, Bauer H. Embryonic development of the CNS microvasculature in the mouse: new insights into the structural mechanisms of early angiogenesis. EXS 1992;61:64–68.
22. Risau W, Sariola H, Zerwes HG, et al. Vasculogenesis and angiogenesis in embryonic-stem-cell-derived embryoid bodies. Development 1988;102:471–478.
23. Wilting J, Brand-Saberi B, Kurz H, Christ B. Development of the embryonic vascular system. Cell Mol Biol Res 1995;41:219–232.
24. Kurz H. Physiology of angiogenesis. J Neuro-Oncol 2000;50:17–35.
25. Hatzopoulos AK, Folkman J, Vasile E, Eiselen GK, Rosenberg RD. Isolation and characterization of endothelial progenitor cells from mouse embryos. Development 1998;125:1457–1468.
26. Vajkoczy P, Blum S, Lamparter M, et al. Multistep nature of microvascular recruitment of ex vivo-expanded embryonic endothelial progenitor cells during tumor angiogenesis. J Exp Med 2003;197:1755–1765.
27. Zadeh G, Guha A. Molecular regulators of angiogenesis in the developing nervous system and adult brain tumors. Int J Oncol 2003;23:557–567.
28. Carmeliet P, Collen D. Molecular analysis of blood vessel formation and disease. Am J Physiol 1997;273:H2091–H2104.

29. Yancopoulos GD, Davis S, Gale NW, Rudge JS, Weigand SJ, Holash J. Vascular-specific growth factors and blood vessel formation. Nature 2000;407:242–248.
30. Gale NW, Thurston G, Davis S, et al. Complementary and coordinated roles of the VEGFs and angiopoietins during normal and pathological vascular formation. Cold Spring Harb Symp Quant Biol 2002;67:267–273.
31. Bikfalvi A, Bicknell R. Recent advances in angiogenesis, anti-angiogenesis and vascular targeting. Trends Pharmacol Sci 2002;23:576–582.
32. Thurston G, Suri C, Smith K, et al. Leakage-resistant blood vessels in mice transgenically overexpressing angiopoietin-1. Science 1999;286:2511–2514.
33. Thurston G, Rudge JS, Ioffe E, et al. Angiopoitin-1 protects the adult vasculature against plasma leakage. Nat Med 2000;6:460–463.
34. Zhang Z, Chopp M. Vascular endothelial growth factor and angiopoietins in focal cerebral ischemia. Trends Cardiovasc Med 2002;12: 62–66.
35. Risau W. Embryonic angiogenesis factors. Pharmacol Ther 1991;51:371–376.
36. Fee D, Grzybicki D, Dobbs, et al. Il-6 promotes vasculogenesis of murine brain microvessel endothelial cells. Cytokine 2000;12: 655–665.
37. Lamszus K, Heese O, Westphal M. Angiogenesis-related growth factors in brain tumors. Cancer Treat Res 2004;117:169–190.
38. Robertson PL, DuBois M, Bowman PD, Goldstein GW. Angiogenesis in developing rat brain: an in vivo and in vitro study. Brain Res 1985;355:219–223.
39. Stump MM, Jordan JL, DeBakey ME, Halpert B. Endothelium grown from circulating blood on isolated intravascular Dacron hub. Am J Pathol 1963;43:361–367.
40. Mackenzie JR, Hackett M, Topuzlu C, Tibbs DJ. Origin of arterial prosthesis lining from circulating blood cells. Arch Surg 1968;97:879–885.
41. Shi Q, Wu MH, Hayashida N, Wechezak AR, Clowes AW, Sauvage LR. Proof of fallout endothelialization of impervious Dacron grafts in the aorta and inferior vena cava of the dog. J Vasc Surg 1994;20:546–556; discussion 556–557.
42. Wu MH, Shi Q, Wechezak AR, Clowes AW, Gordon IL, Sauvage LR. Definitive proof of endothelialization of a Dacron arterial prosthesis in a human being. J Vasc Surg 1995;21:862–867.
43. Quaini F, Urbanek K, Beltrami AP, et al. Chimerism of the transplanted heart. N Engl J Med 2002;346:5–15.
44. Lagaaij EL, Cramer-Knijnenburg GF, van Kemenade FJ, van Es LA, Bruijn JA, van Krieken JH. Endothelial cell chimerism after renal transplantation and vascular rejection. Lancet 2001;6:33–37.
45. Hove WR, van Hoek B, Bajema IM, Ringers J, van Krieken JH, Lagaaij EL. Extensive chimerism in liver transplants: vascular endothelium, bile duct epithelium, and hepatocytes. Liver Transpl 2003;9:552–556.
46. Solovey A, Lin Y, Browne P, Choong S, Wayner E, Hebbel RP. Circulating activated endothelial cells in sickle cell anemia. N Engl J Med 1997;337:1584–1590.
47. Mutin M, Canavy I, Blann A, Bory M, Sampol J, Dignat-George F. Direct evidence of endothelial injury in acute myocardial infarction and unstable angina by demonstration of circulating endothelial cells. Blood 1999;93:2951–2958.
48. Grefte A, van der Giessen M, van Son W. The TH. Circulating Cytomegalovirus (CMV)-infected endothelial cells in patients with an active CMV infection. J Infect Dis 1993;167:270–277.
49. Abbott J, Giordano FJ. Stem cells and cardiovascular disease. J Nucl Cardiol 2003;10:403–412.
50. Verfaillie CM. Adult stem cells: assessing the case for pluripotency. Trends Cell Biol 2002;12:502–508.
51. Verfaillie CM, Schwartz R, Reyes M, Jiang Y. Unexpected potential of adult stem cells. Ann NY Acad Sci 2003;996:231–234.
52. Jiang Y, Jahagirdar BN, Reinhardt RL, et al. Pluripotency of mesenchymal stem cells derived from adult marrow. Nature 2002; 418:41–49.
53. Prockop DJ. Marrow stromal cell as stem cells for nonhematopoietic tissues. Science 1997;276:71–74.
54. Asahara T, Murohara T, Sullivan A, et al. Isolation of putative progenitor endothelial cells for angiogenesis. Science 1997;5302: 964–967.
55. Asahara T, Masuda H, Takahashi T, et al. Bone marrow origin of endothelial progenitor cells responsible for postnatal vasculogenesis in physiological and pathological neovascularization. Circ Res 1999;85:221–228.
56. Lin Y, Weisdorf DJ, Solovey A, Hebbel RP. Origins of circulating endothelial cells and endothelial outgrowth from blood. J Clin Invest 2000;105:71–77.

57. Shi Q, Rafii S, Wu MHD, et al. Evidence for circulating bone marrow-derived endothelial cells. Blood 1998;92:362–367.
58. Gehling UM, Ergun S, Schumacher U, et al. In vitro differentiation of endothelial cells from AC 133-positive progenitor cells Blood 2000;95:3106–3112.
59. Quirici N, Soligo D, Caneva L, Servida F, Bossolasco P, Deliliers DG. Differentiation and expansion of endothelial cells from human bone marrow. British J Hematol 2001;115:186–194.
60. Peichev M, Naiyer AJ, Pereira D, et al. Expression of VEGFR-2 and AC133 by circulating human CD34+ cells identifies a population of functional endothelial precursors. Blood 2000;95:952–958.
61. Rehman J, Li J, Orschell CM, March KL. Peripheral blood endothelial progenitors cells are derived from monocyte/macrophages and secrete angiogenic growth factors. Circulation 2003;107:1164–1169.
62. Schmeisser A, Garlichs CD, Zhang H, et al. Monocytes coexpress endothelial and macrophagocytic lineage markers and form cord-like structures in Matrigel under angiogenic conditions. Cardiovasc Res 2001;49:671–680.
63. Nakul-Aquaronne D, Bayle J, Frelin C. Coexpression of endothelial markers and CD14 by cytokine mobilized CD34+ cells under angiogenic stimulation. Cardiovasc Res 2003;57:816–823.
64. Harraz M, Jiao C, Hanlon HD, Hartley RS. Schatteman GCCD34-blood-derived human endothelial cell progenitors. Stem Cells 2001;19:304–312.
65. Fernandez Pujol B, Licibello FC, Gehling UM, et al. Endothelial-like cells derived from CD14 positive monocytes. Differentiation 2000;65:287–300.
66. Bailey AS, Jiang S, Afentoulis M, et al. Transplanted adult hematopoietic stems cells differentiate into functional endothelial cells. Blood 2004;103:13–19.
67. Havemann K, Pujol BF, Adamkiewicz J. In vitro transformation of monocytes and dendritic cells into endothelial like cells. Adv Exp Med Biol 2003;522:47–57.
68. Glod J, Kobiler D, Noel M, Maric D, Fine HA. Monocytes form a vascular barrier and parcipitate in vessel repair after brain injury. Blood 2005, in press.
69. Fujiyama S, Amano K, Uehira K, et al. Bone marrow monocyte lineage cells adhere to injured endothelium in a monocyte chemoattractant protein-1-dependent manner and accelerate reendothelialization as endothelial progenitor cells. Circ Res 2003;93:980–989.
70. Freedman SB, Isner JM. Therapeutic angiogenesis for ischemic cardiovascular disease. J Mol Cell Cardiol 2001;33:379–393.
71. Zhang ZG, Zhang L, Jiang Q, Chopp M. Bone marrow-derived endothelial progenitor cells participate in cerebral neovascularization after focal cerebral ischemia in the adult mouse. Circ Res 2002;90:284–288.
72. Lyden D, Hattori K, Dias S, et al. Impaired recruitment of bone-marrow-derived endothelial and hematopoietic precursor cells blocks tumor angiogenesis and growth. Nat Med 2001;7:1194–1201.
73. Ferrari N, Glod J, Lee J, Kobiler D, Fine HA. Bone marrow-derived, endothelial progenitor-like cells as angiogenesis-selective gene-targeting vectors. Gene Ther 2003;10:647–656.
74. Machein MR, Renninger S, de Lima-Hahn E, Plate KH. Minor contribution of bone marrow-derived endothelial progenitors to the vascularization of murine gliomas. Brain Pathol 2003;13:582–597.
75. Murayama T, Asahara T. Bone marrow-derived endothelial progenitor cells for vascular regeneration. Curr Opin Mol Ther 2002;4:395–402.
76. Rafii S, Lyden D, Benezra R, Hattori K, Heissig B. Vascular and hematopoietic stem cells: novel targets for anti-angiogenesis therapy. Nat Rev Cancer 2002;2:826–835.
77. Rabbany SY, Heissig B, Hattori K, Rafii S. Molecular pathways regulating mobilization of marrow-derived stem cells for tissue revascularization. Trends Mol Med 2003;9:109–117.
78. Kalka C, Masuda H, Takahashi T, et al. Vascular Endothelial Growth Factor (165) gene transfer augments circulating endothelial progenitor cells in human subjects. Circ Res 2000;86:1198–1202.
79. Kalka C, Tehrani H, Laudenberg B, et al. VEGF gene transfer mobilizes endothelial progenitor cells in patients with inoperable coronary disease. Ann Thorac Surg 2000;70:829–834.
80. Bahlmann FH, DeGroot K, Spandau JM, et al. Erythropoietin regulates endothelial progenitor cells. Blood 2004;103:921–926.
81. Heeschen C, Aicher A, Lehmann R, et al. Erythropoietin is a potent physiologic stimulus for endothelial progenitor cell mobilization. Blood 2003;102:1340–1346.
82. Llevadot J, Murasawa S, Kureishi Y, et al. HMG-CoA reductase inhibitor mobilizes bone marrow-derived endothelial progenitor cells. J Clin Invest 2001;108:399–405.
83. Heissig B, Hattori K, Dias S, et al. Recruitment of stem and progenitor cells from the bone marrow niche requires MMP-9 mediated release of kit-ligand. Cell 2002;109:625–637.

84. Hess DC, Hill WD, Martin-Studdard A, Carroll J, Brailer J, Carothers J. Bone marrow as a source of endothelial cells and NeuN-expressing cells after stroke. Stroke 2002;33:1362–1368.
85. Beck H, Voswinckel R, Wagner S, et al. Participation of bone marrow-derived cells in long-term repair processes after experimental stroke. J Cereb Blood Flow Metab 2003;23:709–717.
86. Moore XL, Lu J, Sun L, Zhu CJ, Tan P, Wong MC. Endothelial progenitor cells' "homing" specificity to brain tumors. Gene Ther 2004;11:811–818.
87. Csaky KG, Baffi JZ, Byrnes GA, et al. Recruitment of marrow-derived endothelial cells to experimental choroidal neovascularization by local expression of vascular endothelial growth factor. Exp Eye Res 2004;78:1107–1116.
88. Espinosa-Heidmann DG, Caicedo A, Hernandez EP, Csaky KG, Cousins SW. Bone-marrow derived progenitor cells contribute to experimental choroidal neovascularization. Invest Ophthalmol Vis Sci 2003;44:4914–4919.
89. Sengupta N, Caballero S, Mames RN, Butler JM, Scott EW, Grant MB. The role of adult bone marrow-derived stem cells in choroidal neovascularization. Invest Opththalmol Vis Sci 2003;44:4908–4913.
90. Vallieres L, Sawchenko PE. Bone marrow-derived cells that populate the adult mouse brain preserve their hematopoietic identity. J Neurosci 2003;23:5197–5207.
91. Szmitko PE, Fedak PWM, Weisel RD, Stewart DJ, Kutryk MJB, Verma S. Endothelial progenitor cells, new hope for a broken heart. Circulation 2003;107:3093–3100.
92. Iihoshi S, Honmou O, Houkin K, Hashi K, Kocsis JD. A therapeutic window for intravenous administration of autologous bone marrow after cerebral ischemia in adult rats. Brain Res 2004;1007:1–9.
93. Mahmood A, Lu D, Wang L, Li Y, Lu M, Chopp M. Treatment of traumatic brain injury in female rats with intravenous administration of bone marrow stromal cells. Neurosurgery 2001;49:1196–1203.
94. Mahmood A, Lu D, Yi L, Chen JL, Chopp M. Intracranial bone marrow transplantation after traumatic brain injury improving functional outcome in adult rats. J Neurosurg 2001;94:683–685.
95. Borlongan CV, Lind JG, Dillon-Carter O, et al. Bone marrow grafts restore cerebral blood flow and blood brain barrier in stroke rats. Brain Res 2004;1010:108–116.
96. Li Y, Chopp M, Chen J, et al. Intrastriatal transplantation of bone marrow nonhematopoietic cells improves functional recovery after stroke in adult mice. J Cereb Blood Flow Metab 2000;20:1311–1319.
97. Chen J, Li Y, Zhang R, et al. Combination therapy of stroke in rats with a nitric oxide donor and human bone marrow stromal cells enhances angiogenesis and neurogenesis. Brain Res 2004;1005: 21–28.
98. Lu D, Li Y, Wang L, Chen J, Mahmood A, Chopp M. Intraarterial administration of marrow stromal cells in a rat model of traumatic brain injury. J Neurotrauma 2001;18:813–819.
99. Lu D, Mahmood A, Wang L, Li Y, Lu M, Chopp M. Adult bone marrow stromal cells administered intravenously to rats after traumatic brain injury migrate into brain and improve neurological outcome. Neuroreport 2001;12:559–563.
100. Chopp M, Li Y. Treatment of neural injury with marrow stromal cells. Lancet Neurol 2002;1:92–100.

18
Microvessel Remodeling in Cerebral Ischemia

Danica B. Stanimirovic, MD, PhD, Maria J. Moreno, PhD, and Arsalan S. Haqqani, PhD

SUMMARY

The neurovascular (NV) unit, broadly defined as a segment of brain vasculature, is composed of functionally integrated cellular (including brain endothelial cells, astrocytes, pericytes, and smooth muscle cells) and acellular elements that form the basement membrane. Brain insults or conditions characterized by tissue hypoxia induce dramatic changes in the NV unit that include the disruption of interendothelial tight junctions, breakdown of the basal lamina, and endothelial proliferation, migration, and reorganization to form new capillaries and microvessels. This NV remodeling is controlled by a complex interplay of numerous mediators originating from cellular components of the NV unit and from breakdown of the extracellular matrix. Several of these mediator families are also involved in neuronal remodeling, thus providing for integration of vascular and neuronal responses to the ischemic state. On the global molecular scale, changes in ischemic cerebral microvessels include temporally controlled regulation of genes and proteins involved in cell cycle, blood–brain barrier phenotypic properties, cell proliferation, motility, and inflammation. The integration/coordination of angiogenic and neuronal remodeling during postischemic brain recovery is an important determinant of the overall neurological outcome.

Key Words: Cerebral ischemia; microvessels; angiogenesis; proteomics; extracellular matrix.

1. NEUROVASCULAR UNIT AND CELL CYCLE

The brain is supplied by an intricate network of vessels that originate from large carotid and basilar arteries and project deep into the brain tissue forming a dense capillary mesh ($>100 \times 10^9$ capillaries with surface area approx 200 m^2) that provides blood supply to virtually every neuron. The neurovascular (NV) unit, broadly defined as a segment of brain vasculature, integrates three principal functionalities: the blood–brain barrier (BBB), regulation of cerebral blood flow (CBF), and neuroimmune interfacing.

BBB is a dynamic physical and functional barrier between the systemic circulation and the brain attributed to the specific features of cerebral endothelial cells (ECs). Cerebral ECs are sealed by tight junctions that restrict paracellular permeability, lack fenestrations, have a reduced number of pinocytic and endocytic vesicles, and express various polarized transporters and receptors that facilitate the uptake of nutrients into the brain, restrict brain access to circulating drugs, neurotoxins, and neurotransmitters, and actively extrude metabolic products and toxins from the brain *(1)*.

Cerebral ECs are demarcated from the surrounding brain cells by a continuous basement membrane (BM) that stabilizes vascular structures. BM is composed of extracellular matrix (ECM) components, including collagens, vitronectin, fibronectin, tenascin, and proteoglycans that undergo a continuous production, deposition, degradation, and remodeling *(2)*. Intact BM

From: *The Cell Cycle in the Central Nervous System*
Edited by: D. Janigro © Humana Press Inc., Totowa, NJ

components act as repressors of cerebral EC cell cycle by inhibiting proliferation and providing an environment that facilitates cell–cell adhesion *(3)*. Under physiological conditions in adult brain, ECs are essentially in a quiescent state and their turnover is exceptionally slow (approximately hundreds of days). The change of quiescent phenotype of EC into an activated, proliferative phenotype is accompanied by BM degradation with the subsequent exposure of cryptic proangiogenic domains of BM components, previously sequestered in the fully assembled BM *(4)*. This indicates that the same BM proteins in different structural configurations can have opposite effects on cerebral EC at different stages of the angiogenic process *(4)*.

Pericytes extend long cytoplasmic processes over the surface of the cerebral EC and contribute to their maturation, proliferation, survival, migration, differentiation, and vascular branching via the secretion of growth factors or modulation of the ECM *(5,6)*. EC–pericyte association stabilizes the vessels and makes them less sensitive to fluctuations in oxygen tension. Recent studies *(7)* have shown that pericytes alone can form functional endothelium-free tubes in early phases of neovascularization. Pericytes also participate in the regulation of BBB transport and vascular permeability *(5)*.

More than 99% of the surface of the cerebral capillary BM is covered by astrocytic endfeet. Interactions between astrocytes and brain ECs have been implicated in induction of tight junctions and specific transporters, EC production of anticoagulant factors, and their proliferation and angiogenesis *(8,9)*.

Neuronal groups in the proximity of cerebral microvessels control the spatial and temporal regulation of brain perfusion in response to brain activation, a phenomenon known as NV coupling *(10)*. The cellular anatomy and organization of the intact NV unit in adult brain is schematically shown in Fig. 1A.

Brain insults (e.g., stroke, trauma) or conditions (e.g., brain tumors) characterized by tissue hypoxia induce dynamic functional changes in the NV unit, including BBB breakdown, vasoparalysis, leukocyte adhesion and infiltration, prothrombotic conversion, and angio- and vasculogenesis *(11)*. At the microanatomy level, alterations include the disruption of inter-endothelial tight junctions, retraction of pericytes from the abluminal surface of the capillary, release of proteases from the activated ECs, breakdown of the basal lamina with transudation of plasma, degradation of the ECM surrounding vessels, EC migration toward angiogenic/chemotactic stimuli and their proliferation, formation of tube-like structures, fusion of the newly formed vessels, and initiation of blood flow (schematically presented in Fig. 1A). At the molecular level this remodeling is accompanied by increased expression of EC-leukocyte adhesion receptors, loss of EC and astrocyte integrin receptors, loss of their matrix ligands, expression of members of several matrix-degrading protease families, and the appearance of receptors associated with angiogenesis and neovascularization *(11,12)*. Brain angiogenesis is a dynamic, multifaceted process, which requires the activation and synchronization of a plethora of angiogenic stimulators and inhibitors (recently reviewed in ref. *13*), but the disruption of one or more of the above steps may be enough to compromise the angiogenic response.

In the brain, angiogenesis is critical for development, and is enhanced during chronic hypoxic conditions and in brain tumors. Recent evidence also suggests that processes of angiogenesis and neurogenesis that participate in brain recovery after injury are closely interdependent and share common mediators.

1.1. EC Activation by Hypoxia

A common denominator of pathologies characterized by angiogenesis is tissue hypoxia. The change of quiescent phenotype of cerebral EC into an activated, proliferative phenotype, commonly termed as an "angiogenic switch," is controlled through a process of gene regulation initiated by the activation of the hypoxia-inducible transcription factors (HIFs). HIF-1 was originally identified as a nuclear factor required for transcriptional regulation of the human erythropoietin *(EPO)* gene *(14)*. HIF-1 is a heterodimeric protein composed of HIF-1α and HIF-1β (ARNT)-subunits both of which belong to the basic helix–loop–helix periodic arylhydrocarbon

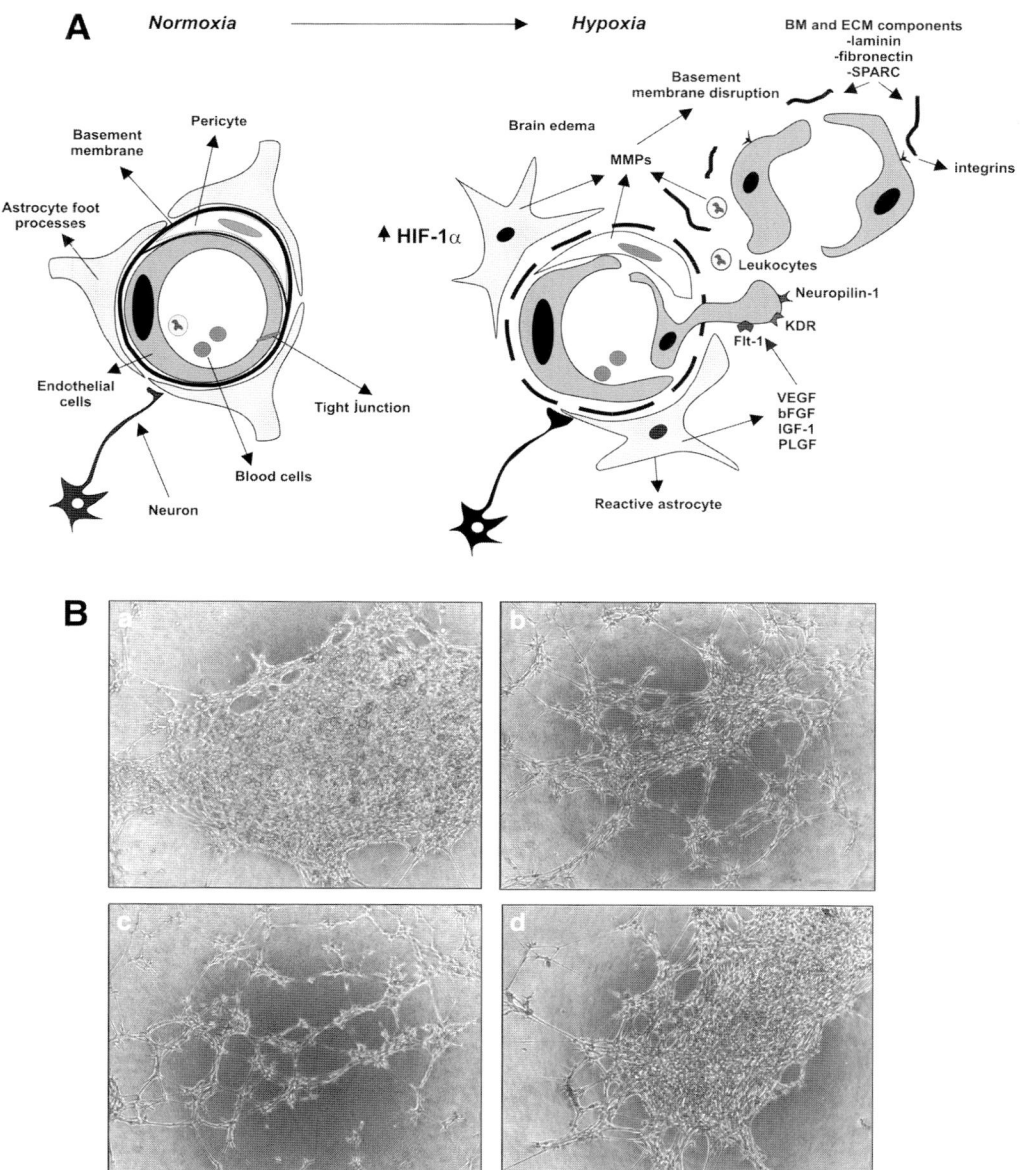

Fig. 1. Hypoxia-induced reorganization of the neurovascular unit. **(A)** Schematic drawing of the neurovascular unit reorganization that takes place in hypoxia/ischemia. Highly structured multicellular anatomy of the NV unit is disrupted, tight junctions are disassembled, and numerous proangiogenic factors and their receptors are upregulated in cerebral ECs, astrocytes, and pericytes. Basement membrane disruption caused by secreted proteases leads to leukocyte infiltration and disinhibition of EC cycle. Several breakdown products of ECM stimulate cell migration and angiogenesis. Proliferating and migrating endothelial cells organize into capillary tubes that subsequently mature into new vessels. **(B)** Capillary tube formation, the final stage of angiogenesis, can be reproduced in vitro in cerebral EC grown in the basement membrane matrix, Matrigel. (a) Mouse cerebromicrovascular endothelial cells (MCECs) in Matrigel under normoxic conditions; (b) MCEC exposed to 10 μM VEGF for 24 h; (c) MCEC isolated from PlGF$^{+/+}$ mice exposed to 6 h hypoxia followed by 18 h reoxygenation reorganized forming an intricate capillary tube network; (d) MCEC derived from PLGF$^{-/-}$ mouse exposed to a 6 h hypoxia followed by 18 h reoxygenation do not form capillary-like tubes.

receptor-simultaneous (bHLH-PAS) family common to a large number of transcription factors *(14)*. Whereas ARNT is only marginally affected by changes in oxygen tension, hypoxia markedly increases HIF-1α protein levels *(15)*. Indirect evidence suggests that HIF-1α induces cell cycle arrest and proapoptotic signaling in EC *(16)*. This proapoptotic effect is counteracted by the simultaneous upregulation of HIF-1-dependent genes that promote cell survival by increasing oxygen delivery (EPO, transferrin, and heme oxygenase), glucose transport (glucose transporter-1), glycolysis (lactate dehydrogenase A), and angiogenesis (vascular endothelial growth factor [VEGF], inducible nitric oxide synthase [iNOS], angiopoietin-2, fibroblast growth factor, and inflammatory cytokines) *(14,17)*. The end result of this gene induction program is tissue adaptation to hypoxia through increased blood and oxygen supply.

VEGF is the major proangiogenic factor induced through HIF-1α activation. VEGF is a multitasking cytokine, which stimulates differentiation, survival, migration, proliferation, tubulogenesis, and vascular permeability in EC *(18)*. It belongs to the family of cystine knot proteins, which includes VEGF-B, VEGF-C, VEGF-D, VEGF-E, and placenta growth factor (PlGF) *(18–19)*. Some of these members also have functional alternative splicing gene variants. Three types of VEGF receptors are tyrosine kinases flt-1 (VEGFR-1), KDR/flk-1 (VEGFR-2), and flt-4 (VEGFR-3), which display a preferential affinity for different VEGF family members *(19)*. Most relevant VEGF functions are mediated via KDR/flk-1, including the control of EC cycle and survival via phosphoinositide 3′-kinase (PI3-K)-dependent activation of the antiapoptotic kinase Akt, that phosphorylates and inhibits the proapoptotic protein Bad and caspase activation *(18–20)*. Direct interactions between VE-cadherin/KDR *(20)* and integrin αvβ3/KDR *(21)* are also important regulators of VEGF-induced endothelial survival.

Neuropilin-1 and -2 (NP-1/NP-2) are nontyrosine-kinase receptors that play a role in axonal guidance by binding class III semaphorins, and in angiogenesis by binding $VEGF_{165}$ and PlGF *(22,23)*. The small size of the NP-1 intracellular domain and the lack of any associated signaling function suggest that NP-1 by itself is not a functional receptor but acts as a coreceptor in a complex with KDR/flk-1 to enhance the binding of VEGF to KDR. Neuropilin-1, with its double capacity to cooperate with major regulator(s) of both angiogenesis and neuronal guidance, is an important link in coordination of the development and/or remodeling of nervous and circulatory systems *(24)*. The balance between guidance molecules and angiogenic factors has been shown to modulate the migration, apoptosis, or survival, and proliferation of both neural progenitor cells *(23)* and ECs *(25)* through neuropilin-1.

PlGF, a ligand of both flt-1 and neuropilin-1, is a dimeric glycoprotein member of the VEGF family expressed in various tissues including brain *(26)*. In contrast to extensive studies on VEGF regulation of normal and abnormal angiogenesis, very little is known about the role of PlGF in this process. Our recent experiment suggested that PlGF and VEGF act synergistically to induce angiogenic switch in human brain EC in vitro *(26)*. Moreover, brain EC derived from ($PlGF^{-/-}$) mice, in contrast to those isolated from $PlGF^{+/+}$ mice, failed to acquire angiogenic phenotype in response to hypoxia (Fig. 1B). Interestingly, PlGF homodimers or PlGF/VEGF heterodimers protect blood vessels from regression in hyperoxic conditions *(26)*.

1.2. Cell–Cell and Cell–ECM Interactions

Following an ischemic insult to the brain, a reorganization of cells and tissues takes place as the surrounding cells attempt to limit the injury, repair the damage, and restore the normal architecture of the brain. This tissue remodeling requires *de novo* synthesis of genes and proteins, which enable cells to actively change their relationship with the existing ECM and with other cells. In this dynamic process, the composition of ECM is also changed to reflect modified secretory status of ECM-producing cells, including astrocytes, microglia, and cerebral ECs *(12)*.

The interactions of cells with ECM components are mediated by specific adhesion molecules, integrins. Integrins are transmembrane receptors composed of α- and β-subunits; these subunits form different heterodimeric combinations that interact with ECM components in a selective and specific manner. Integrins recognize specific motifs, such as arginine–glycine–aspartic acid

(RGD) sequence and are responsible for anchoring cells to the ECM *(12)*. Signaling pathways elicited in brain cells via integrins are responsible for changes in cell morphology and migration within a given ECM environment, processes essential for both acute response to ischemic brain injury (e.g., microglial infiltration and neutrophil infiltration) and late reorganization, stem cell migration, and neurite extension involved in brain repair *(12,27)*. Studies in nonhuman primates *(28)* have shown an early loss of the integrin $\alpha_6\beta_4$ in perivascular astrocytes and the disruption of integrity between astrocytes and microvessel wall after focal brain ischemia. The expression of VEGF, integrin $\alpha_V\beta_3$, and proliferating cell nuclear antigen were highly correlated on the same microvessels in this model *(29)*.

1.2.1. Proteolysis of ECM

Extracellular proteolysis plays a critical role in vascular remodeling *(30)*. Two major systems that modify ECM in the brain are the plasminogen activator (PA) and matrix metalloprotease (MMP) axes *(30,31)*. Deleterious effects of ECM proteolysis by PA-MMP include the disruption of the BBB integrity *(32)*, amplification of inflammatory infiltrates, demyelination, and possibly interruption of cell–cell and cell–matrix interactions that may trigger cell death *(3–32)*. In contrast, these PA-MMP actions may contribute to ECM proteolysis that mediates parenchymal and angiogenic recovery after brain injury *(33)*.

PAs are divided into two groups, intravascular tissue-type PA (tPA), which mainly degrade fibrin clots, and tissue urokinase-type PA (uPA), which participate in cell migration, invasion, and tissue remodeling. Both PA families are involved in catalyzing conversion of plasminogen into plasmin. The active protease plasmin directly digests various components of ECM and/or activates MMPs, which can further degrade ECM *(34)*. uPA is constitutively expressed in EC and is highly upregulated by various angiogenic factors, transforming growth factors-β (TGF-β) and hypoxia *(34)*.

MMPs are a gene family of zinc- and calcium-dependent endopeptidases that cleave components of the ECM and are currently grouped in collagenases, gelatinases, stromelysins, and membrane-type MMPs. Activation of the MMP cascade is triggered by plasmin-mediated conversion and is thus dependent on the activation of tissue PA *(31)*. MMPs can be induced in various cells by cytokines or can be released from storage granules *(31,34)*. Gelatinase A (MMP-2; 72 kDa) and gelatinase B (MMP-9; 92 kDa), also known as type IV collagenases, degrade type IV collagen, fibronectin, and gelatin—the essential components of the BM. Plasminogen activation *(30,31)* and increased activities of MMP-2 and MMP-9 *(35,36)* have been shown in rodent and human brain after focal ischemia, whereas the inhibition of MMP-9 was neuroprotective in rat focal ischemia *(36,37)*. Both PA- and MMP-related proteolytic systems also involve "controlling elements" that limit proteolytic activities and ECM degradation: PA inhibitor(s) (PAIs) and tissue inhibitors of metalloproteases. Therefore, the extent of proteolytic ECM degradation after stroke is dependent on the balance of proteolytic enzymes and their inhibitors produced in injured tissues *(31)*. Hypoxia-stimulated mediator cascades involved in ECM proteolysis and angiogenesis are shown in Fig. 2A.

An ECM protein recently implicated in NV remodeling, a multifunctional 32-kDa glycoprotein secreted protein acidic and rich in cysteine (SPARC), also termed osteonectin or BM-40, is expressed by both brain EC and astrocytes *(38–40)*. SPARC is characterized by three distinct domains: a highly acidic domain, a cysteine-rich domain homologous to a repeated domain in follistatin, and an extracellular Ca^{2+} binding EC domain *(38)*. SPARC can be cleaved by certain members of the MMP family into fragments with sequence motifs that could regulate different aspects of EC functions including angiogenesis *(41)*. SPARC also contributes to cell migration by increasing the production and activity of MMPs *(38)*. The EC domain of SPARC binds vitronectin, an ECM molecule found in high abundance along the vessel BM *(38)*. In addition, SPARC binds to and inhibits PAI-1, thus facilitating uPA-mediated degradation of the ECM. Whereas full length SPARC inhibits the mitogenic effect of VEGF on microvascular EC *(42)*, its cleavage products have proangiogenic properties *(41)*. In the context of brain injury, SPARC

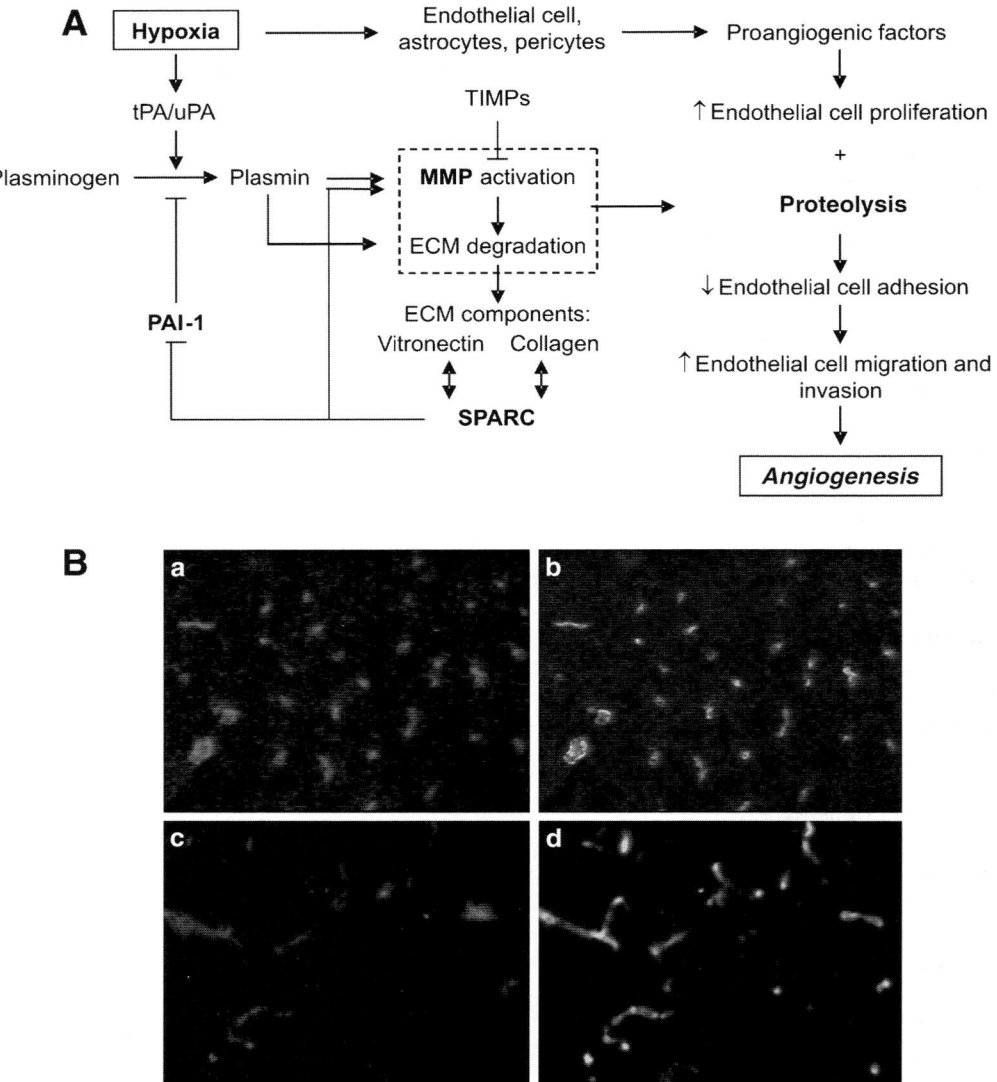

Fig. 2. (**A**) Schematic diagram of processes involved in the extracellular matrix remodeling in response to cerebral ischemia. (**B**) Immunohistochemical staining of brain section for SPARC (red) and ricinus communis agglutinin (green). (a,b) Sham-operated animals (c,d) animals subjected to 10 min global cerebral ischemia and reperfused for 24 h. A profound loss in SPARC-positive vessels was observed in ischemic animals. (Please *see* companion CD for color version of this figure.)

mRNA is induced in mature blood vessels close to the lesion, as well as in blood vessels, which develop following the localized trauma of brain tissue *(38,43)*. Interestingly, our recent studies demonstrated loss of SPARC immunoreactivity from microvascular compartment 6–24 h after global cerebral ischemia (Fig. 2B), suggesting proteolytic degradation/disruption of the BM and potential release of proangiogenic SPARC fragments.

1.2.2. ECM Remodeling and Post-Ischemic Brain Inflammation

Postischemic vascular inflammation is characterized by increased leukocyte adhesion and brain infiltration *(44,45)*. Degradation of BM components by activated MMPs likely facilitates neutrophil movement across the BBB *(46)*. Both neutrophils and brain EC have been shown to express and store various MMPs *(31,33)*. Redistribution and a subsequent loss of tight junction proteins occludin and zonnula occludens-1 from brain EC have been seen in vivo during neutrophil-induced BBB breakdown *(47)*. Adherent neutrophils activate endothelial myosin light chain kinase via calcium transients, resulting in the stimulation of endothelial contractile apparatus and paracellular gap formation that facilitates polymorpho nucleocytes transendothelial migration in response to chemotactic stimuli *(48)*. Moreover, yet unknown soluble factors released from neutrophils have been shown to activate EC MMP-2 *(47)*. ECM microenvironment also elicits remarkable changes in leukocyte molecular make-up and behavior as cells adhere to and migrate across the endothelial barrier *(47,48)*.

In summary, the disruption of BM in ischemic vessels can have several important consequences: (1) BBB permeation and accumulation of plasma components in the perivascular spaces, (2) facilitated infiltration of inflammatory cells, and (3) exposure of pro- and antiangiogenic cues *(3,4)* that can modulate EC proliferation and migration.

1.3. Analyses of Gene and Protein Expression in Brain Vessels During Ischemia

Recent studies attempted to correlate temporal profiles of angiogenesis and microvascular remodeling in ischemia with gene and protein expression. Analyses of 96 genes implicated in angiogenesis in peri-infarct brain tissues using cDNA arrays demonstrated that most of the well-known angiogenic factors, as well as vessel-stabilizing factors such as thrombospondins (TSPs) increased as early as 1 h after middle cerebral artery occlusion (MCAO) *(49)*. To examine molecular changes that occur specifically in cerebral vessels, we recently applied laser capture microdissection (LCM) microscopy technique to dissect microvessels from brain sections *(50)* at different times after a transient global cerebral ischemia model in rat *(51)*. Captured microvessels (Fig. 3A) were analyzed by a sensitive, gel-free isotope coded affinity tag (ICAT) proteomics technique coupled to mass-spectroscopy (MS–MS) protein identification *(52)* (Fig. 3B). The approach enables quantitative comparative analyses of differentially expressed proteins in small tissue samples. Proteins differentially expressed between control and ischemic vessels were identified and functionally categorized. The summary of temporal patterns of four functional protein categories in post-ischemic cerebral vessels is shown in Fig. 3C. Proteins involved in solute and nutrient transport across the BBB, including various subunits of Na K-ATPase and high-affinity glutamate and neutral amino acid transporters, were substantially downregulated in brain microvessels 1 h after a transient 20-min global cerebral ischemia. The loss of transporter proteins coincided with the initial BBB opening measured in the same model *(51)*. Cell cycle-regulating proteins as well as several structural, predominantly cytoskeletal proteins were also profoundly downregulated in early reperfusion. In contrast, proteins involved in inflammation and immunity were upregulated as early as 1 h of reperfusion, reaching maximal expression at 6 h, consistent with dynamics of leukocyte trafficking described in similar ischemia models *(53)*. Twenty four hours after transient cerebral ischemia, a substantial upregulation of majority of identified proteins was detected in the microvascular compartment, including those belonging to inflammatory, cell proliferation and motility, cell cycle, and angiogenic functional categories (Fig. 3C). HIF-1α, TGFβ-1, and laminin-a5 were upregulated at this time suggesting a prolonged hypoxic state that initiates vascular and ECM remodeling. Upregulation of TGFβ-1 expression in the brain after stroke has previously been linked to the pathogenesis of the angiogenic response *(54,55)*. These molecular changes coincided with the second BBB opening described in the same model *(51)*.

In addition to these profiling studies that outlined general patterns of vascular changes, several studies reported angiogenesis-related gene and protein changes in cerebral ischemia using *in situ* hybridization and immunohistochemistry. In the permanent MCAO, the expression of VEGF and VEGF receptors was strongly upregulated in the ischemic border between 6 and 24 h and at 48 h, respectively *(56)*, and followed by the appearance of newly formed vessels at the border of the infarction at 48–72 h of MCAO *(56)*. Angiopoietin-2 mRNA was upregulated in

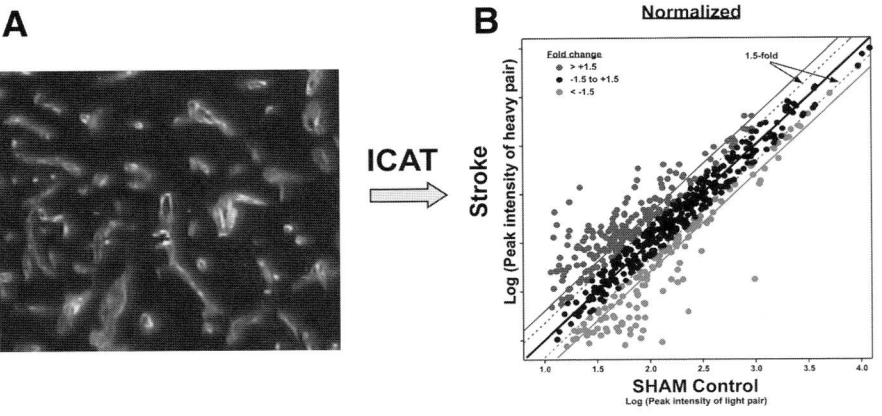

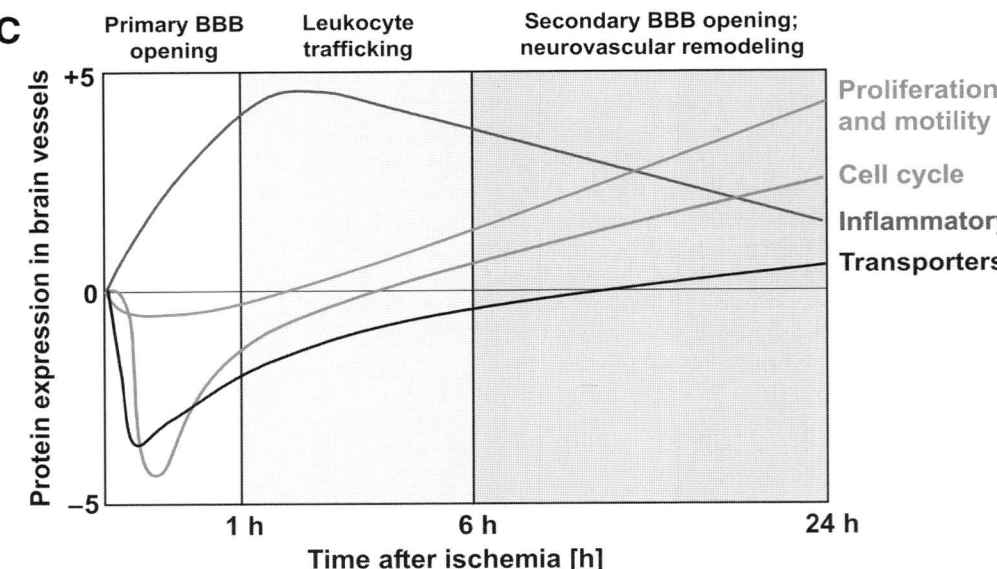

Fig. 3. Analyses of protein expression changes in cerebral microvessels in rats exposed to a 20-min transient global cerebral ischemia. (**A**) Brain vessels were stained with RCA-1 and collected from brain tissue sections 1, 6, or 24 h after ischemia using laser capture microdissection microscopy. (**B**) Ischemic vessels were analyzed using isotope-coded affinity tag (ICAT) proteomics coupled to MS–MS protein sequencing and identification. (**C**) Schematic representation of dynamic changes of various protein categories in rat brain vessels at different times after transient forebrain ischemia. Shaded blocks show temporal correlation between observed protein changes and functional pathology of the NV unit. (Please *see* companion CD for color version of this figure.)

EC cord tips in the peri-infarct and infarct areas 6 h after MCAO (57,58). Decreases in some antiangiogenic mediators, such as angiostatic brain angiogenesis inhibitor 2 (BAI2), have been observed in concert with increases in VEGF in focal cerebral ischemia (59).

Post-ischemic angiogenesis is short-lived and may be completely terminated within a few weeks after ischemic insult (60,61). The molecular mechanisms involved in the dissolution of post-ischemic angiogenic processes are poorly understood. Interestingly, in parallel with the upregulation of proangiogenic mediators, increases in antiangiogenic factors have been found in ischemic brain tissue. TSPs are naturally occurring angiostatic factors that inhibit angiogenesis in vivo (62). A robust expression of TSP-1 and TSP-2 with different cellular temporal profiles has been detected in the ischemic brain *(61)*, suggesting a role for TSPs in spontaneous resolution of post-ischemic angiogenesis.

2. ANGIOGENESIS AFTER STROKE—GOOD OR BAD?

NV remodeling and angiogenesis observed after brain injury and stroke are adaptive processes through which tissue maximizes functional reorganization to respond to disease environment. Vascular and neuronal remodeling take place in concert and are often integrated through the same systems of mediators, such as VEGF–neuropilin–semaphorin system. The concept of integration of vascular and neuronal remodeling after brain injury has been supported by recent studies showing loss of expression of class III semaphorins in neurons in parallel with high expression of NP-1 in vessels, neurons, and astrocytes surrounding the infarct *(63)*. A shift in the balance between semaphorins and VEGF/PLGF, which compete for NP-1 binding, may also determine the timing or nature of angiogenic and neuroprotective effects of this mediator system after ischemia *(63,64)*.

In principle, NV remodeling and angiogenesis are deemed beneficial for brain recovery and favorable neurological outcome. The underlying mechanisms of the beneficial effects of angiogenesis have been debated; some studies suggest a role for ischemia-induced microvessels in facilitating macrophage infiltration and removal of necrotic brain *(65)*, whereas others argue that arteriolar collateral growth and new capillaries facilitate restored perfusion in the ischemic border *(66)*. The later explanation was supported by a magnetic resonance imaging study that mapped increases in regional CBF to the expression of angiogenesis gene products in MCAO model *(67)*.

Potential harmful aspects of the angiogenic process cannot be overlooked. Increased microvascular density in peri-infarct areas is accompanied with compromised BBB permeability and vasogenic edema *(58,68)*. Several studies that have attempted to stimulate the angiogenic process in post-ischemic brain reported that exogenous administration of VEGF after cerebral ischemia results in increased neuronal survival, reduced infarct size, and improved neurological outcome; these effects were produced by both direct neuroprotective actions of VEGF and increased angiogenesis and revascularization *(68,69)*. Chronic intraventricular infusions of $VEGF_{165}$ increased vascular density in a dose-dependent manner; proper dosing was important for achieving a significant increase in vessel density with minimal associated brain edema and ventriculomegaly *(69)*. The timing of VEGF application was also important: whereas early (1 h) post-ischemic administration of $VEGF_{165}$ significantly increased BBB leakage, hemorrhagic transformation, and ischemic lesions, late (48 h) administration of $VEGF_{165}$ to ischemic rats enhanced angiogenesis in the ischemic penumbra, and significantly improved neurological recovery *(68)*. In a recent study *(70)*, low and intermediate doses of VEGF, which did not induce angiogenesis, were significantly neuroprotective in the ischemic brain, whereas high doses of VEGF, which induced angiogenesis, showed no neuroprotection in ischemia and damaged neurons in normal brains. These findings suggested that VEGF-induced neuroprotection does not always occur simultaneously with angiogenesis and that neuroprotection may be compromised by uncontrolled angiogenesis.

An interesting, albeit indirect, evidence of the importance of angiogenesis in overall brain remapping and remodeling after injury originated from experiments showing that physical activity (exercise preconditioning)-induced reduction of stroke damage and improved functional recovery were attributable to angiogenesis and neurotrophin overexpression in brain regions supplied by the MCA following exercise *(71)*. Similarly, peripheral infusion of human bone marrow stromal cells induced both angiogenesis and neurogenesis in the ischemic boundary zone after stroke and improved neurological outcome *(72)*.

3. CONCLUSIONS

In conclusion, NV remodeling after ischemic brain injury is a complex, integrated process that involves cellular and acellular elements of the NV unit and a coordinated regulation of vascular and neuronal responses. The process is orchestrated by complex, multiphasic programs of gene and protein expression involving cross-talk among multiple mediators. As this remodeling is exceptionally important for both anatomical and functional brain recovery, future direct or indirect strategies to induce concurrently both neurogenesis and angiogenesis may provide a breakthrough in treating and managing ischemic brain diseases.

REFERENCES

1. Rapoport SI. Blood-brain barrier in physiology and medicine. New York: Raven Press, 1976.
2. Atwood CS, Bowen RL, Smith MA, Perry G. Cerebrovascular requirement for sealant, anti-coagulant and remodeling molecules that allow for the maintenance of vascular integrity and blood supply. Brain Res Brain Res Rev 2003;43:164–178.
3. Newton LK, Yung WK, Pettigrew LC, Steck PA. Growth regulatory activities of endothelial extracellular matrix: mediation by transforming growth factor-beta. Exp Cell Res 1990;190:127–132.
4. Bix G, Iozzo RV. Matrix revolutions: 'tails' of basement-membrane components with angiostatic functions. Trends Cell Biol 2005;15:52–60.
5. Ramsauer M, Krause D, Dermietzel R. Angiogenesis of the blood-brain barrier in vitro and the function of cerebral pericytes. FASEB J 2002;16:1274–1276.
6. Nehls V, Schuchardt E, Drenckhahn D. The effect of fibroblasts, vascular smooth muscle cells, and pericytes on sprout formation of endothelial cells in a fibrin gel angiogenesis system. Microvasc Res 1994;48:349–63.
7. Ozerdem U, Stallcup WB. Early contribution of pericytes to angiogenic sprouting and tube formation. Angiogenesis 2003;6:241–249.
8. Abbott NJ. Astrocyte-endothelial interactions and blood-brain barrier permeability. J Anat 2002;200:629–638.
9. Stanimirovic DB, Ball R, Wong J, Durkin JP. Evidence for a regulatory role of astrocyte-derived factor(s) and protein kinase C in proliferation of rat cerebromicrovascular endothelial cells. Neurosci Lett 1995;197:219–222.
10. Dirnagl U. Metabolic aspects of neurovascular coupling. Adv Exp Med Biol 1997;413:155–159.
11. del Zoppo GJ, Mabuchi T. Cerebral microvessel responses to focal ischemia. J Cereb Blood Flow Metab 2003;23:879–894.
12. Ellison JA, Barone FC, Feuerstein GZ. Matrix remodeling after stroke. De novo expression of matrix proteins and integrin receptors. Ann NY Acad Sci 1999;890:204–222.
13. Patan S. Vasculogenesis and angiogenesis. Cancer Treat Res 2004;117:3–32.
14. Semenza GL. Hypoxia-inducible factor 1: control of oxygen homeostasis in health and disease. Pediatr Res 2001;49:614–617.
15. Brat DJ, Kaur B, Van Meir EG. Genetic modulation of hypoxia induced gene expression and angiogenesis: relevance to brain tumors. Front Biosci 2003;8:100–116.
16. Iida T, Mine S, Fujimoto H, Suzuki K, Minami Y, Tanaka Y. Hypoxia-inducible factor-1alpha induces cell cycle arrest of endothelial cells. Genes Cells 2002;7:143–149.
17. Stanimirovic DB. Inflammatory activation of brain cells by hypoxia: Transcription factors and signaling pathways. In: Feuerstein G (ed), Inflammation and Stroke. Basel: Birkhauser Verlag AG, 2001, pp. 101–111.
18. Carmeliet P. Mechanisms of angiogenesis and arteriogenesis. Nat Med 2000;6:389–395.
19. Bicknell R, Harris AL. Novel angiogenic signaling pathways and vascular targets. Annu Rev Pharmacol Toxicol 2004;44:219–238.
20. Carmeliet P, Lampugnani MG, Moons L, et al. Targeted deficiency or cytosolic truncation of the VE-cadherin gene in mice impairs VEGF-mediated endothelial survival and angiogenesis. Cell 1999;98(2):147–157.
21. Soldi R, Mitola S, Strasly M, Defilippi P, Tarone G, Bussolino F. Role of alphavbeta3 integrin in the activation of vascular endothelial growth factor receptor-2. EMBO J 1999;18:882–892.
22. Rosenstein JM, Krum JM. New roles for VEGF in nervous tissue—beyond blood vessels. Exp Neurol 2004;187:246–253.
23. Bagnard D, Vaillant C, Khuth ST, et al. Semaphorin 3A-vascular endothelial growth factor-165 balance mediates migration and apoptosis of neural progenitor cells by the recruitment of shared receptor. J Neurosci 2001;21:3332–3341.
24. Emanueli C, Schratzberger P, Kirchmair R, Madeddu P. Paracrine control of vascularization and neurogenesis by neurotrophins. Br J Pharmacol 2003;140:614–619.

25. Miao HQ, Soker S, Feiner L, Alonso JL, Raper JA, Klagsbrun M. Neuropilin-1 mediates collapsin-1/semaphorin III inhibition of endothelial cell motility: functional competition of collapsin-1 and vascular endothelial growth factor-165. J Cell Biol 1999;146:233–242.
26. Autiero M, Waltenberger J, Communi D, et al. Role of PlGF in the intra- and intermolecular cross talk between the VEGF receptors Flt1 and Flk1. Nat Med 2003;9:936–943.
27. Choi BH. Role of the basement membrane in neurogenesis and repair of injury in the central nervous system. Microsc Res Tech 1994;28:193–203.
28. Wagner S, Tagaya M, Koziol JA, Quaranta V, del Zoppo GJ. Rapid disruption of an astrocyte interaction with the extracellular matrix mediated by integrin alpha(6)beta4 during focal cerebral ischemia/reperfusion. Stroke 1997;28:858–865.
29. Abumiya T, Lucero J, Heo JH, et al. Activated microvessels express vascular endothelial growth factor and integrin alpha(v)beta3 during focal cerebral ischemia. J Cereb Blood Flow Metab 1999;19:1038–1050.
30. Lo EH, Wang X, Cuzner ML. Extracellular proteolysis in brain injury and inflammation: Role for plasminogen activators and matrix metalloproteinases. J Neurosci Res 2002;69:1–9.
31. Mun-Bryce S, Rosenberg GA. Matrix-metalloproteinases in cerebrovascular disease. J Cereb Blood Flow Metab 1998;18:1163–1172.
32. Fujimura M, Gasche Y, Morita-Fujimura Y, Massengale J, Kawase M, Chan PH. Early appearance of activated matrix metalloproteinase-9 and blood-brain barrier disruption in mice after focal cerebral ischemia and reperfusion. Brain Res 1999;84:92–100.
33. Iruela-Arispe ML, Diglio CA, Sage EH. Modulation of extracellular matrix proteins by endothelial cells undergoing angiogenesis in vitro. Arterioscler Thromb 1991;11:805–815.
34. Lukes A, Mun-Bryce S, Lukes M, Rosenberg GA. Extracellular matrix degradation by metalloproteinases and central nervous system diseases. Mol Neurobiol 1999;19:267–284.
35. Heo JH, Lucero J, Abumiya T, Koziol JA, Copeland BR, del Zoppo GJ. Matrix-metalloproteinases increase very early during experimental focal cerebral ischemia. J Cereb Blood Flow Metab 1999;19:624–633.
36. Romanic AM, White RF, Arleth AJ, Ohlstein EH, Barone FC. Matrix metalloproteinase expression increases after cerebral focal ischemia in rats: inhibition of matrix-metalloproteinase-9 reduces infarct size. Stroke 1998;29:1020–1030.
37. Jiang X, Namura S, Nagata I. Matrix metalloproteinase inhibitor KB-R7785 attenuates brain damage resulting from permanent focal cerebral ischemia in mice. Neurosci Lett 2001;305:41–44.
38. Bradshaw AD, Sage EH. SPARC, a matricellular protein that functions in cellular differentiation and tissue response to injury. Clin Invest 2001;107:1049–1054.
39. Mendis DB, Malaval L, Brown IR. SPARC, an extracellular matrix glycoprotein containing the follistatin module, is expressed by astrocytes in synaptic enriched regions of the adult brain. Brain Res 1995;67:69–79.
40. Mendis DB, Ivy GO, Brown IR. SPARC/osteonectin mRNA is induced in blood vessels following injury to the adult rat cerebral cortex. Neurochem Res 1998;23:1117–1123.
41. Sage EH, Reed M, Funk SE, et al. Cleavage of the matricellular protein SPARC by matrix metalloproteinase 3 produces polypeptides that influence angiogenesis. J Biol Chem 2003;278:37,849–37,857.
42. Kupprion C, Motamed K, Sage EH. SPARC (BM-40, osteonectin) inhibits the mitogenic effect of vascular endothelial growth factor on microvascular endothelial cells. J Biol Chem 1998;273:29,635–29,640.
43. Mendis DB, Ivy GO, Brown IR. Induction of SC1 mRNA encoding a brain extracellular matrix glycoprotein related to SPARC following lesioning of the adult rat forebrain. Neurochem Res 2000;25:1637–1644.
44. Iadecola C, Alexander M. Cerebral ischemia and inflammation. Curr Opin Neurol 2001;14:89–94.
45. Stanimirovic D, Satoh K. Inflammatory mediators of cerebral endothelium: A role in ischemic brain inflammation. Brain Pathol 2000;10:113–126.
46. Delclaux C, Delacourt C, D'Ortho MP, Boyer V, Lafuma C, Harf A. Role of gelatinase B and elastase in human polymorphonuclear neutrophil migration across basement membrane. Am J Resp Cell Mol Biol 1996;14:288–295.
47. Bolton SJ, Anthony DC, Perry VH. Loss of the tight junction proteins occludin and zonula occludens-1 from cerebral vascular endothelium during neutrophil-induced blood-brain barrier breakdown in vivo. Neuroscience 1998;86:1245–1257.
48. Garcia JGN, Verin AD, Hereniyova M, English D. Adherent neutrophils activate endothelial myosin light chain kinase: role in transendothelial migration. J Appl Physiol 1998;84:1817–1821.

49. Hayashi T, Noshita N, Sugawara T, Chan PH. Temporal profile of angiogenesis and expression of related genes in the brain after ischemia. J Cereb Blood Flow Metab 2003;23:166–180.
50. Mojsilovic-Petrovic J, Nesic M, Pen A, Zhang W, Stanimirovic D. Development of rapid staining protocols for laser-capture microdissection of brain vessels from human and rat coupled to gene expression analyses. J Neurosci Methods 2004;133:39–48.
51. Preston E, Sutherland G, Finsten A. Three openings of the blood-brain barrier produced by forebrain ischemia in the rat. Neurosci Lett 1993;149:75–78.
52. Turecek F. Mass spectrometry in coupling with affinity capture-release and isotope-coded affinity tags for quantitative protein analysis. J Mass Spectrom 2002;37:1–14.
53. Uhl E, Beck J, Stummer W, Lehmberg J, Baethmann A. Leukocyte-endothelium interactions in pial venules during the early and late reperfusion period after global cerebral ischemia in gerbils. J Cereb Blood Flow Metab 2000;20:979–987.
54. Krupinski J, Kumar P, Kumar S, Kaluza J. Increased expression of TGF-beta 1 in brain tissue after ischemic stroke in humans. Stroke 1996;27:852–857.
55. Yamashita K, Gerken U, Vogel P, Hossmann K, Wiessner C. Biphasic expression of TGF-beta1 mRNA in the rat brain following permanent occlusion of the middle cerebral artery. Brain Res 1999;836:139–145.
56. Marti HJ, Bernaudin M, Bellail A, et al. Hypoxia-induced vascular endothelial growth factor expression precedes neovascularization after cerebral ischemia. Am J Pathol 2000;156:965–976.
57. Beck H, Acker T, Wiessner C, Allegrini PR, Plate KH. Expression of angiopoietin-1, angiopoietin-2, and tie receptors after middle cerebral artery occlusion in the rat. Am J Pathol 2000;157:1473–1483.
58. Zhang ZG, Zhang L, Tsang W, et al. Correlation of VEGF and angiopoietin expression with disruption of blood-brain barrier and angiogenesis after focal cerebral ischemia. J Cereb Blood Flow Metab 2002;22:379–392.
59. Kee HJ, Koh JT, Kim MY, et al. Expression of brain-specific angiogenesis inhibitor 2 (BAI2) in normal and ischemic brain: involvement of BAI2 in the ischemia-induced brain angiogenesis. J Cereb Blood Flow Metab 2002;22:1054–1067.
60. Lin TN, Sun SW, Cheung WM, Li F, Chang C. Dynamic changes in cerebral blood flow and angiogenesis after transient focal cerebral ischemia in rats. Evaluation with serial magnetic resonance imaging. Stroke 2002;33:2985–2991.
61. Lin TN, Kim GM, Chen JJ, Cheung WM, He YY, Hsu CY. Differential regulation of thrombospondin-1 and thrombospondin-2 after focal cerebral ischemia/reperfusion. Stroke 2003;34:177–186.
62. Lawler J. Thrombospondin-1 as an endogenous inhibitor of angiogenesis and tumor growth. J Cell Mol Med 2002;6:1–12.
63. Beck H, Acker T, Puschel AW, Fujisawa H, Carmeliet P, Plate KH. Cell type-specific expression of neuropilins in an MCA-occlusion model in mice suggests a potential role in post-ischemic brain remodeling. J Neuropathol Exp Neurol 2002;61:339–350.
64. Zhang ZG, Tsang W, Zhang L, Powers C, Chopp M. Up-regulation of neuropilin-1 in neovasculature after focal cerebral ischemia in the adult rat. J Cereb Blood Flow Metab 2001;21:541–549.
65. Manoonkitiwongsa PS, Jackson-Friedman C, McMillan PJ, Schultz RL, Lyden PD. Angiogenesis after stroke is correlated with increased numbers of macrophages: the clean-up hypothesis. J Cereb Blood Flow Metab 2001;21:1223–1231.
66. Wei L, Erinjeri JP, Rovainen CM, Woolsey TA. Collateral growth and angiogenesis around cortical stroke. Stroke 2001;32:2179–2184.
67. Lin TN, Sun SW, Cheung WM, Li F, Chang C. Dynamic changes in cerebral blood flow and angiogenesis after transient focal cerebral ischemia in rats. Evaluation with serial magnetic resonance imaging. Stroke 2002;33:2985–2991.
68. Zhang ZG, Zhang L, Jiang Q, et al. VEGF enhances angiogenesis and promotes blood-brain barrier leakage in the ischemic brain. J Clin Invest 2000;106:829–838.
69. Harrigan MR, Ennis SR, Masada T, Keep RF. Intraventricular infusion of vascular endothelial growth factor promotes cerebral angiogenesis with minimal brain edema. Neurosurgery 2002;50:589–598.
70. Manoonkitiwongsa PS, Schultz RL, McCreery DB, Whitter EF, Lyden PD. Neuroprotection of ischemic brain by vascular endothelial growth factor is critically dependent on proper dosage and may be compromised by angiogenesis. J Cereb Blood Flow Metab 2004;24:693–702.
71. Ding Y, Li J, Luan X, et al. Exercise pre-conditioning reduces brain damage in ischemic rats that may be associated with regional angiogenesis and cellular overexpression of neurotrophin. Neuroscience 2004;124:583–591.
72. Chen J, Zhang ZG, Li Y, et al. Intravenous administration of human bone marrow stromal cells induces angiogenesis in the ischemic boundary zone after stroke in rats. Circ Res 2003;92:692–699.

19
Vascular and Neuronal Effects of VEGF in the Nervous System

Implications for Neurological Disorders

Lieve Moons, PhD, Peter Carmeliet, MD, PhD, and Mieke Dewerchin, PhD

SUMMARY

Blood vessels and nerves are both vital channels to and from tissues. Recent genetic insights show that they have much more in common than was originally anticipated. They use similar signals and principles to differentiate, grow, and navigate toward their targets. Moreover, the vascular and nervous systems cross talk and, when dysregulated, this contributes to medically important diseases. No factor is better known for its angiogenic effects than vascular endothelial growth factor (VEGF)—this molecule has been implicated in virtually every type of angiogenic disorder, including those associated with cancer, ischemia, and inflammation. Recent studies revealed, however, that VEGF is also involved in neurodegeneration. The role of VEGF in the nervous system is not restricted only to regulating vessel growth: VEGF also has direct effects on different types of neural cells—including even neural stem cells. Furthermore, genetic studies showed that mice with reduced VEGF levels develop adult-onset motor neuron degeneration reminiscent of the human neurodegenerative disorder amyotrophic lateral sclerosis (ALS), and additional genetic studies confirmed that VEGF is a modifier of motor neuron degeneration in humans and in SOD1^{G93A} mice—a model of ALS. Reduced VEGF levels may promote motor neuron degeneration by limiting neural tissue perfusion, by reducing VEGF-dependent neuroprotection and/or by influencing neuroregeneration/neurogenesis. VEGF also affects neuron survival after acute spinal cord or cerebral ischemia and has been implicated in other neurological disorders such as diabetic and ischemic neuropathy, nerve regeneration, Parkinson's disease, Alzheimer's disease, and multiple sclerosis. These findings offer novel opportunities to better decipher the insufficiently understood molecular pathogenesis of many neurodegenerative disorders, and promise to open new avenues for future improved treatment.

Key Words: VEGF; neurotrophic; motor neuron; ALS; neurodegeneration; perfusion; neurogenesis.

1. VEGF: ITS RECEPTORS AND LIGANDS

Over the last decade, vascular endothelial growth factor (VEGF), initially discovered as a "vascular permeability factor" *(1)*, has emerged as a major player of angiogenesis in health and disease *(2,3)*. VEGF, also referred to as VEGF-A, regulates angiogenesis by interacting mainly with two tyrosine kinase receptors, *fms*-like tyrosine kinase (Flt1 or VEGF receptor-1 [VEGFR-1]) and fetal liver kinase (Flk1, also known as VEGF receptor-2 [VEGFR-2] and, in humans, as kinase insert domain-containing receptor, [KDR]). VEGFR-2 has generally been considered to be the main transducer of the VEGF-A-dependent angiogenic signals *(3–5)*. Compared with the strong

From: *The Cell Cycle in the Central Nervous System*
Edited by: D. Janigro © Humana Press Inc., Totowa, NJ

tyrosine phosphorylation of VEGFR-2, VEGFR-1 is only weakly phosphorylated in cultured endothelial cells (ECs) *(6)*. This has cast doubts on whether VEGFR-1 transmits any significant angiogenic signals at all. Indeed, in the embryo, the extracellular ligand-binding domains of VEGFR-1, present as soluble protein, trap VEGF and thereby reduce VEGFR-2-driven angiogenesis *(7,8)*. However, VEGFR-1 also has positive effects. For instance, expression of VEGFR-1 is highly upregulated in disease *(9–11)*, and contributes to the angiogenic switch in pathological conditions *(12)*. Neuropilin-1 (NP-1), expressed on neuronal cells and more recently also identified on ECs, is a specific receptor for the $VEGF_{165}$ isoform and a coreceptor of VEGFR-2 *(13)*. It also binds semaphorin-3A (Sema-3A), a neurorepellant implicated in guidance of axons. Neuropilin-2 (NP-2) binds $VEGF_{165}$ and $VEGF_{145}$, as well as Sema-3C and Sema-3F. The fact that NP-1 and NP-2 bind semaphorins and VEGF suggests that these receptors have roles in both the nervous and cardiovascular system *(14)*.

VEGF is translated from a single gene that is alternatively spliced into various VEGF isoforms with molecular weights of 121, 145, 165, or 189 kDa. Translation of these isoforms is initiated at a classical AUG start codon. A much larger VEGF form (L-VEGF) is generated after translation initiation at an additional CUG codon, in frame with the AUG start codon *(15)*. Proteolytic processing of L-VEGF generates a C-terminal fragment, identical to the secreted AUG-initiated isoforms. Therefore, L-VEGF is assumed to constitute an intracellular store of VEGF. Since the discovery of VEGF-A, several other homologs have been identified, including VEGF-B, VEGF-C, VEGF-D, VEGF-E (also called orf virus VEGF), and placental growth factor (PlGF) (reviewed in ref. *16*). However, much less is known about the specific functions of all these factors. Gene inactivation studies revealed that PlGF and VEGF-B, both binding to VEGFR-1, are not important for physiological angiogenesis. However, in pathological conditions, PlGF activates VEGFR-1 and stimulates the growth and maturation of new vessels via direct signaling through VEGFR-1 and via a cross-talk between the VEGF receptors *(12,17,18)*. The biological function of VEGF-B in vivo remains enigmatic and even controversial *(19–24)*, whereas VEGF-C and VEGF-D primarily stimulate lymphangiogenesis *(25–28)*.

More than 10,000 papers reported on the angiogenic activity of VEGF over the last decennium. However, for a few years now, this growth factor has been assigned a novel role in the nervous system. Indeed, approx 100 recent studies described in vitro or in vivo neural effects of VEGF, providing exciting novel insight and raising strong interest in assessing novel approaches to treat various neurodegenerative disorders. In this chapter, we will overview this promising novel development and discuss the therapeutic potential of VEGF for amyotrophic lateral sclerosis (ALS), stroke, diabetic and ischemic neuropathy, nerve regeneration, Parkinson's disease (PD), Alzheimer's disease (AD), and multiple sclerosis (MS).

2. HYPOXIC UPREGULATION OF VEGF PRODUCTION AND ITS IMPORTANCE IN NEURODEGENERATION

VEGF expression is significantly stimulated by hypoxia. The hypoxic upregulation of VEGF protein levels is controlled at the transcriptional and translational level. Hypoxia-inducible transcription factors (HIFs), HIF1 and HIF2, are members of a basic helix–loop–helix superfamily of transcription factors, which are fundamental in cellular oxygen homeostasis *(29)*. These proteins form heterodimers consisting of class 1, HIF1β, and class 2, for example HIF1α- and HIF2α-subunits. The heterodimers bind a hypoxia-response element (HRE) in various genes. By binding the HRE element in the 5′-flanking region of the *VEGF* gene, they upregulate VEGF transcription in hypoxia *(30)*. On the other hand, hypoxia also affects VEGF mRNA stability via regulating the binding of HuR mRNA-binding protein *(31)* and other factors *(32)* to specific sequences in the 3′-untranslated region *(33)*. A third mechanism of hypoxic induction of VEGF involves the two internal ribosome entry sites (IRESs), IRES-A and -B, located in the 5′-untranslated region of the *VEGF* gene *(34–36)*, which ensure efficient internal initiation of translation in conditions such as hypoxia, when normal translation via ribosome scanning is severely impaired *(37)*.

It was recently shown that HIFs are involved in motor neuron degeneration or survival strategies after injury by the cell-and time-dependent upregulation of HIF1 expression in a rat model of intraspinal axotomy of motoneurons *(38)*. In addition, we recently showed a link between HIFs and VEGF in neuronal survival, by generating knock-in mice with a subtle deletion of the HRE in the VEGF promoter region ($VEGF^{\partial/\partial}$ mice). Quite surprisingly, these mice, which show an impaired hypoxic upregulation of VEGF and have reduced VEGF levels in the nervous system, exhibited signs of motor neuron degeneration, similar to those seen in humans with ALS (*see* Subheading 5.4 and also *see* ref. *39*).

3. ANGIOGENIC AND NEUROGENIC ROLE OF VEGF IN DEVELOPING CENTRAL NERVOUS SYSTEM

Several lines of evidence indicate that VEGF and its receptors importantly determine nervous system development. At the early stages of vascularization of the developing brain, VEGF, expressed in neurons in the subventricular zone *(40)*, induces VEGFR-2-expressing ECs of the perineural vascular plexus and capillaries to invade the neural tissue *(41,42)*. In addition, VEGF, expressed by sensory neurons or Schwann cells, is sufficient to instruct arterial vessels to branch alongside nerves *(43)*. Moreover, restricted inactivation of VEGF in the nervous system results in decreased vessel density, leading to ischemia, severe degeneration, and signs of neuronal acellularity in the cerebral cortex, whereas severe reductions in neural VEGF levels even lead to specific degeneration of the cerebral cortex and neonatal lethality *(44)*. VEGF-A dosage indeed seems to be a critical parameter regulating the density of the vascular plexus in the developing central nervous system (CNS) that is in turn a key determinant in the development and architectural organization of the nervous system *(44)*. Apart from these vascular effects, VEGF also affects neural cells directly. For instance, VEGF competes for binding to NP-1 with Sema-3A, which is a well-known repulsive axonal guidance molecule, and may, thereby, attract growth cones *(45)*. That NP-1 is important during vascular and nervous system development is shown in NP-1 deficient mice, which die at midgestation owing to defects in the heart, vasculature, and nervous system *(46,47)*. NP-2, in conjunction with axonal plexin-A3 receptor, transduces the repulsive axonal guidance effects of Sema-3F, another member of the class III semaphorin family. NP-2-null mice are viable but exhibit multiple axonal path finding defects in the nervous system and have lymphatic vessel abnormalities *(48,49)*. Altogether, these studies underscore the importance of VEGF and its receptors for developing neurons. Importantly, they also contribute to our understanding of the newly emerging link between the vascular and neural system, namely how blood vessels and nerves use common signals and pathways to differentiate, grow, and navigate toward their targets *(50)*.

4. VEGF AFFECTS NEURAL CELLS

4.1. Neurons

VEGF, once considered as the most EC-specific growth factor, has recently been shown to exert direct effects on several neural cell types (Fig. 1). VEGF protects neural stem cell 34 (NSC34) motoneuron cells against hypoxia and oxidative-stress-induced apoptosis by acting through VEGFR-2 and NP-1 *(39)*. VEGF also stimulates axonal outgrowth and promotes the survival of neurons and satellite cells in explant cultures of dorsal root ganglia *(51)*, whereas inhibition of VEGFR-2 signaling blocks axonal outgrowth in response to VEGF *(52)*. In the CNS, VEGF promotes the growth and survival of dopaminergic neurons and astrocytes in ventral mesencephalon explants *(53)*. VEGF seems to exert its neurotrophic effect mainly via signaling through VEGFR-2, phosphatidylinositol 3′-kinase (PI3-K), and Akt. Indeed, this signaling pathway protects HN33 cells (a somatic cell fusion of hippocampal neurons and neuroblastoma cells) against death elicited by deprivation of serum *(54)*, or of oxygen and glucose *(55)*. VEGFR-2 signaling via PI3-K/Akt and methyl ethyl ketone (MEK)/extracellular signal-regulated

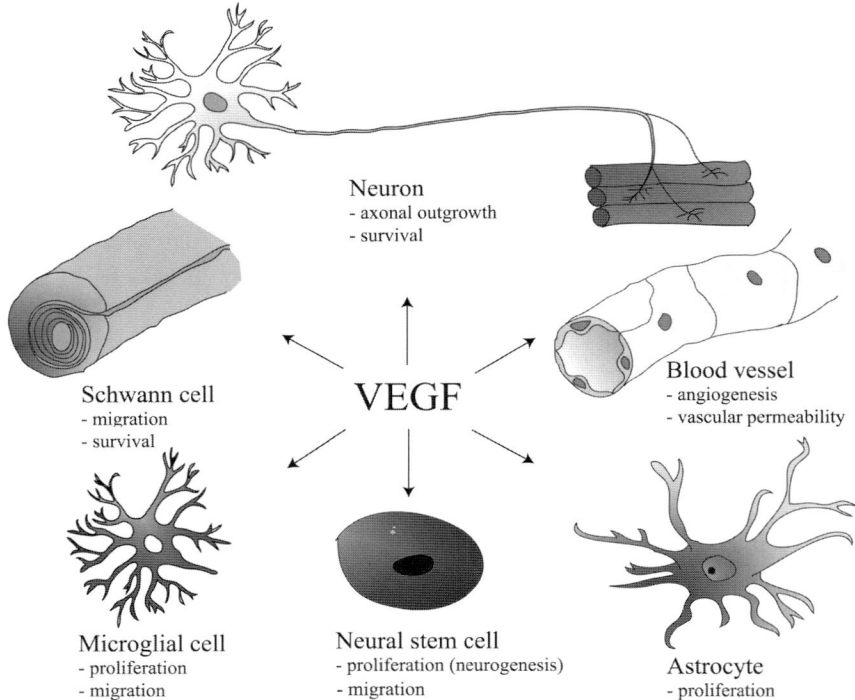

Fig. 1. Pleiotropic effects of VEGF. VEGF, originally considered as an endothelial specific growth factor, was recently shown to have direct effects on different neural cell types including neurons, Schwann cells, astrocytes, neural stem cells, and microglia. The effects of VEGF on the different cell types are indicated. (Reproduced from ref. *164* with permission of Wiley-Liss, Inc., a subsidiary of John Wiley & Sons, Inc.) (Please *see* companion CD for color version of this figure.)

kinase (ERK) also protects cultured hippocampal neurons against glutamate-induced toxicity *(56)* or hypoxic injury, probably via inhibition of excitotoxic processes *(57)*, and stimulates the survival of cerebellar granule neurons following hypoxic preconditioning *(58)*. VEGF/VEGFR-2 signaling also suppresses cell death pathways mediated by caspase-3 in cortical neuron cultures subjected to hypoxia *(59,60)*. Alternatively, VEGF stimulates neuronal survival via regulation of potassium currents, depending on signaling via VEGFR-1, not via VEGFR-2 *(61)*. Indeed, in SH-SY5Y neuroblastoma cells, hypoxia and glucose deprivation upregulate levels of VEGF and the voltage-gated potassium channel Kv1.2, thereby increasing tyrosine phosphorylation of this ion channel. Addition of exogenous VEGF under these conditions promotes cell survival and Kv1.2, tyrosine phosphorylation, whereas inhibition of VEGF has the opposite effects *(62)*. Interestingly, VEGF increases the expression of neuronal microtubule markers TUJ1 and MAP2 when applied to primary cortical neurons, suggesting that VEGF also plays a role in the growth, development, and structural stability of neurons, by upregulating neuronal proteins related to microtubule function *(63)*. All these studies obviously demonstrate that VEGF has direct in vitro neurotrophic effects on several neuronal cell types, including autonomic, sensory, dopaminergic, hippocampal, cerebellar, and cortical neurons.

4.2. Neural Progenitors

Besides its effects on differentiated neurons, VEGF also stimulates neurogenesis in vitro and in vivo. Intracerebroventricular infusion of VEGF in vivo stimulates bromodeoxyuridine incorporation in the rat brain, both in the subventricular zone and in the subgranular layer of

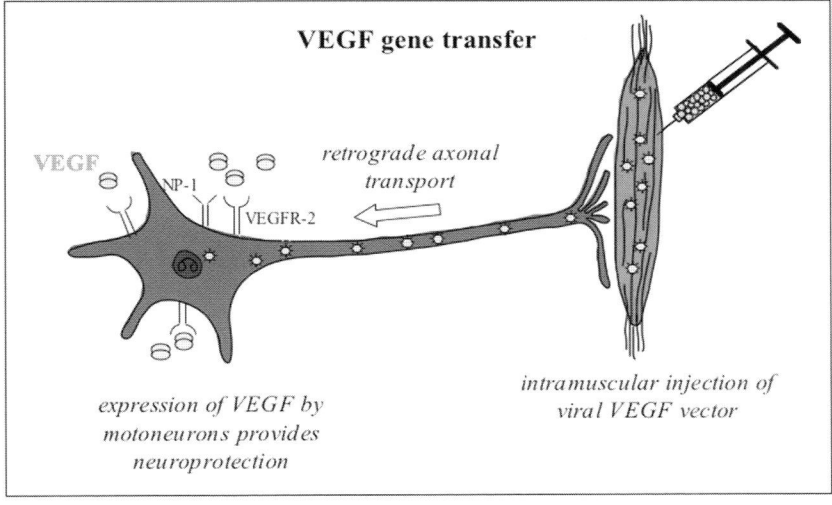
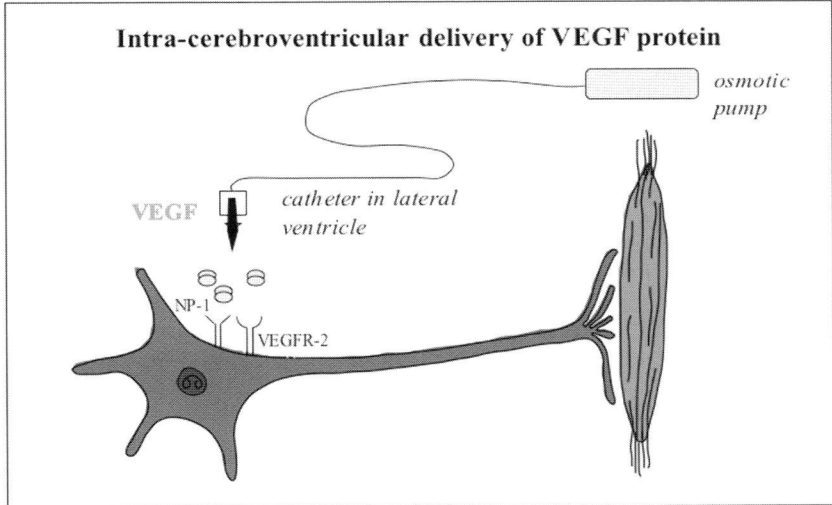

Fig. 3. Possible strategy to deliver VEGF into the brain or spinal cord. Prevention or retardation of motoneuron degeneration in ALS patients will likely require chronic treatment strategies, for instance using gene therapy. The figure illustrates gene transfer of VEGF to the neuronal cell body via intramuscular administration of retrogradely transported recombinant lentivirus. (Reproduced from ref. *165* with permission from the Muscular Dystrophy Association–USA.) (Please *see* companion CD for color version of this figure.)

symptoms in patients with ALS, although IGF-1 delivery has reduced disease progression in one but not in another study *(115,116)*. In SOD1^{G93A} mice, delivery of recombinant LIF improved motor neuron survival without prolonging the life span in one study *(117)*, and had no effect in another study *(118)*. Unfortunately, delivery of recombinant BDNF, CNTF, or IGF-1 was never evaluated in SOD1^{G93A} mice, whereas GDNF was administered only to healthy animals *(119)*. At least part of the failure can be ascribed to the short half-life, immunogenicity, dual effect on neuronal survival vs apoptosis, undesired toxicity, limited ability to cross the blood–brain barrier (BBB) after systemic delivery, or the insufficient penetration from the CSF into the spinal cord parenchyma after central delivery of these proteins *(120,121)*. Indeed, more recently, gene transfer of GDNF but especially of IGF-1, using a retrogradely axon-transported adenoassociated viral vector, was shown to prolong survival of SOD1^{G93A} mice *(122,123)*.

VEGF is able to prevent mutant SOD1-mediated motor neuron cell death in vitro *(124)*. More importantly, intramuscular administration of a VEGF-expressing rabies G-pseudotyped lentivirus, which is retrogradely transported to the neuronal cell body and upregulates VEGF in the spinal motor neurons (Fig. 3), remarkably delayed onset, improved motor performance, and prolonged survival in SOD1^{G93A} mice *(125)*. Initial data using an adenoassociated viral vector expressing VEGF also showed a promising therapeutic effect in ALS mouse models (unpublished observations). Furthermore, intracerebroventricular delivery of recombinant VEGF (Fig. 3) in SOD1^{G93A} rats has similar beneficial therapeutic effects *(126)*. Thus, using various routes and methods of administration, VEGF consistently improved the outcome of ALS in different preclinical animal models.

6.2. Ischemic Brain Injury (Stroke)

VEGF expression is strongly induced after acute focal cerebral ischemia, as early as 1–3 h reaching peak values at 24–48 h after onset of ischemia *(127,128)*. VEGF is upregulated in the ischemic core in which leakage of the BBB occurs *(129)*. Treatment with a soluble VEGF receptor (Flt) chimeric protein, mFlt(1–3)-IgG, which traps endogenous VEGF, reduces the volume of edematous brain tissue and results in a significant sparing of cortical tissue *(130)*. Conversely, intravenous infusion of VEGF$_{165}$ immediately after onset of focal cerebral embolic ischemia induces leakage of the BBB with resultant hemorrhagic transformation of the ischemic lesions *(131)*. However, intraventricular infusion as well as topical application of VEGF on the cerebral cortex immediately after restoration of cerebral blood flow, reduce edema formation and infarct volume *(132,133)*, suggesting that the route of administration importantly determines the effect of VEGF on vascular permeability. Thus, leakage may primarily occur when VEGF is administered via intravascular routes, where it can act directly on ECs.

Apart from its effects on vascular permeability, VEGF may also play a role in neovascularization following focal cerebral ischemia. VEGF is expressed in the penumbra of the infarct (a region of ischemic tissue surrounding the infarct, which is dysfunctional, yet potentially salvageable). Expression starts between 6 and 24 h, persists up to 28 d after onset of ischemia *(128,129)*, and precedes neovascularization of the ischemic border zone *(134)*. Intravenous infusion of VEGF within 48 h after onset of focal embolic cerebral ischemia enhances angiogenesis in the penumbra and improves neural recovery *(131)*. VEGF may thus have favorable effects on the recovery from ischemic brain insult, as it may salvage the viable neural tissue at risk by improving perfusion (via its vasodilatory activity), preventing EC dysfunction and/or stimulating angiogenesis. VEGF could also have direct neuroprotective effects after focal cerebral ischemia, as the VEGF receptors VEGFR-2, VEGFR-1, and NP-1 are also expressed in neurons and glial cells *(129,135,136)*, and their expression is upregulated in ischemic neuronal cells after transient or permanent middle cerebral artery occlusion *(135)*. Furthermore, antisense knockdown of VEGF after ischemic stroke results in an increased number of damaged neurons and an enlarged infarct volume *(137)*. In addition, a neuroprotective effect of intra-arterial infusion of VEGF was observed in a rat brain ischemia model, in the absence of angiogenesis *(138)*. This neuroprotective effect was lost at higher VEGF dosages that also induced angiogenesis, indicating that in vivo neuroprotection by VEGF administration not necessarily correlates with angiogenesis, and underlining the importance of proper VEGF dosage *(138)*. Finally, the therapeutic potential of VEGF in the treatment of stroke is further highlighted by the finding that ICV administration of VEGF 1 d after reperfusion reduces infarct size, improves neurological performance, enhances the survival of newborn neurons in the dentate gyrus and subventricular zone, and stimulates angiogenesis *(106)*.

6.3. Diabetic and Ischemic Neuropathy

Chronic hyperglycemia in diabetes mellitus causes various metabolic defects, which can provoke vasoconstriction, reduced endoneurial blood flow, and nerve hypoxia, further leading to reduced nerve conduction velocities, axonal loss, demyelination, and nerve dysfunction *(139)*.

Increased VEGF expression has been reported in cell bodies and axons of the sciatic nerve and dorsal root ganglia, in a rat model of type I diabetes, exhibiting chronic hyperglycemia for 12 wk. Moreover, application of insulin and/or nerve growth factor, which is known to improve axonal and Schwann cell regeneration, normalized VEGF expression in these diabetic animals *(140)*. On the other hand, intramuscular gene transfer of $VEGF_{165}$ slowed or reversed the development of reduced nerve conduction velocities and sensory nerve action potentials in a rabbit model of ischemic peripheral neuropathy. Furthermore, axonal loss and myelin degeneration were prevented or reversed, whereas neural blood flow was preserved at normal levels in VEGF-treated animals. Finally, VEGF stimulated the migration and prevented hypoxia-induced apoptosis of Schwann cells in vitro, indicating that VEGF could also have direct effects on neural cell integrity *(72)*. In addition, in rats with streptozotocin-induced diabetes, which showed markedly reduced nerve blood flow and developed, in parallel, severe peripheral neuropathy characterized by significant slowing of motor and sensory nerve conduction velocities, intramuscular *VEGF* gene transfer increased vascularity and blood flow in the nerves of treated animals to similar levels as those found in nondiabetic control rats, with a full restoration of large and small fiber peripheral nerve function *(141)*. Moreover, a preliminary clinical study in patients with critical hindlimb ischemia showed improvement in chronic ischemic neuropathy after intramuscular *VEGF* gene therapy *(142)*. A clinical study evaluating the safety and impact of *VEGF* gene transfer on sensory neuropathy in diabetic patients is ongoing *(143)*.

6.4. Spinal Cord Ischemia/Injury and Nerve Regeneration

VEGF is believed to play a role in CNS trauma, as the expression of VEGF and of its receptors VEGFR-2, VEGFR-1, and NP-1 is induced in experimental models of spinal cord injury *(38,144,145)*. In addition, in a rabbit model of transient spinal cord ischemia, expression of $VEGF_{121}$ and $VEGF_{165}$ increases within several hours of reperfusion in large motor neurons and in interneurons in the ventral horn of the spinal cord *(146)*. After spinal cord injury, the acute phase, characterized by hemorrhage, BBB destruction and infiltration of inflammatory cells, is followed by a subacute and chronic phase in which secondary degeneration occurs, involving the formation of a cystic cavity and a glial scar. Local injection of $VEGF_{165}$ immediately after induction of weight drop spinal cord injury in rats improved the behavioral outcome, increased the blood vessel density and survival of spinal cord tissue, and reduced the number of apoptotic cells *(147)*. On the other hand, local injection of VEGF, 72 h after injury, has recently been documented to exacerbate spinal cord injury and lesion volume, as it resulted in increased microvascular permeability and infiltration of leukocytes in spinal cord parenchyma *(148)*. To further investigate the role of VEGF in acute motor neuron death, we have used a mouse model of spinal cord ischemia-reperfusion *(149)*. $VEGF^{\partial/\partial}$ mice (*see* Subheading 5.1.) were found to be unusually sensitive to transient ischemic attacks, as a minor ischemic insult leads to permanent paralysis in $VEGF^{\partial/\partial}$ mice, whereas it only causes transient clinical deficits in wild-type mice *(86)*. Furthermore, addition of exogenous $VEGF_{165}$ protected the at-risk motor neurons of wild-type mice after extended spinal cord ischemia. The neurological outcome of VEGF-treated mice improved more rapidly and histopathological analysis also showed that the loss of motor neurons in the lumbar gray matter was reduced in VEGF-treated mice *(86)*. These data indicate that low VEGF levels render motor neurons more susceptible to acute motor neuron death, whereas VEGF administration can rescue motor neurons from ischemic death. To determine whether VEGF will also be useful to treat spinal cord injury, further research is mandated.

VEGF was also suggested to be implicated in nerve regeneration. Angiogenesis plays an essential role in nerve regeneration, but it remains unclear whether there is a direct interrelationship between vascularization and efficacy of nerve regeneration within a nerve conduit. Sondell and co-workers reported that when cell-free nerve grafts, used to bridge a gap in the sciatic nerve, were pretreated with VEGF, the outgrowth of Schwann cells and blood vessels was stimulated *(150)*. The interdependence between increased vascularization and enhanced nerve regeneration within an acellular conduit was further highlighted by Hobson and colleagues *(151)*, who showed that VEGF significantly augmented blood vessel penetration within silicone nerve regeneration chambers, correlating with an increase in axonal regeneration and Schwann cell migration. Moreover, VEGF also enhanced target organ reinnervation as measured by the recovery of gastrocnemius muscle weights and footpad axonal terminal density. Spinal cord transsection in adult rats results in retrograde degeneration of the corticospinal tract axons that do not regenerate. This phenomenon is believed to be owing to either ischemia from insufficient blood flow or to a deficiency of growth factors *(152)*. In support of this notion, application of exogenous VEGF to the transsected spinal cord resulted in increased vascularization and reduced retrograde degeneration of axons of the corticospinal tract *(153)*. These findings indicate that administration of VEGF to neural scar tissue could be used to promote long-distance regeneration in the adult CNS.

6.5. Other Neurodegenerative Disorders

The possible involvement of VEGF in other neurodegenerative disorders remains enigmatic. Cerebrovascular pathology, including cerebral amyloid angiopathy and small vessel disease, are prominent features of human AD *(154)*. Moreover, the cerebral blood flow is lower in AD cases than in controls, and this cerebral hypoperfusion often precedes the clinical onset of the disease *(155)*. The presence of vascular abnormalities in subjects with AD *(154)* suggests a possible role or involvement of VEGF in this neurodegenerative disease. Interestingly, increased VEGF protein expression has been documented in clusters of reactive astrocytes in the neocortex, in walls of many large intraparenchymal vessels and in diffuse perivascular deposits in AD patients *(156)*. Furthermore, CSF levels of VEGF are significantly enhanced in individuals with AD and vascular dementia *(157)*. However, whether the increased VEGF expression implies compensatory repair mechanisms to counter insufficient vascularity, impaired endothelial transport of nutrients or reduced perfusion, remains to be elucidated.

Furthermore, because if its angiogenic and/or neurotrophic properties VEGF might also exert beneficial effects in PD, an adult-onset neurodegenerative disorder characterized by progressive dopaminergic cell death in the nervous system without a known cause. Evidence for a role of VEGF in PD was recently observed after unilateral implantation of VEGF-producing cells into the striatum of rats, treated with 6-hydroxydopamine, a neurotoxin used to induce dopaminergic neurodegeneration. Administration of VEGF reduced amphetamine-induced rotational behavior, augmented tyrosine hydroxylase-positive neurons and fibers, and increased vascularization and proliferation of glial cells *(158)*. In addition, a single bolus injection of VEGF into the striatum of unilaterally 6-hydroxydopamine lesioned rats, before transplantation of solid ventral mesencephalic grafts into the same striatum, was reported to result in a more homogeneous distribution of small blood vessels throughout the graft and in an accelerated recovery of amphetamine-induced rotational asymmetry *(159)*. Despite these preliminary findings, an in depth analysis of the role and therapeutic potential of VEGF protein or gene delivery for PD awaits future research.

The role of VEGF in MS, a chronic inflammatory disease of the CNS leading to focal destruction of myelin is, at present, more enigmatic, although a vascular component has long been associated with the pathological changes in MS. A possible involvement of VEGF in MS is suggested by the elevated *VEGFR-1* expression, seen in white matter brain tissue from MS patients *(160)*. Using an animal model of experimental allergic encephalomyelitis (EAE), VEGF expression was recently shown to be upregulated during the acute disease peaking at day 26, which was the transition from the acute-inflammatory to chronic-demyelinating phase, before gradually returning to baseline levels *(161)*. VEGF-positive cells with astrocytic morphology were found in association with inflammatory cells. Furthermore, intracerebral infusion of VEGF in animals previously immunized with myelin basic protein induced an inflammatory response in the brain, whereas infusion of vehicle, or infusion of VEGF in naive animals, did not *(162)*. These results suggest that overexpression of VEGF may exacerbate the inflammatory response in autoimmune diseases of the CNS by inducing focal BBB breakdown and migration of inflammatory cells into the lesions. Anyway, it remains to be assessed whether VEGF might enhance MS lesion formation via its proangiogenic and inflammatory activity in the acute phase or whether VEGF might be beneficial by protecting against axon loss and neurodegeneration via its neuroprotective and axon outgrowth-inducing activities.

A primary deficit in neuronal stem cell proliferation, migration or differentiation is hypothesized to contribute to net cell loss and neuronal circuit disruption in neurodegenerative disorders such as AD, PD, and Huntington's disease *(163)*. Therefore, VEGF and other growth factors, capable of stimulating adult neurogenesis *(64,106)*, might also be implicated in these disorders. However, future research will need to assess the role and possible therapeutic potential of VEGF in these neurodegenerative disorders.

7. CONCLUSIONS

Since the discovery of VEGF, its angiogenic activity has been extensively studied. Only recently, however, direct effects of VEGF on neuronal cell types have been described. It is now evident that VEGF is implicated in the human neurodegenerative disorder ALS, but the exact mechanisms via which reduced VEGF levels lead to motor neuron degeneration await unraveling. Further research is also required to define the role of VEGF in other neurodegenerative disorders, including AD, Huntington's disease, PD, and MS. VEGF is also involved in stroke. Importantly, VEGF might have beneficial therapeutic effects, but the optimal delivery modalities need to be further evaluated. Nevertheless, the initial evidence from preclinical studies that *VEGF* gene transfer or protein delivery improves the outcome of ALS and other neurodegenerative disorders is highly promising.

REFERENCES

1. Senger DR, Galli SJ, Dvorak AM, Perruzzi CA, Harvey VS, Dvorak HF. Tumor cells secrete a vascular permeability factor that promotes accumulation of ascites fluid. Science 1983;219:983–985.
2. Carmeliet P. Angiogenesis in health and disease. Nat Med 2003;9:653–660.
3. Ferrara N. Role of vascular endothelial growth factor in regulation of physiological angiogenesis. Am J Physiol Cell Physiol 2001; 280:C1358–C1366.
4. Shibuya M. Structure and function of VEGF/VEGF-receptor system involved in angiogenesis. Cell Struct Funct 2001;26:25–35.
5. Neufeld G, Cohen T, Gengrinovitch S, Poltorak Z. Vascular endothelial growth factor (VEGF) and its receptors. FASEB J 1999;13:9–22.
6. Ferrara N, Davis-Smyth T. The biology of vascular endothelial growth factor. Endocr Rev 1997;18:4–25.
7. Hiratsuka S, Minowa O, Kuno J, Noda, T, Shibuya M. Flt-1 lacking the tyrosine kinase domain is sufficient for normal development and angiogenesis in mice. Proc Natl Acad Sci USA 1998;95:9349–9354.
8. Kendall RL, Wang G, Thomas KA. Identification of a natural soluble form of the vascular endothelial growth factor receptor, FLT-1, and its heterodimerization with KDR. Biochem Biophys Res Commun 1996;226:324–328.
9. Nagura S, Katoh R, Miyagi E, Shibuya M, Kawaoi A. Expression of vascular endothelial growth factor (VEGF) and VEGF receptor-1 (Flt-1) in Graves disease possibly correlated with increased vascular density. Hum Pathol 2001;32:10–17.
10. Luttun A, Tjwa M, Moons L, et al. Revascularization of ischemic tissues by PlGF treatment, and inhibition of tumor angiogenesis, arthritis and atherosclerosis by anti-Flt1. Nat Med 2002;8: 831–840.
11. Autiero M, Luttun A, Tjwa M, Carmeliet P. Placental growth factor and its receptor, vascular endothelial growth factor receptor-1: novel targets for stimulation of ischemic tissue revascularization and inhibition of angiogenic and inflammatory disorders. J Thromb Haemost 2003;1:1356–1370.
12. Carmeliet P, Moons L, Luttun A, et al. Synergism between vascular endothelial growth factor and placental growth factor contributes to angiogenesis and plasma extravasation in pathological conditions. Nat Med 2001;7:575–583.

13. Soker S, Takashima S, Miao HQ, Neufeld G, Klagsbrun M. Neuropilin-1 is expressed by endothelial and tumor cells as an isoform- specific receptor for vascular endothelial growth factor. Cell 1998;92:735–745.
14. Neufeld G, Cohen T, Shraga N, Lange T, Kessler O, Herzog Y. The neuropilins: multifunctional semaphorin and VEGF receptors that modulate axon guidance and angiogenesis. Trends Cardiovasc Med 2002;12:13–19.
15. Huez I, Bornes S, Bresson D, Creancier L, Prats H. New vascular endothelial growth factor isoform generated by internal ribosome entry site-driven CUG translation initiation. Mol Endocrinol 2001;15:2197–2210.
16. Yancopoulos GD, Davis S, Gale NW, Rudge JS, Wiegand SJ, Holash J. Vascular-specific growth factors and blood vessel formation. Nature 2000;407:242–248.
17. Hiratsuka S, Maru Y, Okada A, Seiki M, Noda T, Shibuya M. Involvement of Flt-1 tyrosine kinase (vascular endothelial growth factor receptor-1) in pathological angiogenesis. Cancer Res 2001;61:1207–1213.
18. Autiero M, Waltenberger J, Communi D, et al. Role of PlGF in the intra- and intermolecular cross talk between the VEGF receptors Flt1 and Flk1. Nat Med 2003;9:936–943.
19. Aase K, von Euler G, Li X, et al. Vascular endothelial growth factor-B-deficient mice display an atrial conduction defect. Circulation 2001;104:358–364.
20. Bellomo D, Headrick JP, Silins GU, et al. Mice lacking the vascular endothelial growth factor-B gene (VEGF-B) have smaller hearts, dysfunctional coronary vasculature, and impaired recovery from cardiac ischemia. Circ Res 2000;86:E29–E35.
21. Reichelt M, Shi S, Hayes M, et al. Vascular endothelial growth factor-B and retinal vascular development in the mouse. Clin Experiment Ophthalmol 2003;31:61–65.
22. Wanstall JC, Gambino A, Jeffery TK, et al. Vascular endothelial growth factor-B-deficient mice show impaired development of hypoxic pulmonary hypertension. Cardiovasc Res 2002;55:361–368.
23. Louzier V, Raffestin B, Leroux A, et al. Role of VEGF-B in the lung during development of chronic hypoxic pulmonary hypertension. Am J Physiol Lung Cell Mol Physiol 2003;284:L926–L937.
24. Mould AW, Tonks ID, Cahill MM, et al. Vegfb gene knockout mice display reduced pathology and synovial angiogenesis in both antigen-induced and collagen-induced models of arthritis. Arthritis Rheum 2003;48:2660–2669.
25. Lohela M, Saaristo A, Veikkola T, Alitalo K. Lymphangiogenic growth factors, receptors and therapies. Thromb Haemost 2003;90:167–184.
26. Stacker SA, Hughes RA, Achen MG. Molecular targeting of lymphatics for therapy. Curr Pharm Des 2004;10:65–74.
27. He Y, Karpanen T, Alitalo K. Role of lymphangiogenic factors in tumor metastasis. Biochim Biophys Acta 2004;1654:3–12.
28. Tille JC, Nisato R, Pepper MS. Lymphangiogenesis and tumour metastasis. Novartis Found Symp 2004;256:112–131; discussion 132–116,259–169.
29. Wenger RH. Cellular adaptation to hypoxia: O2-sensing protein hydroxylases, hypoxia-inducible transcription factors, and O2-regulated gene expression. FASEB J 2002;16:1151–1162.
30. Liu Y, Cox SR, Morita T, Kourembanas S. Hypoxia regulates vascular endothelial growth factor gene expression in endothelial cells. Identification of a 5′ enhancer. Circ Res 1995;77:638–643.
31. Levy NS, Chung S, Furneaux H, Levy AP. Hypoxic stabilization of vascular endothelial growth factor mRNA by the RNA-binding protein HuR. J Biol Chem 1998;273:6417–6423.
32. Claffey KP, Shih SC, Mullen A, et al. Identification of a human VPF/VEGF 3′ untranslated region mediating hypoxia-induced mRNA stability. Mol Biol Cell 1998;9:469–481.
33. Levy AP, Levy NS, Goldberg MA. Post-transcriptional regulation of vascular endothelial growth factor by hypoxia. J Biol Chem 1996; 271:2746–2753.
34. Stein I, Itin A, Einat P, Skaliter R, Grossman Z, Keshet E. Translation of vascular endothelial growth factor mRNA by internal ribosome entry: implications for translation under hypoxia. Mol Cell Biol 1998;18:3112–3119.
35. Miller DL, Dibbens JA, Damert A, Risau W, Vadas MA, Goodall GJ. The vascular endothelial growth factor mRNA contains an internal ribosome entry site. FEBS Lett 1998;434:417–420.
36. Huez I, Creancier L, Audigier S, Gensac MC, Prats AC, Prats H. Two independent internal ribosome entry sites are involved in translation initiation of vascular endothelial growth factor mRNA. Mol Cell Biol 1998;18:6178–6190.

37. van der Velden AW, Thomas AA. The role of the 5' untranslated region of an mRNA in translation regulation during development. Int J Biochem Cell Biol 1999;31:87–106.
38. Skold MK, Marti HH, Lindholm T, et al. Induction of HIF1alpha but not HIF2alpha in motoneurons after ventral funiculus axotomy-implication in neuronal survival strategies. Exp Neurol 2004;188:20–32.
39. Oosthuyse B, Moons L, Storkebaum E, et al. Deletion of the hypoxia-response element in the vascular endothelial growth factor promoter causes motor neuron degeneration. Nat Genet 2001;28:131–138.
40. Breier G, Albrecht U, Sterrer S, Risau W. Expression of vascular endothelial growth factor during embryonic angiogenesis and endothelial cell differentiation. Development 1992;114:521–532.
41. Dumont DJ, Fong GH, Puri MC, Gradwohl G, Alitalo K, Breitman ML. Vascularization of the mouse embryo: a study of flk-1, tek, tie, and vascular endothelial growth factor expression during development. Dev Dyn 1995;203:80–92.
42. Millauer B, Wizigmann-Voos S, Schnurch H, et al. High affinity VEGF binding and developmental expression suggest Flk-1 as a major regulator of vasculogenesis and angiogenesis. Cell 1993;72:835–846.
43. Mukouyama YS, Shin D, Britsch S, Taniguchi M, Anderson DJ. Sensory nerves determine the pattern of arterial differentiation and blood vessel branching in the skin. Cell 2002;109:693–705.
44. Haigh JJ, Morelli PI, Gerhardt H, et al. Cortical and retinal defects caused by dosage-dependent reductions in VEGF-A paracrine signaling. Dev Biol 2003;262:225–241.
45. Bagnard D, Vaillant C, Khuth ST, et al. Semaphorin 3A-vascular endothelial growth factor-165 balance mediates migration and apoptosis of neural progenitor cells by the recruitment of shared receptor. J Neurosci 2001;21:3332–3341.
46. Kitsukawa T, Shimizu M, Sanbo M, et al. Neuropilin-semaphorin III/D-mediated chemorepulsive signals play a crucial role in peripheral nerve projection in mice. Neuron 1997;19:995–1005.
47. Kawasaki T, Kitsukawa T, Bekku Y, et al. A requirement for neuropilin-1 in embryonic vessel formation. Development 1999;126: 4895–4902.
48. Chen H, Bagri A, Zupicich JA, et al. Neuropilin-2 regulates the development of selective cranial and sensory nerves and hippocampal mossy fiber projections. Neuron 2000;25:43–56.
49. Yuan L, Moyon D, Pardanaud L, et al. Abnormal lymphatic vessel development in neuropilin 2 mutant mice. Development 2002;129: 4797–4806.
50. Carmeliet P. Blood vessels and nerves: common signals, pathways and diseases. Nat Rev Genet 2003;4:710–720.
51. Sondell M, Lundborg G, Kanje M. Vascular endothelial growth factor has neurotrophic activity and stimulates axonal outgrowth, enhancing cell survival and Schwann cell proliferation in the peripheral nervous system. J Neurosci 1999;19:5731–5740.
52. Sondell M, Sundler F, Kanje M. Vascular endothelial growth factor is a neurotrophic factor which stimulates axonal outgrowth through the flk-1 receptor. Eur J Neurosci 2000;12:4243–4254.
53. Silverman WF, Krum JM, Mani N, Rosenstein JM. Vascular, glial and neuronal effects of vascular endothelial growth factor in mesencephalic explant cultures. Neuroscience 1999;90:1529–1541.
54. Jin KL, Mao XO, Greenberg DA. Vascular endothelial growth factor rescues HN33 neural cells from death induced by serum withdrawal. J Mol Neurosci 2000;14:197–203.
55. Jin KL, Mao XO, Greenberg DA. Vascular endothelial growth factor: direct neuroprotective effect in in vitro ischemia. Proc Natl Acad Sci USA 2000;97:10,242–10,247.
56. Matsuzaki H, Tamatani M, Yamaguchi A, et al. Vascular endothelial growth factor rescues hippocampal neurons from glutamate-induced toxicity: signal transduction cascades. FASEB J 2001;15:1218–1220.
57. Svensson B, Peters M, Konig HG, et al. Vascular endothelial growth factor protects cultured rat hippocampal neurons against hypoxic injury via an antiexcitotoxic, caspase-independent mechanism. J Cereb Blood Flow Metab 2002;22:1170–1175.
58. Wick A, Wick W, Waltenberger J, Weller M, Dichgans J, Schulz JB. Neuroprotection by hypoxic preconditioning requires sequential activation of vascular endothelial growth factor receptor and Akt. J Neurosci 2002;22:6401–6407.
59. Jin K, Mao XO, Batteur SP, McEachron E, Leahy A, Greenberg DA. Caspase-3 and the regulation of hypoxic neuronal death by vascular endothelial growth factor. Neuroscience 2001;108:351–358.
60. Ogunshola OO, Antic A, Donoghue MJ, et al. Paracrine and autocrine functions of neuronal vascular endothelial growth factor (VEGF) in the central nervous system. J Biol Chem 2002;277:11,410–11,415.
61. Xu JY, Zheng P, Shen DH, et al. Vascular endothelial growth factor inhibits outward delayed-rectifier potassium currents in acutely isolated hippocampal neurons. Neuroscience 2003;118:59–67.

62. Qiu MH, Zhang R, Sun FY. Enhancement of ischemia-induced tyrosine phosphorylation of Kv1.2 by vascular endothelial growth factor via activation of phosphatidylinositol 3-kinase. J Neurochem 2003;87:1509–1517.
63. Rosenstein JM, Mani N, Khaibullina A, Krum JM. Neurotrophic effects of vascular endothelial growth factor on organotypic cortical explants and primary cortical neurons. J Neurosci 2003;23: 11,036–11,044.
64. Jin K, Zhu Y, Sun Y, Mao XO, Xie L, Greenberg DA. Vascular endothelial growth factor (VEGF) stimulates neurogenesis in vitro and in vivo. Proc Natl Acad Sci USA 2002;99:11,946–11,950.
65. Fabel K, Tam B, Kaufer D, et al. VEGF is necessary for exercise-induced adult hippocampal neurogenesis. Eur J Neurosci 2003;18:2803–2812.
66. Maurer MH, Tripps WK, Feldmann RE, Jr, Kuschinsky W. Expression of vascular endothelial growth factor and its receptors in rat neural stem cells. Neurosci Lett 2003;344:165–168.
67. Zhu Y, Jin K, Mao XO, Greenberg DA. Vascular endothelial growth factor promotes proliferation of cortical neuron precursors by regulating E2F expression. FASEB J 2003;17:186–193.
68. Zhang H, Vutskits L, Pepper MS, Kiss JZ. VEGF is a chemoattractant for FGF-2-stimulated neural progenitors. J Cell Biol 2003;163:1375–1384.
69. Krum JM, Mani N, Rosenstein JM. Angiogenic and astroglial responses to vascular endothelial growth factor administration in adult rat brain. Neuroscience 2002;110:589–604.
70. Krum JM, Khaibullina A. Inhibition of endogenous VEGF impedes revascularization and astroglial proliferation: roles for VEGF in brain repair. Exp Neurol 2003;181:241–257.
71. Forstreuter F, Lucius R, Mentlein R. Vascular endothelial growth factor induces chemotaxis and proliferation of microglial cells. J Neuroimmunol 2002;132:93–98.
72. Schratzberger P, Schratzberger G, Silver M, et al. Favorable effect of VEGF gene transfer on ischemic peripheral neuropathy. Nat Med 2000;6:405–413.
73. Rowland LP, Shneider NA. Amyotrophic lateral sclerosis. N Engl J Med 2001;344:1688–1700.
74. Hand CK, Rouleau GA. Familial amyotrophic lateral sclerosis. Muscle Nerve 2002;25:135–159.
75. Green SL, Tolwani RJ. Animal models for motor neuron disease. Lab Anim Sci 1999;49:480–487.
76. Yang Y, Hentati A, Deng HX, et al. The gene encoding alsin, a protein with three guanine-nucleotide exchange factor domains, is mutated in a form of recessive amyotrophic lateral sclerosis. Nat Genet 2001;29:160–165.
77. Wilhelmsen KC. Disinhibition-dementia-parkinsonism-amyotrophy complex (DDPAC) is a non-Alzheimer's frontotemporal dementia. J Neural Transm Suppl 1997;49:269–275.
78. Hutton M, Lendon CL, Rizzu P, et al. Association of missense and 5′-splice-site mutations in tau with the inherited dementia FTDP-17. Nature 1998;393:702–705.
79. Rabin BA, Griffin JW, Crain BJ, Scavina M, Chance PF, Cornblath DR. Autosomal dominant juvenile amyotrophic lateral sclerosis. Brain 1999;122(Pt 8):1539–1550.
80. Hosler BA, Siddique T, Sapp PC, et al. Linkage of familial amyotrophic lateral sclerosis with frontotemporal dementia to chromosome 9q21-q22. JAMA 2000;284:1664–1669.
81. Hand CK, Khoris J, Salachas F, et al. A novel locus for familial amyotrophic lateral sclerosis, on chromosome 18q. Am J Hum Genet 2002;70:251–256.
82. Ruddy DM, Parton MJ, Al-Chalabi A, et al. Two families with familial amyotrophic lateral sclerosis are linked to a novel locus on chromosome 16q. Am J Hum Genet 2003;73:390–396.
83. Sapp PC, Hosler BA, McKenna-Yasek D, et al. Identification of two novel loci for dominantly inherited familial amyotrophic lateral sclerosis. Am J Hum Genet 2003;73:397–403.
84. Cleveland DW, Rothstein JD. From Charcot to Lou Gehrig: deciphering selective motor neuron death in ALS. Nat Rev Neurosci 2001;2:806–819.
85. Dal Canto MC, Gurney ME. Neuropathological changes in two lines of mice carrying a transgene for mutant human Cu,Zn SOD, and in mice overexpressing wild type human SOD: a model of familial amyotrophic lateral sclerosis (FALS). Brain Res 1995;676:25–40.
86. Lambrechts D, Storkebaum E, Morimoto M, et al. VEGF is a modifier of amyotrophic lateral sclerosis in mice and humans and protects motoneurons against ischemic death. Nat Genet 2003;34: 383–394.
87. Gros-Louis F, Laurent S, Lopes AA, et al. Absence of mutations in the hypoxia response element of VEGF in ALS. Muscle Nerve 2003;28:774–775.
88. Devos D, Moreau C, Lassalle P, et al. Low levels of the vascular endothelial growth factor in CSF from early ALS patients. Neurology 2004;62:2127–2129.

89. Storkebaum E, Lambrechts D, Manka D, et al. VEGF, a modifier of motor neuron degeneration in SOD1 G93A mice, protects against motor neuron loss after spinal cord ischemia: evidence for a vascular hypothesis. Program No. 602.11. 2003 Abstract Viewer/Itinerary Planner. Washington, DC: Society for Neuroscience, online.
90. Sopher BL, Thomas PS, Jr, LaFevre-Bernt MA, et al. Androgen receptor YAC transgenic mice recapitulate SBMA motor neuronopathy and implicate VEGF164 in the motor neuron degeneration. Neuron 2004;41:687–699.
91. Pramatarova A, Laganiere J, Roussel J, Brisebois K, Rouleau GA. Neuron-specific expression of mutant superoxide dismutase 1 in transgenic mice does not lead to motor impairment. J Neurosci 2001;21:3369–3374.
92. Gong YH, Parsadanian AS, Andreeva A, Snider WD, Elliott JL. Restricted expression of G86R Cu/Zn superoxide dismutase in astrocytes results in astrocytosis but does not cause motoneuron degeneration. J Neurosci 2000;20:660–665.
93. Lino MM, Schneider C, Caroni P. Accumulation of SOD1 mutants in postnatal motoneurons does not cause motoneuron pathology or motoneuron disease. J Neurosci 2002;22:4825–4832.
94. Buijs PC, Krabbe-Hartkamp MJ, Bakker CJ, et al. Effect of age on cerebral blood flow: measurement with ungated two-dimensional phase-contrast MR angiography in 250 adults. Radiology 1998;209: 667–674.
95. Ochiai-Kanai R, Hasegawa K, Takeuchi Y, Yoshioka H, Sawada T. Immunohistochemical nitrotyrosine distribution in neonatal rat cerebrocortical slices during and after hypoxia. Brain Res 1999;847:59–70.
96. Kobari M, Obara K, Watanabe S, Dembo T, Fukuuchi Y. Local cerebral blood flow in motor neuron disease: correlation with clinical findings. J Neurol Sci 1996;144:64–69.
97. Waldemar G, Vorstrup S, Jensen TS, Johnsen A, Boysen G. Focal reductions of cerebral blood flow in amyotrophic lateral sclerosis: a [99mTc]-d,l-HMPAO SPECT study. J Neurol Sci 1992;107:19–28.
98. Heath PR, Shaw PJ. Update on the glutamatergic neurotransmitter system and the role of excitotoxicity in amyotrophic lateral sclerosis. Muscle Nerve 2002;26:438–458.
99. Ramsden DB, Parsons RB, Ho SL, Waring RH. The aetiology of idiopathic Parkinson's disease. Mol Pathol 2001;54:369–380.
100. Kennea NL, Mehmet H. Neural stem cells. J Pathol 2002;197: 536–550.
101. Yamamoto S, Yamamoto N, Kitamura T, Nakamura K, Nakafuku M. Proliferation of parenchymal neural progenitors in response to injury in the adult rat spinal cord. Exp Neurol 2001;172:115–127.
102. Nakatomi H, Kuriu T, Okabe S, et al. Regeneration of hippocampal pyramidal neurons after ischemic brain injury by recruitment of endogenous neural progenitors. Cell 2002;110:429.
103. Ogawa Y, Sawamoto K, Miyata T, et al. Transplantation of in vitro-expanded fetal neural progenitor cells results in neurogenesis and functional recovery after spinal cord contusion injury in adult rats. J Neurosci Res 2002;69:925–933.
104. Craig CG, Tropepe V, Morshead CM, Reynolds BA, Weiss S, der Kooy D. In vivo growth factor expansion of endogenous subependymal neural precursor cell populations in the adult mouse brain. J Neurosci 1996;16:2649–2658.
105. Martens DJ, Seaberg RM, van der Kooy D. In vivo infusions of exogenous growth factors into the fourth ventricle of the adult mouse brain increase the proliferation of neural progenitors around the fourth ventricle and the central canal of the spinal cord. Eur J Neurosci 2002;16:1045–1057.
106. Sun Y, Jin K, Xie L, et al. VEGF-induced neuroprotection, neurogenesis, and angiogenesis after focal cerebral ischemia. J Clin Invest 2003;111:1843–1851.
107. Palmer TD, Willhoite AR, Gage FH. Vascular niche for adult hippocampal neurogenesis. J Comp Neurol 2000;425:479–494.
108. Shah NM, Groves AK, Anderson DJ. Alternative neural crest cell fates are instructively promoted by TGFbeta superfamily members. Cell 1996;85:331–343.
109. Mi H, Haeberle H, Barres BA. Induction of astrocyte differentiation by endothelial cells. J Neurosci 2001;21:1538–1547.
110. Louissaint A, Jr, Rao S, Leventhal C, Goldman SA. Coordinated interaction of neurogenesis and angiogenesis in the adult songbird brain. Neuron 2002;34:945–960.
111. Katoh-Semba R, Asano T, Ueda H, et al. Riluzole enhances expression of brain-derived neurotrophic factor with consequent proliferation of granule precursor cells in the rat hippocampus. FASEB J 2002;16:1328–1330.

112. Miller RG, Mitchell JD, Lyon M, Moore DH. Riluzole for amyotrophic lateral sclerosis (ALS)/motor neuron disease (MND). Amyotroph Lateral Scler Other Motor Neuron Disord 2003;4:191–206.
113. Ochs G, Penn RD, York M, et al. A phase I/II trial of recombinant methionyl human brain derived neurotrophic factor administered by intrathecal infusion to patients with amyotrophic lateral sclerosis. Amyotroph Lateral Scler Other Motor Neuron Disord 2000;1: 201–206.
114. Miller RG, Petajan JH, Bryan WW, et al. A placebo-controlled trial of recombinant human ciliary neurotrophic (rhCNTF) factor in amyotrophic lateral sclerosis. rhCNTF ALS Study Group. Ann Neurol 1996;39:256–260.
115. Lai EC, Felice KJ, Festoff BW, et al. Effect of recombinant human insulin-like growth factor-I on progression of ALS. A placebo-controlled study. The North America ALS/IGF-I Study Group. Neurology 1997;49:1621–1630.
116. Borasio GD, Robberecht W, Leigh PN, et al. A placebo-controlled trial of insulin-like growth factor-I in amyotrophic lateral sclerosis. European ALS/IGF-I Study Group. Neurology 1998;51: 583–586.
117. Azari MF, Lopes EC, Stubna C, et al. Behavioural and anatomical effects of systemically administered leukemia inhibitory factor in the SOD1(G93A G1H) mouse model of familial amyotrophic lateral sclerosis. Brain Res 2003;982:92–97.
118. Feeney SJ, Austin L, Bennett TM, et al. The effect of leukaemia inhibitory factor on SOD1 G93A murine amyotrophic lateral sclerosis. Cytokine 2003;23:108–118.
119. Ramer MS, Bradbury EJ, Michael GJ, Lever IJ, McMahon SB. Glial cell line-derived neurotrophic factor increases calcitonin gene-related peptide immunoreactivity in sensory and motoneurons in vivo. Eur J Neurosci 2003;18:2713–2721.
120. Boyd JG, Gordon T. A dose-dependent facilitation and inhibition of peripheral nerve regeneration by brain-derived neurotrophic factor. Eur J Neurosci 2002;15:613–626.
121. Thorne RG, Frey WH, 2nd. Delivery of neurotrophic factors to the central nervous system: pharmacokinetic considerations. Clin Pharmacokinet 2001;40:907–946.
122. Acsadi G, Anguelov RA, Yang H, et al. Increased survival and function of SOD1 mice after glial cell-derived neurotrophic factor gene therapy. Hum Gene Ther 2002;13:1047–1059.
123. Kaspar BK, Llado J, Sherkat N, Rothstein JD, Gage FH. Retrograde viral delivery of IGF-1 prolongs survival in a mouse ALS model. Science 2003;301:839–842.
124. Li B, Xu W, Luo C, Gozal D, Liu R. VEGF-induced activation of the PI3-K/Akt pathway reduces mutant SOD1-mediated motor neuron cell death. Brain Res Mol Brain Res 2003;111:155–164.
125. Azzouz M, Ralph GS, Storkebaum E, et al. VEGF delivery with retrogradely transported lentivector prolongs survival in a mouse ALS model. Nature 2004;429:413–417.
126. Storkebaum E, Lambrechts D, Dewerchin M, et al. Treatment of motorneuron degeneration by intracerebroventricular delivery of VEGF in a rat model of ALS. Nat Neurosci 2005;8:5–7.
127. Hayashi T, Abe K, Suzuki H, Itoyama Y. Rapid induction of vascular endothelial growth factor gene expression after transient middle cerebral artery occlusion in rats. Stroke 1997;28:2039–2044.
128. Plate KH, Beck H, Danner S, Allegrini PR, Wiessner C. Cell type specific upregulation of vascular endothelial growth factor in an MCA-occlusion model of cerebral infarct. J Neuropathol Exp Neurol 1999;58:654–666.
129. Zhang ZG, Zhang L, Tsang W, et al. Correlation of VEGF and angiopoietin expression with disruption of blood-brain barrier and angiogenesis after focal cerebral ischemia. J Cereb Blood Flow Metab 2002;22:379–392.
130. van Bruggen N, Thibodeaux H, Palmer JT, et al. VEGF antagonism reduces edema formation and tissue damage after ischemia/reperfusion injury in the mouse brain. J Clin Invest 1999;104:1613–1620.
131. Zhang ZG, Zhang L, Jiang Q, et al. VEGF enhances angiogenesis and promotes blood-brain barrier leakage in the ischemic brain. J Clin Invest 2000;106:829–838.
132. Harrigan MR, Ennis SR, Sullivan SE, Keep RF. Effects of intraventricular infusion of vascular endothelial growth factor on cerebral blood flow, edema, and infarct volume. Acta Neurochir (Wien) 2003;145:49–53.
133. Hayashi T, Abe K, Itoyama Y. Reduction of ischemic damage by application of vascular endothelial growth factor in rat brain after transient ischemia. J Cereb Blood Flow Metab 1998;18:887–895.
134. Marti HJ, Bernaudin M, Bellail A, et al. Hypoxia-induced vascular endothelial growth factor expression precedes neovascularization after cerebral ischemia. Am J Pathol 2000;156:965–976.
135. Lennmyr F, Ata KA, Funa K, Olsson Y, Terent A. Expression of vascular endothelial growth factor (VEGF) and its receptors (Flt-1 and Flk-1) following permanent and transient occlusion of the middle cerebral artery in the rat. J Neuropathol Exp Neurol 1998;57:874–882.

136. Beck H, Acker T, Puschel AW, Fujisawa H, Carmeliet P, Plate KH. Cell type-specific expression of neuropilins in an MCA-occlusion model in mice suggests a potential role in post-ischemic brain remodeling. J Neuropathol Exp Neurol 2002;61:339–350.
137. Yang ZJ, Bao WL, Qiu MH, et al. Role of vascular endothelial growth factor in neuronal DNA damage and repair in rat brain following a transient cerebral ischemia. J Neurosci Res 2002;70:140–149.
138. Manoonkitiwongsa PS, Schultz RL, McCreery DB, Whitter EF, Lyden PD. Neuroprotection of ischemic brain by vascular endothelial growth factor is critically dependent on proper dosage and may be compromised by angiogenesis. J Cereb Blood Flow Metab 2004;24:693–702.
139. Veves A, King GL. Can VEGF reverse diabetic neuropathy in human subjects? J Clin Invest 2001;107:1215–1218.
140. Samii A, Unger J, Lange W. Vascular endothelial growth factor expression in peripheral nerves and dorsal root ganglia in diabetic neuropathy in rats. Neurosci Lett 1999;262:159–162.
141. Schratzberger P, Walter DH, Rittig K, et al. Reversal of experimental diabetic neuropathy by VEGF gene transfer. J Clin Invest 2001;107:1083–1092.
142. Simovic D, Isner JM, Ropper AH, Pieczek A, Weinberg DH. Improvement in chronic ischemic neuropathy after intramuscular phVEGF165 gene transfer in patients with critical limb ischemia. Arch Neurol 2001;58:761–768.
143. Isner JM, Ropper A, Hirst K. VEGF gene transfer for diabetic neuropathy. Hum Gene Ther 2001;12:1593–1594.
144. Vaquero J, Zurita M, de Oya S, Coca S. Vascular endothelial growth/permeability factor in spinal cord injury. J Neurosurg 1999; 90:220–223.
145. Skold M, Cullheim S, Hammarberg H, et al. Induction of VEGF and VEGF receptors in the spinal cord after mechanical spinal injury and prostaglandin administration. Eur J Neurosci 2000;12:3675–3686.
146. Hayashi T, Sakurai M, Abe K, Sadahiro M, Tabayashi K, Itoyama Y. Expression of angiogenic factors in rabbit spinal cord after transient ischaemia. Neuropathol Appl Neurobiol 1999;25:63–71.
147. Widenfalk J, Lipson A, Jubran M, et al. Vascular endothelial growth factor improves functional outcome and decreases secondary degeneration in experimental spinal cord contusion injury. Neuroscience 2003;120:951–960.
148. Benton RL, Whittemore SR. VEGF165 therapy exacerbates secondary damage following spinal cord injury. Neurochem Res 2003;28:1693–1703.
149. Lang-Lazdunski L, Matsushita K, Hirt L, et al. Spinal cord ischemia. Development of a model in the mouse. Stroke 2000;31:208–213.
150. Sondell M, Lundborg G, Kanje M. Vascular endothelial growth factor stimulates Schwann cell invasion and neovascularization of acellular nerve grafts. Brain Res 1999;846:219–228.
151. Hobson MI, Green CJ, Terenghi G. VEGF enhances intraneural angiogenesis and improves nerve regeneration after axotomy. J Anat 2000;197(Pt 4):591–605.
152. Silver J, Miller JH. Regeneration beyond the glial scar. Nat Rev Neurosci 2004;5:146–156.
153. Facchiano F, Fernandez E, Mancarella S, et al. Promotion of regeneration of corticospinal tract axons in rats with recombinant vascular endothelial growth factor alone and combined with adenovirus coding for this factor. J Neurosurg 2002;97:161–168.
154. Kalaria RN. Small vessel disease and Alzheimer's dementia: pathological considerations. Cerebrovasc Dis 2002;13:48–52.
155. Nagata K, Kondoh Y, Atchison R, et al. Vascular and metabolic reserve in Alzheimer's disease. Neurobiol Aging 2000;21:301–307.
156. Kalaria RN, Cohen DL, Premkumar DR, Nag S, LaManna JC, Lust WD. Vascular endothelial growth factor in Alzheimer's disease and experimental cerebral ischemia. Brain Res Mol Brain Res 1998;62:101–105.
157. Tarkowski E, Issa R, Sjogren M, et al. Increased intrathecal levels of the angiogenic factors VEGF and TGF-beta in Alzheimer's disease and vascular dementia. Neurobiol Aging 2002;23:237–243.
158. Yasuhara T, Shingo T, Kobayashi K, et al. Neuroprotective effects of vascular endothelial growth factor (VEGF) upon dopaminergic neurons in a rat model of Parkinson's disease. Eur J Neurosci 2004;19:1494–1504.
159. Pitzer MR, Sortwell CE, Daley BF, et al. Angiogenic and neurotrophic effects of vascular endothelial growth factor (VEGF165): studies of grafted and cultured embryonic ventral mesencephalic cells. Exp Neurol 2003:182:435–445.
160. Graumann U, Reynolds R, Steck AJ, Schaeren-Wiemers N. Molecular changes in normal appearing white matter in multiple sclerosis are characteristic of neuroprotective mechanisms against hypoxic insult. Brain Pathol 2002;13:554–573.

161. Kirk SL, Karlik SJ. VEGF and vascular changes in chronic neuroinflammation. J Autoimmun 2003:21:353–363.
162. Proescholdt MA, Jacobson S, Tresser N, Oldfield EH, Merrill MJ. Vascular endothelial growth factor is expressed in multiple sclerosis plaques and can induce inflammatory lesions in experimental allergic encephalomyelitis rats. J Neuropathol Exp Neurol 2002;61:914–925.
163. Armstrong RJ, Barker RA. Neurodegeneration: a failure of neuroregeneration? Lancet 2001;358: 1174–1176.
164. Storkebaum E, Lambrechts D, Carmeliet P. VEGF: once just a specific angiogenic factor, now attractive for neuroprotection? BioEssays 2004;26:943–954.
165. Wahl M. Support for VEGF treatment grows. MDA/ALS Newsmagazine 2004;9(7).

20
Epidermal Growth Factor Receptor in the Adult Brain

Carmen Estrada, MD, PhD and Antonio Villalobo, MD, PhD

SUMMARY

The epidermal growth factor receptor (EGFR) plays a relevant role in brain development. Expression of EGFR and its ligands occurs in those regions in which neuronal and glial cells are generated, and EGFR activation is essential for the proliferation of multipotent neural precursors, as well as the survival, migration, and differentiation of the immature daughter cells. In the adult brain, EGFR is expressed in specific regions in physiological conditions. In the subventricular zone, where neurogenesis persists into adulthood, the EGFR contributes to the maintenance of the progenitor pool, and probably serves functions similar to those observed during development. Expression of both EGFR and its ligands is upregulated in different types of brain lesions where they may participate in brain self-repair by mechanisms that are not yet well understood. Finally, EGFR overexpression, as well as aberrant forms of the receptor, is present in a variety of cerebral tumors. Future research on cerebral EGFR function may lead to new therapeutic strategies in neurodegenerative or ischemic diseases and brain tumors.

Key Words: Brain repair; brain tumors; EGF; EGFR; gliogenesis; HB-EGF; neural stem cells; neurogenesis; reactive astroglia; TGF-α.

1. THE EPIDERMAL GROWTH FACTOR RECEPTOR

The epidermal growth factor receptor (EGFR), also denoted erythroblastosis B1 receptor/human epidermal growth factor receptor 1 (ErbB1/HER1), belongs to a family of tyrosine kinase receptors located at the plasma membrane, which also includes the ErbB2/Neu/HER2, ErbB3/HER3, and ErbB4/HER4 receptors. All these receptors possess a single membrane-spanning domain, an extracellular region containing the ligand-binding site, and a cytosolic region that holds the tyrosine kinase catalytic domain *(1,2)*. The activation of EGFR occurs on binding of a variety of polypeptide ligands including epidermal growth factor (EGF), heparin-binding EGF (HB-EGF), transforming growth factor-α (TGF-α), and amphiregulin among others *(3,4)*. Ligand binding induces receptor homodimerization or heterodimerization with other ErbB family members, the activation of its intrinsic tyrosine kinase, and the trans(auto)phosphorylation of five tyrosine residues located at its C-terminal end *(1–3)*. The phosphorylated tyrosine residues act as recruitment points for a series of adaptor and effector proteins containing Src homology 2 (SH2) and phosphotyrosine-binding (PTB) domains, which initiate the activation of a variety of signaling pathways leading to the progression of the cell cycle and subsequently cell proliferation. Nevertheless, EGFR is also involved in the control of a series of other cellular functions such as cell survival, migration, and differentiation. After signals are transmitted, the EGFR could be subjected to short-term downregulation mediated by regulatory protein kinases *(5,6)* and calmodulin-binding *(7)*, and to long-term downregulation consisting in ligand-induced

From: *The Cell Cycle in the Central Nervous System*
Edited by: D. Janigro © Humana Press Inc., Totowa, NJ

internalization, followed by its degradation into lysosomes or its recycling back to the plasma membrane (8). Recently, it has been shown that the EGFR is transactivated by a variety of G protein-coupled receptors (GPCRs), enlarging the spectrum of physiological functions controlled by EGFR (9,10).

2. EXPRESSION OF EGFR AND ITS LIGANDS IN THE DEVELOPING AND ADULT BRAIN

During embryonic development, EGFR expression starts as early as E11 in the rat brain (11). Both EGFR mRNA and protein appear first in the midbrain and subsequently in the germinal zones, where the expression increases progressively to reach a maximum during the perinatal period (11). The EGFR ligands TGF-α and HB-EGF are also expressed from E12–E14 in different brain regions including the developing cortical plate, hippocampus, cerebellar Purkinje cells, and ventrobasal thalamus (11–13). EGF, however, is very scarce in the brain, with concentrations one or two orders of magnitude lower than those of TGF-α (12,14). It is interesting to note that there is an early presence of both EGFR and its ligands in the midbrain, in a situation that extensively overlaps that of developing dopaminergic neurons (11,15).

At the end of the first week after birth, there is a peak of TGF-α mRNA levels in the mouse striatum, in close proximity with the EGFR-expressing cells that proliferate in the subependymal layer (16). The functional role of the receptor in this system is suggested by the decreased proliferation reported in the TGF-α-deficient mouse, waved-1 (16). TGF-α and EGFR have also been identified in the postnatal cerebellum, where neurogenesis persists for a short period after birth (17). HB-EGF mRNA is expressed in all principal cell layers of the hippocampus during the postnatal period, but becomes restricted to the dentate gyrus (DG) granule cells in adulthood (18).

At all developmental stages, EGFR is expressed by cells that also contain nestin, a marker of neuroepithelial precursor cells (13,19), whereas EGFR ligands have been found in all cell types, including precursors, neuronal, and glial cells (20).

In the adult brain, the presence of EGFR mRNA and protein is restricted almost exclusively to the subventricular zone (SVZ) (16,17,21), an area where neurogenesis persists into adulthood (see Subheading 5.1.). Cells expressing EGFR are distributed particularly in the dorsolateral aspect of the SVZ, and their number, as well as the density of hybridization, diminishes with age (17). Interestingly, the DG of the hippocampus, in which neurogenesis also persists in adult mammals, contains a very low number of EGFR-labeled cells, compared with the SVZ (19). In contrast to the restricted location of EGFR, TGF-α, HB-EGF mRNA, and proteins are widespread in the adult mouse and rat brain, with a distribution similar to that found in postnatal animals (12,14,22,23).

Although in physiological conditions EGFR is not abundant in the adult brain, expression of the receptor, as well as its ligands TGF-α and/or HB-EGF, is upregulated in a wide variety of pathological conditions (24–28). Within the first week after an ischemic episode, EGFR expression increases in astrocytes in the lesioned area (24), microglial cells in the whole hemisphere (24), and undifferentiated precursors in the SVZ (28). At the same time, there is an activation of the HB-EGF gene expression in different brain regions including hippocampus, cerebral cortex, thalamus and cerebellar granule, and Purkinje cell layers (25). In the hippocampus exposed to hypoxic-ischemic injury, both EGFR and HB-EGF are induced in cortical area 3 and DG (25,26), but not in CA1, which is the region most vulnerable to ischemia (27). Kainate-induced excitotoxic seizures induced marked increases in HB-EGF mRNA levels in the hippocampus, as well as in several other cortical and limbic forebrain regions (18). Following radiofrequency lesions in the preoptic area of the hypothalamus, TGF-α mRNA and protein were enhanced in two types of reactive astrocytes: those sending long processes toward the lesion site, and those with stellate shape located more peripherally (29). These effects are not strictly local, because hypoglossal nerve crush results in induction of both EGFR and TGF-α in the hypoglossal nucleus, at a considerable distance from the lesioned area (30).

3. EFFECTS OF EGFR ACTIVATION IN CULTURED CELLS OF NEURAL ORIGIN

In neural cultures derived from the central nervous system (CNS), EGF, TGF-α, HB-EGF, and amphiregulin stimulate the proliferation of multipotent progenitors and astrocytes, and enhance neuron survival *(13,15,19,31–33)*. As shown in vivo, neural cells in culture enhanced their expression of EGFR and its ligands in response to hypoxia *(19)*.

3.1. Neural Stem Cells

Neural stem cells are self-renewing, multipotent cells that reside in the brain germinal zone during embryonic and postnatal development, and give rise to the diverse types of cells in the CNS. Depending on the developmental stage and CNS location, neural stem cells are regulated by various soluble, extracellular matrix- or membrane-bound signals. These signals determine stem cell survival, proliferation, and eventual differentiation toward a specific phenotype.

Neural stem cells can be isolated from embryo germinal zones and from postnatal or adult neurogenic areas, and maintained in culture in the presence of the appropriate growth factors. Data obtained in stem cells from mouse embryos and adult forebrain indicate that initial divisions from quiescent cells can be induced by either fibroblast growth factor 2 (FGF-2) or EGF, but that EGFR stimulation is necessary for the expansion of a daughter-cell population that perpetuates the stem cell-like characteristics and is highly proliferative *(31,34)*. On EGFR stimulation, these precursor cells divide and form neurospheres. When single neurospheres are dissociated, different phenotypes can be found in the progeny, which indicates that they were generated from multipotent cells. Furthermore, some of the individual cells give rise to new neurospheres, thus indicating that the neurosphere-forming cells can be self-renewed *(35,36)*. These results suggest that EGFR activation is an essential step in neural stem cell survival and proliferation.

In addition, EGFR ligands may induce differentiation of neural stem cells in vitro. It has recently been reported that, on treatment with EGF or TGF-α, but not with FGF-2, cultured cells isolated from embryonic brain or adult SVZ, give rise to radial glial cells that produce elongated processes and support neuronal cell migration *(37)*.

3.2. Glial Cells

Profound reorganization of the cytoskeleton has been observed in cultured astrocytes from different origins when they are exposed to EGF or TGF-α *(33,38)*. Cells become elongated with very long, thin processes, and lose the monolayer arrangement to form multilayered cribiform architecture. These effects are inhibited by the EGFR inhibitor tyrophostin AG1478, and are not produced by platelet-derived growth factor (PDGF) or FGF-2 *(33)*. An additional change produced in cultured astrocytes by EGFR stimulation is the upregulation of the glutamate transporter expression *(38,39)*, which may contribute to neuroprotection. EGFR has also a function in microglial cells, in which EGF induces motility owing to both chemotaxis and chemokinesis stimulation *(40)*.

3.3. Neuronal Cells

It has been proposed that HB-EGF has a neuroprotective action, as shown in embryonic hippocampal cell cultures exposed to elevated intracytoplasmic Ca^{2+} concentration produced by kainate toxicity *(18)*. However, the neuron survival effects of EGFR ligands are indirect and require the presence of non-neuronal cells, mainly astroglia, because cells that have initiated differentiation toward the neuronal lineage do not express EGFR *(13,15)*. The mechanism by which HB-EGF or TGF-α promotes neuron survival is not known, although the mitogen-activated protein kinase (MAPK) as well as the Akt-signaling pathways seem to be involved *(13,15)*. Neurite formation is induced by co-culture of neurons with EGF-treated astrocytes. Because EGF activates fibronectin and laminin gene expression in astrocytes, these matrix proteins are probably responsible for neurite outgrowth *(41)*.

4. THE ROLE OF EGFR IN BRAIN DEVELOPMENT

The presence of EGFR and its ligands in strategic locations in the developing brain, together with the consequences of EGFR activation in neural cultures, suggests that this receptor has a relevant function during brain formation. Different models of mice lacking EGFR present smaller, albeit cytoarchitecturally normal, brains at birth. However, during the first week of postnatal life, a massive and progressive neurodegeneration and apoptosis takes place in the cortex, olfactory bulb (OB), and thalamus *(42,43)*. In addition, postnatal TGF-α-deficient mice present a reduction in the number of astrocytes in the striatum, and fewer proliferating cells in the SVZ *(16)*. These data indicate that the functional role of EGFR is prominent during the perinatal period, a time when neurogenesis and gliogenesis are very active.

5. THE ROLE OF EGFR IN THE NORMAL ADULT BRAIN

In the normal adult brain, the presence of EGFR is moderate and restricted to certain specific areas, such as the SVZ, rostral migratory stream, and OB, where it is expressed by neural progenitors; this suggests that EGFR is involved in the neurogenesis that persists in these regions in adult life. Other functions have been proposed for EGFR in the adult brain, such as the stimulation of hypothalamic astrocytes and tanycytes, endowed with the receptor, inducing the secretion of luteinizing hormone-releasing hormone (LHRH) by neighboring neurons *(44)*. However, in this section we will focus on the role of EGFR in cell cycle-controlling events in the adult brain.

5.1. EGFR and Adult Neurogenesis

It is currently well established that neural progenitors with the ability to produce new neurons and glia remain in the adult CNS throughout life, in various mammalian species including humans *(45)*. Quiescent stem cells in the adult, as well as the embryonic, brain can be activated by FGF-2 *(31,34,46)* or other unidentified factor(s) provided by ependymal cells or endothelial basal lamina *(47)*, and divide giving rise to a highly proliferative EGF-responsive transit-amplifying cell population.

Neural stem cells in the adult mammalian brain constantly generate new neurons in two restricted areas: the SVZ, in the lateral walls of the lateral ventricles, and the DG of the hippocampus (reviewed in refs. *48–50*). In the adult SVZ (Fig. 1), periventricular astrocytes (B-cells) have been identified as the neural stem cells *(36)*. The transit-amplifying cells (C-cells), which can be identified by their expression of nestin and the transcription factor Dlx2, and the absence of neuronal- or glial-specific antigens *(51,52)*, give rise to neuroblasts (A-cells), that, after tangential migration through the rostral migratory stream, become inhibitory interneurons in the OB. In the SVZ, EGFR is specifically expressed by the transit-amplifying cells *(52)*. In the adult DG, whereas proliferation is not as massive as in the SVZ, the transit-amplifying cell population is less numerous, in accordance with the scarce presence of EGFR, and the low or absent proliferative response to EGFR stimulation *(19,53)*.

Short intracerebroventricular infusions of EGF produce a rapid increase in the proportion of C-cells and the number of C-cells undergoing mitosis, as well as a reduction in the number and proliferation of neuroblasts in the SVZ *(52)*. Longer infusions (3–14 d) of EGF, TGF-α, or HB-EGF largely increase proliferation of the SVZ undifferentiated precursors and produce profound changes in the fate of postmitotic cells (Fig. 2). Migration toward the OB decreases and, instead, the newly formed cells migrate away from the walls of the lateral ventricles into adjacent brain parenchyma, including the striatum, septum, corpus callosum, and cortex *(19,52,54)*. Infusion of EGF also results in the appearance of *de novo* gliogenesis in the regions of SVZ expansion, with extensions of radial-like VIM-positive processes emanating from the SVZ *(52,53,55)*. Furthermore, EGFR stimulation induces the generation of radial glial cells that express the specific markers rapid cell 2 (RC2), nestin, and brain lipid binding protein, and adopt a radial morphology, with long processes directed toward the adjacent parenchyma *(37)*.

EGFR in the Adult Brain

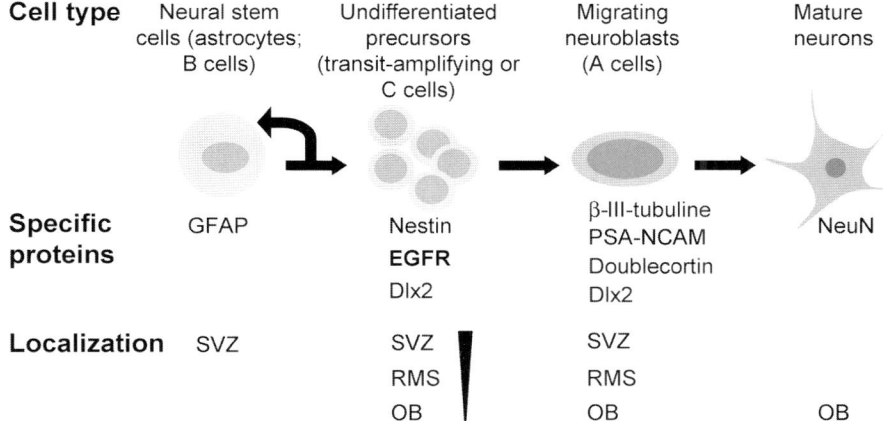

Fig. 1. Cell types involved in adult neurogenesis in the rodent subventricular zone (SVZ). Neural stem cells, derived from apparently mature astrocytes lining the ventricle wall, give rise to EGF-responsive highly proliferative undifferentiated precursors. Part of the precursors generates neuroblasts that initiate a tangential migration through the rostral migratory stream (RMS) to the olfactory bulb (OB), where they become mature interneurons. Given in parentheses is the nomenclature used by Doetsch et al. *(51)*. Proteins expressed by the different cell types, which can be used for their immunohistochemical identification are shown. EGFR is selectively expressed by C-cells.

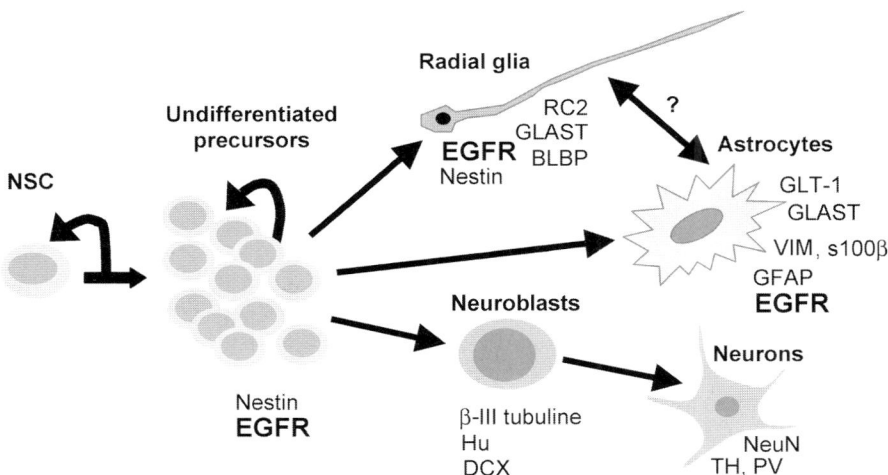

Fig. 2. *De novo* cell generation in the SVZ in brains exposed to intracerebroventricular administration of EGF or TGF-α. There is enhanced proliferation of the undifferentiated precursor population. Newborn glial cells appear, presenting astrocyte or radial glial morphology and expressing specific antigens. Newborn neuroblasts migrate toward the surrounding parenchyma. When EGFR ligands are administered in a lesioned brain, some neuroblasts migrate toward the lesioned area and differentiate as mature neurons whose specific phenotypes depend on the brain region.

Multiple labeling with RC2, Br-deoxyundine (which labels newborn cells), and the immature-neuron antigen doublecortin, indicated that newborn neuronal cells migrated along RC2 fibers toward the striatum *(37)*.

Although the effects of exogenously administered EGFR ligands on adult neurogenesis have been well characterized, there is little information on the physiological role of EGFR in this process. Mice null for TGF-α present reduced cell proliferation in the SVZ and fewer new neurons reaching the OB, indicating that TGF-α is a mitogenic signal necessary for the maintenance of the C-cell population *(56)*. However, this result does not provide full information on the role of the EGFR, because the receptor might be partially activated by redundant EGFR ligands, such as HB-EGF or amphiregulin.

6. EGFR AND BRAIN REPAIR

6.1. EGFR and Its Ligands Are Upregulated in Brain Lesions

Following injury in the CNS, EGFR is upregulated in reactive astrocytes that migrate toward the lesion from adjacent undamaged parenchyma *(24,57)*, and also in undifferentiated precursors in the neurogenic areas *(25,26,28)*. Simultaneously, there is a widespread activation of TGF-α and *HB-EGF* gene expression in both neuronal and glial cells *(18,25,27,38)*.

The mechanisms underlying upregulation of EGFR and its ligands in the injured brain are not well understood. Molecules released during inflammation, hypoxia, or other pathological conditions may directly regulate transcription of the receptor and its ligands, although involvement of indirect transactivation of the EGFR should also be considered. It has been reported that several external stimuli, such as GPCR ligands, cytokines, growth hormones, or steroids may activate intracellular signaling pathways by an indirect mechanism involving the EGFR *(9,10,41,58)*. A major potential mechanism of the cross-communication between GPCR and EGFR is the activation of matrix metalloproteinases (MPs) that cause proteolytic cleavage of the transmembrane precursor pro-HB-EGF, pro-TGF-α, or proamphiregulin, to yield the respective soluble ligands *(58–60)*. Metalloproteinases, which are activated in the injured brain and contribute to tissue remodeling after lesions, might also participate in the control of cell proliferation, neuronal survival, and astrogliosis, by providing the extracellular fluid with soluble EGF ligands (Fig. 3). The transformation of the EGF ligands or EGFR signaling system from physiological to brain-injury conditions is accelerated by the existence of positive feedback mechanisms involved in the regulation of ligand expression (Fig. 3), as it has been reported that activation of EGFR enhances transcription of the *TGF-α* gene and stimulates cleavage of the soluble form from the transmembrane precursor *(44)*.

6.2. The Role of EGFR Activation in Brain Lesions

Upregulation of EGFR and its ligand proteins in nontumoral brain pathologies may be part of the mechanisms intended to minimize neuronal damage or to enhance brain repair. Although there are no experimental results demonstrating a beneficial effect of spontaneous EGFR activation in brain lesions, the effects of intracerebroventricular or local infusions of EGF or TGF-α have been evaluated in several brain lesion models in adult rodents. The data obtained so far suggest that, when applied immediately after the lesion, EGFR activation has a neuroprotective effect *(61)*, although administration over a longer period, once the lesion has been incurred, improves cerebral self-repair mechanisms *(27,28)*.

It is well-known that the adult CNS has a very limited capacity for self-repair or regeneration after excitotoxic, ischemic, or traumatic lesions. However, the demonstration that neurogenesis persists in some areas in the adult brain, and that this process can be stimulated on injury, has opened new research approaches from both biological and therapeutic perspectives. It has been observed that several lesions, such as transient global *(62)* or focal *(27,28,63)* ischemia, seizures *(64)*, or apoptotic degeneration induction *(65)*, stimulate proliferation of neural progenitors in the DG and/or SVZ, and lead to an increased production of new neurons that eventually migrate toward the lesioned area. However, the capacity of neuronal replacement in the adult brain is very low, with estimations between 0.2 and 10% of the lost neurons, depending on the area and the type of lesion *(27,28,66)*.

If endogenous progenitors contribute to the production of new neurons, expansion of their endogenous pool by concomitant administration of growth factors may improve the regenerative

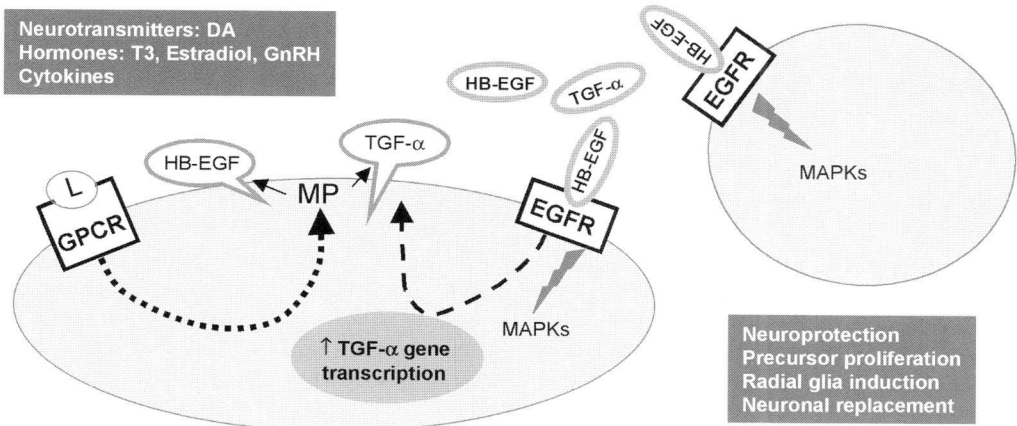

Fig. 3. EGFR-mediated mechanisms by which extracellular signal molecules may affect nervous tissue remodeling. Neurotransmitters, hormones, cytokines, and other molecules (L) acting through G protein-coupled receptors (GPCRs) activate metalloproteases (MPs), which cause proteolytic cleavage of pro-TGF-α and pro-HB-EGF, and release of soluble TGF-α and HB-EGF to the extracellular fluid. Binding of these growth factors to the epidermal growth factor receptor (EGFR) in the same (autocrine) or neighbor (paracrine) cells leads to activation of the mitogen-activated protein kinase (MAPK) and other intracellular pathways. EGFR activation also upregulates transcription of EGFR ligands, thus, generating a positive feedback mechanism that amplifies the initial stimulus. The functional consequences of EGFR activation include neuronal survival and glial reorganization and, eventually, neuronal replacement. DA, dopamine; GnRH, gonadotropin-releasing hormone; T3, tri-iodothyronine.

capacity of the damaged CNS. In this respect, it has been reported that 7-d EGF infusions in animals exposed to transient middle cerebral artery occlusion significantly increased the number of newborn cells that migrated from the SVZ into the infarcted striatum; these cells progressively expressed young- and mature-neuron antigens within the next few weeks *(28)*. Interestingly, EGF also increased the number of astrocytes exhibiting long processes, which formed a network linking the striatum and the SVZ *(28)*. In another model of ischemia, the transient carotid artery occlusion, a selective apoptosis of hippocampal CA1 pyramidal neurons, occurs *(27)*. A few weeks later, a small number of cells expressing young neuron antigens appears in the lesioned layer. Continuous infusion of EGF and FGF-2 during 2–5 d after ischemia did not reduce cell death within the first week, but dramatically increased the number of new neurons, resulting in a recovery of approx 40% of the lost neurons 4 wk later *(27)*. The newly formed neurons acquired a phenotype characteristic of the regenerated region, and received synaptic inputs, suggesting their integration in neuronal circuits. The growth factor-treated animals also improved their performance in the water maze, a test that evaluates the hippocampal function *(27)*.

The origin of the newly formed neurons that initiate cell replacement after lesions, and that are stimulated by EGF, remains elusive, but most authors suggest that they migrate from the walls of the lateral ventricles *(27,28,65)*. Also interesting is the experiment reported by Fallon et al. *(67)*, in which 6-hydroxydopamine (6OHDA) lesions of the substantia nigra–ventral tegmental area, accompanied by intrastriatal infusion of TGF-α, resulted in a massive proliferation of EGFR-expressing cells in the SVZ, followed by an extensive migration of the proliferating precursors toward the TGF-α injection site. After 3–4 wk of infusion, newly born cells expressing dopaminergic neuron-specific antigens were present in the striatum, and there was a partial improvement in the apomorphine-induced rotation test. The directed migration described in these studies could be explained by two alternative or complementary mechanisms: EGFR ligands may exert a chemotactic action on neuronal precursors, and/or the glial transformations induced by EGFR ligands may provide the scaffold and the trophic support necessary for neuroblast migration.

7. THE EGFR IN BRAIN TUMORS

Among the tumors of the brain and associated structures where EGFR expression has been found, we should mention meningiomas *(68)*, ependynomas *(69)*, capillary hemangioblastomas *(70)*, neuroblastomas *(71)*, central neurocytomas *(72)*, and most significantly, tumors of glial origin including oligodendrogliomas, astrocytomas, and glioblastomas *(68,69,73)*. No EGFR expression has been found, however, in pituitary macroadenomas *(74)*, although it is known that the EGFR is present in normal cells of the anterior pituitary gland *(75)*.

Amplifications of the *EGFR* gene, mRNA, and receptor overexpression, as well as mutations including truncations, deletions, and duplications of functional domains, and missense mutations of the receptor are common events in some tumors of the CNS. The frequency of these alterations varies considerably depending on the tumor type and its degree of malignancy *(76)*. Thus, the expression of EGFR in oligodendrogliomas is weak *(77)* with a moderate (approx 50%) incidence *(73)*, and, on average, lower expression than in low-grade astrocytomas *(78)*. Low expression levels (1000–6000 receptors per cell) were also found in ependymomas *(69)*.

Amplification of the *EGFR* gene is very frequent (40–50%) in glioblastomas *(76)* but it is not detected, however, in capillary hemangioblastomas *(70)* or central neurocytomas *(72)*. The amplification of the *EGFR* gene is often associated with the expression of a diversity of mutated receptors resulting from gene rearrangement, together with the wild-type EGFR, which coexist within the same tumor *(76,79)*. Rearrangements usually take place near the 5′-end (exon 1–7 region) and near the 3′-end (exon 22–26 region) of the *EGFR* gene. Most rearrangements near the 3′-end cause the loss of exons 23–25 resulting in a truncated receptor at residue 958 denoted EGFRvV because of a frame shift *(76,80)*. Interestingly, the donor or acceptor rearrangement sites are distinct in different glioblastomas, a process that fails to support a homologous recombination mechanism, as shown for rearrangements occurring near the 3′-end *(80)*.

A common in-frame deletion of exons 2–7 (801 bp) of the *EGFR* gene results in a 140 kDa receptor known as EGFRvIII or just ΔEGFR, with a deletion of 267 amino acids (residues 6–273) at its extracellular region, which is frequently coexpressed with wild-type EGFR in glioblastomas. EGFRvIII lacks the capacity to bind the ligand, has constitutively active tyrosine kinase activity, is inefficiently downregulated, and confers enhanced tumorigenicity because of increased proliferation and inhibition of apoptosis *(79,81–83)*. Related mutations consisting in the deletion of exons 2–7 plus exons 12–13 results in a 125 kDa receptor denoted EGFRvIII/Δ12–13 *(76,79)*, and deletion of residues 6–185, have also been found in glioblastomas *(76)*. An additional deletion, yielding receptors lacking residues 521–603, denoted that EGFRvII is also very frequently found in glioblastomas, and this deletion could also be associated with truncation at residue 958, resulting in a double mutated receptor *(76)*. Furthermore, deletion of residues 959–1030 has also been found in these tumors *(76)*.

Studies with transgenic mice expressing EGFRvIII have shown that those animals have an increased number of astrocytes but do not develop gliomas. In contrast, the cooperation of Ha-Ras and EGFRvIII in transgenic animals expressing both proteins results in decreased survival because of the development of infiltrating oligodendrogliomas and mixed oligodendrogliomas, compared with the development of fibrillary astrocytomas induced by Ha-Ras alone *(84)*. This is consistent with the activation of the Ras/MAPK pathway mediated by Shc and Grb2 induced by this deletion mutant receptor that worsens the prognosis of patients with glioblastoma multiforme *(85,86)*. The tumorigenicity of EGFRvIII in glioblastomas is mediated by its hyperactive tyrosine kinase activity, because the tyrosine phosphatase TC45, a 45 kDa variant of the so-called T-cell protein tyrosine phosphatase (TCPTP), which dephosphorylates this mutant receptor, suppresses receptor-mediated signaling, and the proliferation and anchorage-independent growth of these tumors *(87)*. However, the signaling pathways activated by the hyperactive EGFRvIII vary depending on the cell type, as the proliferative capacity mediated by the MAPK route is operative on both transfected astrocytes and fibroblasts, whereas the phosphoinositide 3-kinase/Akt pathway is only activated in immortalized astrocytes *(88,89)*, a process that induces radioresistance *(88)*. The activation of the phosphoinositide 3-kinase/Akt pathway is also responsible for the downregulation of the cyclin-dependent kinase (CDK) inhibitor p27 and the subsequent increase of CDK2/cyclin A activity in glioblastomas failing to arrest the cell cycle at its G_1 phase in response to serum withdrawal *(89)*.

In contrast, tandem duplication of exons 2–7 results in a 180 kDa receptor named EGFR. Tandem duplication mutant (TDM)/2–7 that has a longer extracellular region, exhibits enhanced basal phosphorylation that is further activated by EGF, and presents impaired downregulation *(79)*. Furthermore, in-frame duplications of exons 18–26 and exons 18–25 result, respectively, in 190- and 185-kDa receptors (EGFR.TDM/18–26 and EGFR.TDM/18–25) that contain the total duplication of the tyrosine kinase domain, and a total or partial duplication of the calcium internalization (CAIN) domain *(90–92)*. The calcium internalization domain, which contains a subdomain denoted calmodulin-like domain (CaM-LD) involved in autoinhibition *(93)*, is required for endocytosis and for raising the cytosolic concentration of free Ca^{2+}. EGFR.TDM/18–26 and EGFR.TDM/18–25 are constitutively autophosphorylated, inefficiently downregulated, and the former has reduced affinity for EGF, although its phosphorylation increases in response to TGF-α, and its internalization appears to occur at the normal rate *(90,91)*.

Potentially useful therapeutic strategies targeting the EGFR have been tested on gliomas that overexpress this receptor. Thus, in a glioblastoma cell line, selective chemical inhibitors of the tyrosine kinase activity, such as the experimental drug ZD1839 (Iressa) that increases the radiosensitivity of these tumor cells, have been used *(94)*. Additionally, neutralizing monoclonal antibodies recognizing specific epitopes of the oncogenic receptor EGFRvIII have been developed, and their potential beneficial effect on glioblastomas and other tumors expressing this variant receptor has been tested *(95,96)*.

ACKNOWLEDGMENTS

Original work in the authors' laboratories was funded by grant (to C.E.) from the Ministerio de Ciencia y Tecnología (Grant No. SAF2002-02131), and grants (to A.V.) from the Ministerio de Ciencia y Tecnología (Grant No. SAF2002-03258), and the Fondo de Investigaciones Sanitarias (Grant No. RTICCC C03/10).

REFERENCES

1. Schlessinger J. Cell signaling by receptor tyrosine kinases. Cell 2002;103:211–225.
2. Jorissen RN, Walker F, Pouliot N, Garrett TPJ, Ward CW, Burgess AW. Epidermal growth factor: mechanisms of activation and signalling. Exp Cell Res 2003;284:31–53.
3. Alroy I, Yarden Y. The ErbB signaling network in embryogenesis and oncogenesis: signal diversification through combinatorial ligand-receptor interactions. FEBS Lett 1997;410:83–86.
4. Harris RC, Chung E, Coffey RJ. EGF receptor ligands. Exp Cell Res 2003;284:2–13.
5. Hunter T, Ling N, Cooper JA. Protein kinase C phosphorylation of the EGF receptor at a threonine residue close to the cytoplasmic face of the plasma membrane. Nature 1984;311:480–483.
6. Countaway JL, Nairn AC, Davis RJ. Mechanism of desensitization of the epidermal growth factor receptor protein-tyrosine kinase. J Biol Chem 1992;267:1129–1140.
7. San José E, Benguría A, Geller P, Villalobo A. Calmodulin inhibits the epidermal growth factor receptor tyrosine kinase. J Biol Chem 1992;267:15,237–15,245.
8. Waterman H, Yarden Y. Molecular mechanisms underlying endocytosis and sorting of ErbB receptor tyrosine kinases. FEBS Lett 2001;490:142–152.
9. Carpenter G. Employment of the epidermal growth factor receptor in growth factor-independent signaling pathways. J Cell Biol 1999;146:697–702.
10. Zwick E, Hackel PO, Prenzel N, Ullrich A. The EGF receptor as central transducer of heterologous signalling systems. Trends Pharm Sci 1999;20:408–412.
11. Kornblum HI, Hussain RJ, Bronstein JM, Gall CM, Lee DC, Seroogy KB. Prenatal ontogeny of the epidermal growth factor receptor and its ligand, transforming growth factor alpha, in the rat brain. J Comp Neurol 1997;380:243–261.

12. Lazar LM, Blum M. Regional distribution and developmental expression of epidermal growth factor and transforming growth factor-α mRNA in mouse brain by a quantitative nuclease protection assay. J Neurosci 1992;12:1666–1 697.
13. Kornblum HI, Zurcher SD, Werb Z, Derynck R, Seroogy KB. Multiple trophic actions of heparin-binding epidermal growth factor (HB-EGF) in the central nervous system. Eur J Neurosci 1999;11:3236–3246.
14. Kaser MR, Lakshmanan J, Fisher DA. Comparison between epidermal growth factor, transforming growth factor-alpha and EGF receptor levels in regions of adult rat brain. Brain Res Mol Brain Res 1992;16:316–322.
15. Farkas LM, Krieglstein K. Heparin-binding epidermal growth factor-like growth factor (HB-EGF) regulates survival of midbrain dopaminergic neurons. J Neural Trans 2002;109:267–277.
16. Weickert CS, Blum M. Striatal TGF-alpha: postnatal developmental expression and evidence for a role in the proliferation of subependymal cells. Brain Res Dev Brain Res 1995;86:203–216.
17. Seroogy KB, Gall CM, Lee DC, Kornblum HI. Proliferative zones of postnatal rat brain express epidermal growth factor receptor mRNA. Brain Res 1995;670:157–164.
18. Opanashuk LA, Mark RJ, Porter J, Damm D, Mattson MP, Seroogy KB. Heparin-binding epidermal growth factor-like growth factor in hippocampus: modulation of expression by seizures and anti-excitotoxic action. J Neurosci 1999;19:133–146.
19. Jin K, Mao XO, Sun Y, et al. Heparin-binding epidermal growth factor-like growth factor: hypoxia-inducible expression in vitro and stimulation of neurogenesis in vitro and in vivo. J Neurosci 2002;22:5365–5373.
20. Nakagawa T, Sasahara M, Hayase Y, et al. Neuronal and glial expression of heparin-binding EGF-like growth factor in central nervous system of prenatal and early-postnatal rat. Brain Res Dev Brain Res 1998;108:263–272.
21. Weickert CS, Webster MJ, Colvin SM, et al. Localization of epidermal growth factor receptors and putative neuroblasts in human subependymal zone. J Comp Neurol 2000;423:359–372.
22. Mishima K, Higashiyama S, Nagashima Y, et al. Regional distribution of heparin-binding epidermal growth factor-like growth factor mRNA and protein in adult rat forebrain. Neurosci Lett 1996;213:153–156.
23. Hayase Y, Higashiyama S, Sasahara M, et al. Expression of heparin-binding epidermal growth factor-like growth factor in rat brain. Brain Res 1998;784:163–178.
24. Planas AM, Justicia C, Soriano MA, Ferrer I. Epidermal growth factor receptor in proliferating reactive glia following transient focal ischemia in the rat brain. Glia 1998;23:120–129.
25. Kawahara N, Mishima K, Higashiyama S, Taniguchi N, Tamura A, Kirino T. The gene for heparin-binding epidermal growth factor-like growth factor is stress-inducible: its role in cerebral ischemia. J Cereb Blood Flow Metab 1999;19:307–320.
26. Tanaka N, Sasahara M, Ohno M, Higashiyama S, Hayase Y, Shimada M. Heparin-binding epidermal growth factor-like growth factor mRNA expression in neonatal rat brain with hypoxic/ischemic injury. Brain Res 1999;827:130–138.
27. Nakatomi H, Kuriu T, Okabe S, et al. Regeneration of hippocampal pyramidal neurons after ischemic brain injury, by recruitment of endogenous neural progenitors. Cell 2002;110:429–441.
28. Teramoto T, Qiu J, Plumier JC, Moskowitz MA. EGF amplifies the replacement of parvalbumin-expressing striatal interneurons after ischemia. J Clin Invest 2003;111:1125–1132.
29. Junier MP, Ma YJ, Costa ME, Hoffman G, Hill DF, Ojeda SR. Transforming growth factor α contributes to the mechanism by which hypothalamic injury induces precocious puberty. Proc Natl Acad Sci USA 1991;88:9743–9747.
30. Lisovoski F, Blot S, Lacombe C, Bellier JP, Dreyfus PA, Junier MP. Transforming growth factor α expression as a response of murine motor neurons to axonal injury and mutation-induced degeneration. J Neuropathol Exp Neurol 1997;56:459–471.
31. Gritti A, Frölichsthal-Schoeller P, Galli R, et al. Epidermal and fibroblast growth factors behave as mitogenic regulators for a single multipotent stem cell-like population from the subventricular region of the adult mouse forebrain. J Neurosci 1999;19:3287–3297.
32. Falk A, Frisen J. Amphiregulin is a mitogen for adult neural stem cells. J Neurosci Res 2002;69:757–762.
33. Liu B, Neufeld AH. Activation of epidermal growth factor receptor causes astrocytes to form cribiform structures. Glia 2004;46:153–168.

34. Tropepe V, Sibilia M, Ciruna BG, Rossant J, Wagner EF, van der Kooy D. Distinct neural stem cells proliferate in response to EGF and FGF in the developing mouse telencephalon. Devel Biol 1999;208:166–188.
35. Reynolds BA, Weiss S. Generation of neurons and astrocytes from isolated cells of the adult mammalian central nervous system. Science 1992;255:1707–1710.
36. Doetsch F, Caille I, Lim DA, García-Verdugo JM, Alvarez-Buylla A. SVZ astrocytes are neural stem cells in the adult mammalian brain. Cell 1999;97:703–716.
37. Gregg C, Weiss S. Generation of functional radial glial cells by embryonic and adult forebrain neural stem cells. J Neurosci 2003; 23:11,587–11,601.
38. Figiel M, Maucher T, Rozyczka J, Bayatti N, Engele J. Regulation of glial glutamate transporter expression by growth factors. Exp Neurol 2003;183:124–135.
39. Zelenaia O, Schlag BD, Gochenauer GE, et al. Epidermal growth factor receptor agonists increase expression of glutamate transporter GLT-1 in astrocytes through pathways dependent on phosphatidylinositol 3- kinase and transcription factor NF-kappaB. Mol Pharmacol 2000;57:667–678.
40. Nolte C, Kirchhoff F, Kettenmann H. Epidermal growth factor is a motility factor for microglial cells in vitro: evidence for EGF receptor expression. Eur J Neurosci 1997;9:1690–1698.
41. Martinez R, Gomes FC. Neuritogenesis induced by thyroid hormone-treated astrocytes is mediated by epidermal growth factor/mitogen-activated protein kinase-phosphatidylinositol 3-kinase pathways and involves modulation of extracellular matrix proteins. J Biol Chem 2002; 277:49,311–49,318.
42. Sibilia M, Steinbach JP, Stingl L, Aguzzi A, Wagner EF. A strain-independent postnatal neurodegeneration in mice lacking the EGF receptor. EMBO J 1998;17:719–731.
43. Kornblum HI, Hussain R, Wiesen J, et al. Abnormal astrocyte development and neuronal death in mice lacking the epidermal growth factor receptor. J Neurosci Res 1998;53:697–717.
44. Junier MP. What role(s) for TGF-α in the central nervous system? Prog Neurobiol 2000;62:443–473.
45. Parati EA, Pozzi S, Ottolina A, Onofrj M, Bez A, Pagano SF. Neural stem cells: an overview. J Endocrinol Invest 2004;27:64–67.
46. Palmer TD, Markakis EA, Willhoite AR, Safar F, Gage FH. Fibroblast growth factor-2 activates a latent neurogenic program in neural stem cells from diverse regions of the adult CNS. J Neurosci 1999;19:8487–8497.
47. Alvarez-Buylla A, Lim DA. For the long run: maintaining germinal niches in the adult brain. Neuron 2004;41:683–686.
48. Gage FH. Mammalian neural stem cells. Science 2000;287: 1433–1438.
49. Temple S. The development of neural stem cells. Nature 2001; 414:112–117.
50. Alvarez-Buylla A, García-Verdugo JM. Neurogenesis in adult subventricular zone. J Neurosci 2002;22:629–634.
51. Doetsch F, García-Verdugo JM, Alvarez-Buylla A. Cellular composition and three-dimensional organization of the subventricular germinal zone in the adult mammalian brain. J Neurosci 1997;17:5046–5061.
52. Doetsch F, Petreanu L, Caille I, García-Verdugo JM, Alvarez-Buylla A. EGF converts transit-amplifying neurogenic precursors in the adult brain into multipotent stem cells. Neuron 2002;36:1021–1034.
53. Kuhn HG, Winkler J, Kempermann G, Thal LJ, Gage FH. Epidermal growth factor and fibroblast growth factor-2 have different effects on neural progenitors in the adult rat brain. J Neurosci 1997;17:5820–5829.
54. Craig CG, Tropepe V, Morshead CM, Reynolds BA, Weiss S, van der Kooy D. In vivo growth factor expansion of endogenous subependymal neural precursor cell populations in the adult mouse brain. J Neurosci 1996;16:2649–2658.
55. Fricker-Gates RA, Winkler C, Kirik D, Rosenblad C, Carpenter MK, Björklund A. EGF infusion stimulates the proliferation and migration of embryonic progenitor cells transplanted in the adult rat striatum. Exp Neurol 2000;165:237–247.
56. Tropepe V, Craig CG, Morshead CM, van der Kooy D. Transforming growth factor-α null and senescent mice show decreased neural progenitor cell proliferation in the forebrain subependyma. J Neurosci 1997;17:7850–7859.
57. Junier MP, Hill DF, Costa ME, Felder S, Ojeda SR. Hypothalamic lesions that induce female precocious puberty activate glial expression of the epidermal growth factor response gene: differential regulation of alternatively spliced transcripts. J Neurosci 1993;13:703–713.

58. Shah BH, Farshori P, Catt KJ. Neuropeptide-induced transactivation of a neuronal epidermal growth factor receptor is mediated by metalloprotease-dependent formation of heparin-binding epidermal growth factor. J Biol Chem 2004;279:414–420.
59. Prenzel N, Zwick E, Daub H, et al. EGF receptor transactivation by G-protein-coupled receptors requires metalloprotease cleavage of proHB-EGF. Nature 1999;402:884–888.
60. Gschwind A, Hart S, Fischer OM, Ullrich A. TACE cleavage of proamphiregulin regulates GPCR-induced proliferation and motility of cancer cells. EMBO J 2003;22:2411–2421.
61. Justicia C, Perez-Asensio FJ, Burguete MC, Salom JB, Planas AM. Administration of transforming growth factor-α reduces infarct volume after transient focal cerebral ischemia in the rat. J Cereb Blood Flow Metab 2001;21:1097–1104.
62. Liu J, Solway K, Messing RO, Sharp FR. Increased neurogenesis in the dentate gyrus after transient global ischemia in gerbils. J Neurosci 1998;18:7768–7778.
63. Jin K, Minami M, Lan JQ, et al. Neurogenesis in dentate subgranular zone and rostral subventricular zone after focal cerebral ischemia in the rat. Proc Natl Acad Sci USA 2001;98:4710–4715.
64. Parent JM, Yu TW, Leibowitz RT, Geschwind DH, Sloviter RS, Lowenstein DH. Dentate granule cell neurogenesis is increased by seizures and contributes to aberrant network reorganization in the adult rat hippocampus. J Neurosci 1997;17:3727–3738.
65. Magavi SS, Blair R, Leavitt BR, Macklis JD. Induction of neurogenesis in the neocortex of adult mice. Nature 2000;405:951–955.
66. Arvidsson A, Collin T, Kirik D, Kokaia Z, Lindvall O. Neuronal replacement from endogenous precursors in the adult brain after stroke. Nat Med 2002;8:963–970.
67. Fallon J, Reid S, Kinyamu R, et al. In vivo induction of massive proliferation, directed migration, and differentiation of neural cells in the adult mammalian brain. Proc Natl Acad Sci USA 2000;97: 14,686–14,691.
68. Baugnet-Mahieu L, Lemaire M, Brotchi J, et al. Epidermal growth factor receptors in human tumors of the central nervous system. Anticancer Res 1990;10:1275–1280.
69. Hall WA, Merrill MJ, Walbridge S, Youle R. Epidermal growth factor receptors on ependymomas and other brain tumors. J Neurosurg 1990;72:641–666.
70. Reifenberger G, Reifenberger J, Bilzer T, Wechsler W, Collins VP. Coexpression of transforming growth factor-α and epidermal growth factor receptor in capillary hemangioblastomas of the central nervous system. Am J Pathol 1995;147:245–250.
71. Meyers MB, Shen WP, Spengler BA, et al. Increased epidermal growth factor receptor in multidrug-resistant human neuroblastoma cells. J Cell Biochem 1988;38:87–97.
72. Tong CY, Ng HK, Pang JC, Hu J, Hui AB, Poon WS. Central neurocytomas are genetically distinct from oligodendrogliomas and neuroblastomas. Histopathology 2000;37:160–165.
73. Reifenberger J, Reifenberger G, Ichimura K, Schmidt EE, Wechsler W, Collins VP. Epidermal growth factor receptor expression in oligodendroglial tumors. Am J Pathol 1996;149:29–35.
74. Bloomer CW, Kenyon L, Hammond E, et al. Cyclooxygenase-2 (COX-2) and epidermal growth factor receptor (EGFR) expression in human pituitary macroadenomas. Am J Clin Oncol 2003;26: S75–S80.
75. Mouihate A, Lestage J. Epidermal growth factor: a potential paracrine and autocrine system within the pituitary. Neuroreport 1995;6: 1401–1404.
76. Frederick L, Wang X-Y, Eley G, James CD. Diversity and frequency of epidermal growth factor receptor mutations in human glioblastomas. Cancer Res 2000;60:1383–1387.
77. Broholm H, Bols B, Heegaard S, Braendstrup O. Immunohistochemical investigation of p53 and EGFR expression of oligodendrogliomas. Clin Neuropathol 1999;18:176–180.
78. Huang H, Okamoto Y, Yokoo H, et al. Gene expression profiling and subgroup identification of oligodendrogliomas. Oncogene 2004;23: 6012–6022.
79. Fenstermaker RA, Ciesielski MJ. Deletion and tandem duplication of exons 2–7 in the epidermal growth factor receptor gene of a human malignant glioma. Oncogene 2000;19:4542–4548.
80. Eley G, Frederick L, Wang X-Y, Smith DI, James CD. 3′ end structure and rearrangements of EGFR in glioblastomas. Genes Chromosomes Cancer 1998;23:248–254.
81. Sugawa N, Ekstrand AJ, James CD, Collins VP. Identical splicing of aberrant epidermal growth factor receptor transcripts from amplified rearranged genes in human glioblastomas. Proc Natl Acad Sci USA 1990;87:8602–8606.
82. Nishikawa R, Ji X-D, Harmon RC, et al. A mutant epidermal growth factor receptor common in human glioma confers enhanced tumorigenicity. Proc Natl Acad Sci USA 1994;91:7727–7731.

83. Sugawa N, Yamamoto K, Ueda S, et al. Function of aberrant EGFR in malignant gliomas. Brain Tumor Pathol 1998;15:53–57.
84. Ding H, Shannon P, Lau N, et al. Oligodendrogliomas result from the expression of an activated mutant epidermal growth factor receptor in a RAS transgenic mouse astrocytoma model. Cancer Res 2003;62:1106–1113.
85. Prigent SA, Nagane M, Lin H, et al. Enhanced tumorigenic behavior of glioblastoma cells expressing a truncated epidermal growth factor receptor is mediated through the Ras-Shc-Grb2 pathway. J Biol Chem 1996;271:25,639–25,645.
86. Feldkamp MM, Lala P, Lau N, Roncari L, Guha A. Expression of activated epidermal growth factor receptors, Ras-guanosine triphosphate, and mitogen-activated protein kinase in human glioblastoma multiforme specimens. Neurosurgery 1999;45:1442–1453.
87. Klingler-Hoffmann M, Fodero-Tavoletti MT, Mishima K, et al. The protein tyrosine phosphatase TCPTP suppresses the tumorigenicity of glioblastoma cells expressing a mutant epidermal growth factor receptor. J Biol Chem 2001;276:46,313–46,318.
88. Li B, Yuan M, Kim I-A, Chang C-M, Bernhard EJ, Shu H-KG. Mutant epidermal growth factor receptor displays increased signaling through the phosphatidylinositol-3 kinase/AKT pathway and promotes radioresistance in cells of astrocytic origin. Oncogene 2004;23:4594–4602.
89. Narita Y, Nagane M, Mishima K, Huang H-JS, Furnari FB, Cavenee WK. Mutant epidermal growth factor receptor signaling down-regulates p27 through activation of the phosphatidylinositol 3-kinase/Akt pathway in glioblastomas. Cancer Res 2002;62:6764–6769.
90. Fenstermaker RA, Ciesielski MJ, Castiglia GJ. Tandem duplication of the epidermal growth factor receptor tyrosine kinase and calcium internalization domains in A-172 glioma cells. Oncogene 1998;16: 3435–3443.
91. Ciesielski MJ, Fenstermaker RA. Oncogenic epidermal growth factor receptor mutants with tandem duplication: gene structure and effects on receptor function. Oncogene 2000;19:810–820.
92. Arjona D, Bello MJ, Alonso ME, et al. Molecular analysis of the erbB gene family calmodulin-binding domain (CaM-BD) and calmodulin-like domain (CaM-LD) in astrocytic gliomas. Int J Oncol 2004;25:1489–1494.
93. Martín-Nieto J, Cusidó-Hita DM, Li H, Benguría A, Villalobo A. Regulation of ErbB receptors by calmodulin. In: Pandalai SG (ed). Recent Research Developments in Biochemistry. Trivandrum: Research Signpost, 2002, Vol 3, Pt I, pp 41–58.
94. Stea B, Falsey R, Kislin K, et al. Time and dose-dependent radiosensitization of glioblastoma multiforme U251 cells by the EGF receptor tyrosine kinase inhibitor ZD1839 ('Iressa'). Cancer Lett 2003;202:43–51.
95. Wikstrand CJ, Hale LP, Batra SK, et al. Monoclonal antibodies against EGFRvIII are tumor specific and react with breast and lung carcinomas and malignant gliomas. Cancer Res 1995;55:3140–3148.
96. Wikstrand CJ, Reist CJ, Archer GE, Zalutsky MR, Bigner DD. The class III variant of the epidermal growth factor receptor (EGFRvIII): characterization and utilization as an immunotherapeutic target. J Neurovirol 1998;4:148–158.

V
CELL CYCLE RE-ENTRY: A MECHANISM OF BRAIN DISEASE?

21
Neurodegeneration and Loss of Cell Cycle Control in Postmitotic Neurons

Randall D. York, PhD, Samantha A. Cicero, and Karl Herrup, PhD

SUMMARY

The concept of cell cycle-related neuronal death (CRND) has now been in the literature for more than a decade. The first evidence linking the loss of cell cycle control with neuronal death arose from studies in transgenic mice in which cell cycle progression was artificially induced in postmitotic neurons during development. Substantial neuronal loss was observed in each instance, suggesting that differentiated neurons make a commitment to maturation that includes a permanent cessation of cell division. In agreement, naturally occurring neuronal death during development involves a re-expression of several cell cycle markers and DNA synthesis. Today, a growing literature supports the view that cell cycle induction in adult neurons may be equally lethal. A large fraction of this literature is based on correlative observations of cell cycle-associated proteins in neurons at risk for death in a variety of neurodegenerative diseases and central nervous system insults. Our intent is not to rehash this information yet again, but instead to discuss the limitations of current approaches and identify gaps in our present knowledge. In doing so, we have attempted to highlight promising avenues of future investigation, which may bring clarity to the realities regarding CRND in multiple neurodegenerative disorders.

Key Words: Neurodegenerative disease; neuronal death; Alzheimer's disease; stroke; ALS; cyclin-dependent kinase; oxidative stress; DNA damage; inflammation; presenilin.

1. INTRODUCTION: CELL CYCLE IN DIFFERENTIATED NEURONS

During the past few years, several excellent reviews have discussed specific aspects of cell cycle regulation in neurodegenerative disease *(1–8)*. In the developing mammalian central nervous system (CNS), neuronal precursors begin to differentiate as they exit the cell cycle and migrate out of the proliferative ventricular zone (for review *see* ref. 9). The coordinate processes of differentiation and migration culminate in proper positioning of neurons in the brain where they exist in a mitotically quiescent state for the remainder of their life. In the normal adult brain, nonneuronal cells proliferate in response to CNS damage. In contrast, differentiated neurons are characteristically postmitotic and show no capacity to replenish lost or injured cells. Neurogenesis may occur in specific regions of the adult CNS, but this involves select neural stem cell pools *(10)*. Nonetheless, the long-enduring dogma that a postmitotic neuron permanently desists from cell cycle processes requires modification.

Generally, differentiated neurons downregulate cell cycle proteins and lose their proliferative response to the multitude of otherwise mitogenic stimuli in the extracellular milieu. On the other hand, an overwhelming body of evidence now indicates that postmitotic neurons retain the capacity to reactivate cell cycle-associated proteins (CCPs) under certain conditions. This aberrant

From: *The Cell Cycle in the Central Nervous System*
Edited by: D. Janigro © Humana Press Inc., Totowa, NJ

neuronal cell cycle re-entry, however, is typically associated with cell death rather than proliferation *(3,11–13)*, which may explain the rare incidence of neuronal tumors *(2)*. This view is clearly illustrated by mouse models in which forced cell cycle re-entry in young neurons of the developing nervous system is intimately linked to apoptotic cell death.

1.1. Mouse Models of Neuronal Cell Cycle Re-Entry and Cell Loss During Development

During the last decade, the ability to make transgenic mice and analyze naturally occurring genetic mutants has provided clues to the role of cell cycle activity in neuronal death during development. For instance, a series of mouse models centered round the Rb/E2F pathway demonstrate the resistance of differentiated neurons to transformation and clearly show that cell cycle reactivation can lead to apoptosis in vivo. It has long been recognized that the retinoblastoma gene product (Rb) and related members of the pocket protein family (p107, p130) function to prevent cell cycle progression and S phase initiation by repressing gene expression through interactions with the E2F family of transcription factors. This repression is accomplished by either blocking the E2F transactivation domain or by recruiting chromatin-modifying enzymes to E2F-binding sites *(12)*. Following mitotic pressure, hyperphosphorylation of Rb liberates E2F, which activates a set of genes required for DNA replication initiation and S phase progression *(14,15)*.

The ability of SV40 large T-antigen (Tag) to sequester and neutralize the function of Rb family members provided the first indication that unscheduled cell cycle re-entry leads to the death of postmitotic neurons in vivo. First, using the opsin promoter to drive expression in rod photoreceptors of transgenic mice, Al-Ubaidi and colleagues showed that SV40 Tag causes degeneration of postmitotic retinal neurons *(16)*. Photoreceptors in these animals express proliferating cell nuclear antigen (PCNA) and synthesize DNA prior to apoptotic death detected by terminal deoxynucleotidyl transferase dUTP nick-end labeling (TUNEL) and DNA laddering *(17)*. A similar induction of DNA synthesis and fragmentation is seen in SV40 Tag-expressing cerebellar Purkinje cells prior to their death *(18,19)*. In support of the proposed role of the Rb/E2F pathway, overexpression of E2F1 in Purkinje cells accelerates SV40 Tag-induced cell loss and ataxia in this model system. Likewise, several labs have shown that direct disruption of Rb itself leads to neuronal death associated with changes in cell cycle regulation *(20–22)*. That is, mice deficient in Rb ($Rb^{-/-}$) display ectopic cell cycle entry and mitotic figures associated with increased neuronal loss. Macleod et al. later showed that the ectopic cell cycle was accompanied by an increased activity of free E2F and activation of the E2F-responsive cell cycle regulator, cyclin E *(23)*. Together, these mouse models indicate that the correlation between unscheduled cell cycle re-entry and cell death occurs commonly in multiple types of differentiated neurons.

In addition to forced cell cycle re-entry, a second series of transgenic mouse models shows that physiological stimuli including chronic depolarization and trophic factor withdrawal during cerebellar development also induce cell cycle-related neuronal death (CRND). In the mouse mutant *staggerer* ($rora^{sg/sg}$), deletion of the orphan nuclear receptor, RORα, leads to the loss of 100% of the cerebellar granule neurons owing to deficient trophic support from Purkinje cells that never fully develop *(24–26)*. In the $rora^{sg/sg}$ cerebellum, differentiated neurons of the internal granule layer (IGL) incorporate bromodeoxyuridine (BrdU) into newly synthesized DNA prior to their loss *(27)*. This DNA synthesis is accompanied by expression of the G_1 and S phase markers cyclin D and PCNA. *Lurcher* mice display a remarkably similar phenotype. In the lurcher mouse, ataxia develops following continuous depolarization and subsequent death of Purkinje cells resulting from a point mutation in the gene encoding an ionotropic glutamate receptor (GRID2) *(28)*. Purkinje cell death begins around the second postnatal week in lurcher mice and continues until about the fourth postnatal week when nearly all Purkinje cells are lost without inducing cell cycle events *(27)*. Progressive Purkinje cell loss is accompanied by the death of most (approx 90%) of the cerebellar granule neurons, which incorporate BrdU and express CCPs prior to dying *(27)*. This model is different from the staggerer mutant in that the cell-autonomous loss of Purkinje cells occurs after completion of development. What is particularly striking is that the developmental stage of Purkinje cell loss does not change the expression of cell cycle proteins and the incorporation of BrdU in granule cells, suggesting that adult neurons may also be vulnerable to CRND.

Together, these mouse models suggest that CRND may play an important role in developmental disorders of the mammalian CNS. However, as others have pointed out, it is worth noting that neuronal DNA synthesis is rarely associated with naturally occurring death during normal development. Therefore, it may be that neurons in healthy individuals repress DNA synthesis, even during apoptosis, and that deregulation of these mechanisms occurs only under pathological conditions.

1.2. Re-Expression of CCPs in Neurodegenerative Disorders

Ample evidence now indicates that cell cycle re-entry precedes apoptosis in neurodegenerative diseases, as well. In Alzheimer's disease (AD) Busser et al. *(29)* have shown that there is increased expression of PCNA, cyclin B_1, cyclin D, and cyclin-dependent kinase (CDK)4 in pyramidal neurons of the hippocampus, in noradrenergic neurons of the locus ceruleus, and in the neurons of the dorsal raphe nucleus from demented patients compared with age-matched controls. These areas of the brain are highly affected by cell loss in AD. In the cerebellum, which is not affected in AD, there is no expression of cell cycle-related proteins. This study indicates that ectopic expression of cell cycle proteins predicts the populations of neurons that are susceptible to death. In addition to the presence of these proteins, Vincent and co-workers have reported an increased activity of the cdc2/cyclin B_1 complex and its activating phosphatase, cdc25, in the AD brain *(30–32)*.

The ectopic expression of cell cycle-related proteins in neurodegenerative disorders is not limited to AD. In fact, the expression of cell cycle markers has also been shown in the brain of patients with Parkinson's disease (PD), amyotrophic lateral sclerosis (ALS), Down's syndrome, Pick's disease, Niemann-Pick syndrome type C, and following focal ischemia and traumatic brain injury *(1,2,5,11,33,34)*. Even though CCPs are re-expressed in a variety of human neurodegenerative diseases, neuronal proliferation is never observed. This suggests that cell cycle re-entry under pathological conditions in the adult may lead to neuronal death rather than proliferation in a manner analogous to aberrant re-entry during development. The mechanisms preventing neurons from entering mitosis in the presence of cell cycle activators is not known.

2. IS IT A BONA FIDE CELL CYCLE?

2.1. Cell Cycle Progression In Vivo

Even though cell cycle reactivation has emerged as a unifying theme for many neurodegenerative diseases, the function of ectopic "cell cycle proteins" is not well understood. Among the most pressing questions that remain unanswered is whether the reactivation of CCPs and DNA synthesis in degenerating neurons is indicative of a bonafide cell cycle. As discussed, a large fraction of the literature is based on correlative observations of CCPs in vulnerable neurons. These include many examples of upregulated G_1, S, and G_2 markers *(29,30,32,35–37)*. In contrast, examples of M phase mitotic markers are limited *(37)* and in vivo evidence for mitotic figures or cytokinesis has not been reported in diseased brains. These patterns suggest the possibility that degenerating neurons are capable of progressing to G_2, but fail to enter mitosis.

However, the mere presence of a protein does not indicate function. Therefore, although these in vivo descriptions are valuable, they do not address whether re-expressed CCPs actually set the cycle in motion. Furthermore, there is a disturbing paucity of reports describing double- and triple-label experiments to address whether the expression of cell cycle proteins is regulated in an orderly fashion. For instance, if G_1, S, and G_2 markers were present in the same cell, this would imply a dysregulated cycle at best. Regardless of molecular signature, all cell cycles are structured around DNA replication (S phase) and division (mitosis). Because mitotic figures are lacking, several groups have focused on whether postmitotic neurons enter S phase by attempting to monitor DNA replication in vivo. One approach has been to inject animals with BrdU and assess incorporation into newly synthesized DNA by immunohistochemistry. Using this technique, convincing evidence for DNA synthesis has been reported for several CRND-associated rodent models of neurodegeneration including target deprivation *(27)*, oxidative stress *(38)*, Parksinonian-like dopaminergic degeneration *(37)*, and focal ischemia *(39)*. These findings clearly demonstrate that DNA synthesis occurs in some cases of CRND. However, the mode of DNA synthesis remains uncertain for each. There is reasonable evidence that normal DNA repair is unlikely to provide the signal intensity observed under the conditions of these experiments. However, an unprecedented and massive DNA repair process specific to neurodegeneration remains possible. In AD, fluorescence *in situ* hybridization (FISH) studies have shown that a significant fraction of the diseased hippocampal pyramidal and basal forebrain neurons fully or partially replicate their genome, suggesting that neurons actually progress to late stages of a cell cycle in vivo *(36)*. The primary limitation of this type of analysis is that it cannot exclude the unlikely possibility that developmentally acquired aneuploidy, as suggested by Chun and colleagues *(40–42)*, coincidentally makes those or neighboring neurons susceptible to re-expression of cell cycle proteins later in life. More recently, FISH analyses have been extended to mouse models of AD *(43)*. Importantly, in these genetic models, the increased frequency of genomic duplications seen late in life is not detected in young adult mice and therefore cannot represent derivatives of aneuploid neuronal precursors. The most parsimonious model resulting from in vivo studies predicts that degenerating neurons re-enter a cell cycle, sometimes progress beyond S phase, and arrest or die prior to entering mitosis.

2.2. Cell Cycle Progression In Vitro

A similar cell cycle re-entry has been shown in primary cortical and cerebellar cultures treated with numerous disease-related stimuli *(44–50)*. In vitro, neurons re-entering a cell cycle appear to die at different phases depending on the inducing stimuli. For example, following excitotoxicity and DNA damaging agents, a significant fraction of cerebellar granule neurons seemingly cross the G_1/S transition and incorporate BrdU prior to dying *(47)*. On the other hand, activity withdrawal, achieved by culturing neurons in low-potassium media, leads to expression of G_1 markers but not BrdU incorporation *(51)*. An apparent progression beyond the G_1/S phase transition has also been detected in vitro following DNA damage and Aβ-stimulated CRND of cortical neurons *(52,53)*. Here, BrdU incorporation indicates DNA synthesis and fluorescence-activated cell sorting (FACS) suggests partial DNA replication and S phase entry. It should be noted that the presumed S phase indicated by a DNA complement ≥2N could theoretically result from the death and DNA fragmentation of G_2 phase or polyploid cells. Despite this caveat, the combination of BrdU detection and fluorescence-activated cell sorting data strongly suggests that DNA is at least partially replicated in these primary culture models of neurodegeneration. Again, the mode of DNA synthesis is not clear. Interestingly, an increase in the fraction of cells with a 4N complement (G_2/M phase) is not observed in vitro, as might be expected from the expression of G_2 markers and FISH data, indicating near-complete duplication of the genome in vivo. Taken together, these findings suggest that death may occur at multiple phases following cell cycle re-entry. It follows that cell type stimulus- and context-dependent differences in the relationship between cell cycle progression and cell death likely contribute to the complexity of human diseases.

2.3. Unconventional Regulation and Targets of CCPs in Degenerating Neurons

To further complicate matters, postmitotic neurons may regulate conventional cell cycle proteins in atypical ways. For example, CCPs are often reported to be "mislocalized" in degenerating neurons *(11,29,30,35,54)*. Given the complications associated with preparing human postmortem tissue, it is often difficult to determine the relevance of these staining patterns to disease. However, for some, similar patterns are also seen in rodent models that allow fixation by perfusion. This aberrant subcellular localization of CCPs may reflect unique functions specific to neuronal death. A well-characterized example involves the mitotic kinase CDK1 (a.k.a. cdc2), which is expressed during G_2/M of a normal cell cycle. Aberrant expression of CDK1 is apparent in ischemic brain and a variety of neurodegenerative diseases including AD, Down's syndrome, and Pick's disease *(11,29,30,55)*. In many cases, CDK1 is localized to the cytoplasm and has been proposed to impart negative effects through the phosphorylation of microtubule-associated proteins such as τ and microtubule associated protein 1B (MAP1b) *(11,29,30,56)*. The BH3-only protein, BCL2 antagonist of cell death (BAD), has recently been identified as an additional cytoplasmic target of CDK1 following activity withdrawal of primary cerebellar granule neurons. In this culture model, CDK1-mediated phosphorylation of BAD disrupts its interaction with 14-3-3 proteins and thereby promotes the apoptotic activity of BAD *(53,57)*. This action of CDK1 is reminiscent of its proposed role upstream of mitochondrial changes in Taxol-induced death of cancer cells *(58–60)*. It remains to be determined whether a similar function of CDK1 contributes to apoptosis following other CRND-inducing stimuli or whether CDK1 targets proapoptotic proteins in vivo. If so, this unusual substrate of CDK1, may explain why neurons that enter G_2 phase die without entering mitosis. It seems such a mechanism could well contribute to relatively rapid death processes such as those in parts of ischemic brain, or following DNA damage and Aβ-induced CRND in vitro *(52,53,61)*. In a chronic neurodegenerative disease such as AD, however, neurons are estimated to express CDK1 and other CCPs for many months prior to their death *(3,35,36)*. The mechanisms that prevent or delay the CDK1/BAD apoptotic pathway in these neurons are not clear. It is worth noting, however, that although CDK1 is indeed found in the cytoplasm in many neurodegenerative diseases, its classical activator, cyclin B, is often detected exclusively in the nucleus *(11,62,63)*.

Abnormally localized cyclins may also contribute to unusual functions in degenerating neurons. For example, nuclear translocation of cyclin D_1 may regulate neuronal death and survival, particularly in AD. In a normally cycling cell, cyclin D_1 accumulates in the nucleus during G_1 and is exported to the cytoplasm during S phase where it is targeted for degradation by proteolysis *(64–66)*. Cyclin D-associated kinase activity is markedly downregulated in differentiated neurons *(67)* and cyclin D_1 levels and associated kinase activities are induced in degenerating neurons undergoing CRND *(1,27,29,49,50,68–70)*. In unstimulated cultures, some cyclin D_1 remains in differentiated neurons where it localizes exclusively to the cytoplasmic fraction *(54,71)*. Interestingly, several CRND-inducing stimuli lead to the nuclear accumulation of cyclin D_1 that is accompanied by phosphorylation of Rb and subsequent death *(50,54,68,70,72)*. Interfering mutants of the cyclin D-dependent kinases (dnCdk4/dnCdk6) prevent Rb phosphorylation in the nucleus, suggesting that nuclear translocation may be important. Finally, overexpression of cyclin D_1 with a fused nuclear localization signal (NLS), but not wild-type cyclin D_1, efficiently kills differentiated cortical neurons *(54)*. Together, these in vitro studies suggest that differentiated neurons normally prevent nuclear accumulation of cyclin D_1 and that neuronal insults abrogate cytoplasmic retention to trigger cell cycle reactivation and cell death via activation of Rb-mediated pathways. Interestingly, induced cyclin D_1 detected in neurodegeneration models in vivo is often localized to the cytoplasm *(27)*. In human patients, much of the induced cyclin D_1 in both AD and mild cognitive impairment is found in the cytoplasm *(35)*. As mentioned in the previous paragraph, cyclin D_1 positive neurons in AD are predicted to remain intact for several months to years prior to dying. It is possible that cytoplasmic retention of the induced cyclin D protects these neurons for some time before additional insults lead to nuclear accumulation, Rb phosphorylation, and death. This hypothesis suggests another example in which the induction of CCPs can be separated in time from apoptosis and is consistent with a multiple hit theory for CRND in neurodegenerative disease. Other CRND-associated cyclins are also found in the cytoplasm, yet little is known about their function and regulation in disease brains. One interesting possibility is that the commonly observed cytoplasmic localization of cyclins allows them to interact with "noncell cycle" proteins resulting in novel functions for each.

3. WHAT TRIGGERS CELL CYCLE RE-ENTRY?

Regardless of the nature of cell cycle progression, understanding the factors initiating cell cycle reactivation and identifying commonalities among diseases seems crucial. Neurodegenerative diseases are defined by the characteristic loss of select populations of neurons in distinct regions of the brain. These patterns of loss may reflect initiating signals unique to each particular disorder. On the other hand, inflammation, oxidative stress, DNA damage, and re-expression of cell cycle proteins are common to many forms of neurodegeneration. In the next section, we briefly consider a hypothesized role of these shared features as points of convergence for disease-specific initiating signals leading to CRND.

3.1. Inflammation

A major interest in our laboratory has been to understand the role of inflammatory responses in neurodegenerative diseases, particularly in AD. AD brains exhibit many hallmarks of inflammation including elevated levels of immune receptors and cell surface proteins, cytokines, and activated microglia (for reviews *see* refs. *73–75*). These observations suggest that a local inflammatory response within the AD brain contributes to the pathophysiology of the disease. This view is strongly supported by persuasive epidemiological studies showing that patient populations treated with nonsteroidal anti-inflammatory drugs (NSAIDs) exhibit a more than 50% decreased risk of AD *(76–78)*. AD patients receiving long-term NSAID therapy exhibit later onset, less severe symptoms, and a slowed rate of cognitive decline *(79)*. NSAID-treated patients also show a 65% reduction in senile plaque-associated reactive microglia *(80)* suggesting that amyloid-β (Aβ) activated microglia may be the principle target of NSAID action. Aβ is sufficient to convert microglia into a reactive phenotype through a unique cell surface receptor complex *(81,82)* that leads to the secretion of a diverse array of bioactive molecules *(83–85)*. Among these secreted molecules, proinflammatory cytokines and mitogens result in a feed-forward activation of microglia and neighboring astrocytes. Ultimately, this local inflammatory environment triggers the death of resident neurons through unresolved mechanisms. Wu et al. demonstrated that mouse cortical neurons exposed to conditioned media from Aβ-activated microglia increased the number of neurons positive for cell cycle markers such as PCNA, cyclin D, and BrdU incorporation *(48)*. Double-labeling with BrdU and terminal deoxyneudeotidyl transferace dUTP nick end labeling suggested that the "cycling" neurons were indeed dying via apoptosis. This study implies that secreted factors from reactive microglia may be the primary source of mitogens that force cell cycle re-entry of postmitotic neurons in AD. This view is supported by the coincidence of inflammatory markers and CCPs in patients at early stages of disease, before the increased deposition of Aβ plaques *(35,86,87)*. The presence of inflammation prior to plaque formation suggests that plaque-associated microglia likely contribute to the progression of disease, but that Aβ oligomers or preplaque fibrils may activate microglia at early stages of disease. It is also plausible that Aβ-independent signals are the initial source of inflammatory stimuli in the earliest stages of AD brains.

The identities of the soluble factor(s) elaborated by the microglia that are responsible for cell cycle re-entry remain unknown. It seems likely that multiple mechanisms may be involved and that local cues will influence the balance between maintained quiescence and cell cycle re-entry. However, as discussed in the following section, the generation of reactive oxygen species (ROS) is likely to be a key factor. Neurotoxic ROS result from Aβ-stimulated NADPH oxidase and subsequent respiratory burst in microglia *(88–90)*. This inflammatory response is not restricted to AD. In fact, there is now substantial evidence that inflammation via microglia activation is a significant event in many CRND-associated neurodegenerative disorders including stroke *(91)*, ALS *(92)*, and PD *(75)*. As is the case for AD, reactive microglia are also observed in several PD mouse models of dopaminergic cell death *(75)*, at least one of which has revealed a critical role of microglial NADPH oxidase *(93)*. Furthermore, ALS mouse models harboring mutations in superoxide dismutase 1 (SOD1) express neuronal CDK4, cyclin D, and phosphorylated Rb that is attenuated by the microgliosis inhibitor, minocycline *(94)*.

3.2. Oxidative Stress

Under normal conditions, the generation of ROS is counteracted by antioxidants (e.g., vitamin E, ascorbic acid, and carotenoids) and detoxifying enzymes such as glutathione peroxidase and catalase. Oxidative stress occurs when the antioxidant capacity of the cell is overcome by the generation of excess ROS leading to damaged lipids, proteins, and DNA *(95)*. Similar to the re-expression of CCPs, markers of oxidative stress including lipid peroxidation and protein or DNA oxidation are found in patients with a variety of neurodegenerative diseases (for reviews see refs. *8,96–99*, and *51*), as well as rodent models of AD *(53,100)* and PD *(101,102)*. Like CCPs, whether the presence of oxidative stress contributes to neuronal death or is merely a consequence of end-stage disease has long been debated. Several in vitro studies have implicated ROS in early stages of neuronal death *(8,103)*, but evidence in vivo had been lacking. The harlequin (*Hq*) mouse mutant now provides convincing evidence to support a causative role of oxidative stress in cell cycle re-entry and neuronal death.

Hq mice display progressive ataxia owing to neurodegeneration of cerebellar granule neurons *(38)*. Unlike the staggerer and lurcher mouse mutants described above, loss of cerebellar granule neurons occurs during aging in *Hq* mice rather than early postnatal development and precedes any loss of Purkinje cells. Aged *Hq* animals also show progressive neurodegeneration in the retina beginning in the ganglion cell layer and resulting in hypoplasia of the optic nerve and further loss of retinal neurons. Using a positional cloning strategy, *Hq* was shown to result from insertion of a murine ecotropic leukemia virus into the apoptosis-inducing factor gene on the X chromosome. Despite its name from overexpression studies, apoptosis-inducing factor may prevent apoptosis under normal conditions by maintaining free radical homeostasis *(104,105)*. Viral insertion causes an 80% reduction in apoptosis-inducing factor activity that correlates with increased levels of lipid peroxidation and catalase activity. Again like CCPs, markers of oxidative stress are found months before the onset of neuronal death, indicating a role early in development or progression of neurodegeneration. Importantly, S phase proteins and BrdU incorporation steadily increase in aging *Hq* cerebellar granule neurons and retinal ganglion, amacrine, and horizontal cells. Most of these cells are positive for activated caspase-3 and display pycnotic nuclei without detectable mitotic figures, suggesting CRND owing to oxidative stress. Interestingly, cortical neurons do not seem to be affected, whereas other neuronal types including retinal photoreceptors and cerebellar Purkinje cells die without expressing CCPs. Clearly, distinct neuronal populations respond differently to oxidative stress. This may be largely owing to intrinsic properties of the different neuronal types because a similar selective sensitivity to oxidative stress is seen among cultured neurons in vitro *(106)*.

Currently, it remains unclear how oxidative stress stimulates some quiescent neurons to exit their G_0 state and re-enter the cell cycle. A number of mitogenic signaling pathways have been implicated downstream of ROS that include activation of growth factor receptors, small GTPases, proteins, and lipid kinase cascades including the mitogen-activated protein kinase and AKT/phosphoinositide 3′-kinase pathways, and transcription factors such as p53 and NFκB (for review, *see* ref. *8*). Klein and Ackerman presented an interesting model based on the ability of oxidative stress to reduce histone deacetylase (HDAC) activity in nonneuronal cells *(107)*. Although a direct connection between oxidative stress and HDAC activity in neurons has not been established, HDAC inhibitors do induce E2F1 activity and a corresponding rise in E2F-responsive cell cycle (cyclin E) and proapoptotic (APAF-1 and capsase-3) proteins in cerebellar granule neurons *(108,109)*. It is therefore quite plausible that oxidative stress-induced inhibition of HDAC activity contributes to the derepression of normally silenced genes in a manner complimentary to hyperphosphorylation of Rb. Examining the activities of HDACs in *Hq* cerebellar granule neurons in culture and in vivo may provide insight into whether the concept of oxidative stress-induced HDAC regulation augments the derepression model proposed by Liu and Greene *(12)*. In any event, the *Hq* model is the first to show both cell loss with respect to age and oxidative stress in vivo and therefore holds great promise for identifying the mechanisms linking ROS to cell cycle re-entry and neuronal death.

3.3. DNA Damage

Many forms of oxidative stress-induced cell damage can lead to apoptosis. However, nucleic acids are likely a major target contributing to neuronal loss in neurodegenerative disease. Evidence for increased DNA oxidation in the form of the modified base, 8-hydroxydeoxyguanosine (8-OHdG), is found in both the mitochondria and nucleus of neurons in AD *(110,111)*, ALS *(112)*, and ischemic brains *(61,113–115)*. Oxidative damage can also lead to changes in DNA structure via strand breaks and rearrangements or deletions, each of which has been documented in degenerating neurons *(116,117)*. In the *Hq* mouse, virtually every cycling neuron expressing S phase markers was also positive for 8-OHdG, correlating oxidative stress-induced DNA damage with CRND. However, not all 8-OHdG-positive cells were in S phase. This is consistent with the hypothesis that oxidative stress-induced DNA damage must reach a threshold that surpasses the repair capacity of the cell for cell cycle reactivation to be triggered. It has been suggested that healthy neurons likely accumulate DNA damage owing to a decrease in DNA repair capacity *(118)*. Interestingly, very little oxidative DNA damage is seen in aged wild-type mouse brains contending with partially deficient repair mechanisms known as the "rodent repairadox" *(119)*. This suggests that these lesions are efficiently repaired in nondiseased brains, despite the relatively low levels of nucleotide excision repair (NER) and other forms of global genome repair in normal differentiated neurons. It should be emphasized that differentiated neurons do indeed retain nucleotide excision repair activity at levels about 10% of their cycling precursors *(118)*. This may in fact be all that is required to maintain genomic integrity in quiescent cells. In mice, there is still no evidence of accumulated neuronal mutations in vivo and methods developed to measure mutation frequencies in aging postmitotic tissue detected marked increases in aging liver, but no change in brain *(120)*. This study is by no means definitive and analysis was limited to select regions of the genome in wild-type mice. However, the view that DNA damage in healthy neurons creates a general susceptibility to death following lost quiescence is not supported by this data. Nor is this view supported by the separation of neuronal DNA synthesis and death in the conditional Rb knockout animals discussed in the following section. Instead, the dual activation of CCPs and induced DNA damage in disease may be necessary to initiate CRND. That is, DNA repair deficiencies may well lead to increased DNA damage in diseased neurons, but this defect may also be part of the disease process. Consistently, AD mouse models show a reduced capacity to remove 8-OHdG lesions compared with wild-type mice *(53)* and human patients show increased damage over age-matched controls *(121)*. Moreover, recent data suggest that DNA damage may itself be the trigger for cell cycle activation. For example, Kruman et al. showed that $A\beta_{1-42}$ and other DNA damaging agents induce CRND in primary cortical neurons *(53)*. Importantly, modified fluorescence-activated cell sorting techniques were used to demonstrate that DNA damage-induced neuronal apoptosis occurs subsequent to DNA synthesis. Neuronal death following S phase re-entry may involve activation of ataxia-telangiectasia-mutated protein (ATM)-dependent checkpoints because neurons from ATM deficient mice ($atm^{-/-}$) are resistant to DNA damage-induced death, whether the lack of ATM effects S phase entry was not reported. Finally, the authors claim that nongenotoxic apoptotic stimuli (e.g., staurosporine and colchicine) do not activate cell cycle reactivation *(53)*. However, close examination of the data clearly shows that staurosporine increases S phase markers and apoptosis following a G_1/S phase transition in the absence of detectable DNA damage, albeit not to the same degree as genotoxic agents. This quantitative rather than qualitative difference suggests that DNA damage, although efficient, may represent only one of many ways of activating CRND.

3.4. Presenilins as Regulators of CRND in AD

Much of this chapter has focused on AD because CRND has been most extensively studied in this disease. Current AD research is dominated by the Aβ hypothesis which states that the deposition of Aβ plaques is the primary cause of neurodegeneration. Despite a lack of spatiotemporal correlation between plaque burden and neuronal cell loss, the focus on Aβ prevails largely owing to the identification of three genes responsible for early-onset familial AD (FAD). The first is the amyloid precursor protein (APP) which is cleaved by the sequential actions of the β- and γ-secretases to form the plaque-forming Aβ peptides. The second two are the presenilins, PS1 and PS2, which are subunits of the γ-secretase. It is likely that PS1-dependent cleavage of amyloid precursor protein results in Aβ production and that FAD mutations in these proteins enhance and modify this process to generate plaques. However, it has recently been shown that genetic ablation of PS1 in the forebrain of *psen2*$^{-/-}$ mice leads to synaptic dysfunction and progressive neurodegeneration that is independent of Aβ production *(122)*. These phenotypes could result from altered proteolytic cleavage of other presenilin targets *(123,124)*, including the liberation of the Notch intracellular domain that translocates to the nucleus to regulate a plethora of physiological responses *(125)*. In addition to proteolytic activity, however, presenilins have also been shown to interact directly with β-catenin *(126–131)*. Nuclear β-catenin binds to members of the T-cell factor/lymphoid enhancer binding factor (TCF/LEF) family of transcription factors to activate a number of target genes *(132)*, including those encoding CCPs. In fact, cyclin D_1 was one of the first identified β-catenin target genes *(133,134)*. Koo and colleagues have shown that mouse embryonic fibroblasts deficient in PS1 accumulate β-catenin that results in TCF/LEF-dependent cyclin D_1 expression and hyperproliferation *(131)*. A similar hyperproliferation is seen in vivo in PS1-null epidermal cells of *psen1*$^{-/-}$ mice rescued with neuron-specific expression of human PS1 *(135)*. Again, increased β-catenin levels are correlated with activation of cyclin D_1 and accelerated G_1/S transitions. In both cases, the effects of PS1 deficiency are reversed by reintroducing wild-type PS1, but not PS1 mutants lacking β-catenin binding (PS1Δcat). Importantly, the PS1Δcat mutant retains normal processing of Notch and APP *(131)*, strengthening the argument that this effect is owing to altered β-catenin interactions. Reintroduction of PS1 harboring two different FAD mutations was also without effect on β-catenin accumulation and cyclin D_1 activation. Together, these findings indicate that PS1 negatively regulates β-catenin levels and that the FAD mutations in PS1 disrupt this function. The speculation would be that, in this situation, the increased neuronal β-catenin drives neurons into an unscheduled cell cycle, thereby killing them.

4. DOES CELL CYCLE RE-ENTRY CAUSE NEURONAL LOSS?

Without doubt, the correlation between cell cycle and neuronal loss has been firmly established in human neurodegenerative diseases. Furthermore, forced cell cycle re-entry by oncogene overexpression or tumor suppressor disruption can clearly lead to neuronal death in some circumstances. However, whether cell cycle reactivation is actually causative or merely associated with neuronal death in neurodegenerative disease is still being addressed.

For instance, we already discussed several studies demonstrating that disruption of the Rb/E2F pathway leads to cell cycle reactivation and subsequent neuronal death, yet recent reports indicate that these are separable processes. First, chimeric mice composed of wild-type and *Rb*$^{-/-}$ cells are viable and fertile with remarkably minor defects *(136,137)*. Importantly, *Rb*$^{-/-}$ cells contribute extensively to all tissues including the CNS, suggesting that apoptosis of Rb-deficient cells is noncell-autonomous. On closer examination, Lipinski et al. have shown that Rb-deficient cells in the CNS of chimeras show comparable levels of ectopic S phase entry, yet continue to survive and differentiate *(138)*. More recently, reports from both the Jacks and Leone labs provide a mechanistic explanation for this noncell-autonomous function of Rb in CNS neuronal death. First, Macpherson et al. describe conditional knockout mice that have Rb deleted from the nervous system, but lack the erythropoiesis defects that lead to CNS hypoxia in Rb-null animals *(139)*. Similar to the chimeras, *Rb*$^{-/-}$ cells in conditional knockouts underwent ectopic S phase entry in the absence of cell death. The authors therefore propose that hypoxia is a necessary cofactor to neuronal death. In support, the Leone lab has recently identified a placental defect in the Rb-null animals that they rescue by either tetraploid aggregation or conditional knockout strategies. In both cases, restoring Rb function to the placenta by supplying wild-type extraembryonic cells suppresses most of the defects seen in Rb-null animals, resulting in normal erythropoeisis *(43,140)*. These animals also show ectopic cell cycle re-entry in the CNS without increased neuronal death. Therefore, Rb appears to have cell-autonomous effects on cell cycle exit and/or maintenance of quiescence, but noncell-autonomous roles in regulating neuronal death.

These studies clearly demonstrate that cell cycle re-entry is not always sufficient to induce neuronal death in rodents. These important findings have prompted many to engage a "multiple-hit" hypothesis to explain CRND in neurodegenerative diseases. This hypothesis posits that loss of quiescence, signified by the re-expression of CCPs and neuronal DNA synthesis, compromises the ability of neurons to suppress apoptosis in stressful environments. In other words, cell cycle reactivation makes neurons vulnerable to secondary death stimuli. To be clear, the multiple-hit hypothesis discussed here is supported by a sophisticated data set and is fundamentally different from a previously proposed "two-hit hypothesis for AD," which is based on the simple and unsurprising observation that more than one signaling pathway is activated in AD brains, even though these could result from a single stimulus *(141)*. Interestingly, in less complex tissue culture models, overexpression of E2F1 alone is sufficient to drive CRND *(142,143)*. These cultures may already be under stress and therefore vulnerable to death as soon as cell cycle is reinitiated. Alternatively, culture systems may lack some of the protective effects of the in vivo environment that must be overcome by additional "hits" in whole animal models. It is possible that these additional stresses simply do not occur within the short life-span and protected environment of laboratory mice. In the case of the conditional RB mutants, the secondary stress appears to be hypoxia. In chronic neurodegenerative diseases, the source of secondary stresses are likely varied and may include transient ischemia, infections, head trauma, and altered cholesterol homeostasis; all of which have been implicated as risk factors for AD *(144–153)*.

Even though cell cycle re-entry can occur without cell death in rodents, CCPs are evidently required for neuronal death in numerous tissue culture and rodent disease models. For example, pharmacological inhibitors of CDKs prevent CRND in a large number of culture and animal models. As always, lack of specificity is clearly a concern when considering the effects of kinase inhibitors and these results should be interpreted with caution. However, there is now corroborating evidence using interfering mutants and expression of endogenous inhibitors for several of these models including kainic-acid-induced neuronal death and stroke models in vivo *(45,68,154)*, as well as DNA damage and Aβ-mediated death in vitro *(50,52,70)*. Notably, by defining the specific kinases involved in each condition, these studies highlight mechanistic differences in CRND among experimental paradigms. Nonetheless, there is strong evidence that CDKs are not only associated with, but often contribute to the death of "cycling" neurons. Moreover, Copani and co-workers have used an antisense approach to show that downregulating specific DNA polymerases not only prevent neuronal DNA replication, but also protects against Aβ-induced neurotoxicity in culture *(155)*. Interestingly, the typical replicative polymerases (polα, δ, or ε) were not required. Instead, Aβ-induced DNA synthesis requires polymerase-β, a classical repair enzyme with limited replicative potential in cycling cells. When contributing to DNA replication in neurons, the low fidelity of polβ-mediated DNA synthesis may create mutations that contribute to the death process. In any event, this study clearly demonstrates that Aβ-induced neuronal DNA synthesis in culture is a novel process involving atypical machinery. It remains to be determined whether a similar requirement for DNA polymerases extends to CRND in vivo.

Although not yet proven in vivo, the suggested requirement for S phase progression has fueled several hypotheses to explain the death of neurons associated with cell cycle reactivation in human neurodegenerative diseases such as AD. Among those receiving considerable attention, two invoke an obligate role of DNA replication. First, Yang et al. *(36)* argues that the chromosomal imbalance in tetraploid cells is itself the fundamental cause of susceptibility to cell loss in AD. In this model, a persistent abnormal gene dose resulting from a G_2-like arrest is the principal defect leading to the slow loss of synapses and atrophy of neuritic processes. A second model is based on the presumed accumulation of unrepaired lesions in the nontranscribed genes of differentiated neurons. This presumption is based on the observations of several groups demonstrating that differentiated neuroblastoma cell lines are less efficient at DNA repair, particularly nucleotide excision repair, than their proliferating precursors *(156)*. These results are consistent with a generally greater sensitivity of neurons to DNA-damaging stimuli than other cell types. Differentiated neurons do retain transcription-coupled repair (TCR) and the specialized differentiation-associated repair (DAR) to maintain the integrity of actively expressed genes. However, Hanawalt and colleagues proposed that differentiated neurons likely accumulate lesions throughout the rest of the genome *(118)*. If true, the prediction is that damaged DNA in neurons undergoing cell cycle re-entry would result in stalled replication forks, double-strand breaks, and activation of apoptotic pathways. Alternatively, if polymerases were able to bypass damaged DNA, previously silent genes may now be expressed with potentially debilitating mutations.

These hypotheses predict that DNA replication is necessary for CRND in vivo. Testing whether this is true will require a means of preventing DNA synthesis without effecting DNA repair or other potential targets effected by CDK inhibitors. Useful strategies to specifically prevent S phase progression or initiation in transformed cells have included inhibiting replication-competent DNA polymerases or preventing assembly of prereplication complexes (PreRCs) that are essential for origin-based recruitment of replicative polymerases. The realization that CRND may be using different enzymes to replicate its DNA cast doubt on the utility of these approaches. It is therefore imperative to first elucidate the basic mechanisms involved in order to design strategies with sufficient specificity to determine whether cell cycle progression, and not just CCP expression, is actually causative or merely associated with neuronal death.

5. CONCLUSIONS

Neurons appear more vulnerable to the loss of cell cycle arrest than previously appreciated. Despite the relatively large literature describing this phenomenon, the nature of cell cycle advancement in postmitotic neurons is only beginning to be addressed. Rather than adhering to the rules established for cell cycle progression in proliferating cells, neurons appear to have adopted unique mechanisms of advancing from G_0 through DNA replication. Whether engagement of these unusual processes is akin to a run-away truck stirring up everything in its path or representative of a highly orchestrated attempt at survival is not clear. Certainly, the primary cause of neuronal demise in CRND remains in question. Even the basic mechanisms of neuronal DNA replication continue to offer surprises. Our next challenge is to further understand the mechanisms by which a neuron stumbles through a "postmitotic cycle" in order to distinguish those processes that are actively contributing to neuronal death from those that represent a failed protective event. This distinction could have decisive effects on strategies to treat neurodegenerative diseases. For instance, if those neurons that achieve a 4N DNA complement are viable for prolonged periods because of a rarely successful protective event; then manipulations that aid S phase progression may prove more advantageous than those aimed at prevention. Although the notion of directing cell cycle progression as opposed to preventing its occurrence would undoubtedly present many obstacles, its success would have profound implications in a broad range of neurodegenerative conditions and the field of neurogenesis in whole. On the other hand, if advancement through S phase proves deleterious, identifying means of specifically preventing this process in neurons without the common side effects of most anticycle drugs would be the therapeutic goal. Before attempting either of these strategies, however, we must first obtain a better understanding of the basic mechanisms involved.

We have highlighted several areas that hold promise for research. These include understanding the neuron-specific localization of CCPs and whether these aberrant locales lead to novel interactions with unique activities. The need to identify CRND-specific targets of CDKs cannot be underestimated and may be aided by the use of pharmacogenetic approaches to primary tissue culture and mouse models, similar to those used for Cdks in yeast and other genetically tractable systems. Although this approach will require considerably more investment in mammalian models, the significant groundwork and direct relevance to multiple disease states make this an attractive and timely endeavor. We have presented several models to facilitate discussion of important lingering issues. We recognize our simplification of these issues and hope only to have ignited more interest in this growing and exciting field. We also acknowledge the many equally exciting ideas we did not elaborate on here, including the role of dedifferentiaton, mitotic and growth factor signaling, and loss of contact inhibition; each of which will undoubtedly continue to incite future research. Finally, it is our hope that by also displaying the shortcomings of the current literature, we will be inspired to tackle the difficult questions regarding the uncertainties that continue to blur our vision.

REFERENCES

1. Copani A, et al. Activation of cell cycle-associated proteins in neuronal death: a mandatory or dispensable path? Trends Neurosci 2001;24(1):25–31.
2. Heintz N. Cell death and the cell cycle: a relationship between transformation and neurodegeneration? Trends Biochem Sci 1993;18(5):157–159.
3. Herrup K, Arendt T. Re-expression of cell cycle proteins induces neuronal cell death during Alzheimer's disease. J Alzheimers Dis 2002;4(3):243–247.
4. Arendt T. Synaptic plasticity and cell cycle activation in neurons are alternative effector pathways: the 'Dr. Jekyll and Mr. Hyde concept' of Alzheimer's disease or the yin and yang of neuroplasticity. Prog Neurobiol 2003;71(2,3):83–248.
5. Becker EB, Bonni A. Cell cycle regulation of neuronal apoptosis in development and disease. Prog Neurobiol 2004;72(1):1–25.
6. Neve RL, et al. Alzheimer's disease: dysfunction of a signalling pathway mediated by the amyloid precursor protein? Biochem Soc Symp 2001;67:37–50.
7. Nguyen MD, et al. Cycling at the interface between neurodevelopment and neurodegeneration. Cell Death Differ 2002;9(12):1294–1306.
8. Klein JA, Ackerman SL. Oxidative stress, cell cycle, and neurodegeneration. J Clin Invest 2003;111(6):785–793.
9. Morest DK, Silver J. Precursors of neurons, neuroglia, and ependymal cells in the CNS: what are they? Where are they from? How do they get where they are going? Glia 2003;43(1):6–18.
10. Gage FH. Neurogenesis in the adult brain. J Neurosci 2002;22(3):612,613.
11. Husseman JW. et al. Mitotic activation: a convergent mechanism for a cohort of neurodegenerative diseases. Neurobiol Aging 2002;21(6):815–828.
12. Liu DX, Greene LA. Regulation of neuronal survival and death by E2F-dependent gene repression and derepression. Neuron 2001;32(3):425–438.
13. Liu DX, Greene LA. Neuronal apoptosis at the G1/S cell cycle checkpoint. Cell Tissue Res 2001;305(2):217–228.
14. Helin K. Regulation of cell proliferation by the E2F transcription factors. Curr Opin Genet Dev 1998;8(1):28–35.
15. Nevins JR. The Rb/E2F pathway and cancer. Hum Mol Genet 2001;10(7):699–703.
16. Al-Ubaidi MR, et al. Photoreceptor degeneration induced by the expression of simian virus 40 large tumor antigen in the retina of transgenic mice. Proc Natl Acad Sci USA 1992;89(4):1194–1198.
17. Al-Ubaidi MR, et al. Unscheduled DNA replication precedes apoptosis of photoreceptors expressing SV40 T antigen. Exp Eye Res 1997;64(4):573–585.
18. Feddersen RM, et al. Disrupted cerebellar cortical development and progressive degeneration of Purkinje cells in SV40 T antigen transgenic mice. Neuron 1992;9(5):955–966.
19. Feddersen RM, et al. In vivo viability of postmitotic Purkinje neurons requires pRb family member function. Mol Cell Neurosci 1995;6(2):153–167.
20. Clarke AR, et al. Requirement for a functional Rb-1 gene in murine development. Nature 1992;359(6393):328–330.
21. Jacks T, et al. Effects of an Rb mutation in the mouse. Nature 1992;359(6393):295–300.
22. Lee EY, et al. Mice deficient for Rb are nonviable and show defects in neurogenesis and haematopoiesis. Nature 1992;359(6393):288–294.
23. Macleod KF, et al. Loss of Rb activates both p53-dependent and independent cell death pathways in the developing mouse nervous system. EMBO J 1996;15(22):6178–6188.
24. Herrup K, Mullen RJ. Regional variation and absence of large neurons in the cerebellum of the staggerer mouse. Brain Res 1979;172(1):1–12.
25. Herrup K. Role of staggerer gene in determining cell number in cerebellar cortex. I. Granule cell death is an indirect consequence of staggerer gene action. Brain Res 1983;313(2):267–274.
26. Herrup K, Sunter K. Numerical matching during cerebellar development: quantitative analysis of granule cell death in staggerer mouse chimeras. J Neurosci 1987;7(3):829–836.
27. Herrup K, Busser JC. The induction of multiple cell cycle events precedes target-related neuronal death. Development 1995;121(8):2385–2395.
28. Zuo J, et al. Neurodegeneration in Lurcher mice caused by mutation in delta2 glutamate receptor gene. Nature 1997;388(6644):769–773.

29. Busser J, et al. Ectopic cell cycle proteins predict the sites of neuronal cell death in Alzheimer's disease brain. J Neurosci 1998;18(8):2801–2807.
30. Vincent I, et al. Aberrant expression of mitotic cdc2/cyclin B1 kinase in degenerating neurons of Alzheimer's disease brain. J Neurosci 1997;17(10):3588–3598.
31. Ding XL, et al. The cell cycle Cdc25A tyrosine phosphatase is activated in degenerating postmitotic neurons in Alzheimer's disease. Am J Pathol 2000;157(6):1983–1990.
32. Vincent I, et al. Constitutive Cdc25B tyrosine phosphatase activity in adult brain neurons with M phase-type alterations in Alzheimer's disease. Neuroscience 2001;105(3):639–650.
33. Jordan-Sciutto KL, et al. Expression patterns of retinoblastoma protein in Parkinson disease. J Neuropathol Exp Neurol 2003;62(1):68–74.
34. Lee SS, et al. Cell cycle aberrations by alpha-synuclein over-expression and cyclin B immunoreactivity in Lewy bodies. Neurobiol Aging 2003;24(5):687–696.
35. Yang Y, et al. Neuronal cell death is preceded by cell cycle events at all stages of Alzheimer's disease. J Neurosci 2003;23(7):2557–2563.
36. Yang Y, et al. DNA replication precedes neuronal cell death in Alzheimer's disease. J Neurosci 2001;21(8):2661–2668.
37. El-Khodor BF, et al. Ectopic expression of cell cycle markers in models of induced programmed cell death in dopamine neurons of the rat substantia nigra pars compacta. Exp Neurol 2003;179(1):17–27.
38. Klein JA, et al. The harlequin mouse mutation downregulates apoptosis-inducing factor. Nature 2002;419(6905):367–374.
39. Jiang W, et al. Cortical neurogenesis in adult rats after transient middle cerebral artery occlusion. Stroke 2001;32(5):1201–1207.
40. Kaushal D, et al. Alteration of gene expression by chromosome loss in the postnatal mouse brain. J Neurosci 2003;23(13):5599–5606.
41. Rehen SK, et al. Chromosomal variation in neurons of the developing and adult mammalian nervous system. Proc Natl Acad Sci USA 2001;98(23):13,361–13,366.
42. Yang AH, et al. Chromosome segregation defects contribute to aneuploidy in normal neural progenitor cells. J Neurosci 2003;23(32):10,454–10,462.
43. Wu L, et al. Extra-embryonic function of Rb is essential for embryonic development and viability. Nature 2003;421(6926):942–947.
44. Freeman RS, et al. Analysis of cell cycle-related gene expression in postmitotic neurons: selective induction of Cyclin D1 during programmed cell death. Neuron 1994;12(2):343–355.
45. Park DS, et al. Cell cycle regulators in neuronal death evoked by excitotoxic stress: implications for neurodegeneration and its treatment. Neurobiol Aging 2000;21(6):771–781.
46. Park DS, et al. Multiple pathways of neuronal death induced by DNA-damaging agents, NGF deprivation, and oxidative stress. J Neurosci 1998;18(3):830–840.
47. Verdaguer E, et al. Kainic acid-induced apoptosis in cerebellar granule neurons: an attempt at cell cycle re-entry. Neuroreport 2002;13(4):413–416.
48. Wu Q, et al. Beta-amyloid activated microglia induce cell cycling and cell death in cultured cortical neurons. Neurobiol Aging 2000;21(6):797–806.
49. Padmanabhan J, et al. Role of cell cycle regulatory proteins in cerebellar granule neuron apoptosis. J Neurosci 1999;19(20):8747–8756.
50. Giovanni A, et al. Involvement of cell cycle elements, cyclin-dependent kinases, pRb, and E2F x DP, in B-amyloid-induced neuronal death. J Biol Chem 1999;274(27):19,011–19,016.
51. Kim JI, et al. Oxidative stress and neurodegeneration in prion diseases. Ann NY Acad Sci 2001;928:182–186.
52. Copani A, et al. Mitotic signaling by beta-amyloid causes neuronal death. FASEB J 1999; 13(15):2225–2234.
53. Kruman II, et al. Cell cycle activation linked to neuronal cell death initiated by DNA damage. Neuron 2004;41(4):549–561.
54. Sumrejkanchanakij P, et al. Role of cyclin D1 cytoplasmic sequestration in the survival of postmitotic neurons. Oncogene 2003;22(54):8723–8730.
55. Bu B, et al. Niemann-Pick disease type C yields possible clue for why cerebellar neurons do not form neurofibrillary tangles. Neurobiol Dis 2002;11(2):285–297.
56. Pei JJ, et al. Up-regulation of cell division cycle (cdc) 2 kinase in neurons with early stage Alzheimer's disease neurofibrillary degeneration. Acta Neuropathol (Berl) 2002;104(4):369–376.

57. Konishi Y, et al. Cdc2 phosphorylation of BAD links the cell cycle to the cell death machinery. Mol Cell 2002;9(5):1005–1016.
58. Ibrado AM, et al. Temporal relationship of CDK1 activation and mitotic arrest to cytosolic accumulation of cytochrome C and caspase-3 activity during Taxol-induced apoptosis of human AML HL-60 cells. Leukemia 1998;12(12):1930–1936.
59. Shen SC, et al. Taxol-induced p34cdc2 kinase activation and apoptosis inhibited by 12-O-tetradecanoylphorbol-13-acetate in human breast MCF-7 carcinoma cells. Cell Growth Differ 1998;9(1):23–29.
60. Yu D, et al. Overexpression of ErbB2 blocks Taxol-induced apoptosis by upregulation of p21Cip1, which inhibits p34Cdc2 kinase. Mol Cell 1998;2(5):581–591.
61. Chan PH. Reactive oxygen radicals in signaling and damage in the ischemic brain. J Cereb Blood Flow Metab 2001;21(1):2–14.
62. Nagy Z, et al. Cell cycle markers in the hippocampus in Alzheimer's disease. Acta Neuropathol (Berl) 1997;94(1):6–15.
63. Smith MZ, et al. Cell cycle-related protein expression in vascular dementia and Alzheimer's disease. Neurosci Lett 1999;271(1):45–48.
64. LaBaer J, et al. New functional activities for the p21 family of CDK inhibitors. Genes Dev 1997;11(7):847–862.
65. Diehl JA, et al. Glycogen synthase kinase-3beta regulates cyclin D1 proteolysis and subcellular localization. Genes Dev 1998;12(22):3499–3511.
66. Alt JR, et al. Phosphorylation-dependent regulation of cyclin D1 nuclear export and cyclin D1-dependent cellular transformation. Genes Dev 2000;14(24):3102–3114.
67. Kranenburg O, et al. Differentiation of P19 EC cells leads to differential modulation of cyclin-dependent kinase activities and to changes in the cell cycle profile. Oncogene 1995;10(1):87–95.
68. Ino H, Chiba T. Cyclin-dependent kinase 4 and cyclin D1 are required for excitotoxin-induced neuronal cell death in vivo. J Neurosci 2001;21(16):6086–6094.
69. Osuga H, et al. Cyclin-dependent kinases as a therapeutic target for stroke. Proc Natl Acad Sci USA 2000;97(18):10,254–10,259.
70. Park DS, et al. Cyclin-dependent kinases participate in death of neurons evoked by DNA-damaging agents. J Cell Biol 1998;143(2):457–467.
71. Ferguson KL, et al. The Rb-CDK4/6 signaling pathway is critical in neural precursor cell cycle regulation. J Biol Chem 2000;275(43):33,593–33,600.
72. Park DS, et al. Cyclin dependent kinase inhibitors and dominant negative cyclin dependent kinase 4 and 6 promote survival of NGF-deprived sympathetic neurons. J Neurosci 1997;17(23):8975–8983.
73. Bamberger ME, Landreth GE. Inflammation, apoptosis, and Alzheimer's disease. Neuroscientist 2002;8(3):276–283.
74. McGeer PL, McGeer EG. Local neuroinflammation and the progression of Alzheimer's disease. J Neurovirol 2002;8(6):529–538.
75. McGeer PL, McGeer EG. Inflammation and neurodegeneration in Parkinson's disease. Parkinsonism Relat Disord 2004;10(Suppl 1):S3–S7.
76. Breitner JC. The role of anti-inflammatory drugs in the prevention and treatment of Alzheimer's disease. Annu Rev Med 1996;47:401–411.
77. Breitner JC. Inflammatory processes and antiinflammatory drugs in Alzheimer's disease: a current appraisal. Neurobiol Aging 1996;17(5):789–794.
78. Stewart WF, et al. Risk of Alzheimer's disease and duration of NSAID use. Neurology 1997;48(3):626–632.
79. Rich JB, et al. Nonsteroidal anti-inflammatory drugs in Alzheimer's disease. Neurology 1995;45(1):51–55.
80. Mackenzie IR, Munoz DG. Nonsteroidal anti-inflammatory drug use and Alzheimer-type pathology in aging. Neurology 1998;50(4):986–990.
81. Bamberger ME, et al. A cell surface receptor complex for fibrillar beta-amyloid mediates microglial activation. J Neurosci 2003;23(7):2665–2674.
82. McDonald DR, et al. beta-Amyloid fibrils activate parallel mitogen-activated protein kinase pathways in microglia and THP1 monocytes. J Neurosci 1998;18(12):4451–4460.
83. Netland EE, et al. Indomethacin reverses the microglial response to amyloid beta-protein. Neurobiol Aging 1998;19(3):201–204.

84. Weldon DT, et al. Fibrillar beta-amyloid induces microglial phagocytosis, expression of inducible nitric oxide synthase, and loss of a select population of neurons in the rat CNS in vivo. J Neurosci 1998;18(6):2161–2173.
85. Kalaria RN. Microglia and Alzheimer's disease. Curr Opin Hematol 1999;6(1):15–24.
86. Tarkowski E, et al. Intrathecal inflammation precedes development of Alzheimer's disease. J Neurol Neurosurg Psychiatry 2003;74(9):1200–1205.
87. Duplan L, et al. Lithostathine and pancreatitis-associated protein are involved in the very early stages of Alzheimer's disease. Neurobiol Aging 2001;22(1):79–88.
88. Meda L, et al. beta-Amyloid(25-35) induces the production of interleukin-8 from human monocytes. J Neuroimmunol 1995;59(1,2):29–33.
89. McDonald DR, et al. Amyloid fibrils activate tyrosine kinase-dependent signaling and superoxide production in microglia. J Neurosci 1997;17(7):2284–2294.
90. Bianca VD, et al. beta-amyloid activates the O-2 forming NADPH oxidase in microglia, monocytes, and neutrophils. A possible inflammatory mechanism of neuronal damage in Alzheimer's disease. J Biol Chem 1999;274(22):15,493–15,499.
91. Allan SM, Rothwell NJ. Inflammation in central nervous system injury. Philos Trans R Soc Lond B Biol Sci 2003;358(1438):1669–1677.
92. Consilvio C, et al. Neuroinflammation, COX-2, and ALS—a dual role? Exp Neurol 2004;187(1): 1–10.
93. Gao HM, et al. Synergistic dopaminergic neurotoxicity of MPTP and inflammogen lipopolysaccharide: relevance to the etiology of Parkinson's disease. FASEB J 2003;17(13):1957–1959.
94. Nguyen MD, et al. Cell cycle regulators in the neuronal death pathway of amyotrophic lateral sclerosis caused by mutant superoxide dismutase 1. J Neurosci 2003;23(6):2131–2140.
95. Contestabile A. Oxidative stress in neurodegeneration: mechanisms and therapeutic perspectives. Curr Top Med Chem 2001;1(6):553–568.
96. Jenner P. Oxidative stress in Parkinson's disease. Ann Neurol 2003;53(Suppl 3);S26–S36; discussion S28–S36.
97. Carri MT, et al. Neurodegeneration in amyotrophic lateral sclerosis: the role of oxidative stress and altered homeostasis of metals. Brain Res Bull 2003;61(4):365–374.
98. Giasson BI, et al. The relationship between oxidative/nitrative stress and pathological inclusions in Alzheimer's and Parkinson's diseases. Free Radic Biol Med 2002;32(12):1264–1275.
99. Albers DS, Beal MF. Mitochondrial dysfunction and oxidative stress in aging and neurodegenerative disease. J Neural Transm Suppl 2000;59:133–154.
100. Matsuoka Y, et al. Fibrillar beta-amyloid evokes oxidative damage in a transgenic mouse model of Alzheimer's disease. Neuroscience 2001;104(3):609–613.
101. Cassarino DS, et al. Elevated reactive oxygen species and antioxidant enzyme activities in animal and cellular models of Parkinson's disease. Biochim Biophys Acta 1997;1362(1):77–86.
102. Sherer TB, et al. An in vitro model of Parkinson's disease: linking mitochondrial impairment to altered alpha-synuclein metabolism and oxidative damage. J Neurosci 2002;22(16):7006–7015.
103. Andersen JK. Oxidative stress in neurodegeneration: cause or consequence? Nat Med 2004;10(Suppl):S18–S25.
104. Lipton SA, Bossy-Wetzel E. Dueling activities of AIF in cell death versus survival: DNA binding and redox activity. Cell 2002;111(2):147–150.
105. Bonni A. Neurodegeneration: A non-apoptotic role for AIF in the brain. Curr Biol 2003; 13(1):R19–R21.
106. White AR, et al. Survival of cultured neurons from amyloid precursor protein knock-out mice against Alzheimer's amyloid-beta toxicity and oxidative stress. J Neurosci 1998;18(16):6207–6217.
107. Rahman I, et al. Oxidative stress and TNF-alpha induce histone acetylation and NF-kappaB/AP-1 activation in alveolar epithelial cells: potential mechanism in gene transcription in lung inflammation. Mol Cell Biochem 2002;234,235(1,2):239–248.
108. Boutillier AL, et al. Selective E2F-dependent gene transcription is controlled by histone deacetylase activity during neuronal apoptosis. J Neurochem 2003;84(4):814–828.
109. Boutillier AL, et al. Constitutive repression of E2F1 transcriptional activity through HDAC proteins is essential for neuronal survival. Ann NY Acad Sci 2002;973:438–442.
110. Gabbita SP, et al. Increased nuclear DNA oxidation in the brain in Alzheimer's disease. J Neurochem 1998;71(5):2034–2040.

111. de la Monte SM, et al. Mitochondrial DNA damage as a mechanism of cell loss in Alzheimer's disease. Lab Invest 2000;80(8):1323–1335.
112. Warita H, et al. Oxidative damage to mitochondrial DNA in spinal motoneurons of transgenic ALS mice. Brain Res Mol Brain Res 2001;89(1,2):147–152.
113. An SJ, et al. Oxidative DNA damage and alteration of glutamate transporter expressions in the hippocampal Ca1 area immediately after ischemic insult. Mol Cells 2002;13(3):476–480.
114. Imai H, et al. Ebselen protects both gray and white matter in a rodent model of focal cerebral ischemia. Stroke 2001;32(9):2149–2154.
115. Zhang WR, et al. Attenuation of oxidative DNA damage with a novel antioxidant EPC-K1 in rat brain neuronal cells after transient middle cerebral artery occlusion. Neurol Res 2001;23(6):676–680.
116. Chen J, et al. Early detection of DNA strand breaks in the brain after transient focal ischemia: implications for the role of DNA damage in apoptosis and neuronal cell death. J Neurochem 1997;69(1):232–245.
117. Gu G, et al. Mitochondrial DNA deletions/rearrangements in parkinson disease and related neurodegenerative disorders. J Neuropathol Exp Neurol 2002;61(7):634–639.
118. Nouspikel T, Hanawalt PC. When parsimony backfires: neglecting DNA repair may doom neurons in Alzheimer's disease. Bioessays 2003;25(2):168–173.
119. Hanawalt PC. Revisiting the rodent repairadox. Environ Mol Mutagen 2001;38(2,3):89–96.
120. Dolle ME, et al. Rapid accumulation of genome rearrangements in liver but not in brain of old mice. Nat Genet 1997;17(4):431–434.
121. Markesbery WR, Carney JM. Oxidative alterations in Alzheimer's disease. Brain Pathol 1999;9(1):133–146.
122. Saura CA, et al. Loss of presenilin function causes impairments of memory and synaptic plasticity followed by age-dependent neurodegeneration. Neuron 2004;42(1):23–36.
123. Kopan R, Goate A. A common enzyme connects notch signaling and Alzheimer's disease. Genes Dev 2000;14(22):2799–2806.
124. Koo EH, Kopan R. Potential role of presenilin-regulated signaling pathways in sporadic neurodegeneration. Nat Med 2004;10(Suppl):S26–S33.
125. Selkoe DJ. Notch and presenilins in vertebrates and invertebrates: implications for neuronal development and degeneration. Curr Opin Neurobiol 2000;10(1):50–57.
126. Zhou J, et al. Presenilin 1 interaction in the brain with a novel member of the Armadillo family. Neuroreport 1997;8(6):1489–1494.
127. Murayama M, et al. Direct association of presenilin-1 with beta-catenin. FEBS Lett 1998;433(1,2):73–77.
128. Yu G, et al. The presenilin 1 protein is a component of a high molecular weight intracellular complex that contains beta-catenin. J Biol Chem 1998;273(26):16,470–16,475.
129. Kang DE, et al. Presenilin 1 facilitates the constitutive turnover of beta-catenin: differential activity of Alzheimer's disease-linked PS1 mutants in the beta-catenin-signaling pathway. J Neurosci 1999;19(11):4229–4237.
130. Stahl B, et al. Direct interaction of Alzheimer's disease-related presenilin 1 with armadillo protein p0071. J Biol Chem 1999;274(14):9141–9148.
131. Soriano S, et al. Presenilin 1 negatively regulates beta-catenin/T cell factor/lymphoid enhancer factor-1 signaling independently of beta-amyloid precursor protein and notch processing. J Cell Biol 2001;152(4):785–794.
132. Behrens J, et al. Functional interaction of beta-catenin with the transcription factor LEF-1. Nature 1996;382(6592):638–642.
133. Shtutman M, et al. The cyclin D1 gene is a target of the beta-catenin/LEF-1 pathway. Proc Natl Acad Sci USA 1999;96(10):5522–5527.
134. Tetsu O, McCormick F. Beta-catenin regulates expression of cyclin D1 in colon carcinoma cells. Nature 1999;398(6726):422–426.
135. Xia X, et al. Loss of presenilin 1 is associated with enhanced beta-catenin signaling and skin tumorigenesis. Proc Natl Acad Sci USA 2001;98(19):10,863–10,868.
136. Williams BO, et al. Extensive contribution of Rb-deficient cells to adult chimeric mice with limited histopathological consequences. EMBO J 1994;13(18):4251–4259.
137. Maandag EC, et al. Developmental rescue of an embryonic-lethal mutation in the retinoblastoma gene in chimeric mice. EMBO J 1994;13(18):4260–4268.

138. Lipinski MM, et al. Cell-autonomous and non-cell-autonomous functions of the Rb tumor suppressor in developing central nervous system. EMBO J 2001;20(13):3402–3413.
139. MacPherson D, et al. Conditional mutation of Rb causes cell cycle defects without apoptosis in the central nervous system. Mol Cell Biol 2003;23(3):1044–1053.
140. de Bruin A, et al. Rb function in extraembryonic lineages suppresses apoptosis in the CNS of Rb-deficient mice. Proc Natl Acad Sci USA 2003;100(11):6546–6551.
141. Zhu X, et al. Differential activation of neuronal ERK, JNK/SAPK and p38 in Alzheimer's disease: the 'two hit' hypothesis. Mech Ageing Dev 2001;123(1):39–46.
142. Hou ST, et al. The transcription factor E2F1 modulates apoptosis of neurons. J Neurochem 2000;75(1):91–100.
143. Smith DS, et al. Induction of DNA replication in adult rat neurons by deregulation of the retinoblastoma/E2F G1 cell cycle pathway. Cell Growth Differ 2000;11(12):625–633.
144. Robinson SR, et al. Challenges and directions for the pathogen hypothesis of Alzheimer's disease. Neurobiol Aging 2004;25(5):629–637.
145. Itzhaki RF, et al. Infiltration of the brain by pathogens causes Alzheimer's disease. Neurobiol Aging 2004;25(5):619–627.
146. Fleminger S, et al. Head injury as a risk factor for Alzheimer's disease: the evidence 10 years on; a partial replication. J Neurol Neurosurg Psychiatry 2003;74(7):857–862.
147. Jellinger KA. Traumatic brain injury as a risk factor for Alzheimer's disease. J Neurol Neurosurg Psychiatry 2004;75(3):511,512.
148. Kalaria RN. The role of cerebral ischemia in Alzheimer's disease. Neurobiol Aging 2000;21(2):321–330.
149. Iadecola C, Gorelick PB. Converging pathogenic mechanisms in vascular and neurodegenerative dementia. Stroke 2003;34(2):335–337.
150. Sadowski M, et al. Links between the pathology of Alzheimer's disease and vascular dementia. Neurochem Res 2004;29(6):1257–1266.
151. Michikawa M. Cholesterol paradox: is high total or low HDL cholesterol level a risk for Alzheimer's disease? J Neurosci Res 2003;72(2):141–146.
152. Puglielli L, et al. Alzheimer's disease: the cholesterol connection. Nat Neurosci 2003;6(4):345–351.
153. Burns M, Duff K. Cholesterol in Alzheimer's disease and tauopathy. Ann NY Acad Sci 2002;977:367–375.
154. O'Hare MJ, Rashidian J, Slack RS, During MJ, Park DS. The role of cyclin dependent kinases in stroke models of neuronal injury. In: Society of Neuroscience 33rd Annual Meeting, Vol. 2003 Abstract Viewer/Itinerary Planner, Program No. 738.734., Washington, DC: Society for Neuroscience, 2003; online.
155. Copani A, et al. Erratic expression of DNA polymerases by beta-amyloid causes neuronal death. FASEB J 2002;16(14):2006–2008.
156. Nouspikel T, Hanawalt PC. Terminally differentiated human neurons repair transcribed genes but display attenuated global DNA repair and modulation of repair gene expression. Mol Cell Biol 2000;20(5):1562–1570.

22
Cell Cycle Activation and the Amyloid-β Protein in Alzheimer's Disease

Katarzyna A. Gustaw, MD, Gemma Casadesus, PhD, Robert P. Friedland, MD, George Perry, PhD, and Mark A. Smith, PhD

SUMMARY

Morphological changes that characterize Alzheimer's disease (AD) are senile plaques, neurofibrillary tangles, and loss of synapses and neurons. At the center of the senile plaque is the polypeptide amyloid-β (Aβ), a product of a transmembrane protein called Aβ protein precursor (AβPP). Aβ formation, deposition, and toxicity have been associated with the cell cycle. Indeed, the mitogenic component appears early in the onset of AD with the reappearance of cell cycle markers and recently has been associated with selective early vulnerability of neurons. Also, because Aβ is mitogenic in vitro, it can induce and maintain cell cycle events in AD. Aβ-mediated cell death in vitro is dependent on the presence of various cell cycle-related elements. Therefore the activation of cell cycle machinery in vivo in the neuron may also similarly mediate its toxic effects. For decades the predominant thinking was that adult neurons do not proliferate. This conclusion resulted from the observation that differentiated neurons fail to divide. In this chapter, it is hypothesized that, in AD, whole populations of non-stem-cell neurons leave their quiescent state and re-enter the cell cycle. However, such neuronal re-entry into the cell cycle is ineffective and eventually leads to neurodegeneration and ultimately AD. Not only Aβ but all of the major genetic and protein elements deregulated in AD are to some extent altered in the nonefficient cell cycle. This group of factors includes AβPP, τ, the presenilins (PSEN1 and PSEN2), and, possibly, apolipoprotein E (ApoE). In addition, AD-related proteins such as ApoE, free radicals, free-radical generators, and antioxidants function also to control the state of the cell cycle. Investigating more defined mechanisms of cell cycle-related activation and arrest in AD involving Aβ and other factors may provide clues to change the natural course of this illness.

Key Words: Alzheimer disease; amyloid-β; cell cycle; dementia; mitogen; neurodegeneration.

1. INTRODUCTION

Alzheimer's disease (AD) is a progressive dementia that results in severe debilitation and finally death. It affects up to 15% of people over the age of 65 yr and nearly half of all individuals by 85 yr *(1)*. The disease process is quickly becoming one of the most serious health problems in the world and has a dehumanizing nature that involves destruction of higher-order brain function *(2)*. In the brain, AD shows selective loss of neurons within the hippocampus, temporal and frontal lobes, and senile plaques, and neurofibrillary tangles (NFTs), which are hallmarks of the disease and thought to be the cause of dementia *(2,3)*. NFTs, which contain a highly phosphorylated form of the microtubule associated protein τ, are the major intracellular

Table 1
Cell Cycle-Associated Proteins Found in Alzheimer's Disease

Substance	Characteristics	Role in the cell cycle	Reference mentioning association with AD
Cyclin A		S to G_2/M	47
Cyclin B		G_2/M	16,47,64
Cyclin C		No known role	47
Cyclin D (D3)		G_0/G_1/lateG_1/S	47,63,64
Cyclin E	Markers	G_1 to G_1/S	
p34 cdc2/CDK1	Markers	Late G_2/M	37,47
CDK4/CDK6	Markers	G_1 to G_1/S	23,43,47
CDK5/p25/p35	Markers	G_2 D_1, D_3 G_1 Cyclins	60
Nclk cdc2-like kinase	Markers	Cyclin A kinase	60
CDK7/MPM2	Markers	CDK-activated kinase	44
Cdc42/rac	Markers	GTPase/cell division	49
p21 ras	Markers	G protein/MAPK	34,72
Ki-67	Markers	Late G_1, S, G_2, M	35,63,64
p105/pRb	Markers	G_2/M TF	43
pCNA	Markers	Non cell cycle-specific	47
p107/pRb	Markers	Cdk2/4/6, checkpoint	43,45 (negative association)
c-myc	Markers	S to G_2 checkpoint	45 (negative association)
ATM	Markers	Checkpoint	45
p44/p42 MAPK (ERK1/2)	Cycle-associated protein	MAP kinase	50
p38 MAPK	Cycle-associated protein	Kinase	48,50
JNK/(SAPK-2/3)-αγ	Cycle-associated protein	Kinase (stress-activated)	50
GSK-3 and β-Catenin	Cycle-associated protein	Proline-dependent protein kinase (PDPK)	60
Sos-1	Cycle-associated protein	Guanine nucleotide exchange factor	46
Grb-2	Cycle-associated protein	Adaptor	46
p16INK4a p18p15p19	Markers	Cyclin D/cdk4/6 inhibitors of M phase	17
p27/Kip1	Markers	Cyclin D and E/cdk7 inhibitor	72 (negative association)

pathology of AD. Senile plaques, on the other hand, are extracellular and primarily composed of amyloid-β (Aβ).

Both Aβ senile plaques and tangles develop preferentially in specific areas of the brain that are associated with cognition such as the hippocampus, entorhinal cortex, and association areas of the neocortex. These areas of the brain show atrophy owing to neuronal death. The mechanisms involved in the formation of these lesions and neuronal death are largely unknown although recent findings indicate a key role for the aberrant re-entry of neurons into the cell cycle.

A growing number of cell cycle-related proteins are associated with the susceptible and vulnerable neurons of AD (Table 1). From their temporal and pathological distribution, such cell cycle changes are indicative of an early and fundamental role in the pathogenesis of AD. The activation of a number of markers, which signal cell growth prior to mitosis including a surge in mitochondrial activity, is seen in AD neurons *(3)* in addition to activation of cyclin-dependent kinases and signal transduction cascades that can lead to τ phosphorylation *(3,4)*. Nonetheless, it is important to note that changes in cell fate ultimately occur through signal transducers that activate specific transcription factors and modulate cell cycle control proteins. These proteins themselves are also regulated in a cell cycle-dependent manner. Therefore, different factors associated with various levels of cell cycle progression should be taken into consideration in delineating the pathogenic mechanisms involved in AD *(5)*. Notably, the possibility exists that senile plaques and NFTs are distal products of these aberrant cell cycle events and may represent neuronal compensations *(3,4)*. Our hypothesis is that Aβ-containing senile plaques are secondary to underlying primary cell cycle-related events in AD.

2. Aβ: THE PATHOLOGICAL HALLMARK OF AD

Pathological formation and deposition of amyloid is a characteristic feature not only of AD *(2)* but also of other diseases, collectively termed amyloidosis, such as Down syndrome *(6,7)*, cerebral amyloid angiopathy *(8)*, multiply myeloma *(9)*, Creutzfeld–Jacob disease *(10)*, or familial amyloid polyneuropathies, and other systemic amyloidoses *(11)*. In each case, a different amyloid protein is responsible for the pathology. For example, in familial amyloid polyneuropathy, the causative deposits consist of mutated transthyretin forming amyloid fibrils, particularly in the peripheral nervous system *(11)*. In AD, on the other hand, the misfolded protein, Aβ, deposits in the central nervous system *(1,2)*. Aβ is derived from a larger precursor, Aβ protein precursor (AβPP), encoded on chromosome 21 *(12)*, which is a cell transmembrane protein, consisting of a large extracellular spanning region and intracellular carboxyterminus. Physiological cleavage results in secretion of the extracellular domain, secreted AβPP, which appears in the medium of cells and CSF. Secreted AβPP has been reported to increase cell survival and adhesion and prevent intracellular calcium accumulation and death of neurons *(5,13)*. Cleavage by α-secretase ends with soluble products. On the other hand, cleavage by β- and γ-secretases generates Aβ40 and Aβ42.

There is a long-standing debate on whether AβPP and Aβ are the cause or the result of neurodegeneration *(2)*. Arguments for a causative role for Aβ in AD include the fact that patients with Down syndrome show Aβ deposition early and often develop dementia by their mid-thirties *(14)*. Moreover, mutations causing AD have been identified on the three separate genes, one of which is in the *AβPP* gene on chromosome 21 and all of which rise to the full spectrum of AD pathology. Likewise, in vitro studies demonstrated that oligomers of Aβ are toxic to neurons by eliciting an inflammatory response from glial cells *(15)*, activating apoptotic cascades and causing microtubule collapse leading to τ protein hyperphosphorylation and depleting presynaptic AβPP with loss of synaptic transmission *(15)*. Nonetheless, how Aβ senile plaques themselves can influence the process of neurodegeneration and whether they are a cause or effect on degeneration remains to be fully determined.

3. THE ASSOCIATION BETWEEN CELL CYCLE REGULATION AND AD

For decades the predominant thinking was adult neurons remain in a nonproliferative state. In AD, however, there is accumulating evidence that susceptible neuronal populations attempt to re-enter into the cell cycle. Quiescence, cell division, and differentiation are states central to regulation of growth and development. Increased growth stimuli, otherwise known as extrinsic mitotic pressure, activate key factors for the progression from G_0 into G_1. These include the complex-forming cyclin-dependent kinases (CDKs), that is, CDKs 4–7 (*see* Table 1), and their cognate-activating cyclins, that is, cyclins D_1, D_3, E, and B1 (*see* Table 1). Such complexes *(16)* are able to regulate DNA replication, cytoskeletal reorganization, and cellular metabolism required for proliferation, development, and cell cycle progression. This unexpected exit from a quiescent state in AD is manifested in several ways, including the ectopic expression of cyclins along with their cognate CDKs and their inhibitors (CDKIs) *(17,18)* and recruitment of mitogenic signal transduction pathway components *(18,19)*. Notably, increased transcriptional activation of a variety of mitosis-related proteins also occurs in AD *(3,19)*. Although the cause of this apparent neuronal re-entry into the cell cycle is not fully known, the consequences for these terminally differentiated cells are disastrous and lead to oxidative stress, cytoskeletal abnormalities, mitochondrial dysfunction, and, ultimately, neuronal death *(3)*. Therefore, in sum, the re-emergence into the cell cycle by neurons accounts for many of the cardinal features of the disease.

It has been argued that a number of the cell cycle-related phenomena found in AD also occur in other cellular processes, such as apoptosis, trophic factor deprivation, and DNA repair *(17–20)*. Furthermore, the signals which lead the potential AD neuron to re-enter the cell cycle remains to be determined. Finally, it is important to note that there is no evidence that a nuclear division has been completed in these vulnerable AD neurons although there is circumstantial evidence of increased DNA in certain vulnerable neuronal populations *(21)*. Additionally, proteins which downregulate cell cycle, such as p16, are unregulated *(22)*. Terminally differentiated neurons in the adult central nervous system are inherently restricted in their ability to divide and, as such, these neurons stop short of actual division and become arrested in an "intermediate state" of the cell cycle. This may account for markers from the different phases of the cell cycle being present in the diseased brain. Indeed, exit from quiescent state is seen by the presence of positive modulators of the cell division cycle such as CDK4 and CDK2 *(23,24)*, which represent the earliest changes of the disease. Therefore, it is unlikely that cell cycle reemergence is epiphenomenal but is rather a direct causative mechanism of neurons death and degeneration *(1)*. In support of this notion, a number of growth factors are known to be elevated in the AD brain, which may drive cell cycle re-entry. Included among such candidate growth factors elevated in the AD brain are neurotrophic factors, nerve growth factor (NGF) *(25)*, transforming growth factor β-1 (TGF-β) *(26)*, platelet-derived growth factor (PDGF), epidermal growth factor (EGF), and basic fibroblast growth factor (bFGF) *(24,27)*. Additionally, insulin-like growth factor-1, which has been shown to mediate transient site-selective increases in τ phosphorylation in primary cortical neurons *(28)*, is involved in axonal growth and development and can mediate the cytoskeletal reorganization that occurs during neurite outgrowth and, perhaps, in aberrant neuronal sprouting *(29)*. This accumulating evidence cements the notion of the early and possibly irreversible fate of these vulnerable neurons. It is fascinating that neurons in phase G_0 for nearly 50–60 yr make compensations that include, among others, an attempted re-entry into the cell cycle. This ectopic cell cycle activation is also seen in mitotically active cells such as those seen during neoplastic transformation and neurogenesis *(1,30)*. The mechanisms of cell cycle activation may differ at many levels, but what is most intriguing is that in AD this re-entry is incomplete and results in stasis, dysfunction, and death. We termed this process as abortosis *(31)*.

4. Aβ AND THE OTHER CELL CYCLE-RELATED FACTORS IN AD

Interestingly, all of the pathological proteins and factors that are seen in AD are sensitive to changes in the cell cycle. The most important are the τ protein *(32)*, AβPP, and PSEN1 and 2 *(33)*.

4.1. AβPP and Aβ

As previously mentioned, Aβ is derived from a larger precursor, AβPP, encoded on chromosome 21 *(12)*. Moreover, familial AD is linked to mutations in the *AβPP* gene. However, even more interesting is the involvement of AβPP in cell cycle-related events. AβPP is upregulated secondary to mitogenic stimulation and AβPP, through the stimulation of Ras-dependent mitogen-activated protein kinases (MAPKs) cascade in vivo is correlated with highly phosphorylated τ *(13)*. Likewise, p21 Ras expression pathway is activated during the post-translational modification of AβPP and τ phosphorylation, which precedes neurofibrillary degeneration and Aβ senile plaque formation *(34)*. Additionally, the presence of p21, highly phosphorylated τ, Ki-67, and cell cycle-associated nuclear antigen protein proliferating cell nuclear antigen (PCNA) may have a role in the production of abnormally phosphorylated τ, which then leads to the formation of cytoskeletal derangements in susceptible neurons *(35)*. This strong link between AβPP and cell cycle events points to cell cycle reactivation and the upstream ectopic expression of cell cycle markers as a critical, and common, early event in AD pathogenesis. Ledoux et al. found that AβPP is upregulated with cell activation *(36)* and activation of mononuclear cells with a cell mitogen led to an increase of AβPP. This suggests that AβPP may participate in the regulation of cell activation in cells and that this could contribute to a circulating pool of AβPP and Aβ. Furthermore, AβPP metabolism is regulated by cell cycle-dependent changes *(37)* and has neurotrophic effects at low (nM) concentrations *(38)* consistent with its mitogenic activity in vitro *(39,40)*. Moreover, similar to AβPP, Aβ itself is upregulated secondary to mitogenic stimulation *(36)*. Presumably, the effect of Aβ is mediated through the MAPK pathway *(41)* and hence could play a direct role in the induction or propagation of cell cycle-mediated events in AD. Therefore, Aβ, along with oxidative stress *(42)* and cell cycle re-entry, may be a part of the same pathological process. However, it is notable that, whereas Aβ-mediated cell death, at least in vitro, is dependent on the presence of various cell cycle-related elements *(43)*, in vivo analysis of the basal nucleus of Meynert and the locus ceruleus, where Aβ is rarely seen, found little or no topographical relationship between Aβ and the ectopic expression of cell cycle markers in diseased brains *(44–50)*. Thus, Aβ may only become toxic in vivo when the neuronal cell cycle machinery is activated or when levels exceed the body's ability to regulate its turnover. Importantly, as highlighted above, cell cycle- and Aβ-related mechanisms also involve other hallmark features of the disease, such as hyperphosphorylated τ.

4.2. τ Phosphorylation

The hyperphosphorylated form of the microtubule associated protein τ forms the core of NFT and is the central intracellular pathological protein seen in AD *(51,52)*. Increased hyperphosphorylation of τ renders it incapable of participation in microtubular dynamics in the neuron and an inability to participate in the stability of microtubules that leads to neuronal dysfunction *(30,53)*. Regarding cell cycle, it is important to note that physiological hyperphosphorylation of τ, driven by CDKs, occurs when cells are mitotically active *(54–58)*. Of note, CDKs, such as CDK2 and CDK5, as well as Cdc kinases and MAP2 kinases, are increased in AD in a topographical manner that completely overlaps with phospho-τ *(59,60)*. Moreover, the same enzymes have been shown to hyperphosphorylate τ in vitro *(30,61–64)*. In addition, CDK7, an age-dependent CDK-activating kinase, is also associated with phospho-τ in AD and may be essential to all other mitotic alterations because CDK7 plays a crucial role as an activator of all the major CDK/cyclin substrates *(44)*.

4.3. Presenilin

Mutations of human presenilin genes, *PSEN1* and *PSEN2*, on chromosomes 14 and 1, are linked to early onset AD *(1)*. Although the functions of the human PSENs are not fully understood to date, their homologues are involved in Notch processing and signaling, and, as such, it is not surprising that these two proteins have redundant roles in cell fate *(65–68)* determination in cytoskeletal anchorage, cell division, early embryonic development *(69)*, and tumorigenesis *(70)*, and interact with centrosomes and kinetochores, thus giving them a significant role during chromosomal segregation and mitosis *(71)*. In this regard, overexpression of PSENs leads to G_1 arrest and this is potentiated by the PSEN2 (N141I) mutation *(72)*. Keeping in mind that PSEN expression arrests cells in the G_1 phase of the cell cycle, changes in the levels of the cell cycle inhibitory proteins p21, p27, and p53 and the cell cycle regulatory proteins c-myc and pRb in PSEN-transfected cell lysates revealed no increases or decreases for any of these proteins suggesting that PSENs mediate cell cycle arrest by mechanisms other than simple changes in the steady-state levels of these cell cycle-related proteins *(72)*. What is interesting is that PSEN overexpression in an adult neuron environment leads to oxidative stress, loss of calcium balance, and also to greater vulnerability to apoptosis *(73)*. Indeed, AD-linked PSENs show greater apoptotic effects *(74)*. Therefore, neurons that are vulnerable to AD have exited from G_0 and entered G_1 but may be blocked from progressing into S phase at the G_1/S boundary by the PSEN (and possibly AβPP) mutations. This G_1 block would lead to an accumulation of cell cycle regulatory proteins and produce the cell cycle stasis seen in AD. Given this, it is not surprising that these mutations yield a contracted time-course of the basic pathophysiology in AD. It is also notable that PSEN and AβPP mutations may lead to the upregulation of CDKs, which can have dual roles in both cell cycle regulation as well as in death signaling *(43,70)*. Indeed, the role of the cell cycle-inhibitory proteins, such as p16, p19, and p21, which also accumulate in AD can act to blunt possible apoptotic stimulation in early AD *(13–15,34)*. A role for the PSENs in cell cycle control may have relevance to both apoptotic and developmental mechanisms owing to the interrelationships of cell death, growth, and differentiation *(71,74)*.

4.4. ApoE

Several genetic studies indicate that AD is associated with *ApoE-ε4 (1,2)* with increased Aβ deposition in patients with AD with an *ApoE4* allele *(75)*. ApoE is a 34-kDa protein associated with triglyceride-rich lipoproteins and high-density lipoproteins (HDL) *(75)*. ApoE may be also involved in cell cycle regulation *(76,77)*. Kothapalli et al. have shown that both high-density lipoproteins and ApoE block S phase entry and cyclin A gene expression and results in G_1-phase arrest *(76)*. The process, however, was found to be indirect *(77)*. Moreover, it was found that neuronal injury induces glial reaction with increased secretion and upregulation *(77)*.

5. CELL CYCLE RE-ENTRY, REDOX IMBALANCE, AND AD

There is abundant evidence that oxidative stress and free radical damage play an essential role in the pathogenesis of AD *(78,79)*. Free radicals, free-radical generators, and antioxidants also act as crucial control parameters of the cell cycle *(80,81)*. It should be noted here that energy is an obligate requirement for dividing cells. Therefore, before mitosis, mitochondrial proliferation is most evident *(82)*. Notably, in AD, increases in mitochondrial DNA, indicative of a premitotic stage, are found in the same neurons that also exhibit cell cycle-related abnormalities and undergo subsequent oxidative damage and cell death *(82)*. Although in a normally mitotic cell, mitochondrial replication is imperative for providing the energy needed for cell division; in AD, in which neuronal cell cycle is interrupted, there is a possibility that neurons incur a stasis with disruption of normal mitochondrial dynamics. Such a mitochondrial imbalance would serve as a potent source of free radicals and cause redox imbalances, especially in redox reactions involving calcium metabolism *(83)*. Thus, cell cycle dysfunction, when mitochondrial mass is highest, has the effect of precipitating an oxidative assault on the cell.

Imbalances in redox homeostasis may be linked to cell cycle control via numerous signal transduction cascades. MEK, ERK1/2, cyclins, cyclin-dependent kinases, and their inhibitors, that is, p16INK4a family, and p21Ras are elevated early in AD and colocalize in pyramidal neurons with NFT *(83)*. Phosphorylation of neuronal ERK, p38, and CREB in AD is induced by nerve growth factor or EGF and is differentially modulated by oxidative and other stresses *(84)*. In support of this notion, compromised mitochondrial function was found to lead to increased cytosolic calcium and to the activation of MAPKs (ERK1/2) *(85)*. Indeed, activation of p38, MAPK, and ERK links τ phosphorylation, oxidative stress, and cell cycle-related events in AD *(48)*.

6. CONCLUSIONS

It is becoming increasingly apparent that an altered cell cycle exists in susceptible neurons in AD. Although these abnormalities may be a partial response to the stress and metabolic imbalance common in degenerating neurons, it is equally likely that such changes are a driving mechanistic force in the disease process. In this regard, dysregulation of the cell cycle fulfills both the sufficient and necessary criteria for the initiation of an oncogenic transformation. Unfortunately, this re-entrance into cell cycle appears to be unsustainable in neurons and leads to stasis in a specific phase of the cell division cycle, with resultant cellular dysfunction and neuronal death. Therapeutic efforts targeted at mediators of such cell cycle alterations may prove extremely efficacious *(86–88)*.

ACKNOWLEDGMENTS

This work was supported in part by the NIH AG1713, Philip Morris USA, the Nickman Family, the Fullerton Foundation, the Kosciuszko Foundation, the GOJO Corporation, the Florence and Joseph Mandel Foundation, and the Alzheimer's Association.

REFERENCES

1. Smith MA. Alzheimer disease. Int Rev Neurobiol 1998;42:1–54.
2. National Institutes of Health, Progress report on Alzheimer's disease. 2001–2002;http://www.alzheimers.org/pr01-02/research.htm.
3. Raina AK, Monteiro MJ, McShea A, Smith MA. The role of cell cycle-mediated events in Alzheimers disease. Int J Exp Pathol 1999;80:71–76.
4. Zhu X, Raina AK, Smith MA. Cell cycle events in neurons. Proliferation or death? Am J Pathol 1999;155:327–329.
5. Stieler JT, Lederer C, Bruckner MK, et al. Impairment of mitogenic activation of peripheral blood lymphocytes in Alzheimer's disease. Neuroreport 2001;12:3969–3972.
6. van Leeuwen FW, de Kleijn DP, van den Hurk HH, et al. Frameshift mutants of beta amyloid precursor protein and ubiquitin-B in Alzheimer's and Down patients. Science 1998;279:242–247.
7. Friedlich AL, Butcher LL. Involvement of free oxygen radicals in beta-amyloidosis: an hypothesis. Neurobiol Aging 1994;15:443–455.
8. Castellani RJ, Smith MA, Perry G, Friedland RP. Cerebral amyloid angiopathy: major contributor or decorative response to Alzheimer's disease pathogenesis. Neurobiol. Aging 2004;25, 599–602; discussion 603–594.
9. Gafumbegete E, Richter S, Jonas L, Nizze H, Makovitzky J. Nonsecretory multiple myeloma with amyloidosis. A case report and review of the literature. Virchows Arch 2004;445:531–536.
10. Goldfarb LG, Brown P. The transmissible spongiform encephalopathies. Annu Rev Med 1995;46:57–65.
11. Sousa MM, Saraiva MJ. Neurodegeneration in familial amyloid polyneuropathy: from pathology to molecular signaling. Prog Neurobiol 2003;71:385–400.
12. Selkoe DJ. Alzheimer's disease results from the cerebral accumulation and cytotoxicity of amyloid beta-protein. J Alzheimers Dis 2001;3:75–80.
13. Greenberg SM, Koo EH, Selkoe DJ, Qiu WQ, Kosik KS. Secreted beta-amyloid precursor protein stimulates mitogen-activated protein kinase and enhances tau phosphorylation. Proc Natl Acad Sci USA 1994;91:7104–7108.
14. Ezquerra M, Ballesta F, Queralt R, et al. Apolipoprotein E epsilon 4 alleles and meiotic origin of non-disjunction in Down syndrome children and in their corresponding fathers and mothers. Neurosci Lett 1998;248:1–4.
15. Clark CM, Karlawish JH. Alzheimer disease: current concepts and emerging diagnostic and therapeutic strategies. Ann Intern Med 2003;138:400–410.
16. Nagy ZS, Smith MZ, Esiri MM, Barnetson L, Smith AD. Hyperhomocysteinaemia in Alzheimer's disease and expression of cell cycle markers in the brain. J Neurol Neurosurg Psychiatry 2000;69:565,566.
17. Arendt T, Rodel L, Gartner U, Holzer M. Expression of the cyclin-dependent kinase inhibitor p16 in Alzheimer's disease. Neuroreport 1996;7:3047–3049.
18. Park DS, Morris EJ, Padmanabhan J, Shelanski ML, Geller HM, Greene LA. Cyclin-dependent kinases participate in death of neurons evoked by DNA-damaging agents. J Cell Biol 1998;143:457–467.
19. Elledge SJ. Cell cycle checkpoints: preventing an identity crisis. Science 1996;274:1664–1672.
20. Padmanabhan J, Park DS, Greene LA, Shelanski ML. Role of cell cycle regulatory proteins in cerebellar granule neuron apoptosis. J Neurosci 1999;19:8747–8756.
21. Yang Y, Mufson EJ, Herrup K. Neuronal cell death is preceded by cell cycle events at all stages of Alzheimer's disease. J Neurosci 2003;23:2557–2563.
22. Bowser R, Smith MA. Cell cycle proteins in Alzheimer's disease: plenty of wheels but no cycle. J Alzheimers Dis 2002;4:249–254.
23. McShea A, Harris PL, Webster KR, Wahl AF, Smith MA. Abnormal expression of the cell cycle regulators P16 and CDK4 in Alzheimer's disease. Am J Pathol 1997;150:1933–1939.
24. Koliatsos VE. Biological therapies for Alzheimer's disease: focus on trophic factors. Crit Rev Neurobiol 1996;10:205–238.
25. Crutcher KA, Scott SA, Liang S, Everson WV, Weingartner J. Detection of NGF-like activity in human brain tissue: increased levels in Alzheimer's disease. J Neurosci 1993;13:2540–2550.
26. van der Wal EA, Gomez-Pinilla F, Cotman CW. Transforming growth factor-beta 1 is in plaques in Alzheimer and Down pathologies. Neuroreport 1993;4:69–72.

27. Stopa EG, Gonzalez AM, Chorsky R, et al. Basic fibroblast growth factor in Alzheimer's disease. Biochem Biophys Res Commun 1990;171:690–696.
28. Lesort M, Johnson GV. Insulin-like growth factor-1 and insulin mediate transient site-selective increases in tau phosphorylation in primary cortical neurons. Neuroscience 2000;99:305–316.
29. Russell JW, Windebank AJ, Schenone A, Feldman EL. Insulin-like growth factor-I prevents apoptosis in neurons after nerve growth factor withdrawal. J Neurobiol 1998;36:455–467.
30. Lindwall G, Cole RD. Phosphorylation affects the ability of tau protein to promote microtubule assembly. J Biol Chem 1984;259:5301–5305.
31. Raina AK, Hochman A, Zhu X, et al. Abortive apoptosis in Alzheimer's disease. Acta Neuropathol (Berl) 2001;101:305–310.
32. An WL, Cowburn RF, Li L, et al. Up-regulation of phosphorylated/ activated p70 S6 kinase and its relationship to neurofibrillary pathology in Alzheimer's disease. Am J Pathol 2003;163:591–607.
33. Raina AK, Zhu X, Rottkamp CA, Monteiro M, Takeda A, Smith MA. Cyclin' toward dementia: cell cycle abnormalities and abortive oncogenesis in Alzheimer disease. J Neurosci Res 2000;61:128–133.
34. Gartner U, Holzer M, Arendt T. Elevated expression of p21ras is an early event in Alzheimer's disease and precedes neurofibrillary degeneration. Neuroscience 1999;91:1–5.
35. Smith TW, Lippa CF. Ki-67 immunoreactivity in Alzheimer's disease and other neurodegenerative disorders. J Neuropathol Exp Neurol 1995;54:297–303.
36. Ledoux S, Rebai N, Dagenais A, et al. Amyloid precursor protein in peripheral mononuclear cells is up-regulated with cell activation. J Immunol 1993;150:5566–5575.
37. Suzuki T, Oishi M, Marshak DR, Czernik AJ, Nairn AC, Greengard P. Cell cycle-dependent regulation of the phosphorylation and metabolism of the Alzheimer amyloid precursor protein. EMBO J 1994;13:1114–1122.
38. Whitson JS, Selkoe DJ, Cotman CW. Amyloid beta protein enhances the survival of hippocampal neurons in vitro. Science 1989;243:1488–1490.
39. McDonald DR, Bamberger ME, Combs CK, Landreth GE. beta-Amyloid fibrils activate parallel mitogen-activated protein kinase pathways in microglia and THP1 monocytes. J Neurosci 1998;18:4451–4460.
40. Pyo H, Jou I, Jung S, Hong S, Joe EH. Mitogen-activated protein kinases activated by lipopolysaccharide and beta-amyloid in cultured rat microglia. Neuroreport 1998;9:871–874.
41. Rapoport M, Ferreira A. PD98059 prevents neurite degeneration induced by fibrillar beta-amyloid in mature hippocampal neurons. J Neurochem 2000;74:125–133.
42. Obrenovich ME, Joseph JA, Atwood CS, Perry G, Smith MA. Amyloid-beta: a (life) preserver for the brain. Neurobiol Aging 2002;23:1097–1099.
43. Giovanni A, Wirtz-Brugger F, Keramaris E, Slack R, Park DS. Involvement of cell cycle elements, cyclin-dependent kinases, pRb, and E2F x DP, in B-amyloid-induced neuronal death. J Biol Chem 1999;274:19,011–19,016.
44. Zhu X, Rottkamp CA, Raina AK, et al. Neuronal CDK7 in hippocampus is related to aging and Alzheimer disease. Neurobiol Aging 2000;21:807–813.
45. McShea A, Wahl AF, Smith MA. Re-entry into the cell cycle: a mechanism for neurodegeneration in Alzheimer disease. Med Hypotheses 1999;52:525–527.
46. McShea A, Zelasko DA, Gerst JL, Smith MA. Signal transduction abnormalities in Alzheimer's disease: evidence of a pathogenic stimuli. Brain Res 1999;815:237–242.
47. Busser J, Geldmacher DS, Herrup K. Ectopic cell cycle proteins predict the sites of neuronal cell death in Alzheimer's disease brain. J Neurosci 1998;18:2801–2807.
48. Zhu X, Rottkamp CA, Boux H, Takeda A, Perry G, Smith MA. Activation of p38 kinase links tau phosphorylation, oxidative stress, and cell cycle-related events in Alzheimer disease. J Neuropathol Exp Neurol 2000;59:880–888.
49. Zhu X, Raina AK, Boux H, Simmons ZL, Takeda A, Smith MA. Activation of oncogenic pathways in degenerating neurons in Alzheimer disease. Int J Dev Neurosci 2000;18:433–437.
50. Zhu X, Castellani RJ, Takeda A, et al. Differential activation of neuronal ERK, JNK/SAPK and p38 in Alzheimer disease: the 'two hit' hypothesis. Mech Ageing Dev 2001;123:39–46.
51. Iqbal K, Zaidi T, Thompson CH, Merz PA, Wisniewski HM. Alzheimer paired helical filaments: bulk isolation, solubility, and protein composition. Acta Neuropathol (Berl) 1984;62:167–177.
52. Grundke-Iqbal I, Iqbal K, Tung YC, Quinlan M, Wisniewski HM, Binder LI. Abnormal phosphorylation of the microtubule-associated protein tau (tau) in Alzheimer cytoskeletal pathology. Proc Natl Acad Sci USA 1986;83:4913–4917.

53. Alonso AC, Grundke-Iqbal I, Iqbal K. Alzheimer's disease hyperphosphorylated tau sequesters normal tau into tangles of filaments and disassembles microtubules. Nat Med 1996;2:783–787.
54. Brion JP, Passarier H, Nunez J, Flament-Durand J. Immunologic determinants of tau protein are present in neurofibrillary tangles of Alzheimer's disease. Arch Biol 1985;95:229–235.
55. Brion JP, Octave JN, Couck AM. Distribution of the phosphorylated microtubule-associated protein tau in developing cortical neurons. Neuroscience 1994;63:895–909.
56. Kanemaru K, Takio K, Miura R, Titani K, Ihara Y. Fetal-type phosphorylation of the tau in paired helical filaments. J Neurochem 1992;58:1667–1675.
57. Goedert M, Jakes R, Crowther RA, et al. The abnormal phosphorylation of tau protein at Ser-202 in Alzheimer disease hrecapitulates phosphorylation during development. Proc Natl Acad Sci USA 1993;90:5066–5070.
58. Pope WB, Lambert MP, Leypold B, et al. Microtubule-associated protein tau is hyperphosphorylated during mitosis in the human neuroblastoma cell line SH-SY5Y. Exp Neurol 1994;126:185–194.
59. Ledesma MD, Correas I, Avila J, Diaz-Nido J. Implication of brain cdc2 and MAP2 kinases in the phosphorylation of tau protein in Alzheimer's disease. FEBS Lett 1992;308:218–224.
60. Baumann K, Mandelkow EM, Biernat J, Piwnica-Worms H, Mandelkow E. Abnormal Alzheimer-like phosphorylation of tau-protein by cyclin-dependent kinases cdk2 and cdk5. FEBS Lett 1993;336:417–424.
61. Arendt T, Holzer M, Grossmann A, Zedlick D, Bruckner MK. Increased expression and subcellular translocation of the mitogen activated protein kinase kinase and mitogen-activated protein kinase in Alzheimer's disease. Neuroscience 1995;68:5–18.
62. Vincent I, Rosado M, Davies P. Mitotic mechanisms in Alzheimer's disease? J Cell Biol 1996;132:413–425.
63. Nagy Z, Esiri MM, Smith AD. Expression of cell division markers in the hippocampus in Alzheimer's disease and other neurodegenerative conditions. Acta Neuropathol (Berl) 1997;93:294–300.
64. Nagy Z, Esiri MM, Cato AM, Smith AD. Cell cycle markers in the hippocampus in Alzheimer's disease. Acta Neuropathol (Berl) 1997;94:6–15.
65. Raina AK, Zhu X, Monteiro M, Takeda A, Smith MA. Abortive oncogeny and cell cycle-mediated events in Alzheimer disease. Prog Cell Cycle Res 2000;4:235–242.
66. Wong PC, Zheng H, Chen H, et al. Presenilin 1 is required for Notch1 and Dll1 expression in the paraxial mesoderm. Nature 1997;387:288–292.
67. Struhl G, Greenwald I. Presenilin is required for activity and nuclear access of Notch in Drosophila. Nature 1999;398:522–525.
68. Ye Y, Lukinova N, Fortini ME. Neurogenic phenotypes and altered Notch processing in Drosophila Presenilin mutants. Nature 1999;398:525–529.
69. Wodarz A, Nusse R. Mechanisms of Wnt signaling in development. Annu Rev Cell Dev Biol 1998;14:59–88.
70. Polakis P. The oncogenic activation of beta-catenin. Curr Opin Genet Dev 1999;9:15–21.
71. Li J, Xu M, Zhou H, Ma J, Potter H. Alzheimer presenilins in the nuclear membrane, interphase kinetochores, and centrosomes suggest a role in chromosome segregation. Cell 1997;90:917–927.
72. Janicki SM, Monteiro MJ. Presenilin overexpression arrests cells in the G1 phase of the cell cycle. Arrest potentiated by the Alzheimer's disease PS2(N141I)mutant. Am J Pathol 1999;155:135–144.
73. Mattson MP, Guo Q, Furukawa K, Pedersen WA. Presenilins, the endoplasmic reticulum, and neuronal apoptosis in Alzheimer's disease. J Neurochem 1998;70:1–14.
74. Wolozin B, Iwasaki K, Vito P, et al. Participation of presenilin 2 in apoptosis: enhanced basal activity conferred by an Alzheimer mutation. Science 1996;274:1710–1713.
75. Rubinsztein DC. Apolipoprotein E: a review of its roles in lipoprotein metabolism, neuronal growth and repair and as a risk factor for Alzheimer's disease. Psychol Med 1995;25:223–229.
76. Kothapalli D, Fuki I, Ali K, et al. Antimitogenic effects of HDL and APOE mediated by Cox-2-dependent IP activation. J Clin Invest 2004;113:609–618.
77. Petegnief V, Saura J, de Gregorio-Rocasolano N, Paul SM. Neuronal injury-induced expression and release of apolipoprotein E in mixed neuron/glia co-cultures: nuclear factor kappaB inhibitors reduce basal and lesion-induced secretion of apolipoprotein E. Neuroscience 2001;104:223–234.
78. Smith MA, Sayre LM, Monnier VM, Perry G. Radical AGEing in Alzheimer's disease. Trends Neurosci 1995;18:172–176.
79. Markesbery WR. Oxidative stress hypothesis in Alzheimer's disease. Free Radic Biol Med 1997;23:134–147.

80. Curcio F, Ceriello A. Decreased cultured endothelial cell proliferation in high glucose medium is reversed by antioxidants: new insights on the pathophysiological mechanisms of diabetic vascular complications. In Vitro Cell Dev Biol 1992;28A:787–790.
81. Barni S, Sciola L, Spano A, Pippia P. Static cytofluorometry and fluorescence morphology of mitochondria and DNA in proliferating fibroblasts. Biotech Histochem 1996;71:66–70.
82. Hirai K, Aliev G, Nunomura A, et al. Mitochondrial abnormalities in Alzheimer's disease. J Neurosci 2001;21:3017–3023.
83. Sousa M, Barros A, Silva J, Tesarik J. Developmental changes in calcium content of ultrastructurally distinct subcellular compartments of preimplantation human embryos. Mol Hum Reprod 1997;3:83–90.
84. Zhang L, Jope RS. Oxidative stress differentially modulates phosphorylation of ERK, p38 and CREB induced by NGF or EGF in PC12 cells. Neurobiol Aging 1999;20:271–278.
85. Luo Y, Bond JD, Ingram VM. Compromised mitochondrial function leads to increased cytosolic calcium and to activation of MAP kinases. Proc Natl Acad Sci USA 1997;94:9705–9710.
86. Bowen RL, Atwood CS, Perry G, Smith MA. Mechanisms involved in gender differences in Alzheimer's disease: the role of leuteinizing and follicle stimulating hormones. In: Legato ML (ed.) Principles of Gender Specific Medicine. San Diego, Academic Press, 2004; pp 1234–1237.
87. Zhu X, Raina AK, Perry G, Smith MA. Alzheimer's disease: the two-hit hypothesis. Lancet Neurol 2004;3:219–226.
88. Webber KM, Bowen R, Casadesus G, Perry G, Atwood CS, Smith MA. Gonadotropins and Alzheimer's disease: the link between estrogen replacement therapy and neuroprotection. Acta Neurobiol Exp (Wars) 2004;64:113–118.

23
Neuronal Precursor Proliferation and Epileptic Malformations of Cortical Development

Jorge A. González-Martínez, MD, PhD, William E. Bingaman, MD, and Imad M. Najm, MD

SUMMARY

The pathogenic mechanisms responsible for the development of cerebral malformations in humans, its epileptogenicity, and its relation with postnatal progenitor cell proliferation and neurogenesis are largely unknown. This chapter summarizes data concerning mechanisms of corticogenesis and postnatal neurogenesis and its possible relation with epileptogenesis in malformations of cortical development.

Key Words: Neurogenesis; epilepsy; apoptosis; stem cell.

1. INTRODUCTION

In recent years, methods of cellular biology and other in vitro approaches have provided unparalleled opportunity to study the molecular mechanisms of neuronal precursor proliferation and its relation with epileptogenicity in some forms of malformations of cortical development (MCD). Here, we present a brief summary of data accumulated over the past four decades concerning vertebrates' neurogenesis in the pre- and postnatal period. We will also summarize recent findings addressing a possible association of neurogenesis with some forms of epileptic MCD.

2. CELL CYCLE IN THE NORMAL MAMMALIAN TELENCEPHALON: CORTICOGENESIS AND POSTNATAL NEUROGENESIS

2.1. Normal Corticogenesis

Recently, several experimental papers have been published about the molecular and cellular processes of cortical mantle development *(1–3)*. The neurons of the cerebral cortex are generated in the ventricular zone, an epithelial layer of progenitor cells that lines the lateral ventricles. Once they have left the proliferative stage, the immature neurons migrate out of the ventricular zone to form the cortical plate, which eventually becomes the gray matter of the cerebral cortex. To reach the cortical plate, these cells migrate under the guidance of radial glial cells, a specialized class of glial cells that preserve contact with both the ventricular and cortical surface. Within the cortical plate, neurons become organized into well-defined layers. The final position of a cortical neuron is correlated precisely with the "birth" of the neuron, or the time at which a dividing precursor undergoes its final round of cell division and gives rise to a postmitotic neuron. Cells that migrate from the ventricular zone and leave the cell cycle at early stages give rise to neurons that settle in the deepest layers of the cortex. In contrast, cells that leave the ventricular zone and exit the cell cycle at progressively later stages migrate over longer distances, past early born neurons, and settle in more superficial layers of the cortex *(4,5)*. Thus, the layering of neurons in the cerebral cortex is established in an inside-first, outside-last manner.

Is the subtype of the newly generated neurons dependent on the cell cycle stage of the progenitor cells? This question has been answered in experiments in which progenitor cells from the ventricular zone of young animals were transplanted into older animals *(4)*. Normally, these cells migrate to deep layers, but in old animals the transplanted cells intermingled, with cells on their way to superficial layers. Remarkably, the destiny of the transplanted cells turns out to depend on the phase of the cell cycle they are in at the time of transplantation. Progenitor cells in S phase at the time of transplantation acquire the fate of neurons generated at later stages of cortical neurogenesis. That is, they give rise to neurons that settle in the superficial layers of the cortex. In contrast, cells that have passed through their final S phase prior to transplantation adopt the normal fate of neurons generated at early stages of cortical neurogenesis. That is, they migrate to layers 5 and 6. Thus young cortical progenitor cells remain sensitive to time-dependent environmental signals that direct their destiny, but as they become postmitotic they also become committed to a specific fate. In contrast, progenitor cells present at later stages are not able to acquire the fate of younger neurons when transplanted in the ventricular zone of a younger host.

The degree of progenitor cell proliferation varies according to the corticogenesis stage. At early stages, progenitors that give rise to neurons of the mature cerebral cortex proliferate rapidly and give rise to additional progenitors, expanding the population of neuronal precursor cells. At later stages, progenitor cells alter their program of cell division and give rise both to neurons and to additional progenitor cells. At a still later stage, they generate only neurons *(5)*.

The number of postmitotic cells during corticogenesis is tailored by a genetically determined regressive mechanism, the physiologically occurring cell death, which is recognized as a prominent developmental event of the vertebrate neurogenesis *(6)*. Based on its time of occurrence during the stages of the central nervous system maturation, two different types of cell death have been indicated *(7)*:

1. "Early or proliferative" cell death, which is observed in the ventricular epithelium during the early stages of development, with simultaneous occurrence of both proliferation and apoptosis. In fact, evidence has been provided that cell death occurs soon after the phase of DNA replication during the S phase, in the G_1 phase of the cycle *(7)*. It was suggested that signals by the local interstitial environment in the ventricular zone may play an epigenetic role on the destiny of precursors cells, either to proliferate or to die. Results from in vitro studies have shown that extrinsic factors from the extracellular environment may regulate cellular proliferation, and they can determine whether replicating cell will differentiate or proceed to the apoptotic pathway *(8)*.
2. "Delayed" programmed cell death has been observed in specific anatomical areas; it has been related to unavailability of functional inputs (e.g., synaptic activity) or even trophic factors *(9)*. Apoptotic cells have been observed both in transient structures of the developing nervous system (such as the proliferative ventricular and subventricular zone), or to a lesser extent in the cortical plate by the end of fetal growth *(10)*. Additionally, during the early stages of the proliferative activity in the ventricular epithelium, low percentage of dying cells are seen, whereas these numbers are increasing at the later stages, when the actual number of proliferating cells is decreasing *(7)*.

2.2. Postnatal Neurogenesis

Adult mammalian neurogenesis has been studied most extensively in the rodent dentate gyrus. In the adult rat, neuronal precursor cells proliferate in clusters in the dentate subgranular zone, located at the border of the granular cell layer and hilus *(11–15)*. Neuronal precursors also persist and continue to proliferate in the adult rodent forebrain subventricular zone (SVZ) *(11,16–18)*. However, unlike in the dentate gyrus, SVZ neuronal progenitors migrate long distances to their final destination in the olfactory bulb *(19,20)*. The immature neurons migrate from the rostral SVZ to the olfactory bulb using a relatively unique form of tangential chain migration *(19–22)* in a restricted forebrain pathway known as the rostral migratory stream (RMS) *(23,24)*. Remarkably, evidence from recent studies related to neurogenesis in the adult primate brain indicates that the capacity for neuronal renewal may not be limited solely to the

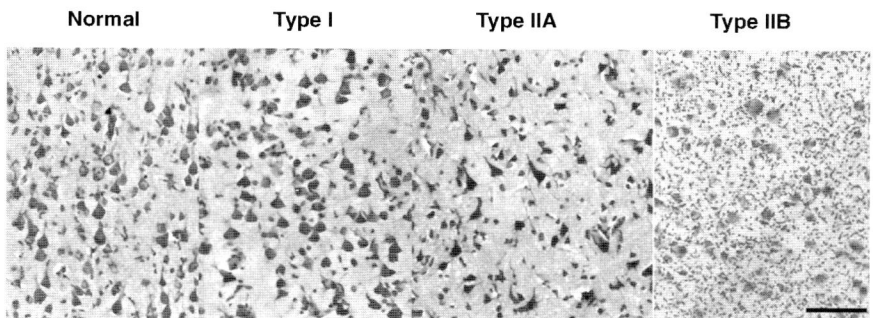

Fig. 1. Histopathological subtypes of MCD as compared with normal tissue. Note the presence of architectural disorganization (type I), dysmorphic neurons (type IIA), and balloon cells (type IIB). Scale bar: 100 μm. (Please *see* companion CD for color version of this figure.)

phylogenetically older olfactory bulb and dentate gyrus brain regions suggested by rodent neurogenesis studies *(25–27)*. Gould and colleagues have shown that the forebrain SVZ of the adult monkey is a proliferative region that generates precursor cells capable of migration through the mature white matter to differentiation into neurons in multiple cortical regions, including neocortical association areas *(28–30)*.

In adult mammals, including humans, postnatal neurogenesis was detected in the dentate gyrus *(31)*. Additionally, the SVZ contains cells that have the characteristics of glial cells, which generate neuronal cells in culture *(2,32–35)*. The systematic decrease in the extent of adult neurogenesis during vertebrate evolution, culminating in primates, may be the result of an adaptation to keep neuronal populations with their accumulative experience for an entire life span *(2)*.

3. MCD AND THE RADIAL UNIT HYPOTHESIS: PATHOGENESIS CONSIDERATIONS

MCD are the pathological substrates of an increasing number of patients with chronic epilepsy *(36–38)*. These lesions are even more frequent in children who are referred for surgical treatment *(39,40)*. As shown in Fig. 1, various histopathological abnormalities have been included under the diagnosis of focal MCD. These include abnormalities of architectural organization (laminar and columnar disorganization) of the cortex, the presence of clearly abnormal cellular elements such as dysmorphic (dysplastics) neurons, large neurons (meganeurons), and balloon cells *(41,42)*. Focal MCD may be restricted to the cortical mantle or may extend to the subcortical white matter and the periventricular regions, clearly representing a departure from the normal cortical cytoarchitectonical organization.

The normal adult neocortex consists of an array of interactive neuronal groups (radial columns) that are interconnected in the vertical dimension, share a common extrinsic connectivity, and have similar functions *(43)*. This hypothesis was based on the observation that postmitotic cells preserve their position during movement by remaining attached to a given radial glial fascicle once they initiate their movement *(4)*. Additionally, the pattern of distribution of radioactive thymidine-labeled cells (a marker for cell proliferation) indicated that the columns likely consist of several clones *(4)*. Therefore, a radial unit consists of cells that originate from several clones but share the same birthplace, migrate along a common pathway, and settle within the same ontogenic column *(4)*. The developing cortical plate consists basically of arrays of such ontogenic columns. Thus, the two-dimensional positional information that is contained within the proliferative zone is transformed into three-dimensional cortical architecture: the x- and y-axis of cell position within the horizontal plane is provided by the site of cell origin, whereas the z-axis along the depth of the cortex is provided by the time of its origin *(4)*.

The radial unit hypothesis has served as a useful concept to explain the cellular and molecular mechanisms involved in the normal and abnormal cortical development. It provides an explanation for a large expansion of cortical surface without the significant increase in thickness that occurred (a) during phylogenetic and ontogenetic development and (b) in pathological abnormalities associated with epilepsy such as MCD. It was proposed that that the gene controlling the number of progenitor cell in S phase at the ventricular surface limits the size of the cortical surface during individual development *(4)*. A relatively minimal change in the timing of developmental cellular events could be associated with important functional consequences: a minimal increase in the length of cell cycles or in the number of cell divisions in the ventricular zone can result in a large increase in the number of progenitor cells that form the radial units *(4)*. Proliferation continues exponentially by the prevalence of symmetric divisions. As a consequence, just an additional round of cell in the proliferative stage doubles the number of progenitor cells and, consequently, the number of radial units. Based on the radial unit hypothesis, the pathogenesis of MCD can be classified into the following three major categories:

1. *MCD in which the number of radial units in the cortex is reduced*, whereas the number of neurons within each ontogenic column remains relatively normal. It can be expected that defects in this category result from an early occurring event that alters the number of proliferative units at the time they are being formed in humans (within the first 6 wk of gestation). Once the number of proliferative units in the ventricular zone is established, each unit can produce a normal or greater number of neurons that become crowded and disorganized in the diminished cerebral vesicle. It could be expected that the cortex would have a smaller surface area despite a normal or enlarged thickness, with the presence of massive neuronal ectopias in the white matter. Identification of human genes associated with the genesis of MCD that may cause a smaller cerebral surface with an absence of convolutions (e.g., pachygyria, hemimegalencephaly, or lissencephaly), will allow new pathogenesis hypothesis for epileptic MCD at the molecular level *(44)*.
2. *MCD in which the number of neurons within radial columns is reduced*, but initial formation of radial units is not affected. The defect in this category should begin after the sixth fetal week, when the normal complement of proliferative units in the human cerebrum already has been established. Such malformations can be caused by interference with cell proliferation through genetic or environmental factors (irradiation or viral infection). Diminished production of neurons in the proliferative units results in fewer neurons within the cortical columns, and the cortex is, therefore, thinner but the cortical surface is relatively preserved.
3. *MCD in which the number of neurons within radial columns is increased.* The number of neurons in radial columns also could be affected by excessive proliferation, diminished apoptosis, and/or a failure of neuronal migration. Here, the "excessive" neurons may fail to reach functional, appropriate loci and often survive in abnormal positions within the cortex or adjacent white matter. Clinical examples in this category are the focal cortical dysplasia, band heterotopia, and periventricular nodular heterotopia.

Finally, it should be emphasized that most MCDs may have features of only one or another category, but in a clinical perspective, most show a mixture of all three categories.

4. EVIDENCE OF POSTNATAL NEURONAL PROGENITOR PROLIFERATION AND INTRINSIC EPILEPTOGENICITY IN MCDs

The literature concerning the association between epilepsy and MCD is vast, complex, unclear, and in need of deliniation and standardization *(45–48)*. Epilepsy has been reported in many different conditions, including congenital encephalopathies, metabolic disorders, anomalies of neuronal migration, sequelae of perinatal brain damage, tumors, infections, pachygyria-agyria syndromes, megalencephalic syndromes, and trauma *(49,50)*. The intrinsic epileptogenicity of MCD was reported previously by Palmini et al., using intraoperative electrocorticography in patients undergoing resective epilepsy surgery. Ictal or continuous epileptogenic discharges mostly were recorded from electrodes overlying dysplastic gyri. Using prolonged subdural grid/video-electroencephalogram monitoring, direct correlations were defined more recently between focal MCD and epileptiform discharges in patients who underwent presurgical invasive evaluation and subsequent focal cortical resection for the management of chronic epilepsy *(51)*. Dysplastic neurons display intrinsic cellular hypoexcitation and hyperexcitation in whole-cell recordings *(52,53)*. Furthermore, the causal relationship between MCD and genesis of seizures has been validated through the correlation of the extent of surgical resection with favorable seizure outcome in patients with chronic epilepsy *(47,54)*, although epileptogenic zone and magnetic resonance imaging-identified dysplastic lesions can be mismatched and partial resection of cortical dysplasias may result in seizure freedom *(55)*. The recent description and characterization of various animal models of MCD and the availability of well-characterized neocortical human tissue have enabled the cellular and molecular characterization of some of the mechanisms underlying epileptogenicity of focal MCDs. Shortly, glutamate receptors have been the primary targets of intense investigation because certain overexcited states were suggested to be pertinent to epilepsy. Overexpression of *N*-methyl-D-asparate (NMDA) receptors, specially the subunits NR2A/B *(56,57)* and α-amino-3-hydroxy-5-methyl-isoxazole-4-propionic acid (AMPA) receptors *(58)* have been correlated with intrinsic epileptogenicity. Other authors *(10,59)* support the hypothesis that the hyperexcitability observed in human MCD may also be a consequence of decreased number of inhibitory neurons, synapses, and receptors subunits. The involvement of postnatal neurogenesis related to MCD's intrinsic epileptogenicity is poorly understood.

Clinical and experimental data in human tissue suggest that some forms of epileptic MCD may biologically behave as nonstatic lesions, growing in the postnatal period. Najm and colleagues have demonstrated that focal MCD with balloon cells may overexpress cellular markers for immaturity as vimentin (immature glial), Tuj1 (immature neurons), and nestin (neuronal committed stem cell) (data not published). Additionally, transcription of genes for nestin and vimentin were enhanced in the balloon cell type cortical dysplasia *(41)*. In a recent study, González and Najm (data not published) demonstrated bromodeoxyuridine (BrdU) uptake (a marker for cell proliferation) in cortical and adjacent white matter samples from patients with MCD. Interestingly, BrdU immunoreactivity was co-immunoexpressed in nestin-positive cells located in the subventricular zone (Fig. 2).

Ying and Najm recently described the overexpression of CD133 (a marker for stem cells) in balloon cells and its coexpression with other intermediate filament proteins (data not published). Balloon cells were found to be immunoreactive to CD133, nestin, Bcl-2, Tuj1, vimentin, and glial fibrillary acidic protein (GFAP). Confocal double-labeling analyses also showed that balloon cells were dual immunopositive for CD133/nestin, CD133/GFAP, CD133/Bcl-2, and nestin/GFAP (Fig. 3).

The expression of CD133 protein in balloon cells may suggest that these cells fail to mature fully and therefore continue to express embryonic genes. Subsequently, these immature cells may lack some of the cellular machinery for migration. This may explain at least in part the localization of the balloon cells in the white matter and gray/white matter junction. Hence, the presence of balloon cells may signify that in this pathological group, the intrauterine insults occurred at earlier stage of embryonic development as compared with other pathological types of cortical dysplasia. Additionally, CD133-positive balloon cells may represent a population of neural stem cells.

5. SEIZURES INDUCE NEURONAL PROGENITAL CELL PROLIFERATION

Several studies have demonstrated that dentate granule cell (DGC) proliferation in adult rats increases in various rodent models of limbic epileptogenesis or acute seizures *(60–64)*. In the kainate and pilocarpine models of temporal lobe epilepsy, chemoconvulsant-induced status epilepticus (SE) increases cell proliferation by approx fivefold to 10-fold in the adult rat dentate gyrus after a latent period of several days *(63)*. A similar effect on DGC proliferation also occurs

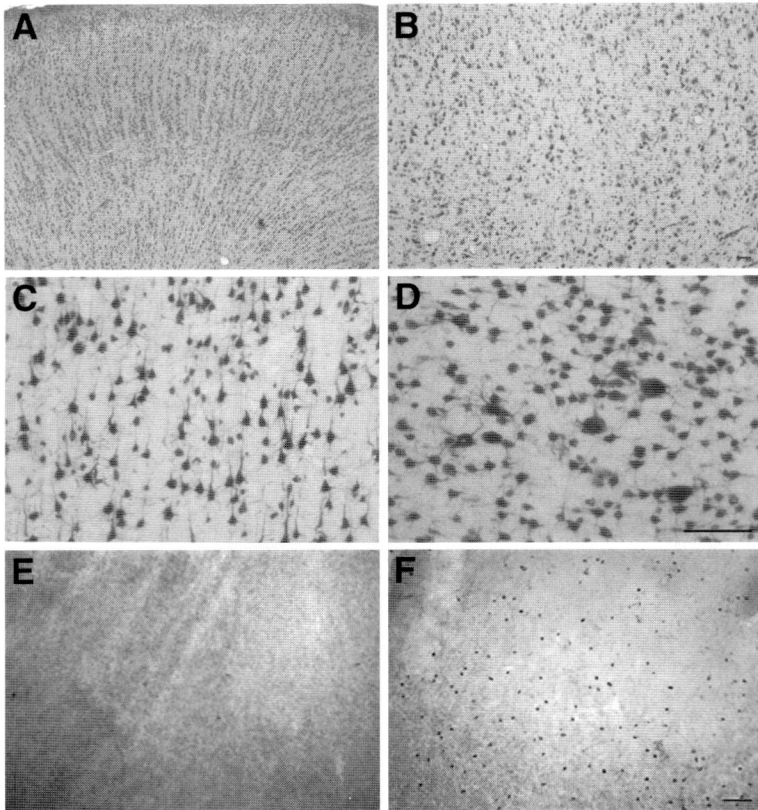

Fig. 2. Difference in BrdU-labeling in normal vs pathological neocortex and adjacent white matter. **(A)** Cresyl-violet (CV) staining from lateral temporal cortex showing normal cytoarchitetonic columnar organization. **(B)** CV staining from dysplastic lateral temporal lobe showing loss of columnar organization and the presence of cytomegalic cells. **(C)** Similar sample depicted in **(A)** showing NeuN expression, demonstrating the normal columnar organization of neocortical neurons. **(D)** Similar sample depicted in **(B)** showing NeuN expression, demonstrating the loss of the columnar organization and the presence of cytomegalic neurons. **(E)** Adjacent white matter (intermediate zone [IZ]) related to the cortical samples depicted in **(A)** and **(C)** showing the absence of BrdU staining. **(F)** IZ related to the cortical samples depicted in **(B)** and **(D)**, demonstrating intense BrdU staining. (Please *see* companion CD for color version of this figure.)

in electrical kindling models of epileptogenesis, including amygdala *(64)*, hippocampal *(65)*, and perforant path *(66)* kindling. Acute seizures in adult rats induced by intermittent perforant path stimulation of brief hippocampal stimulation also accelerate DGC neurogenesis *(65,61)*.

Although these findings suggest the possibility that electrical stimulation directly stimulated neuronal progenitor cell cycle, cell death in the DGC layer also occurs after even brief, discrete seizure-like discharges *(65,67)*. Therefore, seizures may act to increase neurogenesis indirectly through injury leading to cell turnover in the dentate gyrus *(61)*. The molecular link between seizure-induced injury and increased proliferation and/or survival of newly generated DGCs is not known. Among the candidate molecules upregulated by seizure activity are growth factors *(68,69)* and neurotrophins *(70)*.

In order to address potential mechanisms of seizure-induced DGC neurogenesis, Parent et al. studied whether constitutively proliferating precursors are activated by seizures or if, instead, quiescent progenitor cells are recruited to proliferate in the dentate subgranular zone. They labeled the constitutively proliferating cells by systemic BrdU administration 1 d prior to inducing

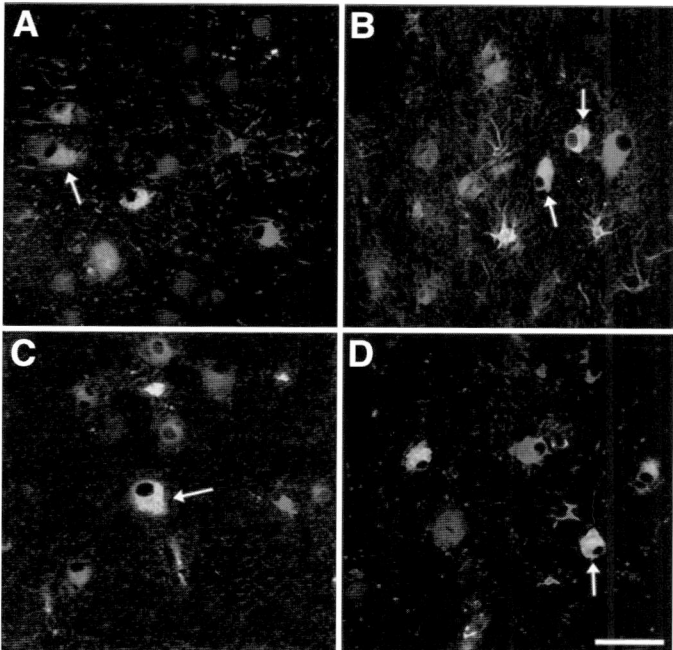

Fig. 3. The colocalization of antibodies against CD133 with antibodies to intermediate filament proteins (nestin and GFAP) were studied in balloon cells by confocal microscopic analysis. Coexpressions of CD133 with either nestin or GFAP were clearly detected in the balloon cells (**A,D**). The presence of singly labeled CD133, nestin, or GFAP were also observed. Coexpression CD133 and Bcl-2 in a subset of balloon cells was shown in (**C**). A population of balloon cells was shown to be immunoreative to both nestin (red) and GFAP (green) (**B**). Double-staining for CD133 and MAP2 failed to reveal any colocolization of these proteins in the balloon cells. (Please *see* companion CD for color version of this figure.)

SE with pilocarpine, and then identified the labeled cells immunohistochemically 2–14 d later *(62)*. After a latent period of 4–7 d, the authors found that SE accelerated the proliferation of cells normally dividing prior to any injury. Although these data do not exclude the involvement of quiescent precursors, it suggests that the mechanisms at least in part relate to accelerated cell division of mitotically active neural precursors. Further studies from the same group also demonstrated the expansion of precursors with migratory cell morphology in the caudal SVZ, infracallosal region, and areas CA1 and CA3 of the hippocampus after seizure-induced injury.

The effect of seizures on the other persistent germinative zone in the adult, the rostral migratory forebrain SVZ, has been relatively unexplored. Using the pilocarpine model of limbic epilepsy, Parent et al. found that 2 h of SE markedly upregulated cell proliferation, as measured by BrdU labeling in the adult rat forebrain SVZ and RMS. Immunostaining for markers of immature neurons revealed that this change in proliferative activity resulted in increased neurogenesis in these same brain regions. The majority of neurons newly generated after seizures migrate through the normal RMS pathway to their appropriate targets in the olfactory bulb. Moreover, a significant proportion of the neuroblasts arising from the SVZ after SE appeared to exit the RMS prematurely and migrate into injured forebrain regions.

6. CONCLUSIONS

The pathogenic mechanisms responsible for the development of cerebral malformations in humans, its epileptogenicity, and its relation with postnatal progenitor cell proliferation and neurogenesis are largely unknown. Morphogenesis of the mature cortical mantle is a consequence of an orchestrated sequence of temporally and spatially organized events that are precisely regulated by genetic and environmental factors. These factors, in a certain moment, determine arrest of neurogenesis. The continuation of neurogenesis in the postnatal period may generate aberrant circuits owing to the incapacity of the newcomer neurons to be adequately incorporated in an already established neural network. Therefore, differently to what occurs in nonhuman primates, our results are in accordance with the hypothesis that postnatal neurogenesis might constitute an undesired process in the adult human brain and, for this reason, is naturally arrested in the postnatal human brain. However, when it continues to occur, it generates pathological conditions, such as epilepsy. Alternatively, we cannot exclude the possibility that postnatal neurogenesis is consequence of the underlying pathological condition. According to this second hypothesis, postnatal neurogenesis could generate inhibitory interneurons, which would attempt to migrate and integrate into the abnormal epileptic circuits in order to suppress its excitatory activity. In this second hypothesis, the quiescent mechanisms of postnatal neurogenesis are requested to be activated, responding to the biological necessity of inhibiting a non-adaptative phenomenon. Lastly, the recent findings of seizure-induced neurogenesis in the adult mammalian forebrain SVZ also raise some other interesting questions: how does seizure activity or seizure-induced injury stimulate SVZ neurogenesis and alter the normal pattern of neuroblast migration in the developing brain? Do differentiating neurons that migrate ectopically into the forebrain survive and integrate into existing networks? Understanding the cellular and molecular mechanisms of neuronal proliferation may help to explain the pathogenesis of the phenotype of previously inexplicable genetic and acquired conditions associated in epileptic MCD, offering new strategies for brain repair or antiepileptic therapies.

REFERENCES

1. Rakic R. Molecular and cellular mechanisms of neuronal migration: relevance to cortical epilepsies. In: Williamson PD, Siegel AM, Roberts DW, Thadani VM, Gazzaniga MS (eds). Advances in Neurology. Philadelphia: Lippincott Williams & Wilkins, 2000, Vol 84.
2. Rakic P. Neurogenesis in the adult primate neocortex: an evolution of the evidence. Nat Rev Neurosci 2002;3:65–71.
3. Rubenstein JLR, Rakic P. Genetic control of cortical development. Cerebral Cortex 1999;9:521–523.
4. Rakic P. A small step for the cell—a giant leap for mankind: a hypothesis of neocortical expansion during evolution. Trends Neurosci 1995b;8:383–388.
5. Rakic P. Differences in the time of origin and in eventual distribution of neurons in areas 17 and 18 of the visual cortex in the rhesus monkey. Exp Brain Res 1976;1(Suppl):244–248.
6. Cowan WM, Fawcett JW, O'Leary DDM. Regressive events in neurogenesis. Science 1984;225:1258–1265.
7. Thomaidou D, Mione MC, Cavanagh JF, Parnavelas JG. Apoptosisand its relation to the cell cycle in the development cerebral cortex. J Neurosci 1997;17:1075–1085.
8. Howard MK, et al. Cell cycle arrest of proliferating neuronal cells by serum deprivation can result in either apoptosis or differentiation. J Neurochem 1993;60(5):1783–1791.
9. Oppenheim RW. Cell death during development of the nervous system. Ann Rev Neurosci 1991;14:453–501.
10. Ferrer L, Soriano E, Del Rio JA, Alcantara S, Auladell C. Cell death and removal in the cerebral cortex during development. Prog Neurobiol 1992;39:11–43.
11. Kaplan MS, Hinds JW. Neurogenesis in adult rat: electron microscopic analysis of light autographs. Science 1977;197:1092–1094.
12. Cameron HA, Gould E. Adult neurogenesis is regulated by adrenal steroids in the dentate gyrus. Neuroscience 1994;61:203–209.
13. Cameron HA, McEwen BS, Gould E. Regulation of adult neurogenesis by excitatory input and NMDA receptor activation in the dentate gyrus. J Neurosci 1995;15:4687–4692.
14. Cameron HA, Woolley CS, McEwen BS, Gould E. Differentiation of newly born neurons and glia in the dentate gyrus of the adult rat. Neuroscience 1993;56:337–344.
15. Kuhn HG, Dickinson-Anson H, Cage FH. Neurogenesis in the dentate gyrus of the adult rat: age-related decrease of neuronal progenitor proliferation. J Neurosci 1996;16:2027–2033.

16. Lois C, Alvarez-Buylla A. Long distance neuronal migration in the adult mammalian brain. Science 1994;264:1145–1148.
17. Thomas LB, Gates MA, Steindler DA. Young neurons from the subependymal zone proliferate and migrate along an astrocyte, extracellular-rich pathway. Glia 1996;17:1–14.
18. Laywell ED, Rakic P, Kukekov VG, Holland EC, Steindler DA. Identification of a multipotent astrocytic stem cell in the immature and adult mouse brain. Proc Natl Acad Sci USA 2000;97:13,883–13,888.
19. Lois C, Alvarez-Buylla A. Proliferating subventricular zone cells in the adult mammalian forebrain can differentiate into neurons and glia. Proc Natl Acad Sci USA 1993;90:2074–2077.
20. Lois C, Garcia-Verdugo, JM, Alvarez-Buylla A. Chain migration of neuronal precursors. Science 1996;271:978–981.
21. Doetsch F, Alvarez-Buylla A. Network of tangential pathways for neuronal migration in adult mammalian brain. Proc Natl Acad Sci USA 1996;93:14,895–14,900.
22. Wichterle H, Garcia-Verdugo JM, Alvarez-Buylla A. Direct evidence of homotypic, glial independent neuronal migration. Neuron 1997;18:779–791.
23. Altman J, Das GD. Autoradiographic and histological evidence of postnatal hippocampal neurogenesis in rats. J Comp Neurol 1965;124:319–335.
24. Kishi K. Golgi studies on the development of granule cells of the rat olfactory bulb with reference to migration in the subependymal layer. J Comp Neurol 1987;258:112–124.
25. Reynolds BA, Weiss S. Generation of neurons and astrocytes from isolated cells of the adult mammalian central nervous system. Science 1992;255:1707–1710.
26. Kornack DR, Rackic P. The generation, migration, and differentiation of olfactory neurons in the adult primate brain. Proc Natl Acad Sci USA 2001;98:4752–4757.
27. Pencea V, Bingaman KD, Freedman LJ, Luskin MB. Neurogenesis in the subventricular zone and rostral migratory stream of the neonatal and adult primate forebrain. Exp Neurol 2001;172:1–6.
28. Gould E, et al. Neurogenesis in the neocortex of adult primates. Science 1999;286:548–552.
29. Gould E, Tanapat P. Lesion-induced proliferation of neuronal progenitors in the dentate gyrus of adult rats. Neuroscience 1997;80:427–436.
30. Gould E, Tanapat P, Rydel T, Hasting N. Regulation of hippocampal neurogenesis in adulthood. Biol Psychiatry 2000;48:715–720.
31. Ericksson PS, et al. Neurogenesis in the adult human hippocampus. Nature Med 1998;4:1313–1317.
32. Nunes MC, et al. Identification and isolation of multipotential neural progenitor cells from the subcortical white matter of the adult human brain. Nature Med 2003;9:439–447.
33. Alvarez-Buylla A, Garcia-Verdugo JM. Neurogenesis in the adult subventricular zone. J Neurosci 2002;22:629–634.
34. Pincus DW, et al. In vitro neurogenesis by adult human epileptic temporal neocortex. Clin Neurosurg 1997;44:17–25.
35. Lie DC, et al. The adult substantia nigra contains progenitor cells with neurogenic potential. J Neurosci 2002;22:6639–6649.
36. Kuzniecky R, Garcia J, Faught E, Morawetz R. Cortical dysplasia in temporal lobe epilepsy: magnetic resonance imaging correlations. Ann Neurol 1991;29:293–298.
37. Li L, et al. High resolution magnetic resonance imaging in adults with partial or secondary generalised epilepsy attending a tertiary referral unit. J Neurol Neurosurg Psychiat 1995;59:384–387.
38. Bartolomei F, et al. Late onset epilepsy associated with regional brain cortical dysplasia. Eur Neurol 1999;42:11–16.
39. Kuzniecky R, et al. Magnetic resonance imaging in childhood intractable partial epilepsies: pathologic correlations. Neurology 1993;43:681–687.
40. Wyllie E, et al. Seizure outcome after epilepsy surgery in children and adolescents. Ann Neurol 1998;44:740–748.
41. Taylor DC, et al. Focal dysplasia of the cerebral cortex in epilepsy. J Neurol Neurosurg Psychiatry 1971;34:369–387.
42. Mischel PS, Nguyen LP, Vinters HV. Cerebral cortical dysplasia associated with pediatric epilepsy. Review of neuropathologic features and proposal for a gradings system. J Neuropathol Exp Neurol 1995;54:137–153.
43. Mountcastle VB. The columnar organization of the neocortex. Brain 1997;20(Pt 4):701–722 (review).
44. Renier O, et al. Isolation of a Miller-Dieker Lissencephaly gene conaining G protein-subunit-like repeats. Nature 1993;364:717–721.

45. Vinters HV, et al. The neuropathology of human symptomatic epilepsy. In: Engel Jr J, (ed). Surgical Treatment of the Epilepsies. New York: Raven Press, 1993, pp 593–608.
46. Palmini A, Andermann F, Olivier A, et al. Focal neuronal migration disorders and intractable partial epilepsy: a study of 30 patients. Ann Neurol 1991;30:741–749.
47. Palmini A, Andermann F, Olivier A, Tampieri D, Robitaille Y. Focal neuronal migration disorders and intractable partial epilepsy: results of surgical treatment. Ann Neurol 1991;30:750–757.
48. Palmini A, Gambardella A, Andermann F, et al. Intrinsic epileptogenicity of human dysplastic cortex as suggested by corticography and surgical results. Ann Neurol 1995;37:476–487.
49. Engel J. Surgery for seizures. N Engl J Med 1996;334:647–652.
50. Raymond A, et al. Abnormalities of gyration, heterotopias, tuberous sclerosis, focal cortical dysplasia, microdysgenesis, dysembryoplastic neuroepithelial tumour and dysgenesis of the archicortex in epilepsy. Clinical, EEG and neuroimaging features in 100 adult patients. Brain 1995;118(Pt 3):629–660. Analysis of 61 cases. Acta Neuropathol 1994;88:166–173.
51. Najm I, et al. NMDA receptor 2A/B subtype differential expression in human cortical dysplasia: Correlation with in situ epileptogenicity. Epilepsia 2000;41:971–976.
52. Avoli M, Bernasconi A, Mattia D, et al. Epileptiform discharges in the human dysplastic neocortex: in vitro physiology and pharmacology. Ann Neurol 1991;46:816–826.
53. Mathern GW, Cepeda C, Hurst RS, et al. Neurons recorded from pediatric epilepsy surgery patients with cortical dysplasia. Epilepsia 2000;41:S162–S167.
54. Hirabayashi S, Binnie C, Janota I, Polkey C. Surgical treatment of epilepsy due to cortical dysplasia: clinical and EEG findings. J Neurol Neurosurg Psychiat 1993;56:765–770.
55. Munari C, Francione S, Kahane P, Tassi L, Hoffman D. Usefulness of stereo-EEG investigations in partial epilepsy associated with cortical dysplastic lesions and gray matter heterotopia. In: Guerrini R, Andemann F, Canaphicci R, Roger J, Zifkin BF, Pfanner P (eds). Dysplasias of Cerebral Cortex and Epilepsy. Philadelphia: Lippincott-Raven, 1996, pp 383–394.
56. Ying Z, Babb TL, Comair YG, et al. Induced expression of NMDAR2 proteins and differential expression of NMDAR1 splice variants in dysplastic neurons of human epileptic neocortex. J Neurpath Exp Neurol 1998;57:47–62.
57. Ying Z, Babb TL, Mikuni N, Najm I, Drazba J, Bingaman W. Selective coexpression of NMDAR2A/B submit proteins in dysplastic neurons of human epileptic cortex. Exp Neurol 1999;159:409–418.
58. Spreafico R, Battaglia G, Arcelli P, et al. Cortical dysplasia: an immunocytochemical study of three patients. Neurology 1998;50:27–36.
59. Crino PB, Duhaime A, Baltuch G. Differential expression of glutamate and GABA-A receptor subunit mRNA in cortical dysplasia. Neurology 2001;56:906–913.
60. Parent JM, Valentin VV, Lowenstein DH. Prolonged seizures increase proliferating neuroblasts in the adult rat subventricular zone-olfactory bulb pathway. J Neurosci 2002;22:3174–3188.
61. Parent JM, et al. Dentate granule cell neurogenesis is increased by seizures and contributes to aberrant network reorganization in the adult rat hippocampus. J Neurosci 1997;17:3727–3738.
62. Parent JM, Tada E, Fike JR, Lowenstein DH. Inhibition of dentate granule cell neurogenesis with brain irradiation does not prevent seizure-induced mossy fiber reorganization in the rat. J Neurosci 1999;19:4508–4519.
63. Gray WP, Sundstrom LE. Kainic acid increases the proliferation of granule cell progenitors in the dentate gyrus of the adult rat. Brain Res 1998;790:52–59.
64. Scott BW, Wang S, Burnham WM, et al. Kindling-induced neurogenesis in the dentate gyrus of the rat. Neurosci Lett 1998;248:73–76.
65. Bengzon J, et al. Apoptosis and proliferation of dentate gyrus neurons after single and intermittent limbic seizures. Proc Natl Acad Sci USA 1997;94:10,432–10,437.
66. Nakagawa E, et al. Enhancement of progenitor cell division in the dentate gyrus triggered by initial limbic seizures in rat models of epilepsy. Epilepsia 2000;41:10–18.
67. Sloviter RS, et al. Apoptosis and necrosis induced in different hippocampal neuron populations by repetitive perforant path stimulation in the rat. J Comp Neurol 1996;366:516–533.
68. Riva MA, Gale K, Mocchetti I. Basic fibroblast growth factor mRNA increases in specific brain regions following convulsive seizures. Mol Brain Res 1992;15:311–318.
69. Humpel C, et al. Fast and widespread increase of basic fibroblast growth factor messenger RNA and in the forebrain after Kainate induced seizures. Neuroscience 1993;57:913–922.
70. Ernfors P, et al. Increased levels of messengers RNAs for neutrophic factors in the brain during epileptogenesis. Neuron 1991;7:165–176.

24
Vascular Differentiation and the Cell Cycle

Luca Cucullo, PhD

SUMMARY

Vascular differentiation is a multi-step process regulated by a complex pattern of cell signaling, soluble factors and local environment, which determine specific functions and the peculiar characteristics the endothelial cells will assume in loco by modulating their ability to respond to different environmental cues. For example, endothelial cells lining the vascular bed at the blood-brain barrier level develop particular characteristics of impermeability toward polar molecules that do not have a specific carrier whereas endothelial cells in other regions of the body allow free diffusion of macromolecules between the vascular and interstitial spaces. What promotes endothelial cells to acquire a distinct barrier phenotype in the central nervous system is a complex pattern of cell signaling regulated by growth factors, surrounding cellular environment (e.g., astrocytes), and mechanical adaptation (e.g., shear stress). Neuronal involvement in vascular differentiation at the BBB level is still unclear; but neurons may play some role through extracellular signaling with astrocytes and from them to the endothelial cells.

Key Words: Blood-brain barrier; angiogenesis; vasculogenesis; cell cycle; endothelial cells; astrocytes; extracellular matrix; shear stress; growth factors; cell signaling.

1. INTRODUCTION

The vasculature of the central nervous system (CNS) is derived from capillary endothelial cells (ECs), which have invaded the early embryonic neuroectoderm from the perineural vascular plexus. The ECs forming the luminal bed of all blood vessels form a structurally and functionally heterogeneous population of cells. The endothelium by itself is considered like a sparse organ system, owing to its vast extension and ability to exert a complex array of specialized functions. Even in the same organ, the endothelium of large and small vessels, veins, and arteries exhibits significant heterogeneity. During embryonic development, they differentiate from a common precursor called angioblast and acquire organ-specific properties. The cell signaling leading to differentiation of angioblast in function-specific ECs may occur through the release of soluble cytokines, cell–cell adhesion, and communication, and the synthesis of matrix proteins on which the endothelium adheres and grows. The acquisition and maintenance of specialized properties by ECs is important in the functional homeostasis of the different organs.

One of the important determinants of EC differentiation and relative expression of different peculiar characteristics is the local environment, especially the interaction with surrounding cells (e.g., astrocytes). There are two processes that lead to the formation of blood vessels.

1. Vasculogenesis (during embryogenesis), whereby a primitive vascular network (such as the dorsal aorta or the cardinal veins) is established from multipotent mesenchymal progenitors that migrate to or differentiate at the location of future vessels, coalesce into cords, differentiate into ECs, and ultimately form patent vessels *(1,2)*.

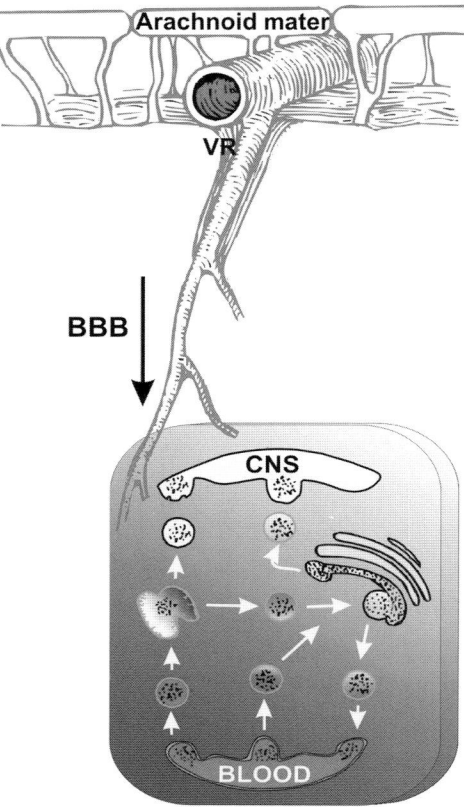

Fig. 1. Passage of solutes from blood to CNS; Virchow-Rodin (VR) space. (Please *see* companion CD for color version of this figure.)

2. Angiogenesis, defined as the formation of new blood vessels by a process of sprouting from pre-existing vessels extending the vascular network. Angiogenesis occurs both during development and in postnatal life *(3,4)*, but recent studies have suggested that endothelial stem cells may persist in adult life, in which they contribute to the formation of new blood vessels *(5,6)*. This suggests that neoangiogenesis in the adult may partially depend on the process of vasculogenesis. Recent studies also demonstrate that the neuronal system plays a fundamental role in the maturation of primitive embryonic vasculature because mutations that disrupt peripheral sensory nerves or Schwann cells prevent proper arteriogenesis, whereas those that disorganize the nerves maintain the alignment of arteries with misrouted axons *(7)*.

2. VASCULAR DIFFERENTIATION AND BLOOD–BRAIN BARRIER

The blood–brain barrier (BBB) is a fundamental component of the CNS. Its functional and structural integrity is vital to maintain the homeostasis of the brain micro-environment. The BBB is present in the majority of the brain capillaries where the ECs display oval or elongated nuclei in the direction of the capillary itself. Inside the subarachnoidal space, the cerebrospinal fluid is interposed between larger-size vessels and the connective tissue of pia. The blood vessels start branching in small capillaries (Fig. 1), whereas the pia disappears and the endothelium acquires the peculiar characteristics of a tight barrier that regulates the exchange of substances between the blood and the brain. Larger vessels (arterioles, small arteries, and veins) differ from capillaries by the presence of smooth muscle cells in their walls. Recent studies also indicate that acquisition

**Table 1
Transport System at the Blood–Brain Barrier**

Transport systems	Substrates
Amine	Choline
Nucleoside	Adenosine
Hexose	D-Glucose
Monocarbolic acid	Lactate
Neutral amino acids	Phenylamine
Basic amino acids	Arginine
Purine	Adenine

of their functional characteristics is at least partially controlled by genetic factors. The expression of a number of Notches (large cell surface receptors important in a wide array of developmental contexts for specifying cell populations) at the endothelial level suggests a requirement for Notch signaling for proper patterning of the vasculature during embryogenesis by modulating the ability of ECs to respond to different environmental cues *(8–11)*. Outside the endothelium, one or two layers of smooth muscle cells are transversely orientated and sandwiched between thick basal laminae. Venules resemble large capillaries and the transition between the two types of vessels is difficult to identify.

At the BBB level the cytoplasm of the EC is of uniform thickness, with very few pinocytotic vesicles, and lacks fenestrations (i.e., openings). The adjacent cell membranes are parallel, and, toward the luminal end, the outer leaflets fuse to form an interendothelial connections structure called tight junctions, which consist of three integral protein types (claudins, occludins, and junctional adhesion molecules). Tight junctions (zonula occludentes) present between the cerebral ECs form a diffusion barrier, which selectively excludes most blood-borne substances from entering the brain *(12,13)*. Therefore, transit across the BBB involves translocation through the capillary endothelium by carrier-mediated transport systems into the internal cytoplasmic domain, and then through the ablumenal membrane, pericyte, and/or basal lamina. Water-soluble (electrolytes) substances of all sizes or polar molecules in general cross the barrier with great difficulty, whereas lipid-soluble substances such as alcohol, narcotics, and anticonvulsants pass with ease. Compounds bound to plasma protein (steroid and thyroxin) take longer than the unbound form. Specific transport systems are responsible for the passage of certain water-soluble but biologically important substances such as D-glucose or phenylamine. The most important of them are summarized in Table 1.

The endothelial cytoplasm is richly endowed with enzymes, including adenosine triphosphatase, nicotinamide adenine dinucleotide, monoamine oxidase, acid and alkaline phosphatases, various dehydrogenases, 3,4-dihydroxy-phenylalanine (DOPA) decarboxylase, and glutamyl transpeptidase. The membrane localization of these enzymes is indicative of the polarity of the endothelial functions in the control of the blood–brain interface. The ECs are also characterized by very high density of mitochondria (almost five times higher than other endothelial phenotypes), denoting high metabolic activity. The periendothelial accessory structures of the BBB include basal membrane, pericytes, and astrocytes. The basal lamina is produced at the BBB level by perivascular astrocytes; it is approx 40–50 nm thick and composed of type IV collagen, heparan sulfate proteoglycan, laminin, fibronectin, and tenascin. The basal lamina or mature extracellular matrix (ECM) is produced at the BBB level by perivascular astrocytes (Fig. 2); it has several functions such as mechanical support for cell adhesion and migration with a mechanism involving transmembrane receptors (integrins) that bridge the cytoskeletal elements of a cell to the ECM *(14)*. The basal lamina also regulates the communication between cells and represents an additional barrier to the passage of macromolecules between the cells and the vascular system *(15)*.

The association of pericytes (represented by a morphologically heterogeneous cell population ranging from circular smooth muscle to elongated fibroblast-like morphology) to blood

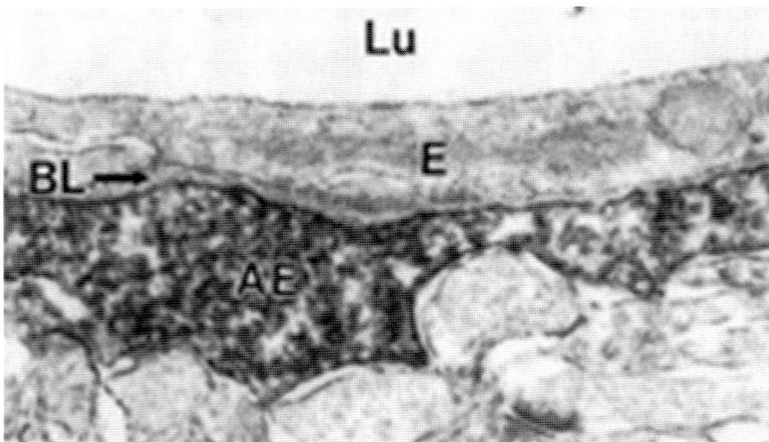

Fig. 2. A cross-section of a cerebral capillary of the BBB. Shown are the astrocytic endfeet (AE), basal lamina (BL), endothelial cell (E), luminal compartment (Lu); *see* ref. *51*.

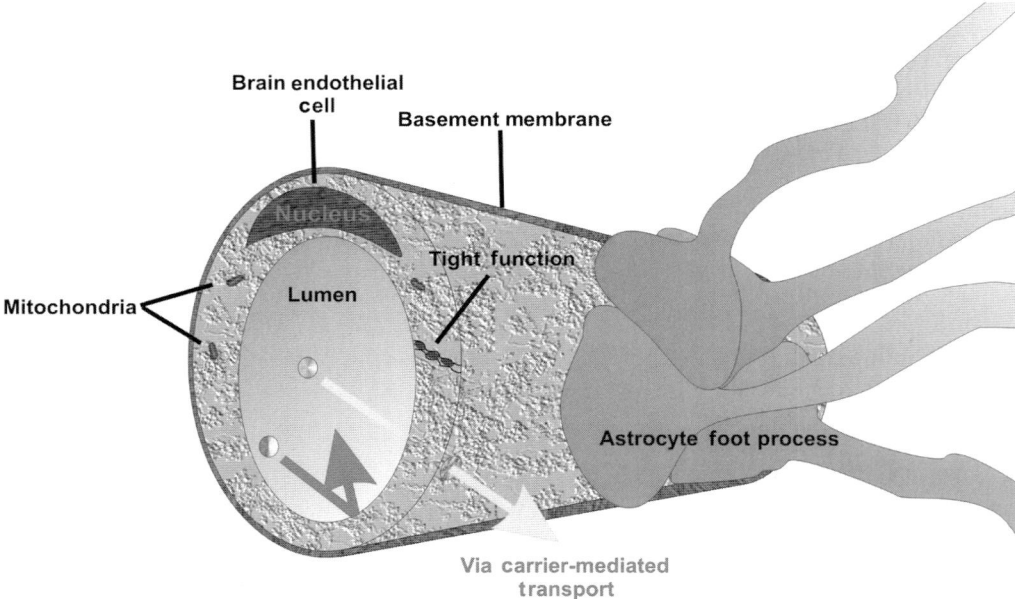

Fig. 3. Schematic representation of the blood–brain barrier with associated astrocytes. (Please *see* companion CD for color version of this figure.)

vessels has been suggested to regulate EC proliferation, survival, migration, differentiation, and vascular branching (as well as other metabolic functions). They are also involved in the regulation of blood flow (via contractility) *(16)*. Whereas pericytes in the periphery are flat, undifferentiated, contractile connective tissue cells which develop around capillary walls, microvascular pericytes have shown to lack the α-actin isoform, typical of contractile cells *(17)*. Pericytes are completely surrounded by a duplication of the basement membrane; their cellular projections penetrate the basal lamina covering approx 20–30% of the microvascular circumference. Additionally, there is some evidence that pericytes are able to mimic astrocyte ability to induct BBB "tightness" *(18)*. Astrocytes are adjacent to the EC; with their endfeet sharing the basal lamina, they envelop more than 99% of the BBB endothelium (Fig. 3). Astrocyte interaction with the cerebral endothelium determines BBB function, regulates protein expression, modulates endothelium differentiation, and appears to be critical for the induction and maintenance of the tight junctions and BBB properties *(19)*. A novel "BBB-protective" role of astrocytes has been recently described, including NO-mediated interleukin-6 (IL-6) release. inter leukin-6 has been shown to trigger production of matrix metalloproteinase (MMP) inhibitors, such as α2-macroglobulins released by perivascular astrocytes *(20–22)*. Glial cells also provide extensive structural and physiological support to neurons by phagocytizing neuronal debris, maintaining the extracellular levels of ions (one of the main functions attributed to astrocytes is the buffering of extracellular [K^+]), neurotransmitter uptake, controlling synaptogenesis, synapse number, function, and plasticity *(23–27)*. They also guide neurons to their proper place during development and direct vessels of the BBB. However, neuronal involvement in the BBB formation is still unclear. In vitro studies support the hypothesis that the neural micro-environment plays a key role in inducing BBB function in capillary ECs *(28)*. Coculture experiments using cerebral capillary ECs and neurons have shown a dose-dependent increases in γ-glutamyl transpeptidase activity (an enzyme responsible for amino acid transport across the BBB), indicating an inductive effect of neurons *(29)*.

3. THE CELL CYCLE

A number of neurodegenerative disorders, such as stagger, lurcher in mice *(30)*, and Pick's disease, intractable temporal lobe epilepsy, progressive supranuclear palsy, Lewy body disease, and Parkinson's disease in humans *(31,32)*, are associated with an anomalous cell cycle, as manifested by a re-entrant mitotic phenotype. The cell cycle is, in fact, a process highly regulated with numerous checks and balances, which ensures a homeostatic balance between cell proliferation and cell death in the presence of appropriate environmental signals. In proliferating cells, the cell cycle consists of five phases. Gap 1 (G_1) is the interval between mitosis and DNA replication that is characterized by cell growth. The transition that occurs at the restriction point (R) in G_1 commits the cell to the proliferative cycle. If the conditions that signal this transition are not present, the cell exits the cell cycle and enters gap 0 (G_0), a nonproliferative phase during which growth, differentiation, and apoptosis occur. Replication of DNA occurs during the synthesis (S) phase, which is followed by a second gap phase (G_2), during which growth and preparation for cell division occurs. Mitosis and the production of two daughter cells occur in M phase. The passage of a cell through the cell cycle is regulated by a class of nuclear enzymes called cyclin-dependent kinases (CDKs) and includes a G_1 CDK (CDK4), an S-phase CDK (CDK2), and the M-phase CDK (CDK1). These enzymes are expressed constitutively but are inactive in the absence of their cyclin partners: G_1 cyclin (cyclin D), S-phase cyclins (cyclins E and A), and mitotic cyclins (cyclins B and A). The cyclin/CDK complex is activated by the sequential phosphorylation and dephosphorylation of the key residues of the complex, located principally on the CDK subunits. The stimulatory effects of cyclin binding are offset by CDK-inhibitory proteins (CKIs), and the activation state of a CDK depends on the relative level of its cyclin and CDK-inhibitory proteins. The tight regulation of cell cycle control intrinsically requires an intimate coupling of the cell death mechanism in order to remove defective cells during the cell cycle, such that cell cycle checkpoints appear to link the cell cycle to apoptosis *(33)*. Cells can exit the cell cycle to enter and stay at G_0 phase, which is the case in terminally differentiated cells. The expression/activation of cyclin D-CDK4,6 complex, triggered by mitotic growth factors (GFs), controls the re-entry of resting G_0 cells into the G_1 phase of cell cycle. Thereafter, the G_1/S transition is controlled by the activation of the cyclin E-CDK2 complex such that absence of cyclin E and/or the inhibition of the cyclin E-CDK2 complex by p21, p27, and p53 will cause the cell cycle to be arrested at the G_1 checkpoint. There are several proteins that can inhibit the cell cycle in G_1. For example if DNA damage has occurred, p53 accumulates in the cell and induces the p21-mediated inhibition of cyclin D-CDK complex. The DNA replication in the S phase and the transition to the G_2 phase is regulated by the activation of cyclin A-CDK2 complex and proliferating cell nuclear antigen *(34)*. The key regulator of the G_2/M transition is

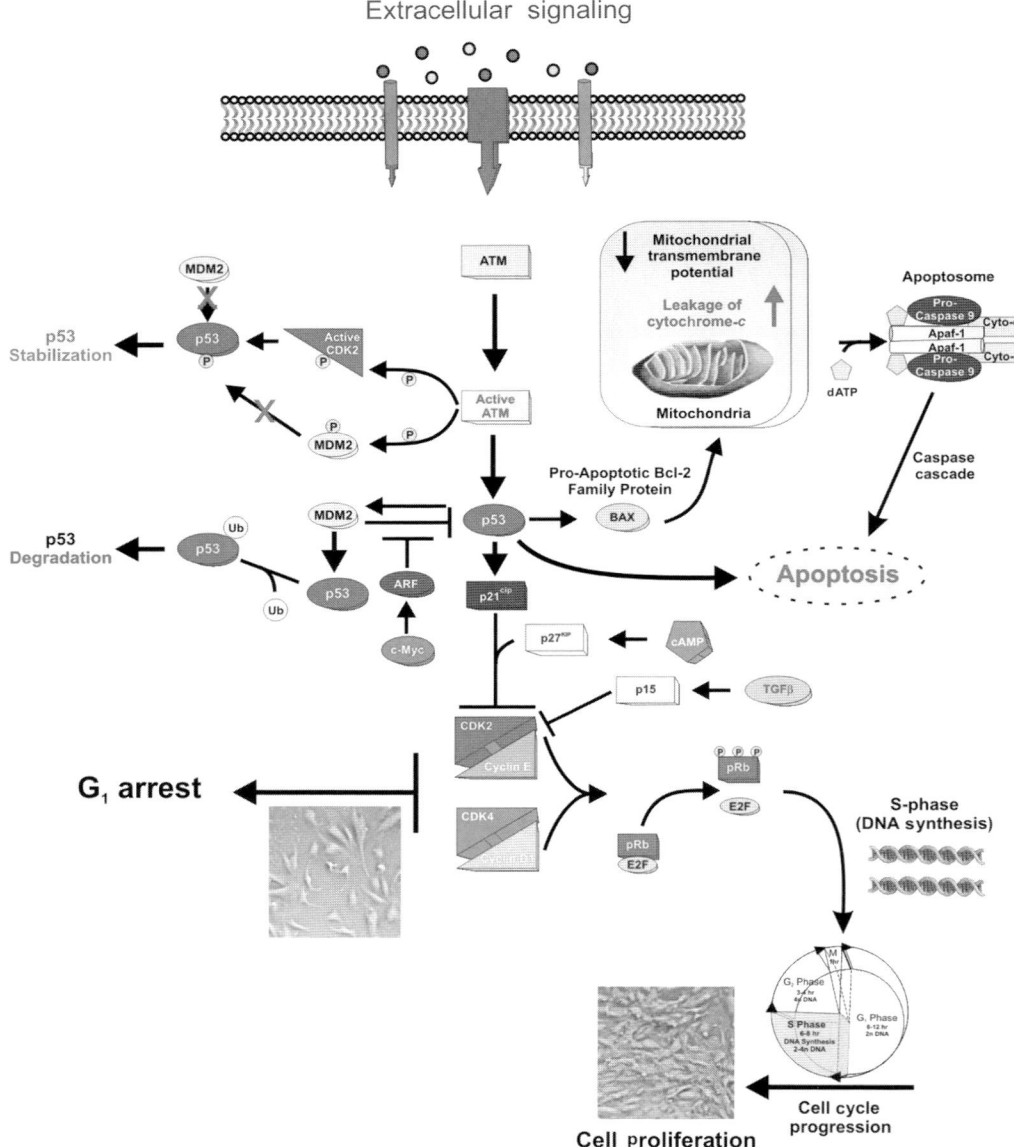

Fig. 4. Diagrammatic representation of cell cycle-related pathways in response to extracellular signaling. (Please *see* companion CD for color version of this figure.)

the cyclin B–cell division cycle 2 complex *(34)*. Cells arrested at G_2/M checkpoint do not have the ability to redifferentiate and die via an apoptotic pathway. Similarly, activation of TGF-β receptors induces the inhibition of cyclin D/CDK by p15, whereas cyclic-adenosine monophosphate inhibits the cyclin D/CDK complex via p27 (Fig. 4). When the cyclin D/CDK complex is inhibited, retinoblastoma protein (Rb) is in a state of low phosphorylation and is tightly bound to the transcription factor E2F, inhibiting its activity. ECs are quiescent in normal blood vessels but undergo rapid bursts of proliferation after vascular injury and during angiogenesis. Recent studies have shown that the activity of CDK2, which regulates G_1 and S phases of the cell cycle, is increased in proliferating ECs but not in ECs that are contact-inhibited for growth *(35)*. Despite these differences in kinase activity, the protein levels of CDK2 and cyclin E do not appear to be modulated by growth conditions or cell contact. The CDK inhibitor p27 is highly expressed in contact-inhibited (nonproliferating) ECs, in contrast to cyclin A, which is preferentially expressed in proliferating ECs.

4. CELL CYCLE REGULATION: ROLE OF ECM

Cells interact with the ECM primarily through integrins, which by binding to their extracellular ligands are responsible for the downstream effects of the matrix on cell function. The ECM plays a key role in angiogenesis by modulating capillary endothelial (CE) cell sensitivity to soluble GFs and, thereby, switching cells between proliferation, differentiation, and involution in the local tissue micro-environment. In the CNS, various ECM components have been identified, which are strongly expressed during development and are downregulated in most areas of the brain during maturation concurrently with changes in their composition. Many molecules, however, appear to be replaced by members of the same protein family, permitting the maintenance of the overall structural organization. The adhesion of ECs to the ECM is mediated by a family of $\alpha\beta$ heterodimeric transmembrane glycoproteins cell surface protein receptors (integrins) that interact with extracellular or cell surface molecules and cytoplasmic molecules, including cytoskeletal and catalytic signaling proteins (e.g., fibronectin). Integrins contain a large extracellular domain (responsible for ligand binding), a single transmembrane domain, and a cytoplasmic tail *(36–38)*. Recent studies suggest that integrins not only receive signals from the ECM but also actively transmit "inside-out" signals in promoting integrin ligand-binding affinity, assembly of ECM, and cell adhesion *(39)*. The interaction between ECM and integrins induces several cytoplasmic signal transduction cascades that regulate cell cycle and cell differentiation, leading to stimulation of mitogenic events associated with the G_0/G_1 transition, including expression of immediate early growth response genes *(40–43)*. Thus, it is also possible that adhesion of ECs to ECM exerts its effects on growth not only via direct integrin receptor signaling mechanisms but also indirectly through associated integrin-dependent changes in cytoskeletal tension and associated changes in cell shape and cytoskeletal structure *(44)*.

5. CELL CYCLE REGULATION: ROLE OF GFS

The development of a functioning vascular network requires a remarkable coordination between different cell types undergoing complex changes, and is mainly dependent on signals exchanged between these cell types. The vascular endothelial GF (VEGF) provides the first example of a GF specific for the vascular endothelium (Fig. 5). The specificity of VEGF for the vascular endothelium results from the specific distribution of VEGF receptors to these cells. VEGF mediates its effects by binding with high affinity to two different tyrosine kinase receptors: VEGF receptor-1 (Fms-like tyrosine kinase 1) and VEGF receptor-2 (kinase domain receptor). VEGF receptor-1 is expressed on ECs and monocytes and mediates cell motility. The proliferative and mitogenic activities of VEGF, as well as vascular permeability, are mediated primarily through VEGF receptor-2 *(45,46)* which also seems to help promote ECs to form tubule-like structures during the earliest stages of vasculogenesis *(47,48)*. Its mitogenic effect is counteracted by the VEG inhibitor, produced by ECs. This protein is a member of the tumor necrosis factor family and seems to be a potent inhibitor of EC proliferation, angiogenesis, and tumor growth. Its mechanism of inhibition is based on an early G_1 arrest in G_0/G_1 cells responding to growth stimuli, and on programmed death in proliferating cells. In vitro studies have shown that VEG inhibitor also inhibits the formation of capillary-like structures by ECs in collagen gels, and the growth of capillaries induced by either basic fibroblast GF or vascular ECGF. However, the complexity of cellular interaction involved during vascular development suggests that other GFs such as angiopoietin-Tie, EphB2-Eph4B, and fibral growth factor (which, like VEGF, requires activation of protein kinase) are also involved and might cooperate with the VEGF system to establish the dynamic blood vessel structures. Similar to VEGF, the specificity of the angiopoietins for the vascular endothelium results from the restricted distribution of the

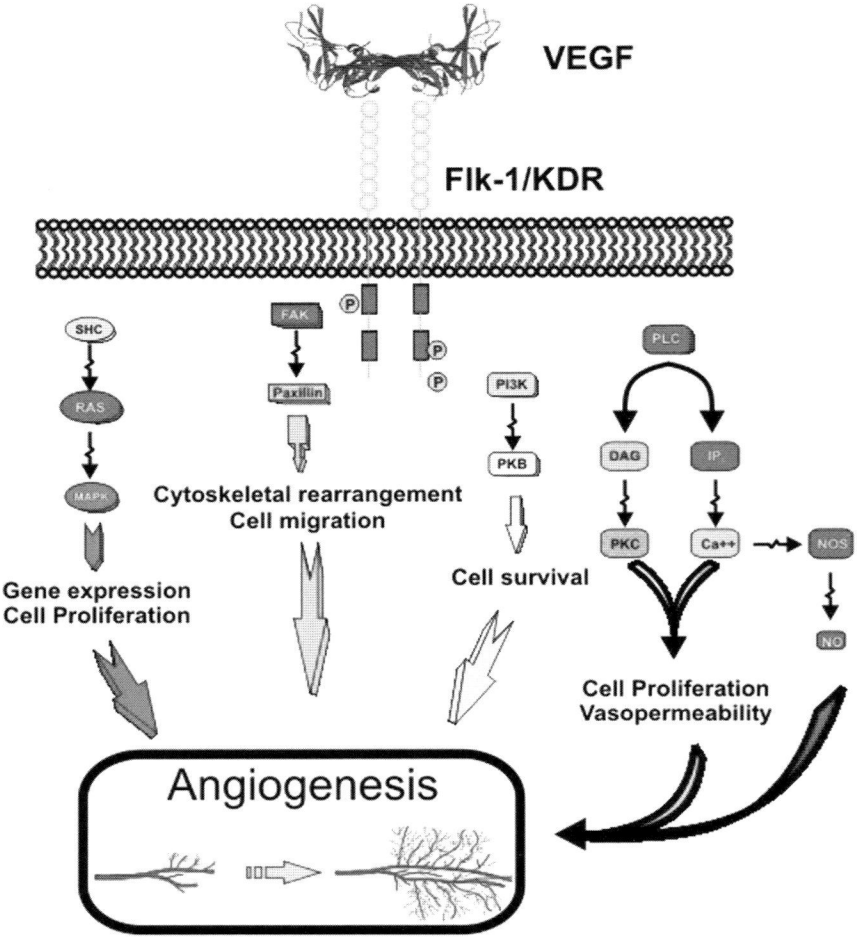

Fig. 5. Diagrammatic representation signaling pathways activated by VEGF. (Please *see* companion CD for color version of this figure.)

angiopoietin receptors Tie1 and Tie2 (tyrosine kinase receptors) to these cells. Their action appears to be different from those of VEGF because they do not seem to induce tubule like structure formation or mitogenic response but rather in vitro EC sprouting *(49)*. The expression of VEGF and angiopoietin receptors is high during brain angiogenesis but low in adult BBB endothelium. They are required for the proper development of a vascular system, and particularly Tie2 is necessary for brain angiogenesis. Signal transduction by these receptors regulates EC growth, permeability, and differentiation. For example, at the BBB level endothelial adherent junctions are a downstream target of VEGF receptor signaling, and this suggests that tyrosine phosphorylation of its components may be involved in the loosening of cell–cell contacts in established vessels to modulate transendothelial permeability and to allow sprouting and cell migration during angiogenesis *(50)*.

6. CONCLUSIONS

Vascular differentiation is a multistep process regulated by a complex pattern of cell signaling, soluble factors, and local environment, which determine specific functions and the peculiar characteristics the ECs will assume *in loco* by modulating their ability to respond to different environmental cues. For example, we have seen that ECs lining the vascular bed at the BBB level are characterized by the presence of tight intercellular junctions, the lack of fenestrations, and a paucity of pinocytotic vesicles that confer them specific characteristics of impermeability toward polar molecules that do not have a specific carrier. Specialization of the apical and basolateral EC membranes also occurs through the preferential expression and distribution of membrane transporters and enzymes that function to protect, as well as promote, substrate delivery to the brain parenchyma. ECs in other regions of the body differ in that because they lack tight junctions and subsequently allow for free diffusion of macromolecules between the vascular and interstitial spaces. However, what promotes ECs to acquire a distinct barrier phenotype in the CNS is a complex pattern of cell signaling regulated by GFs, surrounding cellular environment, and mechanical adaptation. It is known, for example, that exposure of the apical membrane to shear stress promotes growth inhibition and differentiation of ECs. It also serves to induce metabolic changes which limit the oxygen and substrate consumption of such cells and allows for trafficking of metabolic fuels to the brain. Exposure of these cells to "permissive" or "promoting" factors such as VEGF or angiopoietin and their binding to the appropriate receptors regulates the proliferative and mitogenic activities, or activates a specific patter of cellular differentiation. Cellular cross talk also plays an important role in cell differentiation. For example, astrocytes have been demonstrated to promote a variety of changes in gene expression in ECs, as well as phenotypic changes including segregation of transporters and enzymes. Neuronal involvement in vascular differentiation at the BBB level is still unclear; but it is possible that neurons may play some role through extracellular signaling with astrocytes and from them to the ECs.

REFERENCES

1. Risau W, Flamme I. Vasculogenesis. Annu Rev Cell Dev Biol 1995;11:73–91.
2. Risau W. Angiogenesis and endothelial cell function. Arzneimittelforschung 1994;44:416–417.
3. Watt SM, Gschmeissner SE, Bates PA. PECAM-1: its expression and function as a cell adhesion molecule on hemopoietic and endothelial cells. Leuk Lymphoma 1995;17:229–244.
4. Reyes M, Lund T, Lenvik T, Aguiar D, Koodie L, Verfaillie CM. Purification and ex vivo expansion of postnatal human marrow mesodermal progenitor cells. Blood 2001;98:2615–2625.
5. Asahara T, Murohara T, Sullivan A, et al. Isolation of putative progenitor endothelial cells for angiogenesis. Science 1997;275:964–967.
6. Gehling UM, Ergun S, Schumacher U, et al. In vitro differentiation of endothelial cells from AC133-positive progenitor cells. Blood 2000;95:3106–3112.
7. Mukouyama YS, Shin D, Britsch S, Taniguchi M, Anderson DJ. Sensory nerves determine the pattern of arterial differentiation and blood vessel branching in the skin. Cell 2002;109:693–705.
8. Lawson ND, Weinstein BM. Arteries and veins: making a difference with zebrafish. Nat Rev Genet 2002;3:674–682.
9. Lawson ND, Vogel AM, Weinstein BM. Sonic hedgehog and vascular endothelial growth factor act upstream of the Notch pathway during arterial endothelial differentiation. Dev Cell 2002;3:127–136.
10. Shawber CJ, Das I, Francisco E, Kitajewski J. Notch signaling in primary endothelial cells. Ann NY Acad Sci 2003;995:162–170.
11. Shawber CJ, Kitajewski J. Notch function in the vasculature: insights from zebrafish, mouse and man. Bioessays 2004;26:225–234.
12. Huber JD, Witt KA, Hom S, Egleton RD, Mark KS, Davis TP. Inflammatory pain alters blood-brain barrier permeability and tight junctional protein expression. Am J Physiol Heart Circ Physiol 2001;280:H1241–H1248.
13. Huber JD, Egleton RD, Davis TP. Molecular physiology and pathophysiology of tight junctions in the blood-brain barrier. Trends Neurosci 2001;24:719–725.
14. Tilling T, Engelbertz C, Decker S, Korte D, Huwel S, Galla HJ. Expression and adhesive properties of basement membrane proteins in cerebral capillary endothelial cell cultures. Cell Tissue Res 2002;310:19–29.
15. Rosenberg GA, Kornfeld M, Estrada E, Kelley RO, Liotta LA, Stetler-Stevenson WG. TIMP-2 reduces proteolytic opening of blood-brain barrier by type IV collagenase. Brain Res 1992;576:203–207.

16. Hellstrom M, Gerhardt H, Kalen M, et al. Lack of pericytes leads to endothelial hyperplasia and abnormal vascular morphogenesis. J Cell Biol 2001;153:543–553.
17. Nehls V, Drenckhahn D. Heterogeneity of microvascular pericytes for smooth muscle type alpha-actin. J Cell Biol 1991;113:147–154.
18. Minakawa T, Bready J, Berliner J, Fisher M, Cancilla PA. In vitro interaction of astrocytes and pericytes with capillary-like structures of brain microvessel endothelium. Lab Invest 1991;65:32–40.
19. Liberto CM, Albrecht PJ, Herx LM, Yong VW, Levison SW. Pro-regenerative properties of cytokine-activated astrocytes. J Neurochem 2004;89:1092–1100.
20. Aschner M. Immune and inflammatory responses in the CNS: modulation by astrocytes. Toxicol Lett 1998;102–103:283–287.
21. Aschner M. Astrocytes as mediators of immune and inflammatory responses in the CNS. Neurotoxicology 1998;19:269–281.
22. Cucullo L, Marchi N, Marroni M, Fazio V, Namura S, Janigro D. Blood-brain barrier damage induces release of α2-macroglobulin. Mol Cell Proteomics 2003;2:234–241.
23. Privat A. Astrocytes as support for axonal regeneration in the central nervous system of mammals. Glia 2003;43:91–93.
24. Araque A, Sanzgiri RP, Parpura V, Haydon PG. Astrocyte-induced modulation of synaptic transmission. Can J Physiol Pharmacol 1999;77:699–706.
25. Araque A, Parpura V, Sanzgiri RP, Haydon PG. Tripartite synapses: glia, the unacknowledged partner. Trends Neurosci 1999;22:208–215.
26. Lin SC, Bergles DE. Synaptic signaling between neurons and glia. Glia 2004;47:290–298.
27. Theodosis DT, Piet R, Poulain DA, Oliet SH. Neuronal, glial and synaptic remodeling in the adult hypothalamus: functional consequences and role of cell surface and extracellular matrix adhesion molecules. Neurochem Int 2004;45:491–501.
28. Bauer HC, Bauer H. Neural induction of the blood-brain barrier: still an enigma. Cell Mol Neurobiol 2000;20:13–28.
29. Tontsch U, Bauer HC. Glial cells and neurons induce blood-brain barrier related enzymes in cultured cerebral endothelial cells. Brain Res 1991;539:247–253.
30. Herrup K, Busser JC. The induction of multiple cell cycle events precedes target-related neuronal death. Development 1995;121:2385–2395.
31. Nagy Z, Esiri MM, Cato AM, Smith AD. Cell cycle markers in the hippocampus in Alzheimer's disease. Acta Neuropathol (Berl) 1997;94:6–15.
32. Nagy Z, Esiri MM, Smith AD. Expression of cell division markers in the hippocampus in Alzheimer's disease and other neurodegenerative conditions. Acta Neuropathol (Berl) 1997;93:294–300.
33. Zhu X, Raina AK, Smith MA. Cell cycle events in neurons. Proliferation or death? Am J Pathol 1999;155:327–329.
34. Grana X, Reddy EP. Cell cycle control in mammalian cells: role of cyclins, cyclin dependent kinases (CDKs), growth suppressor genes and cyclin-dependent kinase inhibitors (CKIs). Oncogene 1995;11:211–219.
35. Chen D, Walsh K, Wang J. Regulation of cdk2 activity in endothelial cells that are inhibited from growth by cell contact. Arterioscler Thromb Vasc Biol 2000;20:629–635.
36. Akiyama SK. Integrins in cell adhesion and signaling. Hum Cell 1996;9:181–186.
37. Albelda SM. Differential expression of integrin cell-substratum adhesion receptors on endothelium. EXS 1992;61:188–192.
38. Albelda SM, Buck CA. Integrins and other cell adhesion molecules. FASEB J 1990;4:2868–2880.
39. Ruoslahti E. Integrin signaling and matrix assembly. Tumour Biol 1996;17:117–124.
40. Schwartz MA, Ingber DE. Integrating with integrins. Mol Biol Cell 1994;5:389–393.
41. Akhta N, Carlso S, Pesarini A, Ambulos N, Passaniti A. Extracellular matrix-derived angiogenic factor(s) inhibit endothelial cell proliferation, enhance differentiation, and stimulate angiogenesis in vivo. Endothelium 2001;8:221–234.
42. Schwartz MA, Schaller MD, Ginsberg MH. Integrins: emerging paradigms of signal transduction. Annu Rev Cell Dev Biol 1995;11:549–599.
43. Dike LE, Ingber DE. Integrin-dependent induction of early growth response genes in capillary endothelial cells. J Cell Sci 1996;109 (Pt 12):2855–2863.
44. Huang S, Chen CS, Ingber DE. Control of cyclin D1, p27(Kip1), and cell cycle progression in human capillary endothelial cells by cell shape and cytoskeletal tension. Mol Biol Cell 1998;9:3179–3193.

45. Clauss M, Weich H, Breier G, et al. The vascular endothelial growth factor receptor Flt-1 mediates biological activities. Implications for a functional role of placenta growth factor in monocyte activation and chemotaxis. J Biol Chem 1996;271:17,629–17,634.
46. Yancopoulos GD, Davis S, Gale NW, Rudge JS, Wiegand SJ, Holash J. Vascular-specific growth factors and blood vessel formation. Nature 2000;407:242–248.
47. Ferrara N, Bunting S. Vascular endothelial growth factor, a specific regulator of angiogenesis. Curr Opin Nephrol Hypertens 1996;5:35–44.
48. Carmeliet P, Ferreira V, Breier G, et al. Abnormal blood vessel development and lethality in embryos lacking a single VEGF allele. Nature 1996;380:435–439.
49. Koblizek TI, Weiss C, Yancopoulos GD, Deutsch U, Risau W. Angiopoietin-1 induces sprouting angiogenesis in vitro. Curr Biol 1998;8:529–532.
50. Esser S, Wolburg K, Wolburg H, Breier G, Kurzchalia T, Risau W. Vascular endothelial growth factor induces endothelial fenestrations in vitro. J Cell Biol 1998;140:947–959.

25
Adult Neurogenesis and Central Nervous System Cell Cycle Analysis

Novel Tools for Exploration of the Neural Causes and Correlates of Psychiatric Disorders

Amelia J. Eisch, PhD and Chitra D. Mandyam, PhD

SUMMARY

Once thought to be restricted to development and the early postnatal period, neurogenesis is now known to be an ongoing event in the brain of adult mammals. One brain region that is a site for adult neurogenesis, the hippocampus, is implicated in a variety of psychiatric disorders: depression, addiction, and schizophrenia. The goal of this chapter is to provide an overview of recent evidence suggesting that altered hippocampal neurogenesis is evident in psychiatric disorders. Given the global topic of this volume, additional emphasis is placed on the use of and potential for cell cycle protein analysis to reveal additional information about the links between adult hippocampal neurogenesis and psychiatric disorders. In this way, it is hoped that this chapter will encourage cell cycle researchers to turn their attention, and hopefully efforts, to exploring these links, as well as encourage researchers of adult neurogenesis to employ cell cycle analysis in their efforts to explore stem and progenitor cell biology in the adult mammalian brain.

Key Words: Adult neurogenesis; hippocampus; subgranular zone; subependymal zone; subventricular zone; cell cycle; depression; addiction; schizophrenia.

1. INTRODUCTION

Psychiatric disorders, such as depression and addiction, pose an enormous economic and emotional burden on all societies in the modern world. The etiology of mental disorders is exceptionally complex. Social, environmental, and genetic factors are all postulated to play roles in the predisposition or development of psychiatric disorders. However, research has linked abnormalities in brain structures to a variety of psychiatric disorders. For example, the hippocampus, a brain region involved with spatial memory and regulation of affect and emotion, is decreased in size and/or presents abnormal morphology in myriad psychiatric disorders: depression, posttraumatic stress disorder, dementia, schizophrenia, and drug dependence *(1–5)*. Congenital abnormalities in brain structure are perhaps not surprising in disorders that are diagnosed in early life, such as autism. However, the connection between abnormal hippocampal structure and adult brain function has drawn much attention, given the distinction of the hippocampus as one of a few brain structures that can make new neurons throughout adulthood. This connection raises questions: Do alterations in adult hippocampal neurogenesis contribute to or correlate with the manifestation of psychiatric symptoms? Will normalization of hippocampal neurogenesis in such disorders contribute to improvement of symptoms?

The primary goal of this chapter is to highlight recent research on adult mammalian neurogenesis relevant to understanding the relationship between psychiatric disorders and adult

From: *The Cell Cycle in the Central Nervous System*
Edited by: D. Janigro © Humana Press Inc., Totowa, NJ

neurogenesis. Much progress has been made in this pursuit; Eisch's 2002 review on this topic lists many goals that have already been accomplished (6). Although the fast pace of research in this field has generated numerous reviews on psychiatric disorders and neurogenesis (7–10), three aspects of this chapter distinguish it from earlier reviews. First, little has been written in regards to using cell cycle analysis to assess the connection between psychiatric disorders and neurogenesis. The emphasis of this book being on investigation of cell cycle in the central nervous system (CNS), this chapter will attempt to fill that important niche by detailing what is known about endogenous cell cycle proteins in relation to adult mammalian neurogenesis. Second, although the links between neurogenesis and certain psychiatric and neurological disorders, such as depression or brain injury, have been discussed in recent reviews, little has been written about advances made in regards to other disorders, such as addiction and schizophrenia. Therefore, this chapter will detail what is known about addiction and schizophrenia, and will attempt to provide an overview of the current assessment of the putative link between adult neurogenesis and all psychiatric disorders. Third, as the nature of this book volume is to present diverse research on the CNS cell cycle, a background on the controversies and recent advances in the field of adult neurogenesis research will precede discussion of the relationship between neurogenesis and specific psychiatric disorders. It is hoped that this background will allow cell cycle researchers outside of the field to turn their attention, and hopefully their expertise, to addressing the complex issue of whether, and how, alterations in adult neurogenesis may be involved in the etiology or progression of these devastating disorders.

2. "NEW NEURONS IN OLD BRAINS": PAST AND CURRENT CONTROVERSIES

The birth of new neurons in the adult mammalian brain was reported almost 50 yr ago (11–16), but it took 40 more years for their occurrence in primates and humans to be generally accepted. Beginning in 1950s and 1960s and extending through early 1990s, intraperitoneal injection of the S phase marker [^{3}H]thymidine into dogs, cats, rats, and mice revealed labeled cells with neuronal morphology in discrete regions of the adult brain. These studies established the occurrence of dividing cells with neurogenic potential in the subgranular zone (SGZ) of the hippocampus and the more anterior subependymal zone (SEZ; also called the subventricular zone [SVZ]) bordering the caudate-putamen and lateral ventricles (17–21). Dividing cells in the SGZ migrate a short distance into the granule cell layer (GCL) and mature into hippocampal granule cells. In contrast, dividing cells in the SEZ migrate a significant distance by entering the aptly named rostral migratory stream (RMS), and eventually become olfactory bulb interneurons. Distinct differences exist between resident progenitors of the SGZ and SEZ (22), and given the more prominent putative role for the hippocampus in psychiatric disorders, the SGZ will be the main focus of this chapter. However, it is notable that neurons coming from both the SGZ and SEZ functionally integrate into hippocampal or olfactory bulb circuitry, respectively (23–25). This underscores the fact that adult mammalian neurogenesis is not a vestigial phenomena; new neurons have the capacity to impact the structure of the adult mammalian brain (26,27).

The function of adult-generated neurons has been more challenging to define (28). Much has been made of the fact that the production of new neurons is increased or decreased by manipulations thought to be "positive" or "negative" influences on brain function. For example, stress, age, and drugs of abuse decrease the number of new cells with neurogenic potential (29–31). In contrast, voluntary exercise, hippocampal-dependent learning, and antidepressant treatment increase the number of new cells with neurogenic potential (32–34). The discrete regional occurrence and the regulation of neurogenesis is intriguing in that both the olfactory bulb and hippocampus are linked to memory function. In fact, correlative evidence that "new neurons lead to better brain function" is recently complemented by attempts to make causal links between adult neurogenesis and brain function. For example, attenuation of new cell birth in the hippocampus leads to diminished hippocampal function in vivo and in vitro (35,36). Might the new neurons in the adult brain somehow contribute to memory formation or retrieval *(37)*? No published work to date provides definitive, causative (as opposed to correlative) evidence, but the results are intriguing enough to encourage more sophisticated exploration, such as with inducible transgenic approaches in which neurogenesis might be "turned off."

Adult-generated neurons positively impact the structure, and perhaps the function, of the brain. Basic researchers have long been able to probe olfaction and memory function in laboratory animals. However, a long-standing controversy in the field is whether adult neurogenesis occurs at sufficiently high levels in the primate or human brain to be of importance to structure and/or function. Early work with primates identified few, if any, [^{3}H]thymidine-labeled cells in the young adult brain. This distinction in the occurrence of neurogenesis in rats and mice but not in primates was blurred in late 1990s with the advent of monoclonal antibody technology, and the increased sensitivity of this method to detect dividing cells. This approach allowed several researchers to visualize new neurons by injecting the thymidine analog bromodeoxyuridine (BrdU), and using a BrdU antibody *(38)*. Several publications appeared in rapid succession detailing evidence of adult neurogenesis in the primate and human brain, raising the question of whether these new neurons in the human brain did indeed contribute to hippocampal function, such as learning and memory. However, skepticism was raised about the appropriateness of this question, given (a) the potential for BrdU to label cells undergoing cell death or DNA repair, not just cell division *(39)* and (b) the much lower level of magnitude and transient existence of neurons in human and primate relative to rats and mice *(40,41)*.

The first concern—is it all DNA repair? *(39)*—has been addressed in several publications, with the majority of work showing that appropriate BrdU administration conditions reveal cells that were in S phase, not undergoing DNA repair, at the time of BrdU injection *(42)*. In addition, evidence suggests that the labeling is not owing to RNA synthesis, because treatment with RNase prior to staining does not alter the number of labeled cells *(43)*. Concerns still exist about whether BrdU labeling in models of brain injury reveal dividing or repairing cells. (*See*, for example, ref. *44)*. This issue is relevant to the current chapter, because animal models of depression and addiction, for example, may result in a mild form of brain injury. As such, this issue will be addressed in detail in later in this chapter.

The second concern is also of great relevance: humans and nonhuman primates have a low number of adult-generated neurons relative to other mammals, and therefore, adult neurogenesis may not of significant consequence in terms of brain function. If this concern is valid, then attempts at understanding the regulation of adult neurogenesis in the laboratory rat and mouse with the goal of linking adult neurogenesis to psychiatric disorders are without merit. Support for the decreased importance of adult-generated neurons to the human brain came from recent work reporting that although dividing cells are seen in the adult human SEZ, no chain migration through the Rostral migratory system is evident in the human brain. Might adult-generated neurons have no impact in the adult primate or human brain? Several points counterbalance this concern *(45)*, and two will be stressed here. First, neuroscience research emphasizes that the brain, in many ways, does not function as a democracy. For example, adult-generated neurons integrate into one of the main entry portals of the hippocampus, the dentate gyrus, and this portal is a site of convergence of many discrete inputs into the hippocampus. Therefore, even the addition of a few adult-generated hippocampal neurons may be well positioned to influence hippocampal function. Thus, although the field moves forward with care and forethought, sufficient evidence exists to pursue the intriguing connection between adult-generated hippocampal cells and hippocampal function. Second, the number or permanence of adult-generated neurons is reliant on the sensitivity of detection techniques. A key component in the delay in recognizing adult neurogenesis as a phenomenon common to all mammals was the relative difficulty of identifying [^{3}H]thymidine-labeled cells. Given that even BrdU is diluted after each cell division, it is possible that more adult-generated neurons persist undetectable even via the more sensitive immunohistochemical labeling of BrdU cells. This latter point emphasizes the need to establish new and more sensitive ways to detect cells within the cell cycle and within various stages of cell development in the adult brain; this is the focus of the following section.

3. VISUALIZATION OF NEW CELLS IN THE ADULT MAMMALIAN BRAIN

The most common method to detect adult neurogenesis has been via immunocytochemistry against exogenous markers such as BrdU or [^{3}H]thymidine. The advantages of exogenous markers are clear: the ability to "birth date" and examine the cell cycle kinetics of dividing cells *(38,46,47)*. The disadvantages of exogenous markers, however, are significant: dilution of the marker over time, low efficiency of labeling cells with very long cell cycles, potential interference with DNA function, and inability to use exogenous markers for study of postmortem tissue or neurogenesis in wild, nonlaboratory animal populations. These disadvantages emphasize the increased need for alternative approaches to visualizing adult-generated neurons in the brain. Retroviral infection has been widely used to study neurogenesis during pre- and postnatal development *(48,49)*. One of the benefits of retroviral infection is the lack of dilution over multiple cell divisions *(50)*. This will prove valuable in future efforts to characterize the clonal history of dividing cells in the adult brain. Two additional approaches will be discussed here: labeling of endogenous proteins and utilization of transgenic mouse models.

3.1. Endogenous Proteins for Detection of Adult-Generated Neurons

A complimentary and alternate approach to using exogenous markers of the cell cycle is to label endogenous cell cycle proteins that have distinct expression patterns in certain phases of the cell cycle. As discussed in this section, many antibodies are available to visualize endogenous cell cycle proteins, some of which have been used to visualize cells in the neurogenic regions of the mammalian brain. They include retinoblastoma protein (Rb), proliferating cell nuclear antigen (PCNA), Ki-67, cyclin-dependent kinase 1 (CDK1), histone-H3, and doublecortin (DCX) (*see* Fig. 1). In addition to labeling for single proteins, labeling for multiple proteins has proven useful in tracking newborn cells in the adult brain through the cell cycle *(51,52)*. In addition, combining endogenous cell cycle protein detection with BrdU detection allows birth dating of newborn cells *(51)*. Here we will give an overview of endogenous cell cycle markers and discuss how they have been used to study questions related to adult neurogenesis in the mammalian brain.

3.1.1. Rb Protein

Rb is a nuclear phosphoprotein that was initially discovered for its association with transcription factors (e.g., the E2F family of transcription factors) that play a role in cell cycle progression *(53–57)*. Rb protein has functional domains where transcription factors bind and form complexes *(58)*. The relationship between Rb, transcription factors, and CDK1 (discussed in Subheading 3.1.4) is dynamic, and has been studied both in vitro and in non-neural tissue *in situ* to elucidate movement of dividing cells through the cell cycle. For example, cells stained for hypophosphorylated Rb are in the late G_1 phase. Subsequent phosphorylation of Rb in late G_1 by CDK1 (and other kinases, including CDK4/CDK6 and CDK2) releases factors, such as E2F, which promote the cell to enter into S phase of the cell cycle *(58)*.

Rb's expression and distribution in the developing *(59)* and adult *(60)* mammalian brain and its relationship to neurogenesis have been assessed. Knockout studies have revealed an important role for Rb in neurogenesis *(61,62)*, and 75–80% of newly born cells in the adult rat SEZ and SGZ express Rb protein *(60)*. However, little work has been done to examine expression of phosphorylated Rb in G_1, or to relate Rb expression to the expression of other cell cycle protein *in situ*, such as E2F. For example, it is possible that colabeling with Rb and E2F may provide a cell cycle phase-specific marker of late G_1 and early S. In addition, given that some studies suggest a role of Rb in quiescent (G_0) cells, work needs to be done to characterize RB expression levels and protein interactions in quiescent cells.

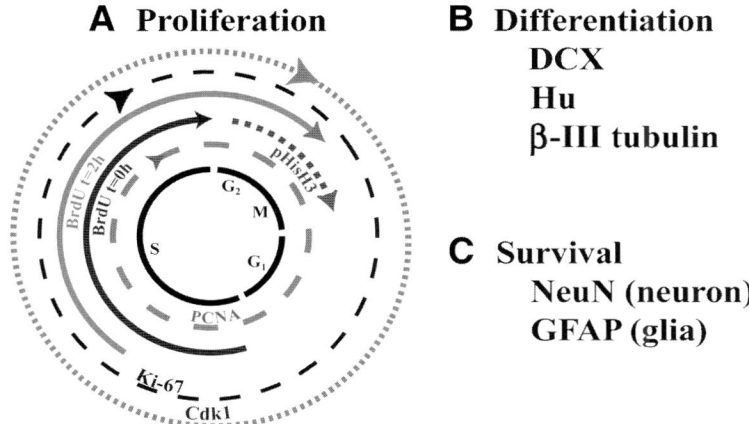

Fig. 1. Endogenous and exogenous markers available to detect proliferation (**A**), differentiation/immature phenotype (**B**), and survival/mature phenotype (**C**). Schematic in (**A**) depicts mouse SGZ cell cycle indicating the proportion of the length of S phase, G_2/M phase and G_1 phase compared with the entire cell cycle *(46)*. Endogenous and exogenous cell cycle markers are shown spanning the putative phase(s) of the cell cycle during which the marker is detectable via immunocytochemistry of adult brain. *See* Subheading 3.1.

3.1.2. Proliferating Cell Nuclear Antigen

Proliferating cell nuclear antigen (PCNA), also known as cyclin, is a catalytic nuclear protein associated with DNA polymerase δ *(63,64)*. PCNA's association with this DNA polymerase explains the colocalization of PCNA protein throughout the cell cycle with markers of DNA.

PCNA protein level within a cell is most abundant during late G_1 and early S, and is expressed at lower levels during G_2 and M *(65,66)*. In the adult mammalian brain, PCNA expression is well correlated with BrdU incorporation *(67,68)*, supporting the expression of PCNA during S phase. However, the expression of PCNA is detectable throughout all four stages of the cell cycle in vitro, and therefore PCNA immunohistochemistry (IHC) is often used to detect proliferating cells throughout the cell cycle, regardless of fluctuating levels of total PCNA protein *(69)*.

Several papers have used PCNA to label proliferating cells *in situ* and in brain tissue *(67,68,70–72)* and it is clear that PCNA labeling is closely associated with BrdU labeling. However, it is notable that few quantitative studies on PCNA *in situ* expression exist. For example, can PCNA be used as a marker of the entire cell cycle, and BrdU be used as a marker of the S phase? If true, the ratio of the number of BrdU+ cells to PCNA+ cells would reveal a ratio of the S phase to the entire cell cycle. Such an approach would allow an approximation of length of S phase in dividing cells in the adult mammalian brain without using multiple injections of S phase markers previously needed *(46,47)*. Preliminary work from our laboratory suggests that this is the case *(52)*.

Given PCNA's intimate interaction with DNA, it is not surprising that PCNA expression may also be evident during DNA repair and cell death. Little work has been done *in situ* to address the proportion of PCNA+ cells that are not in the cell cycle (e.g., are undergoing either DNA repair or cell death). Therefore, although the ratio of PCNA+ to BrdU+ cells holds the potential to reveal cell cycle kinetics in adult mammalian brain, clearly the ability of PCNA to mark *in situ* cycling cells vs damaged or dying cells remains to be clarified prior to acceptance of PCNA immunostaining as a strict marker of proliferating cells.

3.1.3. Ki-67

Ki-67 is a nuclear, nonhistone protein whose presence in a cell is significantly correlated with cell proliferation *(73,74)*. Antisense oligonucleotides targeting Ki-67 inhibit cell proliferation in vivo and in vitro *(74,75)*, further supporting the role of Ki-67 in progression through the cell cycle. There are several splice variants of the 30,000-base-pairs-long *Ki-67* gene, and only two proteins encoded by two open reading frames are transcribed at detectable levels *(74)*. The primary structure of Ki-67 has potential sites for phosphorylation by various kinases including protein kinase C, casein kinase II, tyrosine kinase, and CDK1 (discussed below) *(74)*. The presence of multiple posttranslational modification sites suggest that Ki-67 is a target for multiple second messenger pathways, and can be modulated via a variety of protein–protein interactions. Phosphorylation and subsequent dephosphorylation of Ki-67 by the regulatory complex cyclin B/CDK1 appears essential for proliferating cells to enter and exit mitosis *(76)*.

Ki-67 protein level is most abundant during late G_1 and early S, and is expressed at lower levels during G_2 and M *(77)*. Ki-67 expression is well correlated with BrdU incorporation *(78,79)*, supporting the expression of Ki-67 during S phase, and this correlation has also been quantified: nearly all newly born BrdU+ cells express Ki-67. Via IHC, Ki-67 is reliably detected in S, G_2, and M phases of the cell cycle. Unlike PCNA and Rb, which are expressed in all phases of the cell cycle, Ki-67 expression in G_1 phase of the cell cycle is dependent on a given cell's division history. For example, in vitro data suggest that proliferating cells entering early G_1 after M express Ki-67, but cells entering G_1 from G_0 do not express Ki-67 *(77)*. Although this remarkable aspect of Ki-67 IHC has been used in vitro to distinguish cells, in the recent history of dividing cells, such an application has not been applied to *in situ* research in regards to adult neurogenesis.

3.1.4. Cyclin-Dependent Kinase 1

CDK1, or p34/cdc2, was initially discovered as the mitosis-promoting factor *(80,81)*. Although CDK1 is expressed throughout the cell cycle as a complex with cyclin B *(82–84)*, the phosphorylation state of CDK1 varies throughout the cell cycle. In its inactive state, the CDK1/cyclin B complex is hyperphosphorylated at Thr 14, Tyr 15, and Thr 161. This hyperphosphorylated state is maintained by the protein Myt1, a Wee1 family protein kinase *(85)*. During G_2 phase of the cell cycle, several protein kinases are upregulated, preparing the cell to enter M phase. The most important kinases include Akt (also known as protein kinase B) and Cdc25C *(84)*. Akt inhibits the hyperphosphorylation of CDK1/cyclin B complex by Myt1, thereby allowing Cdc25C to dephosphorylate the CDK1/cyclin B complex. At the onset of mitosis, dephosphorylation of CDK1 occurs at Thr 14 and Tyr 15 by Cdc25C. The active, dephosphorylated CDK1/cyclin B complex remains in that state until the cells have divided *(81,86)*.

Given the putative expression of CDK1 throughout the cell cycle, it is interesting that only 33% of proliferating cells in the adult rat SEZ and SGZ have been shown to express CDK1 *(60)*. This contradictory result could have several explanations, such as CDK1 having a different cell cycle expression pattern relative in adult brain relative to in vitro systems, or the antibody used to detect CDK1 in that study was against an epitope masked by CDK1's interaction with cyclin B. Regardless, more work needs to be done to clarify the expression of CDK1 *in situ*—in either its inactive, hyperphosphorylated state in most of the cell cycle, or its active, dephosphorylated state in late G_2 and throughout M phase—in regards to cell cycle phase specificity.

3.1.5. Histone-H3

Histones, the core of the nucleosome, have long been appreciated for their ability to regulate movement through the cell cycle *(87)*. The basic unit of the chromatin, the nucleosome, is made up of an octamer of core histones (two of each histone-H2A, histone-H2B, histone-H3, and histone-H4) that are wrapped around two superhelical turns of DNA *(88)*. All the histones have been implicated in chromosome condensation and chromosome assembly. Phosphorylation and subsequent dephosphorylation of histone-H3 *(87)* at serine 10 (S10) is thought to be one of several steps necessary for chromosome condensation and successful exit from mitosis *(89–91)*. Histone-H3 is phosphorylated at S10 from early G_2 to the end of M *(92)*. Although the role of pHis-H3 in cell division has been extensively studied in vitro *(91,93–96)*, given the ability of pHis-H3 to label cells in a distinct portion of the cell cycle—G_2/M—it is notable that few reports of pHis-H3 IHC in adult neural tissue have been published.

3.1.6. Doublecortin

DCX was discovered as a brain-specific protein, whose gene was found mutated in patients with a "double cortex" and lissencephaly *(97)*. DNA sequence analysis of DCX indicates multiple phosphorylation sites by mitogen-activated protein kinase and several other kinases. Although the precise function of DCX is unknown, its signaling is thought to play an important role in neuronal migration *(97)*. DCX immunoreactivity is evident in neurogenic regions of the adult mammalian brain, namely SGZ and SEZ *(67,98)*, and has been used to establish the presence of adult-generated neurons in typically nonneurogenic regions of the adult brain after injury *(99)*. DCX labeling is seen in proliferating cells that are postmitotic and in young developing neurons *(100,101)*, suggesting that DCX is a marker for immature neurons *(102)*. Although DCX is thought to be a postmitotic maker, recent work suggests that many cells in the cell cycle also express DCX *(98)*. In addition, DCX labeling is also seen in regions, such as the piriform cortex, where no adult-generated neurons have been detected. These studies emphasize that it is not yet clear what DCX labeling means, and encourage caution in interpreting DCX immunostaining until the details of this protein and its expression in the adult brain are resolved.

Summing all these, there are numerous benefits of studying endogenous cell cycle proteins in adult brain tissue: verification of active proliferation without introduction of exogenous marker (via staining for PCNA or Ki-67), identification of cell cycle phase (via staining for pHis-H3 or pCDK1 for G_2/M phase, or Rb/E2F for late G_1 and early S phase), division history (via staining for Ki-67), age and fate of newborn cells (DCX), and the potential to track cell cycle dynamics (via a combination of staining for BrdU, PCNA, and pHis-H3). Although much work has been done to identify the expression of endogenous cell cycle proteins in vitro, little work has been done to study their expression in adult neural tissue *in situ*. Whereas the close association of cell cycle and cell death cascades encourages caution in interpretation of data resulting from endogenous cell cycle studies *(103)*, several approaches can be used to more clearly define cells as cycling or dying: regional expression, morphological analysis, multiple labeling with different cell cycle, and differentiation markers. With these caveats in hand, examination of endogenous cell cycle markers is likely to benefit most researchers in the adult neurogenesis field, including those tackling the challenge of assessing the relationship between adult neurogenesis and psychiatric disorders and the potential importance of adult neurogenesis in the human brain *(104,105)*.

3.2. Transgenic Mouse Models for Detection of Adult-Generated Neurons

Recent advancements in transgenic mouse technology have produced mouse lines to visualize proliferating cells without the aid of exogenous or endogenous markers. For example, the transgenic approach has proven useful in verifying glial fibrillary acidic protein (GFAP)-positive cells in the adult SGZ as one source of hippocampal stem cells *(106,107)*. However, given the immense lack of knowledge about fundamental aspects of adult-generated neurons, of particular value have been reporter mice, in which fluorescent reporter genes allow in vitro and in vivo analysis of the newly born cells without the limitations of exogenous markers *(108)*. Two lines of reporter mice will be highlighted here to illustrate their impact on the field of adult neurogenesis.

3.2.1. Nestin-eGFP Mouse

The nestin-eGFP mouse was initially developed for examination of stem cells in the developing brain. Preliminary studies with the mouse showed that the nestin promoter was sufficient to drive GFP in neurogenic regions and in stem cells of the embryonic *(109)* and early postnatal *(110)* mouse. Recent studies have used this mouse for study in the adult brain *(110–112)*. Detailed analysis of GFP+ cells reveals a heterogeneous population of cells, each with a distinct morphology and labeling of other markers, such as GFAP and DCX. This heterogeneity has been used by some groups to examine the impact of external factors (such as voluntary exercise) on stem, progenitor, and committed progenitor cells. That nestin eGFP mice can be used to distinguish cells at different stages of stem and progenitor cell differentiation makes this an extremely valuable tool for assessment of the relationship between adult neurogenesis and psychiatric disorders.

In the adult mouse, all proliferating cells in the adult SGZ and SEZ appear to express eGFP *(51)*. However, only a fraction of eGFP+ cells colabel with BrdU. This emphasizes that the nestin promoter is active throughout the cell cycle, not just in S, G_2, and M phases. Therefore, it is feasible to use the nestin-eGFP mouse to assess all cells in the cell cycle, and to combine this expression with phase-specific markers of the cell cycle, for example pHis-H3 and Ki-67, to determine the respective proportion that each phase contributes to the cell cycle of a proliferative population *(51)*. BrdU will still be needed to birth-date cells, and therefore will be needed to assess cell cycle kinetics. However, the time resolution of gene promoters may improve soon to the extent in which a transgenic mouse, for example, expresses a destabilized fluorescent protein that turns on during anaphase, and is degraded by the completion of telophase.

3.2.2. Propiomelanocortin-eGFP Mouse

Unlike the nestin-eGFP line that is used to assess the proliferating population, the propiomelanocortin (POMC)-eGFP line can be used to assess differentiation of newborn cells in the SGZ. In the POMC-eGFP mouse, differentiating cells express eGFP under the control of POMC promoter *(113)*. Injection of POMC-eGFP mice with BrdU showed a gradual increase in BrdU+ and GFP+ cells. Very few GFP+ cells were BrdU+ at early proliferating time-points (3 d) and maximal colabeling was seen at 11–12 d *(113)*. This emphasizes the utility of the POMC-eGFP mouse to track cells that have left the cell cycle and have begun to differentiate and perhaps migrate to their mature destination. More work needs to be done with the POMC-eGFP line to determine the "age" of the proliferating/differentiating cells. This includes labeling GFP with a series of endogenous markers—PCNA, to determine if POMC promoter is active in all phases of the cell cycle; pHis-H3, in G_2 and M phases of the cell cycle; and DCX, to determine if the newborn cells are going to become immature neurons.

Although exceptionally promising for enabling visualization of stem and progenitor cells throughout the life of the animal, the expression of a reporter molecule in cells early in their development may interfere with normal function of the cells. Therefore, the challenge remains to use transgenic technology to visualize adult-generated neurons via an inducible transgenic system in which a scientist has temporal control over expression of a reporter molecule or removal of a given gene *(114,115)*. The inducible approach, and the growing knowledge of progenitor-selective and region-selective promoters, likely will enable researchers to turn off or turn on cytogenesis or neurogenesis in a particular brain region of adult animals. Inducible transgenic systems will be essential for evaluating the hypothesis that more neurogenesis equals better memory and less neurogenesis equals impaired memory. One recent paper used a conditional, forebrain-specific knockout of the *presenilin-1* gene to address the relationship between neurogenesis and memory *(116)*. The authors conclude that neurogenesis may be essential for clearance of memories from the hippocampus, suggesting a novel relationship between neurogenesis and memory. However, significant technical differences exist between this and earlier studies *(32,35)*, such as the BrdU injection paradigm used and the timing of the behavioral analysis after BrdU injection *(116)*. This emphasizes not only the need for additional adult neurogenesis studies utilizing conditional transgenic mice, but also for the consensus of general standards of cytogenesis and neurogenesis *(6,7)*.

The remainder of this chapter will provide an overview of reports in the past 3 yr, which are relevant to the question at hand: Are psychiatric disorders marked by abnormal adult hippocampal neurogenesis? The majority of these studies have been done in animal models, not in human or postmortem human tissue. Scientists can probe myriad aspects of stress, depression, and addiction in laboratory animals, and the bulk of information does suggest a correlation: more new adult-generated hippocampal neurons, better hippocampal function and improvement of "psychiatric-related symptoms" in the laboratory animal. However, scientists are cautious about directly interpreting such correlative results to the human condition, especially because the extent and importance of adult hippocampal neurogenesis in the human brain is still unclear. In this regard, the endogenous cell cycle analysis proposed above should prove useful in determining the role and regulation of adult-generated cells in the human brain. In addition, scientists are beginning to move beyond simply *counting* the number of newly generated cells and instead probing the mechanism underlying their regulation. In spite of these technical limitations, the following sections will emphasize the intriguing connection that exists between adult hippocampal neurogenesis and stress-related disorder, substance abuse, and schizophrenia.

4. STRESS-RELATED DISORDERS

4.1. Stresses in Adulthood

In humans, psychosocial, environmental, or physical stress is implicated in the onset or relapse of a variety of psychiatric disorders, including mood disorders, substance abuse, and schizophrenia *(117–121)*. Stress has also been linked to decreased hippocampal volume. Interestingly, one of the most potent inhibitors of adult hippocampal neurogenesis is stress. In laboratory animals, stress can be applied by exposure to a dominant member of the same or predatory species (or even to their odor alone), by physically isolating or restraining the animal, or by a several-week exposure to a variety of mild environmental or psychosocial stressors. These stressors have all been shown to decrease the number of newly born neurons in the adult mammalian SGZ *(122–127)*. Interestingly, stress does not appear to negatively impact neurogenesis in SEZ *(125)*, emphasizing nascent differences between the progenitor cells in the adult SGZ and SEZ. In the human condition, the duration or intensity of stress appears to be important in stress-induced relapse to depression or substance abuse. This point is mirrored in recent research with animal models of stress. For example, chronic, but not acute, physical restraint results in diminished adult hippocampal neurogenesis and volume of granule cell layer *(128)*. In addition to decreasing adult neurogenesis, stress decreases the performance on learning and memory tasks and decreases hippocampal volume *(129–131)*. Although correlative, these findings are part of the growing body of evidence that supports a role for new neurons in hippocampal function.

Adrenal glucocorticoids (GCs), such as corticosterone, are hypothesized to mediate the negative impact of stress on hippocampal neurogenesis. This is supported by the negative impact of systemic corticosterone on adult neurogenesis *(30,132)* and the positive impact of removal of the source of endogenous GCs by adrenalectomy (ADX) *(30,133)*. A corticosterone replacement regimen that mimics the circadian fluctuation of the hormone is effective in normalizing neurogenesis in ADX rats *(43,134)*, emphasizing the potential importance of diurnal and nocturnal cycles of hormones in regulating cell birth. Different receptors mediate the impact of stress hormones: the low-affinity mineralocorticoid (MR) vs high-affinity GC receptor. Although no direct evidence has yet been found for location of either receptor on progenitor cells in the adult brain *(135)*, several pieces of evidence implicate MR and GC receptors in regulation of neurogenesis. First, disruption of the MC receptor interferes with neurogenesis in the adult hippocampus *(136)*. Second, new neurons appear to be sensitive to MR or GC stimulation during different stages of their proliferation, differentiation, and maturation *(137)*. This suggests a dynamic expression of stress-hormone receptors over the life span of adult-generated neurons. Finally, the balance between cell birth and cell death in the hippocampus is influenced by stimulation of MR or GC receptors *(138)*. Taken together, these studies indicate that GCs are an important mediator of the stress-induced decrease of adult neurogenesis.

Although stress and stress-related hormones are widely embraced as inhibitors of adult hippocampal neurogenesis, several recent papers highlight situations in which stress and stress-related hormones do not appear to mediate alterations in adult neurogenesis. For example, although dominant males in a social dominance hierarchy have higher levels of neurogenesis compared with subordinates and controls, the groups do not differ in basal or stress levels of corticosterone *(139)*. Likewise, as discussed in more detail below, chronic opiate exposure inhibits hippocampal neurogenesis independent of corticosterone *(43)*. Finally, some groups find no correlation between cell birth and corticosterone levels or stress response parameters

(140). These findings, and others, suggest that in some instances variables other than stress—or compounds other than corticosterone—may be responsible for the decreases in adult neurogenesis. For example, neurosteroids, glutamate, serotonin, estrogen, and testosterone, are all viable candidates for mediating stress-induced alterations in neurogenesis *(141–145)*. In addition, stress clearly causes other changes in the hippocampus, which could be responsible for the stress-induced reduction in hippocampal volume *(146)*. Therefore, although stress and corticosterone can inhibit adult neurogenesis and this may lead to stress-induced reduction in hippocampal volume, other factors should be examined in addition to stress.

4.2. Pre- and Postnatal Stress

Stress in early life increases susceptibility to psychiatric disorders in adulthood and causes long-lasting changes in hippocampal structure and function *(147–150)*. In recent years, many studies have shown that prenatal or early postnatal stress, including dietary deficiency or food restriction, also leads to decreased hippocampal cell proliferation in adulthood *(151–153)*, and that this decrease in newly born cells in the adult SGZ is correlated with impaired performance on behavioral tasks *(151)*. For example, rats separated from their mother in early life have decreased hippocampal progenitor proliferation and immature neuron production as adults *(154)*. Interestingly, the postnatal stressed rats did not show stress-induced inhibition of adult neurogenesis *(154)*, and they displayed other markers of abnormal responsivity to stress hormones in adulthood. Such findings have been replicated in primates using prenatal stress *(155)*. When exposed to stress either early or late in pregnancy and examined during adulthood, stressed primates had decreased hippocampal neurogenesis and volume, abnormal stress-hormone responsivity, and altered emotionality relative to controls. Given the gender imbalance of certain stress-related psychiatric disorders, it is interesting to note that early stress-induced inhibition of neurogenesis is greater in females than in males *(156)*. Lastly, it is intriguing that the relatively novel antidepressant treatment of acupuncture can normalize the decreased cell proliferation seen after maternal separation *(157)*. Taken together, these studies provide evidence that early stress can inhibit adult neurogenesis and alter the ability of the adult hippocampus to respond to stress, and that these changes may respond favorably to antidepressant treatment in adulthood. Clearly, more studies are warranted to examine whether such early stress-induced alterations in neurogenesis are associated with predisposition to psychiatric disorders.

4.3. Mood Disorders

Mood disorders, including major depression and bipolar disorder, are among the most prevalent causes of morbidity, disability and suicide throughout the world *(158)*. Given that patients with mood disorders have alterations in effect, learning, and memory, researchers have postulated that the hippocampus plays a role in the pathophysiology of mood disorders *(159)*. In support of this hypothesis, many studies have found volume and cell loss and abnormal morphology of the hippocampus and related brain regions in patients with mood disorders *(2,160–167)*. Such structural evidence from patients in turn support a two-pronged hypothesis promoted by many basic scientists and psychiatrists *(117,168–170)*. First, mood disorders are marked by decreased hippocampal neurogenesis. Second, reversal of mood disorder-induced decrease of adult hippocampal neurogenesis will alleviate symptoms or otherwise contribute to improvement of the patient. This two-pronged hypothesis has received much attention, given recent high-profile publications and debates. As such, an overview of the most recent evidence for and against each prong of the hypothesis will be provided here. Readers are referred to other in-depth reviews of this timely topic *(9,171–178)*, for additional perspectives.

Direct evidence for the first prong of the hypothesis—are mood disorders marked by decreased adult hippocampal neurogenesis?—is scant. A decrease in hippocampal volume in the depressed brain may be owing to decreased adult hippocampal neurogenesis, or it may be owing to loss of neuronal processes, increased cell death, and so on. Provision of direct evidence for decreased neurogenesis in the depressed brain, such as via brain imaging, is difficult owing to both the lack of markers for in vivo examination of human progenitor cells and the exceptionally high resolution needed to visualize dividing cells in the living brain. The ability to label proliferating cells in the postmortem human brain via endogenous cell cycle markers has been applied to neurological disorders, such as Huntington's, Parkinson's, and Alzheimer's diseases *(179)*. However, this approach has not yet been extended to postmortem analysis of the depressed or bipolar brain. One unique challenge of working with postmortem tissue from mood disorder patients is the high percentage of patients who die via suicide, thus complicating the interpretation of the resulting data. Challenges common to working with postmortem tissue from all neurological disorders and psychiatric disorders also pose problems. For example, some cell cycle markers are also implicated in DNA repair or cell death *(180)*. Therefore, a decrease in a cell division protein, such as PCNA, in the hippocampus of a postmortem brain can only be considered as a decrease in progenitor proliferation after exclusion of the possibilities of PCNA labeling damaged or dying cells. In addition, because many disorders may be marked by decreased hippocampal volume, decreases in endogenous cell cycle markers must be considered in the broader context of hippocampal cell density and volume. For example, fewer PCNA cells in a single section from the hippocampus of a depressed patient may simply reflect the smaller volume of the hippocampus in this disorder. Therefore, direct evidence of decreased hippocampal neurogenesis in mood disorders awaits rigorous application of stereological and immunohistochemical analysis of the postmortem brain as well as significant technical advances for imaging neurogenesis in the living brain.

There are a few pieces of indirect evidence for mood disorders being marked by decreased adult hippocampal neurogenesis. First, clinical studies suggest that depression is marked by altered proliferation of non-CNS cells *(181)*. Assessment of proliferation in skin fibroblasts or neurogenesis in olfactory epithelium of depressed patients *(182,183)* would allow expansion of this hypothesis to include nonimmune and CNS cells. Second, basic research studies have used animal models of depression to assess whether decreased neurogenesis correlates with appearance of a depressive-like phenotype. Such studies have yielded conflicting results. One report showed that exposure to inescapable shock, which results in a state of behavioral despair, decreases hippocampal cell proliferation, and that this effect can be reversed by antidepressant treatment *(184)*. However, inescapable shock typically only produces a state of behavioral despair in a portion of animals. Another study exploited this interanimal variability and showed that development of behavioral despair does not correlate with decreased hippocampal proliferation *(185)*. It remains to be seen whether development of behavioral despair correlates with other states of new neuron development and maturation. Finally, if adult neurogenesis were to play a causal role in mood disorders, one might hypothesize that decreased adult neurogenesis would lead to depressive-like symptoms in laboratory animals. In fact, several studies have shown that decreased neurogenesis itself does not produce depressive-like symptoms *(185,186)*. Although these indirect studies are not as clinically relevant as human studies, much can still be learnt from additional studies with animal models of mood disorders in terms of evaluating the hypothesis of whether mood disorders are marked by decreased neurogenesis.

Direct evidence for the second prong of the hypothesis—will reversal of mood disorder-induced decrease of adult hippocampal neurogenesis alleviate symptoms of the disorder?—also suffers from the challenges of working with the human brain. Medications for mood disorders can lead to the normalization of brain and hippocampal volume *(187)*. However, it is difficult to say whether this normalization is owing to increased adult neurogenesis, sprouting of processes of mature cells, or attenuation of cell death.

Although direct evidence is lacking, indirect evidence for the ability of effective mood disorder treatments to increase adult neurogenesis is ample. In general, clinically effective treatments for mood disorders increase the number of new cells and new neurons in the adult mammalian hippocampus. This is true for several classes of antidepressant medications, including selective reuptake inhibitors for serotonin or for norepinephrine, inhibitors of monoamine oxidase or phosphodiesterase IV, and lithium *(188–194)*. The specificity of these agents to increase cell proliferation is supported by the lack of effect of other neuroactive drugs, such as the antipsychotic haloperidol *(186,188)*. Recent work has also shown the ability of other putative antidepressants to increase hippocampal cell proliferation: α-amino-3-hydroxy-5-methyl-isoxazole-4-propionic acid (AMPA) glutamate receptor potentiators *(195)*, substance P antagonists *(196)*, and the endogenous steroid dehydroepiandrosterone *(197)*. Notably, the in vivo time-course of many of these agents to increase hippocampal proliferation is similar to their therapeutic time-course *(186,188)*. These studies suggest that effective treatments for mood disorders may mediate some of their therapeutic actions via increases in adult neurogenesis.

One exception to the growing list of antidepressant treatments that increase hippocampal cell proliferation is transcranial magnetic stimulation (TMS) *(126)*. The inability of TMS itself to increase proliferation may be owing to key anatomical differences between humans and laboratory animals used in this study. However, TMS appears to normalize the stress-induced decrease in hippocampal cell proliferation. This raises a critical point: a confound of most studies evaluating the impact of mood disorder treatments on hippocampal cell proliferation is that they are performed in naive, "nondepressed" animals. Antidepressant treatments and lithium have, in general, no effect on control subjects. Therefore, a challenge facing the field is to directly demonstrate that mood disorders decrease, and effective antidepressant treatments normalize, adult hippocampal neurogenesis. To this end, several studies have found that inescapable shock and stress- or stress-hormone-induced decreases in hippocampal cell proliferation can be ameliorated by chronic treatment with antidepressants *(126,184,186,198,199)*. In most of these studies, normalization of hippocampal proliferation was accompanied by improvements in behavioral measures of depression-like symptoms, such as grooming or coat state. Far from the clinical situations, these studies provide additional correlative evidence for the link between animal models of depression, neurogenesis, and improvement of both by antidepressant treatment.

Two recent papers have attempted to more closely examine the correlative relationship between effective treatments for mood disorders and neurogenesis. One paper found that lithium, a common and effective treatment for bipolar disorder, enhanced a cellular form of hippocampal learning prior to its enhancement of hippocampal cell proliferation *(200)*. This apparent disconnection between impact on cellular function and neurogenesis was explained by lithium's more immediate action on other signaling molecules that might themselves increase neurogenesis. The second paper studied an animal model of antidepressant action and concluded that adult hippocampal neurogenesis was required for the behavioral impact of antidepressant treatment *(186)*. This paper was welcomed for its forward thinking and thoroughness in regards to experimental design, but weaknesses identified in the technology used to "knock out" neurogenesis in this paper encourage further exploration of whether, indeed, antidepressant efficacy requires new hippocampal neurons *(103,201–205)*.

Recent work has provided greater insight into the mechanism underlying antidepressant-induced increases in hippocampal cell proliferation *(173)*. Neurochemicals implicated include growth factors *(206–208)*, glutamate and serotonin action at distinct receptor subtypes *(195,209)*, and choline *(210)*. Given that glia may be one source of new neurons in the adult hippocampus *(211)*, it is also interesting that antidepressants appear to influence glia proliferation in the hippocampus *(190,212)*. One intracellular cascade common to both neurons and glia, and that many of these aforementioned factors may converge on to regulate hippocampal cell proliferation, is the cyclic adenosine mono phosphate pathway *(213)*. Key components of the cyclic adenosine mono phosphate pathway have recently been identified in hippocampal progenitor cells, and alterations of these components are being investigated for their ability to contribute to the antidepressant-induced increase in hippocampal neurogenesis *(192,214,215)*. Importantly, broad proteomic and genomic approaches have yielded several additional candidates for mediating the neuroadaptive changes that occur in the hippocampus after effective treatment of mood disorders *(216–218)*. All of these studies found candidates related to cell proliferation and maturation, underscoring the potential for neurogenesis to be an important component of effective treatment for mood disorders.

5. SUBSTANCE ABUSE

Like stress and mood disorders, abuse of illicit drugs is widely recognized as a disease of the brain *(219,220)*, albeit one compounded by myriad genetic and social factors. All drugs of abuse studied to date acutely activate the brain's natural "reward" pathway. Although other regions of the brain are undoubtedly involved in the maintenance of or relapse to drug-taking, the reward pathway has been the focus of most efforts to understand the neurobiological underpinnings of addiction. Recently, the hippocampus has received attention for its potential role in the development and maintenance of, and relapse to, addiction *(221,222)*. The hippocampus is activated during drug-taking *(223,224)*, and it is a site involved in both drug-taking *(225,226)* and relapse to drug-taking *(227)*. In addition, drug-taking impairs hippocampal function *(228–233)* and results in hippocampal and limbic pathology *(1,233–235)*. In addition, drug-induced reduction in hippocampal volume appears to normalize with abstinence. Therefore, although the reward pathway remains an important focus for addiction researchers, understanding how the hippocampus is involved in drug addiction, recovery, and relapse will likely improve our understanding of addiction in general.

In addition to stimulating the reward pathway, recent research suggests that all drugs of abuse have another commonality: drugs of abuse inhibit the hippocampal cell proliferation. As discussed in detail in Subheadings 5.1.–5.5., chronic, not acute, exposure to opiates, stimulants, tetrahydrocannabinoid (THC), nicotine, and ethanol decrease adult neurogenesis. This similarity is striking given that each drug has a distinct mechanism of action. For example, opiates primarily act via the μ opiate receptors (MORs), although stimulants act to block or reverse the dopamine transporter. How is it that drugs with such diverse neuromechanisms of action all result in decreased hippocampal proliferation? Does the decreased hippocampal proliferation underlie drug-induced deficits in hippocampal function, or is it merely coincidental? Examination of recent progress in the impact of drugs of abuse on adult mammalian neurogenesis reveals some progress toward answering these questions *(236)*.

5.1. Opiates

Opiates, including heroin and morphine, were the first class of drugs of abuse shown to inhibit hippocampal cell proliferation in the adult mammal *(43)*. As with stress, opiates act to inhibit SGZ proliferation but not SEZ proliferation. This underscores the distinct regulation of these two populations of neural stem cells in the adult brain, and emphasizes the lack of general toxicity of chronic morphine treatment to adult progenitor cells. It is notable that chronic administration of opiates is necessary to decrease neurogenesis, because acute treatment has no effect *(43)*. This suggests that chronic opiate exposure leads to adaptive changes that mediate the inhibition of adult neurogenesis. Inhibition of adult neurogenesis now joins the growing list of neural adaptations that occur after chronic exposure to opiates *(220)*.

One unique aspect of morphine-induced decrease in adult hippocampal neurogenesis is that it is not mediated by stress hormones, like corticosterone. This distinction also emphasizes that comprehension of the mechanisms underlying morphine-induced decrease in adult hippocampal neurogenesis may reveal novel information about adult hippocampal progenitor cells.

In regards to the mechanism, it is known that chronic opiates act to inhibit neurogenesis via their action at the MOR *(43)*. However, it is unclear whether MOR are on progenitor cells themselves or on afferents to the SGZ *(237,238)*. Opiates have been shown to act on a variety of second messenger cascades to alter proliferation in vitro *(239,240)*. Once the location of MOR involved in morphine-induced decrease in adult hippocampal neurogenesis is identified, it will be important to assess whether morphine induces expected alterations in these second messenger cascades in the MOR containing cells. In addition, although overall cell death is not detectably increased in the hippocampus after chronic morphine *(43)*, in vitro research shows that neurons and microglia can undergo cell death after morphine exposure *(241–243)*. Recent data show that chronic morphine induces premature mitosis of proliferating cells in the adult mouse SGZ *(51)*. Therefore, it will be interesting to see if these premature mitotic cells seen after chronic morphine then move onto die; the numbers are low enough that they have previously escaped detection with more global markers of hippocampal cell death *(43)*. Although MOR and cell death may play a role in morphine-induced decreased in neurogenesis, it is also likely that morphine induces other changes in the SGZ, perhaps to make it less neurogenic *(244)*. Additional research needs to be done to clarify how the neurogenic environment in the SGZ is altered after chronic exposure to opiates.

Although the functional implications of opiate-induced inhibition of adult neurogenesis remain unclear, it is reasonable to speculate that this inhibition may contribute to the cognitive deficits seen after chronic opiate use *(228)*. Opioid peptides play an important role in memory formation *(245,246)*. Clinically, heroin users have poorer performance on attention, verbal fluency, and memory tasks than controls *(228)*. In animals, chronic opiate exposure interferes with the acquisition of the radial maze task without disrupting recall of a previously learned task *(229)*. In contrast, chronic administration of naltrexone improves the rate of learning of a new task *(247)*. Recent work has shown that chronic morphine treatment can alter a form of cellular plasticity in the adult hippocampus *(248,249)*, adding further support to the idea that chronic opiate exposure may diminish hippocampal function.

The extent to which opiate-induced decreases in adult neurogenesis are selective to morphine and heroin is unclear. Although methadone is a safe and highly effective treatment for opiate addiction *(250)*, there is some evidence of cognitive impairment after methadone treatment *(251,252)*, suggesting that examination of the effect of methadone on adult hippocampal neurogenesis is warranted. Comprehension of how opiates inhibit adult neurogenesis will provide important insight into how proliferating cells in the adult brain are regulated.

5.2. Stimulants

Administration of stimulants in the prenatal or early postnatal period has long-lasting effects on brain structure and function *(253)*. Less well studied is the impact of stimulants on neurogenesis and hippocampal function in the adult brain. Acute exposure to methamphetamine decreases the number of newly born cells in the adult gerbil SGZ *(254)*. In contrast, acute exposure to amphetamine does not change the number of newly born cells in the adult rat SGZ *(255)*. It has been suggested that these results differ because of the longer half-life and more potent pharmacological effects of methamphetamine *(255)*. However, comparison of these studies is difficult given the difference in species, dose, and regimen of BrdU used, and method of quantitation of proliferation. It is notable that acute amphetamine reportedly alters cytogenesis in the adult rat striatum, presumably affecting non-neural progenitor cells *(255)*.

In contrast to acute studies, few studies have assessed the effect of chronic stimulant exposure on adult neurogenesis. We have found that chronic exposure to a low dose of cocaine does not alter the number of proliferating cells in the adult SGZ *(6)*. Given that astrocytes in the adult hippocampus are one putative source of neural progenitor cells *(256)*, it is interesting that astrocytes show a time-dependent alteration in expression in the hippocampus *(257)*: acute and repeated cocaine injections produce increases in GFAP immunoreactivity but chronic exposure does not. Other stimulants, such as caffeine, amphetamine, or 3,4-methylenedioxymethamphetamine (ecstasy) are known to have significant effects on behavior, brain structure, and proliferation in a developmental or immune context *(258,259)*, and several laboratories are currently investigating these drugs in regards to hippocampal cell proliferation. Recent study shows increased cell proliferation after administration of a dopamine agonist selective for one of the dopamine receptor subtypes *(260)*. Clearly more work needs to be done to evaluate the impact of stimulants on adult hippocampal neurogenesis.

5.3. Nicotine

Nicotine has a well-documented stimulatory effect on proliferative cells in the developing brain *(261)*. A recent report examines nicotine's effect on the newly born cells in the SGZ of the adult rat (262). Rats trained to self-administer nicotine for 6 wk show dose-dependent decreases in newly born cells in the SGZ relative to controls. Given that increased adult neurogenesis has previously been correlated with learning and memory, it is interesting that chronic treatment with nicotine produces a decrease in newly born cells in the SGZ. Nicotine produces long-lasting improvement in performance on attention and memory tasks (263), and nicotinic mechanisms influence forms of synaptic plasticity thought to underlie learning and memory (264,265). Although this exact nicotine administration paradigm has not been evaluated for its effect on cognitive function, it is intriguing that these data show that a stimulus generally known to improve memory can decrease newly born cells in the adult SGZ.

5.4. Ethanol

Two recent studies have examined the impact of alcohol on proliferating cells in the adult brain. Adult rats with prolonged ethanol exposure have altered hippocampal progenitor proliferation (266,267). As shown with many antidepressants and with opiates, ethanol's impact appears restricted to the hippocampal progenitors, as the progenitors in the SEZ/SVZ are not altered. Although correlative, these data fit with ethanol's negative impact on hippocampal function (268). Interestingly, these changes can be prevented by administration of an antioxidant (267). Adult hippocampal progenitor cells are known to be positively influenced by antioxidants (269–272), and this protective role of antioxidants against ethanol-induced decrease in hippocampal neurogenesis warrants additional research.

5.5. Tetrahydrocannabinoid

THC, the active ingredient in marijuana, is a compound with significant effects on hippocampal morphology and function. Although THC has significant inhibitory effects on proliferation of immune (273–276) and other cells (277,278), its impact on hippocampal neurogenesis was only recently explored. Using both in vivo and in vitro approaches, a recent study showed the inhibitory effect of a cannabinoid agonist on proliferating cells in the adult brain (279). This study also made important strides to exploring the intracellular molecules involved in the cannabinoid-induced decrease in new cells; perhaps not surprisingly, they are some of the same intracellular cascades implicated in the neurogenic effect of treatments for mood disorders (171).

Summing all these, exposure to drugs of abuse appears to inhibit hippocampal progenitor cells without altering SEZ/SVZ progenitor cells and, in most cases, without causing overt damage to mature hippocampal neurons. This relative selectivity of drugs of abuse suggests alterations in neurogenesis can be added to the growing list of specific, identifiable alterations caused by prolonged exposure to drugs of abuse. This is in sharp contrast to the widespread media campaign of past decades depicting an egg frying in a hot pan with the statement "this is your brain on drugs." That drugs can cause such discrete changes in brain structure encourages more detailed research into the mechanisms underlying these alterations in neurogenesis. For example, how is it that all drugs with such distinct mechanisms of action lead to decreased hippocampal neurogenesis? Opiates appear to inhibit neurogenesis without involvement of corticosterone. Will all drugs of abuse show this independence from GCs? Opiates also appear to cause premature mitosis of dividing cells. Will this be a common impact of all drugs of abuse? One way in which addiction research in particular will benefit from such additional studies on drug-induced regulation of adult neurogenesis is that unlike other psychiatric disorders, such as mood disorders and schizophrenia, addiction is easily modeled in a laboratory setting. Animals will readily self-administer most drugs of abuse, and behavioral and cognitive changes induced by the addicted state can be assessed using a variety of tests. Therefore, there exists a battery of well-defined tools with which to assess functional impact of drug-induced alterations in adult neurogenesis. Combined with the endogenous cell cycle protein analyses and transgenic mouse approaches stressed earlier in this chapter, these tools will be especially useful in examining whether a relationship exists between the newly born cells in the hippocampus and the acquisition or maintenance of the addictive phenotype, or with the memory component of addiction *(246,280,281)*. Postmortem analyses using endogenous cell cycle markers also hold tremendous potential for elucidating the impact of addiction. These advances in our knowledge hold promise not only to help us identify the underlying cause of drug abuse and dependence, but also to aid the development of effective treatment strategies.

6. SCHIZOPHRENIA

Schizophrenia is one of the most complex, multifaceted psychiatric disorders, making it both one of the most challenging and important disorders to study from both a clinical and basic research standpoint *(4)*. Several facets of schizophrenia suggest hippocampal involvement in the disorder: symptoms of disrupted attention, learning, memory, and information processing *(282,283)*; alterations in hippocampal structure *(284–286)*, some of which are there prior to the onset of psychosis *(287)*. Genetic alterations associated with schizophrenia have strengthened the proposed role of altered hippocampal function and structure in development of schizophrenia. For example, the nonmutated form of one gene found to be mutated in some forms of schizophrenia (*disrupted-in-schizophrenia 1*) is enriched in the adult mammalian hippocampus *(288)*. Most postmortem studies of schizophrenic brains show no gliosis or other evidence of neurodegeneration or neural injury *(289,290)*. Therefore, researchers have explored whether more subtle alteration in hippocampal structure, such as alterations in neurogenesis, are evident in the schizophrenic brain.

To date, postmortem study on schizophrenic brains and neurogenesis are lacking. Hopefully, this will be amended in the near future given the utility of endogenous cell cycle markers to assess hippocampal neurogenesis in the postmortem brain. In lieu of these postmortem studies, two other approaches have been used: animal models and administration of antipsychotic agents.

The developmental hypothesis of schizophrenia has gained ground in recent years *(291)*. Schizophrenic brains do have abnormalities in neurodevelopmental measures, such as dysregulation of cellular adhesion molecules and related signaling pathways *(292–295)*. This evidence has encouraged the use of animal models of schizophrenia thought to mimic in part gestational or early postnatal trauma or hippocampal lesion *(296–298)*. Many of these previous models were unsuitable for studying adult neurogenesis because the hippocampal lesion made examination of adult-generated cells difficult to interpret. Some of the more recent studies use a relatively more specific approach: inhibition of neurogenesis during gestation and examination of hippocampal structure and function in the adult *(299,300)*. Other animal models of schizophrenia include treatment with methamphetamine or *N*-methyl-D-aspartate receptor antagonists, pharmacological agents already known to alter adult cytogenesis *(141,254)*. Further investigation of whether adult neurogenesis is altered in schizophrenia will require either new animal models of the disorder or examination of postmortem schizophrenic tissue, perhaps using one of the alternatives to mitotic markers suggested in this chapter.

An alternative approach to animal models of schizophrenia is to assess the impact of clinically effective antipsychotic medications on adult progenitor proliferation. Acute haloperidol increases proliferation in the SEZ/SVZ in the gerbil *(301)*. This effect may either be transient or species-specific, however, because chronic systemic injections of haloperidol in the rat *(188)* or mouse *(186)*, or chronic oral administration of haloperidol in the rat *(6)*, did not alter proliferation in either the SEZ or the SGZ. Antipsychotic medication varies in their mechanism of action; therefore, it is interesting that a recent study found that chronic administration of atypical antipsychotics increased SEZ/SVZ proliferation in the adult rat *(302)*. This study is also one of the few to use an endogenous cell cycle protein, PCNA, to underscore their results. This increased number of new potential olfactory bulb cells seen after chronic antipsychotic administration may have implications for altered olfaction-related proliferation seen in schizophrenics

(183). Further work needs to be done to evaluate whether antipsychotic medication has any effect on adult hippocampal progenitor proliferation.

Postmortem studies are clearly necessary to assess whether adult hippocampal neurogenesis is altered in schizophrenia. Appropriate interpretation of endogenous cell cycle protein expression in the schizophrenic brain has a great potential in this regard. Such studies may reveal new treatment strategies for schizophrenia, and will be useful in evaluating whether alterations in adult neurogenesis contribute to the neurobiological underpinnings of schizophrenia.

7. CONCLUSIONS

A direct link between adult hippocampal neurogenesis and psychiatric disorders or the improvement in symptoms with effective treatment remains lacking, but the links are compelling to warrant additional vigorous research. Should neurogenesis be shown to be an important mechanism in any of these disorders and their treatments, an understanding of the underlying molecular and cellular mechanisms involved could offer novel approaches to the treatment of mood disorders. However, the findings uncovered along the way will also have important implications for an understanding of adult neural stem cell biology. By using complementary and alternate approaches to endogenous proliferation markers, such as incorporating cell cycle markers, future studies will be able to get beyond simply counting new cells and further explore the mechanisms underlying the alterations in cell birth.

ACKNOWLEDGMENTS

This work was supported by grants to A.J.E. from the National Institute on Drug Abuse, the National Institute of Mental Health, the National Institute of Aging, and the National Alliance for Research on Schizophrenia and Depression, and by a postdoctoral fellowship to C.D.M. from the National Institute on Drug Abuse.

REFERENCES

1. Bartzokis G, Beckson M, Lu PH, et al. Age-related brain volume reductions in amphetamine and cocaine addicts and normal controls: implications for addiction research. Psychiat Res 2000;98:93–102.
2. Sheline YI, Wang PW, Gado MH, Csernansky JG, Vannier MW. Hippocampal atrophy in recurrent major depression. Proc Natl Acad Sci USA 1996;93:3908–3913.
3. McEwen BS, Sapolsky RM. Stress and cognitive function. Curr Opin Neurobiol 1995;5:205–216.
4. Benes FM, Berretta S. Amygdalo-entorhinal inputs to the hippocampal formation in relation to schizophrenia. Ann NY Acad Sci 2000;911:293–304.
5. Carlen PL, Wilkinson DA. Reversibility of alcohol-related brain damage: clinical and experimental observations. Acta Med Scand Suppl 1987;717:19–26.
6. Eisch AJ. Adult neurogenesis: implications for psychiatry. Prog Brain Res 2002;138:315–342.
7. Eisch AJ, Nestler EJ. To be or not to be: adult neurogenesis and psychiatry. Clin Neurosci Res 2002;2:93–108.
8. Duman RS. Structural alterations in depression: cellular mechanisms underlying pathology and treatment of mood disorders. CNS Spectr 2002;7:140–142.
9. Jacobs BL. Adult brain neurogenesis and depression. Brain Behav Immun 2002;16:602–609.
10. Kempermann G, Kronenberg G. Depressed new neurons—adult hippocampal neurogenesis and a cellular plasticity hypothesis of major depression. Biol Psychiat 2003;54:499–503.
11. Greenough WT, Cohen NJ, Juraska JM. New neurons in old brains: learning to survive? Nat Neurosci 1999;2:203–205.
12. Messier B, Leblond CP, Smart IH. Presence of DNA synthesis and mitosis in the brain of young adult mice. Exp Cell Res 1958;14:224–226.
13. Messier B, Leblond CP. Cell proliferation and migration as revealed by radioautography after injection of thymidine-H3 into male rats and mice. Am J Anat 1960;106:247–285.
14. Smart IH. The subependymal layer of the mouse brain and its cell production as shown by radioautography after thymidine-H3 injection. J Comp Neurol 1961;116:325–348.

15. Altman J, Das GD. Autoradiographic and histological evidence of postnatal hippocampal neurogenesis in rats. J Comp Neurol 1965;124:319–335.
16. Altman J, Das GD. Autoradiographic and histological studies of postnatal neurogenesis. I. A longitudinal investigation of the kinetics, migration and transformation of cells incorporating tritiated thymidine in neonate rats, with special reference to postnatal neurogenesis in some brain regions. J Comp Neurol 1966;126:337–389.
17. Kaplan MS, Hinds JW. Neurogenesis in the adult rat: electron microscopic analysis of light radioautographs. Science 1977;197:1092–1094.
18. Bayer SA. 3H-thymidine-radiographic studies of neurogenesis in the rat olfactory bulb. Exp Brain Res 1983;50:329–340.
19. Margolis FL, Verhaagen J, Biffo S, Huang FL, Grillo M. Regulation of gene expression in the olfactory neuroepithelium: a neurogenetic matrix. Prog Brain Res 1991;89:97–122.
20. Alvarez-Buylla A, Herrera DG, Wichterle H. The subventricular zone: source of neuronal precursors for brain repair. Prog Brain Res 2000;127:1–11.
21. Hastings NB, Tanapat P, Gould E. Neurogenesis in the adult mammalian brain. Clin Neurosci Res 2001;1:175–182.
22. Seaberg RM, van der Kooy D. Adult rodent neurogenic regions: the ventricular subependyma contains neural stem cells, but the dentate gyrus contains restricted progenitors. J Neurosci 2002;22:1784–1793.
23. van Praag H, Schinder AF, Christie BR, Toni N, Palmer TD, Gage FH. Functional neurogenesis in the adult hippocampus. Nature 2002;415:1030–1034.
24. Corotto FS, Henegar JR, Maruniak JA. Odor deprivation leads to reduced neurogenesis and reduced neuronal survival in the olfactory bulb of the adult mouse. Neuroscience 1994;61:739–744.
25. Gheusi G, Cremer H, McLean H, Chazal G, Vincent JD, Lledo PM. Importance of newly generated neurons in the adult olfactory bulb for odor discrimination. Proc Natl Acad Sci USA 2000;97:1823–1828.
26. Rakic P. Adult neurogenesis in mammals: an identity crisis. J Neurosci 2002;22:614–618.
27. Gould E, Gross CG. Neurogenesis in adult mammals: some progress and problems. J Neurosci 2002;22:619–623.
28. Kempermann G. Why new neurons? Possible functions for adult hippocampal neurogenesis. J Neurosci 2002;22:635–638.
29. Gould E, Cameron HA, Daniels DC, Woolley CS, McEwen BS. Adrenal hormones suppress cell division in the adult rat dentate gyrus. J Neurosci 1992;12:3642–3650.
30. Cameron HA, Gould E. Adult neurogenesis is regulated by adrenal steroids in the dentate gyrus. Neuroscience 1994;61:203–209.
31. Gould E, McEwen BS, Tanapat P, Galea LA, Fuchs E. Neurogenesis in the dentate gyrus of the adult tree shrew is regulated by psychosocial stress and NMDA receptor activation. J Neurosci 1997;17:2492–2498.
32. Gould E, Beylin A, Tanapat P, Reeves A, Shors TJ. Learning enhances adult neurogenesis in the hippocampal formation. Nat Neurosci 1999;2:260–265.
33. Nilsson M, Perfilieva E, Johansson U, Orwar O, Eriksson PS. Enriched environment increases neurogenesis in the adult rat dentate gyrus and improves spatial memory. J Neurobiol 1999;39:569–578.
34. van Praag H, Christie BR, Sejnowski TJ, Gage FH. Running enhances neurogenesis, learning, and long-term potentiation in mice. Proc Natl Acad Sci USA 1999;96:13,427–13,431.
35. Shors TJ, Miesegaes G, Beylin A, Zhao M, Rydel T, Gould E. Neurogenesis in the adult is involved in the formation of trace memories. Nature 2001;410:372–376.
36. Snyder JS, Kee N, Wojtowicz JM. Effects of adult neurogenesis on synaptic plasticity in the rat dentate gyrus. J Neurophysiol 2001;85:2423–2431.
37. Prickaerts J, Koopmans G, Blokland A, Scheepens A. Learning and adult neurogenesis: survival with or without proliferation? Neurobiol Learn Mem 2004;81:1–11.
38. Miller MW, Nowakowski RS. Use of bromodeoxyuridine-immunohistochemistry to examine the proliferation, migration and time of origin of cells in the central nervous system. Brain Res 1988;457:44–52.
39. Cooper-Kuhn CM, Kuhn HG. Is it all DNA repair? Methodological considerations for detecting neurogenesis in the adult brain. Brain Res Dev Brain Res 2002;134:13–21.
40. Kornack DR, Rakic P. Continuation of neurogenesis in the hippocampus of the adult macaque monkey. Proc Natl Acad Sci USA 1999;96:5768–5773.
41. Gould E, Vail N, Wagers M, Gross CG. Adult-generated hippocampal and neocortical neurons in macaques have a transient existence. Proc Natl Acad Sci USA 2001;98:10,910–10,917.

42. Parent JM, Tada E, Fike JR, Lowenstein DH. Inhibition of dentate granule cell neurogenesis with brain irradiation does not prevent seizure-induced mossy fiber synaptic reorganization in the rat. J Neurosci 1999;19:4508–4519.
43. Eisch AJ, Barrot M, Schad CA, Self DW, Nestler EJ. Opiates inhibit neurogenesis in the adult rat hippocampus. Proc Natl Acad Sci USA 2000;97:7579–7584.
44. Pandey S, Wang E. Cells en route to apoptosis are characterized by the upregulation of c-fos, c-myc, c-jun, cdc2, and RB phosphorylation, resembling events of early cell-cycle traverse. J Cell Biochem 1995;58:135–150.
45. Shors TJ. Memory traces of trace memories: neurogenesis, synaptogenesis and awareness. Trends Neurosci 2004;27:250–256.
46. Hayes NL, Nowakowski RS. Dynamics of cell proliferation in the adult dentate gyrus of two inbred strains of mice. Brain Res Dev Brain Res 2002;134:77–85.
47. Cameron HA, McKay RD. Adult neurogenesis produces a large pool of new granule cells in the dentate gyrus. J Comp Neurol 2001;435:406–417.
48. Luskin MB. Restricted proliferation and migration of postnatally generated neurons derived from the forebrain subventricular zone. Neuron 1993;11:173–189.
49. Morshead CM, Craig CG, van der Kooy D. In vivo clonal analyses reveal the properties of endogenous neural stem cell proliferation in the adult mammalian forebrain. Development 1998;125:2251–2261.
50. Arsenijevic Y, Villemure JG, Brunet JF, et al. Isolation of multipotent neural precursors residing in the cortex of the adult human brain. Exp Neurol 2001;170:48–62.
51. Mandyam CD, Norris RD, Eisch AJ. Chronic morphine induces premature mitosis of proliferating cells in the adult mouse subgranular zone. J Neurosci Res 2004;76:783–794.
52. Eisch AJ, Mandyam CD. Beyond BrdU: Basic and clinical implications for analysis of endogenous cell cyle proteins. In: Focus on Stem Cell Research. Erik V. Greer, ed. Nova Science, New York, 2004;pp. 111–142.
53. Bagchi S, Weinmann R, Raychaudhuri, P. The retinoblastoma protein copurifies with E2F-I, an E1A-regulated inhibitor of the transcription factor E2F. Cell 1991;65:1063–1072.
54. Bandara LR, Adamczewski JP, Hunt T, La Thangue NB. Cyclin A and the retinoblastoma gene product complex with a common transcription factor. Nature 1991;352:249–251.
55. Chellappan SP, Hiebert S, Mudryj M, Horowitz JM, Nevins JR. The E2F transcription factor is a cellular target for the RB protein. Cell 1991;65:1053–1061.
56. Defeo-Jones D, Huang PS, Jones RE, et al. Cloning of cDNAs for cellular proteins that bind to the retinoblastoma gene product. Nature 1991;352:251–254.
57. Kaelin WG, Jr, Pallas DC, DeCaprio JA, Kaye FJ, Livingston DM. Identification of cellular proteins that can interact specifically with the T/E1A-binding region of the retinoblastoma gene product. Cell 1991;64:521–532.
58. Yoshikawa K. Cell cycle regulators in neural stem cells and postmitotic neurons. Neurosci Res 2000;37:1–14.
59. Jiang W, Gu W, Brannstrom T, Rosqvist R, Wester P. Cortical neurogenesis in adult rats after transient middle cerebral artery occlusion. Stroke 2001;32:1201–1207.
60. Okano HJ, Pfaff DW, Gibbs RB. RB and Cdc2 expression in brain: correlations with 3H-thymidine incorporation and neurogenesis. J Neurosci 1993;13:2930–2938.
61. Ferguson KL, Slack RS. The Rb pathway in neurogenesis. Neuroreport 2001;12:N55–N62.
62. Lee DW, Miyasato LE, Clayton NS. Neurobiological bases of spatial learning in the natural environment: neurogenesis and growth in the avian and mammalian hippocampus. Neuroreport 1998;9:R15–R27.
63. Bravo R, Frank R, Blundell PA, Macdonald-Bravo H. Cyclin/PCNA is the auxiliary protein of DNA polymerase-delta. Nature 1987;326:515–517.
64. Prelich G, Tan CK, Kostura M, et al. Functional identity of proliferating cell nuclear antigen and a DNA polymerase-delta auxiliary protein. Nature 1987;326:517–520.
65. Kawabe T, Suganuma M, Ando T, Kimura M, Hori H, Okamoto T. Cdc25C interacts with PCNA at G_2/M transition. Oncogene 2002;21:1717–1726.
66. Takahashi T, Caviness VS, Jr. PCNA-binding to DNA at the G_1/S transition in proliferating cells of the developing cerebral wall. J Neurocytol 1993;22:1096–1102.
67. Jin K, Minami M, Lan JQ, et al. Neurogenesis in dentate subgranular zone and rostral subventricular zone after focal cerebral ischemia in the rat. Proc Natl Acad Sci USA 2001;98:4710–4715.
68. Wildemann B, Schmidmaier G, Ordel S, Stange R, Haas NP, Raschke M. Cell proliferation and differentiation during fracture healing are influenced by locally applied IGF-I and TGF-beta1: Comparison of two proliferation markers, PCNA and BrdU. J Biomed Mater Res 2003;65B:150–156.

69. Celis JE, Celis A. Cell cycle-dependent variations in the distribution of the nuclear protein cyclin proliferating cell nuclear antigen in cultured cells: subdivision of S phase. Proc Natl Acad Sci USA 1985;82:3262–3266.
70. Belvindrah R, Rougon G, Chazal G. Increased neurogenesis in adult mCD24-deficient mice. J Neurosci 2002;22:3594–3607.
71. Ino H, Chiba T. Expression of proliferating cell nuclear antigen (PCNA) in the adult and developing mouse nervous system. Brain Res Mol Brain Res 2000;78:163–174.
72. Nagai R, Tsunoda S, Asada H, Urabe N. Proliferating cell nuclear antigen positive cells in the hippocampal subgranular zone decline after irradiation in a rodent model. Neurol Res 2002;24: 517–520.
73. Gerdes J, Li L, Schlueter C, et al. Immunobiochemical and molecular biologic characterization of the cell proliferation-associated nuclear antigen that is defined by monoclonal antibody Ki-67. Am J Pathol 1991;138:867–873.
74. Schluter C, Duchrow M, Wohlenberg C, et al. The cell proliferation-associated antigen of antibody Ki-67: a very large, ubiquitous nuclear protein with numerous repeated elements, representing a new kind of cell cycle-maintaining proteins. J Cell Biol 1993;123:513–522.
75. Kausch I, Lingnau A, Endl E, et al. Antisense treatment against Ki-67 mRNA inhibits proliferation and tumor growth in vitro and in vivo. Int J Cancer 2003;105:710–716.
76. Endl E, Gerdes J. Posttranslational modifications of the KI-67 protein coincide with two major checkpoints during mitosis. J Cell Physiol 2000;182:371–380.
77. Gerdes J, Lemke H, Baisch H, Wacker HH, Schwab U, Stein H. Cell cycle analysis of a cell proliferation-associated human nuclear antigen defined by the monoclonal antibody Ki-67. J Immunol 1984;133:1710–1715.
78. Dayer AG, Ford AA, Cleaver KM, Yassaee M, Cameron HA. Short-term and long-term survival of new neurons in the rat dentate gyrus. J Comp Neurol 2003;460:563–572.
79. Kee N, Sivalingam S, Boonstra R, Wojtowicz JM. The utility of Ki-67 and BrdU as proliferative markers of adult neurogenesis. J Neurosci Methods 2002;115:97–105.
80. Draetta G, Beach D. Activation of cdc2 protein kinase during mitosis in human cells: cell cycle-dependent phosphorylation and subunit rearrangement. Cell 1998;54:17–26.
81. Nurse P. Universal control mechanism regulating onset of M-phase. Nature 1990;344:503–508.
82. Duckworth BC, Weaver JS, Ruderman JV. G2 arrest in Xenopus oocytes depends on phosphorylation of cdc25 by protein kinase A. Proc Natl Acad Sci USA 2002;99:16,794–16,799.
83. Nigg EA. Cell cycle regulation by protein kinases and phosphatases. Ernst Schering Res Found Workshop 2001;34:19–46.
84. Okumura E, Fukuhara T, Yoshida H, et al. Akt inhibits Myt1 in the signalling pathway that leads to meiotic G2/M-phase transition. Nat Cell Biol 2002;4:111–116.
85. Nigg EA. Mitotic kinases as regulators of cell division and its checkpoints. Nature Rev Mol Cell Biol 2001;2:21–32.
86. Steinmann KE, Belinsky GS, Lee D, Schlegel R. Chemically induced premature mitosis: differential response in rodent and human cells and the relationship to cyclin B synthesis and p34cdc2/cyclin B complex formation. Proc Natl Acad Sci USA 1991;88:6843–6847.
87. Bradbury EM, Cary PD, Crane-Robinson C, Rattle HW. Conformations and interactions of histones and their role in chromosome structure. Ann NY Acad Sci 1973;222:266–289.
88. Arents G, Burlingame RW, Wang BC, Love WE, Moudrianakis, EN. The nucleosomal core histone octamer at 3.1 A resolution: a tripartite protein assembly and a left-handed superhelix. Proc Natl Acad Sci USA 1991;88:10,148–10,152.
89. Guo XW, Th'ng JP, Swank RA, et al. Chromosome condensation induced by fostriecin does not require p34cdc2 kinase activity and histone H1 hyperphosphorylation, but is associated with enhanced histone H2A and H3 phosphorylation. EMBO J 1995;14:976–985.
90. Hendzel MJ, Bazett-Jones DP. Fixation-dependent organization of core histones following DNA fluorescent in situ hybridization. Chromosoma 1997;106:114–123.
91. Wei Y, Yu L, Bowen J, Gorovsky MA, Allis CD. Phosphorylation of histone H3 is required for proper chromosome condensation and segregation. Cell 1999;97:99–109.
92. Crosio C, Fimia GM, Loury R, et al. Mitotic phosphorylation of histone H3: spatio-temporal regulation by mammalian Aurora kinases. Mol Cell Biol 2002;22:874–885.
93. Ahmad K, Henikoff S. Histone-H3 variants specify modes of chromatin assembly. Proc Natl Acad Sci USA 2002;99(Suppl 4):16,477–16,484.

94. Cobb J, Miyaike M, Kikuchi A, Handel MA. Meiotic events at the centromeric heterochromatin: histone H3 phosphorylation, topoisomerase II alpha localization and chromosome condensation. Chromosoma 1999;108:412–425.
95. Schmiesing JA, Gregson HC, Zhou S, Yokomori K. A human condensin complex containing hCAP-C-hCAP-E and CNAP1, a homolog of Xenopus XCAP-D2, colocalizes with phosphorylated histone H3 during the early stage of mitotic chromosome condensation. Mol Cell Biol 2000;20:6996–7006.
96. Wei Y, Mizzen CA, Cook RG, Gorovsky MA, Allis CD. Phosphorylation of histone H3 at serine 10 is correlated with chromosome condensation during mitosis and meiosis in Tetrahymena. Proc Natl Acad Sci USA 1998;95:7480–7484.
97. Allen KM, Walsh CA. Genes that regulate neuronal migration in the cerebral cortex. Epilepsy Res 1999;36:143–154.
98. Brown JP, Couillard-Despres S, Cooper-Kuhn CM, Winkler J, Aigner L, Kuhn HG. Transient expression of doublecortin during adult neurogenesis. J Comp Neurol 2003;467:1–10.
99. Magavi SS, Leavitt BR, Macklis JD. Induction of neurogenesis in the neocortex of adult mice. Nature 2000;405:951–955.
100. Friocourt G, Koulakoff A, Chafey P, et al. Doublecortin functions at the extremities of growing neuronal processes. Cereb Cortex 2003;13:620–626.
101. Rao MS, Shetty AK. Efficacy of doublecortin as a marker to analyse the absolute number and dendritic growth of newly generated neurons in the adult dentate gyrus. Eur J Neurosci 2004;19:234–246.
102. Kempermann G, Gast D, Kronenberg G, Yamaguchi M, Gage FH. Early determination and long-term persistence of adult-generated new neurons in the hippocampus of mice. Development 2003;130:391–399.
103. Limoli CL, Giedzinski E, Rola R, Otsuka S, Palmer TD, Fike JR. Radiation response of neural precursor cells: linking cellular sensitivity to cell cycle checkpoints, apoptosis and oxidative stress. Radiat Res 2004;161:17–27.
104. Bedard A, Parent A. Evidence of newly generated neurons in the human olfactory bulb. Brain Res Dev Brain Res 2004;151:159–168.
105. Jordan-Sciutto KL, Dorsey R, Chalovich EM, Hammond RR, Achim CL. Expression patterns of retinoblastoma protein in Parkinson disease. J Neuropathol Exp Neurol 2003;62:68–74.
106. Garcia AD, Doan NB, Imura T, Bush TG, Sofroniew MV. GFAP-expressing progenitors are the principal source of constitutive neurogenesis in adult mouse forebrain. Nat Neurosci 2004;7:1233–1241.
107. Sofroniew MV, Bush TG, Blumauer N, Lawrence K, Mucke L, Johnson MH. Genetically-targeted and conditionally-regulated ablation of astroglial cells in the central, enteric and peripheral nervous systems in adult transgenic mice. Brain Res 1999;835:91–95.
108. Sawamoto K, Yamamoto A, Kawaguchi A, et al. Okano H. Direct isolation of committed neuronal progenitor cells from transgenic mice coexpressing spectrally distinct fluorescent proteins regulated by stage-specific neural promoters. J Neurosci Res 2001;65:220–227.
109. Kawaguchi A, Miyata T, Sawamoto K, et al. Nestin-EGFP transgenic mice: visualization of the self-renewal and multipotency of CNS stem cells. Mol Cell Neurosci 2001;17:259–273.
110. Yamaguchi M, Saito H, Suzuki M, Mori K. Visualization of neurogenesis in the central nervous system using nestin promoter-GFP transgenic mice. Neuroreport 2000;11:1991–1996.
111. Kronenberg G, Reuter K, Steiner B, et al. Subpopulations of proliferating cells of the adult hippocampus respond differently to physiologic neurogenic stimuli. J Comp Neurol 2003;467:455–463.
112. Filippov V, Kronenberg G, Pivneva T, et al. Subpopulation of nestin-expressing progenitor cells in the adult murine hippocampus shows electrophysiological and morphological characteristics of astrocytes. Mol Cell Neurosci 2003;23:373–382.
113. Overstreet LS, Hentges ST, Bumaschny VF, et al. A transgenic marker for newly born granule cells in dentate gyrus. J Neurosci 2004;24:3251–3259.
114. Chen J, Kelz MB, Zeng G, et al. Transgenic animals with inducible, targeted gene expression in brain. Mol Pharmacol 1998;54:495–503.
115. Feil R, Wagner J, Metzger D, Chambon P. Regulation of Cre recombinase activity by mutated estrogen receptor ligand-binding domains. Biochem Biophys Res Commun 1997;237:752–757.
116. Feng R, Rampon C, Tang YP, et al. Deficient neurogenesis in forebrain-specific presenilin-1 knockout mice is associated with reduced clearance of hippocampal memory traces. Neuron 2001;32:911–926.
117. Duman RS, Heninger GR, Nestler EJ. A molecular and cellular theory of depression. Arch Gen Psychiat 1997;54:597–606.

118. Monroe SM, Roberts JE, Kupfer DJ, Frank E. Life stress and treatment course of recurrent depression: II Postrecovery associations with attrition, symptom course, and recurrence over 3 years. J Abnorm Psychol 1996;105:313–328.
119. Monroe SM, Kupfer DJ, Frank E. Life stress and treatment course of recurrent depression: 1. Response during index episode. J Consult Clin Psychol 1992;60:718–724.
120. Paykel ES, Tanner J. Life events, depressive relapse and maintenance treatment. Psychol Med 1976;6:481–485.
121. Paykel ES. Life events, social support and depression. Acta Psychiatr Scand Suppl 1994;377:50–58.
122. Gould E, Cameron HA. Early NMDA receptor blockade impairs defensive behavior and increases cell proliferation in the dentate gyrus of developing rats. Behav Neurosci 1997;111:49–56.
123. Gould E, Tanapat P, Rydel T, Hastings N. Regulation of hippocampal neurogenesis in adulthood. Biol Psychiat 2000;48:715–720.
124. Gould E, Tanapat P, McEwen BS, Flugge G, Fuchs E. Proliferation of granule cell precursors in the dentate gyrus of adult monkeys is diminished by stress. Proc Natl Acad Sci USA 1998;95:3168–3171.
125. Tanapat P, Hastings NB, Rydel TA, Galea LA, Gould E. Exposure to fox odor inhibits cell proliferation in the hippocampus of adult rats via an adrenal hormone-dependent mechanism. J Comp Neurol 2001;437:496–504.
126. Czeh B, Welt T, Fischer AK, et al. Chronic psychosocial stress and concomitant repetitive transcranial magnetic stimulation: effects on stress hormone levels and adult hippocampal neurogenesis. Biol Psychiat 2002;52:1057–1065.
127. Alonso R, Griebel G, Pavone G, Stemmelin J, Le Fur G, Soubrie P. Blockade of CRF(1) or V(1b) receptors reverses stress-induced suppression of neurogenesis in a mouse model of depression. Mol Psychiat 2004;9:278–286, 224.
128. Pham K, Nacher J, Hof PR, McEwen BS. Repeated restraint stress suppresses neurogenesis and induces biphasic PSA-NCAM expression in the adult rat dentate gyrus. Eur J Neurosci 2003;17:879–886.
129. Fuchs E, Flugge G. Modulation of binding sites for corticotropin-releasing hormone by chronic psychosocial stress. Psychoneuroendocrinology 1995;20:33–51.
130. Ohl F, Michaelis T, Vollmann-Honsdorf GK, Kirschbaum C, Fuchs E. Effect of chronic psychosocial stress and long-term cortisol treatment on hippocampus-mediated memory and hippocampal volume: a pilot-study in tree shrews. Psychoneuroendocrinology 2000;25:357–363.
131. Kitraki E, Kremmyda O, Youlatos D, Alexis M, Kittas C. Spatial performance and corticosteroid receptor status in the 21-day restraint stress paradigm. Ann NY Acad Sci 2004;1018:323–327.
132. Gould E, Woolley CS, Cameron HA, Daniels DC, McEwen BS. Adrenal steroids regulate postnatal development of the rat dentate gyrus: II Effects of glucocorticoids and mineralocorticoids on cell birth. J Comp Neurol 1991;313:486–493.
133. Cameron HA, McKay RD. Restoring production of hippocampal neurons in old age. Nat Neurosci 1999;2:894–897.
134. Rodriguez JJ, Montaron MF, Petry KG, et al. Complex regulation of the expression of the polysialylated form of the neuronal cell adhesion molecule by glucocorticoids in the rat hippocampus. Eur J Neurosci 1998;10:2994–3006.
135. Cameron HA, Woolley CS, Gould E. Adrenal steroid receptor immunoreactivity in cells born in the adult rat dentate gyrus. Brain Res 1993;611:342–346.
136. Gass P, Kretz O, Wolfer DP, et al. Genetic disruption of mineralocorticoid receptor leads to impaired neurogenesis and granule cell degeneration in the hippocampus of adult mice. EMBO Rep 2000;1:447–451.
137. Yu IT, Lee SH, Lee YS, Son H. Differential effects of corticosterone and dexamethasone on hippocampal neurogenesis in vitro. Biochem Biophys Res Commun 2004;317:484–490.
138. Montaron MF, Piazza PV, Aurousseau C, Urani A, Le Moal M, Abrous DN. Implication of corticosteroid receptors in the regulation of hippocampal structural plasticity. Eur J Neurosci 2003;18:3105–3111.
139. Kozorovitskiy Y, Gould E. Dominance hierarchy influences adult neurogenesis in the dentate gyrus. J Neurosci 2004;24:6755–6759.
140. Heine VM, Maslam S, Joels M, Lucassen PJ. Prominent decline of newborn cell proliferation, differentiation, and apoptosis in the aging dentate gyrus, in absence of an age-related hypothalamus-pituitary-adrenal axis activation. Neurobiol Aging 2004;25:361–375.
141. Cameron HA, McEwen BS, Gould E. Regulation of adult neurogenesis by excitatory input and NMDA receptor activation in the dentate gyrus. J Neurosci 1995;15:4687–4692.

142. Tanapat P, Hastings NB, Reeves AJ, Gould E. Estrogen stimulates a transient increase in the number of new neurons in the dentate gyrus of the adult female rat. J Neurosci 1999;19:5792–5801.
143. Galea LA, McEwen BS. Sex and seasonal differences in the rate of cell proliferation in the dentate gyrus of adult wild meadow voles. Neuroscience 1999;89:955–964.
144. Nacher J, Rosell DR, Alonso-Llosa G, McEwen BS. NMDA receptor antagonist treatment induces a long-lasting increase in the number of proliferating cells, PSA-NCAM-immunoreactive granule neurons and radial glia in the adult rat dentate gyrus. Eur J Neurosci 2001;13:512–520.
145. Smith MT, Pencea V, Wang Z, Luskin MB, Insel TR. Increased number of BrdU-labeled neurons in the rostral migratory stream of the estrous prairie vole. Horm Behav 2001;39:11–21.
146. Reagan LP, Rosell DR, Wood GE, et al. Chronic restraint stress up-regulates GLT-1 mRNA and protein expression in the rat hippocampus: reversal by tianeptine. Proc Natl Acad Sci USA 2004;101: 2179–2184.
147. Meaney MJ, Aitken DH, van Berkel C, Bhatnagar S, Sapolsky RM. Effect of neonatal handling on age-related impairments associated with the hippocampus. Science 1988;239:766–768.
148. Meaney MJ, Aitken DH, Bhatnagar S, Sapolsky RM. Postnatal handling attenuates certain neuroendocrine, anatomical, and cognitive dysfunctions associated with aging in female rats. Neurobiol Aging 1991;12:31–38.
149. Vaido AI, Shiryaeva NV, Vshivtseva VV. Effect of prenatal stress on proliferative activity and chromosome aberrations in embryo brain in rats with different excitability of the nervous system. Bull Exp Biol Med 2000;129:452–455.
150. Teicher MH, Andersen SL, Polcari A, Anderson CM, Navalta CP. Developmental neurobiology of childhood stress and trauma. Psychiat Clin North Am 2002;25:397–426.
151. Lemaire V, Koehl M, Le Moal M, Abrous DN. Prenatal stress produces learning deficits associated with an inhibition of neurogenesis in the hippocampus. Proc Natl Acad Sci USA 2000;97: 11,032–11,037.
152. Craciunescu CN, Brown EC, Mar MH, Albright CD, Nadeau MR, Zeisel SH. Folic Acid deficiency during late gestation decreases progenitor cell proliferation and increases apoptosis in fetal mouse brain. J Nutr 2004;134:162–166.
153. Akman C, Zhao Q, Liu X, Holmes GL. Effect of food deprivation during early development on cognition and neurogenesis in the rat. Epilepsy Behav 2004;5:446–454.
154. Mirescu C, Peters JD, Gould E. Early life experience alters response of adult neurogenesis to stress. Nat Neurosci 2004;7:841–846.
155. Coe CL, Kramer M, Czeh B, et al. Prenatal stress diminishes neurogenesis in the dentate gyrus of juvenile rhesus monkeys. Biol Psychiat 2003;54:1025–1034.
156. Schmitz C, Rhodes ME, Bludau M, et al. Depression: reduced number of granule cells in the hippocampus of female, but not male, rats due to prenatal restraint stress. Mol Psychiatr 2002;7: 810–813.
157. Park HJ, Lim S, Lee HS, et al. Acupuncture enhances cell proliferation in dentate gyrus of maternally-separated rats. Neurosci Lett 2002;319:153–156.
158. Greden JF. The burden of disease for treatment-resistant depression. J Clin Psychiat 2001;62:26–31.
159. Sheline YI. Neuroimaging studies of mood disorder effects on the brain. Biol Psychiat 2003;54: 338–352.
160. Sheline YI. Hippocampal atrophy in major depression: a result of depression-induced neurotoxicity? Mol Psychiat 1996;1:298–299.
161. Sheline YI, Sanghavi M, Mintun MA, Gado MH. Depression duration but not age predicts hippocampal volume loss in medically healthy women with recurrent major depression. J Neurosci 1999;19:5034–5043.
162. Bremner JD, Narayan M, Anderson ER, Staib LH, Miller HL, Charney DS. Hippocampal volume reduction in major depression. Am J Psychiat 2000;157:115–118.
163. Ongur D, Drevets WC. Price JL. Glial reduction in the subgenual prefrontal cortex in mood disorders. Proc Natl Acad Sci USA 1998;95:13,290–13,295.
164. Sheline YI, Gado MH, Price JL. Amygdala core nuclei volumes are decreased in recurrent major depression. Neuroreport 1998;9:2023–2028.
165. Drevets WC. Neuroimaging and neuropathological studies of depression: implications for the cognitive-emotional features of mood disorders. Curr Opin Neurobiol 2001;11:240–249.
166. Sheline YI. 3D MRI studies of neuroanatomic changes in unipolar major depression: the role of stress and medical comorbidity. Biol Psychiat 2000;48:791–780.

167. Dowlatshahi D, MacQueen G, Wang JF, et al. Increased hippocampal supragranular Timm staining in subjects with bipolar disorder. Neuroreport 2000;11:3775–3778.
168. Duman RS, Malberg J, Nakagawa S, D'Sa C. Neuronal plasticity and survival in mood disorders. Biol Psychiat 2000;48:732–739.
169. Jacobs BL, Praag H, Gage FH. Adult brain neurogenesis and psychiatry: a novel theory of depression. Mol Psychiat 2000;5:262–269.
170. Duman RS, Malberg J, Nakagawa S. Regulation of adult neurogenesis by psychotropic drugs and stress. J Pharmacol Exp Ther 2001;299:401–407.
171. Manji HK, Moore GJ, Chen G. Clinical and preclinical evidence for the neurotrophic effects of mood stabilizers: implications for the pathophysiology and treatment of manic-depressive illness. Biol Psychiat 2000;48:740–754.
172. Kempermann G. Regulation of adult hippocampal neurogenesis—implications for novel theories of major depression. Bipolar Disord 2002;4:17–33.
173. Coyle JT, Duman RS. Finding the intracellular signaling pathways affected by mood disorder treatments. Neuron 2003;38:157–160.
174. Dremencov E, Gur E, Lerer B, Newman ME. Effects of chronic antidepressants and electroconvulsive shock on serotonergic neurotransmission in the rat hippocampus. Prog Neuropsychopharmacol Biol Psychiat 2003;27:729–739.
175. Henn FA, Vollmayr B. Neurogenesis and depression: etiology or epiphenomenon? Biol Psychiat 2004;56:146–150.
176. Duman RS. Depression: a case of neuronal life and death? Biol Psychiat 2004;56:140–145.
177. Sapolsky RM. Is impaired neurogenesis relevant to the affective symptoms of depression? Biol Psychiat 2004;56:137–139.
178. Malberg JE. Implications of adult hippocampal neurogenesis in antidepressant action. J Psychiat Neurosci 2004;29:196–205.
179. Curtis MA, Penney EB, Pearson AG, et al. Increased cell proliferation and neurogenesis in the adult human Huntington's disease brain. Proc Natl Acad Sci USA 2003;100:9023–9027.
180. Yang Y, Mufson EJ, Herrup K. Neuronal cell death is preceded by cell cycle events at all stages of Alzheimer's disease. J Neurosci 2003;23:2557–2563.
181. Irwin M. Immune correlates of depression. Adv Exp Med Biol 1999;461:1–24.
182. Mahadik SP, Mukherjee S, Laev H, Reddy R, Schnur DB. Abnormal growth of skin fibroblasts from schizophrenic patients. Psychiat Res 1991;37:309–320.
183. Feron F, Perry C, Hirning MH, McGrath J, Mackay-Sim A. Altered adhesion, proliferation and death in neural cultures from adults with schizophrenia. Schizophr Res 1999;40:211–218.
184. Malberg JE, Duman RS. Cell proliferation in adult hippocampus is decreased by inescapable stress: Reversal by fluoxetine treatment. Neuropsychopharmacology 2003;28:1562–1571.
185. Vollmayr B, Simonis C, Weber S, Gass P, Henn F. Reduced cell proliferation in the dentate gyrus is not correlated with the development of learned helplessness. Biol Psychiat 2003;54:1035–1040.
186. Santarelli L, Saxe M, Gross C, et al. Requirement of hippocampal neurogenesis for the behavioral effects of antidepressants. Science 2003;301:805–809.
187. Moore GJ, Bebchuk JM, Wilds IB, Chen G, Manji HK. Lithium-induced increase in human brain grey matter. Lancet 2000;356:1241–1242.
188. Malberg JE, Eisch AJ, Nestler EJ, Duman RS. Chronic antidepressant treatment increases neurogenesis in adult rat hippocampus. J Neurosci 2000;20:9104–9110.
189. Madsen TM, Treschow A, Bengzon J, Bolwig TG, Lindvall O, Tingstrom A. Increased neurogenesis in a model of electroconvulsive therapy. Biol Psychiat 2000;47:1043–1049.
190. Manev R, Uz T, Manev H. Fluoxetine increases the content of neurotrophic protein S100beta in the rat hippocampus. Eur J Pharmacol 2001;420:R1–R2.
191. Manev H, Uz T, Smalheiser NR, Manev R. Antidepressants alter cell proliferation in the adult brain in vivo and in neural cultures in vitro. Eur J Pharmacol 2001;411:67–70.
192. Nakagawa S, Kim JE, Lee R, et al. Regulation of neurogenesis in adult mouse hippocampus by cAMP and the cAMP response element-binding protein. J Neurosci 2002;22:3673–3682.
193. Scott BW, Wojtowicz JM, Burnham WM. Neurogenesis in the dentate gyrus of the rat following electroconvulsive shock seizures. Exp Neurol 2000;165:231–236.
194. Chen G, Rajkowska G, Du F, Seraji-Bozorgzad N, Manji HK. Enhancement of hippocampal neurogenesis by lithium. J Neurochem 2000;75:1729–1734.

195. Bai F, Bergeron M, Nelson DL. Chronic AMPA receptor potentiator (LY451646) treatment increases cell proliferation in adult rat hippocampus. Neuropharmacology 2003;44:1013–1021.
196. Rupniak NM. Elucidating the antidepressant actions of substance P (NK1 receptor) antagonists. Curr Opin Invest Drugs 2002;3:257–261.
197. Karishma KK, Herbert J. Dehydroepiandrosterone (DHEA) stimulates neurogenesis in the hippocampus of the rat, promotes survival of newly formed neurons and prevents corticosterone-induced suppression. Eur J Neurosci 2002;16:445–453.
198. Hellsten J, Wennstrom M, Mohapel P, Ekdahl CT, Bengzon J, Tingstrom A. Electroconvulsive seizures increase hippocampal neurogenesis after chronic corticosterone treatment. Eur J Neurosci 2002;16:283–290.
199. Alfonso J, Pollevick GD, Van Der Hart MG, Flugge G, Fuchs E, Frasch AC. Identification of genes regulated by chronic psychosocial stress and antidepressant treatment in the hippocampus. Eur J Neurosci 2004;19:659–666.
200. Son H, Yu IT, Hwang SJ, et al. Lithium enhances long-term potentiation independently of hippocampal neurogenesis in the rat dentate gyrus. J Neurochem 2003;85:872–881.
201. Monje ML, Mizumatsu S, Fike JR, Palmer TD. Irradiation induces neural precursor-cell dysfunction. Nat Med 2002;8:955–962.
202. Monje ML, Palmer T. Radiation injury and neurogenesis. Curr Opin Neurol 2003;16:129–134.
203. Madsen TM, Kristjansen PE, Bolwig TG, Wortwein G. Arrested neuronal proliferation and impaired hippocampal function following fractionated brain irradiation in the adult rat. Neuroscience 2003;119:635–642.
204. Rola R, Raber J, Rizk A, et al. Radiation-induced impairment of hippocampal neurogenesis is associated with cognitive deficits in young mice. Exp Neurol 2004;188:316–330.
205. Raber J, Rola R, LeFevour A, et al. Radiation-induced cognitive impairments are associated with changes in indicators of hippocampal neurogenesis. Radiat Res 2004;162:39–47.
206. Russo-Neustadt A. Brain-derived neurotrophic factor, behavior, and new directions for the treatment of mental disorders. Semin Clin Neuropsychiat 2003;8:109–118.
207. Khawaja X, Xu J, Liang JJ, Barrett JE. Proteomic analysis of protein changes developing in rat hippocampus after chronic antidepressant treatment: Implications for depressive disorders and future therapies. J Neurosci Res 2004;75:451–460.
208. Castren E. Neurotrophic effects of antidepressant drugs. Curr Opin Pharmacol 2004;4:8–64.
209. Banasr M, Hery M, Printemps R, Daszuta A. Serotonin-induced increases in adult cell proliferation and neurogenesis are mediated through different and common 5-HT receptor subtypes in the dentate gyrus and the subventricular zone. Neuropsychopharmacology 2004;29:450–460.
210. Sartorius A, Neumann-Haefelin C, Vollmayr B, Hoehn M, Henn FA. Choline rise in the rat hippocampus induced by electroconvulsive shock treatment. Biol Psychiat 2003;53:620–623.
211. Alvarez-Buylla A, Garcia-Verdugo JM, Tramontin AD. A unified hypothesis on the lineage of neural stem cells. Nat Rev Neurosci 2001;2:287–293.
212. Manev H, Uz T, Manev R. Glia as a putative target for antidepressant treatments. J Affect Disord 2003;75:59–64.
213. Nibuya M, Nestler EJ, Duman RS. Chronic antidepressant administration increases the expression of cAMP response element binding protein (CREB) in rat hippocampus. J Neurosci 1996;16:2365–2372.
214. Yang EJ, Ahn YS, Chung KC. Protein kinase Dyrk1 activates cAMP response element-binding protein during neuronal differentiation in hippocampal progenitor cells. J Biol Chem 2001;22:22.
215. Nakagawa S, Kim JE, Lee R, et al. Localization of phosphorylated cAMP response element-binding protein in immature neurons of adult hippocampus. J Neurosci 2002;22:9868–9876.
216. Newton SS, Collier EF, Hunsberger J, et al. Gene profile of electroconvulsive seizures: induction of neurotrophic and angiogenic factors. J Neurosci 2003;23:10,841–10,851.
217. Altar CA, Laeng P, Jurata LW, et al. Electroconvulsive seizures regulate gene expression of distinct neurotrophic signaling pathways. J Neurosci 2004;24:2667–2677.
218. Ogden CA, Rich ME, Schork NJ, et al. Candidate genes, pathways and mechanisms for bipolar (manic-depressive) and related disorders: an expanded convergent functional genomics approach. Mol Psychiat 2004;9:1007–1029.
219. Kreek MJ. Drug addictions. Molecular and cellular endpoints. Ann NY Acad Sci 2001;937:27–49.
220. Nestler EJ. Molecular basis of long-term plasticity underlying addiction. Nat Rev Neurosci 2001;2:119–128.
221. Nestler EJ. Neurobiology. Total recall-the memory of addiction. Science 2001;292:2266–2267.

222. Rogers RD, Robbins TW. Investigating the neurocognitive deficits associated with chronic drug misuse. Curr Opin Neurobiol 2001;11:250–257.
223. Sell LA, Morris JS, Bearn J, Frackowiak RS, Friston KJ, Dolan RJ. Neural responses associated with cue evoked emotional states and heroin in opiate addicts. Drug Alcohol Depend 2000; 60:207–216.
224. Kilts CD, Schweitzer JB, Quinn CK, et al. Neural activity related to drug craving in cocaine addiction. Arch Gen Psychiat 2001;58:334–341.
225. Wise RA. Opiate reward: sites and substrates. Neurosci Biobehav Rev 1989;13:129–133.
226. McBride WJ, Murphy JM, Ikemoto S. Localization of brain reinforcement mechanisms: intracranial self- administration and intracranial place-conditioning studies. Behav Brain Res 1999;101:129–152.
227. Vorel SR, Liu X, Hayes RJ, Spector JA, Gardner EL. Relapse to cocaine-seeking after hippocampal theta burst stimulation. Science 2001;292:1175–1178.
228. Guerra D, Sole A, Cami J, Tobena A. Neuropsychological performance in opiate addicts after rapid detoxification. Drug Alcohol Depend 1987;20:261–270.
229. Spain JW, Newsom GC. Chronic opioids impair acquisition of both radial maze and Y-maze choice escape. Psychopharmacology 1991;105:101–106.
230. Parrott AC, Lasky J. Ecstasy (MDMA) effects upon mood and cognition: before, during and after a Saturday night dance. Psychopharmacology (Berl) 1998;139:261–268.
231. Selby MJ, Azrin RL. Neuropsychological functioning in drug abusers. Drug Alcohol Depend 1998;50:39–45.
232. Ornstein TJ, Iddon JL, Baldacchino AM, et al. Profiles of cognitive dysfunction in chronic amphetamine and heroin abusers. Neuropsychopharmacology 2000;23:113–126.
233. Bowden SC, Crews FT, Bates ME, Fals-Stewart W, Ambrose ML. Neurotoxicity and neurocognitive impairments with alcohol and drug-use disorders: potential roles in addiction and recovery. Alcohol Clin Exp Res 2001;25:317–321.
234. Robinson TE, Kolb B. Alterations in the morphology of dendrites and dendritic spines in the nucleus accumbens and prefrontal cortex following repeated treatment with amphetamine or cocaine. Eur J Neurosci 1999;11:1598–1604.
235. Liu RS, Lemieux L, Shorvon SD, Sisodiya SM, Duncan JS. Association between brain size and abstinence from alcohol. Lancet 2000;355:1969–1970.
236. Eisch AJ, Mandyam CD. Drug dependence and addiction, II: Adult neurogenesis and drug abuse. Am J Psychiat 2004;161:426.
237. Drake CT, Chang PC, Harris JA, Milner TA. Neurons with mu opioid receptors interact indirectly with enkephalin- containing neurons in the rat dentate gyrus. Exp Neurol 2002;176:254–261.
238. Drake CT, Milner TA. Mu opioid receptors are in discrete hippocampal interneuron subpopulations. Hippocampus 2002;12:119–136.
239. Hauser KF, Stiene-Martin A, Mattson MP, Elde RP, Ryan SE, Godleske CC. mu-Opioid receptor-induced Ca^{2+} mobilization and astroglial development: morphine inhibits DNA synthesis and stimulates cellular hypertrophy through a $Ca(2+)$-dependent mechanism. Brain Res 1996;720:191–203.
240. Persson AI, Thorlin T, Bull C, et al. Mu- and delta-opioid receptor antagonists decrease proliferation and increase neurogenesis in cultures of rat adult hippocampal progenitors. Eur J Neurosci 2003;17:1159–1172.
241. Yin D, Mufson RA, Wang R, Shi Y. Fas-mediated cell death promoted by opioids. Nature 1999;397:218.
242. Hu S, Sheng WS, Lokensgard JR, Peterson PK. Morphine induces apoptosis of human microglia and neurons. Neuropharmacology 2002;42:829–836.
243. Mao J, Sung B, Ji RR, Lim G. Neuronal apoptosis associated with morphine tolerance: evidence for an opioid-induced neurotoxic mechanism. J Neurosci 2002;22:7650–7661.
244. Song P, Zhao ZQ. The involvement of glial cells in the development of morphine tolerance. Neurosci Res 2001;39:281–286.
245. Mauk MD, Warren JT, Thompson RF. Selective, naloxone-reversible morphine depression of learned behavioral and hippocampal responses. Science 1982;216:434–436.
246. White NM. Addictive drugs as reinforcers: multiple partial actions on memory systems. Addiction 1996;91:921–949; discussion 951–965.
247. Spain JW, Newsom GC. Chronic naltrexone enhances acquisition of the radial maze task in rats. Proc West Pharmacol Soc 1989;32:141–142.
248. Pu L, Bao GB, Xu NJ, Ma L, Pei G. Hippocampal long-term potentiation is reduced by chronic opiate treatment and can be restored by re-exposure to opiates. J Neurosci 2002;22:1914–1921.

249. Harrison JM, Allen RG, Pellegrino MJ, Williams JT, Manzoni OJ. Chronic morphine treatment alters endogenous opioid control of hippocampal mossy fiber synaptic transmission. J Neurophysiol 2002;87:2464–2470.
250. Kreek MJ. Opiates opioids and addiction. Mol Psychiat 1996;1:232–254.
251. Darke S, Sims J, McDonald S, Wickes W. Cognitive impairment among methadone maintenance patients. Addiction 2000;95:687–695.
252. Curran HV, Kleckham J, Bearn J, Strang J, Wanigaratne S. Effects of methadone on cognition, mood and craving in detoxifying opiate addicts: a dose-response study. Psychopharmacology (Berl) 2001;154:153–160.
253. Harvey JA. Cocaine effects on the developing brain: current status. Neurosci Biobehav Rev 2004;27:751–764.
254. Teuchert-Noodt G, Dawirs RR, Hildebrandt K. Adult treatment with methamphetamine transiently decreases dentate granule cell proliferation in the gerbil hippocampus. J Neural Transm 2000;107:133–143.
255. Mao L, Wang JQ. Gliogenesis in the striatum of the adult rat: alteration in neural progenitor population after psychostimulant exposure. Brain Res Dev Brain Res 2001;130:41–51.
256. Alvarez-Buylla A, Seri B, Doetsch F. Identification of neural stem cells in the adult vertebrate brain. Brain Res Bull 2002;57:751–758.
257. Fattore L, Puddu MC, Picciau S, et al. Astroglial in vivo response to cocaine in mouse dentate gyrus: a quantitative and qualitative analysis by confocal microscopy. Neuroscience 2002;110:1–6.
258. Marret S, Gressens P, Van-Maele-Fabry G, Picard J, Evrard P. Caffeine-induced disturbances of early neurogenesis in whole mouse embryo cultures. Brain Res 1997;773:213–216.
259. Connor TJ, Kelly JP, Leonard BE. An assessment of the acute effects of the serotonin releasers methylenedioxymethamphetamine, methylenedioxyamphetamine and fenfluramine on immunity in rats. Immunopharmacology 2000;46:223–235.
260. Van Kampen JM, Hagg T, Robertson HA. Induction of neurogenesis in the adult rat subventricular zone and neostriatum following dopamine D3 receptor stimulation. Eur J Neurosci 2004;19:2377–2387.
261. Opanashuk LA, Pauly JR, Hauser KF. Effect of nicotine on cerebellar granule neuron development. Eur J Neurosci 2001;13:48–56.
262. Abrous DN, Adriani W, Montaron MF, et al. Nicotine self-administration impairs hippocampal plasticity. J Neurosci 2002;22:3656–3662.
263. Rezvani AH, Levin ED. Cognitive effects of nicotine. Biol Psychiat 2001;49:258–267.
264. Fujii S, Sumikawa K. Nicotine accelerates reversal of long-term potentiation and enhances long-term depression in the rat hippocampal CA1 region. Brain Res 2001;894:340–346.
265. Dani JA, Ji D, Zhou FM. Synaptic plasticity and nicotine addiction. Neuron 2001;31:349–352.
266. Nixon K, Crews FT. Binge ethanol exposure decreases neurogenesis in adult rat hippocampus. J Neurochem 2002;83:1087–1093.
267. Herrera DG, Yague AG, Johnsen-Soriano S, et al. Selective impairment of hippocampal neurogenesis by chronic alcoholism: protective effects of an antioxidant. Proc Natl Acad Sci USA 2003;100:7919–7924.
268. Paula-Barbosa MM, Brandao F, Madeira MD, Cadete-Leite A. Structural changes in the hippocampal formation after long-term alcohol consumption and withdrawal in the rat. Addiction 1993;88:237–247.
269. Ciaroni S, Cuppini R, Cecchini T, et al. Neurogenesis in the adult rat dentate gyrus is enhanced by vitamin E deficiency. J Comp Neurol 1999;411:495–502.
270. Ferri P, Cecchini T, Ciaroni S, et al. Vitamin E affects cell death in adult rat dentate gyrus. J Neurocytol 2003;32:1155–1164.
271. Cecchini T, Ciaroni S, Ferri P, et al. Alpha-tocopherol, an exogenous factor of adult hippocampal neurogenesis regulation. J Neurosci Res 2003;73:447–455.
272. Cuppini R, Ciaroni S, Cecchini T, et al. Tocopherols enhance neurogenesis in dentate gyrus of adult rats. Int J Vitam Nutr Res 2002;72:170–176.
273. Klein TW, Newton CA, Widen R, Friedman H. The effect of delta-9-tetrahydrocannabinol and 11-hydroxy-delta-9-tetrahydrocannabinol on T-lymphocyte and B-lymphocyte mitogen responses. J Immunopharmacol 1985;7:451–466.
274. Pross S, Klein T, Newton C, Friedman H. Differential effects of marijuana components on proliferation of spleen, lymph node and thymus cells in vitro. Int J Immunopharmacol 1987;9:363–370.
275. Schwarz H, Blanco FJ, Lotz M. Anadamide an endogenous cannabinoid receptor agonist inhibits lymphocyte proliferation and induces apoptosis. J Neuroimmunol 1994;55:107–115.
276. Bhargava HN, House RV, Thorat SN, Thomas PT. Cellular immune function in mice tolerant to or abstinent from l-trans-delta 9-tetrahydrocannabinol. Pharmacology 1996;52:271–282.

277. Luthra YK, Esber HJ, Lariviere DM, Rosenkrantz H. Assessment of tolerance to immunosuppressive activity of delta 9-tetrahydrocannabinol in rats. J Immunopharmacol 1980;2:245–256.
278. Green LG, Stein JL, Stein GS. A decreased influence of cannabinoids on macromolecular biosynthesis and cell proliferation in human cells which metabolize polycyclic hydrocarbon carcinogens. Anticancer Res 1983;3:211–217.
279. Rueda D, Navarro B, Martinez-Serrano A, Guzman M, Galve-Roperh I. The endocannabinoid anandamide inhibits neuronal progenitor cell differentiation through attenuation of the RAP1/B-RAF/ERK pathway. J Biol Chem 2002;16:16.
280. Koob GF, Le Moal M. Drug addiction, dysregulation of reward, and allostasis. Neuropsychopharmacology 2001;24:97–129.
281. Ungless MA, Whistler JL, Malenka RC, Bonci A. Single cocaine exposure in vivo induces long-term potentiation in dopamine neurons. Nature 2001;411:583–587.
282. Kuperberg G, Heckers S. Schizophrenia and cognitive function. Curr Opin Neurobiol 2000;10:205–210.
283. Lewis DA, Gonzalez-Burgos G. Intrinsic excitatory connections in the prefrontal cortex and the pathophysiology of schizophrenia. Brain Res Bull 2000;52:309–317.
284. Becker T, Elmer K, Schneider F, et al. Confirmation of reduced temporal limbic structure volume on magnetic resonance imaging in male patients with schizophrenia. Psychiat Res 1996;67:135–143.
285. Altshuler LL, Bartzokis G, Grieder T, et al. An MRI study of temporal lobe structures in men with bipolar disorder or schizophrenia. Biol Psychiat 2000;48:147–162.
286. Gur RE, Turetsky BI, Cowell PE, et al. Temporolimbic volume reductions in schizophrenia. Arch Gen Psychiat 2000;57:769–775.
287. Matsumoto H, Simmons A, Williams S, Pipe R, Murray R, Frangou S. Structural magnetic imaging of the hippocampus in early onset schizophrenia. Biol Psychiat 2001;49:824–831.
288. Austin CP, Ky B, Ma L, Morris JA, Shughrue PJ. Expression of Disrupted-In-Schizophrenia-1, a schizophrenia-associated gene, is prominent in the mouse hippocampus throughout brain development. Neuroscience 2004;124:3–10.
289. Bachus SE, Kleinman JE. The neuropathology of schizophrenia. J Clin Psychiat 1996;57:72–83.
290. Arnold SE. Neurodevelopmental abnormalities in schizophrenia: insights from neuropathology. Dev Psychopathol 1999;11:439–456.
291. Weinberger DR. Implications of normal brain development for the pathogenesis of schizophrenia. Arch Gen Psychiat 1987;44:660–669.
292. Barbeau D, Liang JJ, Robitalille Y, Quirion R, Srivastava LK. Decreased expression of the embryonic form of the neural cell adhesion molecule in schizophrenic brains. Proc Natl Acad Sci USA 1995;92:2785–2789.
293. Cotter D, Kerwin R, al-Sarraji S, et al. Abnormalities of Wnt signalling in schizophrenia—evidence for neurodevelopmental abnormality. Neuroreport 1998;9:1379–1383.
294. van Kammen DP, Poltorak M, Kelley ME, et al. Further studies of elevated cerebrospinal fluid neuronal cell adhesion molecule in schizophrenia. Biol Psychiat 1998;43:680–686.
295. Vawter MP. Dysregulation of the neural cell adhesion molecule and neuroPD. Eur J Pharmacol 2000;405:385–395.
296. Lillrank SM, Lipska BK, Weinberger DR. Neurodevelopmental animal models of schizophrenia. Clin Neurosci 1995;3:98–104.
297. Lipska BK, Weinberger DR. To model a psychiatric disorder in animals: schizophrenia as a reality test. Neuropsychopharmacology 2000;23:223–239.
298. Lipska BK. Using animal models to test a neurodevelopmental hypothesis of schizophrenia. J Psychiat Neurosci 2004;29:282–286.
299. Flagstad P, Mork A, Glenthoj BY, Van Beek J, Michael-Titus AT, Didriksen M. Disruption of neurogenesis on gestational day 17 in the rat causes behavioral changes relevant to positive and negative schizophrenia symptoms and alters amphetamine-induced dopamine release in nucleus accumbens. Neuropsychopharmacology 2004;9:2052–2064.
300. Gourevitch R, Rocher C, Pen GL, Krebs MO, Jay TM. Working memory deficits in adult rats after prenatal disruption of neurogenesis. Behav Pharmacol 2004;15:287–292.
301. Dawirs RR, Hildebrandt K, Teuchert-Noodt G. Adult treatment with haloperidol increases dentate granule cell proliferation in the gerbil hippocampus. J Neural Transm 1998;105:317–327.
302. Wakade CG, Mahadik SP, Waller JL, Chiu FC. Atypical neuroleptics stimulate neurogenesis in adult rat brain. J Neurosci Res 2002;69:72–79.

26
Neurogenesis in Alzheimer's Disease
Compensation, Crisis, or Chaos?

Gemma Casadesus, PhD, Xiongwei Zhu, PhD, Hyoung-gon Lee, PhD, Michael W. Marlatt, MS, Robert P. Friedland, MD, Katarzyna A. Gustaw, MD, George Perry, PhD, and Mark A. Smith, PhD

SUMMARY

The potential of hippocampal neurogenesis, the innate capacity of "self" stem cells in the hippocampus to generate new neurons throughout our lifespan, is rapidly gaining importance not only as a potential therapeutic avenue (i.e., replacement of damaged neurons) but also as a potential pathogenic mechanism for disease (i.e., inability to generate new neurons) for various disease processes, most notably neurodegenerative diseases such as Alzheimer's disease (AD) and Parkinson's disease. In this review, we focus on the role of neurogenesis in AD in which it is notable that conditions that improve cognitive output and increase neurogenesis are associated with a decreased incidence of AD, whereas, conversely, conditions that lead to declines in behavioral output and neurogenesis are associated with increases in incidence. These parallels suggest that changes in hippocampal neurogenesis may play a significant role in the development of AD and that modulating this process may provide a mechanistic and therapeutic inroad to the disease. However, contradictions within the neurogenesis literature itself and across studies examining this process in animal models of AD have led to a confusing state of affairs regarding the role of neurogenesis in AD. Herein, we critically examine these contradicting reports and provide additional/alternate insight into the function of hippocampal neurogenesis in AD.

Key Words: Aging; Alzheimer's disease; cell cycle; cognition; hippocampal neurogenesis; memory.

1. INTRODUCTION

Recent advances in health care and preventive medicine have lead to dramatic increases in life expectancy, rising from 50 yr at the beginning of last century to approx 80 yr today. Unfortunately, increases in life expectancy inevitably bring a higher incidence of age-related illnesses. In this regard, AD is the most common cause of dementia and its prevalence is directly related to age *(1)*. More than 4×10^6 individuals are currently affected with the disease in the United States alone and this number is projected to increase to 14×10^6 by 2050 *(2)*. Of concern, despite valiant effort by the scientific field to understand the molecular underpinnings of this insidious disease, little progress has been made with regards to mechanisms, diagnostic tests, and treatments.

The area of the brain that has attracted most attention with respect to its involvement in AD is the hippocampus, an area that is associated with early deposition of neurofibrillary tangles and senile plaques, the main hallmarks of the disease *(3,4)*. Notably, the hippocampus plays

From: *The Cell Cycle in the Central Nervous System*
Edited by: D. Janigro © Humana Press Inc., Totowa, NJ

a key role in the modulation of spatial and episodic memory which shows the earliest and most severe declines in AD *(5)*. Likely, not coincidentally, the hippocampus is also the region of the brain that uniquely grows new neurons in adulthood *(6–8)*. This latter aspect—neurogenesis—together with an awaking interest in the regenerative capabilities of stem cells, has opened another hopeful avenue for the investigation of AD-associated mechanisms and, more importantly, for the development of new therapeutic treatments. The aim of this review is to summarize and critically examine the relevance of current research efforts in hippocampal neurogenesis regarding AD and how these may be linked to known AD-associated mechanisms and therapeutic strategies.

2. SIGNIFICANCE OF HIPPOCAMPAL NEUROGENESIS

Most neurons in the central neurons system do not retain the capacity to be replaced after birth or during early postnatal stages. Neurons are often terminally differentiated and exist throughout the life of the organism; therefore there has been a long-held notion that when a neuron dies it cannot be replaced. However, through the use of stereological methods, we now know that the dentate gyrus of the hippocampus *(6–8)*, the olfactory bulb *(8,9)*, and certain areas of the cortex *(10–13)* retain the capacity to produce neurons during adulthood and that such neurons can integrate into the brain regions and be functional.

In the adult dentate gyrus, cells divide continuously in the subgranular zone, a region located between the granular cell layer and the hilus of the hippocampus *(14,15)*. This cellular process, termed neurogenesis, is a comprehensive word used to define the proliferation, survival, migration, and differentiation of hippocampal precursor cells. Whereas originally thought of as only playing a role in development, neurogenesis is now known to also occur in adults and has been reported in a variety of vertebrate species including mice *(16)*, rats *(17)*, tree shrews *(18)*, primates *(19)*, and humans *(20)* among others. The study of neurogenesis in vivo is possible using the proliferation marker, bromodeoxyuridine (BrdU), which incorporates into the DNA of dividing cells and allows the identification of these cells in a time-dependent and location-specific manner (*see also* Chapter 2). After their birth, many of these progenitor cells migrate to the granular cell layer and differentiate into mature granule cells whereas a smaller portion develops glial phenotypes *(21)*. In this regard, Kempermann and colleagues *(22)*, using BrdU labeling, showed that hippocampal cells divide daily and that, after a period of 4 wk, 67% of the initially counted cells remains. Of these remaining cells, about 50% go on to develop a neuronal phenotype and migrate to the granule cell layer, 15% differentiate into glial cells, and the remaining 35% do not differentiate. It is apparent, as outlined below, that newly differentiated cells are functional and play important roles in health and disease. Indeed, newly born neurons, once they have incorporated into the granule cell layer of the dentate gyrus, become indistinguishable from mature granule cells in so much as they extend axons, make connections with the hippocampal CA3 region *(23–25)*, receive synaptic input *(23)*, and demonstrate functional properties in vivo *(26)* and in vitro *(25)*. Therefore, based on the evidence summarized previously, cells are born in the dentate gyrus of the hippocampus throughout the life span of an animal; they can survive, differentiate into mature phenotypes, and acquire functional properties. As such, it is reasonable to assume that neurogenesis, the intrinsic capacity of the brain to generate new neurons, may play a fundamental role in the various functional aspects of the hippocampus including cognitive behavior.

2.1. Neurogenesis and Hippocampally Related Cognition

The mechanisms by which neurogenesis in the hippocampus takes place and its direct significance to learning and memory is presently being established *(27,28)*. Certainly, the fact that neurogenesis occurs throughout the life span of many species makes this process highly susceptible to environmental/experience-dependent structural changes. To this end, various recent studies have identified several factors that can regulate the production and survival of hippocampal

Table 1
Various Factors Known to Modulate Neurogenesis, Either Positively (↑) or Negatively (↓), Also Modulate Risk of AD, Either Increasing (↑) or Decreasing (↓) it

Factor/condition	Neurogenesis	AD
Chronic aspects		
Education/environmental complexity	↑	↓
Exercise	↑	↓
Estrogen	↑	↓[a]
Depression	↓	↑
Epileptic seizures	↓	↑
Stress	↓	↑
Cholinergic denervation	↓	↑[b]
Aging	↓	↑
Acute aspects		
Ischemia	↑[c] ↓	↑
Head injury	↑[c] ↓	↑

[a]Following Women's Health Initiative, the role of estrogen has been questioned as a treatment strategy. However, the majority of studies show protection from developing disease (see refs. 137 and 138 for further discussion).

[b]Cholinergic denervation is a classic consequence of AD and the only one targeted with currently available FDA-approved therapeutics.

[c]Effects on neurogenesis may be transient. Long-term consequences likely lead to downregulation.

neurons developed during rodent adulthood. Some, like estrogen (29–31), environmental complexity (16,17,32–36), exercise (37–39), N-methyl-D-aspartate (NMDA)-related excitatory input (40–42), head injury, and transient global and focal ischemia (43–46), positively regulate neurogenesis, whereas others such as depression (47–49), stress (18,50–55), cholinergic denervation (56), epileptic seizures (57) and aging (14,33,36,58) decrease the levels of neurogenesis (Table 1).

Importantly, the aforementioned factors that affect neurogenesis correlate with changes in behavioral performance, including learning and memory (28,59–64), indicating a direct relationship between neurogenesis and cognition. For example, placing rodents in an enriched housing environment (16,17,32–36) or exercise (37–39) increases neurogenesis by increasing the survival of progenitor cells and also leads to enhanced performance in the water-maze task. Conversely, other factors such as stress, which are linked to a decrease in neurogenesis, impair behavioral performance (18,50–55). Therefore neurogenesis appears to be, at least in part, involved in modulating behavioral output.

Linking changes in newly differentiated neurons in adult rats to impairment in hippocampus-dependent behavior provides a direct connection between hippocampal neurogenesis and learning and memory. To this end, Shors and colleagues (28) demonstrated that in adult rats, substantial reduction in the number of newly generated neurons directly impaired hippocampus-dependent trace conditioning, a task in which an animal must associate stimuli that are separated in time, but had no effect on learning during hippocampus-independent tasks. Therefore neurogenesis appears to be directly involved in modulating hippocampally related cognitive output, and as such, may prove important in degenerative diseases associated with impairments in this region such as AD.

3. NEUROGENESIS IN AD

The process of hippocampal neurogenesis is rapidly gaining importance not only as a potential therapeutic avenue (i.e., replacement of damaged neurons) but also as a potential pathogenic mechanism (i.e., inability to generate new neurons) for disease. In this regard, factors that increase neurogenesis and that are associated with cognitive improvement such as enriched environment (16,17,32–36), learning (28,59–64), and estrogen levels (29–31) have all been associated with a decreased incidence of AD *(65–71)*. Likewise, factors which are commonly associated with decreased cognition reduce neurogenesis, such as depression *(47–49)*, stress *(18,50–55)*, epileptic seizures *(57)*, cholinergic denervation *(56)*, and, most importantly, aging *(14,33,36,58)*, are all associated with increased incidence of AD *(1,72–77)*. These striking parallels suggest that AD may indeed be associated with declines in hippocampal neurogenesis (Table 1). One obvious place to begin examining this possibility is in the established animal models of this disease.

Linkage studies show that mutations in at least three genes, the amyloid-β precursor protein (AβPP) gene located on chromosome 21, and the two homologous genes, presenilin 1 and 2 (PS1 and PS2) located on chromosome 14 and 1, respectively, are associated with early-onset AD *(78)*. Based on these findings, several transgenic mouse models of AD have been generated to mimic the most extensively studied pathological hallmark of AD, namely Aβ plaques *(79)*. The creation of transgenic lines expressing mutated AβPP, a type 1 membrane protein with a large extracellular domain and a short cytoplasmic domain belonging to a larger AβPP family that also includes amyloid precursor-like protein 1 (APLP1) and APLP2 *(80)*, or mutations in the genes encoding PS1 and PS2, which are polytopic proteins with six-to-eight-transmembrane domains that play a central role in intramembrane proteolysis (δ cleavage) of a number of cell surface proteins including AβPP, and which give rise to Aβ peptide *(81)*, have been used to establish these models as they share the common feature that they all alter γ-secretase cleavage of AβPP to increase the production of Aβ42, the primary component of amyloid plaques. Likewise, the fact that the penetrance of the mutations is almost 100% suggests they play critical roles in neurodegeneration and perhaps, given the parallels between modulators of neurogenesis and incidence factors in AD, these mutations or the overexpression of Aβ can also affect neurogenesis.

3.1. Declines in Hippocampal Neurogenesis Are Present in Animal Models of AD

Three studies now suggest that animals carrying the PS1 familial AD mutation show impaired adult neurogenesis as shown by declines in precursor proliferation *(82,83)* and survival *(83)* as well as various degrees of memory impairment *(82,83)*. Interestingly, however, these differences are only visible when the animals are placed in an enriched environment, an intervention that would maximize differences when present, as it is a powerful positive modulator of neurogenesis.

Similar studies using transgenic animals exhibiting mutations in the AβPP protein show parallel results. In this regard, two groups of investigators have demonstrated that animals carrying AβPP mutation show declines in neurogenesis affecting both proliferation and survival *(84–86)* of precursor cells. However, the studies contradict each other regarding the age at which these animals show such declines. That is, whereas Dong and colleagues *(86)* show declines in neurogenesis as early as 3 mo well before the deposition of Aβ plaques begins, suggesting that declines in neurogenesis may be independent of Aβ senile plaque formation, Hughey and colleagues *(84)* show that the declines in neurogenesis are restricted to the period when Aβ plaque deposition begins (12–14 mo) as they do not find that younger animals (3 mo) exhibit such declines, thus suggesting that these declines are associated and perhaps dependent of plaque deposition. Nonetheless, together, these data indicate that in AD, prior to Aβ deposition or during later stages of the disease when Aβ deposition is extensive, would lead to declines in hippocampal neurogenesis which could then turn into deficits in hippocampally associated learning and memory that are readily observed in this disease. Nevertheless, recent studies contradict such assumption and suggest a self-repair mechanism to be responsible for increases rather than decreases in hippocampal neurogenesis in AD.

3.2. Increased Neurogenesis Not Associated With Cognitive Improvement is Associated with Increased Incidence of AD

Several studies demonstrate that factors such as ischemic *(43–46)* or seizure-induced brain insults *(87–93)* lead to increased neurogenesis in the absence of improvements in cognition.

(*See also* Chapter 2.) These findings have lent the field to consider neurogenesis as a possible endogenous repair mechanism that the brain utilizes to diminish damage from trauma, blood flow deprivation insults, and/or overactivity *(93)*. Because both brain trauma and ischemic episodes are associated with increased incidence of AD, this capacity of the brain to seemingly repair itself from these insults has been proposed as major grounds for the use of stem cells in AD *(94–96)*. In addition, the accumulation of pathological agents such as Aβ or phosphorylated τ, which are in most cases described as toxic *(97,98)*, have led to the supposition that these pathologies, like during ischemia or brain trauma, could drive the brain to attempt to self-regenerate by inducing neurogenesis. Therefore under this assumption, AD would lead not to reduced neurogenesis, as previously described, but to enhanced neurogenesis as suggested in the literature summarized below.

3.3. AD and Aβ Treatment Lead to Increases in Neurogenesis

Various recent studies "invoke" the brain-repair theory to explain paradoxical findings concerning the presence of increased neurogenesis in AD *(99)*, animal models of the disease *(100)*, and after in vitro treatment with Aβ *(101)*. To this end, a recent study demonstrates that when different concentrations of the Aβ peptide are added into neural stem cell (NSC) cultures of hippocampi from postnatal day 0 Bl6 mice, the total number of neurons increases in a dose-dependent manner *(101)*. However, this report contradicts previous findings demonstrating that treatment of neural stem cells with Aβ leads to reduced neurogenesis and increased apoptosis of these cells in vitro *(85)*. Several differences are apparent that could account for the contradictions between studies such as differences in cell lines (postnatal hippocampal cultures vs prenatal cortical cultures) or types of Aβ peptides (Aβ42 vs Aβ25–35). However, given that postnatal hippocampal cultures which include mature neurons in addition to glia, as those used in Lopez-Toledano and Shelanski study, more closely resemble the hippocampal environment in vivo and the fact that Aβ42 is the peptide produced in the animal models of the disease, one would predict the that first study *(101)* mimicked more closely the findings in the animal models, however, the opposite occurs. One possible explanation for these puzzling results is the fact that in vitro studies have demonstrated that AβPP and its proteolytic fragments (i.e., Aβ peptide and sAβPP) have mitogenic properties *(102–106)*, hence these results may merely reflect an in-vitro specific event. Nevertheless, corroboration for the in vitro data is found in two recent studies demonstrating that there is increased hippocampal neurogenesis in the brains of patients with AD, as shown by increased numbers of immature but not mature neuronal markers compared with aged-matched controls *(99)* and in animals carrying the Indiana and Swedish mutations in which both young (3 mo) and older (12 mo) transgenic mice *(100)* show enhanced BrdU labeling in dentate gyrus, directly contradicting the previously mentioned reports *(84–86)*. These findings indeed point to a possible repair-driven strategy of the AD brain to protect itself from possible pathology-associated damage, however, in the case of the human study *(99)* no correlations between the number of immature cells and Aβ deposition were made and therefore such a deduction remains unfeasible. Moreover, in the transgenic mouse study, the authors discuss that a self-repair strategy does not need to be Aβ dependent as young animals, with no Aβ deposition, show increases neurogenesis and suggest that repair is targeted toward synaptic and neurotransmission type deficits also present in AD animal models.

Whereas a self-repair mechanism is a possibility, the incapacity of hippocampal neurogenesis increases to yield any significant improvements as shown by the progressive nature of this disease point to the fact that other mechanisms may be at play in these findings. Regarding the human study, one possibility that remains unexplored is that indeed the brain is attempting self-repair but not from AD specifically but from the ischemic-like insults that occur late in the course of the disease *(107)*, as mentioned before, a factor demonstrated to result in increased neurogenesis *(43–46)*. Also worthy of consideration, and as suggested by Jin and colleagues, is that self-repair is not Aβ-dependent but rather is dependent on other mechanisms. One such mechanism extensively described to play a major role in AD is oxidative stress *(108–113)*. In addition, various studies now suggest an antioxidant role for Aβ *(114–116)* and several reports demonstrate that antioxidant treatment *(117–120)* or antioxidant-related mechanisms such as caloric restriction *(121)* promote hippocampal neurogenesis. Therefore, a factor that could account for the increases in neurogenesis in these in vivo studies is that Aβ was indeed responsible for promotion of neurogenesis by serving an antioxidant function.

Finally, another possibility that should also be considered is that a repair strategy is indeed implemented but not specifically by promoting hippocampal neurogenesis but by using a parallel but more general mechanism such as trying to reactivate the cell cycling machinery in vulnerable cell populations of the diseases (*see also* Chapter 16). This molecular event has now been suggested as a possible mechanism for AD pathogenesis and has been reviewed elsewhere *(122–124)*. In this regard, the absence of colabeling between BrdU and immature cell populations in the human AD study *(99)* allows this theory to remain a possibility.

3.4. A Clear or Not-So-Clear Link Between Hippocampal Neurogenesis and AD

From the literature summarized above in Subheadings 3.1.–3.3., conclusively linking hippocampal neurogenesis and AD remains controversial. In addition to the contradictions across similar studies many other inconsistencies complicate this matter further. For example, the fact that brain injury, which is associated with increased incidence of AD, leads to increased hippocampal neurogenesis and the fact that caloric restriction which increases neurogenesis *(121)*, and is thought to be beneficial to normal subjects and to protect from many risk factors associated with AD (i.e., oxidative stress, inflammation, etc.), is detrimental or even lethal to the AβPP transgenic animals *(125)* contradicts why animal models of the disease would show deceased neurogenesis. More importantly, a recent report demonstrating that AD transgenic animals that live in an enriched environment, an intervention that increases neurogenesis and improves learning and memory *(28,59–64)*, develop higher rates of amyloid plaques *(126)*, a condition that, based on some of the animal model studies, is supposed to negatively impact cognition and decrease hippocampal neurogenesis, is certainly puzzling and suggest that an Aβ deposition focused explanation for the declines of neurogenesis is questionable or even that Aβ could indeed have a protective role. Moreover, the fact that a recent study demonstrates that aged animals with higher levels of new cells in the hippocampus show lower cognitive performance than those animals with fewer cells *(127)* only creates further confusion by suggesting that increases in neurogenesis, at times, may in fact represent a response to deleterious events. In other words, neurogenesis may act as a surrogate marker of brain and cognitive health.

These findings and those described throughout the course of the manuscript bring to light the generalized confusion over the field of hippocampal neurogenesis in general and with relation to AD, and bring to question the validity of the established animal models of this disease. In this regard, it is important to note that the genes associated with the mutations that lead to Aβ generation and that are used in these animal models also have powerful roles in neurodevelopment *(128)*. For example, mutant mice lacking all three *AβPP* genes display early lethality as well as high incidence of cortical dysplasia, suggesting a critical role in neural development *(129)*. Moreover, PS1 is involved in γ-cleavage of the Notch receptor, which plays a critical role in determining cell fate during early development *(130)*, and is also involved in degradation *(131,132)*, such that PS1-deficient mice display profound developmental abnormalities *(133,134)* and transgenic mice that express a constitutively active β-catenin in neuroepithelial precursor cells develop enlarged brains because a greater proportion of neural precursors re-enter cell cycle after mitosis *(135)*. Therefore, because both AβPP and PS1 seem to be tightly involved in cell cycle regulation during development they are also likely to play a role in hippocampal neurogenesis in animal models of AD independently of any AD-related event, namely, overexpression of Aβ.

4. CONCLUSIONS

The conflicts stated in this chapter suggest that rather than critically examining the true role of hippocampal neurogenesis in a disease such as AD, the field seems to use "increases" or "decreases" in neurogenesis as superficial "utensil" to fit the hypothesis that Aβ is detrimental in AD. If neurogenesis is increased it is because it is associated with self-repair after Aβ toxicity *(101)* and if is decreased it is associated with declining cognition owing to Aβ deposition *(84,85)*. However, the fact that no correlations exist between Aβ and neurogenesis in any of these studies in addition to the other mitogenic or antioxidant properties of this molecule, both of which capable of altering neurogenesis, make such assumptions premature. As such, no compensation or crisis but rather chaotic state of affairs seems to be present in the literature regarding the role of neurogenesis in AD and judging from the disparity in results across the neurogenesis-AD studies the jury is still out on what the exact function of neurogenesis is, whether increases or decreases in neurogenesis would dictate benefit or detriment in AD, and more importantly whether the dogmatic detrimental powers of Aβ are a major player in this game *(115,136–138)*.

ACKNOWLEDGMENT

Work in the authors' laboratories is supported by Philip Morris USA Inc. and Philip Morris International.

REFERENCES

1. Katzman R. Alzheimer's disease as an age-dependent disorder. Ciba Found Symp 1988;134:69–85.
2. Larson EB, Kukull WA, Katzman RL. Cognitive impairment: dementia and Alzheimer's disease. Annu Rev Public Health 1992;13:431–449.
3. Selkoe DJ. Images in neuroscience. Alzheimer's disease: from genes to pathogenesis. Am J Psychiat 1997;154:1198.
4. Avila J, Lim F, Moreno F, Belmonte C, Cuello AC. Tau function and dysfunction in neurons: its role in neurodegenerative disorders. Mol Neurobiol 2002;25:213–231.
5. de Toledo-Morrell L, Dickerson B, Sullivan MP, Spanovic C, Wilson R, Bennett DA. Hemispheric differences in hippocampal volume predict verbal and spatial memory performance in patients with Alzheimer's disease. Hippocampus 2000;10:136–142.
6. Altman J, Das GD. Post-natal origin of microneurones in the rat brain. Nature 1965;207:953–956.
7. Rakic P. DNA synthesis and cell division in the adult primate brain. Ann NY Acad Sci 1985;457:193–211.
8. Suhonen JO, Peterson DA, Ray J, Gage FH. Differentiation of adult hippocampus-derived progenitors into olfactory neurons in vivo. Nature 1996;383:624–627.
9. Luskin MB. Restricted proliferation and migration of postnatally generated neurons derived from the forebrain subventricular zone. Neuron 1993;11:173–189.
10. Kriegstein AR. Cortical neurogenesis and its disorders. Curr Opin Neurol 1996;9:113–117.
11. Perez-Canellas MM, Garcia-Verdugo JM. Adult neurogenesis in the telencephalon of a lizard: a [3H]thymidine autoradiographic and bromodeoxyuridine immunocytochemical study. Brain Res Dev Brain Res 1996;93:49–61.
12. Gould E, Reeves AJ, Graziano MS, Gross CG. Neurogenesis in the neocortex of adult primates. Science 1999;286:548–552.
13. Magavi SS, Macklis JD. Induction of neuronal type-specific neurogenesis in the cerebral cortex of adult mice: manipulation of neural precursors in situ. Brain Res Dev Brain Res 2002;134:57–76.
14. Kuhn HG, Dickinson-Anson H, Gage FH. Neurogenesis in the dentate gyrus of the adult rat: age-related decrease of neuronal progenitor proliferation. J Neurosci 1996;16:2027–2033.
15. Gage FH, Kempermann G, Palmer TD, Peterson DA, Ray J. Multipotent progenitor cells in the adult dentate gyrus. J Neurobiol 1998;36:249–266.
16. Kempermann G, Brandon EP, Gage FH. Environmental stimulation of 129/SvJ mice causes increased cell proliferation and neurogenesis in the adult dentate gyrus. Curr Biol 1998;8:939–942.
17. Nilsson M, Perfilieva E, Johansson U, Orwar O, Eriksson PS. Enriched environment increases neurogenesis in the adult rat dentate gyrus and improves spatial memory. J Neurobiol 1999;39: 569–578.
18. Gould E, McEwen BS, Tanapat P, Galea LA, Fuchs E. Neurogenesis in the dentate gyrus of the adult tree shrew is regulated by psychosocial stress and NMDA receptor activation. J Neurosci 1997;17:2492–2498.

19. Gould E, Reeves AJ, Fallah M, Tanapat P, Gross CG, Fuchs E. Hippocampal neurogenesis in adult old world primates. Proc Natl Acad Sci USA 1999;96:5263–5267.
20. Eriksson PS, Perfilieva E, Bjork-Eriksson T, et al. Neurogenesis in the adult human hippocampus. Nat Med 1998;4:1313–1317.
21. Bayer SA, Altman J. Hippocampal development in the rat: cytogenesis and morphogenesis examined with autoradiography and low-level X-irradiation. J Comp Neurol 1974;158:55–79.
22. Kempermann G, Kuhn HG, Gage FH. Genetic influence on neurogenesis in the dentate gyrus of adult mice. Proc Natl Acad Sci USA 1997;94:10,409–10,414.
23. Markakis EA, Gage FH. Adult-generated neurons in the dentate gyrus send axonal projections to field CA3 and are surrounded by synaptic vesicles. J Comp Neurol 1999;406:449–460.
24. Hastings NB, Gould E. Rapid extension of axons into the CA3 region by adult-generated granule cells. J Comp Neurol 1999;413:146–154.
25. Song HJ, Stevens CF, Gage FH. Neural stem cells from adult hippocampus develop essential properties of functional CNS neurons. Nat Neurosci 2002;5:438–445.
26. van Praag H, Schinder AF, Christie BR, Toni N, Palmer TD, Gage FH. Functional neurogenesis in the adult hippocampus. Nature 2002;415:1030–1034.
27. Shors TJ. Can new neurons replace memories lost? Sci Aging Knowledge Environ 2003;49:pe35.
28. Shors TJ, Miesegaes G, Beylin A, Zhao M, Rydel T, Gould E. Neurogenesis in the adult is involved in the formation of trace memories. Nature 2001;410:372–376.
29. Tanapat P, Hastings NB, Reeves AJ, Gould E. Estrogen stimulates a transient increase in the number of new neurons in the dentate gyrus of the adult female rat. J Neurosci 1999;19:5792–5801.
30. Saravia F, Revsin Y, Lux-Lantos V, Beauquis J, Homo-Delarche F, De Nicola AF. Oestradiol restores cell proliferation in dentate gyrus and subventricular zone of streptozotocin-diabetic mice. J Neuroendocrinol 2004;16:704–710.
31. Perez-Martin M, Azcoitia I, Trejo JL, Sierra A, Garcia-Segura LM. An antagonist of estrogen receptors blocks the induction of adult neurogenesis by insulin-like growth factor-I in the dentate gyrus of adult female rat. Eur J Neurosci 2003;18:923–930.
32. Brown J, Cooper-Kuhn CM, Kempermann G, et al. Enriched environment and physical activity stimulate hippocampal but not olfactory bulb neurogenesis. Eur J Neurosci 2003;17:2042–2046.
33. Kempermann G, Gast D, Gage FH. Neuroplasticity in old age: sustained fivefold induction of hippocampal neurogenesis by long-term environmental enrichment. Ann Neurol 2002;52:135–143.
34. Kempermann G, Gage FH. Experience-dependent regulation of adult hippocampal neurogenesis: effects of long-term stimulation and stimulus withdrawal. Hippocampus 1999;9:321–332.
35. Kempermann G, Kuhn HG, Gage FH. More hippocampal neurons in adult mice living in an enriched environment. Nature 1997;386:493–495.
36. Kempermann G, Kuhn HG, Gage FH. Experience-induced neurogenesis in the senescent dentate gyrus. J Neurosci 1998;18:3206–3212.
37. Kempermann G, van Praag H, Gage FH. Activity-dependent regulation of neuronal plasticity and self repair. Prog Brain Res 2000;127:35–48.
38. van Praag H, Kempermann G, Gage FH. Running increases cell proliferation and neurogenesis in the adult mouse dentate gyrus. Nat Neurosci 1999;2:266–270.
39. Kitamura T, Mishina M, Sugiyama H. Enhancement of neurogenesis by running wheel exercises is suppressed in mice lacking NMDA receptor epsilon 1 subunit. Neurosci Res 2003;47:55–63.
40. Arvidsson A, Kokaia Z, Lindvall O. N-methyl-D-aspartate receptor-mediated increase of neurogenesis in adult rat dentate gyrus following stroke. Eur J Neurosci 2001;14:10–18.
41. Cameron HA, McEwen BS, Gould E. Regulation of adult neurogenesis by excitatory input and NMDA receptor activation in the dentate gyrus. J Neurosci 1995;15:4687–4692.
42. Nacher J, Rosell DR, Alonso-Llosa G, McEwen BS. NMDA receptor antagonist treatment induces a long-lasting increase in the number of proliferating cells, PSA-NCAM-immunoreactive granule neurons and radial glia in the adult rat dentate gyrus. Eur J Neurosci 2001;13:512–520.
43. Yagita Y, Kitagawa K, Ohtsuki T, et al. Neurogenesis by progenitor cells in the ischemic adult rat hippocampus. Stroke 2001;32:1890–1896.
44. Liu J, Solway K, Messing RO, Sharp FR. Increased neurogenesis in the dentate gyrus after transient global ischemia in gerbils. J Neurosci 1998;18:7768–7778.
45. Kee NJ, Preston E, Wojtowicz JM. Enhanced neurogenesis after transient global ischemia in the dentate gyrus of the rat. Exp Brain Res 2001;136:313–320.

46. Jin K, Minami M, Lan JQ, et al. Neurogenesis in dentate subgranular zone and rostral subventricular zone after focal cerebral ischemia in the rat. Proc Natl Acad Sci USA 2001;98:4710–4715.
47. Malberg JE, Eisch AJ, Nestler EJ, Duman RS. Chronic antidepressant treatment increases neurogenesis in adult rat hippocampus. J Neurosci 2000;20:9104–9110.
48. Malberg JE, Duman RS. Cell proliferation in adult hippocampus is decreased by inescapable stress: reversal by fluoxetine treatment. Neuropsychopharmacology 2003;28:1562–1571.
49. Jacobs BL. Adult brain neurogenesis and depression. Brain Behav Immun 2002;16:602–609.
50. Tanapat P, Galea LA, Gould E. Stress inhibits the proliferation of granule cell precursors in the developing dentate gyrus. Int J Dev Neurosci 1998;16:235–239.
51. McEwen BS. Stress and hippocampal plasticity. Annu Rev Neurosci 1999;22:105–122.
52. Czeh B, Welt T, Fischer AK, et al. Chronic psychosocial stress and concomitant repetitive transcranial magnetic stimulation: effects on stress hormone levels and adult hippocampal neurogenesis. Biol Psychiatry 2002;52:1057–1065.
53. Pham K, Nacher J, Hof PR, McEwen BS. Repeated restraint stress suppresses neurogenesis and induces biphasic PSA-NCAM expression in the adult rat dentate gyrus. Eur J Neurosci 2003;17:879–886.
54. Lemaire V, Koehl M, Le Moal M, Abrous DN. Prenatal stress produces learning deficits associated with an inhibition of neurogenesis in the hippocampus. Proc Natl Acad Sci USA 2000;97:11,032–11,037.
55. Lemaire V, Aurousseau C, Le Moal M, Abrous DN. Behavioural trait of reactivity to novelty is related to hippocampal neurogenesis. Eur J Neurosci 1999;11:4006–4014.
56. Cooper-Kuhn CM, Winkler J, Kuhn HG. Decreased neurogenesis after cholinergic forebrain lesion in the adult rat. J Neurosci Res 2004;77:155–165.
57. Holmes GL. Effects of early seizures on later behavior and epileptogenicity. Ment Retard Dev Disabil Res Rev 2004;10:101–105.
58. Bizon JL, Gallagher M. Production of new cells in the rat dentate gyrus over the lifespan: relation to cognitive decline. Eur J Neurosci 2003;18:215–219.
59. Gould E, Tanapat P, Hastings NB, Shors TJ. Neurogenesis in adulthood: a possible role in learning. Trends Cogn Sci 1999;3:186–192.
60. Gould E, Beylin A, Tanapat P, Reeves A, Shors TJ. Learning enhances adult neurogenesis in the hippocampal formation. Nat Neurosci 1999;2:260–265.
61. Kempermann G. Why new neurons? Possible functions for adult hippocampal neurogenesis. J Neurosci 2002;22:635–638.
62. Drapeau E, Mayo W, Aurousseau C, Le Moal M, Piazza PV, Abrous DN. Spatial memory performances of aged rats in the water maze predict levels of hippocampal neurogenesis. Proc Natl Acad Sci USA 2003;100:14,385–14,390.
63. Prickaerts J, Koopmans G, Blokland A, Scheepens A. Learning and adult neurogenesis: survival with or without proliferation? Neurobiol Learn Mem 2004;81:1–11.
64. Schmidt-Hieber C, Jonas P, Bischofberger J. Enhanced synaptic plasticity in newly generated granule cells of the adult hippocampus. Nature 2004;429:184–187.
65. Pope SK, Shue VM, Beck C. Will a healthy lifestyle help prevent Alzheimer's disease? Annu Rev Public Health 2003;24:111–132.
66. Callahan CM, Hall KS, Hui SL, Musick BS, Unverzagt FW, Hendrie HC. Relationship of age, education, and occupation with dementia among a community-based sample of African Americans. Arch Neurol 1996;53:134–140.
67. Glatt SL, Hubble JP, Lyons K, et al. Risk factors for dementia in Parkinson's disease: effect of education. Neuroepidemiology 1996;15:20–25.
68. Fritsch T, McClendon MJ, Smyth KA, Ogrocki PK. Effects of educational attainment and occupational status on cognitive and functional decline in persons with Alzheimer-type dementia. Int Psychogeriatr 2002;14:347–363.
69. Tang MX, Jacobs D, Stern Y, et al. Effect of oestrogen during menopause on risk and age at onset of Alzheimer's disease. Lancet 1996;348:429–432.
70. Stoppe G. Dementia: risk and protective factors with special consideration of gender and hormone replacement therapy. Z Arztl Fortbild Qualitatssich 2000;94:217–222.
71. Geerlings MI, Ruitenberg A, Witteman JC, et al. Reproductive period and risk of dementia in postmenopausal women. JAMA 2001;285:1475–1481.
72. Green RC, Cupples LA, Kurz A, et al. Depression as a risk factor for Alzheimer disease: the MIRAGE study. Arch Neurol 2003;60:753–759.

73. Speck CE, Kukull WA, Brenner DE, et al. History of depression as a risk factor for Alzheimer's disease. Epidemiology 1995;6:366–369.
74. Wilson RS, Evans DA, Bienias JL, Mendes de Leon CF, Schneider JA, Bennett DA. Proneness to psychological distress is associated with risk of Alzheimer's disease. Neurology 2003;61: 1479–1485.
75. Aronson MK, Ooi WL, Geva DL, Masur D, Blau A, Frishman W. Dementia. Age-dependent incidence, prevalence, and mortality in the old old. Arch Intern Med 1991;151:989–992.
76. Gavrilova SI, Bratsun AL. Epidemiology and risk factors of Alzheimer's disease. Vestn Ross Akad Med Nauk 1999;1:39–46.
77. Gaitatzis A, Carroll K, Majeed A, W Sander J. The epidemiology of the comorbidity of epilepsy in the general population. Epilepsia 2004;45:1613–1622.
78. Hardy J. Amyloid, the presenilins and Alzheimer's disease. Trends Neurosci 1997;20:154–159.
79. Selkoe DJ. Alzheimer's disease: genes, proteins, and therapy. Physiol Rev 2001;81:741–766.
80. Coulson EJ, Paliga K, Beyreuther K, Masters CL. What the evolution of the amyloid protein precursor supergene family tells us about its function. Neurochem Int 2000;36:175–184.
81. Selkoe DJ, Schenk D. Alzheimer's disease: molecular understanding predicts amyloid-based therapeutics. Annu Rev Pharmacol Toxicol 2003;43:545–584.
82. Feng R, Rampon C, Tang YP, et al. Deficient neurogenesis in forebrain-specific presenilin-1 knockout mice is associated with reduced clearance of hippocampal memory traces. Neuron 2001;32:911–926.
83. Wang R, Dineley KT, Sweatt JD, Zheng H. Presenilin 1 familial Alzheimer's disease mutation leads to defective associative learning and impaired adult neurogenesis. Neuroscience 2004;126:305–312.
84. Haughey NJ, Liu D, Nath A, Borchard AC, Mattson MP. Disruption of neurogenesis in the subventricular zone of adult mice, and in human cortical neuronal precursor cells in culture, by amyloid beta-peptide: implications for the pathogenesis of Alzheimer's disease. Neuromol Med 2002;1:125–135.
85. Haughey NJ, Nath A, Chan SL, Borchard AC, Rao MS, Mattson MP. Disruption of neurogenesis by amyloid beta-peptide, and perturbed neural progenitor cell homeostasis, in models of Alzheimer's disease. J Neurochem 2002;83:1509–1524.
86. Dong H, Goico B, Martin M, Csernansky CA, Bertchume A, Csernansky JG. Modulation of hippocampal cell proliferation, memory, and amyloid plaque deposition in APPsw (Tg2576) mutant mice by isolation stress. Neuroscience 2004;127:601–609.
87. Parent JM, Yu TW, Leibowitz RT, Geschwind DH, Sloviter RS, Lowenstein DH. Dentate granule cell neurogenesis is increased by seizures and contributes to aberrant network reorganization in the adult rat hippocampus. J Neurosci 1997;17:3727–3738.
88. Scott BW, Wojtowicz JM. Burnham WM. Neurogenesis in the dentate gyrus of the rat following electroconvulsive shock seizures. Exp Neurol 2000;165:231–236.
89. Wang YL, Sun RP, Lei GF, Wang JW, Guo SH. Neurogenesis of dentate granule cells following kainic acid induced seizures in immature rats. Zhonghua Er Ke Za Zhi 2004;42:621–624.
90. Cha BH, Akman C, Silveira DC, Liu X, Holmes GL. Spontaneous recurrent seizure following status epilepticus enhances dentate gyrus neurogenesis. Brain Dev 2004;26:394–397.
91. Mohapel P, Ekdahl CT, Lindvall O. Status epilepticus severity influences the long-term outcome of neurogenesis in the adult dentate gyrus. Neurobiol Dis 2004;15:196–205.
92. Hellsten J, Wennstrom M, Bengzon J, Mohapel P, Tingstrom A. Electroconvulsive seizures induce endothelial cell proliferation in adult rat hippocampus. Biol Psychiat 2004;55:420–427.
93. Parent JM. The role of seizure-induced neurogenesis in epileptogenesis and brain repair. Epilepsy Res 2002;50:179–189.
94. Gage FH. Brain, repair yourself. Sci Am 2003;289:46–53.
95. Reinoso Suarez F. Adult neurogenesis and stem cells. Functional capacity. An R Acad Nac Med (Madr) 2002;119:507–521; discussion 521–508.
96. Domanska-Janik K. Stem cells—potential therapeutic use in neurological diseases. Neurol Neurochir Pol 2002;36(Suppl 1):107–117.
97. Bateman DA, Chakrabartty A. Interactions of Alzheimer amyloid peptides with cultured cells and brain tissue, and their biological consequences. Biopolymers 2004;76:4–14.
98. Walsh DM, Klyubin I, Fadeeva JV, Rowan MJ, Selkoe DJ. Amyloid-beta oligomers: their production, toxicity and therapeutic inhibition. Biochem Soc Trans 2002;30:552–557.
99. Jin K, Peel AL, Mao XO, et al. Increased hippocampal neurogenesis in Alzheimer's disease. Proc Natl Acad Sci USA 2004;101:343–347.

100. Jin K, Galvan V, Xie L, et al. Enhanced neurogenesis in Alzheimer's disease transgenic (PDGF-APPSw, Ind) mice. Proc Natl Acad Sci USA 2004;101:13,363–13,367.
101. Lopez-Toledano MA, Shelanski ML. Neurogenic effect of beta-amyloid peptide in the development of neural stem cells. J Neurosci 2004;24:5439–5444.
102. Milward EA, Papadopoulos R, Fuller SJ, et al. The amyloid protein precursor of Alzheimer's disease is a mediator of the effects of nerve growth factor on neurite outgrowth. Neuron 1992;9:129–137.
103. Schubert D, Cole G, Saitoh T, Oltersdorf T. Amyloid beta protein precursor is a mitogen. Biochem Biophys Res Commun 1989;162:83–88.
104. Copani A, Condorelli F, Canonico PL, Nicoletti F, Sortino MA. Cell cycle progression towards Alzheimer's disease. Funct Neurol 2001;16:11–15.
105. Hoffmann J, Twiesselmann C, Kummer MP, Romagnoli P, Herzog V. A possible role for the Alzheimer amyloid precursor protein in the regulation of epidermal basal cell proliferation. Eur J Cell Biol 2000;79:905–914.
106. Schmitz A, Tikkanen R, Kirfel G, Herzog V. The biological role of the Alzheimer amyloid precursor protein in epithelial cells. Histochem Cell Biol 2002;117:171–180.
107. Jellinger KA, Mitter-Ferstl E. The impact of cerebrovascular lesions in Alzheimer disease—a comparative autopsy study. J Neurol 2003;250:1050–1055.
108. Smith MA, Sayre LM, Monnier VM, Perry G. Radical Ageing in Alzheimer's disease. Trends Neurosci 1995;18:172–176.
109. Smith MA, Perry G, Richey PL, et al. Oxidative damage in Alzheimer's. Nature 1996;382:120–121.
110. Perry G, Castellani RJ, Hirai K, Smith MA. Reactive oxygen species mediate cellular damage in Alzheimer disease. J Alzheimers Dis 1998;1:45–55.
111. Perry G, Smith MA. Is oxidative damage central to the pathogenesis of Alzheimer disease? Acta Neurol Belg 1998;98:175–179.
112. Smith MA, Rottkamp CA, Nunomura A, Raina AK, Perry G. Oxidative stress in Alzheimer's disease. Biochim Biophys Acta 2000;1502:139–144.
113. Nunomura A, Perry G, Aliev G, et al. Oxidative damage is the earliest event in Alzheimer disease. J Neuropathol Exp Neurol 2001;60:759–767.
114. Joseph J, Shukitt-Hale B, Denisova NA, Martin A, Perry G, Smith MA. Copernicus revisited: amyloid beta in Alzheimer's disease. Neurobiol Aging 2001;22:131–146.
115. Rottkamp CA, Atwood CS, Joseph JA, Nunomura A, Perry G, Smith MA. The state versus amyloid-beta: the trial of the most wanted criminal in Alzheimer disease. Peptides 2002;23:1333–1341.
116. Smith MA, Atwood CS, Joseph JA, Perry G. Ill-fated amyloid-beta vaccine. J Neurosci Res 2002;69:285.
117. Herrera DG, Yague AG, Johnsen-Soriano S, et al. Selective impairment of hippocampal neurogenesis by chronic alcoholism: protective effects of an antioxidant. Proc Natl Acad Sci USA 2003;100:7919–7924.
118. Cecchini T, Ciaroni S, Ferri P, et al. Alpha-tocopherol, an exogenous factor of adult hippocampal neurogenesis regulation. J Neurosci Res 2003;73:447–455.
119. Cuppini R, Ciaroni S, Cecchini T, et al. Tocopherols enhance neurogenesis in dentate gyrus of adult rats. Int J Vitam Nutr Res 2002;72:170–176.
120. Shen LH, Zhang JT. Ginsenoside Rg1 promotes proliferation of hippocampal progenitor cells. Neurol Res 2004;26:422–428.
121. Lee J, Duan W, Long JM, Ingram DK, Mattson MP. Dietary restriction increases the number of newly generated neural cells, and induces BDNF expression, in the dentate gyrus of rats. J Mol Neurosci 2000;15:99–108.
122. Raina AK, Zhu X, Rottkamp CA, Monteiro M, Takeda A, Smith MA. Cyclin toward dementia: cell cycle abnormalities and abortive oncogenesis in Alzheimer disease. J Neurosci Res 2000;61:128–133.
123. Raina AK, Zhu X, Smith MA. Alzheimer's disease and the cell cycle. Acta Neurobiol Exp (Wars) 2004;64:107–112.
124. Zhu X, Raina AK, Perry G, Smith MA. Alzheimer's disease: the two-hit hypothesis. Lancet Neurol 2004;3:219–226.
125. Mattson MP. Neuroprotective signaling and the aging brain: take away my food and let me run. Brain Res 2000;886:47–53.

126. Jankowsky JL, Xu G, Fromholt D, Gonzales V, Borchelt DR. Environmental enrichment exacerbates amyloid plaque formation in a transgenic mouse model of Alzheimer disease. J Neuropathol Exp Neurol 2003;62:1220–1227.
127. Bizon JL, Lee HJ, Gallagher M. Neurogenesis in a rat model of age-related cognitive decline. Aging Cell 2004;3:227–234.
128. Bothwell M, Giniger E. Alzheimer's disease: neurodevelopment converges with neurodegeneration. Cell 2000;102:271–273.
129. Koo EH. The beta-amyloid precursor protein (APP) and Alzheimer's disease: does the tail wag the dog? Traffic 2002;3:763–770.
130. Selkoe D, Kopan R. Notch and presenilin: regulated intramembrane proteolysis links development and degeneration. Annu Rev Neurosci 2003;26:565–597.
131. Kang DE, Soriano S, Xia X, et al. Presenilin couples the paired phosphorylation of beta-catenin independent of axin: implications for beta-catenin activation in tumorigenesis. Cell 2002;110:751–762.
132. Willert K, Nusse R. Beta-catenin: a key mediator of Wnt signaling. Curr Opin Genet Dev 1998;8:95–102.
133. Shen J, Bronson RT, Chen DF, Xia W, Selkoe DJ, Tonegawa S. Skeletal and CNS defects in Presenilin-1-deficient mice. Cell 1997;89:629–639.
134. Wong PC, Zheng H, Chen H, et al. Presenilin 1 is required for Notch1 and Dll1 expression in the paraxial mesoderm. Nature 1997;387:288–292.
135. Chenn A, Walsh CA. Regulation of cerebral cortical size by control of cell cycle exit in neural precursors. Science 2002;297:365–369.
136. Smith MA, Casadesus G, Joseph JA, Perry G. Amyloid-beta and tau serve antioxidant functions in the aging and Alzheimer brain. Free Radic Biol Med 2002;33:1194–1199.
137. Webber KM, Bowen R, Casadesus G, Perry G, Atwood CS, Smith MA. Gonadotropins and Alzheimer's disease: the link between estrogen replacement therapy and neuroprotection. Acta Neurobiol Exp (Wars) 2004;64:113–118.
138. Casadesus G, Zhu X, Atwood CS, et al. Beyond estrogen: targeting gonadotropin hormones in the treatment of Alzheimer's disease. Curr Drug Targets CNS Neurol Disord 2004;3:281–285.

VI
THE BIOLOGY OF GLIOMAS

27
p53 and Multidrug Resistance Transporters in the Central Nervous System

Shirley Teng, PhD and Micheline Piquette-Miller, PhD

SUMMARY

Multidrug resistance (MDR) is a complication that is often seen during the treatment of cancer and epilepsy. This resistance has been attributed to an increased expression of ATP-binding cassette (ABC)-membrane efflux transporters in brain. Expression of these transporters, in particular P-glycoprotein (P-gp) and multidrug resistance-associated protein 1 (MRP1), have been shown to be regulated by the tumor suppressor gene *p53* in that p53 mutations can lead to transporter induction. On the other hand, increased p53 expression in the brain is associated with neuronal cell death via apoptosis. Epileptic brain has been shown to regulate the MDR genes in a manner similar to tumor cells. Seizures have also been shown to precipitate a generalized acute inflammatory response and cytokine release within the central nervous system (CNS). The overall phenomenon of cell cycle arrest, apoptosis, and induction of MDR in epilepsy may be linked through common signaling pathways activated by these stimuli. Complexities in the development and regulation of MDR within the CNS underscore the need to delineate the molecular pathways involved. Ultimately, this knowledge may be used in the development of novel therapeutics in refractory epilepsy or malignancies.

Key Words: Multidrug resistance; P-glycoprotein; multidrug resistance associated transporter; epilepsy; p53; NF-κB; cytokine.

1. INTRODUCTION

Pharmacotherapy for the treatment of disorders of the central nervous system (CNS) requires that sufficient drug levels should be achieved at target sites within the brain. Factors such as drug permeation, cellular distribution, and clearance of drug from the CNS influence achievable brain concentrations. Tight junctions of the blood–brain barrier (BBB) and the blood–cerebrospinal fluid (CSF) barrier form very effective obstacles to the free diffusion of many solutes into the brain. Moreover, expression of efflux transporters at the site of these membrane barriers effectively exports substrates from the CNS back into the capillary lumen, thus limiting brain accumulation of potential toxins and metabolic byproducts. To date, the principle active efflux transporters found in these membranes include the ATP-binding cassette (ABC) transporters. The overexpression of these active efflux transporters at the BBB and in cells of the CNS is thought to play an important role in the development of drug resistance in CNS disorders such as epilepsy and brain tumors.

2. MULTIDRUG RESISTANCE TRANSPORTERS

Despite appropriate drug therapy, approx 30% of epileptic patients fail to respond to antiepileptic drugs *(1)*. Moreover, patients who are resistant to one type of drug are often also refractory to other classes of anticonvulsant drugs, which act via alternate mechanisms. The phenomenon of

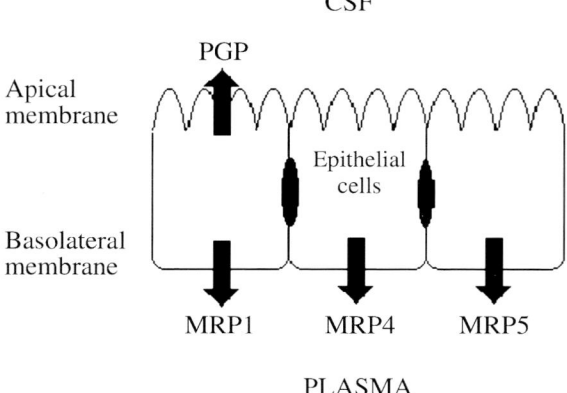

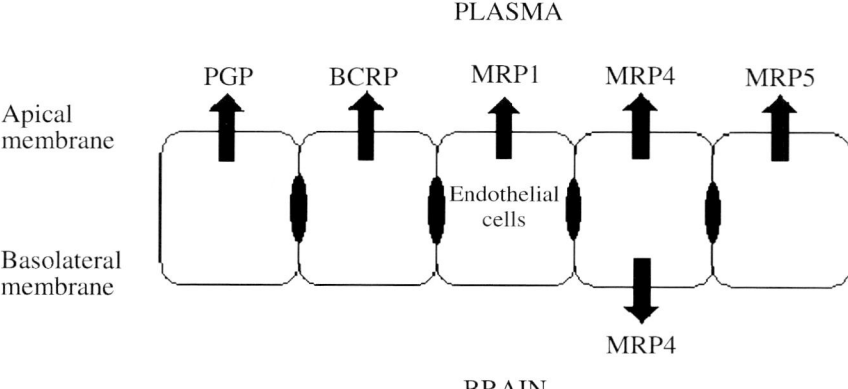

Fig. 1. Expression of multidrug resistance transporters at the blood–CSF barrier (**A**) and the blood–brain barrier (**B**).

multidrug resistance (MDR)—defined as the simultaneous resistance to a number of unrelated drugs—is well recognized. First described in 1972 by Dano *(2)* in tumor cells that were resistant to both vinca alkaloids and anthracyclines, this resistant phenotype has been studied extensively and has been observed in many different cancers both clinically and in vitro in tumor-derived cell lines *(3)*. Many tumors have been shown to either inherently express numerous genes associated with the drug resistant phenotype or to develop alterations in these genes following exposure to chemotherapy. Tumor-inherent and -acquired resistance are seen with many structurally and functionally unrelated chemotherapeutics, subsequently resulting in loss of drug efficacy and poor clinical outcome. Several mechanisms of resistance have been studied including increased drug metabolism, decreased drug uptake, increased repair of damaged drug targets, and prevention of cell cycle arrest or apoptosis. However, increased expression of drug efflux transporters has largely been accepted as the predominant mechanism of MDR in tumors. In particular, much attention has been focused on the role of two members of the ABC superfamily of transport proteins, P-glycoprotein (P-gp) and the MDR-associated proteins (MRP). The molecular and cellular mechanisms that lead to drug resistance in epilepsy are still unknown. However, these transporters are expressed at the major barriers to drug absorption in the CNS, the BBB, and the blood–CSF barrier of the choroid plexus (Fig. 1). Furthermore, overexpression of P-gp and MRP are seen in brain tissue of refractory epileptic patients, suggesting that these transporters contribute to drug resistance in epilepsy.

2.1. P-Glycoprotein

P-gp is a 170-kDa apical membrane efflux transporter which is encoded by the MDR genes *mdr1* in humans and *mdr1a* and *mdr1b* in rodents. P-gp was first identified in 1976 by Juliano and Ling as a protein that modified colchicine absorption in Chinese hamster ovary cells *(4)*. Overexpression of P-gp in tumors arising from a variety of cell types has been shown to impart resistance to numerous antineoplastic agents *(3)*. The importance of P-gp in drug absorption, distribution, and clearance in normal tissues has also become well established. It is estimated that up to 50% of drug candidates may be substrates for P-gp. Apart from the fact that many of these compounds are highly lipophilic, few structural similarities exist. Major substrates include anticancer drugs, immunosuppressive agents, protease inhibitors, cardiovascular agents, opioids, as well as several antiepileptic drugs *(5)*.

P-gp is expressed in a variety of cells and secretory organs such as the liver, kidney, intestine, and placenta, thus it serves as an important protective barrier against potential toxins. The CNS is another major site of P-gp expression, as it is highly expressed in endothelial cells of the BBB and in epithelial cells, which make up the choroid plexus *(6,7)*. The capillary endothelial cells of the BBB express P-gp at the apical/luminal membrane, thereby transporting compounds into the blood and away from brain parenchyma. P-gp has also been detected at the apical membrane of choroid plexus epithelial cells *(7)*. In addition to these two sites, P-gp expression has been found in neighboring astrocyte foot processes of isolated human brain capillaries *(8)*, as well as in primary cultures of rat astrocytes and microglia *(9,10,110,134)*. The importance of P-gp in restricting the entry of drugs into the CNS has been made apparent with the generation and characterization of P-gp knockout mice. Schinkel et al. *(11)* first demonstrated elevated levels of the neurotoxin ivermectin and the anticancer drug vinblastine in the brain of *mdr1a*-null mice, which was correlated with dramatically increased sensitivity to these compounds. Mouse models lacking *mdr1a*, *mdr1b*, or both genes have also been shown to have higher brain concentrations of many other drugs including quinidine *(12)*, morphine *(13,14)*, paclitaxel *(15)*, and HIV protease inhibitors *(16)*. Clearly, P-gp-mediated drug transport in the brain can significantly impact the accumulation and thus the efficacy of CNS-targeted drug therapeutics.

2.2. MDR-Associated Protein

The MRP were first identified in 1992 when Cole et al. *(17)* cloned a 190 kDa basolateral transporter, termed MRP1, from doxorubicin-resistant human lung cancer cells. Since then, several additional members of the MRP family of transporters have been identified. To date, the MRP family consists of nine members, termed MRPs 1–9. Substrate specificity for the MRPs is very broad, including anticancer drugs such as cisplatin, doxorubicin, and etoposide, many organic anions and bile acids as well as glutathione, glucuronide, and sulfate conjugates. Whereas some overlap exists between MRP family members in terms of substrate selectivity, each transporter has also been found to have distinct structural requirements for transport. For example, whereas MRP2 primarily transports glutathione conjugates, MRP3 has a much higher affinity for glucuronide conjugates. On the other hand, the major substrates of MRP4 and MRP5 are nucleotide analogs *(18)*. Overexpression of MRP1 has been frequently reported both clinically in drug-resistant tumors and in vitro in resistant carcinoma cell lines *(19–22)*, thus implicating this transporter in tumor drug resistance. To date, the majority of studies have focused on MRP1, however the importance of other MRP transporters in conferring drug resistance is slowly emerging. For example, it is now believed that MRP2 plays a role in conferring resistance to cisplatin *(23)*, whereas overexpression of MRP3 and MRP4 are believed to play a role in conferring resistance to the anticancer agents methotrexate *(24)* and topotecan *(25)*, respectively. Moreover, the role of MRP transporters in CNS drug distribution is becoming more apparent with the recent generation of *MRP* knockout mice. For instance, as compared with wild type, *MRP4* (–/–) deficient mice demonstrated dramatic increases in the accumulation of topotecan in brain tissue and CSF *(25)*. Similarly, etoposide levels in the CSF were 10-fold higher in triple knockout mice deficient in *MRP1*, *mdr1a*, and *mdr1b* *(26)*.

Table 1
ABC Transporters in the CNS

Transporter	Tissue localization	Source	Location in the CNS							References
			BBB	CP	AS	MG	N	G	WB	
P-gp										
mdr1	Apical: liver, kidney, gut, pancreas, placenta, testis, and heart	Human brain	++	++	++/−		−	−	+	(6,7,27–30)
		Primary culture	+							
mdr1a		Rat brain	++	+					+	(9,10,31–35)
		Primary culture	+++/+		+	++				
		Cell lines	+			−				
mdr1b		Rat brain	−	+		+			+	(9,10,31–35)
		Primary culture	+++		++	+				
		Cell lines	+/+++							
MRP1	BBB: apical Basolateral: Ubiquitous expression but highest in lung, testis, and kidney	Human brain	+++/−	+	++		−	−	++/−	(6,7,9,10, 26–29, 31–33, 35–46)
		Primary culture	++							
		Rat brain	+/++	+++	++	++		+	++	
		Primary culture	+			+/−			+	
		Cell lines	+/++/−	+++						
		Mouse brain	+++							
		Bovine brain	++						++	
MRP2	Apical: Liver, gut, kidney, and placenta	Human brain	++/−	+/−	−	−	−	−	−/++	(31,35,36,39 42,43,45,46)
		Rat brain								
		Primary culture								
		Bovine primary culture	−							
MRP3	Basolateral: Liver, gut, placenta, kidney, pancreas, gall bladder, and adrenal	Rat brain		+	+	+	+	−	+	(31,35,36, 42–45)
		Primary culture		+						
		Mouse brain								
		Bovine brain	−							
		Primary culture	+/−							

MRP4	BBB: apical and basolateral Apical: kidney Basolateral: prostate	Human brain Rat brain Primary culture MLS-9 cell line Mouse brain Bovine primary culture	++ ++ ++	+	+/++ +	+	+	(9,25,31, 43,45,47)
MRP5	BBB: apical basolateral: ubiquitous expression but highest in muscle, heart, brain, and urogenital tissue	Rat brain Primary culture MLS-9 cell line Mouse isolated cells Bovine primary culture	+++ +++ ++ +	+	++/− +	+	++	(9,31,35, 43,45,47)
MRP6	Basolateral: Liver, kidney, gut, skin, retina, and heart[a]	Rat brain Primary culture Mouse brain Bovine primary culture	++ +	+	+ −	− +	+	(31,35, 43,45)
MRP7	Colon, spleen, heart, liver, muscle, and kidney[b]	Human brain Rat primary culture Bovine primary culture	−				+/−	(35,48)
MRP8	Apical: Breast, testis, liver, and placenta	Human brain					+	(49)
BCRP	Apical: Placenta, gut, colon, liver, and breast	Human brain Mouse primary culture Rat primary culture Porcine primary culture	++ +++ +++ +++	+				(6,50–52)

[a] Reports on tissue localization between humans and rodents are conflicting.
[b] Expression of MRP7 is very low in most tissues. Only tissues with relatively higher transcripts in the rat are indicated. Membrane localization is uncertain.

BBB, blood–brain barrier; CP, choroid plexus; AS, astrocytes; MG, microglia; N, neurons; G, glial cells; WB, whole brain homogenate.

Note: Two different expression levels separated by a slash ("/") indicates that there are conflicting results in the literature. "Brain" levels were measured in isolated cells or tissues from the indicated species. "Primary culture" denotes a culturing of the cell type indicated; brain capillary endothelial cells and epithelial cells of the choroid plexus are cultured for BBB and CP, respectively.

To date the majority of studies examining expression of MRP transporters have been done in rodents using whole brain homogenates, isolated capillaries, or cultures enriched for specific cell types (i.e., neurons, astrocytes, and microglia). Expression results from these studies are summarized in Table 1. Both BBB and choroid plexus are the major sites of MRP expression. MRP1, MRP2, and MRP4 are expressed on the luminal surface of brain capillary endothelium of different species including mouse, rat, and human. Studies in cultured bovine brain microvessel endothelial cells suggest that other MRP isoforms, including *MRP1*, *MRP4*, *MRP5*, and *MRP6* are also expressed on the luminal surface of the BBB *(36,53)*. Epithelial cells of the choroid plexus express high levels of *MRP1*, *MRP4*, *MRP5*, and *MRP6* in the basolateral membrane whereas *MRP2* and *MRP3* are expressed to a lesser extent. Of note are recent reverse transcription-polymerase chain reaction (RT-PCR) results demonstrating a greater mRNA expression of *MRP1*, *MRP4*, and *MRP5* in the choroid plexus of rats as compared with levels seen in kidney, liver, or intestine *(31)*, organs which have a major role in drug excretion. Hence these MRP transporters are likely to play a more prominent role in the distribution and secretion of drugs within the CNS than in the whole body. Moreover, this suggests a role of the MRPs in the normal physiological transport of endogenous compounds found primarily in the CNS. In astrocyte and microglial cultures, *MRP1* and *MRP5* are most highly expressed, whereas significant levels of *MRP3*, *MRP4*, and *MRP6* have also been detected. There are very few studies examining the expression of *MRP7* and *MRP8* in the CNS, although these two transporters have been detected in whole brain homogenates *(48,49)*. On the other hand, numerous conflicting reports regarding the expression of MRP transporters exist in the literature. For example whereas some studies have reported *MRP2* expression in whole brain homogenates as well as in isolated endothelial cells and choroid plexus *(31,39,42)*, other groups did not detect *MRP2* in these tissues *(43,45)*. Furthermore, although a high expression of MRP1 has been found in rat whole brain homogenates *(42)*, low levels of *MRP1* expression are seen in isolated endothelial cell fractions and primary cultures *(32,41)*. Many of these discrepancies likely arise from experimental differences in obtaining enriched cell fractions or in culture techniques. Indeed, it has been shown that culturing affects the expression of several transporters *(40,54)*. For example, culturing of human microvessels was found to result in the induction of *MRP1* and suppression of P-gp mRNA and protein as compared with levels seen in freshly isolated cells *(30)*. In addition, co-cultures of glia and brain capillary endothelial cells express much higher levels of P-gp than solo cultures of glia or brain capillaries *(35)*. Hence identification of significant levels of expression in cultures may overrepresent levels which would be seen in vivo, under normal physiological conditions.

2.3. Breast Cancer Resistance Protein

A more recent addition to the ABC transporter family is the breast cancer resistance protein (BCRP) (ABCG2), which is also known as the mitoxantrone resistance protein (MXR). BCRP is an ATP-dependent 72.1 kD "half transporter," which was independently cloned from placenta *(55)*, as well as drug-resistant MCF-7/AdrVp breast carcinoma cells *(56)* and the mitoxantrone-resistant colon carcinoma cells S1-M1-80 *(57)*. BCRP is involved in the extrusion of substrates such as mitoxantrone, topotecan, flavopiridol, or methotrexate from cells, thereby reducing intracellular concentrations and imparting drug resistance. As the importance of BCRP in drug transport has only been recently recognized, the majority of substrates have yet to be identified. However, it is felt that considerable overlaps with P-gp and MRP substrates will be found with BCRP. Moreover, tissue distribution of BCRP closely resembles that of P-gp, with the greatest expression seen in placenta, intestine, colon, liver, kidney, and brain *(58)*. Altered absorption and excretion of BCRP substrates has been reported in *BCRP* (–/–) knockout mice *(59–61)*. In the brain, BCRP is primarily expressed on the luminal surface of the human, murine, and porcine BBB *(50,51,62)*. Consequently, it has been speculated that BCRP poses a barrier to drug access to the brain. This may be of particular importance in the treatment of AIDS, as BCRP has been found to confer cellular resistance to the nucleoside reverse transcriptase inhibitors *(63)*. BCRP mRNA

Table 2
Antiepileptic Drugs That Are Substrates of P-gp or MRP

Drug	P-gp	MRP	References
Carbamazepine	±	+	(66,67)
Felbamate	+		(68)
Lamotrigine	+		(68)
Phenobarbital	+		(68)
Phenytoin	+	+	(69)
Topiramate	+		(70)
Valproic acid		+	(71)

levels at the BBB are three times higher in *mdr1a*-null mice as compared with wild types and this is associated with increased CNS efflux of BCRP substrates such as prazosin and mitoxantrone *(52)*. This suggests that BCRP expression is induced to compensate for the absence of functional P-gp, and thus serves as an additional barrier to protect the brain from potentially harmful xenobiotics. Whether BCRP induction would be seen in situations whereby P-gp is intentionally suppressed or inhibited remains to be determined. On the other hand, induction of both P-gp and BCRP have been reported in numerous human cancers including glioblastomas *(6,64)*. Although a recent investigation comparing expression in epileptic and nonepileptic brain tissues did not detect differences in BCRP expression *(65)*, the involvement of BCRP in drug resistant epilepsy is still relatively unknown.

3. INVOLVEMENT OF MDR TRANSPORTERS IN REFRACTORY EPILEPSY

Treatment of epilepsy is generally effective in controlling seizures. However, a large proportion (30%) of patients fail to respond or are inadequately responsive to treatment with anticonvulsant drugs *(1)*. Several anticonvulsant drugs have been demonstrated to be substrates of P-gp and/or MRP (Table 2). Thus activity of P-gp and MRP likely play a major role in this resistance. Indeed, overexpression of *MDR1*/P-gp has been frequently detected in brain tissues of patients with drug-resistant epilepsy *(27,29,72,73)*. Moreover, genetic polymorphisms of *MDR1* have been linked to drug-resistant epilepsy, although data in this study suggest involvement of multiple factors *(74)*. Microarray analysis of RNA obtained from endothelial cells isolated from temporal lobe blood vessels of patients with refractory epilepsy have demonstrated an induction of *MDR1*, *MRP2*, and *MRP5* mRNA in addition to increased levels of immunoreactive MRP1 *(73)*, suggesting involvement of multiple ABC transporters in drug resistance. Likewise, increased levels of immunoreactivity to P-gp, MRP1, and MRP2 were seen in the hippocampus isolated from patients with drug resistant temporal lobe epilepsy *(27,29)*. Interestingly, observed specificity in the induction of transporters in different cell types, including neurons and astrocytic foot end processes, suggests that each transporter may play a unique role in drug resistance, affecting drug penetration at different levels. Likewise, high levels of P-gp and MRP1 have also been detected in abnormal neurons, astrocytes, and microglia of pediatric patients with refractory epilepsy *(75)*.

Experimental animal models of epilepsy such as amygdala-kindled rats are also associated with increases in *mdr1a*/P-gp expression *(76–79)*. Using the amygdala-kindling model of temporal lobe epilepsy, Potschka et al. *(80)* demonstrated dramatic increases in P-gp expression in the kindled focus of phenytoin-resistant as compared with sensitive rats. In a separate study by these authors, inhibition of MRP2 in kindled rats by probenecid significantly increased the efficacy of phenytoin *(69)*. Moreover, phenytoin exerted a higher anticonvulsant effect in MRP2-deficient rats. Overall this indicates that P-gp and MRP2 contribute to the eventual activity of antiepileptic drugs by restricting their accumulation in the brain, and are thus likely to contribute to drug resistance in epilepsy. On the other hand, CNS expression of other MRP transporters in experimental models of epilepsy has not yet been examined. Clearly, there is a need to further elucidate their role and involvement in conferring drug resistance.

One question that must be considered when discussing involvement of the ABC transporters in epilepsy is whether the overexpression occurs because of the chronic exposure to antiepileptic drug therapy, or because of the events resulting from episodes of epileptic seizures and/or owing to underlying pathophysiological changes in epileptic patients. Moreover, whether or not MDR/MRP overexpression is a factor in the etiology of epilepsy could also be contemplated. Evidence from experimental animal models of epilepsy suggests that overexpression of the MDR genes is a complex phenomenon, influenced by exposure to both chemotherapeutics and seizure activity. Induction of MDR does not appear to require prior drug exposure. Acute seizures in genetic epilepsy-prone, drug-naive rats elevated expression of the *mdr1a* gene for more than 7 d *(81)*. Studies in mice also observed increases in *mdr1* expression within hours following kainic-acid induced seizures, whereas repeated doses of phenytoin or carbamazepine were not found to affect *mdr1* expression *(77)*. Likewise, Wang et al. *(82)* observed significant increases in P-gp expression in kindled rats, which had not received antiepileptic drug therapy. Although it could be argued that only transient elevations in *mdr*/P-gp occur in these animal models, it is plausible that frequent and/or repetitive seizures may lead to chronic changes in P-gp/*mdr* regulation. Indeed, induction of *mdr1* mRNA levels is seen in limbic areas of rats after chronic epileptic activity *(77)*. On the other hand, chronic exposure to drug therapy is also likely to contribute to drug resistance as kindled rats receiving antiepileptic drug therapy expressed higher levels of P-gp than drug naive controls *(82)*. Thus multiple factors including both chronic drug exposure and chronic seizure activity likely play a role in P-gp overexpression.

Reports indicating overexpression of MRP1 and MRP2 transporters in human epileptic tissue are not surprising as these MDR genes have many overlapping regulatory pathways with MDR1. However, whereas we can surmise that seizure and/or drug exposure are likely involved in MRP1 and MRP2 induction, little is known about the molecular pathways involved. As apoptosis and cell cycling have been thought to potentially play an important role in the multidrug resistant phenotype of tumors and in both P-gp and MRP induction, much attention has recently been focused on the involvement of these pathways in epilepsy.

4. RELATIONSHIP BETWEEN p53 EXPRESSION AND MDR

It has been proposed that similar to tumor cells, re-entry into the cell cycle and abnormalities in the p53 signaling pathways are triggers for overexpression of the MDR genes in epilepsy *(83,110)*. Apoptosis or programmed cell death is an essential mechanism to eliminate potentially dangerous cells. The tumor suppressor gene, *p53*, is considered to be a key transcription factor in cell apoptosis whereby activation of *p53* is able to regulate the expression of numerous genes involved in either apoptosis or cell cycle arrest *(84)*. Under normal homeostatic conditions, the expression and activity of *p53* is low; however, it becomes highly elevated on exposure to stress stimuli such as hypoxia, DNA damage, and oncogene activation *(84)*. In this manner, *p53* plays a vital role in activating signaling pathways aimed at repairing or preventing further damage, which could result in tumorigenesis. Indication of a role for p53 in the mechanism of drug resistance was first brought forth by Lowe et al. *(85)* who showed that immuno-compromised mice injected with tumors lacking the *p53* gene exhibited resistance to treatment with adriamycin whereas tumors carrying the wild-type *p53* gene were responsive and underwent apoptosis. Numerous carcinoma cell lines expressing a nonfunctional or mutant *p53* have since been demonstrated to lack sensitivity to chemotherapeutic drugs *(86,87)*. Moreover, mutations or absence of functional *p53* have been correlated with clinical findings of tumor resistance *(83)*. Overall, these studies provide firm evidence that the onset of drug resistance is dependent at least in part on *p53* status or function.

Promoter construct studies have demonstrated that transfection of cells with mutant *p53* induces *mdr1* and confers resistance to P-gp substrates, whereas transfection with wild-type *p53* represses *mdr1* promoter activity *(88–90)*. Using *p53* knockout mice, Bush and Li *(91)* found that basal expression of *mdr1a* and *mdr1b* mRNA in various tissues were upregulated as compared with wild-type mice, which could be attenuated by inclusion of a wild-type *p53* expression plasmid. P-gp induction was correlated with increased resistance to the P-gp substrates doxorubicin and vincristine. Clinical findings are also in agreement with the in vitro data *(92,93)*. Studies aimed at elucidating the signaling pathways involved have examined the impact of p53 mutation on interaction with *mdr1* promoter constructs. Induction of *mdr1* by mutant *p53* was found to occur via interactions of *p53* with distinct regions of the *mdr1* promoter. Studies have demonstrated that wild-type *p53* inhibition of reporter gene activity could be linked to the promoter regions between –2 to +133 and –189 to +133, which could be activated and inhibited by mutant p53, respectively *(94,95)*. Furthermore, the activity of the mutant p53 was dependent on its interaction with the transcription factor ETS-1 *(96)*. This suggests a gain of function mutation that activates transcription via the downstream promoter of *mdr1*. It has also been observed that the majority of *p53* mutants with gain of function properties have mutations within the DNA-binding domain *(97)*, which could therefore affect recognition of site-specific binding sequences on the *mdr1* gene, resulting in modulation of *mdr1* expression. Indeed, the transcriptional activation and oligomerization functions of p53 are required for *mdr1* induction by mutant p53, further indicating that mutations lead to a gain of function *(98,99)*. However the impact of p53 on the expression of P-gp is still somewhat controversial. Whereas the majority of studies have demonstrated an association between loss of p53 function with *mdr1* induction, Goldsmith et al. *(100)* reported activation of *mdr1* transcription by wild-type *p53*, whereas mutant *p53* slightly suppressed activity. The reason for these contrasting result is not clear, but may be owing to differences in the cell lines or the specific *p53* mutation examined. Indeed, Strauss and Haas *(94)* showed that only some *p53* mutants were able to activate *mdr1* transcription at a specific promoter site, thus indicating that individual mutations can function differently.

The expression of MRP1 has also been shown to be affected by *p53* status in a manner similar to that regulating P-gp. Numerous studies have reported that wild-type *p53* suppresses human and rodent MRP1 promoter activity *(101,102)*, whereas mutant *p53* was associated with induced MRP1 expression *(103–105)*. On the other hand, no effect of mutant p53 on MRP1 expression was found in the human *p53*-null cell lines Saos-2 (osteosarcoma) and Caco-2 (colon carcinoma) *(96)*. The reason for this discrepancy remains unclear. Wang and Beck *(101)* showed that transactivation of *MRP1* by the transcription factor Sp1 could be attenuated by wild-type *p53*, suggesting that interaction between Sp1 and p53 occurs through a negative transcriptional regulatory mechanism. Hence, decreases in p53 expression or activity during disease may therefore contribute to *MRP1* induction through increases in Sp1 activity. With regards to oxidative stress, p53 status has been found to play a role in the induction of MRP1 and MRP3, but not MRP2 by pro-oxidants *(106)*. On the other hand, the role of p53 in MRP2 or MRP3 transcriptional regulation has not been reported. Clearly, further investigation into the role of p53 in the regulation of the MRP transporters is needed.

5. RELATIONSHIP BETWEEN p53 EXPRESSION AND NEURONAL VIABILITY IN EPILEPSY

The response of neurons to injury is highly dependent on the expression and activity of *p53*. Whereas induction of *p53* has been correlated with neuronal injury *(107)*, the absence or inhibition of *p53* has been shown to confer neuroprotection to seizure-induced apoptosis *(108,109)*. Furthermore, there is a possibility that modulation of *p53* and subsequently the expression of MDR transporters in neurons and astrocytes could be a result of the underlying brain dysfunction. Clearly, p53 plays a major role in deciding the fate of a neuron and therefore the health of the brain. Modulation of the expression or function of p53 leads to the induction of P-gp and MRP, which contributes to the onset of MDR. Pharmacoresistant epilepsy is also seen in patients with pre-existing brain malformations such as cortical dysplasia *(135,136)* and may be linked to p53 modulation. Intriguingly, Marroni et al. *(110)* showed that induction of P-gp and MRP1 protein in human epileptic astrocytes is coupled with a loss of p53 in the astrocytes, but paradoxically, epilepsy is associated with an induced expression of p53 in other cell types resulting in neuron damage and apoptosis *(107,111,112)*. This apparent contradiction in the role of p53 in the brain may be owing to a cell type-dependent differential effect of seizures on p53 expression. According to the model proposed by Marroni et al. *(110)*, the seizure-induced increase in neuronal p53 levels contributes to neuronal apoptosis and furthers the epileptic phenotype. On the other hand, the concurrent loss of p53 in epileptic astrocytes prevents the onset of apoptosis and contributes to the MDR phenotype, with the possibility of the initiation of tumorigenesis. Thus, attempts to treat epilepsy by modulating p53 expression face a predicament in that increased p53 levels contribute to seizure-induced neuronal apoptosis whereas the loss of p53 in glial cells may lead to drug resistance.

6. OTHER REGULATORY PATHWAYS

Over the last decade, evidence that inflammatory processes are involved in seizures and epilepsy has accrued. Infectious diseases and inflammatory disorders are associated with an increased incidence of seizures in humans. In experimental animal models, systemic exposure to bacterial endotoxin has been shown to enhance seizure susceptibility *(113)*. Increased levels of proinflammatory cytokines such as interleukin (IL)-1β, IL-6, and tumor necrosis factor-α (TNF-α) are found in the CSF of patients and in experimental models of epilepsy *(114–116)*. Therefore one may contemplate whether cytokines play a role in the etiology or progression of epilepsy. Whereas activation of the cytokine network has been clearly associated with stimulation of febrile seizures *(115,117)*, epileptic seizures themselves provoke an acute phase response resulting in a cascade of cytokine release *(118)*. Moreover, inflammatory mediators released during this acute phase response include IL-1β, which has been linked to epileptic activity. Whereas significant induction of IL-1β mRNA are seen in the amygdala and hippocampus of kindled rats within hours after seizure *(119)*, intrahippocampal injections of IL-1β have been shown to have proconvulsant effects in experimental models *(116)*. Hence IL-1β may be involved in altering seizure threshold and induction. Underlying brain malformations in epileptic patients may also contribute to cytokine activation.

Cytokine induction in epileptic seizures may contribute to the development of MDR. In human hepatomas and rat hepatocytes, proinflammatory cytokines such as IL-1β, IL-6, and TNF-α have been shown to modulate expression of P-gp and the MRP transporters *(120–123)*. Likewise, induction of either CNS-localized or systemic inflammation alters the expression of P-gp in the brain *(124)*. Whereas pathways for cytokine-mediated changes in the expression of the MDR genes may occur distinct from p53-mediated routes, it is interesting that cytokine-activated transcription factors such as nuclear factor-κB (NF-κB) play an important role in the regulation of both p53-dependent and -independent signaling pathways. NF-κB, which is activated by numerous stimuli including cytokines and environmental toxins, plays an essential role in the regulation of genes which govern apoptosis, inflammation, cell cycle transition, and carcinogenesis. The relationship between NF-κB and apoptosis has been extensively examined over many years *(125)*. To date, evidence exists which suggests an antiapoptotic effect of NF-κB in neurons, which may prevent neuronal degeneration in Alzheimer's and ischemic diseases *(126–128)*. Moreover, observations of cross-competition between NF-κB and p53 have been reported *(129–131)*. Of note is that response elements for NF-κB have been located in the promoter region of human *mdr1 (132)*. The presence of an NF-κB-binding site together with an adjacent p53-response element in the rat *mdr1b* promoter is also required for basal promoter activity *(133)*. Whereas it is highly plausible that activation of NF-κB in epilepsy may play an early role in the development of drug-resistance through cross-competition with p53 and MDR induction, the detailed interconnections among NF-κB, apoptosis, cell cycle and MDR are still undefined. Although more studies are clearly needed, one could hypothesize that inhibition of NF-κB in astrocytes through specific inhibitors or antisense technologies could prevent or circumvent the development of drug resistance in epilepsy.

7. CONCLUSIONS

As the involvement of P-gp and the MRP transporters in drug resistant epilepsy is becoming more apparent, further studies examining their role in the transport of antiepileptic drugs are warranted. A relationship between epilepsy and transporter regulation by p53 and NF-κB has been established, and this has led to a much greater understanding of the complex interplay between disease, gene regulation, apoptosis, and drug resistance. Although a clearer picture of the mechanism of MDR in epilepsy is emerging, more research is needed to determine ways of modulating the expression and/or activity of transporters in order to overcome resistance. Knowledge of transporter regulation can significantly contribute to the development of specific transporter modulators, which can be used to optimize antiepilepsy drug therapy.

REFERENCES

1. Regesta G, Tanganelli P. Clinical aspects and biological bases of drug-resistant epilepsies. Epilepsy Res 1999;34:109–122.
2. Dano K. Cross resistance between vinca alkaloids and anthracyclines in Ehrlich ascites tumor in vivo. Cancer Chemother Rep 1972;56:701–708.
3. Ling V. Multidrug resistance: molecular mechanisms and clinical relevance. Cancer Chemother Pharmacol 1997;40:S3–S8.
4. Juliano RL, Ling V. A surface glycoprotein modulating drug permeability in Chinese hamster ovary cell mutants. Biochim Biophys Acta 1976;455:152–162.
5. Litman T, Druley TE, Stein WD, Bates SE. From MDR to MXR: new understanding of multidrug resistance systems, their properties and clinical significance. Cell Mol Life Sci 2001;58:931–959.
6. Zhang W, Mojsilovic-Petrovic J, Andrade MF, Zhang H, Ball M, Stanimirovic DB. Expression and functional characterization of ABCG2 in brain endothelial cells and vessels. FASEB J 2003;17: 2085–2087.
7. Rao VV, Dahlheimer JL, Bardgett ME, et al. Choroid plexus epithelial expression of MDR1 P glycoprotein and multidrug resistance-associated protein contribute to the blood–cerebrospinal-fluid drug-permeability barrier. Proc Natl Acad Sci USA 1999;96:3900–3905.
8. Golden PL, Pardridge WM. P-Glycoprotein on astrocyte foot processes of unfixed isolated human brain capillaries. Brain Res 1999;819:143–146.
9. Ballerini P, DiIorio P, Ciccarelli R, et al. Glial cells express multiple ATP binding cassette proteins which are involved in ATP release. Neuroreport 2002;13:1789–1792.
10. Lee G, Schlichter L, Bendayan M, Bendayan R. Functional expression of P-glycoprotein in rat brain microglia. J Pharmacol Exp Ther 2001;299:204–212.
11. Schinkel AH, Smit JJ, van Tellingen O, et al. Disruption of the mouse mdr1a P-glycoprotein gene leads to a deficiency in the blood–brain barrier and to increased sensitivity to drugs. Cell 1994;77:491–502.
12. Kusuhara H, Suzuki H, Terasaki T, Kakee A, Lemaire M, Sugiyama Y. P-Glycoprotein mediates the efflux of quinidine across the blood–brain barrier. J Pharmacol Exp Ther 1997;283:574–580.
13. Xie R, Hammarlund-Udenaes M, de Boer AG, de Lange EC. The role of P-glycoprotein in blood–brain barrier transport of morphine: transcortical microdialysis studies in mdr1a (–/–) and mdr1a (+/+) mice. Br J Pharmacol 1999;128:563–568.
14. Thompson SJ, Koszdin K, Bernards CM. Opiate-induced analgesia is increased and prolonged in mice lacking P-glycoprotein. Anesthesiology 2000;92:1392–1399.
15. Gallo JM, Li S, Guo P, Reed K, Ma J. The effect of P-glycoprotein on paclitaxel brain and brain tumor distribution in mice. Cancer Res 2003;63:5114–5117.
16. Polli JW, Jarrett JL, Studenberg SD, et al. Role of P-glycoprotein on the CNS disposition of amprenavir (141W94), an HIV protease inhibitor. Pharm Res 1999;16:1206–1212.
17. Cole SP, Bhardwaj G, Gerlach JH, et al. Overexpression of a transporter gene in a multidrug-resistant human lung cancer cell line. Science 1992;258:1650–1654.
18. Haimeur A, Conseil G, Deeley RG, Cole SP. The MRP-related and BCRP/ABCG2 multidrug resistance proteins: biology, substrate specificity and regulation. Curr Drug Metab 2004;5:21–53.
19. Roelofsen H, Vos TA, Schippers IJ, et al. Increased levels of the multidrug resistance protein in lateral membranes of proliferating hepatocyte-derived cells. Gastroenterology 1997;112:511–521.
20. Filipits M, Malayeri R, Suchomel RW, et al. Expression of the multidrug resistance protein (MRP1) in breast cancer. Anticancer Res 1999;19:5043–5049.

21. Meijer GA, Schroeijers AB, Flens MJ, et al. Increased expression of multidrug resistance related proteins Pgp, MRP1, and LRP/MVP occurs early in colorectal carcinogenesis. J Clin Pathol 1999; 52:450–454.
22. Young LC, Campling BG, Cole SP, Deeley RG, Gerlach JH. Multidrug resistance proteins MRP3, MRP1, and MRP2 in lung cancer: correlation of protein levels with drug response and messenger RNA levels. Clin Cancer Res 2001;7:1798–1804.
23. Itoh Y, Tamai M, Yokogawa K, et al. Involvement of multidrug resistance-associated protein 2 in in vivo cisplatin resistance of rat hepatoma AH66 cells. Anticancer Res 2002;22:1649–1653.
24. Zeng H, Bain LJ, Belinsky MG, Kruh GD. Expression of multidrug resistance protein-3 (multispecific organic anion transporter-D) in human embryonic kidney 293 cells confers resistance to anticancer agents. Cancer Res 1999;59:5964–5967.
25. Leggas M, Adachi M, Scheffer GL, et al. Mrp4 confers resistance to topotecan and protects the brain from chemotherapy. Mol Cell Biol 2004;24:7612–7621.
26. Wijnholds J, deLange EC, Scheffer GL, et al. Multidrug resistance protein 1 protects the choroid plexus epithelium and contributes to the blood–cerebrospinal fluid barrier. J Clin Invest 2000;105:279–285.
27. Aronica E, Gorter JA, Jansen GH, et al. Expression and cellular distribution of multidrug transporter proteins in two major causes of medically intractable epilepsy: focal cortical dysplasia and glioneuronal tumors. Neuroscience 2003;118:417–429.
28. Spiegl-Kreinecker S, Buchroithner J, Elbling L, et al. Expression and functional activity of the ABC-transporter proteins P-glycoprotein and multidrug-resistanc protein 1 in human brain tumor cells and astrocytes. J Neuro-Oncol 2002;57:27–36.
29. Aronica E, Gorter JA, Ramkema M, et al. Expression and cellular distribution of multidrug resistance-related proteins in the hippocampus of patients with mesial temporal lobe epilepsy. Epilepsia 2004;45:441–451.
30. Seetharaman S, Barrand MA, Maskell L, Scheper RJ. Multidrug resistance-related transport proteins in isolated human brain microvessels and in cells cultured from these isolates. J Neurochem 1998;70:1151–1159.
31. Choudhuri S, Cherrington NJ, Li N, Klaassen CD. Constitutive expression of various xenobiotic and endobiotic tranporter mRNAs in the choroid plexus of rats. Drug Metab Disp 2003;31:1337–1345.
32. Decleves X, Regina A, Laplanche JL, et al. Functional expression of P-glycoprotein and multidrug resistance-associated protein (Mrp1) in primary cultures of rat astrocytes. J Neurosci Res 2000;60:594–601.
33. Hosoya KI, Takashima T, Tetsuka K, et al. mRNA expression and transport characterization of conditionally immortalized rat brain capillary endothelial cell lines; a new in vitro BBB model for drug targeting. J Drug Target 2000;8:357–370.
34. Brady JM, Cherrington NJ, Hartley DP, Buist SC, Li N, Klaassen CD. Tissue distribution and chemical induction of multiple drug resistance genes in rats. Drug Metab Disp 2002;30:838–844.
35. Berezowski V, Landry C, Dehouck MP, Cecchelli R, Fenart L. Contribution of glial cells and pericytes to the mRNA profiles of P-glycoprotein and multidrug-resistance associated proteins in an in vitro model of the blood–brain barrier. Brain Res 2004;1018:1–9.
36. Zhang Y, Han H, Elmquist WF, Miller DW. Expression of various multidrug resistance-associated protein (MRP) homologues in brain microvessel endothelial cells. Brain Res 2000;876:148–153.
37. Dallas S, Zhu X, Baruchel S, Schlichter L, Bendayan R. Functional expression of the multidrug resistance protein 1 in microglia. J Pharmacol Exp Ther 2003;307:282–290.
38. Nishino J, Suzuki H, Sugiyama D, et al. Transepithelial transport of organic anions across the choroid plexus: possible involvement of organic anion transporter and multidrug resistance-associated protein. J Pharmacol Exp Ther 1999;290:289–294.
39. Miller DS, Nobmann SN, Gutmann H, Toeroek M, Drewe J, Fricker G. Xenobiotic transport across isolated brain microvessels studied by confocal microscopy. Mol Pharmacol 2000;58:1357–1367.
40. Regina A, Koman A, Piciotti M, et al. Mrp1 multidrug resistance-associated protein and P-glycoprotein expression in rat brain microvessel endothelial cells. J Neurochem 1998;71:705–715.
41. Sugiyama Y, Kusuhara H, Suzuki H. Kinetic and biochemical analysis of carrier-mediated efflux of drugs through the blood–brain and blood–cerebrospinal fluid barriers: importance in the drug delivery to the brain. J Control Release 1999;62:179–186.
42. Cherrington NJ, Hartley DP, Li N, Johnson DR, Klaassen CD. Organ distribution of multidrug resistance proteins 1, 2, and 3 (Mrp 1, 2, and 3) mRNA and hepatic induction of Mrp3 by constitutive androstane receptor activators in rats. J Pharmacol Exp Ther 2002;300:97–104.

43. Lee YJ, Kusuhara H, Sugiyama Y. Do multidrug resistance-associated protein-1 and -2 play any role in the elimination of estradiol-17 beta-glucuronide and 2,4-dinitrophenyl-S-glutathione across the blood–cerebrospinal fluid barrier? J Pharm Sci 2004;93:99–107.
44. Hirrlinger J, Konig J, Keppler D, Lindenau J, Schulz JB, Dringen R. The multidrug resistance protein MRP1 mediates the release of glutathione disulfide from rat astrocytes during oxidative stress. J Neurochem 2001;76:627–636.
45. Hirrlinger J, Konig J, Dringen R. Expression of mRNAs of multidrug resistance proteins (Mrps) in cultured rat astrocytes, oligodendrocytes, microglial cells and neurones. J Neurochem 2002;82: 716–719.
46. Sugiyama D, Kusuhara H, Lee YJ, Sugiyama Y. Involvement of multidrug resistance associated protein 1 (Mrp1) in the efflux transport of 17beta estradiol-D-17beta-glucuronide (E217betaG) across the blood–brain barrier. Pharm Res 2003;20:1394–1400.
47. Dallas S, Schlichter L, Bendayan, R. Multidrug resistance protein (MRP) 4- and MRP 5-mediated efflux of 9-(2-phosphonylmethoxyethyl)adenine by microglia. J Phramacol Exp Ther 2004;309: 1221–1229.
48. Hopper E, Belinsky MG, Zeng H, Tosolini A, Testa JR, Kruh GD. Analysis of the structure and expression pattern of MRP7 (ABCC10), a new member of the MRP subfamily. Cancer Lett 2001;162:181–191.
49. Bera TK, Lee S, Salvator G, Lee B, Pastan I. MRP8, a new member of ABC transporter superfamily, identified by EST database mining and gene prediction program, is highly expressed in breast cancer. Mol Med 2001;7:509–516.
50. Cooray HC, Blackmore CG, Maskell L, Barrand MA. Localisation of breast cancer resistance protein in microvessel endothelium of human brain. Neuroreport 2002;13:2059–2063.
51. Eisenblatter T, Huwel S, Galla HJ. Characterisation of the brain multidrug resistance protein (BMDP/ABCG2/BCRP) expressed at the blood–brain barrier. Brain Res 2003;971:221–231.
52. Cisternino S, Mercier C, Bourasset F, Roux F, Scherrmann JM. Expression, up-regulation, and transport activity of the multidrug-resistance protein Abcg2 at the mouse blood–brain barrier. Cancer Res 2004;64:3296–3301.
53. Zhang Y, Schuetz JD, Elmquist WF, Miller DW. Plasma membrane localization of multidrug resistance-associated protein (MRP) homologues in brain capillary endothelial cells. J Pharmacol Exp Ther 2004;311:449–455.
54. Torok M, Huwyler J, Gutmann H, Fricker G, Drewe J. Modulation of transendothelial permeability and expression of ATP-binding cassette transporters in cultured brain capillary endothelial cells by astrocytic factors and cell-culture conditions. Exp Brain Res 2003;153:356–365.
55. Allikmets R, Schriml LM, Hutchinson A, Romano-Spica V, Dean M. A human placenta-specific ATP-binding cassette gene (ABCP) on chromosome 4q22 that is involved in multidrug resistance. Cancer Res 1998;58:5337–5339.
56. Doyle LA, Yang W, Abruzzo LV, et al. A multidrug resistance transporter from human MCF-7 breast cancer cells. Proc Natl Acad Sci USA 1998;95:15,665–15,670.
57. Miyake K, Mickley L, Litman T, et al. Molecular cloning of cDNAs which are highly overexpressed in mitoxantrone-resistant cells: demonstration of homology to ABC transport genes. Cancer Res 1999;59:8–13.
58. Doyle LA, Ross DD. Multidrug resistance mediated by the breast cancer resistance protein BCRP (ABCG2). Oncogene 2003;22:7340–7358.
59. Jonker JW, Buitelaar M, Wagenaar E, et al. The breast cancer resistance protein protects against a major chlorophyll-derived dietary phototoxin and protoporphyria. Proc Natl Acad Sci USA 2002;99:15,649–15,654.
60. Mizuno N, Suzuki M, Kusuhara H, et al. Impaired renal excretion of 6-hydroxy-5,7-dimethyl-2-methylamino-4-(3-pyridylmethyl) benzothiazole (E3040) sulfate in breast cancer resistance protein (BCRP1/ABCG2) knockout mice. Drug Metab Dispos 2004;32:898–901.
61. Van Herwaarden AE, Jonker JW, Wagenaar E, et al. The breast cancer resistance protein (Bcrp1/Abcg2) restricts exposure to the dietary carcinogen 2-amino-1-methyl-6-phenylimidazo[4,5-b]pyridine. Cancer Res 63:6447–6452.
62. Maliepaard M, Scheffer GL, Faneyte IF, et al. Subcellular localization and distribution of the breast cancer resistance protein transporter in normal human tissues. Cancer Res 2001;61:3458–3464.
63. Wang X, Nitanda T, Shi M, et al. Induction of cellular resistance to nucleoside reverse transcriptase inhibitors by the wild-type breast cancer resistance protein. Biochem Pharmacol 2004;68:1363–1370.

64. Tews DS, Nissen A, Kulgen C, Gaumann AK. Drug resistance-associated factors in primary and secondary glioblastomas and the precursor tumors. J Neuro-Oncol 2000;50:227–237.
65. Sisodiya SM, Martinian L, Scheffer GL, et al. Major vault protein, a marker of drug resistance, is upregulated in refractory epilepsy. Epilepsia 2003;44:1388–1396.
66. Potschka H, Fedrowitz M, Loscher W. P-glycoprotein and multidrug resistance-associated protein are involved in the regulation of extracellular levels of the major antiepileptic drug carbamazepine in the brain. Neuroreport 2001;12:3557–3560.
67. Owen A, Pirmohamed M, Tettey JN, Morgan P, Chadwick D, Park BK. Carbamazepine is not a substrate for P-glycoprotein. Br J Clin Pharmacol 2001;51:345–349.
68. Potschka H, Fedrowitz M, Loscher W. P-Glycoprotein-mediated efflux of phenobarbital, lamotrigine, and felbamate at the blood–brain barrier: evidence from microdialysis experiments in rats. Neurosci Lett 2002;327:173–176.
69. Potschka H, Fedrowitz M, Loscher W. Multidrug resistance protein MRP2 contributes to blood–brain barrier function and restricts antiepileptic drug activity. J Pharmacol Exp Ther 2003;306:124–131.
70. Sills GJ, Kwan P, Butler E, de Lange EC, van den Berg DJ, Brodie MJ. P-glycoprotein-mediated efflux of antiepileptic drugs: preliminary studies in mdr1a knockout mice. Epilepsy Behav 2002;3:427–432.
71. Gibbs JP, Adeyeye MC, Yang Z, Shen DD. Valproic acid uptake by bovine brain microvessel endothelial cells: role of active efflux transport. Epilepsy Res 2004;58:53–66.
72. Tishler DM, Weinberg KI, Hinton DR, Barbaro N, Annett GM, Raffel C. MDR1 gene expression in brain of patients with medically intractable epilepsy. Epilepsia 1995;36:1–6.
73. Dombrowski SM, Desai SY, Marroni M, et al. Overexpression of multiple drug resistance genes in endothelial cells from patients with refractory epilepsy. Epilepsia 2001;42:1501–1506.
74. Siddiqui A, Kerb R, Weale ME, et al. Association of multidrug resistance in epilepsy with a polymorphism in the drug-transporter gene ABCB1. N Engl J Med 2003;348:1442–1448.
75. Lazarowski A, Massaro M, Schteinschnaider A, Intruvini S, Sevlever G, Rabinowicz A. Neuronal MDR-1 gene expression and persistent low levels of anticonvulsants in a child with refractory epilepsy. Ther Drug Monit 2004;26:44–46.
76. Zhang L, Ong WY, Lee T. Induction of P-glycoprotein expression in astrocytes following intracerebroventricular kainate injections. Exp Brain Res 1999;126:509–516.
77. Rizzi M, Caccia S, Guiso G, et al. Limbic seizures induce P-glycoprotein in rodent brain: functional implications for pharmacoresistance. J Neurosci 2002;22:5833–5839.
78. Lazarowski A, Ramos AJ, Garcia-Rivello H, Brusco A, Girardi E. Neuronal and glial expression of the multidrug resistance gene product in an experimental epilepsy model. Cell Mol Neurobiol 2004;24:77–85.
79. Volk HA, Potschka H, Loscher W. Increased expression of the multidrug transporter P-glycoprotein in limbic brain regions after amygdala-kindled seizures in rats. Epilepsy Res 2004;58:67–79.
80. Potschka H, Volk HA, Loscher W. Pharmacoresistance and expression of multidrug transporter P-glycoprotein in kindled rats. Neuroreport 2004;15:1657–1661.
81. Kwan P, Sills GJ, Butler E, Gant TW, Meldrum BS, Brodie MJ. Regional expression of multidrug resistance genes in genetically epilepsy-prone rat brain after a single audiogenic seizure. Epilepsia 2002;43:1318–1323.
82. Wang Y, Zhou D, Wang B, et al. A kindling model of pharmacoresistant temporal lobe epilepsy in Sprague-Dawley rats induced by Coriaria lactone and its possible mechanism. Epilepsia 2003;44:475–488.
83. Bush JA, Li G. Cancer chemoresistance: the relationship between p53 and multidrug transporters. Int J Cancer 2002;98:323–330.
84. Oren M. Decision making by p53: life, death and cancer. Cell Death Differ 2003;10:431–442.
85. Lowe SW, Bodis S, McClatchey A, et al. p53 status and the efficacy of cancer therapy in vivo. Science 1994;266:807–810.
86. Keshelava N, Zuo JJ, Chen P, et al. Loss of p53 function confers high-level multidrug resistance in neuroblastoma cell lines. Cancer Res 2001;61:6185–6193.
87. Perego P, Giarola M, Righetti SC, et al. Association between cisplatin resistance and mutation of p53 gene and reduced bax expression in ovarian carcinoma cell systems. Cancer Res 1996;56:556–562.
88. Chin KV, Ueda K, Pastan I, Gottesman MM. Modulation of activity of the promoter of the human MDR1 gene by Ras and p53. Science 1992;255:459–462.
89. Zastawny RL, Salvino R, Chen J, Benchimol S, Ling V. The core promoter region of the P-glycoprotein gene is sufficient to confer differential responsiveness to wild-type and mutant p53. Oncogene 1993;8:1529–1535.

90. Thottassery JV, Zambetti GP, Arimori K, Schuetz EG, Schuetz JD. p53-dependent regulation of MDR1 gene expression causes selective resistance to chemotherapeutic agents. Proc Natl Acad Sci USA 1997;94:11,037–11,042.
91. Bush JA Li G. Regulation of the Mdr1 isoforms in a p53-deficient mouse model. Carcinogenesis 2002;23:1603–1607.
92. de Kant E, Heide I, Thiede C, Herrmann R, Rochlitz CF. MDR1 expression correlates with mutant p53 expression in colorectal cancer metastases. J Cancer Res Clin Oncol 1996;122:671–675.
93. Schneider J, Rubio MP, Barbazan MJ, Rodriguez-Escudero FJ, Seizinger BR, Castresana JS. P-glycoprotein, HER-2/neu, and mutant p53 expression in human gynecologic tumors. J Natl Cancer Inst 1994;86:850–855.
94. Strauss BE, Haas M. The region 3′ to the major transcriptional start site of the MDR1 downstream promoter mediates activation by a subset of mutant P53 proteins. Biochem Biophys Res Commun 1995;217:333–340.
95. Strauss BE, Shivakumar C, Deb SP, Deb S, Haas M. The MDR1 downstream promoter contains sequence-specific binding sites for wild-type p53. Biochem Biophys Res Commun 217:825–831.
96. Sampath J, Sun D, Kidd VJ, et al. Mutant p53 cooperates with ETS and selectively up-regulates human MDR1 not MRP1. J Biol Chem 2001;276:39,359–39,367.
97. Pfeifer GP. P53 mutational spectra and the role of methylated CpG sequences. Mutat Res 2000;450:155–166.
98. Lin J, Teresky AK, Levine AJ. Two critical hydrophobic amino acids in the N-terminal domain of the p53 protein are required for the gain of function phenotypes of human p53 mutants. Oncogene 1995;10:2387–2390.
99. Lanyi A, Deb D, Seymour RC, Ludes-Meyers JH, Subler MA, Deb S. "Gain of function" phenotype of tumor-derived mutant p53 requires the oligomerization/nonsequence-specific nucleic acid-binding domain. Oncogene 1998;18:3169–3176.
100. Goldsmith ME, Gudas JM, Schneider E, Cowan KH. Wild type p53 stimulates expression from the human multidrug resistance promoter in a p53-negative cell line. J Biol Chem 1995;270:1894–1898.
101. Wang Q, Beck WT. Transcriptional suppression of multidrug resistance-associated protein (MRP) gene expression by wild-type p53. Cancer Res 1998;15:5762–5769.
102. Muredda M, Nunoya K, Burtch-Wright RA, Kurz, EU, Cole SP, Deeley RG. Cloning and characterization of the murine and Rat mrp1 promoter regions. Mol Pharmacol 2003;64:1259–1269.
103. Sullivan GF, Yang JM, Vassil A, Yang J, Bash-Babula J, Hait WN. Regulation of expression of the multidrug resistance protein MRP1 by p53 in human prostate cancer cells. J Clin Invest 2000;105: 1261–1267.
104. Fukushima Y, Oshika Y, Tokunaga T, et al. Multidrug resistance-associated protein (MRP) expression is correlated with expression of aberrant p53 protein in colorectal cancer. Eur J Cancer 1999;35:935–938.
105. Tsang WP, Chau SP, Fung KP, Kong SK, Kwok TT. Modulation of multidrug resistance-associated protein 1 (MRP1) by p53 mutant in Saos-2 cells. Cancer Chemother Pharmacol 2003;51:161–166.
106. Lin-Lee YC, Tatebe S, Savaraj N, Ishikawa T, Kuo MT. Differential sensitivities of the MRP gene family and gamma-glutamylcysteine synthetase to prooxidants in human colorectal carcinoma cell lines with different p53 status. Biochem Pharmacol 2001;61:555–563.
107. Sakhi S, Bruce A, Sun N, Tocco G, Baudry M, Schreiber SS. p53 induction is associated with neuronal damage in the central nervous system. Proc Natl Acad Sci USA 1994;91:7525–7529.
108. Culmsee C, Zhu X, Yu QS, et al. A synthetic inhibitor of p53 protects neurons against death induced by ischemic and excitotoxic insults, and amyloid beta-peptide. J Neurochem 2001;77:220–228.
109. Xiang H, Hochman DW, Saya H, Fujiwara T, Schwartzkroin PA, Morrison RS. Evidence for p53-mediated modulation of neuronal viability. J Neurosci 1996;16:6753–6765.
110. Marroni M, Agrawal ML, Kight K, et al. Relationship between expression of multiple drug resistance proteins and p53 tumor suppressor gene proteins in human brain astrocytes. Neuroscience 2003;121:605–617.
111. Tan Z, Sankar R, Shin D, et al. Differential induction of p53 in immature and adult rat brain following lithium-pilocarpine status epilepticus. Brain Res 2002;928:187–193.
112. Chopp M, Li Y, Zhang ZG, Freytag SO. p53 expression in brain after middle cerebral artery occlusion in the rat. Biochem Biophys Res Commun 1992;14:1201–1207.
113. Sayyah M, Javad-Pour M, Ghazi-Khansari M. The bacterial endotoxin lipopolysaccharide enhances seizure susceptibility in mice: involvement of proinflammatory factors: nitric oxide and prostaglandins. Neuroscience 2003;122:1073–1080.

114. Peltola J, Hurme M, Miettinen A, Keranen T. Elevated levels of interleukin-6 may occur in cerebrospinal fluid from patients with recent epileptic seizures. Epilepsy Res 1998;31:129–133.
115. Virta M, Hurme M, Helminen M. Increased plasma levels of pro- and anti-inflammatory cytokines in patients with febrile seizures. Epilepsia 2002;43:920–923.
116. Vezzani A, Moneta D, Richichi C, et al. Functional role of inflammatory cytokines and anti-inflammatory molecules in seizures and epileptogenesis. Epilepsia 2002;43:30–35.
117. Haspolat S, Mihci E, Coskun M, et al. Interleukin-1beta, tumor necrosis factor-alpha, and nitrite levels in febrile seizures. J Child Neurol 2002;17:749–751.
118. Peltola J, Laaksonen J, Haapala AM, Hurme M, Rainesalo S, Keranen T. Indicators of inflammation after recent tonic-clonic epileptic seizures correlate with plasma interleukin-6 levels. Seizure 2002;11:44–46.
119. Plata-Salaman CR, Ilyin SE, Turrin NP, et al. Kindling modulates the IL-1beta system, TNF-alpha, TGF-beta1, and neuropeptide mRNAs in specific brain regions. Brain Res Mol Brain Res 2000;75:248–258.
120. Lee G, Piquette-Miller M. Influence of IL-6 on MDR and MRP-mediated multidrug resistance in human hepatoma cells. Can J Physiol Pharmacol 2001;79:876–884.
121. Lee G, Piquette-Miller M. Cytokines alter the expression and activity of the multidrug resistance transporters in human hepatoma cell lines; analysis using RT-PCR and cDNA microarrays. J Pharm Sci 2003;92:2152–2163.
122. Sukhai M, Yong A, Pak A, Piquette-Miller M. Decreased expression of P-glycoprotein in interleukin-1beta and interleukin-6 treated rat hepatocytes. Inflamm Res 2001;50:362–370.
123. Sukhai M, Yong A, Kalitsky J, Piquette-Miller, M. Inflammation and interleukin-6 mediate reductions in the hepatic expression and transcription of the mdr1a and mdr1b Genes. Mol Cell Biol Res Commun 2000;4:248–256.
124. Goralski KB, Hartmann G, Piquette-Miller M, Renton KW. Downregulation of mdr1a expression in the brain and liver during CNS inflammation alters the in vivo disposition of digoxin. Br J Pharmacol 2003;139:35–48.
125. Chen F, Castranova V, Shi X. New insights into the role of nuclear factor-kappaB in cell growth regulation. Am J Pathol 2001;159:387–397.
126. Bales KR, Dy Y, Dodel RC, Yan GM, Hamilton-Byrd E, Paul SM. The NF-kappaB/Rel family of proteins mediates Abeta-induced neurotoxicity and glial activation. Brain Res Mol Brain Res 1998;57:63–72.
127. Guo Q, Robinson N, Mattson MP. Secreted beta-amyloid precursor protein counteracts the proapoptotic action of mutant presenilin-1 by activation of NF-kappaB and stabilization of calcium homeostasis. J Biol Chem 1998;273:12,341–12,351.
128. Yu Z, Zhou D, Bruce-Keller AJ, Kindy MS, Mattson MP. Lack of the p50 subunit of nuclear factor-kappaB increases the vulnerability of hippocampal neurons to excitotoxic injury. J Neurosci 1999;19:8856–8865.
129. Shao J, Fujiwara T, Kadowaki Y, et al. Overexpression of the wild-type p53 gene inhibits NF-kappaB activity and synergizes with aspirin to induce apoptosis in human colon cancer cells. Oncogene 2000;19:726–736.
130. Ikeda A, Sun X, Li Y, et al. p300/CBP-dependent and -independent transcriptional interference between NF-kappaB RelA and p53. Biochem Biophys Res Commun 2000;272:375–379.
131. Pise-Masison CA, Mahieux R, Jiang H, Ashcroft M, et al. Inactivation of p53 by human T-cell lymphotropic virus type 1 Tax requires activation of the NF-kappaB pathway and is dependent on p53 phosphorylation. Mol Cell Biol 2000;20:3377–3386.
132. Labialle S, Gayet L, Marthinet E, Rigal D, Baggetto LG. Transcriptional regulators of the human multidrug resistance 1 gene: recent views. Biochem Pharmacol 2002;64:943–948.
133. Sukhai M, Piquette-Miller M. Regulation of the multidrug resistance genes by stress signals. J Pharm Pharm Sci 2000;3:268–280.
134. Marchi N, Hallene KL, Kight KM, et al. Significance of MDR1 and multiple drug resistance in refractory human epileptic brain. BMC Med 2004;2:37.
135. Aronica E, Gorter JA, Jansen GH, et al. Expression and cellular distribution of multidrug transporter proteins in two major causes of medically intractable epilepsy: focal cortical dysplasia and glioneuronal tumors. Neuroscience 2003;118:417–429.
136. Sisodiya SM, Lin WR, Harding BN, Squier MV, Thom M. Drug resistance in epilepsy: expression of drug resistance proteins in common causes of refractory epilepsy. Brain 2002;125:22–31.

28
Signaling Modules in Glial Tumors and Implications for Molecular Therapy

Gurpreet S. Kapoor, PhD and Donald M. O'Rourke, MD

SUMMARY

Gliomas (ependymomas are also considered gliomas) are the most common primary central nervous system tumors. They are graded on a scale of I–IV, based on their degree of malignancy as judged by variable histological features. Genetic and biochemical evidence have proven that gliomagenesis involves a stepwise accumulation of genetic lesions affecting either signal transduction pathways activated by receptor tyrosine kinases (RTKs) or cell cycle growth arrest pathways. Many of these observed molecular alterations are now being used to compliment clinical diagnosis. Genetic alterations affecting RTK signaling results in the activation of several downstream pathways, such as the phosphatidylinositol 3-kinase (PI3-K)/Akt and Ras/Raf/MEK/MAPK pathways, which provide a number of novel targets for glioma therapy. This chapter aims to present a broad understanding of the RTK signaling networks involved in gliomagenesis. Molecular classification of primary glial tumors and elucidation of cooperative interactions between different genetic lesions will eventually allow us to target distinct glioma subsets and will provide a more rational approach to adjuvant therapies.

Key Words: Glioma; receptor tyrosine kinase; epidermal growth factor receptor; platelet-derived growth factor receptor; Ras; PI3-kinase; Akt; PTEN; TGF-α; mitogen-activated protein kinase.

1. INTRODUCTION

Receptor tyrosine kinases (RTKs), which play a critical role in normal cell proliferation and differentiation, have been extensively studied for their possible role in gliomagenesis. RTKs constitute a family of at least 20 members containing an extracellular ligand-binding domain, a single-transmembrane domain, and an intracellular cytoplasmic domain with intrinsic tyrosine kinase activity *(1)*. Many studies have shown aberrant signaling by RTKs in a variety of cancers, including human brain tumors. Constitutive activation of RTKs is one of the important features of aberrant signaling leading to malignant transformation and tumor proliferation, and can occur by several mechanisms *(2)*. Deregulated RTK signaling can occur via gene amplification, overexpression, and activating mutations, including deletions in the extracellular domain or alterations in the RTK cytoplasmic domain. Another mechanism of aberrant RTK signaling involves the activation of autocrine growth factor/receptor loops *(3)*. In central nervous system (CNS) tumors, in particular astrocytomas, the classical examples of autocrine growth factor/receptor loops involve production of platelet-derived growth factor (PDGF), epidermal growth factor (EGF), and transforming growth factor-α (TGF-α) and their receptors *(4)*. Therefore, it has become imperative to focus on mitogenic and transforming signaling cascades generated by PDGF/PDGFR, EGF/EGFR, and TGF-α/EGFR autocrine loops, in order to understand the

From: *The Cell Cycle in the Central Nervous System*
Edited by: D. Janigro © Humana Press Inc., Totowa, NJ

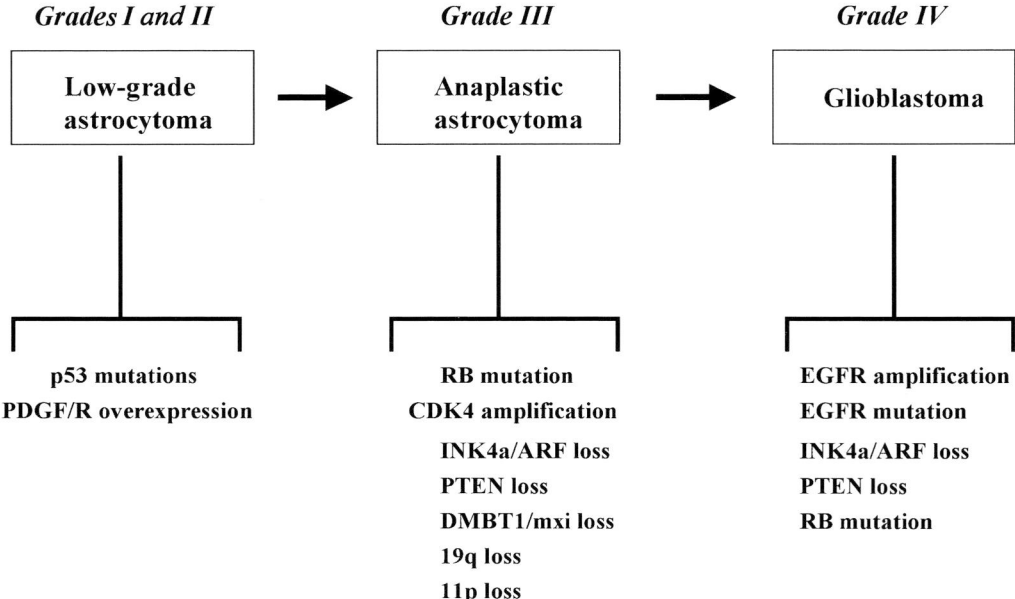

Fig. 1. Stepwise accumulation of genetic alterations in gliomagenesis. PDGF/R, platelet-derived growth factor/receptor; RB, retinoblastoma; CDK4, cyclin-dependent kinase 4; INK4a, inhibitor of cyclin-dependent kinase on chromosome 4; ARF, alternative reading frame on INK4a locus; PTEN, phosphatase/tensin homolog on chromosome 10; DMBT1, deleted in malignant brain tumors 1; EGFR, epidermal growth factor receptor.

molecular mechanisms underlying glioma formation, and to use these signaling modules as novel therapeutic targets.

2. PDGF/PDGFR SIGNALING IN HUMAN AND MOUSE GLIAL TUMORS

The accumulated evidence has suggested that PDGF and PDGFR play a significant role in glial development and lineage commitment *(5)*. In cell culture, PDGF functions to block differentiation and promote proliferation of the O-2A glial progenitors that give rise to either oligodendrocytes or type-2 astrocytes *(6,7)*. Although coexpression of PDGF and PDGFR has been shown in all brain tumor stages, including low-grade astrocytomas, anaplastic astrocytomas, and glioblastoma multiforme (GBM) *(4,8–10)*, PDGF/PDGFR overexpression has been most commonly observed in low-grade astrocytomas in association with loss of function of the p53 tumor suppressor *(11,12)*. These observations suggest a cooperative relationship between the PDGF/PDGFR and p53 signaling pathways (Fig. 1). A recent study using cell culture and a transgenic mouse model showed that overexpression of PDGF in neural progenitors induced the formation of oligodendrogliomas, whereas PDGF transfer into differentiated astrocytes induced the formation of either oligodendrogliomas or mixed oligoastrocytomas *(13)*. The observed histologies of these glial tumors were consistent with low-grade neoplasms. Collectively, these reports suggest that the PDGF mitogenic signaling loop may be an early or initiating event in driving neural precursors or differentiated glial cells to low-grade astrocytomas and/or oligodendrogliomas, before malignant transformation of low-grade clones. It therefore appears that PDGF/R alterations are most commonly observed in "secondary GBM" or those malignant gliomas that arise from lower-grade tumors *(9)*.

The PDGF family consists of four members, PDGF-A, -B, -C, and -D, which transduce signals through the PDGF-α and -β receptors. Biosynthesis and processing of PDGFs involve the formation of dimers PDGF-AA, -BB, -CC, and -DD and the heterodimer PDGF-AB *(14)*. Numerous studies have reported the expression of PDGF-A and -B ligands in glioblastomas and indicate that autocrine signaling by these isoforms is required for cell survival *(15–17)*. Moreover, PDGF-A, -B, and -C, and PDGFR have also been implicated in medulloblastomas and ependymomas *(18,19)*. Very recently, it was reported that PDGF-B enhances glioma angiogenesis by stimulating vascular endothelial growth factor (VEGF) expression in tumor endothelia and promoting pericyte recruitment to neovessels *(20)*. A recent study in glioma cell lines and primary glioblastoma tissues using quantitative reverse transcriptase-polymerase chain reaction (PCR) implicated PDGF-C and -D ligands in the formation of brain tumors and confirmed the existence of autocrine signaling by PDGF-A and -B in brain tumors *(21)*. PDGF-AA and -CC selectively bind to PDGFR-α, whereas PDGF-DD preferentially binds to PDGF-β, with PDGF-BB displaying affinity for both receptors *(14,22–24)*. Binding of PDGF stabilizes PDGF receptor dimerization, which is followed by autophosphorylation of tyrosine residues, leading to increased tyrosine kinase activity *(25,26)* and formation of docking sites for signal relay molecules containing src homology 2 (SH2) domains. A large number of SH2 domain-containing enzymes, such as phosphatidylinositol 3-kinase (PI3-K), phospholipase C (PLC)-γ, the src family of tyrosine kinases, the protein tyrosine phosphatase (PTP) SHP-2, and a GTPase activating protein (GAPs) for Ras have been shown to bind to particular SH2 sites on PDGFR-α- and β-receptors and modulate different signaling pathways. PDGFRs bind to other molecules such as Grb2, Grb7, Nck, and Shc, which lack enzymatic activities and have adapter functions, linking the activated receptor to downstream signaling molecules and distinct pathways. PDGFRs also bind to transcription factors belonging to the signal transducer and activator of transcription (STAT) family, which translocate to the nucleus to directly activate the transcription of several genes *(27)*.

3. EGFR/ERBB SIGNALING CASCADES IN HUMAN AND EXPERIMENTAL ASTROCYTOMAS

EGFR belongs to the ErbB family of type-I RTKs, based on structural homology to the v-erbB oncogene carried by the avian erythroblastosis virus *(28)*. The family includes four members: EGFR (also termed erbB1/HER1), neu (erbB2 or HER2), erbB3 (HER3), and erbB4 (HER4). In high-grade astrocytomas or GBMs, the majority of gene-amplification events involve EGFR *(29–31)*, and it has been reported that approx 50% of GBMs and only a small percentage of anaplastic astrocytomas express high levels of EGFR *(31)*. These observations suggest that EGFR overexpression and/or gene alteration is a late event in gliomagenesis and is frequently observed in "primary" or "*de novo*" glioblastomas occurring in older patients (Fig. 1) *(29–31)*. Furthermore, sequencing of the amplified EGFR genes revealed that frequent gene rearrangements result in essentially seven classes of variant EGFR transcripts *(32)*. The most common rearrangement is a genomic deletion of exons 2–7, resulting in an in-frame deletion of 801 bp of the coding sequence to generate a mutant receptor called de2-7 EGFR, ΔEGFR, or EGFRvIII, which cannot bind ligand because of the truncated extracellular domain, but is constitutively active *(32–38)*. This mutant receptor has also been detected in cancers of the lung, breast, and prostate *(39,40)*, but not in normal tissues *(35)*. Amplification of EGFR genes has been implicated in poor prognosis of patients with GBM *(41)* and it has been demonstrated that patients with EGFRvIII-positive GBMs have shorter life expectancies *(42)*. Unlike wtEGFR, EGFRvIII transforms NIH3T3 cells *(43)* and strongly enhances the tumorigenicity of human gliomas in nude mice *(36,44)*. Overexpression of mutant EGFRs in astrocytes or their precursors in transgenic mice has been shown to promote the development of glioblastoma *(45)*. However, the novel glycine residue resulting from gene rearrangement creates a new epitope at the splice site and the tumor-specific expression of EGFRvIII makes this mutant a potential tumor-specific target for therapy in gliomas and in other cancers *(46–48)*. Moreover, expression of EGFRvIII in gliomas provides resistance to cisplatin, a commonly used chemotherapeutic agent *(49)*, which suggests a need for EGFRvIII-targeted inhibition in combination with chemotherapy.

Enhanced EGFR signaling has also been reported to cooperate with other alterations in the development of GBM (Fig. 1). The most common alterations are those that disrupt cell cycle arrest and include deletion of p16^{INK4a}/p19^{ARF} *(50,51)*, deletion of RB *(52,53)*, loss of function of p53 *(52)*, and amplification of CDK4 *(54,55)*, CDK6 *(56)*, cyclin D$_1$ *(57)*, and MDM2 *(51)*. However, the associated genetic lesions that cooperate with enhanced EGFR signaling in the development of GBM have not been completely characterized. In a recent study, it was demonstrated that expression of EGFRvIII, but not wtEGFR, in mouse astrocytes harboring activated oncogenic Ras resulted in the formation of oligodendroglioma and mixed oligoastrocytoma tumors *(58)*. Many reports evaluating human tumor tissues have shown that combined loss of p16^{INK4a} and p19^{ARF}, but not of either p53, p16^{INK4a}, or p19^{ARF} alone, is associated with EGFR activation in GBM *(12,59,60)*. On the other hand, a study using a mouse model system showed that human telomerase catalytic component (hTERT) overexpressed in normal human astrocytes cooperates with p53/pRb inactivation and Ras pathway activation, but not PI3-K/Akt pathway or EGFR activation, to allow the formation of intracranial tumors strongly resembling p53/pRb pathway-deficient, telomerase-positive, Ras-activated human grade-III anaplastic astrocytomas *(61)*. Also, p53 loss in combination with NF1 (neurofibromatosis type-I) loss leads to astrocytomas and glioblastoma formation in mice *(62)*. Furthermore, a study in mice showed that combined activation of Ras and Akt in neural progenitors induces the formation of glioblastoma *(63)*. A more recent study showed that p16^{INK4a}/p19^{ARF} loss cooperates with Ras and Akt activation in astrocyte precursors and neural progenitors to generate glioblastomas of various morphologies *(64)*, indicating that gliomagenesis occurs through an intricate cooperativity between genetic alterations often involving RTK signaling pathways and cell cycle-regulatory molecules.

Elevated levels of activated Akt have also been associated with loss of PTEN (phosphatase/tensin homolog on chromosome 10), a tumor suppressor, in many glioblastomas (Fig. 1) *(65–67)*. Collectively, these studies suggest that robust signaling by overexpressed EGFR and/or loss of functional PTEN and other genes leads to increased activation of Akt in GBMs, and thus increased transformation. However, it is unclear whether the specific state of glial cell differentiation plays a restrictive role in glioma progression. Recent work in mice has demonstrated that deregulation of specific genetic pathways (i.e., Ink4a/Arf inactivation and EGFR activation) may be more important than the distinct neural cell of origin in dictating the emergence and phenotype of malignant gliomas *(68)*.

Notably, other erbB proteins have been implicated in other CNS tumors. Both erbB2 and erbB4 expression levels have been shown to predict prognosis of childhood medulloblastoma and ependymoma *(69,70)*. In addition to their pathological role in brain tumors, erbB proteins have been implicated in many systemic human cancers such as colon, head and neck, pancreas, lung, breast, kidney, ovary, and bladder *(71)*. Studies in these cancers have linked prognosis to excessive receptor kinase activity, receptor overexpression, ligand-independent constitutive activation of receptor mutants, and/or autocrine stimulation *(72–75)*. Increased receptor activity leads to increased downstream signaling and enhanced cell transformation. Studies on erbB family members have demonstrated that homodimerization and heterodimerization are the initial events in a variety of cellular signals required for cell growth and differentiation of many cell types under physiological conditions *(76)*. Thus, on EGF stimulation, EGFR either forms a homodimer and/or heterodimer with other family members, which leads to auto-/transphosphorylation and activation of RTK activity, recruitment of various signaling relay molecules, and initiation of variety of a intracellular signaling cascades including The mitogen-activated protein kinase (MAPK), PI3-K/Akt, PLC-γ, and STATs. Among these, MAPK and PI3-K/Akt pathways have been more extensively studied in glial tumors.

4. TGF-α/EGFR AUTOCRINE LOOP IN BRAIN TUMORS

TGF-α is a member of the EGF family and a potent mitogen for a number of cell types in culture. Mature TGF-α is a 5.5 kDa peptide *(77)* sharing 30% structural homology with EGF. TGF-α binds to EGFR and activates RTK activity *(78–80)*. Binding of TGF-α to EGFR initiates receptor dimerization and autophosphorylation, followed by recruitment of src-homology 2 (SH2) domain containing molecules, which link EGFR to similar intracellular pathways initiated by the EGF/EGFR loop (1,81). Several tumors and tumor cell lines have been shown to coexpress EGFR and TGF-α (82), indicating the existence of autocrine activation loop driving tumor growth. High levels of TGF-α have also been reported in human glioma (59,83,84). Increased TGF-α levels have been observed in many primary human glioblastomas and anaplastic astrocytomas (84). The highest level of TGF-α expression was found in recurrent tumors, which apparently had undergone transformation from lower grade to high-grade malignant anaplastic tumors. Elevated expression of TGF-α has also been reported in primitive neuroectodermal brain tumors including medulloblastomas (PNET/MB) (85).

It has been reported previously that induction of TGF-α via a tetracycline-on/off system in a glioma cell line, U1242MG, resulted in increased motility at the single cell level, suggesting that coexpression of EGFR and TGF-α was capable of forming an independent autocrine locomotory loop and that TGF-α may form an important component of the glioma cell invasion machinery (86). TGF-α has also been shown to decrease GFAP (marker for astrocytic differentiation) mRNA levels in U373MG glioblastoma cells. On the other hand, TGF-α upregulates gene expression for nestin, a marker for undifferentiated astrocytic precursors, without affecting vimentin gene transcription. These changes in the gene expression of intermediate filament proteins were correlated with increased motility and less stellate morphology, suggesting that TGF-α may be required to induce dedifferentiation in glial tumors in order to promote motility (87). In a very recent study, U1242MG cells containing a TGF-α inducible transgene were used to determine the effect of autocrine TGF-α on cell proliferation in vitro and on subcutaneous tumor growth in nude mice (88). This study showed that induction of TGF-α expression in the absence of tetracycline resulted in increased cell proliferation in vitro, which was inhibited by EGFR-blocking monoclonal antibody C225 and an EGFR-specific tyrosine kinase inhibitor (RTKI). Similarly, U1242MG clones expressing TGF-α developed into tumors of varied sizes in mice not fed with tetracycline. Moreover, mice injected with the TGF-α-expressing clones developed larger tumors than those injected with the control clones. These studies clearly indicate that the TGF-α/EGFR autocrine loop plays an important role in glial tumor progression.

5. MAPK SIGNALING CASCADE IN GLIOMAS

MAPKs are proline directed serine–threonine (Ser–Thr) kinases, which are activated by a variety of cellular stimuli and regulate a variety of cellular processes, such as proliferation, differentiation, development, and tumorigenesis. The MAPK superfamily has been clearly divided into three subgroups: the extracellularly responsive kinases (p42/44^{MAPK} or Erk1/2); the c-jun N-terminal kinases (JNKs) (p46/54^{JNK}), which are also known as the stress-activated protein kinases (SAPKs); and p38^{MAPK} (also known as RK, Mxi-2, CSBP1/2, or HOG-1-related kinases). Although the MAPK families are structurally related, they are generally activated by distinct extracellular stimuli, thus comprising a series of separate MAPK cascades. Genetic and biochemical analyses have identified the universally conserved Ras/Raf/MEK/ p42/44^{MAPK} pathway (89,90) as one of the most ubiquitous MAPK pathways stimulated by various growth factor receptors including PDGFR, EGFR, and other RTKs. In general, receptor–ligand interaction leads to recruitment of adapter proteins such as Grb2, which brings the guanine nucleotide exchange factor Sos to the receptor to form a stable complex which is required for the activation of membrane bound Ras, a small monomeric G protein, by the exchange of guanosine diphosphate for triphosphate (Fig. 2) (91). In addition, EGF-induced activation of Ras may be transduced via another adapter protein, Shc, which binds to activated EGFR and becomes phosphorylated, creating an additional binding site for Grb2 (92). Recent work indicates that the protein tyrosine phosphatase SHP-2 gets recruited to the EGFR-bound adapter Gab1 and facilitates Sos relocation to guanosine triphosphate-bound Ras (93,94). A recent study in U87MG glioblastoma cells showed that activation of Ras/Raf/MEK/p42/44^{MAPK} pathway by EGFRvIII was blocked by PI3-K inhibitors, wortmannin and LY294002, whereas wild-type EGFR-induced Erk activation was

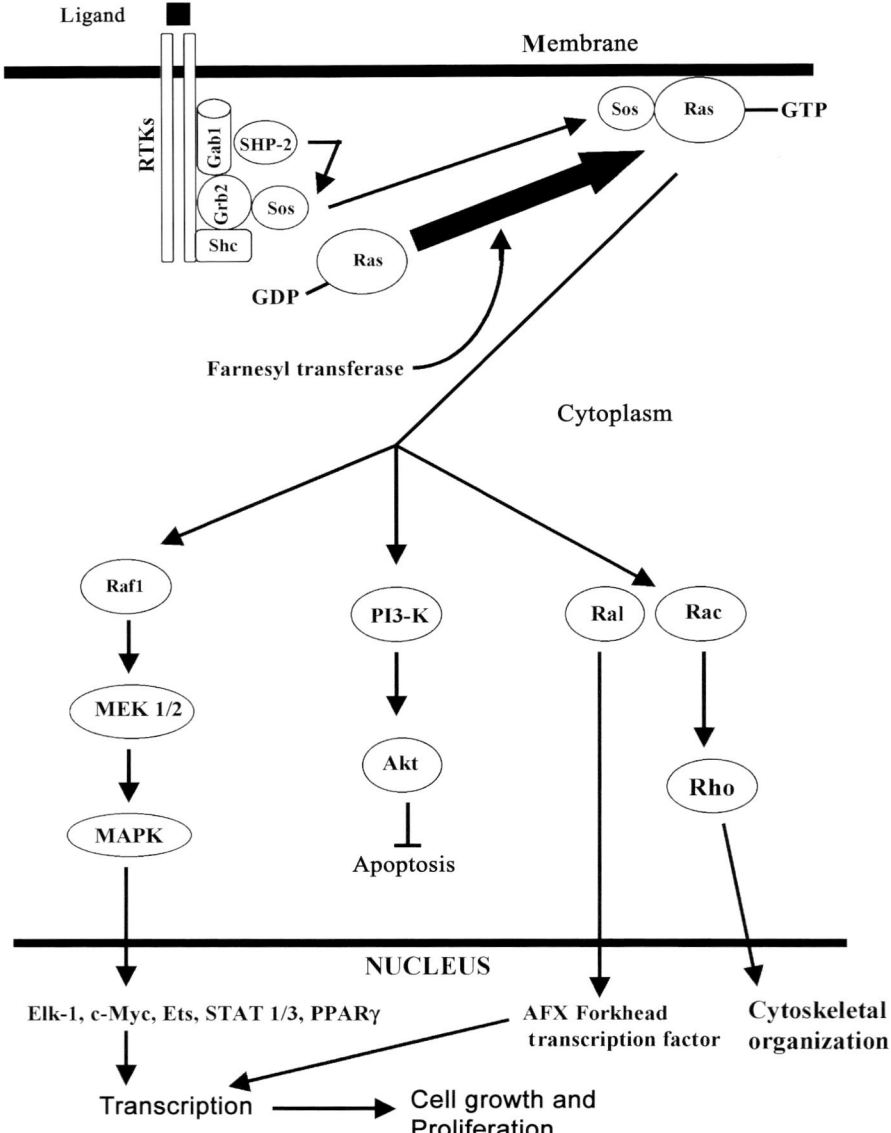

Fig. 2. Receptor tyrosine kinase (RTK)-induced activation of Ras/Raf/MEK/MAPK pathway. RTK, Grb2, growth factor receptor 3-binding protein; Gab1, Grb2 associated binder1; Sos, son-of-sevenless (*Drosophila* homolog of ras); Ras, the protein product of c-ras; Raf1, a kinase, the protein product of c-raf; MEK, MAP kinase kinase; MAPK, mitogen-activated protein serine–threonine kinases; PPAR, peroxisome proliferator-activated receptor.

largely unaffected, suggesting that EGFRvIII and EGFR wild-type preferentially use different signaling pathways to activate Ras/Raf/MEK/p42/ 44^{MAPK} pathways *(95)*.

One of the best-characterized effectors of Ras are the Raf Ser–Thr kinases, which are required for the activation of MEK/p42/44^{MAPK} pathway and are felt to be critical for Ras-mediated cell transformation (Fig. 2) *(91,96,97)*. The other three effector pathways that have demonstrated roles in Ras transformation are those mediated by Ras activation of PI3-K, by guanine nucleotide exchange factors (GEFs) for the Ras-related small GTPase Ral (RalGDS) and by Rac (Tiam 1). In the classical pathway, Ras-activated Raf stimulates downstream mitogen-activated protein kinase (MEK), which in turn phosphorylates MAPKs (also named ERKs) *(98–101)*. MAPKs or ERKs can translocate to the nucleus to phosphorylate and activate several transcription factors for activation of growth-inducing genes (Fig. 2). A recent study showed that normal human astrocytes undergo a p16^{INK4a}-associated senescence-like growth arrest in response to sustained activation of the Ras/Raf/MEK/p42/44^{MAPK} pathway. However, high-grade glioma cells that have dismantled p16^{INK4a}-associated senescence-like growth arrest pathways are potentially regulated by a second p21^{cip1}-dependent growth arrest pathway in response to sustained Ras/Raf/MEK/p42/44^{MAPK} pathway activation *(102)*.

Mutated and constitutively activated forms of Ras are found in approx 50% of all human metastatic tumors *(103)*. In gliomas, specific mutations affecting Ras have not been observed. However, high levels of Ras-GTP have been documented in high-grade astrocytomas *(104,105)*. It was recently demonstrated that a region on chromosome 10 (10p13), which carries a Ras suppressor, RSU-1, was deleted in 30% of the high-grade gliomas tested and two out of three oligodendrogliomas, but not in other CNS tumors, bladder, or colon tumors, or normal tissue *(106)*. Ectopic expression of RSU-1 inhibited tumorigenesis of a glioblastoma cell line. Similarly, a novel Ras-related protein, Rig (Ras-related inhibitor of cell growth) has been shown to express at high levels in normal cardiac and neural tissue. However, expression of Rig protein was frequently lost or downregulated in neural tumor-derived cell lines and in primary human astrocytomas. Moreover, ectopic Rig expression in human astrocytomas suppressed cell growth *(107)*. These studies suggest that high levels of Ras-GTP in high-grade astrocytomas may be owing in part to a gradual accumulation of Ras suppressor protein mutations in primary glial tumors, as well as from activated RTK signaling.

Ras transformation of mouse NIH3T3 cells results in elevated levels of cyclin D$_1$ and accelerated G$_1$ progression *(108)*. Thus, increased Ras activation leads to increased levels of the G$_1$-specific protein complex CDK4/cyclin D in gliomas, which drives cells into S phase and mitosis. In neurofibromatosis type-I (NF1), Ras is constitutively active because of the mutation in the NF1 gene, which encodes the Ras GAP-related protein neurofibromin required to convert active GTP-bound Ras to inactive GDP-bound Ras *(109,110)*. Activation of other small G proteins such as Rap1, which also activates the Raf1/MEK/MAPK pathway, may result in increased cell proliferation in astrocytomas *(111)*. Tuberin, a tuberous sclerosis complex 2 (TSC2) gene product, regulates the activity of Rap1, via a GAP-related domain at its C-terminus *(112)*. Analysis of sporadic astrocytomas and ependymomas demonstrated either increased Rap1 or reduced/absent tuberin protein expression in 50–60% of different cohorts of gliomas, compared with a small percentage of schwannomas and meningiomas, and none of the oligodendrogliomas studied *(111)*, suggesting that alterations in Rap1 signaling may play important role in the development of certain sporadic human gliomas.

RalGEFs have been associated with Ras-induced transformation in very few cell lines *(113,114)*, but their role in the development of brain tumors and other human cancers is still unclear but may be more significant than originally thought *(111)*. Activated forms of Rac have been reported to induce survival in Rat1 fibroblasts and M14 melanoma cells *(115,116)*, but their role in glial cell survival has not been studied extensively. A recent study with primary glioma and astrocyte cell cultures showed that Rac1 regulates a major survival pathway in most glioma cells, and that suppression of Rac1 activity stimulates death in virtually all glioma cells, regardless of the mutational status *(117)*. However, normal astrocytes were not affected, suggesting that Rac survival pathway is specific to transformed glial cells and could be used as a potential therapeutic target for the treatment of malignant gliomas.

The SAPKs, in particular JNKs, have been implicated in human tumorigenesis *(118,119)*. It has been reported that overexpression of EGFRvIII in mouse NIH3T3 cells leads to constitutive activation of the JNK pathway, which correlates with enhanced transformation by EGFRvIII *(120)*. Moreover, a recent study showed constitutively active forms of JNK isoforms in primary glial tumors, indicating a possible association between EGFRvIII and JNKs in the development of GBMs *(121)*. Previously, we showed that overexpression of transforming ErbB receptor complexes leads to constitutive activation of Erk MAPK and particularly JNK MAPKs, including JNK2, enhanced transformation and resistance to apoptosis in primary human glioblastoma cells *(122)*. A single study in T98G glioblastoma cells showed that JNK2 is required for growth of T98G cells in nonstress conditions and that $p21^{cip1/waf1}$ may contribute to the sustained growth arrest of JNK2-depleted T98G cultures *(123)*. Very recently, we have observed that JNK pathway activation may play an important and specific role in cell migration and motility in GBMs (Kapitonov and O'Rourke, manuscript submitted). These observations clearly indicate that JNK pathway events may play a critical role in glial tumor development and progression.

6. ROLE OF PI3-K/AKT AXIS IN GLIOMAGENESIS

The PI3-K/Akt pathway is one of the major survival pathways in epithelial cells. In general, activation of the PI3-K/Akt pathway begins with receptor activation, rapid stimulation of phosphoinositol metabolism *(124,125)*, and coupling of PI3-K to phosphorylated docking proteins, such as Gab1 *(126)*. PI3 kinase is a phospholipid kinase comprising of a regulatory subunit, p85, which contains two SH2 and one SH3 domains, and a catalytic subunit designated p110. We have recently documented that coupling of the SHP-2 protein tyrosine phosphatase to Gab1 is essential for EGFR-mediated PI3-K activation in glioblastoma cells *(127–129)*. Activated PI3-K phosphorylates phosphatidylinositol 4,5-biphosphate (PIP_2) or $PtdIns(4,5)P_2$ via its p110 subunit to generate second messengers $PtdIns(3,4)P_2$ (PIP_2) and $PtdIns(3,4,5)P_3$ (PIP_3). PIP_3 mediates membrane translocation of several signaling proteins, such as the Ser–Thr kinases PDK1 and Akt, the docking protein Gab1, and PLC-γ *(124,125)*. PDK1 phosphorylates Akt at Thr308 *(130)*. It has been proposed that an unidentified protein kinase (a hypothetical PDK2) is responsible for phosphorylation of Akt at Ser473, leading to its complete activation. Activated Akt phosphorylates and inhibits several proapoptotic proteins such as Bad *(131)*, the Forkhead transcription factor *(132)*, and glycogen synthase kinase-3 (GSK-3) *(133)* to promote cell survival. It also promotes protein synthesis by activating p70S6 kinase via mammalian target of rapamycin (mTOR) (Fig. 3).

Interestingly, approx 80% of all human GBMs express activated Akt, which might account for increased cell survival and a worse prognosis for the patients with these tumors *(67)*. Overexpression of constitutively activated Akt has been shown to convert anaplastic astrocytomas to GBM in a human astrocyte glioma model *(61)*. A previous study demonstrated an increased association of PI3-K with focal adhesion kinase (FAK) with sustained PI3-K activity, PIP_3 levels, and Akt phosphorylation in glioblastoma and breast cancer cells expressing a PTEN phosphatase-inactivate mutant, suggesting a possible role for PI3-K in cell migration, invasion, spreading, and focal adhesions *(134)* (reviewed in ref. *135*). Interestingly, another study in PTEN mutant C6 glioma cells showed a positive association between increased PI3-K/Akt signaling and increased invasiveness and gelatinase activity, again suggesting that unchecked PI3-K/Akt activation in gliomas not only serves as a survival signal but might also participate in tumor motility and infiltration *(136)*. Previously, we reported that EGFR transcriptionally upregulates vascular VEGF in human glioblastoma cells via PI3-K-dependent pathway *(137)*. We recently observed that PTEN mutation in human glioblastoma cells cooperates with EGFR activation to upregulate VEGF expression in a PI3-K/Akt-dependent manner, suggesting that PI3-K/Akt axis constitutes an important component of the angiogenic cascade required for development of GBM *(138)*.

It has been well documented that activation of PI3-K/Akt axis is regulated by the lipid phosphatase activity of the PTEN tumor suppressor, which blocks Akt activation by converting PIP_3 to PIP_2, thereby mediating cell cycle arrest and/or apoptosis *(65,139–141)*. Interestingly, various groups have reported that a region including the PTEN locus on the long arm of chromosome 10 is deleted in many tumors, including glioblastomas, breast, prostrate, endometrial carcinoma, and melanoma *(142–144)*, suggesting a pathogenic role for the constitutively active Akt survival pathway in these cancers. Interestingly, it has been shown that PTEN suppresses

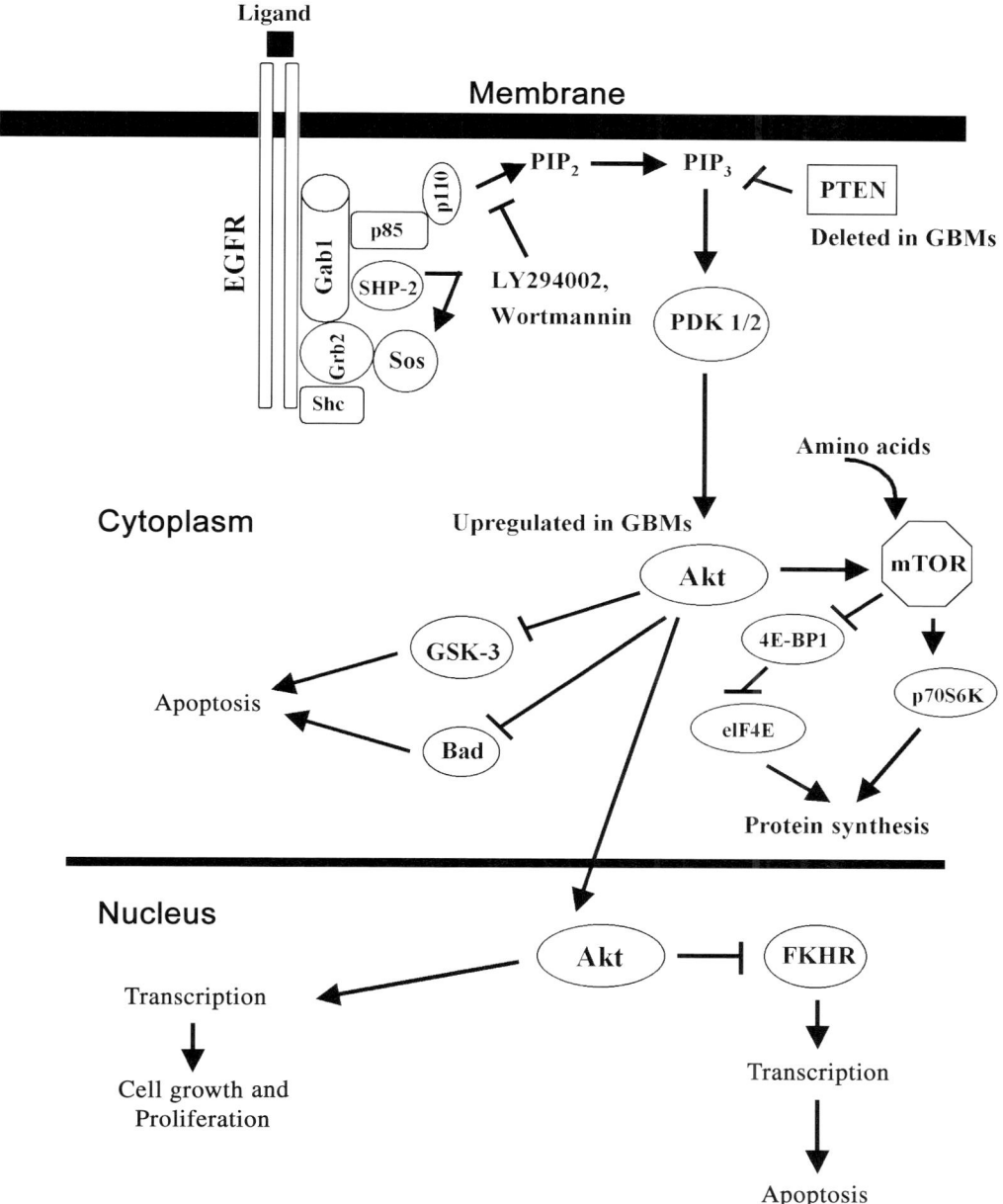

Fig. 3. EGFR-mediated activation and regulation of PI3-K/Akt pathway. PDK1/2, phosphoinositide-dependent kinase 1 and 2; GSK-3, glycogen synthase kinase-3; mTOR, mammalian target of rapamycin; FKHR, forkhead/winged helix protein; 4E-BP1, eIF4E-binding protein 1; eIF4E, eukaryotic initiation factor 4E; p70S6K, ribosomal protein S6 kinase.

growth of U87MG glioblastoma cells by blocking cell cycle progression through G_1, and this was correlated to a significant accumulation of the cell cycle kinase inhibitor p27^{kip1} *(140)*, suggesting that loss of function of PTEN is one of the prerequisites for GBM development. It has also been observed that PTEN mutations are more frequent in primary or *de novo* GBMs, but not in secondary GBMs, which arise from low-grade (grade II) or anaplastic astrocytomas (grade III)

(145). The observation that secondary glioblastomas contain p53 mutations as a genetic hallmark but rare PTEN mutations suggests that primary and secondary GBMs develop through distinct genetic pathways involving RTK signaling *(145,146)*. Approximately 40% of the GBMs are associated with deletions of the PTEN locus on chromosome 10 *(147–149)*. Earlier studies using minichromosome transfer experiments showed that introduction of human chromosome 10q into glioblastoma cells suppressed the growth in soft agar or formation of tumors in nude mice *(150)*. Overexpression of PTEN in glioblastoma cells via adenovirus-mediated gene delivery generated similar results in soft agar and in nude mice *(151)*. Furthermore, we reported previously that overexpression of signal regulatory protein, SIRPα1, in glioblastoma cell lines negatively regulate EGFR-mediated PI3-K/Akt signaling and led to reduced transformation, reduced cell migration and cell spreading, and enhanced apoptosis following DNA damage *(127)*. Studies are under way to establish the mechanism by which SIRPα1 proteins mediate their effects on PI3-K/Akt activation and also to determine the factors regulating the expression of these proteins in normal astrocytes and GBMs under normal physiological conditions.

7. PLC-γ PATHWAY AND GLIOMA

PLC-γ is a phosphoinositide-specific phospholipase, which plays an important role in tumor cell migration occurring via PDGF and EGFR by an undefined mechanism *(152)*. A single study showed that inhibition of PLC-γ activation by a pharmacological inhibitor or a dominant negative PLC-γ (PLCz)-blocked glioma cell motility and invasion of fetal rat brain aggregates, suggesting PLC-γ as a potential target for antiinvasive therapy for GBMs *(153)*. On receptor activation, PLC-γ is rapidly recruited to the receptor through the binding of its SH2 domains to pTyr sites in adapter proteins such as Gab1 *(154)*. Coupling to the receptor activates PLC-γ, which hydrolyzes its substrate PIP_2 to generate two secondary messengers, diacylglycerol (DAG) and inositol 1,4,5-triphosphate (IP_3) *(155)*. IP_3 binds to specific intracellular receptors and stimulates the release of intracellular Ca^{2+}. Ca^{2+} then binds to calmodulin, which in turn activates a family of Ca^{2+}/calmodulin-dependent protein kinases (CaMKs). DAG and Ca^{2+} also activate members of the protein kinase C (PKC) family *(156,157)*. CaMKs and PKCs in turn exert both stimulatory and inhibitory effects on downstream Ras/Raf/MEK/MAPK signaling *(158)* (Fig. 4). Several in vitro studies have supported the involvement of PKCs in glioma cell proliferation and invasion *(159–163)*. A recent article reviewed that PKC-α and -ε cooperates with EGFR in the induction of ornithine decarboxylase (ODC) to increase glioma cell proliferation *(164)*. Very recently, it was shown that a scaffolding protein receptor for activated C-kinase1 (RACK1) mediates interaction between integrin-β chain and phorbol myristate acetate (PMA)-activated PKC-ε, resulting in increased focal adhesion and lamellipodia formation, indicating that PKC-ε positively regulates integrin-dependent adhesion, spreading, and motility of human glioma cells *(165)*. Another study showed that PKC-ε differentially activated Erks at focal adhesions and was required for PMA-induced adhesion and migration of human glioma cell *(166)*. Use of PKC inhibitors suggest that PKCs also play fundamental role in regulating cell cycle and participating in cell survival mechanism. A recent study in human glioma cells reported that PKC-τ and PKC-β-II phosphoylates cyclin-dependent kinase activating kinase (CAK), suggesting their role in cell cycle regulation *(167)*. Interestingly, overexpression of PKC-η has been shown to increase proliferative capacity of glioblastoma cells and block ultraviolet- and γ-irradiation-induced apoptosis by inhibiting caspase-9 activation. These studies suggest that broad spectrum targeting of PKC isoforms may form a basis for future glioma therapy.

8. SIGNAL TRANSDUCER AND ACTIVATOR TRANSCRIPTIONS

It has been shown that EGFRs activate three STAT forms, STAT1, STAT3, and STAT5 *(168)*, whereas PDGFR binds and activates only STAT5 *(169)*. On receptor stimulation, STATs can be activated by phosphorylation via Janus kinase (JAK)-dependent or JAK-independent pathways *(168,170)*. Activated STATs bind to homotypic or heterotypic STATs through their SH2

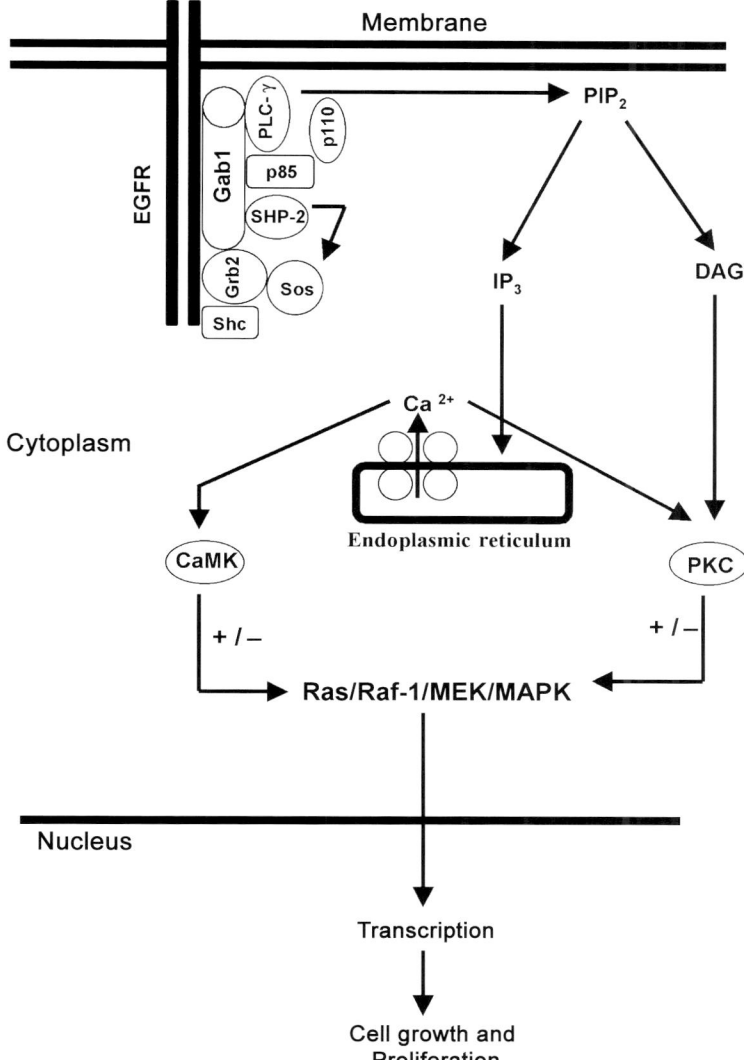

Fig. 4. EGFR-induced PLC-γ pathway participates in activation of different pathways. PLC-γ, phosphoinositide-specific phospholipase C-γ; DAG, diacylglycerol; IP_3, inositol 1,4,5-triphosphate; CaMK, Ca^{2+}/calmodulin-dependent protein kinases.

domains to form homodimers or heterodimers and translocate into the nucleus to bind to sequence-specific STAT-responsive elements on DNA, and activate the transcription of specific target genes such as $p21^{CIP1}$, cyclin D_1, myc, Bcl-2, Bcl-xL, and caspase 1. Collectively, these target genes can regulate cell cycle progression and apoptosis (Fig. 5).

Of the three STATs, STAT1 and STAT3 have been shown to be involved in EGFR-mediated cell cycle regulation *(168)*. STAT1 has been implicated as a negative regulator of cell cycle progression and a promoter of apoptosis *(171)*, whereas STAT3 acts as a positive regulator of cell cycle progression with antiapoptotic activities *(172,173)*. A constitutively activated STAT3α has been reported in large percentage of gliomas and medulloblastomas, indicating that STAT3 may play an important role in EGFR-mediated oncogenesis *(174)*.

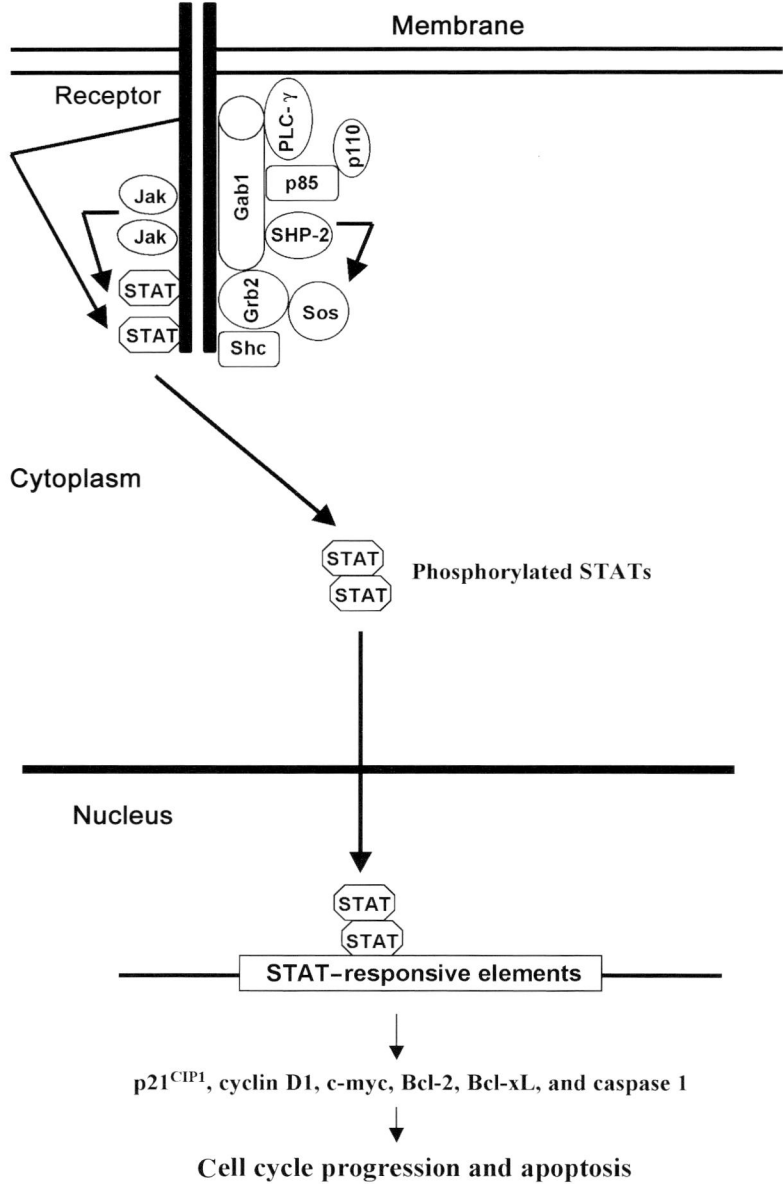

Fig. 5. Activation of STAT pathway by EGFR or PDGFR. STAT, signal transducers and activators of transcription; JAK, Janus kinase.

9. TARGETED MOLECULAR THERAPY OF GLIOMAS

Increased knowledge of the structure and activating mechanisms of RTKs and distinct downstream signaling modules have substantially improved our understanding of the cellular machinery that mediates gliomagenesis and maintains the malignant phenotype of transformed glia. There has been a constant search for new approaches to target specific steps in the pathogenesis of high-grade gliomas because treatment with conventional cytotoxic agents has shown very little progress (175). More importantly, the heterogeneous nature of malignant gliomas owing to different genetic lesions makes a compelling argument for more rational targeted therapies.

The molecular and pharmacotherapeutic approaches to gliomas can be broadly divided into immunotherapy (monoclonal antibodies [MAbs] and tumor vaccination), antisense oligonucleotides, gene therapy, and small molecules such as tyrosine kinase inhibitors (TKIs) and farnesyltransferase inhibitors. In this section, we will focus mainly on small molecule and immunotherapeutic approaches that target RTK signaling with a brief overview on antiangiogenesis therapy (Fig. 6).

9.1. Immunotherapy: MAbs in Glioma Therapy

As approx 50% of GBMs coexpress high levels of EGFR and a mutant EGFR receptor, EGFRvIII *(31,35)*, generation of MAbs directed against the EGFR and/or EGFRvIII mutant may be useful in blocking tumor progression mediated by aberrant EGFR signaling. MAbs directed against the extracellular domain of erbB family RTKs have proven to be an effective strategy to kill tumor cells derived from systemic epithelial cancers *(176)*. Well-known examples include herceptin (trastuzumab), against HER2 or erbB2, and IMC-C225 (cetuximab or erbitux, ImClone) against erbB1 receptor. The United States Food and Drug Administration (FDA) has approved herceptin for breast cancer treatment *(177–179)* whereas C225 is in phase III clinical trials and has been shown to promote growth inhibitory effects in variety of tumors including pancreatic, colorectal, renal, and breast carcinomas *(180,181)*. According to the data presented at American Association of Clinical Oncology (ASCO), 2003, 329 patients with metastatic colorectal cancer were enrolled in randomized phase II trial to receive either cetuximab or a combination of cetuximab and irinotecan (CPT-11), the standard chemotherapeutic agent. The study showed that tumors shrank in 22.9% of the patients receiving two-drug regimen as compared with 10.8% of those receiving cetuximab alone (reviewed in ref. *182*). In addition, two-drug regimen patients had no signs of tumor progression for a median of 4 mo whereas patients on cetuximab alone had no tumor progression for a median of 1.5 mo. This indicates that anti-tumor efficacy can be enhanced in most cases when MAbs are used in combination with cytotoxic agents.

A number of MAbs have also been generated against EGFRvIII *(35,183)*. The most effective MAb was the murine IgG$_{2a}$ MAb Y10, which recognizes both human EGFRvIII and a murine homolog of this mutation *(184)*. It was also shown that incubation of EGFRvIII-expressing cells with Y10 inhibited DNA synthesis and cell proliferation leading to cell death. In addition, Y10 was able to mediate cell death of EGFRvIII-positive cells in the presence of complement, as well as with both murine and human cells bearing Fc receptors *(184)*. Intraperitoneal injection of Y10 led to increased survival in the mice bearing subcutaneous EGFRvIII-expressing tumors. However, it failed to increase survival in animals with intracerebral tumors because of its inability to cross blood–brain barrier. Similarly, another MAb directed against EGFRvIII, MAb 806 showed reduced tumor volume and increased survival of mice bearing xenografts of U87MG.EGFRvIII, LN-Z308.EGFRvIII, or A1207. EGFRvIII gliomas, but is ineffective with mice bearing U87MG tumors, suggesting the specificity of this antibody *(185,186)*. In principle, MAb have been shown to promote receptor internalization, resulting in attenuation of receptor phosphorylation and downstream signaling *(37,185)*.

Alternatively, MAbs can be "armed" with toxins or radionuclides *(187)*. An anti-EGFRvIII antibody fused to pseudomonas exotoxin A generates cytotoxicity in mouse fibroblasts and human glioblastoma cells expressing EGFRvIII without effecting parental cells, suggesting that "armed" MAbs can be more specific than "unarmed" or "naked" antibodies *(47)*. MAbs against EGFR and EGFRvIII have also been used to deliver ^{125}I to GBM cells in animal xenografts and in patients *(188–191)*. However, iodinated antibodies may pose a risk of killing nonneoplastic cells. A relatively new approach involves the use of immunoliposomes, in which liposomes are attached to antibody fragments, to deliver a variety of cytotoxic agents, toxins, or even genes for therapy *(187,192–194)*. A practical limitation for clinical use of full-length MAbs in the treatment of brain tumors is the requirement for these large molecules to permeate the blood–brain barrier. This limitation is the rationale behind generation of smaller MAb fragments to treat cancers *(37,184)*.

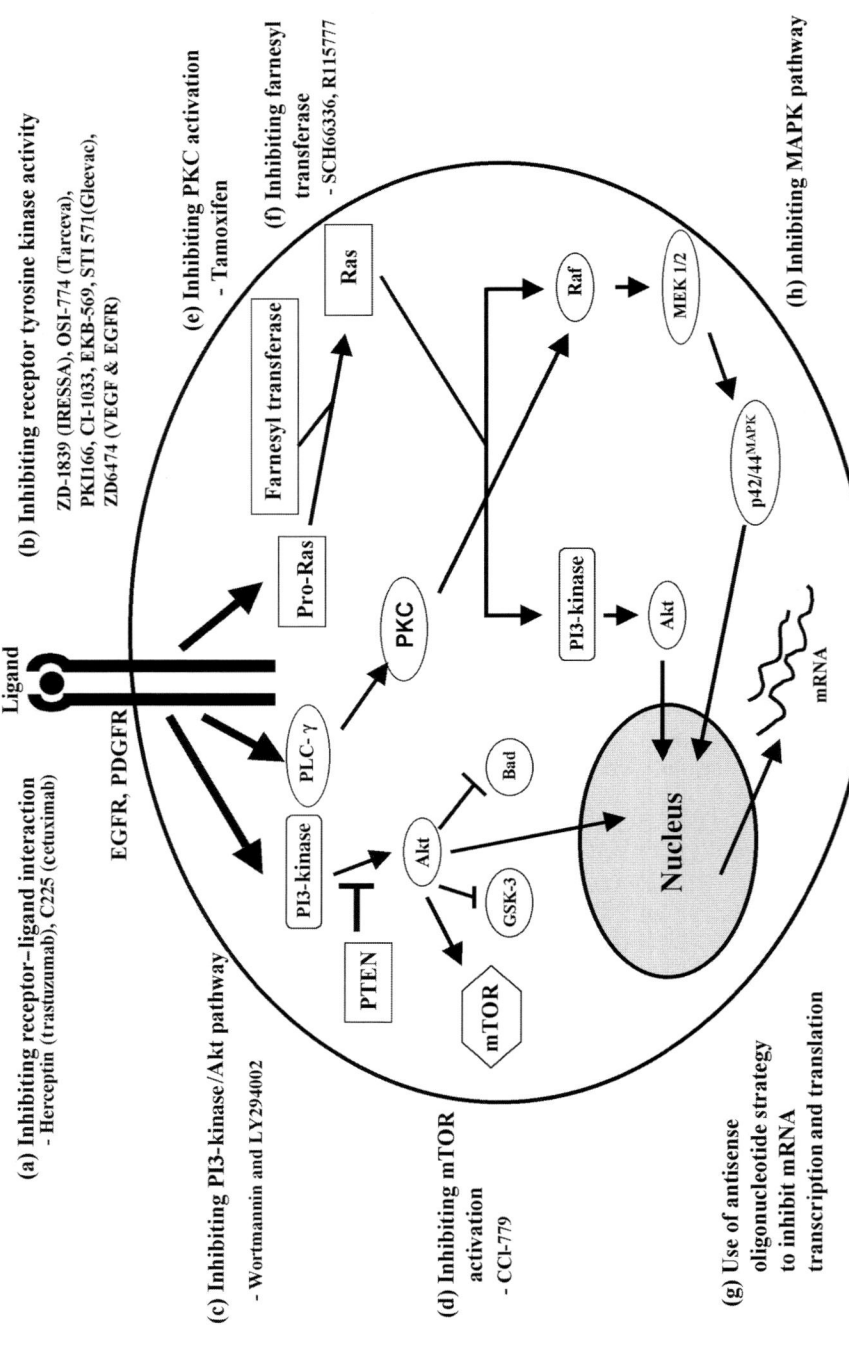

Fig. 6. Novel strategies to inhibit aberrant RTK signaling in glial tumor cells by inhibiting receptor–ligand interactions, the TK domain of RTK, PI3-kinase/Akt pathway, farnesyl transferase, MAPK pathway, or using antisense oligonucleotides to inhibit the translation of key proteins. PDGFR, platelet-derived growth factor receptor; mRNA, messenger ribonucleic acid.

9.2. Immunotherapy: Tumor Vaccination for Gliomas

Recent advances in basic research on antigen presentation, use of specific cytokines, and T-cell-mediated cytotoxicity have given new directions for glioma immunotherapy. One of the recent approaches is the use of adoptive immunotherapy, which can broadly grouped into two categories.

9.2.1. Vaccination with Dendritic Cells Pulsed With Tumor Antigens

Antigen presentation by antigen presenting cells (APCs) is a pivotal step in inducing antigen-specific immunity and is essential for tumor vaccine design. Dendritic cells have always been considered most potent antigen presenting cells for initiating an immune response. Dendritic cells are bone marrow-derived cells capable of presenting antigens in human leukocyte antigen (HLA)-restricted manner. The ability to efficiently prime CD4 T-helper cells and to generate CD8 cytotoxic T-cells (CTLs) make dendritic cells attractive in vaccine strategies for glioma therapy. An earlier study showed that immunizing tumor cells mixed with syngeneic spleen-derived dendritic cells significantly prolonged mean survival for rats harboring pre-established intracranial tumors *(195)*. Another group also reported a prolonged survival in rats harboring pre-established intracranial 9L gliomas after vaccination with bone marrow-derived dendritic cells pulsed with acid-eluted protein from 9L glioma cells *(196)*. Similar results were demonstrated by vaccinating mice with dendritic cells pulsed with Semliki Forest virus-mediated glioma complementary DNA *(197)* and dendritic cells fused to glioma cells *(198)*. These observations in animal glioma models have prompted a number of vaccine clinical trials in patients with glioma. In the phase I clinical trial, patients receiving an autologous peripheral blood dendritic cells pulsed with peptides from autologous glioma cell surface showed a prolonged survival time with an increased intratumoral cytotoxicity and memory for T-cell infiltration *(199)*. However, another phase I clinical trial involving patients vaccinated with a novel fusion product of autologous dendritic and glioma cells did not show statistically significant treatment-associated response to therapy *(200)*. However, there were no serious adverse autoimmune responses observed in this study, indicating that dendritic cell-based immunotherapy can be used safely in humans as an adjunct to currently available glioma therapies *(201)*.

9.2.2. Gene Technique-Based Glioma Vaccination

Genetic manipulation of tumor cells to express certain cytokines such as interferon (IFN)-γ, granulocyte-macrophage colony-stimulating factor (GM-CSF), or interleukin (IL)-12 has been shown to stimulate a potent immunity against tumors with the brain and provides a basis for gene-based immunotherapy *(202)* (reviewed in ref. *203*). Vaccination of allogeneic pre-B-cells expressing EGFRvIII has been shown to produce a systemic immune response against autologous intracranial tumor expressing the same antigen *(204)*. Vaccination with genetically engineered glioma cells expressing antisense molecules that block specific gene expression, such as glioma-derived immunosuppressive factors TGF-β2 and insulin-like growth factor-1 (IGF-1), has also been shown to suppress intracranial tumor growth *(205)*. Two approaches have been used to introduce cytokines intracranially. The first approach involves the direct intracranial implantation of genetically engineered tumor cells, which secrete cytokines. Interleukin (IL)-2, IL-4, GM-CSF, TNF-α, and IFN-γ have all shown to demonstrate a significant survival advantage in animal models *(202,206)*. The second approach involves direct *in situ* cytokine gene transfection. Genetically engineered adenovirus and herpes simplex virus expressing various cytokines have been tested and showed definite survival advantage when used in experimental brain tumor models *(207–209)*. Despite these encouraging results, there have been very few clinical trial results reported. A study involving single GBM patient showed that repeated immunization of autologos tumor cells and a genetically modified fibroblast that secretes IL-2 promoted an antitumor response, mediated in part by cytotoxic T-lymphocytes *(210)*. Similarly, another phase I clinical trial involving 11 GBM patients immunized with autologous tumor cells modified with Newcastle disease virus showed noticeable peripheral immune responses, but no survival advantage over patients who received conventional combination treatment of surgery, radiotherapy, and chemotherapy *(211)*.

9.3. Small Molecule Therapy for RTK Signaling Pathways

One promising approach to inhibit aberrant RTK signaling is the generation of small-molecule drugs that selectively interfere with intrinsic tyrosine kinase activity and thereby inhibit receptor autophosphorylation and downstream signaling cascades *(212)*. RTK signaling can be targeted at three main levels: the receptor itself, the PI3-K/Akt, and/or Ras/MAPK signaling modules (Fig. 4).

9.3.1. Targeting at the Level of the Receptor: EGFR, EGFRvIII, and PDGFR

One of the reasonable approaches to inhibit RTK signaling is to generate small molecule drugs that interfere with the function of the receptor. A number of synthetic small molecules that are TKIs have been generated, particularly for the EGFR/ErbB family. The most advanced of these are ATP analog of the quinazoline and pyridopyrimidine family that compete with ATP for ATP-binding sites and thus block RTK activation *(213)*. Reversible inhibitors ZD-1839 (IRESSA, AstraZeneca, Boston, MA) and OSI-774 (Tarceva, Roche/OSI, Melville, NJ) are two potent EGFR inhibitors currently being tested in clinical trials against solid tumors *(214,215)*. OSI-774 has been approved for use in phase III clinical trials alone or in combination with conventional chemotherapy, for nonsmall-cell lung and pancreatic cancers. Similarly, another reversible ErbB inhibitor, PKI166 (Novartis, Cambridge, MA), and the irreversible inhibitors CI1033 (Pfizer/Warner-Lambert, Cambridge, MA/Morris Plains, NJ) and EKB-569 (Wyeth-Ayerst, Philadelphia, PA) also inhibit EGFR signaling by competing with ATP for binding to the receptor *(216,217)*. CI1033 has also been reported to block EGFRvIII activation *(217)*. The FDA has approved STI 571 (Gleevec, Novartis), another TKI, which inhibits the non-RTK bcr-abl kinase, and two RTKs, PDGFR and c-Kit, for the treatment of chronic myeloid leukemia *(218)*. Gleevec also inhibits the growth of GBM xenografts in vivo *(219)* and is currently in clinical trials for malignant gliomas.

9.3.2. Targeting the PI3-K/Akt Pathway

Biallelic PTEN loss with consequent deregulation of the PI3-K/Akt pathway is one of the major hallmarks of most of the GBMs *(67,149)*. Because this pathway is so crucial for glioblastoma cell proliferation and survival, the PTEN/PI3-kinase/Akt pathway serves as a potential target for therapy. PI3-K inhibitors such as wortmannin and LY294002 are effective, but toxic, and it has been difficult to generate specific PI3-K/Akt inhibitors. An alternative approach is to inhibit FKBP12-rapamycin-associated protein/mammalian target of rapamycin (FRAP/mTOR) kinase, a downstream target of the PI3-K/Akt pathway *(220)*, with less toxicity *(221)*. The immunosuppressant drug rapamycin forms a complex with the immunophilin FK506 binding protein 12 (FKB12) that binds specifically to mTOR and inhibits its kinase activity *(221,222)*. An ester analog of rapamycin, CCI-779 (Wyeth Ayerst), showed growth inhibition of PTEN-deficient mouse as well as human cancer cells, indicating the possibility of using mTOR inhibitors for treating PTEN-null human cancers, including glioblastomas *(221)*. CCI-779 is currently being evaluated in phase I testing in human malignant gliomas. The main goal is to determine the efficacy of CCI-779 in these trials and to determine whether patient sensitivity correlates with PTEN loss and PI3-K/Akt activation. It is also important to determine whether EGFR and/or EGFRvIII expression are associated with tumor response to therapy.

9.3.3. Inhibition of the Ras/MAPK Pathway

As discussed earlier, specific mutations affecting Ras have not been identified in human gliomas. Rather, constitutive Ras activation in GBMs mainly results from overactivation of EGFR, EGFRvIII, and/or PDGFR signaling. For signal transduction, Ras must attach to the plasma membrane, which depends on the posttranslational addition of a farnesyl group to its C-terminal end by the enzyme farnesyl transferase. Thus, farnesyl transferase inhibitors (FTIs) may be helpful in blocking Ras-mediated signaling, and there has been a great deal of effort into the development of FTIs as cancer therapeutics. FTIs may also target other pathways such as Rho-mediated signaling and the PI3-K/Akt pathway, which require farnesylation *(223)*. Two synthetic

FTIs, SCH66336 (Schering-Plough, Kenilworth, NJ) *(224)* and R115777 (Zarnestra, Janssen Research Foundation, Spring House, PA), have shown promising results in preclinical models, including inhibition of growth of GBM cell lines *(225–227)*. Moreover, preliminary findings suggest that EGFR overexpresssion in GBM cells can confer increased sensitivity to SCH66336 *(226)*. R115777 is in phase I/II clinical trial evaluation involving patients with recurrent glioma *(228)*.

We recently observed in our laboratory that overexpression of mutant EGFRvIII in human glioblastoma cells leads to constitutive activation of the Erk MAPK pathway in GBM cells and pharmacological inhibition of this pathway in turn downregulates EGFRvIII phosphorylation and transformation *(229)*, indicating that the use of inhibitors of the MAPK pathway may be an alternative approach to treat EGFRvIII-containing malignant gliomas. There are several ongoing efforts to inhibit constituents of the Ras/Raf/MEK/MAPK pathway, which may eventually achieve enhanced efficacy with reduced toxicity in the management of malignant gliomas that are often refractory to conventional therapy.

9.4. Antiangiogenesis Therapy

Malignant gliomas are among the most highly vascularized solid human tumors *(230)*. The invasive nature of gliomas and accompanying neovasculature results from the expression of stimulatory angiogenic factors, which induce endothelial cell proliferation and promote complex interaction between cancer cell and extracellular matrix (ECM). The most extensively studied factors include vascular VEGFs and VEGF receptors, integrins, and metalloproteinases. In recent years, targeting these molecules to generate effective antiangiogenic drugs has become one of the major thrusts in the field of glioma therapy. VEGF was shown to be differentially overexpressed in gliomas when compared with normal tissue *(231)* (reviewed in ref. *232*). Furthermore, expression of VEGF has been strongly correlated to microvessel density in gliomas and meningiomas *(233)*, indicating their potential as a target for drug development.

Antibodies against VEGF and VEGF receptors are currently being tested in phase II/III clinical trials. It was recently reported that a recombinant humanized monoclonal antibody against VEGF, bevacizumab (AVASTIN, Genentech, San Francisco, CA), given with conventional chemotherapy significantly improved survival of patients with metastatic colorectal cancers, suggesting that bevacizumab can be effective in treatment of VEGF-overexpressing cancers *(234)*. Similarly, small molecule inhibitors of VEGFR tyrosine kinases such as SU5416, SU6668, PTK787, and ZD4190 are already under clinical investigation (reviewed in ref. *235*).

The second category of target molecules includes integrins and metalloproteinases 2 and 9 (MMP-2 and MMP-9), which participate in ECM attachment and degradation to promote neovascularization. Integrins interact with ECM constituents through specific peptide motifs such as Arg-Gly-Asp (RGD) present on ECM proteins to facilitate attachment. Synthetic analogs of RGDs are now being considered as therapeutic tools to inhibit tumor growth and invasion. It was recently demonstrated in an in vivo mouse model and by in vitro experiments that an α v-integrin antagonist EMD 121974, a cyclic RGD pentapeptide, was able to induce apoptosis in α v-integrin-expressing U87MG glioblastoma and DAOY medulloblastoma cell lines, suggesting a potential of using RGDs for glioma therapy *(236)*. Also, conditionally replicative adenoviruses (CRAds) carrying RGD motifs are being tried in preclinical studies to treat various cancers, including gliomas *(237,238)*. Furthermore, antibodies against a variety of integrins have been shown to inhibit matrigel invasion of glioma cell lines and primary cultures *(239)*. A recent study showed that intraperitoneal administration of IS20I, a specific inhibitor of α(v)β-3 integrin, reduces tumor growth in nude mouse bearing intracranial and subcutaneous glioma tumors *(240)*. Synthetic MMP inhibitors such as Marimastat, Batimastat (BB-94), AG3340, and Bay12-9566 have already been used in phase II and III clinical trials in several cancers including malignant gliomas (reviewed in ref. *235*).

Only recently, a large number of naturally occurring endogenous inhibitors of angiogenesis, such as platelet-derived factor-4 (PF-4), thrombospondin-1, metallospondin, tissue inhibitors of metalloproteinases (TIMPs), plasminogen activator/inhibitor, angiopoietin-2 (Ang-2), angiostatin (cleaved from plasminogen), and endostatin (cleaved from collagen XVIII) have been discovered, which in theory might be exploited in antiangiogenesis therapy. Among them, angiostatin and endostatin are already in phase I trials. Previous studies showed that systemic administration of angiostatin successfully treated subcutaneous and intracerebral gliomas of rat and human origin in nude mouse model *(241,242)*. In addition, retroviral or adenoviral transduction of the angiostatin gene into established brain tumors in mice inhibited tumor growth effectively *(243–245)*. Similarly, endostatin has been successfully used systemically or by gene transfer methods to treat a variety of cancers, including fibrosarcoma, melanoma, hemangioendothelioma, prostate, renal, mammary, lung, and colon carcinomas *(246)*. A recent study involving immunological analysis of frozen tissues from 51 patients with astrocytic tumors (grades II–IV) showed increased levels of tissue endostatin, suggesting a definite role of the proteins in glioma formation *(247)*. Overexpression of endostatin has been shown to reduce tumor growth rate by 90% in a rat C6 glioma model *(248)*. Similarly, endostatin was shown to reduce vascularization, blood flow, and tumor growth in rat gliosarcoma model *(249)*. Furthermore, combined administration of endostatin and a PKC-α-inhibiting DNA enzyme improved survival rate in rats bearing intracranial malignant glioma BT(4)C, suggesting that combined treatment may represent an attractive therapeutic strategy against malignant gliomas *(250)*. A study in ovarian cancer cells showed that combined application of angiostatin and endostatin produced synergistic effects on tumor suppression and suggested that the molecular targets differ for the two inhibitory molecules *(251)*. Taken together, these studies indicated that the angiogenic machinery involves a plethora of interacting molecules, which could be exploited as potential candidates for antiangiogenesis drug development.

10. FUTURE DIRECTIONS

Over the past decade tremendous progress has been made in the elucidation of the molecular pathogenesis of malignant gliomas. Empirical observations and experimental studies led to emerging consensus that malignant progression of gliomas occurs through either abnormal activation of signal transduction pathways downstream of RTKs and/or disruption of cell cycle-arrest pathways. The elucidation of the signaling pathways activation owing to different genetic alterations is now being translated into glial tumor treatment strategies. Initial efforts have been focused on the use of single agents directed at specific target molecules. However, the complexity and cross talk between different signaling cascades may limit potential efficacy of targeting a single molecule. Therefore, it has become imperative to consider combinatorial treatment regimens involving either multiple inhibitors targeting different pathways or combination of these inhibitors with traditional cytotoxic drugs. Furthermore, efforts should also be made to design new strategies to determine the optimal dose and effectiveness of these agents by evaluating their molecular effects as an end point rather than assessing traditional maximally tolerated dose and/or overall tumor response. Finally, the molecular heterogeneity of GBMs possesses a great challenge for any targeted molecular therapy. Therefore, it is essential to develop an expanded molecular subclassification of GBMs. This can be achieved by using genetic approaches such as gene profiling coupled with careful biological validation and development of new biochemical and clinical markers. Future treatment strategies for malignant glial tumors will involve a fine integration between a broad-spectrum drug regimen, advances in molecular profiling of tumors, and protocol design. The potential benefit of these strategies offers a hope for significant improvement in the prognosis for patients with malignant glial brain tumors.

ACKNOWLEDGMENTS

Our research was supported by grants to D.M.O. from the National Institutes of Health, the Department of Veterans Affairs (Merit Review Program), and The Brain Tumor Society.

REFERENCES

1. Ullrich A, Schlessinger J. Signal transduction by receptors with tyrosine kinase activity. Cell 1990;61:203–212.

2. Kolibaba KS, Druker BJ. Protein tyrosine kinases and cancer. Biochim Biophys Acta 1997;1333:F217–F248.
3. Hanahan D, Weinberg RA. The hallmarks of cancer. Cell 2000;100: 57–70.
4. Maher EA, Furnari FB, Bachoo RM, et al. Malignant glioma: genetics and biology of a grave matter. Genes Dev 2001;15:1311–1333.
5. Yeh YL, Kang YM, Chaibi MS, Xie JF, Graves DT. IL-1 and transforming growth factor-beta inhibit platelet-derived growth factor-AA binding to osteoblastic cells by reducing platelet- derived growth factor-alpha receptor expression. J Immunol 1993;150:5625–5632.
6. Noble M, Murray K, Stroobant P, Waterfield MD, Riddle P. Platelet-derived growth factor promotes division and motility and inhibits premature differentiation of the oligodendrocyte/type-2 astrocyte progenitor cell. Nature 1988;333:560–562.
7. Richardson WD, Pringle N, Mosley MJ, Westermark B, Dubois-Dalcq M. A role for platelet-derived growth factor in normal gliogenesis in the central nervous system. Cell 1988;53:309–319.
8. Claesson-Welsh L. Signal transduction by the PDGF receptors. Prog Growth Factor Res 1994;5:37–54.
9. Heldin CH, Westermark B. Platelet-derived growth factor: mechanism of action and possible in vivo function. Cell Regul 1990;1:555–566.
10. Westermark B, Heldin CH, Nister M. Platelet-derived growth factor in human glioma. Glia 1995;15:257–263.
11. Chung R, Whaley J, Kley N, et al. TP53 gene mutations and 17p deletions in human astrocytomas. Genes Chromosomes Cancer 1991;3:323–331.
12. von Deimling A, Eibl RH, Ohgaki H, et al. p53 mutations are associated with 17p allelic loss in grade II and grade III astrocytoma. Cancer Res 1992;52:2987–2990.
13. Dai C, Celestino JC, Okada Y, Louis DN, Fuller GN, Holland EC. PDGF autocrine stimulation dedifferentiates cultured astrocytes and induces oligodendrogliomas and oligoastrocytomas from neural progenitors and astrocytes in vivo. Genes Develop 2001;15:1913–1925.
14. Betsholtz C, Karlsson L, Lindahl P. Developmental roles of platelet-derived growth factors. Bioessays 2001;23:494–507.
15. Guha A, Dashner K, Black PM, Wagner JA, Stiles CD. Expression of PDGF and PDGF receptors in human astrocytoma operation specimens supports the existence of an autocrine loop. Int J Cancer 1995;60:168–173.
16. Hermanson M, Funa K, Hartman M, et al. Platelet-derived growth factor and its receptors in human glioma tissue: expression of messenger RNA and protein suggests the presence of autocrine and paracrine loops. Cancer Res 1992;52:3213–3219.
17. Vassbotn FS, Ostman A, Langeland N, et al. Activated platelet-derived growth factor autocrine pathway drives the transformed phenotype of a human glioblastoma cell line. J Cell Physiol 1994;158:381–389.
18. Black P, Carroll R, Glowacka D. Expression of platelet-derived growth factor transcripts in medulloblastomas and ependymomas. Pediatr Neurosurg 1996;24:74–78.
19. Andrae J, Molander C, Smits A, Funa K, Nister M. Platelet-derived growth factor-B and -C and active alpha-receptors in medulloblastoma cells. Biochem Biophys Res Commun 2002;296:604–611.
20. Guo P, Hu B, Gu W, et al. Platelet-derived growth factor-B enhances glioma angiogenesis by stimulating vascular endothelial growth factor expression in tumor endothelia and by promoting pericyte recruitment. Am J Pathol 2003;162:1083–1093.
21. Lokker NA, Sullivan CM, Hollenbach SJ, Israel MA, Giese NA. Platelet-derived growth factor (PDGF) autocrine signaling regulates survival and mitogenic pathways in glioblastoma cells: evidence that the novel PDGF-C and PDGF-D ligands may play a role in the development of brain tumors. Cancer Res 2002;62:3729–3735.
22. Bergsten E, Uutela M, Li X, et al. PDGF-D is a specific, protease-activated ligand for the PDGF beta-receptor. Nat Cell Biol 2001;3:512–516.
23. Gilbertson DG, Duff ME, West JW, et al. Platelet-derived growth factor C (PDGF-C), a novel growth factor that binds to PDGF alpha and beta receptor. J Biol Chem 2001;276:27,406–27,414.
24. LaRochelle WJ, Jeffers M, McDonald WF, et al. PDGF-D, a new protease-activated growth factor. Nat Cell Biol 2001;3:517–521.
25. Fantl WJ, Escobedo JA, Williams LT. Mutations of the platelet-derived growth factor receptor that cause a loss of ligand-induced conformational change, subtle changes in kinase activity, and impaired ability to stimulate DNA synthesis. Mol Cell Biol 1989;9:4473–4478.

26. Kazlauskas A, Cooper JA. Autophosphorylation of the PDGF receptor in the kinase insert region regulates interactions with cell proteins. Cell 1989;58:1121–1133.
27. Heldin CH, Ostman A, Ronnstrand L. Signal transduction via platelet-derived growth factor receptors. Biochim Biophys Acta 1998;1378:F79–F113.
28. Salomon DS, Brandt R, Ciardiello F, Normanno N. Epidermal growth factor-related peptides and their receptors in human malignancies. Crit Rev Oncol Hematol 1995;19:183–232.
29. Bigner SH, Humphrey PA, Wong AJ, et al. Characterization of the epidermal growth factor receptor in human glioma cell lines and xenografts. Cancer Res 1990;50:8017–8022.
30. Libermann TA, Nusbaum HR, Razon N, et al. Amplification, enhanced expression and possible rearrangement of EGF receptor gene in primary human brain tumours of glial origin. Nature 1985;313:144–147.
31. Wong AJ, Bigner SH, Bigner DD, Kinzler KW, Hamilton SR, Vogelstein B. Increased expression of the epidermal growth factor receptor gene in malignant gliomas is invariably associated with gene amplification. Proc Natl Acad Sci USA 1987;84:6899–6903.
32. Rasheed BK, Wiltshire RN, Bigner SH, Bigner DD. Molecular pathogenesis of malignant gliomas. Curr Opin Oncol 1999;11:162–167.
33. Ekstrand AJ, Sugawa N, James CD, Collins VP. Amplified and rearranged epidermal growth factor receptor genes in human glioblastomas reveal deletions of sequences encoding portions of the N- and/or C-terminal tails. Proc Natl Acad Sci USA 1992;89:4309–4313.
34. Wong AJ, Ruppert JM, Bigner SH, et al. Structural alterations of the epidermal growth factor receptor gene in human gliomas. Proc Natl Acad Sci USA 1992;89:2965–2969.
35. Wikstrand CJ, Reist CJ, Archer GE, Zalutsky MR, Bigner DD. The class III variant of the epidermal growth factor receptor (EGFRvIII): characterization and utilization as an immunotherapeutic target. J Neurovirol 1998;4:148–158.
36. O'Rourke DM, Nute EJ, Davis JG, et al. Inhibition of a naturally occurring EGFR oncoprotein by the p185neu ectodomain: implications for subdomain contributions to receptor assembly. Oncogene 1998;16:1197–1207.
37. Kuan CT, Wikstrand CJ, Bigner DD. EGF mutant receptor vIII as a molecular target in cancer therapy. Endocr Relat Cancer 2001;8:83–96.
38. Nagane M, Lin H, Cavenee WK, Huang HJ. Aberrant receptor signaling in human malignant gliomas: mechanisms and therapeutic implications. Cancer Lett 2001;162(Suppl):S17–S21.
39. Moscatello DK, Holgado-Madruga M, Godwin AK, et al. Frequent expression of a mutant epidermal growth factor receptor in multiple human tumors. Cancer Res 1995;55:5536–5539.
40. Olapade-Olaopa EO, Moscatello DK, MacKay EH, et al. Evidence for the differential expression of a variant EGF receptor protein in human prostate cancer. Br J Cancer 2000;82:186–194.
41. Schlegel J, Stumm G, Brandle K, et al. Amplification and differential expression of members of the erbB-gene family in human glioblastoma. J Neuro-Oncol 1994;22:201–207.
42. Feldkamp MM, Lala P, Lau N, Roncari L, Guha A. Expression of activated epidermal growth factor receptors, Ras-guanosine triphosphate, and mitogen-activated protein kinase in human glioblastoma multiforme specimens. Neurosurgery 1999;45:1442–1453.
43. Moscatello DK, Montgomery RB, Sundareshan P, McDanel H, Wong MY, Wong AJ. Transformational and altered signal transduction by a naturally occurring mutant EGF receptor. Oncogene 1996;13:85–96.
44. Nishikawa R, Ji XD, Harmon RC, et al. A mutant epidermal growth factor receptor common in human glioma confers enhanced tumorigenicity. Proc Natl Acad Sci USA 1994;91:7727–7731.
45. Holland EC, Hively WP, DePinho RA, Varmus HE. A constitutively active epidermal growth factor receptor cooperates with disruption of G1 cell cycle arrest pathways to induce glioma-like lesions in mice. Genes Dev 1998;12:3675–3685.
46. Hills D, Rowlinson-Busza G, Gullick WJ. Specific targeting of a mutant, activated FGF receptor found in glioblastoma using a monoclonal antibody. Int J Cancer 1995;63:537–543.
47. Lorimer IA, Keppler-Hafkemeyer A, Beers RA, Pegram CN, Bigner DD, Pastan I. Recombinant immunotoxins specific for a mutant epidermal growth factor receptor: targeting with a single chain antibody variable domain isolated by phage display. Proc Natl Acad Sci USA 1996;93:14,815–14,820.
48. Lorimer IA. Mutant epidermal growth factor receptors as targets for cancer therapy. Curr Cancer Drug Targets 2002;2:91–102.
49. Nagane M, Levitzki A, Gazit A, Cavenee WK, Huang HJ. Drug resistance of human glioblastoma cells conferred by a tumor-specific mutant epidermal growth factor receptor through modulation of Bcl-XL and caspase-3-like proteases. Proc Natl Acad Sci USA 1998;95:5724–5729.

50. Giani C, Finocchiaro G. Mutation rate of the CDKN2 gene in malignant gliomas. Cancer Res 1994;54:6338–6339.
51. Reifenberger G, Liu L, Ichimura K, Schmidt EE, Collins VP. Amplification and overexpression of the MDM2 gene in a subset of human malignant gliomas without p53 mutations. Cancer Res 1993;53:2736–2739.
52. Henson JW, Schnitker BL, Correa KM, et al. The retinoblastoma gene is involved in malignant progression of astrocytomas. Ann Neurol 1994;36:714–721.
53. Newcomb EW, Alonso M, Sung T, Miller DC. Incidence of p14ARF gene deletion in high-grade adult and pediatric astrocytomas. Hum Pathol 2000;31:115–119.
54. Holland EC, Hively WP, Gallo V, Varmus HE. Modeling mutations in the G1 arrest pathway in human gliomas: overexpression of CDK4 but not loss of INK4a-ARF induces hyperploidy in cultured mouse astrocytes. Genes Dev 1998;12:3644–3649.
55. Rollbrocker B, Waha A, Louis DN, Wiestler OD, von Deimling A. Amplification of the cyclin-dependent kinase 4 (CDK4) gene is associated with high cdk4 protein levels in glioblastoma multiforme. Acta Neuropathol (Berl) 1996;92:70–74.
56. Costello JF, Plass C, Arap W, et al. Cyclin-dependent kinase 6 (CDK6) amplification in human gliomas identified using two-dimensional separation of genomic DNA. Cancer Res 1997;57:1250–1254.
57. Buschges R, Weber RG, Actor B, Lichter P, Collins VP, Reifenberger G. Amplification and expression of cyclin D genes (CCND1, CCND2 and CCND3) in human malignant gliomas. Brain Pathol 1999;9:435–442; discussion 432–433.
58. Ding H, Shannon P, Lau N, et al. Oligodendrogliomas result from the expression of an activated mutant epidermal growth factor receptor in a RAS transgenic mouse astrocytoma model. Cancer Res 2003;63:1106–1113.
59. Ekstrand AJ, James CD, Cavenee WK, Seliger B, Pettersson RF, Collins VP. Genes for epidermal growth factor receptor, transforming growth factor alpha, and epidermal growth factor and their expression in human gliomas in vivo. Cancer Res 1991;51:2164–2172.
60. Hayashi Y, Ueki K, Waha A, Wiestler OD, Louis DN, von Deimling A. Association of EGFR gene amplification and CDKN2 (p16/MTS1) gene deletion in glioblastoma multiforme. Brain Pathol 1997;7:871–875.
61. Sonoda Y, Ozawa T, Aldape KD, Deen DF, Berger MS, Pieper RO. Akt pathway activation converts anaplastic astrocytoma to glioblastoma multiforme in a human astrocyte model of glioma. Cancer Res 2001;61:6674–6678.
62. Reilly KM, Loisel DA, Bronson RT, McLaughlin ME, Jacks T. Nf1;Trp53 mutant mice develop glioblastoma with evidence of strain-specific effects. Nat Genet 2000;26:109–113.
63. Holland EC, Celestino J, Dai C, Schaefer L, Sawaya RE, Fuller GN. Combined activation of Ras and Akt in neural progenitors induces glioblastoma formation in mice. Nat Genet 2000;25:55–57.
64. Uhrbom L, Dai C, Celestino JC, Rosenblum MK, Fuller GN, Holland EC. Ink4a-Arf loss cooperates with KRas activation in astrocytes and neural progenitors to generate glioblastomas of various morphologies depending on activated Akt. Cancer Res 2002;62:5551–5558.
65. Davies MA, Lu Y, Sano T, et al. Adenoviral transgene expression of MMAC/PTEN in human glioma cells inhibits Akt activation and induces anoikis. Cancer Res 1998;58:5285–5290.
66. Fujisawa H, Kurrer M, Reis RM, Yonekawa Y, Kleihues P, Ohgaki H. Acquisition of the glioblastoma phenotype during astrocytoma progression is associated with loss of heterozygosity on 10q25-qter. Am J Pathol 1999;155:387–394.
67. Haas-Kogan D, Shalev N, Wong M, Mills G, Yount G, Stokoe D. Protein kinase B (PKB/Akt) activity is elevated in glioblastoma cells due to mutation of the tumor suppressor PTEN/MMAC. Curr Biol 1998;8:1195–1198.
68. Bachoo RM, Maher EA, Ligon KL, et al. Epidermal growth factor receptor and Ink4a/Arf: convergent mechanisms governing terminal differentiation and transformation along the neural stem cell to astrocyte axis. Cancer Cell 2002;1:269–277.
69. Gilbertson RJ, Bentley L, Hernan R, et al. ERBB Receptor Signaling Promotes Ependymoma Cell Proliferation and Represents a Potential Novel Therapeutic Target for This Disease. Clin Cancer Res 2002;8:3054–3064.
70. Gilbertson RJ, Perry RH, Kelly PJ, Pearson AD, Lunec J. Prognostic significance of HER2 and HER4 coexpression in childhood medulloblastoma. Cancer Res 1997;57:3272–3280.
71. Lui VW, Grandis JR. EGFR-mediated cell cycle regulation. Anticancer Res 2002;22:1–11.

72. Chow NH, Liu HS, Lee EI, et al. Significance of urinary epidermal growth factor and its receptor expression in human bladder cancer. Anticancer Res 1997;17:1293–1296.
73. Grandis JR, Melhem MF, Gooding WE, et al. Levels of TGF-alpha and EGFR protein in head and neck squamous cell carcinoma and patient survival. J Natl Cancer Inst 1998;90:824–832.
74. Turkeri LN, Erton ML, Cevik I, Akdas A. Impact of the expression of epidermal growth factor, transforming growth factor alpha, and epidermal growth factor receptor on the prognosis of superficial bladder cancer. Urology 1998;51:645–649.
75. Zwick ME, Cutler DJ, Chakravarti A. Patterns of genetic variation in Mendelian and complex traits. Annu Rev Genom Hum Genet 2000;1:387–407.
76. Carraway KL, III, Cantley LC. A neu acquaintance for erbB3 and erbB4: a role for receptor heterodimerization in growth signaling. Cell 1994;78:5–8.
77. Derynck R, Roberts AB, Winkler ME, Chen EY, Goeddel DV. Human transforming growth factor-alpha: precursor structure and expression in E. coli. Cell 1984;38:287–297.
78. Massague J. Epidermal growth factor-like transforming growth factor. II. Interaction with epidermal growth factor receptors in human placenta membranes and A431 cells. J Biol Chem 1983;258: 13,614–13,620.
79. Shum L, Reeves SA, Kuo AC, Fromer ES, Derynck R. Association of the transmembrane TGF-alpha precursor with a protein kinase complex. J Cell Biol 1994;125:903–916.
80. Yarden Y, Ullrich A. Growth factor receptor tyrosine kinases. Annu Rev Biochem 1988;57:443–478.
81. Tang P, Steck PA, Yung WK. The autocrine loop of TGF-alpha/EGFR and brain tumors. J Neuro-Oncol 1997;35:303–314.
82. Derynck R, Goeddel DV, Ullrich A, et al. Synthesis of messenger RNAs for transforming growth factors alpha and beta and the epidermal growth factor receptor by human tumors. Cancer Res 1987;47:707–712.
83. Schlegel U, Moots PL, Rosenblum MK, Thaler HT, Furneaux HM. Expression of transforming growth factor alpha in human gliomas. Oncogene 1990;5:1839–1842.
84. Yung WK, Zhang X, Steck PA, Hung MC. Differential amplification of the TGF-alpha gene in human gliomas. Cancer Commun 1990;2:201–205.
85. Huber H, Eggert A, Janss AJ, et al. Angiogenic profile of childhood primitive neuroectodermal brain tumours/medulloblastomas. Eur J Cancer 2001;37:2064–2072.
86. El-Obeid A, Bongcam-Rudloff E, Sorby M, Ostman A, Nister M, Westermark B. Cell scattering and migration induced by autocrine transforming growth factor alpha in human glioma cells in vitro. Cancer Res 1997;57:5598–5604.
87. Zhou R, Skalli O. TGF-alpha differentially regulates GFAP, vimentin, and nestin gene expression in U-373 MG glioblastoma cells: correlation with cell shape and motility. Exp Cell Res 2000;254:269–278.
88. El-Obeid A, Hesselager G, Westermark B, Nister M. TGF-alpha-driven tumor growth is inhibited by an EGF receptor tyrosine kinase inhibitor. Biochem Biophys Res Commun 2002;290:349–358.
89. Garrington TP, Johnson GL. Organization and regulation of mitogen-activated protein kinase signaling pathways. Curr Opin Cell Biol 1999;11:211–218.
90. Wilkinson MG, Millar JB. Control of the eukaryotic cell cycle by MAP kinase signaling pathways. FASEB J 2000;14:2147–2157.
91. Cox AD, Der CJ. Ras family signaling: therapeutic targeting. Cancer Biol Ther 2002;1:599–606.
92. Schlessinger J. Cell signaling by receptor tyrosine kinases. Cell 2000;103:211–225.
93. Li W, Nishimura R, Kashishian A, et al. A new function for a phosphotyrosine phosphatase: linking GRB2-Sos to a receptor tyrosine kinase. Mol Cell Biol 1994;14:509–517.
94. Yart A, Laffargue M, Mayeux P, et al. A critical role for phosphoinositide 3-kinase upstream of Gab1 and SHP2 in the activation of ras and mitogen-activated protein kinases by epidermal growth factor. J Biol Chem 2001;276:8856–8864.
95. Lorimer IA, Lavictoire SJ. Activation of extracellular-regulated kinases by normal and mutant EGF receptors. Biochim Biophys Acta 2001;1538:1–9.
96. Hamad NM, Elconin JH, Karnoub AE, et al. Distinct requirements for Ras oncogenesis in human versus mouse cells. Genes Dev 2002;16:2045–2057.
97. Katz ME, McCormick F. Signal transduction from multiple Ras effectors. Curr Opin Genet Dev 1997;7:75–79.
98. Boulton TG, Nye SH, Robbins DJ, et al. ERKs: a family of protein-serine/threonine kinases that are activated and tyrosine phosphorylated in response to insulin and NGF. Cell 1991;65:663–675.

99. Dent P, Haser W, Haystead TA, Vincent LA, Roberts TM, Sturgill TW. Activation of mitogen-activated protein kinase kinase by v-Raf in NIH 3T3 cells and in vitro. Science 1992;257:1404–1407.
100. Howe LR, Leevers SJ, Gomez N, Nakielny S, Cohen P, Marshall CJ. Activation of the MAP kinase pathway by the protein kinase raf. Cell 1992;71:335–342.
101. Kyriakis JM, Force TL, Rapp UR, Bonventre JV, Avruch J. Mitogen regulation of c-Raf-1 protein kinase activity toward mitogen-activated protein kinase-kinase. J Biol Chem 1993;268: 16,009–16,019.
102. Fanton CP, McMahon M, Pieper RO. Dual growth arrest pathways in astrocytes and astrocytic tumors in response to Raf-1 activation. J Biol Chem 2001;276:18,871–18,877.
103. Chambers AF, Tuck AB. Ras-responsive genes and tumor metastasis. Crit Rev Oncol 1993;4:95–114.
104. Guha A, Feldkamp MM, Lau N, Boss G, Pawson A. Proliferation of human malignant astrocytomas is dependent on Ras activation. Oncogene 1997;15:2755–2765.
105. Gutmann DH, Giordano MJ, Mahadeo DK, Lau N, Silbergeld D, Guha A. Increased neurofibromatosis 1 gene expression in astrocytic tumors: positive regulation by p21-ras. Oncogene 1996;12: 2121–2127.
106. Chunduru S, Kawami H, Gullick R, Monacci WJ, Dougherty G, Cutler ML. Identification of an alternatively spliced RNA for the Ras suppressor RSU-1 in human gliomas. J Neuro-Oncol 2002;60: 201–211.
107. Ellis CA, Vos MD, Howell H, Vallecorsa T, Fults DW, Clark GJ. Rig is a novel Ras-related protein and potential neural tumor suppressor. Proc Natl Acad Sci USA 2002;99:9876–9881.
108. Liu JJ, Chao JR, Jiang MC, Ng SY, Yen JJ, Yang-Yen HF. Ras transformation results in an elevated level of cyclin D1 and acceleration of G1 progression in NIH 3T3 cells. Mol Cell Biol 1995;15:3654–3663.
109. Reed N, Gutmann DH. Tumorigenesis in neurofibromatosis: new insights and potential therapies. Trends Mol Med 2001;7:157–162.
110. Zhu Y, Parada LF. A particular GAP in mind. Nat Genet 2001; 27:354–355.
111. Gutmann DH, Saporito-Irwin S, DeClue JE, Wienecke R, Guha A. Alterations in the rap1 signaling pathway are common in human gliomas. Oncogene 1997;15:1611–1616.
112. Jin F, Wienecke R, Xiao GH, Maize JC, Jr, DeClue JE, Yeung RS. Suppression of tumorigenicity by the wild-type tuberous sclerosis 2 (Tsc2) gene and its C-terminal region. Proc Natl Acad Sci USA 1996;93:9154–9159.
113. Lu Z, Hornia A, Joseph T, et al. Phospholipase D and RalA cooperate with the epidermal growth factor receptor to transform 3Y1 rat fibroblasts. Mol Cell Biol 2000;20:462–467.
114. Ward Y, Wang W, Woodhouse E, Linnoila I, Liotta L, Kelly K. Signal pathways which promote invasion and metastasis: critical and distinct contributions of extracellular signal-regulated kinase and Ral-specific guanine exchange factor pathways. Mol Cell Biol 2001;21:5958–5969.
115. Nishida K, Kaziro Y, Satoh T. Anti-apoptotic function of Rac in hematopoietic cells. Oncogene 1999;18:407–415.
116. Ruggieri R, Chuang YY, Symons M. The small GTPase Rac suppresses apoptosis caused by serum deprivation in fibroblasts. Mol Med 2001;7:293–300.
117. Senger DL, Tudan C, Guiot MC, et al. Suppression of Rac activity induces apoptosis of human glioma cells but not normal human astrocytes. Cancer Res 2002;62:2131–2140.
118. Behrens A, Jochum W, Sibilia M, Wagner EF. Oncogenic transformation by ras and fos is mediated by c-Jun N-terminal phosphorylation. Oncogene 2000;19:2657–2663.
119. Chen N, Nomura M, She QB, et al. Suppression of skin tumorigenesis in c-Jun NH(2)-terminal kinase-2- deficient mice. Cancer Res 2001;61:3908–3912.
120. Antonyak MA, Moscatello DK, Wong AJ. Constitutive activation of c-Jun N-terminal kinase by a mutant epidermal growth factor receptor. J Biol Chem 1998;273:2817–2822.
121. Tsuiki H, Tnani M, Okamoto I, et al. Constitutively active forms of c-Jun NH_2-terminal kinase are expressed in primary glial tumors. Cancer Res 2003;63:250–255.
122. Wu CJ, Qian X, O'Rourke DM. Sustained mitogen-activated protein kinase activation is induced by transforming erbB receptor complexes. DNA Cell Biol 1999;18:731–741.
123. Potapova O, Gorospe M, Bost F, et al. c-Jun N-terminal kinase is essential for growth of human T98G glioblastoma cells. J Biol Chem 2000;275:24,767–24,775.
124. Czech MP. Lipid rafts and insulin action. Nature 2000;407:147–148.
125. Rameh LE, Cantley LC. The role of phosphoinositide 3-kinase lipid products in cell function. J Biol Chem 1999;274:8347–8350.

126. Rodrigues GA, Falasca M, Zhang Z, Ong SH, Schlessinger J. A novel positive feedback loop mediated by the docking protein Gab1 and phosphatidylinositol 3-kinase in epidermal growth factor receptor signaling. Mol Cell Biol 2000;20:1448–1459.
127. Wu CJ, Chen Z, Ullrich A, Greene MI, O'Rourke DM. Inhibition of EGFR-mediated phosphoinositide-3-OH kinase (PI3-K) signaling and glioblastoma phenotype by signal-regulatory proteins (SIRPs). Oncogene 2000;19:3999–4010.
128. Wu CJ, O'Rourke DM, Feng GS, Johnson GR, Wang Q, Greene MI. The tyrosine phosphatase SHP-2 is required for mediating phosphatidylinositol 3-kinase/Akt activation by growth factors. Oncogene 2001;20:6018–6025.
129. Kapoor GS, Zhan Y, Johnson GR, O'Rourke DM. Distinct domains in the SHP-2 phosphatase differentially regulate epidermal growth factor receptor/NF-kappaB activation through Gab1 in glioblastoma cells. Mol Cell Biol 2004;24:823–836.
130. Alessi DR, Andjelkovic M, Caudwell B, et al. Mechanism of activation of protein kinase B by insulin and IGF-1. EMBO J 1996;15:6541–6551.
131. Cardone MH, Roy N, Stennicke HR, et al. Regulation of cell death protease caspase-9 by phosphorylation. Science 1998;282:1318–1321.
132. Brunet A, Bonni A, Zigmond MJ, et al. Akt promotes cell survival by phosphorylating and inhibiting a Forkhead transcription factor. Cell 1999;96:857–868.
133. Cross DA, Alessi DR, Cohen P, Andjelkovich M, Hemmings BA. Inhibition of glycogen synthase kinase-3 by insulin mediated by protein kinase B. Nature 1995;378:785–789.
134. Tamura M, Gu J, Danen EH, Takino T, Miyamoto S, Yamada KM. PTEN interactions with focal adhesion kinase and suppression of the extracellular matrix-dependent phosphatidylinositol 3-kinase/Akt cell survival pathway. J Biol Chem 1999;274:20,693–20,703.
135. Yamada KM, Araki M. Tumor suppressor PTEN: modulator of cell signaling, growth, migration and apoptosis. J Cell Sci 2001;114:2375–2382.
136. Kubiatowski T, Jang T, Lachyankar MB, et al. Association of increased phosphatidylinositol 3-kinase signaling with increased invasiveness and gelatinase activity in malignant gliomas. J Neurosurg 2001;95:480–488.
137. Maity A, Pore N, Lee J, Solomon D, O'Rourke DM. Epidermal growth factor receptor transcriptionally up-regulates vascular endothelial growth factor expression in human glioblastoma cells via a pathway involving phosphatidylinositol 3(-kinase and distinct from that induced by hypoxia. Cancer Res 2000;60:5879–5886.
138. Pore N, Liu S, Haas-Kogan DA, O'Rourke DM, Maity A. PTEN mutation and epidermal growth factor receptor activation regulate vascular endothelial growth factor (VEGF) mRNA expression in human glioblastoma cells by transactivating the proximal VEGF promoter. Cancer Res 2003;63:236–241.
139. Furnari FB, Lin H, Huang HS, Cavenee WK. Growth suppression of glioma cells by PTEN requires a functional phosphatase catalytic domain. Proc Natl Acad Sci USA 1997;94:12,479–12,484.
140. Li DM, Sun H. PTEN/MMAC1/TEP1 suppresses the tumorigenicity and induces G1 cell cycle arrest in human glioblastoma cells. Proc Natl Acad Sci USA 1998;95:15,406–15,411.
141. Tian XX, Pang JC, To SS, Ng HK. Restoration of wild-type PTEN expression leads to apoptosis, induces differentiation, and reduces telomerase activity in human glioma cells. J Neuropathol Exp Neurol 1999;58:472–479.
142. Cantley LC, Neel BG. New insights into tumor suppression: PTEN suppresses tumor formation by restraining the phosphoinositide 3-kinase/AKT pathway. Proc Natl Acad Sci USA 1999;96:4240–4245.
143. Di Cristofano A, Pandolfi PP. The multiple roles of PTEN in tumor suppression. Cell, 2000; 100:387–390.
144. Rasheed BK, Fuller GN, Friedman AH, Bigner DD, Bigner SH. Loss of heterozygosity for 10q loci in human gliomas. Genes, Chromosomes Cancer 1992;5:75–82.
145. Tohma Y, Gratas C, Biernat W, et al. PTEN (MMAC1) mutations are frequent in primary glioblastomas (de novo) but not in secondary glioblastomas. J Neuropathol Exp Neurol 1998;57:684–689.
146. Peraud A, Watanabe K, Plate KH, Yonekawa Y, Kleihues P, Ohgaki H. p53 mutations versus EGF receptor expression in giant cell glioblastomas. J Neuropathol Exp Neurol 1997;56:1236–1241.
147. Liu W, James CD, Frederick L, Alderete BE, Jenkins RB. PTEN/MMAC1 mutations and EGFR amplification in glioblastomas. Cancer Res 1997;57:5254–5257.
148. Schmidt EE, Ichimura K, Goike HM, Moshref A, Liu L, Collins VP. Mutational profile of the PTEN gene in primary human astrocytic tumors and cultivated xenografts. J Neuropathol Exp Neurol 1999;58:1170–1183.

149. Smith JS, Tachibana I, Passe SM, et al. PTEN mutation, EGFR amplification, and outcome in patients with anaplastic astrocytoma and glioblastoma multiforme. J Natl Cancer Inst 2001;93:1246–1256.
150. Pershouse MA, Stubblefield E, Hadi A, Killary AM, Yung WK, Steck PA. Analysis of the functional role of chromosome 10 loss in human glioblastomas. Cancer Res 1993;53:5043–5050.
151. Cheney IW, Johnson DE, Vaillancourt MT, et al. Suppression of tumorigenicity of glioblastoma cells by adenovirus-mediated MMAC1/PTEN gene transfer. Cancer Res 1998;58:2331–2334.
152. Chen P, Xie H, Sekar MC, Gupta K, Wells A. Epidermal growth factor receptor-mediated cell motility: phospholipase C activity is required, but mitogen-activated protein kinase activity is not sufficient for induced cell movement. J Cell Biol 1994;127:847–857.
153. Khoshyomn S, Penar PL, Rossi J, Wells A, Abramson DL, Bhushan A. Inhibition of phospholipase C-gamma1 activation blocks glioma cell motility and invasion of fetal rat brain aggregates. Neurosurgery 1999;44:568–577; discussion 577,578.
154. Gual P, Giordano S, Williams TA, Rocchi S, Van Obberghen E, Comoglio PM. Sustained recruitment of phospholipase C-gamma to Gab1 is required for HGF-induced branching tubulogenesis. Oncogene 2000;19:1509–1518.
155. Falasca M, Logan SK, Lehto VP, Baccante G, Lemmon MA, Schlessinger J. Activation of phospholipase C gamma by PI 3-kinase-induced PH domain-mediated membrane targeting. EMBO J 1998;17:414–422.
156. Karin M, Hunter T. Transcriptional control by protein phosphorylation: signal transmission from the cell surface to the nucleus. Curr Biol 1995;5:747–757.
157. Hunter T. Signaling—2000 and beyond. Cell 2000;100:113–127.
158. Luttrell LM, Daaka Y, Lefkowitz RJ. Regulation of tyrosine kinase cascades by G-protein-coupled receptors. Curr Opin Cell Biol 1999;11:177–183.
159. Sharif TR, Sharif M. Overexpression of protein kinase C epsilon in astroglial brain tumor derived cell lines and primary tumor samples. Int J Oncol 1999;15:237–243.
160. Couldwell WT, Antel JP, Yong VW. Protein kinase C activity correlates with the growth rate of malignant gliomas: Part II. Effects of glioma mitogens and modulators of protein kinase C. Neurosurgery 1992;31:717–724; discussion 724.
161. da Rocha AB, Mans DR, Lenz G, et al. Protein kinase C-mediated in vitro invasion of human glioma cells through extracellular-signal-regulated kinase and ornithine decarboxylase. Pathobiology 2000;68:113–123.
162. Chintala SK, Kyritsis AP, Mohan PM, et al. Altered actin cytoskeleton and inhibition of matrix metalloproteinase expression by vanadate and phenylarsine oxide, inhibitors of phosphotyrosine phosphatases: modulation of migration and invasion of human malignant glioma cells. Mol Carcinog 1999;26:274–285.
163. Cho KK, Mikkelsen T, Lee YJ, Jiang F, Chopp M, Rosenblum ML. The role of protein kinase Calpha in U-87 glioma invasion. Int J Dev Neurosci 1999;17:447–461.
164. da Rocha AB, Mans DR, Regner A, Schwartsmann G. Targeting protein kinase C: new therapeutic opportunities against high- grade malignant gliomas? Oncologist 2002;7:17–33.
165. Besson A, Wilson TL, Yong VW. The anchoring protein RACK1 links protein kinase Cepsilon to integrin beta chains. Requirements for adhesion and motility. J Biol Chem 2002;277:22,073–22,084.
166. Besson A, Davy A, Robbins SM, Yong VW. Differential activation of ERKs to focal adhesions by PKC epsilon is required for PMA-induced adhesion and migration of human glioma cells. Oncogene 2001;20:7398–7407.
167. Acevedo-Duncan M, Patel R, Whelan S, Bicaku E. Human glioma PKC-iota and PKC-betaII phosphorylate cyclin-dependent kinase activating kinase during the cell cycle. Cell Prolif 2002;35:23–36.
168. Bromberg J, Chen X. STAT proteins: signal tranducers and activators of transcription. Methods Enzymol 2001;333:138–151.
169. Valgeirsdottir S, Paukku K, Silvennoinen O, Heldin CH, Claesson-Welsh L. Activation of Stat5 by platelet-derived growth factor (PDGF) is dependent on phosphorylation sites in PDGF beta-receptor juxtamembrane and kinase insert domains. Oncogene 1998;16:505–515.
170. Ihle JN, Kerr IM. Jaks and Stats in signaling by the cytokine receptor superfamily. Trends Genet 1995;11:69–74.
171. Bromberg JF, Fan Z, Brown C, Mendelsohn J, Darnell JE Jr. Epidermal growth factor-induced growth inhibition requires Stat1 activation. Cell Growth Differ 1998;9:505–512.
172. Coqueret O, Gascan H. Functional interaction of STAT3 transcription factor with the cell cycle inhibitor p21WAF1/CIP1/SDI1. J Biol Chem 2000;275:18,794–18,800.
173. Bienvenu F, Gascan H, Coqueret O. Cyclin D1 represses STAT3 activation through a Cdk4-independent mechanism. J Biol Chem 2001;276:16,840–16,847.

174. Schaefer LK, Ren Z, Fuller GN, Schaefer TS. Constitutive activation of Stat3alpha in brain tumors: localization to tumor endothelial cells and activation by the endothelial tyrosine kinase receptor (VEGFR-2). Oncogene 2002;21:2058–2065.
175. Stewart J. Modulation of the subjective and physiological effects of drugs by contexts and expectations—the search for mechanisms: comment on Alessi, Roll, Reilly, and Johanson (2002). Exp Clin Psychopharmacol 2002;10:96–98; discussion 101–103.
176. Fan Z, Mendelsohn J. Therapeutic application of anti-growth factor receptor antibodies. Curr Opin Oncol 1998;10:67–73.
177. Pegram MD, Lipton A, Hayes DF, et al. Phase II study of receptor-enhanced chemosensitivity using recombinant humanized anti-p185HER2/neu monoclonal antibody plus cisplatin in patients with HER2/neu-overexpressing metastatic breast cancer refractory to chemotherapy treatment. J Clin Oncol 1998;16:2659–2671.
178. Cobleigh MA, Vogel CL, Tripathy D, et al. Multinational study of the efficacy and safety of humanized anti-HER2 monoclonal antibody in women who have HER2-overexpressing metastatic breast cancer that has progressed after chemotherapy for metastatic disease. J Clin Oncol 1999;17:2639–2648.
179. Slamon DJ, Leyland-Jones B, Shak S, et al. Use of chemotherapy plus a monoclonal antibody against HER2 for metastatic breast cancer that overexpresses HER2. New Engl J Med 2001;344:783–792.
180. Overholser JP, Prewett MC, Hooper AT, Waksal HW, Hicklin DJ. Epidermal growth factor receptor blockade by antibody IMC-C225 inhibits growth of a human pancreatic carcinoma xenograft in nude mice. Cancer 2000;89:74–82.
181. Prewett M, Rothman M, Waksal H, Feldman M, Bander NH, Hicklin DJ. Mouse-human chimeric anti-epidermal growth factor receptor antibody C225 inhibits the growth of human renal cell carcinoma xenografts in nude mice. Clin Cancer Res 1998;4:2957–2966.
182. McCarthy M. Antiangiogenesis drug promising for metastatic colorectal cancer. Lancet 2003;361:1959.
183. Wikstrand CJ, Hale LP, Batra SK, et al. Monoclonal antibodies against EGFRvIII are tumor specific and react with breast and lung carcinomas and malignant gliomas. Cancer Res 1995;55:3140–3148.
184. Sampson JH, Crotty LE, Lee S, et al. Unarmed, tumor-specific monoclonal antibody effectively treats brain tumors. Proc Natl Acad Sci USA 2000;97:7503–7508.
185. Mishima K, Johns TG, Luwor RB, et al. Growth suppression of intracranial xenografted glioblastomas overexpressing mutant epidermal growth factor receptors by systemic administration of monoclonal antibody (mAb) 806, a novel monoclonal antibody directed to the receptor [erratum appears in Cancer Res 2001;61(20):7703–7705]. Cancer Res 2001;61:5349–5354.
186. Luwor RB, Johns TG, Murone C, et al. Monoclonal antibody 806 inhibits the growth of tumor xenografts expressing either the de2-7 or amplified epidermal growth factor receptor (EGFR) but not wild-type EGFR. Cancer Res 2001;61:5355–5361.
187. Carter P. Improving the efficacy of antibody-based cancer therapies. Nat Rev Cancer 2001;1:118–129.
188. Brady LW, Miyamoto C, Woo DV, et al. Malignant astrocytomas treated with iodine-125 labeled monoclonal antibody 425 against epidermal growth factor receptor: a phase II trial. Int J Radiat Oncol Biol Phys 1992;22:225–230.
189. Foulon CF, Reist CJ, Bigner DD, Zalutsky MR. Radioiodination via D-amino acid peptide enhances cellular retention and tumor xenograft targeting of an internalizing anti-epidermal growth factor receptor variant III monoclonal antibody. Cancer Res 2000;60L:4453–4460.
190. Brady LW. A new treatment for high grade gliomas of the brain. Bull Memoires Acad Roy Med Bel 1998;153:255–261; discussion 261,262.
191. Emrich JG, Brady LW, Quang TS, et al. Radioiodinated (I-125) monoclonal antibody 425 in the treatment of high grade glioma patients: ten-year synopsis of a novel treatment. Am J Clin Oncol 2002;25:541–546.
192. Park JW, Hong K, Kirpotin DB, et al. Anti-HER2 immunoliposomes: enhanced efficacy attributable to targeted delivery. Clin Cancer Res 2002;8:1172–1181.
193. Siwak DR, Tari AM, Lopez-Berestein G. The potential of drug-carrying immunoliposomes as anticancer agents. Commentary re: Park JW, et al. Anti-HER2 immunoliposomes: enhanced efficacy due to targeted delivery. Clin Cancer Res 2002;8:955–956, 1172–1181.
194. Zhang Y, Zhu C, Pardridge WM. Antisense gene therapy of brain cancer with an artificial virus gene delivery system. Mol Ther J Am Soc Gene Ther 2002;6:67–72.

195. Siesjo P, Visse E, Sjogren HO. Cure of established, intracerebral rat gliomas induced by therapeutic immunizations with tumor cells and purified APC or adjuvant IFN-gamma treatment. J Immunother Emphasis Tumor Immunol 1996;19:334–345.
196. Liau LM, Black KL, Prins RM, et al. Treatment of intracranial gliomas with bone marrow-derived dendritic cells pulsed with tumor antigens. J Neurosurg 1999;90:1115–1124.
197. Yamanaka R, Zullo SA, Tanaka R, Blaese M, Xanthopoulos KG. Enhancement of antitumor immune response in glioma models in mice by genetically modified dendritic cells pulsed with Semliki forest virus-mediated complementary DNA. J Neurosurg 2001;94:474–481.
198. Akasaki Y, Kikuchi T, Homma S, Abe T, Kofe D, Ohno T. Antitumor effect of immunizations with fusions of dendritic and glioma cells in a mouse brain tumor model [comment]. J Immunother 2001;24:106–113.
199. Yu JS, Wheeler CJ, Zeltzer PM, et al. Vaccination of malignant glioma patients with peptide-pulsed dendritic cells elicits systemic cytotoxicity and intracranial T-cell infiltration. Cancer Res 2001;61:842–847.
200. Kikuchi T, Akasaki Y, Irie M, Homma S, Abe T, Ohno T. Results of a phase I clinical trial of vaccination of glioma patients with fusions of dendritic and glioma cells. Cancer Immunol Immunother 2001;50:337–344.
201. Graf MR, Prins RM, Hawkins WT, Merchant RE. Irradiated tumor cell vaccine for treatment of an established glioma. I. Successful treatment with combined radiotherapy and cellular vaccination. Cancer Immunol Immunother 2002;51:179–189.
202. Sampson JH, Ashley DM, Archer GE, et al. Characterization of a spontaneous murine astrocytoma and abrogation of its tumorigenicity by cytokine secretion. Neurosurgery 1997;41:1365–1372; discussion 1372–1373.
203. Liu MA. DNA vaccines: a review. J Int Med 2003;253:402–410.
204. Ashley DM, Sampson JH, Archer GE, Batra SK, Bigner DD, Hale LP. A genetically modified allogeneic cellular vaccine generates MHC class I-restricted cytotoxic responses against tumor-associated antigens and protects against CNS tumors in vivo. J Neuroimmunol 1997;78:34–46.
205. Resnicoff M, Sell C, Rubini M, et al. Rat glioblastoma cells expressing an antisense RNA to the insulin-like growth factor-1 (IGF-1) receptor are nontumorigenic and induce regression of wild-type tumors. Cancer Res 1994;54:2218–2222.
206. Yu JS, Burwick JA, Dranoff G, Breakefield XO. Gene therapy for metastatic brain tumors by vaccination with granulocyte-macrophage colony-stimulating factor-transduced tumor cells. Hum Gene Ther 1997;8:1065–1072.
207. Liu Y, Ehtesham M, Samoto K, et al. In situ adenoviral interleukin 12 gene transfer confers potent and long-lasting cytotoxic immunity in glioma. Cancer Gene Ther 2002;9:9–15.
208. Parker JN, Gillespie GY, Love CE, Randall S, Whitley RJ, Markert JM. Engineered herpes simplex virus expressing IL-12 in the treatment of experimental murine brain tumors. Proc Natl Acad Sci USA 2000;97:2208–2213.
209. Andreansky S, He B, van Cott J, et al. Treatment of intracranial gliomas in immunocompetent mice using herpes simplex viruses that express murine interleukins. Gene Ther 1998;5:121–130.
210. Sobol RE, Fakhrai H, Shawler D, et al. Interleukin-2 gene therapy in a patient with glioblastoma. Gene Ther 1995;2:164–167.
211. Schneider T, Gerhards R, Kirches E, Firsching R. Preliminary results of active specific immunization with modified tumor cell vaccine in glioblastoma multiforme. J Neuro-Oncol 2001;53:39–46.
212. Levitzki A. Protein tyrosine kinase inhibitors as novel therapeutic agents. Pharmacol Ther 1999;82:231–239.
213. Noonberg SB, Benz CC. Tyrosine kinase inhibitors targeted to the epidermal growth factor receptor subfamily: role as anticancer agents. Drugs 2000;59:753–767.
214. Ciardiello F, Caputo R, Bianco R, et al. Antitumor effect and potentiation of cytotoxic drugs activity in human cancer cells by ZD-1839 (Iressa), an epidermal growth factor receptor-selective tyrosine kinase inhibitor. Clin Cancer Res 2000;6:2053–2063.
215. Hidalgo M, Siu LL, Nemunaitis J, et al. Phase I and pharmacologic study of OSI-774, an epidermal growth factor receptor tyrosine kinase inhibitor, in patients with advanced solid malignancies. J Clin Oncol 2001;19:3267–3279.
216. Shawver LK, Slamon D, Ullrich A. Smart drugs: tyrosine kinase inhibitors in cancer therapy. Cancer Cell 2002;1:117–123.
217. Traxler P, Bold G, Buchdunger E, et al. Tyrosine kinase inhibitors: from rational design to clinical trials. Med Res Rev 2001;21:499–512.

218. Druker BJ, Sawyers CL, Kantarjian H, et al. Activity of a specific inhibitor of the BCR-ABL tyrosine kinase in the blast crisis of chronic myeloid leukemia and acute lymphoblastic leukemia with the Philadelphia chromosome. N Engl J Med 2001;344:1038–1042.
219. Kilic T, Alberta JA, Zdunek PR, et al. Intracranial inhibition of platelet-derived growth factor-mediated glioblastoma cell growth by an orally active kinase inhibitor of the 2-phenylaminopyrimidine class. Cancer Res 2000;60:5143–5150.
220. Sekulic A, Hudson CC, Homme JL, et al. A direct linkage between the phosphoinositide 3-kinase-AKT signaling pathway and the mammalian target of rapamycin in mitogen-stimulated and transformed cells. Cancer Res 2000;60:3504–3513.
221. Neshat MS, Mellinghoff IK, Tran C, et al. Enhanced sensitivity of PTEN-deficient tumors to inhibition of FRAP/mTOR. Proc Natl Acad Sci USA 2001;98:10,314–10,319.
222. Vivanco I, Sawyers CL. The phosphatidylinositol 3-Kinase AKT pathway in human cancer. Nat Rev Cancer 2002;2:489–501.
223. Jiang K, Coppola D, Crespo NC, et al. The phosphoinositide 3-OH kinase/AKT2 pathway as a critical target for farnesyltransferase inhibitor-induced apoptosis. Mol Cell Biol 2000;20:139–148.
224. Adjei AA. Farnesyltransferase inhibitors. Cancer Chemother Biol Res Mod 2001;19:149–164.
225. Karp JE, Kaufmann SH, Adjei AA, Lancet JE, Wright JJ, End DW. Current status of clinical trials of farnesyltransferase inhibitors. Curr Opin Oncol 2001;13:470–476.
226. Glass TL, Liu TJ, Yung WK. Inhibition of cell growth in human glioblastoma cell lines by farnesyltransferase inhibitor SCH66336. Neuro-Oncology 2000;2:151–158.
227. Feldkamp MM, Lau N, Guha A. The farnesyltransferase inhibitor L-744,832 inhibits the growth of astrocytomas through a combination of antiproliferative, antiangiogenic, and proapoptotic activities. Ann NY Acad Sci 1999;886:257–260.
228. Cloughesy TF, Filka E, Nelson G, et al. Irinotecan treatment for recurrent malignant glioma using an every-3-week regimen. Am J Clin Oncol 2002;25:204–208.
229. Zhan Y, O'Rourke DM. SHP-∂-dependent mitogen-activated protein kinase activation regulates EGFRvIII but not wild-type epidermal growth factor receptor phosphorylation and glioblastoma cell survival. Cancer Res 2004;64:8292–8298.
230. Brem S, Cotran R, Folkman J. Tumor angiogenesis: a quantitative method for histologic grading. J Natl Cancer Inst 1972;48:347–356.
231. Plate KH, Breier G, Weich HA, Risau W. Vascular endothelial growth factor is a potential tumour angiogenesis factor in human gliomas in vivo. Nature 1992;359:845–848.
232. Veikkola T, Alitalo K. VEGFs, receptors and angiogenesis. Sem Cancer Biol 1999;9:211–220.
233. Samoto K, Ikezaki K, Ono M, et al. Expression of vascular endothelial growth factor and its possible relation with neovascularization in human brain tumors. Cancer Res 1995;55:1189–1193.
234. Fernando NH, Hurwitz HI. Inhibition of vascular endothelial growth factor in the treatment of colorectal cancer. Sem Oncol 2003;30:39–50.
235. Kirsch M, Santarius T, Black PM, Schackert G. Therapeutic anti-angiogenesis for malignant brain tumors. Onkologie 2001;24:423–430.
236. Taga T, Suzuki A, Gonzalez-Gomez I, et al. alpha v-Integrin antagonist EMD 121974 induces apoptosis in brain tumor cells growing on vitronectin and tenascin. Int J Cancer 2002;98:690–697.
237. Lamfers ML, Grill J, Dirven CM, et al. Potential of the conditionally replicative adenovirus Ad5-Delta24RGD in the treatment of malignant gliomas and its enhanced effect with radiotherapy. Cancer Res 2002;62:5736–5742.
238. Bauerschmitz GJ, Barker SD, Hemminki A. Adenoviral gene therapy for cancer: from vectors to targeted and replication competent agents (review). Int J Oncol 2002;21:1161–1174.
239. Paulus W, Tonn JC. Basement membrane invasion of glioma cells mediated by integrin receptors. J Neurosurg 1994;80:515–519.
240. Bello L, Lucini V, Giussani C, et al. IS20I, a specific alphavbeta3 integrin inhibitor, reduces glioma growth in vivo. Neurosurgery 2003;52:177–185; discussion 185,186.
241. Kirsch M, Strasser J, Allende R, Bello L, Zhang J, Black PM. Angiostatin suppresses malignant glioma growth in vivo. Cancer Res 1998;58:4654–4659.
242. Griscelli F, Li H, Bennaceur-Griscelli A, et al. Angiostatin gene transfer: inhibition of tumor growth in vivo by blockage of endothelial cell proliferation associated with a mitosis arrest. Proc Natl Acad Sci USA 1998;95:6367–6372.
243. Tanaka T, Cao Y, Folkman J, Fine HA. Viral vector-targeted antiangiogenic gene therapy utilizing an angiostatin complementary DNA. Cancer Res 1998;58:3362–3369.

244. Ma HI, Guo P, Li J, et al. Suppression of intracranial human glioma growth after intramuscular administration of an adeno-associated viral vector expressing angiostatin. Cancer Res 2002;62: 756–763.
245. Ma HI, Lin SZ, Chiang YH, et al. Intratumoral gene therapy of malignant brain tumor in a rat model with angiostatin delivered by adeno-associated viral (AAV) vector. Gene Ther 2002;9:2–11.
246. Boehm T, Folkman J, Browder T, O'Reilly MS. Antiangiogenic therapy of experimental cancer does not induce acquired drug resistance. Nature 1997;390:404–407.
247. Morimoto T, Aoyagi M, Tamaki M, et al. Increased levels of tissue endostatin in human malignant gliomas. Clin Cancer Res 2002;8:2933–2938.
248. Peroulis I, Jonas N, Saleh M. Antiangiogenic activity of endostatin inhibits C6 glioma growth. Int J Cancer 2002;97:839–845.
249. Sorensen DR, Read TA, Porwol T, et al. Endostatin reduces vascularization, blood flow, and growth in a rat gliosarcoma. Neuro-Oncology 2002;4:1–8.
250. Sorensen DR, Leirdal M, Iversen PO, Sioud M. Combination of endostatin and a protein kinase Calpha DNA enzyme improves the survival of rats with malignant glioma. Neoplasia (NY) 2002;4: 474–479.
251. Yokoyama Y, Dhanabal M, Griffioen AW, Sukhatme VP, Ramakrishnan S. Synergy between angiostatin and endostatin: inhibition of ovarian cancer growth. Cancer Res 2000;60:2190–2196.

29
Detection of Proliferation in Gliomas by Positron Emission Tomography Imaging

Alexander M. Spence, MD, David A. Mankoff, MD, PhD, Joanne M. Wells, MS, Mark Muzi, MS, John R. Grierson, PhD, Janet F. Eary, MD, S. Finbarr O'Sullivan, PhD, Jeanne M. Link, PhD, Daniel L. Silbergeld, MD, and Kenneth A. Krohn, PhD

SUMMARY

This chapter is a review of positron emission tomography (PET) imaging of cellular proliferation in brain tumors. PET with [C-11]-thymidine or [F-18]-fluorothymidine is in the developmental stages. Estimation of proliferation requires (a) injecting one of these tracers intravenously followed by (b) collecting emission data from the tumor with the tomograph, (c) sampling arterial blood radioactivity during imaging, (d) analyzing plasma metabolites of the tracers, and (e) mathematically modeling all of the data. The calculated specific retention of the tracers as estimates of proliferation is expressed as a flux constant with units of milliliters per minute per gram. The blood–brain barrier (BBB) limits the transport, and ultimately the uptake, of both of these tracers. This complicates assessing proliferation in regions of tumors located behind an intact BBB and requires the estimation of the transport rate in addition to flux in regions where the BBB is broken down. Despite these challenges, early data show that estimations of proliferation by PET correlate well with tumor grade. However, more work is necessary to evaluate how reliably these tracers can be used to assess response to treatment interventions.

Key Words: Brain tumor; glioma; PET; thymidine; fluorothymidine; FLT; proliferation.

1. INTRODUCTION

This chapter reviews positron emission tomography (PET) imaging of biosynthesis of DNA in brain tumors. PET imaging of pathophysiology in vivo requires an understanding of the biology of both tumors and their host organs at a molecular level. The influence of the tumor host organ milieu, viz., brain, on tumor cellular proliferation cannot be assessed in vitro, whereas with PET imaging, proliferation can now be estimated *in situ*. Whole tumors can be analyzed at multiple time-points in the patient's clinical course, an advantage over invasive biopsy approaches that sample limited regions of a tumor and cannot be repeated easily over time. PET imaging is assuming an increasingly important role in the design of cancer treatments as well as monitoring therapy results earlier and more reliably.

For gliomas, standard noninvasive imaging procedures, X-ray computed tomography (CT) and magnetic resonance imaging (MRI), provide excellent anatomical precision and sensitivity. Unfortunately, their specificity in distinguishing neoplastic disease from vascular or inflammatory processes can be problematic in individual cases. Treatment effects including surgical trauma, corticosteroid-induced reduction of edema and contrast enhancement, and radionecrosis

From: *The Cell Cycle in the Central Nervous System*
Edited by: D. Janigro © Humana Press Inc., Totowa, NJ

cannot always reliably be distinguished from tumor recurrence or response to therapy. As gliomas grow and infiltrate the normal cellular milieu, tumor cell density relative to the normal elements increases. MRI T2 and FLAIR sequences reliably estimate the margins of this advancing wave of pathology, but within a given abnormal volume the ratio of tumor to normal elements is not well defined. Magnetic resonance spectroscopy can profile some of the main chemical constituents of tumor and normal tissue but has limitations in that the volume of resolution is large relative to that provided by conventional MRI. As our understanding of pathology moves increasingly to the molecular level, especially in the brain where tissue sampling in vivo carries significant risks, it is imperative that we develop methods to measure molecular pathological processes as they progress over time. PET adds this capability to our clinical and research armamentarium and, if performed rigorously, yields quantitative estimates of metabolic rates. MRI and CT cannot estimate growth rate in a single setting, whereas PET can provide quantitative measures of this with tracers of DNA biosynthesis.

There are several potential applications for proliferation imaging with PET, although a great deal of additional work, such as the following, is needed before these become part of practice in neuro-oncology:

1. Delineating tumors from non-neoplastic processes.
2. Grading and estimating prognosis.
3. Localizing the optimum site for biopsy.
4. Distinguishing recurrence from frank radionecrosis seen with CT and/or MRI contrast enhancement or from nonenhancing radiation-induced white matter changes.
5. Assessing response to therapy.
6. Determining malignant degeneration in low-grade glioma.
7. Designing therapy.

2. [C-11]-THYMIDINE

The most direct measure of tumor growth and proliferation is the rate of DNA synthesis *(1–3)*. Of the four nucleosides used in DNA synthesis, only thymidine (TdR) is used in DNA and not RNA; therefore, most tracer approaches for measuring tumor growth have used a labeled form of TdR or an analog *(1,3)*. Although TdR labeled with H-3 or C-14 has been used for over 40 yr to measure cellular proliferation in vitro, more recently, positron-emitting versions of TdR labeled with C-11 and several TdR analogs labeled with F-18 have been developed *(4,5)*.

The application of cellular proliferation imaging to brain tumors is supported by several investigations of the S phase fraction (SPF) in astrocytic gliomas that established a significant correlation between labeling index (LI) and histological grade *(6–10)*. Glioblastoma SPF averages about 8%, anaplastic astrocytoma 4%, and low grade 1–2%, with normal brain at or close to zero *(9,10)*. Consequently, tracers of DNA synthesis potentially provide high contrast between tumor and normal brain in proportion to the grade and proliferation rate. Following radiotherapy (RT) the LI's in gliomas fall by roughly one-half such that cellular proliferation imaging may reliably measure response to treatment early after the intervention *(9,11)*.

The metabolic pathways traced by TdR and [F-18]-3′-deoxy-3′-fluorothymidine (FLT), the chief analog of TdR used in PET, are illustrated in Fig. 1 and structures in Fig. 2. Circulating TdR, extrinsic to the cell, is taken up and incorporated into DNA via the exogenous or salvage pathway. TdR enters cells by facilitated, non-energy-dependent transporters or by active, Na^+-dependent carriers *(12,13)*. Although TdR delivery from the blood to the cell is rapid in most somatic tissues, limited transport of TdR across the blood–brain barrier (BBB) makes transport a potentially rate-limiting step in brain tumor imaging *(14,15)*.

Once inside the cell, TdR is rapidly phosphorylated three times to TdR triphosphate (TTP) before being incorporated into DNA *(1)*. The initial rate-limiting step is phosphorylation by TdR kinase-1 (TK1), a cytosolic enzyme, to TdR monophosphate (TMP). This supplements the production of TMP via the *de novo* pathway from deoxyuridine MP catalyzed by thymidylate synthase. The ultimate rate-limiting step is the incorporation of TTP into DNA by DNA polymerase *(1,16)*. TK1 is upregulated several-fold as cells pass from G_1 to the S phase of the cell

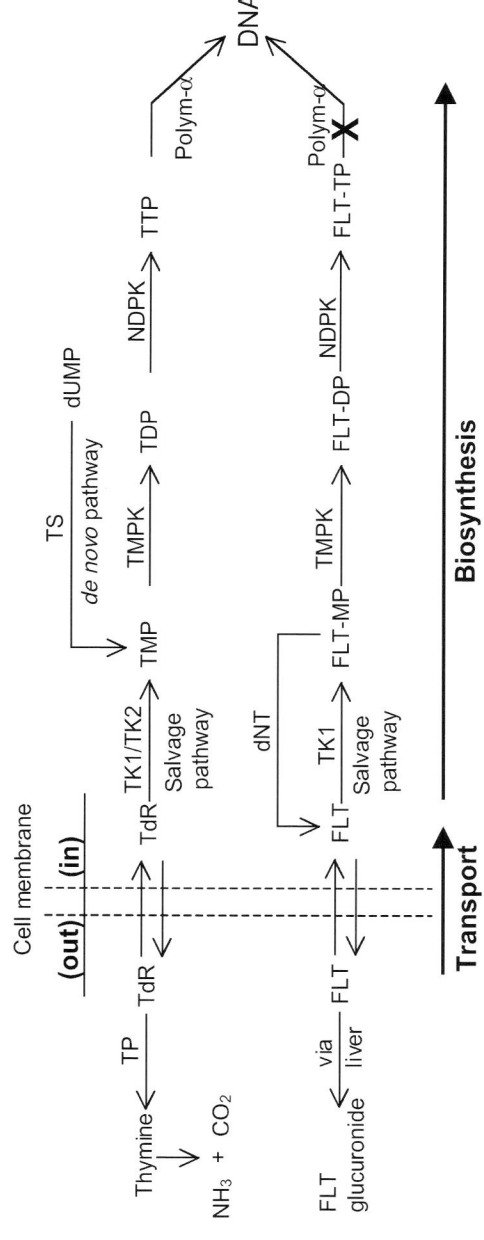

dNT: deoxynucleotidase
dUMP: deoxyuridine monophosphate
NDPK: nucleoside diphosphate kinase
Polym-α: DNA polymerase-α
TK1: thymidine kinase 1
TMPK: thymidylate kinase
TP: thymidine phosphorylase
TS: thymidylate synthase

Fig. 1. The metabolic pathways followed intracellularly by thymidine leading to incorporation in DNA and by FLT through phosphorylation steps in the exogenous (salvage) pathway. PET images measure tracer uptake via the exogenous pathway.

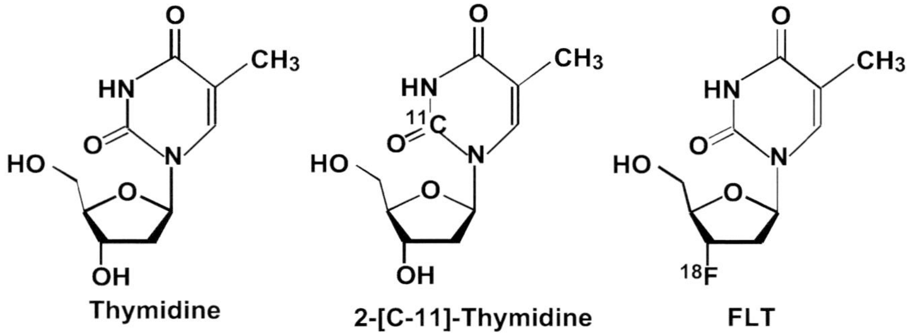

Fig. 2. Chemical structures of 2-[C-11]-thymidine (half-life 20 min) and [F-18]-3′-deoxy-3′-fluorothymidine (half-life 110 min) along side thymidine to show the positions of the radiolabels. These are the main tracers studied thus far for proliferation imaging with PET in gliomas.

cycle *(17)*. TdR also reacts with the mitochondrial isozyme, TdR kinase 2 (TK2) which is used for mitochondrial DNA replication and repair, but TK2 is not regulated by the cell cycle and has low expression in most tissues except liver. Thus, retention of exogenous TdR tracers from TMP to DNA reflects DNA synthesis. For some TdR analogs such as FLT, which are not incorporated into DNA, retention of tracer reflects the activity of *TK1*.

The activity of TK1 and flux of TdR through the salvage pathway are closely related under most circumstances. However, in the cell, the salvage pathway competes with *de novo* synthesis from intracellular deoxyuridine, and therefore the rate of TdR uptake is affected by the relative utilization of the salvage and *de novo* pathways *(1,16)*. Tumors wherein the *de novo* pathway dominates DNA synthesis may be poorly imaged with PET tracers that are limited to following the salvage pathway *(18)*.

TdR labeling with C-11 in either the methyl or 2-position produced the earliest PET tracers for imaging cellular proliferation *(19–22)*. The use of these tracers is complicated by rapid metabolism, yielding [C-11]-CO_2 from 2-[C-11]-TdR and other labeled metabolites from methyl-[C-11]-TdR *(23)*. The main reason for this is the presence of large pools of TdR phosphorylase (TP) in blood, liver, and spleen that rapidly degrade TdR. This limits its incorporation into DNA and produces a large pool of radiolabeled blood metabolites.

Semiquantitative images can be produced by injecting these tracers intravenously and collecting the emission data in the field of view of the PET scanner for up to 60 min *(24–26)*. The signal in the PET images includes not only the radioactivity of TdR bound in the DNA synthetic pathway but also the labeled unmetabolized TdR in the tumor plus that in the blood as well as circulating labeled metabolites, especially [C-11]-CO_2, some of which re-enter the brain and tumor. The image as such is a composite of the bound TdR in the DNA synthetic pathway plus labeled precursor TdR and metabolic byproducts that contaminate the image. Importantly, this approach has limited usefulness because it does not distinguish the proportion of unmetabolized TdR in the tissue that accumulates owing to transport via a disrupted BBB versus the proportion that actually enters and binds in the DNA synthesis pathway.

Despite these confounding issues, a limited number of patient studies collected in this manner have been reported. Vander Borght et al. imaged 13 glioma-bearing patients with 2-[C-11]-TdR *(26)*. There were 10 high-grade, 3 low-grade, 8 untreated, and 5 recurrent tumors. They were imaged for 60 min but blood metabolites were not measured. The tumor-to-cortex uptake ratio was 1.27 ± 0.23 (standard deviation) and no correlation was found between total C-11 uptake and tumor grade. A later study with methyl-[C-11]-TdR showed increased C-11 in eight of ten gliomas *(25)*. However, studies by Lonneux et al. with methyl-[C-11]-TdR showed that the metabolism of this molecule, which results in labeled thymine and several other labeled degradation products, yielded uninterpretable images *(27)*.

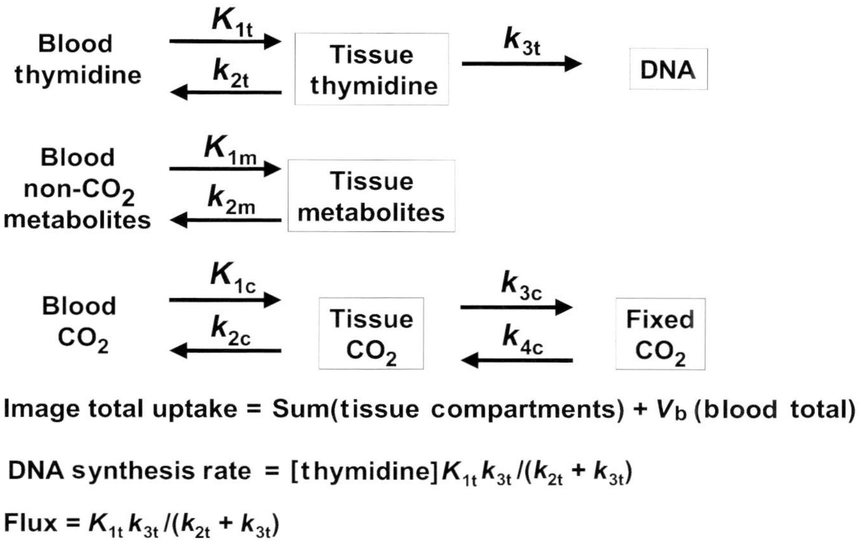

```
      K₁ₜ
Blood  ───▶  Tissue      k₃ₜ
thymidine ◀── thymidine  ───▶  DNA
      k₂ₜ
```

Image total uptake = Sum(tissue compartments) + V_b (blood total)

DNA synthesis rate = [thymidine]$K_{1t} k_{3t} / (k_{2t} + k_{3t})$

Flux = $K_{1t} k_{3t} / (k_{2t} + k_{3t})$

Fig. 3. Schematic of the compartmental model for kinetic analysis of [C-11]-CO_2 and 2-[C-11]-thymidine images.

Our early efforts with methyl-[C-11]-TdR yielded similar limited results, so we developed and validated [C-11]-TdR labeled in the ring-2 position building on the work of Vander Borght (Fig. 2) *(26)*. Although this derivative is metabolized in vivo at the same rate as the labeled methyl derivative, the dominant metabolic product carrying the label is [C-11]-CO_2 whose pharmacokinetics are well described by existing models for imaging brain tumors *(28,29)*.

As stated above, a PET TdR image is the sum of the TdR metabolic processes for delivery, uptake, and retention in DNA by the tumor and degradative metabolism in the whole patient. In the brain this metabolism causes a special set of image analysis challenges. [C-11]-CO_2 rapidly crosses the BBB and distributes throughout the brain and tumor. Blood metabolite analysis and kinetic modeling of dynamic plasma concentrations and imaging data are required to separate the contributions of [C-11]-TdR and [C-11]-CO_2. TdR itself is poorly transported across the BBB causing concern that TdR imaging may not be able to separate permeability in regions of BBB disruption from authentic DNA synthesis in proliferating tumor. This can be accomplished as we have shown in the initial validation of our 2-[C-11]-TdR methods for use in patient imaging studies *(24)*. In the first three patients reported, the studies included [C-11]-CO_2 injections and imaging preceeding the injections of 2-[C-11]-TdR. Analysis of the plasma and tissue kinetics of the [C-11]-CO_2 and 2-[C-11]-TdR data combined with compartmental modeling yielded parameters from which a flux constant for exchange of TdR into DNA was calculated as follows (Fig. 3):

TdR flux constant = $K_{TdR} = K_{1t} k_{3t} / (k_{2t} + k_{3t})$

In this formulation, K_{TdR} represents flux of [C-11]-TdR into the DNA synthetic pathway and K_{1t} represents transport of [C-11]-TdR from blood to the intracellular space which may in large part be owing to disruption of the BBB.

CO_2 rapidly crosses the BBB, whereas TdR has more limited transport *(14,28,29)*. Examination of the TdR-summed image collected from 20–60 min postinjection in Fig. 4B shows high background in the normal cortex, most likely owing to [C-11]-CO_2 that has accumulated from TdR metabolism elsewhere in the body. The image in Fig. 4C demonstrates the application of the kinetic model and mixture analysis to the dynamic TdR images *(30,31)*. It is a calculated image of TdR flux, which indicates the rate of TdR incorporation into DNA with the

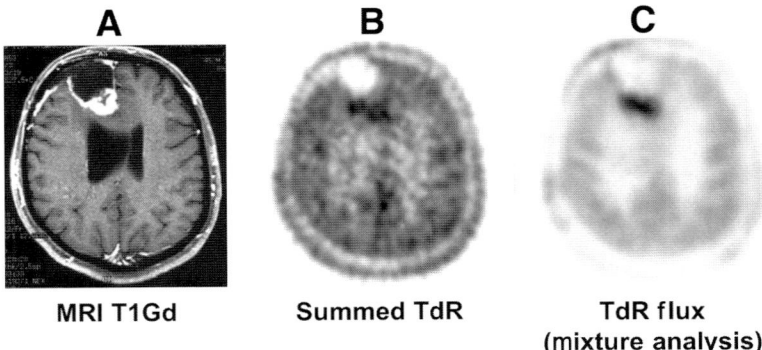

Fig. 4. 2-[C-11]-thymidine images of a patient with a recurrent right frontal glioma. The MRI (**A**) shows distinct contrast enhancement. The summed TdR image (**B**) shows uptake in the tumor but also shows the effect of [C-11]-CO_2 uptake degrading the image. The TdR flux (mixture analysis) image (**C**) shows the focus most clearly in the posterior aspect of the resection cavity. (Reproduced with permission from ref. *24*.)

model correction for labeled metabolites. Comparison of the TdR-summed and TdR flux images from mixture analysis (Fig. 4B vs Fig. 4C) shows the suppression of normal brain CO_2 background and significant enhancement of tumor contrast in the calculated flux image, resulting in much better definition of the tumor.

More recently, detailed kinetic analysis of 2-[C-11]-TdR PET images in a group of patients with a broad range of brain tumors was reported *(15,32)*. Validation of the modeling approach was achieved in that PET studies with sequential injections of [C-11]-CO_2 followed by 2-[C-11]-TdR provided dynamic images that allowed mathematical modeling (Fig. 3) to estimate K_{TdR} (retention) and K_{1t} (transport) with standard errors of 10% and 15%, respectively *(15)*. The positive correlation between tumor grade and TdR flux into tumor DNA suggested that kinetic analysis could separate the effects of altered transport, owing to BBB disruption, from TdR retention in DNA.

In more detail, this study encompassed 20 patients with either primary or recurrent gliomas *(32)*. A small number of patients were studied more than once or had more than one lesion so the total number of kinetic data sets was 26. There were eight oligodendrogliomas of grade II, six mixed gliomas of grade II, one astrocytoma of grade II, two low-grade gliomas not otherwise specified, two anaplastic astrocytomas, and one anaplastic mixed glioma. Fourteen of the tumors had been treated and 10 of the lesions were contrast-enhancing on MRI. One patient had four scans over the course of tumor progression and treatment. An additional patient was included with pure left anterior temporal radionecrosis owing to fast neutron RT of a nasopharyngeal carcinoma.

An example of the images and type of dynamic data are shown in Fig. 5. This patient had a recurrent right parietal anaplastic mixed glioma. For tumor, K_{TdR} (retention) and K_{1t} (transport) were 0.059 and 0.081 mL/min/g, respectively, and for the whole brain 0.016 and 0.011 mL/min/g, respectively.

In the whole patient population there was an association between transport (K_{1t}) and contrast enhancement on gadolinium contrast enhanced MRI in the expected direction of increased TdR transport into the tumors with contrast enhancement as an indication of BBB damage ($p < 0.001$) (Fig. 6A). There was overlap between the estimated retention (K_{TdR}) for contrast-enhancing and -nonenhancing lesions, which suggested that the flux constant was indeed measuring retention, not simply reflecting transport (Fig. 6B). TdR retention in untreated tumors ($n = 9$ for low grade, $n = 3$ for high grade) was significantly greater in the high-grade than low-grade tumors despite low population numbers ($p < 0.02$) (Fig. 7A). Previously treated patients with high-grade tumors tended to show lower TdR retention than those not yet

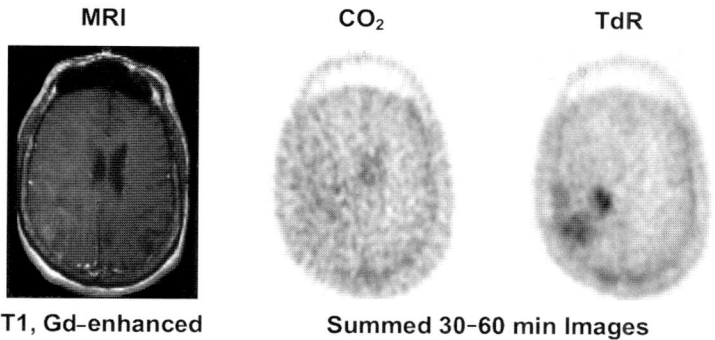
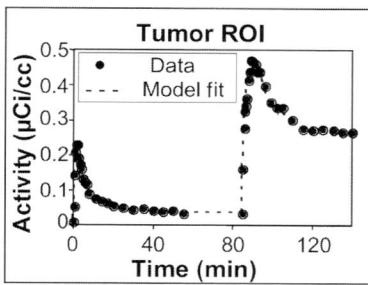
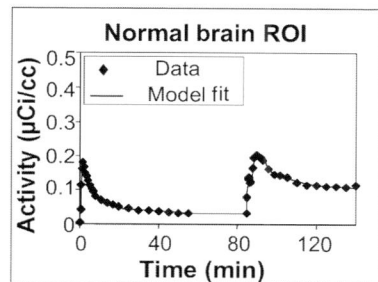

Fig. 5. Summed [C-11]-CO_2 and [C-11]-TdR images from a recurrent anaplastic mixed glioma with corresponding tissue time-activity curves representing the tissue dynamic PET data in the tumor region of interest (ROI) and whole brain ROI. [C-11]-CO_2 was injected at $t = 0$ min and [C-11]-TdR was injected at $t = 85$ min. The initial peaks are from the [C-11]-CO_2 injection and the second peaks are owing to the injection of [C-11]-TdR. (Reproduced with permission from ref. 32.)

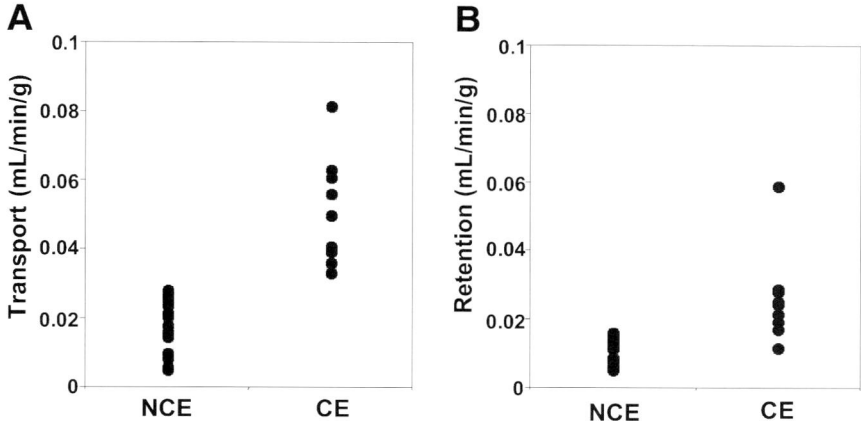

Fig. 6. Relationship between contrast enhancement on MRI and thymidine transport (K_{1t}) (**A**) and thymidine retention (K_{TdR}) (**B**). CE, contrast enchancement; NCE, nonenhanced. (Reproduced with permission from ref. 32.)

treated. The lack of significance was likely owing to the small number of cases ($n = 3$ for untreated, $n = 4$ for treated) ($p > 0.2$) (Fig. 7B). Also, there was no significant difference between patients with low-grade tumors that had received or had not received treatment ($p > 0.7$) (Fig. 7B).

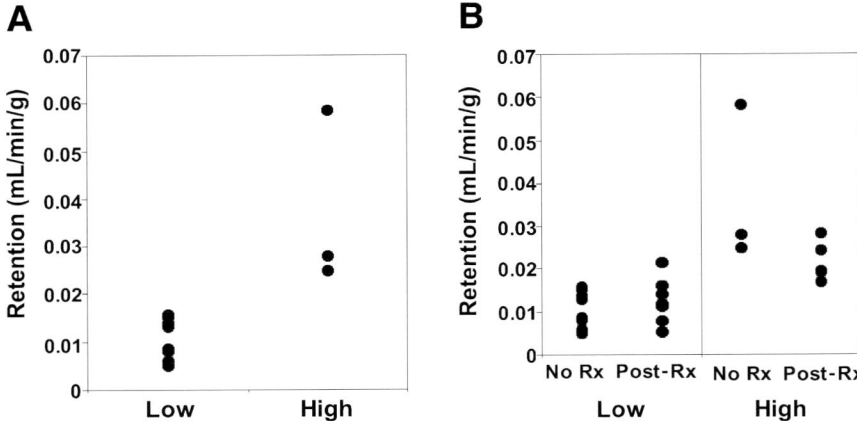

Fig. 7. Thymidine retention against tumor grade for untreated tumors (**A**) and tumor grade and treatment status (**B**). (Reproduced with permission from ref. *32*.)

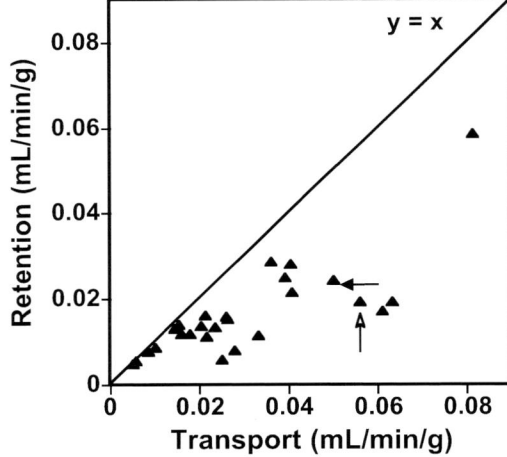

Fig. 8. Relationship between K_{TdR} (retention) and K_{1t} (transport). Closed arrow is a patient with an extremely aggressive tumor; open arrow is the patient with pure radionecrosis. (Reproduced with permission from ref. *32*.)

The relationship between K_{TdR} and K_{1t} for all lesions studied is shown in Fig. 8. At low retention and transport rates (K_{TdR} and $K_{1t} < 0.01$ mL/min/g), at the level of normal brain, K_{1t} and K_{TdR} closely correlate, which may indicate that tracer uptake is transport limited. Above these levels, retention and transport were not proportional, suggesting that they can be measured independently. Interestingly, the patient shown with the closed arrow had an extremely aggressive tumor, and the model estimated a high retention and a high transport rate. The patient shown with the open arrow had pure radionecrosis resulting from neutron RT of an extracranial nasopharyngeal carcinoma. Although the model estimated a nonzero K_{TdR}, it is in the region of the graph, where K1t is much greater than K_{TdR}, suggesting a predominant effect of transport rather than retention in DNA.

One patient was scanned four times over the course of tumor progression and treatment. The time-course of changes in K_{TdR} and K_{1t} is shown in Fig. 9. The tumor was a mixed grade II glioma at the time of the first scan when the model estimated a low level of retention. At the time of the

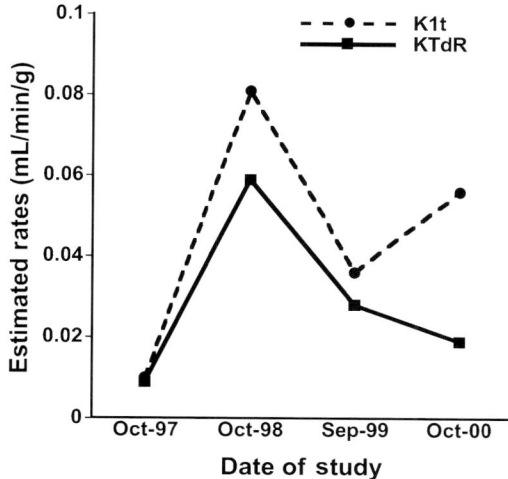

Fig. 9. Estimated flux constants and transport rates for a single patient scanned four times over course of tumor progression and treatment: Oct-97, grade II mixed glioma diagnosed by biopsy; Oct-98, transformed to grade III followed by surgery, chemotherapy and RT; Sep-99, after therapy when retention and transport both were reduced; and Oct-00, increased contrast enhancement on MRI raising question of active tumor vs radionecrosis, but the kinetic analysis showed a decreased retention and increased transport rates consistent with a radionecrotic effect. (Reproduced with permission from ref. *32*.)

second scan, MRI suggested that the tumor had transformed into a high-grade lesion, and the K_{TdR} was also much higher. Partial resection confirmed a mixed grade III glioma. After treatment by RT and chemotherapy (third scan), the model estimated a lower retention level, corresponding with residual contrast enhancement on MRI and a clinical course suggesting treated, but viable, tumor. One year later the patient was clinically well but had increased contrast enhancement on MRI, leading to the clinical question of active tumor versus radionecrosis. The kinetic analysis showed a decreased retention and increased transport rates, suggesting radionecrosis. The patient did well for 1 yr without additional treatment or evidence of tumor progression.

3. [F-18]-3′-DEOXY-3′-FLUOROTHYMIDINE

An alternate approach to imaging brain tumor cellular proliferation is to label TdR analogs with F-18 that are resistant to degradation by TP. This eliminates the background of labeled metabolites and provides a longer-lived and more convenient tracer *(33)*. The leading compound for this is [F-18]-fluoro-L-thymidine (FLT) *(34,35)*. FLT is a selective substrate for TK1, which is used for nuclear DNA replication, but only goes as far as the triphosphate along the DNA synthesis pathway (Fig. 1). FLT undergoes relatively little degradation after injection aside from production of the glucuronide in the liver. *TP* in the blood does not break it down. However, as with TdR the uptake of FLT is restricted by the BBB.

Several studies of tumors exposed in vitro to FLT have validated that its uptake correlates positively with TK1 activity in cycling cells *(18,36)*. A study of the short-term metabolic fate of FLT in the DNA salvage pathway in exponentially growing A549 tumor cells has provided additional validation of the usefulness of FLT for imaging cellular proliferation (Fig. 10) *(37)*. TK1 activity produced FLT-monophosphate (FLTMP) which dominated the labeled nucleotide pool. Subsequent phosphorylations by thymidylate kinase (TMPK) and nucleotide diphosphate kinase (NDPK) led to FLT-triphosphate (FLTTP) which comprised about 30% of the metabolic pool after 1 h. A putative deoxynucleotidase (dNT), which degrades FLTMP to FLT, provided the primary mechanism for tracer efflux from cells. In contrast, FLTTP was resistant to degradation and highly retained.

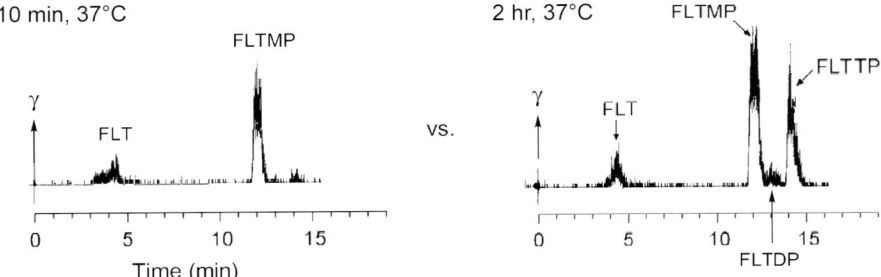

Fig. 10. [F-18]-FLT metabolism in A549 cell cultures evaluated after 10 min and 2 h exposure by ion exchange analysis of activity accumulated by A549 cells. All three FLT nucleotides were detected in cell extracts. Initial metabolism was dominated by monophosphate (FLTMP) synthesis, whereas exposure for 2 h led to an additional significant triphosphate (FLTTP) fraction. Only trace amounts of the diphosphate (FLTDP) were detected. Metabolites were exclusively nucleotides. (Reproduced with permission from ref. *37*.)

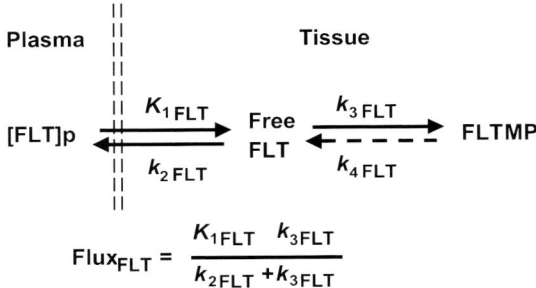

Fig. 11. Schematic of a two-compartment, rate-constant-4 model for kinetic analysis of FLT-PET images.

The mathematical model describing the uptake of FLT is much simpler than that for TdR because the only metabolite in the plasma that needs to be measured is the glucuronide conjugate (Fig. 11). Similar to TdR, retention of FLT is reflected in the flux constant, K_{FLT}, and transport in K_{1FLT}

$$\text{Flux constant} = K_{FLT} = K_{1FLT} \, k_{3FLT} / (k_{2FLT} + k_{3FLT})$$

Few brain tumor cases have been studied with FLT-PET. Sloan and co-workers reported 29 patients with gliomas at presentation or recurrence *(38–40)*. Tracer uptake was quantified as standard uptake value (SUV) which equals the tissue concentration of tracer normalized to the injected dose of tracer and body weight, that is, SUV = tissue concentration of tracer/ (injected tracer dose/body weight). The value of the pixel in the image with the highest SUV is the SUV_{max}. From their results SUV_{max} of FLT correlated with tumor grade and MIB-1 immunolabeling for proliferation. SUV_{max} in normal brain was 0.32, low-grade gliomas, 0.49–0.79; intermediate grade, 0.85–1.46; and high grade, 2.04–3.90. Uptake in areas of radionecrosis was low with SUV_{max} less than 1.0. FLT uptake was seen exclusively in areas which showed contrast enhancement in MRI images *(40)*. These investigators also reported that brain tumor tissue biopsied shortly after FLT injection contained the labeled species largely in the form of phosphorylated FLT *(38)*.

Our preliminary work is illustrated in Table 1 and Figs. 12–14. In Table 1 and Fig. 12 the data and images are shown for a 67-yr-old man who had a bicentric gliosarcoma of the right frontal and temporal lobes. The right temporal lesion was resected and the frontal lesion was left

Table 1
FLT Modeling Results on the Patient Shown in Fig. 12

Parameter	Pre-RT Right frontal tumor	Post-RT Right frontal tumor
ROI volume of increased uptake (cc)	19.0	33.8
Tumor volume encompassed by increased uptake (cc)	48.4	73.9
Flux (mL/min/g)	0.027	0.015
K1 (mL/min/g)	0.066	0.059

See text for additional explanation.

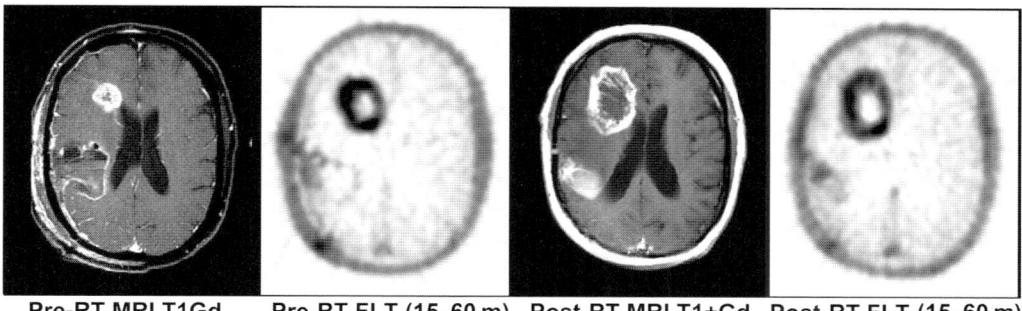

Pre-RT MRI T1Gd   Pre-RT FLT (15–60 m)   Post-RT MRI T1+Gd   Post-RT FLT (15–60 m)

Fig. 12. Images from a 65-yr-old man with a bicentric gliosarcoma in the right temporal lobe (resected) and the right frontal lobe (unresected). Labels under the images identify their sources and times.

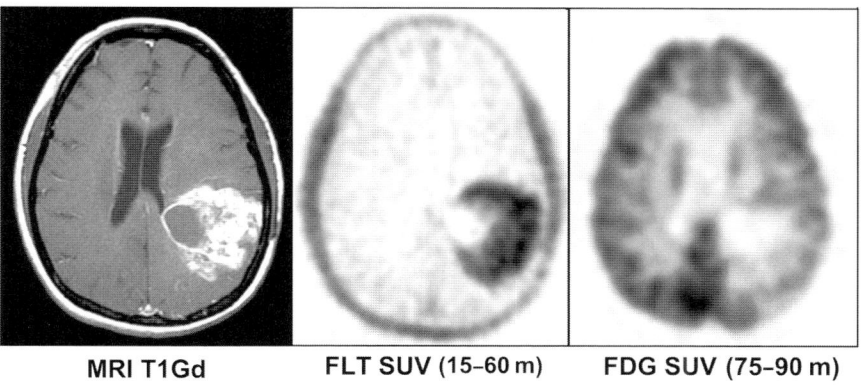

MRI T1Gd   FLT SUV (15–60 m)   FDG SUV (75–90 m)

Fig. 13. Gadolinium contrast-enhanced MRI, FLT, and FDG images of an anaplastic astrocytoma of the left parietal lobe. Uptake of FLT is in the tumor region that showed BBB breakdown on gadolinium contrast enhanced MRI, but not outside this region.

untouched. FLT images were obtained both 2.5 wk before and 2.5 wk after 63 Gy of fractionated external beam RT. The K_{FLT} before RT for the frontal lesion was 0.030 mL/min/g and the tumor volume encompassed by increased FLT uptake was 48.4 cc. The images taken after RT show that the volume had decidedly increased as assessed by gadolinium contrast enhanced MRI and the FLT volumes (73.9 cc), consistent with tumor progression by conventional standards *(41)*. However, the K_{FLT} was roughly halved after RT (0.017 mL/min/g), suggesting that

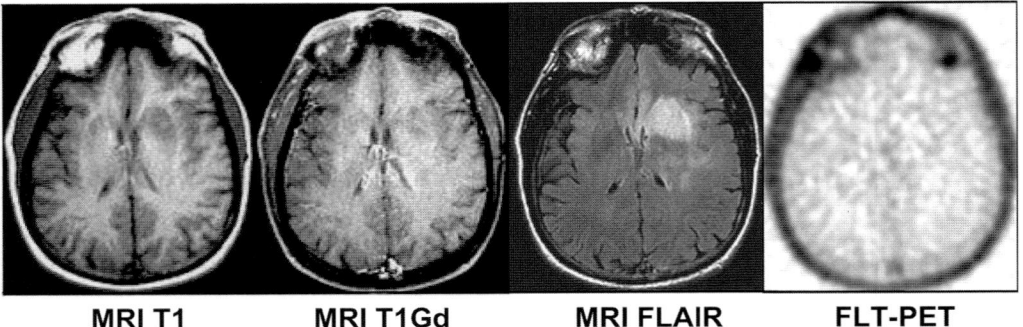

Fig. 14. Images from a 61-yr-old man who had a bifrontal anaplastic astrocytoma. The MRI T1 and T1Gd show no contrast enhancement, that is, no detectable breakdown of the BBB. The FLAIR image shows abnormality in the deep medial left frontal lobe. Other slices showed similar findings in the right frontal lobe. The FLT image shows no detectable uptake even though the microscopic examination of the tumor showed a MIB-1 labeling level of 10%.

growth rate had been reduced by the RT, but this was not enough to stabilize or shrink the tumor volume. The FLT-PET images clearly show an additional dimension to assessing response to therapy achievable by this approach. With additional therapy the patient survived a total of 10 mo from diagnosis.

The images shown in Fig. 13 are those of a 44-yr-old woman who had an anaplastic astrocytoma of the left parietal lobe biopsied in 2001. She received 59.4 Gy of external beam RT. In May and October of 2003, her gadolinium contrast enhanced MRI scans raised the question of recurrence because of increased volume. However, her neurological examination showed minimal findings despite the presence of a very large lesion on MRI. This raised the question of radionecrosis and lead to successive PET studies with FLT and 2-[F-18]-fluoro-2-deoxyglucose (FDG). There was uptake of FLT in the tumor in the volume that showed BBB breakdown on gadolinium contrast enhanced MRI, but not outside this volume. The tumor and brain K_{1FLT} transport levels were 0.058 and 0.005 mL/min/g, respectively, whereas the K_{FLT} flux constants were 0.017 and 0.003 mL/min/g, respectively. The level of FDG uptake was greater than white matter but less than cortex consistent with recurrence of the tumor but this interpretation is associated with a sensitivity of 86% and a specificity of only 22% *(42)*. All of the clinical, MRI, and PET data together favored recurrence over radionecrosis.

Another illustrative case is shown in Fig. 14. This 61-yr-old man had a bifrontal anaplastic astrocytoma that completely lacked contrast enhancement on gadolinium contrast-enhanced MRI at the time of clinical presentation. Biopsy preceeded the FLT-PET by 20 d. This malignant glioma had a high MIB-1 level of 10% but no breakdown of the BBB and little, if any, uptake of FLT. The tumor and brain K_{1FLT} transport levels were 0.006 and 0.011 mL/min/g, respectively, whereas the K_{FLT} flux constants were both 0.003 mL/min/g. These numbers suggest very low transport into tumor and brain. The unexpectedly low flux value for tumor is likely a reflection of transport limitation because the tumor MIB-1 was 10%. As with TdR, low transport may limit tracer uptake and impair the ability to estimate flux.

4. CONCLUSIONS

Future studies will clarify the capacity of PET imaging of cellular proliferation to measure early response to treatment. Pilot studies in somatic tumors have demonstrated efficacy for this *(43,44)*. In brain tumors proliferation imaging with FLT shows significant promise but has not been exploited thoroughly enough to allow judgement of its potential benefit to the practice of neuro-oncology. Further studies will be necessary to determine how significantly the BBB exclusion of proliferation tracers limits the usefullness of this imaging technique to estimate flux. Additional tracers also need to be developed which circumvent the limitation imposed by the BBB.

ACKNOWLEDGMENT

This work was supported by Grants nos. CA42045 and S10RR17229.

REFERENCES

1. Cleaver JE. Thymidine metabolism and cell kinetics. Front Biol 1967;6:43–100.
2. Livingston RB, Ambus U, George SL, Freireich EJ, Hart JS. In vitro determination of thymidine-[H-3] labeling index in human solid tumors. Cancer Res 1974;34:1376–1380.
3. Tannock IF, Hill RP (eds). The Basic Science of Oncology. New York: McGraw-Hill, 1992.
4. Krohn KA, Mankoff DA, Eary JF. Imaging cellular proliferation as a measure of response to therapy. J Clin Pharmacol Suppl 2001;4:S96–S103.
5. Mankoff DA, Dehdashti F, Shields AF. Characterizing tumors using metabolic imaging: PET imaging of cellular proliferation and steroid receptors. Neoplasia 2000;2:71–88.
6. Coons SW, Johnson PC, Pearl DK. Prognostic significance of flow cytometry deoxyribonucleic acid analysis of human astrocytomas. Neurosurgery 1994;35:119–125.
7. Hoshino T, Ahn D, Prados MD, Lamborn K, Wilson CB. Prognostic significance of the proliferative potential of intracranial gliomas measured by bromodeoxyuridine labeling. Int J Cancer 1993;53:550–555.
8. Lamborn KR, Prados MD, Kaplan SB, Davis RL. Final report on the University of California-San Francisco experience with bromodeoxyuridine labeling index as a prognostic factor for the survival of glioma patients. Cancer 1999;85:925–935.
9. Matsutani M. Cell kinetics, In: Berger MS, Wilson CB (eds). The Gliomas. Philadelphia: WB Saunders Co., 1999.
10. Shibuya M, Ito S, Davis RL, Wilson CB, Hoshino T. A new method for analyzing the cell kinetics of human brain tumors by double labeling with bromodeoxyuridine in situ and with iododeoxyuridine in vitro. Cancer 1993;71:3109–3113.
11. Fujimaki T, Matsutani M, Takakura K. Analysis of BUdR (bromodeoxyuridine) labeling indices of cerebral glioblastomas after radiation therapy. J Jpn Soc Ther Radiol Oncol 1990;2:263–273.
12. Damaraju VL, Damaraju S, Young JD, et al. Nucleoside anticancer drugs: the role of nucleoside transporters in resistance to cancer chemotherapy. Oncogene 2003;22:7524–7536.
13. Young JD, Cheeseman CI, Mackey JR, Cass CE, Baldwin SA. Gastrointestinal Transport, Molecular Physiology, In: Fambrough D, Benos D, Barrett K, Domowitz M (eds). Current Topics in Membranes. San Diego, CA: Academic Press, 2000, pp 329–378.
14. Cornford EM, Oldendorf WH. Independent blood–brain barrier transport systems for nucleic acid precursors. Biochim Biophys Acta 1975;394:211–219.
15. Wells JM, Mankoff DA, Muzi M, et al. Kinetic analysis of 2-[11C] thymidine PET imaging studies of malignant brain tumors: compartmental model investigation and mathematical analysis. Mol Imaging 2002;1:151–159.
16. Mankoff DA, Shields AF, Graham MM, Link JM, Eary JF, Krohn KA. Kinetic analysis of 2-[carbon-11]thymidine PET imaging studies: compartmental model and mathematical analysis. J Nucl Med 1998;39:1043–1055.
17. Sherley JL, Kelly TJ. Regulation of human thymidine kinase during the cell cycle. J Biol Chem 1988;263:8350–8358.
18. Schwartz JL, Tamura Y, Jordan R, Grierson JR, Krohn KA. Monitoring tumor cell proliferation by targeting DNA synthetic processes with thymidine and thymidine analogs. J Nucl Med 2003;44:2027–2032.
19. Christman D, Crawford EJ, Friedkin M, Wolf AP. Detection of DNA synthesis in intact organisms with positron-emitting (methyl- 11 C)thymidine. Proc Natl Acad Sci USA 1972;69:988–992.
20. Link JM, Grierson J, Krohn K. Alternatives in the synthesis of 2-[C-11]-thymidine. J Label Comp Radiopharm 1995;37:610–612.
21. Sundoro-Wu BM, Schmall B, Conti PS, Dahl JR, Drumm P, Jacobsen JK. Selective alkylation of pyrimidyldianions: synthesis and purification of 11C labeled thymidine for tumor visualization using positron emission tomography. Int J Appl Radiat Isot 1984;35:705–708.

22. Vander Borght T, Labar D, Pauwels S, Lambotte L. Production of [2-11C]thymidine for quantification of cellular proliferation with PET. Int J Rad Appl Instrum [A] 1991;42:103–104.
23. Shields AF, Lim K, Grierson J, Link J, Krohn KA. Utilization of labeled thymidine in DNA synthesis: studies for PET. J Nucl Med 1990;31:337–342.
24. Eary JF, Mankoff DA, Spence AM, et al. 2-[C-11]thymidine imaging of malignant brain tumors. Cancer Res 1999;59:615–621.
25. De Reuck J, Santens P, Goethals P, et al. [Methyl-11C]thymidine positron emission tomography in tumoral and non-tumoral cerebral lesions. Acta Neurol Belg 1999;99:118–125.
26. Vander Borght T, Pauwels S, Lambotte L, et al. Brain tumor imaging with PET and 2-[carbon-11]thymidine. J Nucl Med 1994;35:974–982.
27. Lonneux M, Labar D, Bol A, Jamar F, Pauwels S. Uptake of 2-11-Cthymidine in colonic and bronchial tumors. In: Paans AMS, Pruim J, Franssen EJF, Vaalburg W, (eds), Metabolic Imaging of Cancer: Proceedings of the European Conference on Research and Application of Positron Emission Tomography in Oncology. Groningen, the Netherlands: PET-Centrum AZG, 1996.
28. Brooks DJ, Lammertsma AA, Beaney RP, et al. Measurement of regional cerebral pH in human subjects using continuous inhalation of $11CO_2$ and positron emission tomography. J Cereb Blood Flow Metab 1984;4:458–465.
29. Buxton RB, Wechsler LR, Alpert NM, Ackerman RH, Elmaleh DR, Correia JA. Measurement of brain pH using $11CO_2$ and positron emission tomography. J Cereb Blood Flow Metab 1984;4:8–16.
30. O'Sullivan F. Metabolic images from dynamic positron emission tomography studies. Stat Methods Med Res 1994;3:87–101.
31. O'Sullivan F, Muzi M, Graham MM, Spence AM. Parametric imaging by mixture analysis in 3-D: Validation for dual-tracer glucose studies. In: Myers R, Cunningham V, Bailey D, Jones T, (eds), Quantitation of Brain Function Using PET. London: Academic Press, Inc, 1996, pp 297–300.
32. Wells JM, Mankoff DA, Eary JF, et al. Kinetic analysis of 2-[11C]thymidine PET imaging studies of malignant brain tumors: preliminary patient results. Mol Imaging 2002;1:145–150.
33. Shields AF, Grierson JR, Kozawa SM, Zheng M. Development of labeled thymidine analogs for imaging tumor proliferation. Nucl Med Biol 1996;23:17–22.
34. Grierson JR, Shields AF. Radiosynthesis of 3′-deoxy-3′-[(18)F] fluorothymidine: [(18)F]FLT for imaging of cellular proliferation in vivo. Nucl Med Biol 2000;27:143–156.
35. Shields AF, Grierson JR, Dohmen BM, et al. Imaging proliferation in vivo with [F-18]FLT and positron emission tomography. Nat Med 1998;4:1334–1336.
36. Rasey JS, Grierson JR, Wiens LW, Kolb PD, Schwartz JL. Validation of FLT uptake as a measure of thymidine kinase-1 activity in A549 carcinoma cells. J Nucl Med 2002;43:1210–1217.
37. Grierson JR, Schwartz JL, Muzi M, Jordan R, Krohn KA. Metabolism of 3′-deoxy-3′-[F-18]fluorothymidine (FLT) in proliferating A549 cells: validations for positron emission tomography (PET). Nuc Med Biol 2004;31:829–837.
38. Bendaly EA, Sloan AE, Dohmen BM, et al. Use of 18F-FLT-PET to assess the metabolic activity of primary and metastatic brain tumors (abstract). J Nucl Med 2002;3:111P.
39. Sloan AE, Bendaly EA, Dohman BM, et al. Use of 18F-FLT-PET to assess the metabolic activity of primary, recurrent and metastatic brain tumors (abstract). Neuro-Oncology 2002;4:363.
40. Sloan AE, Shields AF, Kupsky W, et al. Superiority of [F-18]FLT-PET compared to FDG PET in assessing proliferative activity and tumor physiology in primary and recurrent intracranial gliomas (abstract). Neuro-Oncology 2001;3:345.
41. Macdonald DR, Cascino TL, Schold SCJ, Cairncross JG. Response criteria for phase II studies of supratentorial malignant glioma. J Clin Oncol 1990;8:1277–1280.
42. Ricci PE, Karis JP, Heiserman JE, Fram EK, Bice AN, Drayer BP. Differentiating recurrent tumor from radiation necrosis: time for re-evaluation of positron emission tomography? AJNR Am J Neuroradiol 1998;19:407–413.
43. Shields AF, Mankoff DA, Link JM, et al. Carbon-11-thymidine and FDG to measure therapy response. J Nucl Med 1998;39:1757–1762.
44. Vesselle H, Grierson J, Muzi M, et al. In vivo validation of 3′deoxy-3′-[(18)F]fluorothymidine ([(18)F]FLT) as a proliferation imaging tracer in humans: correlation of [(18)F]FLT uptake by positron emission tomography with Ki-67 immunohistochemistry and flow cytometry in human lung tumors. Clin Cancer Res 2002;8: 3315–3323.

30
Transformation of Normal Astrocytes Into a Tumor Phenotype

Sean E. Aeder, PhD and Isa M. Hussaini, PhD

SUMMARY

Genetic alterations leading to changes in cell cycle regulation and growth factor signaling transform astrocytes into a tumor phenotype. This chapter describes four grades of astrocytic tumors, discusses how genetic alterations lead to astrocyte transformation, and stresses how growth factor and cell cycle signaling networks contribute to this transformation. Finally we discuss cell lines and animal models pertinent to the study of astrocytic tumors.

Key Words: Astrocyte; cell cycle; growth factor; invasion; PKC; astrocytoma; glioblastoma; RB; p53; transformation.

1. INTRODUCTION

Differentiation of neural stem cells into protoplasmic and fibrillary astrocytes is largely controlled by signals from growth factors and cytokines produced during development and throughout life. These signals include the epidermal growth factor (EGF), cytokines of the ciliary neurotrophic factor (CNTF)/leukemia inhibitory factor (LIF) family, and bone morphogenetic proteins (BMPs). After differentiation, astrocytes provide a wide variety of functions throughout the central nervous system (CNS). They are responsible for regulating neuronal growth and survival *(1)*, guiding neuron migration during development *(2–7)*, promoting synapse formation and modulating synaptic transmission *(8–11)*, and mediating inflammatory responses during infection and injury *(12,13)*.

With a retained capacity for cell division, neural stem cells and astrocytes are especially susceptible to transformation. This susceptibility is enhanced following high doses of radiation, such as those given during prophylactic therapy for acute lymphoblastic leukemia or for pituitary tumors *(14–16)*. Additionally, syndromes predisposing patients to these tumors include Li-Fraumeni syndrome, neurofibromatosis 1 and 2, tuberous sclerosis, and von Hippel-Lindau and Turcot's syndrome. These disorders involve genetic mutations that increase the risk for a variety of cancers in addition to astrocytomas *(17)*. Apart from rare hereditary syndromes and ionizing radiation, no major environmental or lifestyle factors that increase the incidence of astrocytomas have been identified. However, many studies have shown that unregulated cell cycle and stimulation of signaling pathways similar to those that are activated during astrocyte differentiation synergistically enhance astrocyte transformation.

2. CLASSIFICATION OF ASTROCYTIC TUMORS

Astrocytic tumors are the largest, most complex, and most diverse group of neuroectodermal tumors. The World Health Organization (WHO) classifies astrocytic tumors into four malignancy grades based on their histology, with grade I the least aggressive and grade IV being the most aggressive. Grade I pilocytic astrocytomas are generally well-circumscribed, slow-growing tumors found in the optic nerve, the optic chiasm/hypothalamus, thalamus, and basal ganglia, cerebral hemispheres, cerebellum, brainstem, or spinal cord of children and young adults. Grades II–IV tumors have average rates of survival of 7, 2, and less than 1 yr, respectively *(18,19)*. These tumors are more common in adult males and most are located in the cerebral hemispheres. Less studied astrocytic tumors include subependymal giant cell astrocytoma (SEGA) and pleomorphic xanthoastrocytoma (PXA), which correspond to WHO grades I and II, respectively. Two tumors beyond the scope of this chapter include grade II oligoastrocytomas and grade III anaplastic oligoastrocytomas. These tumors are made up of two populations of cells, oligodendroglioma cells and astrocytic tumor cells. Just as with astrocytic tumors, genetic alterations in these cells affect growth factor receptor and cell cycle signaling pathways *(20,21)*.

2.1. Pilocytic Astrocytoma

Pilocytic astrocytomas are WHO grade I tumors, and are unusual in their ability to maintain their relatively benign status over many years. With a lack of an invasive phenotype and a slow rate of proliferation, they are generally nonaggressive and can be completely resected when the regional anatomy permits. A limited number of studies looking at the underlying genetic abnormalities of pilocytic astrocytomas have found allelic losses on chromosomes 17p and 17q. These chromosomes include the p53 and neurofibromatosis I (*NF1*) genes, whose protein products regulate the cell cycle and Ras activity, respectively.

2.2. Diffuse Astrocytoma

Cells of grades II–IV astrocytic tumors constitute a range of neoplasms that share molecular and genetic abnormalities, a propensity for diffuse microinvasion of surrounding tissues, and a tendency for local recurrence. Additionally, sequential acquisitions of genetic alterations enable these tumors to increase in grade with time.

2.2.1. WHO Grade II

Diffuse astrocytomas classified as WHO grade II are the lowest grade of common adult astrocytic tumors. Cells from these tumors are highly differentiated, slow-growing, and infiltrate into neighboring brain structures, making surgical resection difficult. They show little nuclear atypia, an increase in cell size, and a cytoplasm that extends into fine processes. Mutated p53 is common with diffuse astrocytomas, though the frequency of these mutations does not increase with malignant progression. Increased genomic instability resulting from the loss of wild-type p53 activity provides a mechanism for progression of these grade II tumors to anaplastic astrocytomas and glioblastoma multiforme (GBMs) *(22–24)*. Other mutations found in diffuse astrocytomas include an increase in platelet-derived growth factor receptor-α (PDGFR-α) and PDGF-α mRNA *(25)*, a gain of chromosome 7q, and amplification of 8q *(26,27)*.

2.2.2. WHO Grade III: Anaplastic Astrocytoma

WHO grade III anaplastic astrocytomas are made up of diffusely infiltrating and highly proliferative tumor cells. They can arise from lower grade astrocytomas and have a notorious tendency to escalate in malignancy to GBM (grade IV). Anaplastic astrocytomas show increased cellularity, they are more pleomorphic than those found in grade II astrocytic tumors, and display distinct nuclear atypia along with increased mitotic activity. Like grade II diffuse astrocytomas, anaplastic astrocytomas display a high frequency of p53 mutations. Additional genetic changes in anaplastic astrocytomas include a specific inhibitor of cyclin-dependent kinase 4 (INK4A) deletions, retinoblastoma (Rb) alterations, p14^{ARF} deletion, cyclin-dependent kinase 4 (CDK4) amplification, phosphatase and tensin homolog deleted on chromosome 10 (PTEN) mutations, loss of heterozygosity (LOH) on chromosome 10q, LOH on chromosome 19q, LOH on chromosome 22q, and deletion of chromosome 6. Unlike grade IV tumors, less than 10% of anaplastic astrocytomas exhibit epidermal growth factor receptor (EGFR) amplification.

2.2.3. WHO Grade IV: GBM

GBMs are the most frequent malignant primary brain tumor, accounting for 12–15% of all intracranial neoplasms and 50–60% of astrocytic tumors *(28)*. They are malignant astrocytic tumors, made up of heterogeneous populations of poorly differentiated neoplastic astrocytes, hyperproliferative endothelial cells, macrophages, and trapped portions of normal brain structures. Cellular polymorphism, nuclear atypia, and high-mitotic activity are found in these grade IV tumors. Tumor cell infiltration is extensive with GBMs, often extending into the adjacent cortex, the basal ganglia, and the contralateral hemisphere. Prominent microvascular proliferation and extensive pseudopalisading necrosis, which may comprise 80% of the tumor mass, are hallmarks of this tumor.

The vast majority (>80%) of GBMs arise in a *de novo* manner with no evidence of a previous lesion or lower malignancy (primary GBMs); however, some develop from progression of grade II and III astrocytic tumors (secondary GBMs). Even though they arise through distinct mechanisms, primary and secondary GBMs show disruptions in regulatory elements of the same genetic pathways and share a common phenotypic end point. Common genetic mutations in primary GBMs include amplification and overexpression of EGFR, amplification and overexpression of human double minute 2 (HDM2), INK4A deletion, LOH of 10p and 10q, PTEN mutations, and Rb alterations. Unlike primary GBMs, secondary GBMs appear to develop genetic alterations in an almost stepwise manner, with the progression from grade II diffuse astrocytomas to GBMs taking anywhere from less than 1 yr to more than 10 yr *(29)*. Initially, differentiated astrocytes or precursor cells become low-grade astrocytomas with p53 mutations and PDGFR-α overexpression. Then, the low-grade astrocytomas further differentiate into anaplastic astrocytomas with frequent mutations including LOH 19q and Rb alterations. Finally, genetic alterations that may accompany the progression to secondary GBMs include LOH 10q, PTEN mutations, loss of deleted in colorectal cancer (DCC) expression, and PDGFR-α amplification. In addition to mutations and losses of genes, silencing by aberrant promoter hypermethylation has been identified as an important mechanism for inactivating some genes in GBMs *(30–32)*. These silenced genes can include p53, Rb, and p14^{ARF}.

2.3. Pleomorphic Xanthoastrocytoma

PXAs are rare (>1% of astrocytic neoplasms) WHO grade II astrocytic tumors that may recur as a malignant diffuse astroctyoma. These tumors occur most commonly in the temporal lobe of patients that have a long-standing history of chronic epilepsy. Unlike diffuse astrocytomas, PXAs show little invasion into adjacent brain and contain areas of necrosis only when undergoing progression to an anaplastic form. The limited numbers of studies looking at genetic alterations in PXA cells have indicated that p53 mutations, EGFR amplification, and HDM2 overexpression may play a role in this tumor *(33,34)*. Interestingly, LOH of 10q, which is common in diffuse astrocytomas, has not been reported in PXAs. This suggests that genetic alterations involved in PXA tumor formation and progression may be significantly different from those involved in other astrocytic tumors.

2.4. Subependymal Giant-Cell Astrocytoma

SEGA, corresponding to WHO grade I, is a circumscribed, slowly growing tumor usually found in the wall of the lateral ventricles and is made up of large ganglioid astrocytes. Immunohistochemically, these tumors are positive for both glial and neuronal markers *(35,36)*. Nearly all patients with SEGA have the autosomal dominant disorder tuberous sclerosis complex (TSC), though only 5% of TSC patients have SEGA. TSC is caused by mutations in the TSC1 and TSC2 tumor suppressor genes, which encode hamartin and tuberin, respectively. In

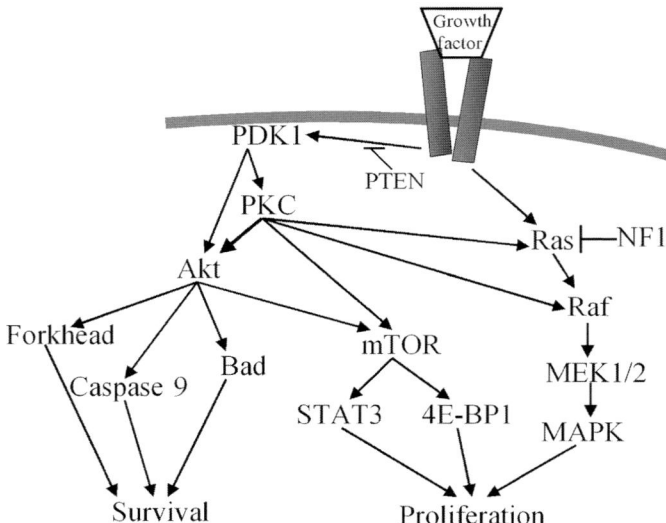

Fig. 1. Growth factor signaling mediates the activation of multiple signaling pathways.

normal cells, hamartin and tuberin form a complex that inhibits mammalian target of rapamycin (mTOR) from enhancing translation and cell growth by functioning as a guanosine triphosphate against Rheb *(37,38)*. Unlike most diffuse astrocytomas, where mTOR activity is increased by protein kinase C (PKCs), mutated PTEN, and tyrosine kinase receptor signaling, mTOR activation in SEGAs is primarily as a result of Rheb activity.

3. GROWTH-FACTOR SIGNALING

After binding a growth factor, receptor tyrosine kinases undergo dimerization, autophosphorylation, and recruit adaptor proteins that interact with various downstream signaling molecules. Normal astrocytes use growth factor signals for survival and proliferation; however, growth factor signaling networks also promote induction and increase malignancy of astrocytic tumors. This occurs through overexpression of the receptors and their ligands or as a consequence of genetic alterations of receptors or their downstream signaling molecules. Most transformed astrocytes express EGFR ligands *(39)*, permitting autocrine stimulation as a mechanism for constitutive signaling. Alternatively, tumors with EGFR amplifications often have mutations and rearrangements of EGFR genes *(40)*. These mostly intragenic mutations and rearrangements generate constitutively active proteins with ligand-independent growth-factor signaling *(41–44)*. Signaling from these receptors is constant because they are unsusceptible to ligand-mediated downregulation, a normal avenue for attenuating EGFR signaling. With its overexpression and amplification in most primary GBMs, and genetic alterations of downstream signaling networks occurring in every grade of astrocytic tumor, EGFR has become a major target of therapeutic intervention and research.

Though EGFR alterations are mainly found in primary GBMs, expression of PDGF ligands and their receptors are equally elevated in all grades of astrocytic tumors. This suggests that autocrine activation of PDGFR is important in the initial stages of transformation. Experiments in a mouse model have supported this hypothesis, as overexpression of PDGF in mouse brains can induce astrocytic tumor formation *(45)*. EGFR and PDGFR stimulate several pathways that are commonly activated in astrocytic tumors, including the Ras, Akt, and PKC signaling cascades (*see* Fig. 1) *(46–49)*. Although some effectors are specific for one of these pathways, converging signals downstream of Ras, Akt, and PKC signaling cascades likely provide redundant mechanisms for astroctye transformation. Selectively targeting activation of EGFR and its downstream effectors is likely important for any successful GBM therapy.

3.1. Ras

The GTP-binding protein Ras is an important downstream effector of receptor tyrosine kinase signaling in many cancers. Once activated, Ras can induce proliferation, invasion, and inhibit apoptosis through the Ras/Raf/MEK/MAPK pathway. In a study examining 20 GBMs, Ras activity was elevated in every instance *(46)*. Because Ras mutations are rare in astrocytomas, the increased Ras activity can mainly be attributed to signaling from receptor tyrosine kinases or the loss of function of the Ras GTPase-homologous neuro fibromatiosis 1 genes *(50)*.

In mouse models, there have been conflicting reports indicating whether or not activation of Ras alone is able to induce GBMs. Mice expressing an active form of Ras or the Ras effector Src under an astrocyte-specific promoter have been shown to develop GBMs *(51,52)*. However, mice with germline genetic modifications resulting in oncogenic Ras or v-Src expression develop normally and only acquire GBMs from a secondary mutation. Likewise, neither active K-Ras nor active Akt expressing central nervous system progenitors induce GBM formation *(49)*, but GBMs develop in mice when both of these genes are expressed in their active forms. These studies suggest that Ras activation in combination with other mutations can give rise to astrocytic tumors.

3.2. Akt and PTEN

Like Ras, activation of the serine/threonine kinase Akt is frequently elevated in many cancers. Downstream of growth factor receptors, phosphatidylinositol 3-kinase (PI3-K)-mediated activation of Akt is normally attenuated by the tumor suppressor protein/phosphatase PTEN. However, LOH on PTEN-encoding chromosome 10 occurs in most primary GBMs and is a pivotal determining factor for progression of a lower grade astrocytic tumor to a secondary GBM *(53)*. As a result of excessive growth factor signaling and/or inactivating PTEN mutations *(40,48)*, Akt activation is particularly high in astrocytic tumors. In fact, expression of activated Akt raises the malignancy level of tumors in mouse models from anaplastic astroctyomas to those that recapitulate the human GBM phenotype *(54)*.

An in vivo study using tissue microarrays from 45 GBMs demonstrated that loss of PTEN or expression of the mutant EGF receptor EGFR vIII correlates with activation of Akt *(55)*. Moreover, Akt can be activated downstream of several different PKC isozymes. Activated Akt has been shown to contribute to astrocyte tumor progression by stimulating a number of pathways leading to cell survival, growth, and proliferation. Phosphorylation-dependent activation of Akt leads to the activation of mTOR, Forkhead transcription factor family members, S6, and inactivation of proapoptotic Bad and Caspase-9 *(55)*. Additionally, translational regulation by Akt and Ras effectors contributes to recruitment of specific mRNAs involved in growth, transcriptional regulation, cell–cell interactions, and morphology to polysomes of transforming astrocytes *(56)*.

3.3. Protein Kinase C

PKC family of calcium and/or lipid-activated serine-threonine kinases functions downstream of most membrane-associated signal transduction pathways, including EGFR and PDGF. Activities of PKC family members are associated with cell growth, differentiation, gene expression, hormone secretion, and motility *(57–62)*. Activation and expression of PKCs are higher in malignant astrocytomas than in non-neoplastic astrocytes *(63–65)*, indicating a role in astrocyte transformation. Interestingly, cell culture studies focused on the roles of PKCs have demonstrated that differential expression of these kinases regulates transformation in an isotype-specific manner. For example, PKC-α, PKC-ε, and PKC-η have been shown to enhance proliferation and inhibit apoptosis, whereas PKC-δ has been shown to reduce proliferation and promote apoptosis *(66–70)*. PKC-mediated proliferation of GBM cells involves activation of the MAPK

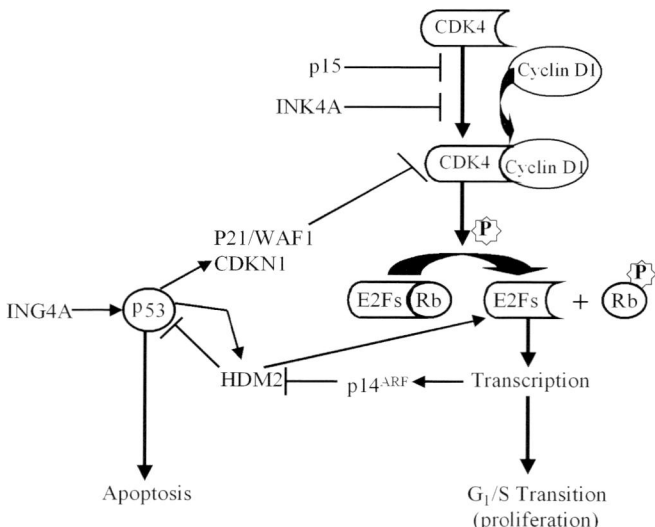

Fig. 2. Rb and p53 signaling pathways regulate the cell cycle and apoptosis. Expression and activity of many proteins in this figure (CDK4, cyclin D1, RB, p53, HDM2, p14^{ARF}) are commonly altered in astrocytic tumors. Interactions between CDK6 and cyclin D3, which behave in a similar manner to CDK4 and cyclin D1, have been omitted for the purpose of simplicity. Figure modified from Collins *(126)*.

and Akt/mTOR pathways, whereas survival appears to be as a result of caspase-9 inhibition or upregulation of Bcl-2.

In addition to functioning as mediators of proliferation, cell cycle, and survival, activation of PKCs has been shown to increase the in vitro invasion of GBM cells by regulating the expression of matrix metalloproteinases (MMPs), tissue inhibitor of metalloproteinases (TIMPs), and vascular endothelial growth factor (VEGF) *(71–74)*. VEGF, which binds to VEGF receptor-1 and VEGF receptor-2, may be the most important regulator of vascular functions in GBM angiogenesis. VEGF receptors are typically not expressed in quiescent endothelium, but are upregulated in proliferating tumor vessels *(25,75–80)*. With matrix metalloproteinases, tissue inhibitor of metalloproteinases, and VEGF expressed by invading GBM cells, tumor angiogenesis can occur through a paracrine mechanism (astrocytoma cell–endothelial cell) *(25,81,82)*. In fact, inhibition of PKC-β, a known regulator of VEGF signaling *(83)*, leads to decreased angiogenesis in GBM xenografts *(84)*.

4. CELL CYCLE

Expression of a constitutively active EGFR alone does not induce astrocyte transformation, but must be accompanied by a disruption of cell cycle arrest pathways to have an oncogenic effect *(44)*. Genetic alterations contributing to unrestricted proliferation are frequent in every grade of astrocytic tumor, from pilocytic astrocytomas to GBMs. Almost 25% of low-grade astrocytomas show mutations in genes involved in cell cycle pathways, whereas about 66% of anaplastic astrocytomas have such mutations, and cell cycle mutations have been found in nearly every GBM analyzed *(85)*. Without a mechanism for cell cycle arrest or apoptosis, astrocytes are at a high risk for transformation. Two pathways regulating cell cycle and apoptosis that are commonly mutated in astrocytic tumors are the Rb and p53 pathways (Fig. 2).

4.1. Retinoblastoma

The tumor suppressor Rb functions as a key regulator of the G_1/S cell cycle checkpoint. In nondividing cells, members of the E2F family of transcription factors form complexes with Rb.

Phosphorylation of Rb by the CDK4/CDK6-cyclin D signaling complex induces Rb to release the E2F transcription factor, thereby allowing the activation of genes necessary for proliferation *(86)*. In nearly all astrocytic tumors, there are genetic alterations in one or more components of the INK4A/CDK4/Rb/E2F pathway *(87,88)*. For example, INK4A in nonneoplastic astrocytes exerts growth control by inhibiting the cyclin-dependent kinases CDK4 and CDK6, reducing phosphorylation of Rb, and thereby inhibiting G_1/S phase transition *(86)*. In high-grade astrocytic tumors, the gene encoding INK4A is commonly deleted from chromosome 9p21. This prevents INK4A from inhibiting CDK4 and CDK6, thereby allowing the G_1/S transition. Additionally, the CDK4 gene is amplified in many high-grade astrocytomas *(89,90)*, particularly in those without CDKN2A/CDKN2B alterations. This indicates the balance between the expression of CDK4 and its inhibitor INK4A can result in a growth advantage that facilitates tumor progression *(89)*.

4.2. p53

Disruptions in the p53 pathway can contribute to genomic alterations in transformed astrocytes. The transcription factor p53 plays a crucial role in the cell response to DNA damage and stress. Normal cells undergoing stress will either arrest their replication or undergo apoptosis via p53 signaling. However, p53, found on chromosome 17p, is mutated in most cancers *(91)*. Mutations on the DNA binding domain of this protein can confer new dominant-negative or gain-of-function properties. Loss of chromosome 17p or p53 mutations are found in one-third of all adult astrocytomas, suggesting that inactivation of p53 is important for the formation of grade II tumors. In contrast, p53 mutations are associated with more advanced stages of malignancy in highly prevalent cancers *(92)*. However, just as with expression of active EGFR mutants, loss of p53 alone is not enough to initiate astrocyte transformation in an animal model *(93,94)*.

p53 inactivation results in increased genomic instability, which appears to play a role in both the formation of low-grade disease and progression toward secondary GBM. In response to DNA damage or other cellular perturbations, p53 degradation is relieved by p53 phosphorylation, disrupting p53 binding to HDM2 or p14^{ARF} binding/sequestering of HDM2 *(95)*. Phosphorylated p53 becomes stabilized in the nucleus, where it transcriptionally upregulates the cyclin/cyclin-dependent kinase p21^{WAF1}. The p21^{WAF1} then binds to and inhibits CDK/cyclin complexes, resulting in G_1/S phase arrest *(95)*. This response pathway decreases the risk of carcinogenesis by reducing the occurrence of mutations. Studies on patients with anaplastic astrocytomas who had multiple biopsies demonstrated that a p53 mutation was already present in more than 90% of patients with preceding low-grade biopsies *(96)*, suggesting that p53 mutations are an early event in this tumor. Additional evidence that loss of p53 is an early event in astrocytic tumor is that patients with Li-Fraumeni syndrome, a familial cancer characterized by the presence of a germline mutation in p53, are predisposed to develop astrocytomas *(24,97)*.

4.2.1. HDM2

In normal cells, the HDM2 regulates p53 by an autoregulatory feedback loop. The HDM2 protein inhibits p53 transcriptional activity on binding p53 or by promoting ubiquitin-mediated degradation of p53 *(98–100)*. On the other hand, transcription of the HDM2 gene is induced by wild-type p53 *(100)*. Therefore, amplification or overexpression of HDM2 is an alternative mechanism for escaping p53-regulated control of cell growth. Interestingly, amplification and overexpression of HDM2 is found in primary GBMs with wild-type p53 and not in secondary GBMs in which p53 mutations are common. In addition to its role in the p53 pathway, HDM2 interferes with Rb-mediated inhibition of cell cycle progression *(101)*.

4.2.2. The CDKN2A Locus

The p53-regulatory protein p14^{ARF} and the Rb-regulatory protein INK4A are encoded on alternative reading frames of the CDKN2A locus *(102)*. Expression of p14^{ARF} suppresses growth by inducing p53-dependent/INK4A-independent cell cycle arrest *(103–106)*. p14^{ARF} prevents HDM2-induced degradation and transactivational silencing of p53 by blocking nucleocytoplasmic shuttling of the HDM2 gene product *(107)*. Conversely, p14^{ARF} expression is negatively regulated by p53 and inversely correlates with p53 function in tumor cell lines *(104)*. Thus, distinct tumor suppressors (INK4A and p14^{ARF}) encoded by a single genetic locus regulate the Rb and p53 pathways, both of which are essential for cell cycle and regulation of apoptosis. In secondary GBMs, p53 is directly mutated, whereas in primary GBMs loss of p14^{ARF} or upregulation of HDM2 disrupt the p53 pathway. In a mouse model, p14^{ARF}-deficient mice spontaneously develop astrocytomas *(108)*.

Genetic alterations of the CDKN2A locus are common in primary GBMs and rare in secondary GBMs *(109)*. Because this locus encodes proteins that mediate both the Rb and p53 pathways, it is thought that simultaneous disruption of the Rb and p53 pathways by deletion of the CDKN2A locus may help explain why primary GBMs develop so much faster than secondary GBMs *(110)*. Comparing mouse studies that removed neuro fibromatosis 1 and p53 in a stepwise manner or removed them simultaneously demonstrated that only concurrent disruption resulted in high-grade astrocytic tumors *(111)*. These mouse studies support the concept that concurrent, rather than stepwise, loss of key growth-regulatory pathways is a more favorable mechanism for inducing the transformation of astrocytes to a highly malignant phenotype.

4.2.3. Inhibitor of Growth 4

The tumor suppressor protein ING4 is a ubiquitously expressed member of the inhibitor of growth (ING) family. Like other ING family members, ING4 forms transcriptional complexes with p53 that are required for the activation of p53-responsive genes *(112)*. Overexpression of ING4A negatively regulates cell growth through upregulation of the p53 target p21^{WAF1}. Moreover, ING4A physically interacts with the p65 (RelA) subunit of NF-κB *(113)*. This interaction regulates astrocytoma angiogenesis through transcriptional repression of NF-κB-responsive genes. Accordingly, ING4 mRNA is two to three times lower in grade II and III astrocytic tumors, and six times lower in GBMs than in tissue derived from normal brain regions.

5. DEVELOPING A MODEL OF ASTROCYTE TRANSFORMATION

High-grade astrocytic tumors can arise from stepwise accumulations of genetic mutations that increase the malignancy of lower grade tumors over a period of time or, more commonly, after a set of mutations transforms cells in a more accelerated manner. Though less prevalent than primary GBMs, studies of secondary GBMs have indicated which mutations are important for marginally malignant phenotypes and which combinations of mutations are required for transformation of astrocytes into high-grade tumor cells. Genetic alterations found in both low- and high-grade tumors are likely involved in early phases of tumor formation. Conversely, genetic alterations that are disrupted in only high-grade astrocytic tumors are likely important for tumor progression. Because similar pathways are disrupted in primary and secondary GBMs, we can speculate that the sequences of events that give rise to primary GBMs occur at a more accelerated and effective manner. Though much is known about the major signaling pathways described in this chapter, effectors of these pathways need to be identified as targets for therapies. Taking into consideration results from clinical cell culture, a model for astrocyte transformation into a tumor phenotype is shown in Fig. 3.

5.1. Cell Lines and Clinical Specimens

Many studies of astrocytic tumors have been initiated using cell culture or tumor specimens to determine genetic alterations and their corresponding phenotypes. Cell lines derived from GBMs include U87-MG, U-373-MG, U251-MG, SKMG-3, and U1242-MG. Possibly as a result of clonal selection from repeated passages, each of these lines is relatively homogenous. Some of the known genetic alterations of these established human GBM cell lines are shown in Table 1. These tumor cell lines, and others, have proven to be useful in vitro and in vivo for characterizing phenotypes that arise from genetic alterations in transformed astrocytes. Additionally, these cell lines have been used as the initial tools to study the effects of potential therapeutic treatments.

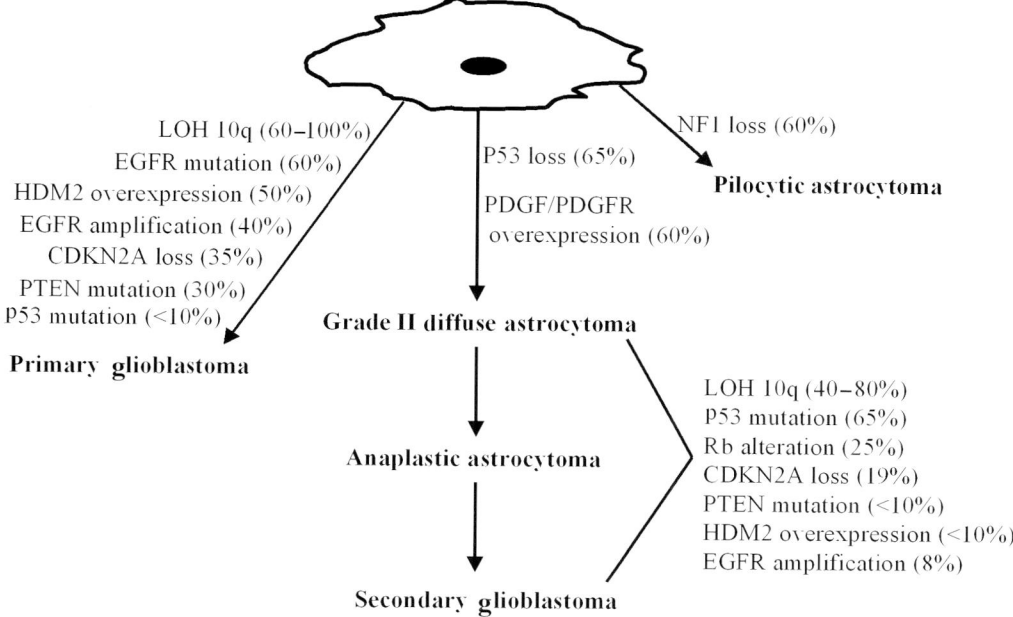

Fig. 3. Astrocyte transformation involves a number of different genetic alterations. Primary GBMs arise from concurrent genetic alterations, whereas secondary glioblastomas develop from a step-wise accumulation of alterations. Figure modified from Kleishues and Ohgaki (85).

Table 1
Glioblastoma Cell Lines Display a Variety of Mutations

	PTEN	p53	RB	INK4A	Others
U-87MG	Mutation	wt	wt	Deletion	HDM2 loss
U-373MG	Deletion	Mutation	Loss	wt	
U-251MG	Mutation	Mutation	wt	Deletion	
SKMG-3	Deletion	Mutation			EGFR amplification

This table illustrates known genetic alterations of several glioblastoma cell lines.

5.2. Animal Models of Astrocytic Tumors

In order to further our understanding of astrocyte transformation, we must develop animal models that better recapitulate in vivo growth of human astrocytic tumors. Our current animal models do not effectively simulate invasive astrocytic tumor growth; however, a number of them have provided information that could not have been successfully obtained in a cell culture system. Detailed descriptions of the current mouse models can be found in several excellent reviews *(114–118)*.

Currently, most animal models use astrocytes as the tumor origin and allow tumor growth for a sufficient amount of time so efficacies of therapies can be determined. However, not all animal models use orthotopic implantation, others vary in their ability to grow like human tumors, and some models have immunological issues that interfere with their usefulness. Even though they have provided researchers with invaluable information, the microinvasive characteristics of human astrocytic tumors are poorly replicated in every model *(119)*. Understanding the microinvasive phenotype is imperative because migratory GBM cells not only "seed" new tumors, but also the invasive cells have been shown to have a reduced susceptibility to apoptosis *(120)*.

The choice of cells to be used in an animal model is of the utmost importance. Presently, researchers use cell lines derived from human astroctytic tumors *(121)*, manipulated normal human astrocytes from embryos *(54)*, mice with transgenic constructs that specifically express in astrocytes *(49)*, and chemically induced astroctyoma cell lines from rats *(122)*. Models using rat cell lines derived from chemically induced tumors, such as rat C6 cells, should be avoided. Implanted C6 cells only partially reproduce invasion, display a rate of proliferation that causes death within weeks, and were produced as a result of nonspecific mutations *(123,124)*. On the other hand, the models that best recapitulate astrocytic tumor formation use xenografts of human astrocytoma cells in the brain of immunosuppressed rodents *(125)*. These models are excellent for clinical trials and studies looking at tumor progression in vivo. However, because the cells are completely transformed prior to implantation, these models fall short in their ability to identify pathways controlling tumor induction. Additionally, studies using immunodeficient rodents do not recapitulate the immunological interactions between tumor and host. To study the induction of tumors, the best models use normal astrocytes with specific genetic alterations. These models include xenografts of astrocytes derived from human embryos placed in the brains of immunocompromised rodents, or transgenic mouse models with genes that exclusively express in astrocytes. Though they do not demonstrate microinvasion, both of these models have been instrumental in helping us understand combinations of signaling pathways that transform normal astrocytes into a tumor phenotype *(49,54)*. Moreover, targeting the same genes or pathways that are commonly disrupted in the corresponding human tumor can specifically generate different types of astrocytic tumors. Though researchers should be wary of their limitations, the above-described models should continue providing evidence of causal roles and molecular mechanisms of genetic pathways involved in astrocyte transformation. Additionally, they will provide preclinical tools for drug screens and hope for patients with these tumors.

REFERENCES

1. Arenander A, de Vellis J. Early response gene induction in astrocytes as a mechanism for encoding and integrating neuronal signals. Prog Brain Res 1992;94:177–188.
2. Bentivoglio M, Mazzarello P. The history of radial glia. Brain Res Bull 1999;49:305–315.
3. Hatten ME, Mason CA. Mechanisms of glial-guided neuronal migration in vitro and in vivo. Experientia 1990;46:907–916.
4. Komuro H, Rakic P. Distinct modes of neuronal migration in different domains of developing cerebellar cortex. J Neurosci 1998;18:1478–1490.
5. Mason CA, Sretavan DW. Glia, neurons, and axon pathfinding during optic chiasm development. Curr Opin Neurobiol 1997;7:647–653.
6. Powell EM, Meiners S, DiProspero NA, Geller HM. Mechanisms of astrocyte-directed neurite guidance. Cell Tissue Res 1997;290: 385–393.
7. van den Pol AN, Spencer DD. Differential neurite growth on astrocyte substrates: interspecies facilitation in green fluorescent protein-transfected rat and human neurons. Neuroscience 2000;95: 603–616.
8. Araque A, Sanzgiri RP, Parpura V, Haydon PG. Astrocyte-induced modulation of synaptic transmission. Can J Physiol Pharmacol 1999;77:699–706.
9. Bacci A, Verderio C, Pravettoni E, Matteoli M. The role of glial cells in synaptic function. Philos Trans R Soc Lond B Biol Sci 1999;354:403–409.
10. Pfrieger FW, Barres BA. New views on synapse-glia interactions. Curr Opin Neurobiol 1996;6:615–621.
11. Vesce S, Bezzi P, Volterra A. The active role of astrocytes in synaptic transmission. Cell Mol Life Sci 1999;56:991–1000.
12. Aschner M. Immune and inflammatory responses in the CNS: modulation by astrocytes. Toxicol Lett 1998;102–103:283–287.
13. Montgomery DL. Astrocytes: form, functions, and roles in disease. Vet Pathol 1994;31:145–167.
14. Walter AW, Hancock ML, Pui CH, et al. Secondary brain tumors in children treated for acute lymphoblastic leukemia at St Jude Children's Research Hospital. J Clin Oncol 1998;16:3761–3767.

15. Neglia JP, Meadows AT, Robison LL, et al. Second neoplasms after acute lymphoblastic leukemia in childhood. N Engl J Med 1991;325:1330–1336.
16. Brada M, Ford D, Ashley S, et al. Risk of second brain tumour after conservative surgery and radiotherapy for pituitary adenoma. BMJ 1992;304:1343–1346.
17. Louis DN, von Deimling A. Hereditary tumor syndromes of the nervous system: overview and rare syndromes. Brain Pathol 1995;5:145–151.
18. Simpson JR, Horton J, Scott C, et al. Influence of location and extent of surgical resection on survival of patients with glioblastoma multiforme: results of three consecutive Radiation Therapy Oncology Group (RTOG) clinical trials. Int J Radiat Oncol Biol Phys 1993;26:239–244.
19. McCormack BM, Miller DC, Budzilovich GN, Voorhees GJ, Ransohoff J. Treatment and survival of low-grade astrocytoma in adults—1977–1988. Neurosurgery 1992;31:636–642; discussion 642.
20. Ding H, Shannon P, Lau N, et al. Oligodendrogliomas result from the expression of an activated mutant epidermal growth factor receptor in a RAS transgenic mouse astrocytoma model. Cancer Res 2003;63:1106–1113.
21. Mueller W, Hartmann C, Hoffmann A, et al. Genetic signature of oligoastrocytomas correlates with tumor location and denotes distinct molecular subsets. Am J Pathol 2002;161:313–319.
22. Hartwell L. Defects in a cell cycle checkpoint may be responsible for the genomic instability of cancer cells. Cell 1992;71:543–546.
23. Lane DP. Cancer. p53, guardian of the genome. Nature 1992; 358:15,16.
24. Malkin D, Li FP, Strong LC, et al. Germ line p53 mutations in a familial syndrome of breast cancer, sarcomas, and other neoplasms. Science 1990;250:1233–1238.
25. Hermanson M, Funa K, Hartman M, et al. Platelet-derived growth factor and its receptors in human glioma tissue: expression of messenger RNA and protein suggests the presence of autocrine and paracrine loops. Cancer Res 1992;52:3213–3219.
26. Nishizaki T, Ozaki S, Harada K, et al. Investigation of genetic alterations associated with the grade of astrocytic tumor by comparative genomic hybridization. Genes Chromosomes Cancer 1998;21:340–346.
27. Schrock E, Blume C, Meffert MC, du Manoir S, Bersch W, Kiessling M. Recurrent gain of chromosome arm 7q in low-grade astrocytic tumors studied by comparative genomic hybridization. Genes Chromosomes Cancer 1996;15:199–205.
28. Zulch K. Brain Tumors. Their Biology and Pathology. 3rd ed. Berlin, Heidelberg: Springer-Verlag; 1986.
29. Ohgaki H, Watanabe K, Peraud A, et al. A case history of glioma progression. Acta Neuropathol (Berl) 1999;97:525–532.
30. Gonzalez-Gomez P, Bello MJ, Arjona D, et al. Promoter hypermethylation of multiple genes in astrocytic gliomas. Int J Oncol 2003;22:601–608.
31. Nakamura M, Watanabe T, Yonekawa Y, Kleihues P, Ohgaki H. Promoter methylation of the DNA repair gene MGMT in astrocytomas is frequently associated with G:C $\geq$ A:T mutations of the TP53 tumor suppressor gene. Carcinogenesis 2001;22:1715–1719.
32. Nakamura M, Watanabe T, Klangby U, Asker C, Wiman K, Yonekawa Y. p14ARF deletion and methylation in genetic pathways to glioblastomas. Brain Pathol 2001;11:159–168.
33. Matsumoto K, Suzuki SO, Fukui M, Iwaki T. Accumulation of MDM2 in pleomorphic xanthoastrocytomas. Pathol Int 2004;54:387–391.
34. Paulus W, Lisle DK, Tonn JC, et al. Molecular genetic alterations in pleomorphic xanthoastrocytoma. Acta Neuropathol (Berl) 1996;91:293–297.
35. Sharma M, Ralte A, Arora R, Santosh V, Shankar SK, Sarkar C. Subependymal giant cell astrocytoma: a clinicopathological study of 23 cases with special emphasis on proliferative markers and expression of p53 and retinoblastoma gene proteins. Pathology 2004;36:139–144.
36. Kim SK, Wang KC, Cho BK, et al. Biological behavior and tumorigenesis of subependymal giant cell astrocytomas. J Neurooncol 2001;52:217–225.
37. Plank TL, Yeung RS, Henske EP. Hamartin, the product of the tuberous sclerosis 1 (TSC1) gene, interacts with tuberin and appears to be localized to cytoplasmic vesicles. Cancer Res 1998;58: 4766–4770.
38. Inoki K, Li Y, Xu T, Guan KL. Rheb GTPase is a direct target of TSC2 GAP activity and regulates mTOR signaling. Genes Dev 2003;17:1829–1834.
39. Ekstrand AJ, James CD, Cavenee WK, Seliger B, Pettersson RF, Collins VP. Genes for epidermal growth factor receptor, transforming growth factor alpha, and epidermal growth factor and their expression in human gliomas in vivo. Cancer Res 1991;51:2164–2172.

40. Frederick L, Wang XY, Eley G, James CD. Diversity and frequency of epidermal growth factor receptor mutations in human glioblastomas. Cancer Res 2000;60:1383–1387.
41. Wong AJ, Ruppert JM, Bigner SH, et al. Structural alterations of the epidermal growth factor receptor gene in human gliomas. Proc Natl Acad Sci USA 1992;89:2965–2969.
42. Ekstrand AJ, Sugawa N, James CD, Collins VP. Amplified and rearranged epidermal growth factor receptor genes in human glioblastomas reveal deletions of sequences encoding portions of the N- and/or C-terminal tails. Proc Natl Acad Sci USA 1992;89:4309–4313.
43. Sugawa N, Ekstrand AJ, James CD, Collins VP. Identical splicing of aberrant epidermal growth factor receptor transcripts from amplified rearranged genes in human glioblastomas. Proc Natl Acad Sci USA 1990;87:8602–8606.
44. Holland EC, Hively WP, DePinho RA, Varmus HE. A constitutively active epidermal growth factor receptor cooperates with disruption of G_1 cell-cycle arrest pathways to induce glioma-like lesions in mice. Genes Dev 1998;12:3675–3685.
45. Uhrbom L, Hesselager G, Nister M, Westermark B. Induction of brain tumors in mice using a recombinant platelet-derived growth factor B-chain retrovirus. Cancer Res 1998;58:5275–5279.
46. Guha A, Feldkamp MM, Lau N, Boss G, Pawson A. Proliferation of human malignant astrocytomas is dependent on Ras activation. Oncogene 1997;15:2755–2765.
47. Guha A. Ras activation in astrocytomas and neurofibromas. Can J Neurol Sci 1998;25:267–281.
48. Haas-Kogan D, Shalev N, Wong M, Mills G, Yount G, Stokoe D. Protein kinase B (PKB/Akt) activity is elevated in glioblastoma cells due to mutation of the tumor suppressor PTEN/MMAC. Curr Biol 1998;8:1195–1198.
49. Holland EC, Celestino J, Dai C, Schaefer L, Sawaya RE, Fuller GN. Combined activation of Ras and Akt in neural progenitors induces glioblastoma formation in mice. Nat Genet 2000;25:55–57.
50. Bos JL. ras oncogenes in human cancer: a review. Cancer Res 1989;49:4682–4689.
51. Weissenberger J, Steinbach JP, Malin G, Spada S, Rulicke T, Aguzzi A. Development and malignant progression of astrocytomas in GFAP-v-src transgenic mice. Oncogene 1997;14:2005–2013.
52. Ding H, Roncari L, Shannon P, et al. Astrocyte-specific expression of activated p21-ras results in malignant astrocytoma formation in a transgenic mouse model of human gliomas. Cancer Res 2001;61:3826–3836.
53. Tohma Y, Gratas C, Biernat W, et al. PTEN (MMAC1) mutations are frequent in primary glioblastomas (de novo) but not in secondary glioblastomas. J Neuropathol Exp Neurol 1998;57:684–689.
54. Sonoda Y, Ozawa T, Hirose Y, et al. Formation of intracranial tumors by genetically modified human astrocytes defines four pathways critical in the development of human anaplastic astrocytoma. Cancer Res 2001;61:4956–4960.
55. Choe G, Horvath S, Cloughesy TF, et al. Analysis of the phosphatidylinositol 3′-kinase signaling pathway in glioblastoma patients in vivo. Cancer Res 2003;63:2742–2746.
56. Rajasekhar VK, Viale A, Socci ND, Wiedmann M, Hu X, Holland EC. Oncogenic Ras and Akt signaling contribute to glioblastoma formation by differential recruitment of existing mRNAs to polysomes. Mol Cell 2003;12:889–901.
57. Nishizuka Y. The molecular heterogeneity of protein kinase C and its implications for cellular regulation. Nature 1988;334:661–665.
58. Nishizuka Y. The role of protein kinase C in cell surface signal transduction and tumour promotion. Nature 1984;308:693–698.
59. Gescher A. Towards selective pharmacological modulation of protein kinase C—opportunities for the development of novel antineoplastic agents. Br J Cancer 1992;66:10–19.
60. Basu A. The potential of protein kinase C as a target for anticancer treatment. Pharmacol Ther 1993;59:257–280.
61. Blobe GC, Obeid LM, Hannun YA. Regulation of protein kinase C and role in cancer biology. Cancer Metastasis Rev 1994;13:411–431.
62. Choe Y, Jung H, Khang I, Kim K. Selective roles of protein kinase C isoforms on cell motility of GT1 immortalized hypothalamic neurones. J Neuroendocrinol 2003;15:508–515.
63. Couldwell WT, Uhm JH, Antel JP, Yong VW. Enhanced protein kinase C activity correlates with the growth rate of malignant gliomas in vitro. Neurosurgery 1991;29:880–886; discussion 886–887.
64. Todo T, Shitara N, Nakamura H, Takakura K, Ikeda K Immunohistochemical demonstration of protein kinase C isozymes in human brain tumors. Neurosurgery 1991;29:399–403; discussion 403–404.
65. Benzil DL, Finkelstein SD, Epstein MH, Finch PW. Expression pattern of alpha-protein kinase C in human astrocytomas indicates a role in malignant progression. Cancer Res 1992;52:2951–2956.

66. da Rocha AB, Mans DR, Lenz G, et al. Protein kinase C-mediated in vitro invasion of human glioma cells through extracellular-signal-regulated kinase and ornithine decarboxylase. Pathobiology 2000; 68:113–123.
67. Mandil R, Ashkenazi E, Blass M, et al. Protein kinase Calpha and protein kinase Cdelta play opposite roles in the proliferation and apoptosis of glioma cells. Cancer Res 2001;61:4612–4619.
68. Aeder SE, Martin PM, Soh JW, Hussaini IM. PKC-eta mediates glioblastoma cell proliferation through the Akt and mTOR signaling pathways. Oncogene 2004;23(56):9062–9069.
69. Hussaini IM, Carpenter JE, Redpath GT, Sando JJ, Shaffrey ME, Vandenberg SR. Protein kinase C-eta regulates resistance to UV- and gamma-irradiation-induced apoptosis in glioblastoma cells by preventing caspase-9 activation. Neurooncol 2002;4:9–21.
70. Hussaini IM, Karns LR, Vinton G, et al. Phorbol 12-myristate 13-acetate induces protein kinase ceta-specific proliferative response in astrocytic tumor cells. J Biol Chem 2000;275:22,348–22,354.
71. Tsai JC, Teng LJ, Chen CT, Hong TM, Goldman CK, Gillespie GY. Protein kinase C mediates induced secretion of vascular endothelial growth factor by human glioma cells. Biochem Biophys Res Commun 2003;309:952–960.
72. Park MJ, Park IC, Hur JH, et al. Protein kinase C activation by phorbol ester increases in vitro invasion through regulation of matrix metalloproteinases/tissue inhibitors of metalloproteinases system in D54 human glioblastoma cells. Neurosci Lett 2000;290:201–204.
73. Park MJ, Park IC, Hur JH, et al. Modulation of phorbol ester-induced regulation of matrix metalloproteinases and tissue inhibitors of metalloproteinases by SB203580, a specific inhibitor of p38 mitogen-activated protein kinase. J Neurosurg 2002;97:112–118.
74. Shih SC, Mullen A, Abrams K, Mukhopadhyay D, Claffey KP. Role of protein kinase C isoforms in phorbol ester-induced vascular endothelial growth factor expression in human glioblastoma cells. J Biol Chem 1999;274:15,407–15,414.
75. Plate KH, Breier G, Farrell CL, Risau W. Platelet-derived growth factor receptor-beta is induced during tumor development and upregulated during tumor progression in endothelial cells in human gliomas. Lab Invest 1992;67:529–534.
76. Plate KH, Breier G, Risau W. Molecular mechanisms of developmental and tumor angiogenesis. Brain Pathol 1994;4:207–218.
77. Plate KH, Breier G, Weich HA, Mennel HD, Risau W. Vascular endothelial growth factor and glioma angiogenesis: coordinate induction of VEGF receptors, distribution of VEGF protein and possible in vivo regulatory mechanisms. Int J Cancer 1994;59:520–529.
78. Plate KH, Risau W. Angiogenesis in malignant gliomas. Glia 1995;15:339–347.
79. Hatva E, Kaipainen A, Mentula P, et al. Expression of endothelial cell-specific receptor tyrosine kinases and growth factors in human brain tumors. Am J Pathol 1995;146:368–378.
80. Chan AS, Leung SY, Wong MP, et al. Expression of vascular endothelial growth factor and its receptors in the anaplastic progression of astrocytoma, oligodendroglioma, and ependymoma. Am J Surg Pathol 1998;22:816–826.
81. Berkman RA, Merrill MJ, Reinhold WC, et al. Expression of the vascular permeability factor/vascular endothelial growth factor gene in central nervous system neoplasms. J Clin Invest 1993;91: 153–159.
82. Plate KH, Breier G, Weich HA, Risau W. Vascular endothelial growth factor is a potential tumour angiogenesis factor in human gliomas in vivo. Nature 1992;359:845–848.
83. Yoshiji H, Kuriyama S, Ways DK, et al. Protein kinase C lies on the signaling pathway for vascular endothelial growth factor-mediated tumor development and angiogenesis. Cancer Res 1999;59: 4413–4418.
84. Teicher BA, Menon K, Alvarez E, Galbreath E, Shih C, Faul M. Antiangiogenic and antitumor effects of a protein kinase Cbeta inhibitor in human T98G glioblastoma multiforme xenografts. Clin Cancer Res 2001;7:634–640.
85. Kleihues P, Ohgaki H. Primary and secondary glioblastomas: from concept to clinical diagnosis. Neuro-Oncol 1999;1:44–51.
86. Serrano M, Hannon GJ, Beach D. A new regulatory motif in cell-cycle control causing specific inhibition of cyclin D/CDK4. Nature 1993;366:704–707.
87. James CD, He J, Carlbom E, Nordenskjold M, Cavenee WK, Collins VP. Chromosome 9 deletion mapping reveals interferon alpha and interferon beta-1 gene deletions in human glial tumors. Cancer Res 1991;51:1684–1688.
88. Olopade OI, Jenkins RB, Ransom DT, et al. Molecular analysis of deletions of the short arm of chromosome 9 in human gliomas. Cancer Res 1992;52:2523–2529.

89. Nishikawa R, Furnari FB, Lin H, et al. Loss of P16INK4 expression is frequent in high grade gliomas. Cancer Res 1995;55:1941–1945.
90. Reifenberger G, Reifenberger J, Ichimura K, Meltzer PS, Collins VP. Amplification of multiple genes from chromosomal region 12q13-14 in human malignant gliomas: preliminary mapping of the amplicons shows preferential involvement of CDK4, SAS, and MDM2. Cancer Res 1994;54:4299–4303.
91. Guimaraes DP, Hainaut P. TP53: a key gene in human cancer. Biochimie 2002;84:83–93.
92. Kinzler KW, Vogelstein B. Lessons from hereditary colorectal cancer. Cell 1996;87:159–170.
93. Jacks T, Remington L, Williams BO, et al. Tumor spectrum analysis in p53-mutant mice. Curr Biol 1994;4:1–7.
94. Donehower LA, Harvey M, Slagle BL, et al. Mice deficient for p53 are developmentally normal but susceptible to spontaneous tumours. Nature 1992;356:215–221.
95. Sherr CJ, McCormick F. The RB and p53 pathways in cancer. Cancer Cell 2002;2:103–112.
96. Watanabe K, Sato K, Biernat W, et al. Incidence and timing of p53 mutations during astrocytoma progression in patients with multiple biopsies. Clin Cancer Res 1997;3:523–530.
97. Srivastava S, Zou ZQ, Pirollo K, Blattner W, Chang EH Germ-line transmission of a mutated p53 gene in a cancer-prone family with Li-Fraumeni syndrome. Nature 1990;348:747–749.
98. Haupt Y, Maya R, Kazaz A, Oren M. Mdm2 promotes the rapid degradation of p53. Nature 1997;387:296–299.
99. Kubbutat MH, Jones SN, Vousden KH. Regulation of p53 stability by Mdm2. Nature 1997;387:299–303.
100. Zauberman A, Flusberg D, Haupt Y, Barak Y, Oren M. A functional p53-responsive intronic promoter is contained within the human mdm2 gene. Nucleic Acids Res 1995;23:2584–2592.
101. Xiao ZX, Chen J, Levine AJ, et al. Interaction between the retinoblastoma protein and the oncoprotein MDM2. Nature 1995;375:694–8.
102. Mao L, Merlo A, Bedi G, et al. A novel p16INK4A transcript. Cancer Res 1995;55:2995–2997.
103. Quelle DE, Cheng M, Ashmun RA, Sherr CJ. Cancer-associated mutations at the INK4a locus cancel cell cycle arrest by p16INK4a but not by the alternative reading frame protein p19ARF. Proc Natl Acad Sci USA 1997;94:669–673.
104. Stott FJ, Bates S, James MC, et al. The alternative product from the human CDKN2A locus, p14(ARF), participates in a regulatory feedback loop with p53 and MDM2. EMBO J 1998;17:5001–5014.
105. Kamijo T, Zindy F, Roussel MF, et al. Tumor suppression at the mouse INK4a locus mediated by the alternative reading frame product p19ARF. Cell 1997;91:649–659.
106. Arap W, Knudsen E, Sewell DA, et al. Functional analysis of wild-type and malignant glioma derived CDKN2Abeta alleles: evidence for an RB-independent growth suppressive pathway. Oncogene 1997;15:2013–2020.
107. Tao W, Levine AJ. P19(ARF) stabilizes p53 by blocking nucleo-cytoplasmic shuttling of Mdm2. Proc Natl Acad Sci USA 1999;96:6937–6941.
108. Kamijo T, Bodner S, van de Kamp E, Randle DH, Sherr CJ. Tumor spectrum in ARF-deficient mice. Cancer Res 1999;59:2217–2222.
109. Biernat W, Tohma Y, Yonekawa Y, Kleihues P, Ohgaki H. Alterations of cell cycle regulatory genes in primary (de novo) and secondary glioblastomas. Acta Neuropathol (Berl) 1997;94:303–309.
110. Zhu Y, Parada LF. The molecular and genetic basis of neurological tumours. Nat Rev Cancer 2002;2:616–626.
111. Reilly KM, Loisel DA, Bronson RT, McLaughlin ME, Jacks T. Nf1;Trp53 mutant mice develop glioblastoma with evidence of strain-specific effects. Nat Genet 2000;26:109–113.
112. Shiseki M, Nagashima M, Pedeux RM, et al. p29ING4 and p28ING5 bind to p53 and p300, and enhance p53 activity. Cancer Res 2003;63:2373–2378.
113. Garkavtsev I, Kozin SV, Chernova O, et al. The candidate tumour suppressor protein ING4 regulates brain tumour growth and angiogenesis. Nature 2004;428:328–332.
114. Holland EC. Gliomagenesis: genetic alterations and mouse models. Nat Rev Genet 2001;2:120–129.
115. Reilly KM, Jacks T. Genetically engineered mouse models of astrocytoma: GEMs in the rough? Semin Cancer Biol 2001;11:177–191.
116. Begemann M, Fuller GN, Holland EC. Genetic modeling of glioma formation in mice. Brain Pathol 2002;12:117–132.
117. Hesselager G, Holland EC. Using mice to decipher the molecular genetics of brain tumors. Neurosurgery 2003;53:685–694; discussion 695.

118. Gutmann DH, Baker SJ, Giovannini M, Garbow J, Weiss W. Mouse models of human cancer consortium symposium on nervous system tumors. Cancer Res 2003;63:3001–3004.
119. Finkelstein SD, Black P, Nowak TP, Hand CM, Christensen S, Finch PW. Histological characteristics and expression of acidic and basic fibroblast growth factor genes in intracerebral xenogeneic transplants of human glioma cells. Neurosurgery 1994;34:136–143.
120. Joy AM, Beaudry CE, Tran NL, Ponce FA, Holz DR, Demuth T. Migrating glioma cells activate the PI3-K pathway and display decreased susceptibility to apoptosis. J Cell Sci 2003;116: 4409–4417.
121. Bernstein JJ, Goldberg WJ, Laws ER, Jr. Human malignant astrocytoma xenografts migrate in rat brain: a model for central nervous system cancer research. J Neurosci Res 1989;22:134–143.
122. Grobben B, De Deyn PP, Slegers H. Rat C6 glioma as experimental model system for the study of glioblastoma growth and invasion. Cell Tissue Res 2002;310:257–270.
123. Barth RF. Rat brain tumor models in experimental neuro-oncology: the 9L, C6, T9, F98, RG2 (D74), RT-2 and CNS-1 gliomas. J Neurooncol 1998;36:91–102.
124. Peterson DL, Sheridan PJ, Brown WE, Jr. Animal models for brain tumors: historical perspectives and future directions. J Neurosurg 1994;80:865–876.
125. Guillamo JS, Lisovoski F, Christov C, et al. Migration pathways of human glioblastoma cells xenografted into the immunosuppressed rat brain. J Neurooncol 2001;52:205–215.
126. Collins VP. Brain tumours: classification and genes. J Neurol Neurosurg Psychiatry 2004; 75(Suppl 2):ii2–ii11.

31
Mechanisms of Gliomagenesis

Wei Zhang, MD, PhD and Howard A. Fine, MD

SUMMARY

Malignant brain tumors are quickly become the leading cause of cancer-related deaths in children and young adults, and the incidence of the disease has increased many folds in the elderly over the last decade. Currently, the rapid development of molecular technology provides great promise for understanding the mechanisms of gliomagenesis. A number of genetic abnormalities have been implicated in this malignant process, which are involved in signal transduction, cell cycle control, cell growth, proliferation, apoptosis, and differentiation. We discus here the genes and pathways believed to be critical in glioma formation and progression.

Key Words: Genetic pathways; glial differentiation; gliomagenesis; growth factors; signal transduction.

1. INTRODUCTION

Oncogenesis, the development of a tumor cells, may entail the accumulation of a series of somatic mutations resulting in the breakdown of normal cell growth control and other alterations of the phenotype. There are three classes of genes, proto-oncogenes, tumor-suppressor genes, and mutator genes, involved in oncogenesis of all tissues including those within the central nervous system (CNS). Cancer cells become malignant as a consequence of activating mutations, overexpression of one or more cellular proto-oncogenes, inactivating mutations, and decreased expression of one or more tumor-suppressor genes. Most tumor-suppressor genes and oncogenes are components of signal transduction pathways that control crucial cellular functions, including cell cycle entry/exit.

Primary CNS tumors are classified based on their predominant cell type arising from the embryonic neural tube (neuroectoderm), according to the most well-developed and accepted World Health Organization (WHO) classification system for brain tumors (1). Brain tumors are among the most aggressive and intractable types of cancer, and the tumors historically thought to arise from glial cells (gliomas) make up the most common group of primary brain tumors. Grade 1 gliomas generally behave in a benign fashion and, in many cases, might even be circumscribed, whereas grade II–IV gliomas are diffusely infiltrate throughout the brain. The prognosis of patients with these tumors varies depending on the grade of the tumor, but malignant gliomas (grades III and IV), despite therapy, continue to have median survival measured in months with few patients surviving beyond 5 yr. Gliomas are composed of tumor cells that may histologically resemble astrocytes, oligodendrocytes, ependymocytes, cells seen during embryogenesis, or may be so atypical that it is difficult to compare them with any normal cell type.

The pathogenesis of brain tumors has long been enigmatic, but in recent years considerable progress has been made, particularly in the molecular genesis of gliomas. The past decade has

From: *The Cell Cycle in the Central Nervous System*
Edited by: D. Janigro © Humana Press Inc., Totowa, NJ

dramatically increased our knowledge of genetic and molecular alterations in human CNS tumors. Several of these steps have recently been identified in human brain tumors. As the genomic maps improve, increasingly more genes are identified as being involved in both the development and progression of gliomas. Here, we will deal primarily with the genetic development of astrocytic tumors, as which have been best studied. A summary of the genes and pathways believed to be involved in glioma formation and progression is discussed in this chapter.

2. MOLECULAR PATHWAYS TO GLIOMA FORMATION

Diffuse gliomas constitute the most common type of intracranial malignant neoplasm, which are classified histologically as astrocytomas, oligodendrogliomas, or oligoastrocytomas, the tumors with morphological features of both cell types. The grade IV astrocytoma or glioblastoma multiforme (GBM) are both the most common and the most malignant, and consequently, have been studied the most. Based on clinical characteristics, GBMs can be divided into those that develop over time from a lower grade astrocytoma and those that appear to develop *de novo* without a pre-existing low-grade lesion *(2)*. Genetic and biochemical evidences have proven that there are distinct genetic pathways involved in the initiation and progression of these tumors, with a stepwise accumulation of genetic lesions *(3)*. A number of genetic alterations have been correlated with human gliomas, and generally affect either signal transduction pathways activated by receptor tyrosine kinases (RTKs) or cell cycle-arrest pathways involving regulators.

2.1. Signaling Pathway Mediated by Growth Factor RTKs

RTKs play a critical role in normal cell proliferation, differentiation, and cell transformation *(4)*. Aberrant signaling by RTKs has been found in a variety of cancers, including brain tumors. RTK is a group of transmembrane proteins with at least 20 members and activates growth factor-mediated signal transduction. After binding to a growth factor, RTKs undergo receptor dimerization, autophosphorylation, and recruitment of adaptor proteins that interact with and activate various downstream effectors, leading to malignant transformation and tumor cell proliferation by several mechanisms *(5)*. One of the mechanisms of aberrant RTKs signaling involves the activation of autocrine growth factor/receptor loops *(6)*. In brain tumors, in particular astrocytomas, such loops include platelet-derived growth factor/receptor (PDGF/PDGFR), epidermal growth factor/receptor (EGF/EGFR), and transforming growth factor-α/ receptor (TGFα/EGFR), PDGF and EGF play important roles in glial development: EGF in neural stem cell proliferation and survival, and PDGF in glial cell differentiation. On the other hand, deregulated RTK signaling can occur via gene amplification, overexpression, and activating mutations, including deletions in the extracellular domain or alterations in the RTK cytoplasmic domain. RTKs activity results in activation of several downstream pathways. The small GTP-binding protein, Ras, is an important downstream effector of the growth factor-RTK signaling pathway. Ras can activate at least three downstream cascades: Raf-MEK (mitogen-activated protein kinase kinase)-MAPK (mitogen-activated protein kinase), phosphatidylinositol 3-kinase (PI3-K)-Akt, and GTP-binding protein, cell division cycle 42 (CDC42)-Rac-Rho *(7)* (Fig. 1).

2.1.1. PDGF/PDGFR Signaling

The PDGF family consists of four members, PDGF-A, -B, -C, and -D, with formation of dimers PDGF-AA, -BB, -CC, and -DD, as well as the heterodimer PDGF-AB *(8)*. PDGFs transduce signals through their receptors PDGFR-α and -β. The PDGF dimers, PDGF-AA and -CC selectively bind to PDGFR-α, PDGF-DD binds to PDGFR-β, whereas PDGF-BB has affinity to both receptors *(9–11)*. Binding of PDGF stabilizes PDGF receptor dimerization, followed by autophosphorylation of tyrosine residues, leading to increased tyrosine kinase activity *(12)* and formation of docking sites for signal relay molecules containing Src homology (SH2). A large number of SH2 domain-containing enzymes, such as PI3-K, phospholipase C-γ, the Src family

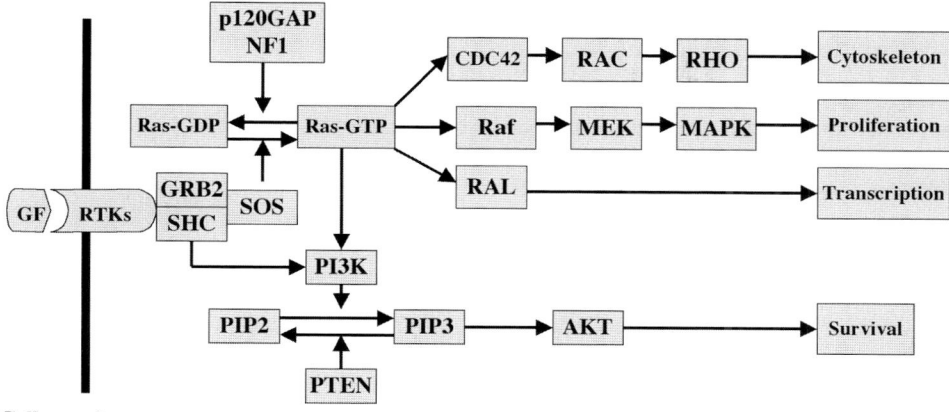

Fig. 1. Growth factors/receptor tyrosine kinases signal pathways in gliomas. Activated by ligands (such as EGF and PDGF) binding, RTKs undergo dimerization and autophosphorylation, and recruit the adaptor proteins (such as SHC and GRB2). These adaptor proteins then activate Ras- or PI3-K-mediated signaling cascades. Ras is activated by guanosine exchange factors (GEFs, such as SOS) and inactivated by GTPase-activating proteins (GAPs, such as p120GAP) and neurofibromin (the NF1 gene product), when the GTP is hydrolysed to GDP. Activated Ras positively regulates the downstream effectors, such as Raf, CDC42, and PI3-K, which mediate signaling cascades further for cell proliferation, differentiation, and survival. PI3-K, which converts phosphatidylinositol-4,5-bisphosphate (PIP2) to phosphatidylinositol-3,4,5-trisphosphate (PIP3), activates AKT-mediated signaling pathway for cell survival. The tumor-suppressor PTEN inhibits the activation of AKT by converting PIP3 back to PIP2. The GF/RTK pathways are frequently modified in gliomas, primarily through amplification and/or overexpression, and also by loss of the activity of negative regulators, such as PTEN and NF1. (Please *see* companion CD for color version of this figure.)

of tyrosine kinases and a guanosine triphosphatase (GTPase)-activating protein for Ras, have been shown to bind SH2 sites on PDGF receptors α and β, and modulate different signaling pathways *(4)*.

Numerous studies have reported the expression of PDGF-A and PDGF-B ligands in glioblastoma and indicate that autocrine signaling by these isoforms is required for cell survival *(13–15)*. Using quantitative reverse transcriptase polymerase chain reaction (PCR), the existence of autocrine signaling by PDGF-A and -B in brain tumors has been confirmed and the PDGF-C and -D ligands have been implicated in the formation of brain tumors *(16)*. PDGF/PDGFR are often coexpressed in the same tumor cells, indicating an autocrine stimulatory loop in astrocytoma cells *(13,14,17)*, which is essential for astrocytoma formation. In addition, PDGF-A, -B, and -C, as well as PDGFR have been found to be implicated in medulloblastomas and ependymomas *(18,19)*.

Although overexpression of PDGF and its receptor (PDGFR) are the common events in all grades of astrocytomas *(20)*, PDGF/PDGFR overexpression has been most commonly observed in low-grade astrocytomas in association with loss-of-function of the p53 tumor suppressor *(21)*. However, there are no reports so far that overexpression of PDGF induces low-grade astrocytomas. It has recently been reported that overexpression of PDGF in neural progenitors induced the formation of oligodendrogliomas, whereas PDGF transfer into differentiated astrocytes induced the formation of either oligodendrogliomas or mixed oligoastrocytomas *(22)*. Mice transduced with a retroviral vector expressing PDGF frequently develop highly invasive gliomas (40%) with a varied histology, many characteristics of GBMs *(23)*. These tumors also expressed PDGFR, supporting a model of autocrine stimulation in tumorigenesis. However, the precise role of PDGF/PDGFR signaling either in glial cell development or in glioma formation remains unclear and needs to be further studied.

2.1.2. EGF/EGFR Signaling Cascades

The involvement of elevated and/or aberrant EGFR activity in human cancers is well established, especially in glioblastomas. EGFR is one of the four members of ErbB family of type 1 RTKs: EGFR (erbB1 or HER1 [human EGFR 1]), neu (erbB2 or HER2), erbB3 (HER3), and erbB4 (HER4). As a transmembrane receptor, EGFR binds extracellular ligands such as EGF and TGFα and transduces an intracellular signal. The EGFR signaling pathway is necessary for sustained proliferation and perhaps survival of the neural stem cell compartment. Peak EGFR expression coincides with the peak of gliogenesis in the embryonic and early perinatal period.

The majority of GBMs have aberrant EGFR activity through EGFR overexpression, amplification or mutation *(24)*. In GBMs, amplification of the EGFR gene and in-frame deletion of exons 2–7 of the gene, EGFRvIII (ΔEGFR or del2-7EGFR), which results in a constitutively active form of EGFR, activate several signaling pathways including the MAPKs, PI3-K/Akt, and Jak/Stat pathways. Amplification of EGFR locus is found in approx 40% of primary GBMs but is rarely found in secondary GBMs *(20,24–26)*. Moreover, approx 80% of patients with EGFR amplification concurrently show gene rearrangement and other mutations, which most often result in constitutively activated EGFR. In vivo, however, expression of the EGFR mutant in neonatal mouse brain failed to induce tumor formation *(27)*. The mutated EGFR induced astrocytoma-like tumors only in the presence of other mutations. Overexpression of mutant EGFRs in astrocytes or their precursors in transgenic mice (Ink4a/Arf –/–) has been shown to promote the development of glioblastoma *(27,28)*.

2.1.3. Ras/Raf/MEK/MAPK Signaling Pathway

The Ras/Raf/MEK/MAPK pathway, one of the best-studied signal-transduction pathways and also one of the most important pathways stimulated by PDGFR, EGFR, and other RTKs, involves in a variety of cellular processes such as proliferation, differentiation, development, and tumorigenesis *(29)*. MAPKs can translocate to the nucleus to phosphorylate and activate several factors for activation of growth-inducing genes. Ras-GTP operates downstream of growth factors at a major signal transduction crossroad, translating extrinsic messages into the Raf-MEK-MAPK, the PI3-K-Akt, or the Rac-Rho pathways. Specific mutations affecting Ras have not been identified in human gliomas, although such Ras mutations are found in approx 30% of human cancers *(29)*. Rather, constitutive Ras activation in GBMs mainly results from overactivation of EGFR, EGFRvIII, and/or PDGFR signaling. It has been reported that Ras activity was elevated in all 20 GBMs analyzed *(30)*. Overexpression of mutant EGFRvIII in human GBM cells leads to constitutive activation of the MAPK pathway and pharmacological inhibition of this pathway in turn downregulates EGFRvIII phosphorylation and transformation. In a mouse transgenic model, overexpression of oncogenic Ras in astrocytes leads to the development of astrocytoma, providing direct evidence that Ras-pathway activation may be important for astrocytoma formation *(31)*. However, Ras may only participate in the transformation of mature astrocytes immortalized with human telomerase (hTERT) when both p53 and pRb are inactivated *(32)*, and further study is needed.

2.1.4. PTEN Loss and PI3-K/AKT Pathway

The tumor-suppressor gene, phosphatase and tensin homolog (PTEN), also known as mutated in multiple advanced cancers (MMAC), or transforming growth-factor-β-regulated and epithelial-cell-enriched phosphatase 1 (TEP1), is located on chromosome 10q23 and encode a protein that plays an important role in the regulation of cell proliferation, apoptosis, and tumor invasion *(33)*. PTEN protein can function as both a protein and lipid phosphatase, and its activity seems to be essential for tumor suppression as many mutations are found in its phosphatase domain. The observation that some mutant forms of the PTEN protein still retain protein phosphatase activity indicates that the ability to dephosphorylate lipids might be more important for tumor suppression. The tumor-suppressor properties of PTEN are closely related to its inhibitory effect on the PI3-K-dependent activation of protein kinase B (Akt), a well-known growth-control signaling pathway *(34)*.

PI3-Ks are lipid kinases that phosphorylate phosphatidylinositol (4,5)-bisphosphate (PIP$_2$) to produce phosphatidylinositol (3,4,5)-triphosphate (PIP$_3$), a second messenger that regulates its downstream effectors including activation of Akt. As a serine-threonine kinase, Akt plays an important role in PI3-K-mediated tumorigenesis. Activated Akt signaling pathways are important for neoplasia, including cell proliferation, adhesion, survival, and motility *(34)*. Nuclear translocation of Akt has been shown to be an important step in Akt-mediated cell proliferation and antiapoptosis. Akt can also influence cell survival by means of indirect effects on two central regulator of cell death-nuclear factor of κB (NF-κB) and p53. Akt can induce degradation of the NF-κB inhibitor, IκB by phosphorylation and activation of IκB kinase (IKK), to exert a positive effect on NF-κB function. In addition, Akt can influence the activity of the proapoptotic tumor suppressor, p53, through phosphorylation of the p53-binding protein murine double mutant 2 (MDM2). MDM2 is a negative regulator of p53 function that targets p53 for degradation by the proteasome through its ubiquitin E3 ligase activity. After being phosphorylated by Akt, MDM2 translocates more efficiently to the nucleus, where it can bind p53, resulting in enhanced p53 degradation. Furthermore, p53 can positively regulate the PTEN promoter. By dephosphorylation of PIP$_3$, PTEN inhibits the PI3-K-Akt pathway *(35)*. Biallelic loss of PTEN with consequent deregulation of the PI3-K/Akt pathway is one of the major hallmarks of many GBMs *(36,37)*. In order to determine if PI3-Ks are genetically altered in tumorigenesis, Samuels and co-workers sequenced PI3-K genes in human cancers and corresponding normal tissues. PIK3CA, which encodes the p110α catalytic subunit, was found to be the only gene with tumor-specific mutations in 35 colorectal cancers examined. PIK3CA mutations generally arise late in tumorigenesis, just prior to or coincident with invasion, and the positions of the mutations within PIK3CA imply that they are likely to increase kinase activity. Subsequent sequence analysis of all coding exons of PIK3CA in 199 additional colorectal cancers revealed mutations in a total of 74 tumors (32%). Mutations in PIK3CA were also identified in 4 of 15 (27%) GBMs, suggesting that the mutant PIK3CA is likely to function as an oncogene in human cancers, including gliomas *(38)*.

GBM formation is driven by a combination of hyperactive PI3-K/Akt and Ras/MAPK pathways together with frequent inactivation of PTEN and INK4a-ARF tumor-suppressor genes *(25,39)*. Most recently, activation of both Akt and NFκB has been reported in diffuse gliomas, with a strong positive correlation between the activation status of Akt and NFκB and glioma biological grade *(40)*. Activation of the PI3-K/Akt pathway and high levels of FoxG1 in glioblastoma cells cooperate to prevent p21^{Cip1} induction and cytostasis by the TGF-β/Smad-FoxO pathway *(41)*. By attenuating PI3-K and FoxG1 functions, the cytostatic affects of TGF-β to GBM cells could be restored. Mutation or loss of PTEN leads to constitutively activated Akt, which in turn phosporylates and inactivates Bad, caspase 9, and Forkhead transcription factors to inhibit apoptosis and to promote cell survival *(42–44)*.

2.2. Aberration of Cell Cycle-Regulatory Pathways

The second group of mutations found in gliomas is the pathways involved in cell cycle arrest. Cell cycle regulation and cell proliferation have been extensively studied in the past decade, with a paradigm of cell cycle regulation developed *(45)*. Cell cycle progression is controlled by serine/threonine kinases known as cyclin-dependent kinases (CDKs) that are activated by cyclins (D$_1$, D$_2$, D$_3$, and E) binding and inhibited by CDK inhibitors, the INK (p16^{INK4a}, p15^{INK4b}, p18^{INK4c}, and p19^{INK4d}) and CIP/KIP (p21^{CIP1}, p27^{KIP1}, and p57^{KIP2}) families *(46)*.

There are two major cell cycle-regulatory pathways involved in gliomagenesis: one controls the phosphorylation of the retinoblastoma protein (pRb) and the other regulates p53 activity. Both pRb and p53 activities dictate whether the cell progresses past a G$_1$ restriction point into S phase and mitosis (Fig. 2). Abnormalities in the cell cycle play a major role in the majority of human neoplasms *(45,47)*. Increased evidence demonstrate that neoplastic cells display alterations in the progression of the normal cell cycle *(45,48)*. Cancer cells become malignant as a consequence of activating mutations and/or increased expression of one or more cellular

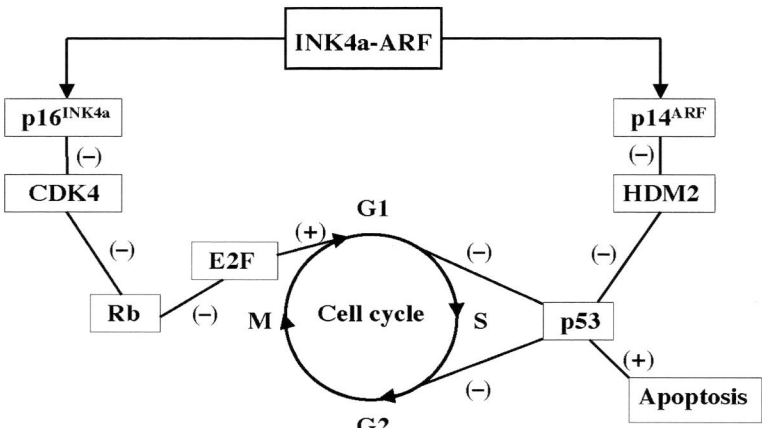

Fig. 2. INK4a-ARF and cell cycle-arrest pathways in gliomas. The INK4a-ARF gene (*CDKN2A*) encodes p16^{INK4a} and p14^{ARF}, which control the activity of retinoblastoma (Rb) and p53, two proteins critically for cell cycle regulation. Rb promotes cell cycle arrest in G_1 and regulates entry into the S phase of the cell cycle through its effects on E2F, whereas p53 promotes G_1 and G_2 arrest. p53 also promotes apoptosis and loss of p53 function causes genomic instability. In human gliomas, both Rb-p16^{INK4a}-CDK4 and p53-p14^{ARF}-HDM2 pathways are frequently altered. (–) or (+), the negative or positive regulations.

proto-oncogenes, and/or decreased expression of one or more tumor-suppressor genes. In contrast to normal cells, tumor cells are unable to stop at predetermined points of the cell cycle, so-called "checkpoints." These pauses in the cell cycle are necessary to verify the integrity of the genome before cells advance to the next phase of the cell cycle *(49)*. A major role of tumor-suppressor genes is the regulation of these checkpoints.

2.2.1. Rb-p16^{INK4a}-Cyclin D-CDK4 Pathway

Rb are pocket proteins that sequester E2F transcription factors, preventing them from activating critical genes in cell proliferation. In addition, pRb/E2F binds to histone deacetylase to form complexes that act as transcriptional repressors *(50)*. In quiescent cells, pRbs are hypophosphorylated and sequester E2F transcription factors. Proliferation occurs when pRb is phosphorylated and inactivated by CDKs. In response to proliferation signals, pRb is phosphorylated by CDK4 and/or CDK6 complexes during G_1 phase and CDK2 at G_1/S interphase. The hyperphosphorylated pRb release E2F, which in turn activate the expression of a set of genes that are required for G_1/S transition and the initiation of DNA synthesis. Most human tumors have abnormalities in some component of the Rb pathway owing either to hyperactivation of CDKs as a result of amplification/overexpression of positive cofactors, cyclins/CDKs, or to downregulation of negative factors, endogenous CDK inhibitors or mutation in the Rb gene product. The tumor suppressor p16^{INK4a} is a selective inhibitor of cyclin D/CDK4/6 *(45)*. Heterozygosity at the p16^{INK4a} locus combines with mutations in the remaining p16 allele to promote glial tumorigenesis *(51)*. These aberrations promote deregulated S phase progression in a way that ignores growth factor signals, with loss of G_1 checkpoints *(45,47)*. Inactivation of pRb in glial fibrillary acidic protein (GFAP)-positive astrocytes in transgenic mice contributed to the development of tumors similar to high-grade astrocytomas *(52)*. However, overexpression of CDK4 was found to be insufficient to induce glial tumorigenesis *(53)*.

Approximately 25% of low-grade gliomas show mutations in genes that encode proteins in the cell cycle-regulatory pathways, particularly loss of the Rb locus, whereas about 66% of anaplastic gliomas have such mutations, suggesting an important role of pRb in glioma formation.

2.2.2. The p53-p14^{ARF}-MDM2 Pathway

The p53 tumor suppressor is a short-lived transcription factor that induces either G_1 and G_2 arrest or apoptosis in response to many external insults such as DNA damage and oncogenic mutations *(54)*. p53 activates p21, which is a "universal" inhibitor of cyclin-CDK complexes. However, p53 alteration does not play a major role in cell cycle arrest/deregulation because the expression of p21 in gliomas does not correlate with proliferative activity *(55)*. p53 inactivation is primarily important in abrogating apoptosis. Loss of normal p53 function has been linked to tumor formation in genetic models and human tumor specimens. *TP53* (the gene encoding p53 in human) mutation or loss is frequently seen in childhood gliomas, low-grade gliomas, and secondary GBMs *(21)*.

In a population-based study for genetic pathways to glioblastoma, 170 TP53 mutations were observed in 126 of 402 GBMs analyzed, which are early and frequent genetic alterations in the pathway leading to secondary GBMs *(56)*. And in another large group of GBMs analyzed, nearly all tumors had mutations in this pathway, most of which were INK4a-ARF deletion or inactivation *(57)*. The loss of INK4a-ARF results in inactivation of both Rb and p53 involved cell cycle-arrest pathways, as the INK4a-ARF locus codes for both tumor-suppressor proteins p16^{INK4a} and p14^{ARF}. p16^{INK4a} prevents Rb phosphorylation by inhibition of the CDK4 or 6-cyclin D_1 complex, whereas p14^{ARF} protects p53 from MDM2-mediated degradation. Deletion of INK4a-ARF can be either homozygous or heterozygous. However, even tumors with heterozygous deletions mostly do not express the remaining wild-type allele of INK4a-ARF, possibly by methylation of its promoter or by other unknown mechanisms *(57)*. It is noteworthy that neither p16^{INK4a} nor p14^{ARF} is produced in about 60% of GBMs, whereas loss-of-function mutations in the gene TP53 have been found in the remaining part of GBMs (about 40%), in combination with either Rb deletion or CDK4 amplification. In the cases whether p14^{ARF} or p53 is directly affected, p53 may still be inactivated by MDM2 amplification and overexpression *(58,59)*. By these mechanisms, both the cell cycle-arrest pathways mediated by Rb and p53 are disrupted in the majority of gliomas, indicating the importance of cell cycle-arrest disruption in gliomagenesis.

3. CELLULAR ORIGIN OF GLIOMAGENESIS

As discussed in Subheading 2, two GBM subtypes have been identified clinically *(1,60)*. Primary GBM typically presents in older patients as an aggressive, highly invasive tumor, usually without any evidence of prior clinical disease. Secondary GBM has a very different clinical history, usually presenting in younger patients with a history of low-grade astrocytoma that transforms into GBM within 5–10 yr of the initial diagnosis, regardless of prior therapy. These GBM subtypes have identified differences in their genetic profiles, predominantly in the penetrance of specific genetic mutations. It has generally been thought that astrocytomas arise from astrocytic precursors, whereas the oligodendrogliomas arise from oligodendrocytic precursors, and the mixed gliomas arise from progenitors of both astrocytes and oligodendrocytes. However, malignant primary brain tumors, like many cancer types, resemble undifferentiated cells in their gene expression and phenotypic characteristics. The cell type that gives rise to gliomas remains unclear, although there has been a significant amount of speculation on this point *(39)*. An important issue for truly understanding gliomagenesis, therefore, is to explore the cellular origins of this neoplasm.

3.1. Neural Stem Cells

Recent studies suggest that brain tumor cells share with neural stem cells the capacity for self-renewal and multilineage differentiation *(61)*. The neural stem cells have the ability to undergo self-renewal and generate neuronal and glial progenitors, which subsequently give rise to all three CNS cell types, neurons, oligodendrocytes, and astrocytes *(62)*. Specific signal transduction pathways have been shown to stimulate such a process and control the differentiation of precursor cells into mature glia. In vitro, PDGF causes the proliferation of an oligodendroglial progenitor population and cooperates with FGF2 to prevent the further differentiation of that population into mature oligodendrocytes *(63,64)*, whereas EGF and ciliary neurotrophic factor (CNTF) force glia progenitors toward astrocytic and oligodendrocytic differentiation *(65)*. However, EGF, but not PDGF can stimulate the proliferation of neural stem cells. Specific inactivation of PTEN in mouse neural stem cells caused increased proliferation, at least in part, by shortening the cell cycle of neural stem cells *(66)*. Such unique roles of EGF signaling and PTEN in neural stem cell proliferation suggest that primary GBM might derive from neural stem cells. Adult neural stem cells that localize predominantly in the subventricular zone and dentate gyrus of the hippocampus have been identified in various species, including humans and rodents *(67,68)*, often expressing the astrocytic marker GFAP *(69)*. In addition, radial glial cells, traditionally considered to be astrocyte precursors, show characteristics of neural stem cells both in vivo and in vitro *(70,71)*. It has been shown in mice that neural stem cells are more susceptible to transformation than are differentiated astrocytes *(26,72)*. These observations indicate a close link between the astrocytic lineage and adult neural stem cells, with animal studies demonstrating the capacity of adult neural stem cells to transform into GBM.

3.2. Dedifferetiation of Glial Cells

The primary GBMs arise mainly in older patients (>50 yr) but the neural stem cell activity reduces with age *(67)*, suggesting an alternative origin of glioma cells. Another possible source of transformed glia with stem cell-like properties is the mature astrocyte or oligodendrocyte that may be induced to dedifferentiate in response to a genetic mutation. Mature astrocytes can dedifferentiate into radial glia, cells that are among the first to appear in the developing CNS, not only at the site of the experimental inoculation of embryonic cells but also at a distance from the site of implantation *(73)*. The diffusible factors acting across long distances can change the fate of a cell and the mature astrocytes retain the ability to respond to these factors. The differentiation of radial glia into astrocytes takes place once all neuronal and glial cells have migrated out of the subventricular zone. Oligodendrocyte precursor cells can also be reprogrammed to become neural stem cells in response to certain exogenous growth factors *(74)*.

The oncogenic alterations that induce glioma formation might cause mature astrocytes and oligodendrocytes to dedifferentiation, thereby allowing these differentiated cells to serve as the cellular origin for gliomagenesis. It has recently been reported that combined loss of INK4a and ARF, but not INK4a, ARF, or p53 alone, results in dedifferentiation of neonatal mouse astrocytes into neural stem cells in response to EGFR activation. When the cells have been transplanted into mice brains, INK4a/Arf (–/–) astrocytes, and INK4a/Arf (–/–) neural stem cells, can lead to the development of high-grade astrocytomas in response to EGF *(28)*.

Taken together, gliomas can arise from neural stem cells, which either exist in the adult brain or can be dedifferentiated from more differentiated cell types (astrocytes or oligodendrocytes) in response to oncogenic mutations (such as loss of INK4a/Arf).

4. INITIATION AND MALIGNANT PROGRESSION OF GLIOMAS

Primary and secondary GBMs may develop from the mutation of different genes that affect the same cellular pathways. Oncogenes (EGFR, PDGF, and its receptors) and tumor-suppressor genes ($p16^{INK4a}$, $p14^{ARF}$, PTEN, RB1, and TP53) are involved in glioma formation and progression. Loss of heterozygosity (LOH) at 10q is the most frequent genetic alteration in both subtypes of GBM (Fig. 3).

4.1. Pathways Involved in Astrocytoma Initiation and Progression

Secondary GBMs develop from low-grade gliomas and thus can be identified with distinct initiation and progression genetic pathways. Certain genetic alterations that present in both low- and high-grade astrocytomas are involved in early phases of tumor formation, whereas those

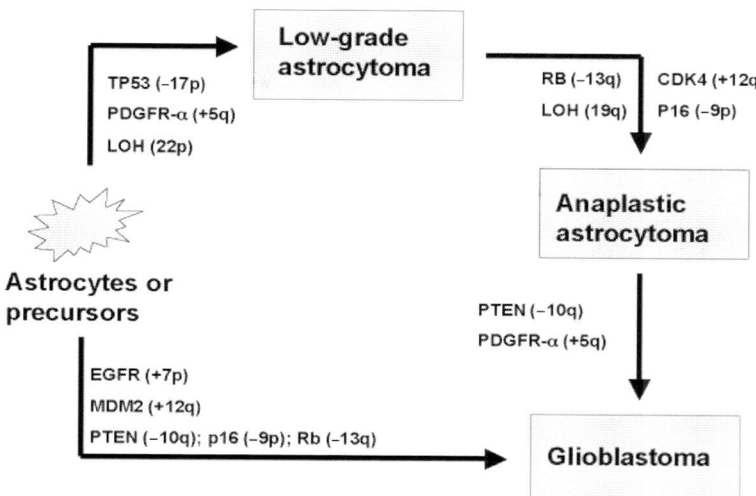

Fig. 3. Genetic alterations in glioma initiation and progression. For secondary glioblastoma tumorigenesis, inactivation of *TP53* gene and overexpression of PDGF/PDGFR play a crucial role in the initial steps, whereas inactivated p16 and Rb, as well as the loss of PTEN, are responsible for tumor progression (upper right road). For primary glioblastoma, the overexpression/amplification of EGFR and MDM2, and the loss of PTEN and p16 play critical roles (low left road). LOH, loss of heterozygosity; (−) or (+), loss or gain of a particular number/part of chromosome. (Please *see* companion CD for color version of this figure.)

specifically disrupted in high-grade but not low-grade astrocytomas are considered to be involved in tumor progression.

The genetic changes that have been described in grade II astrocytomas include inactivation of the *p53* gene, PDGF and related receptor amplification, and loss of chromosome 22q. Over 60% of astrocytomas have loss of alleles on 17q, including TP53 locus, and the retained TP53 allele is mutated in the majority of cases *(56,75)*. The most common inactivation of *p53* gene is a point mutation of the highly conserved domains 5, 7, 8 in one allele, coupled with loss of the remaining allele. However, loss of TP53 alone is insufficient to initiate astrocytoma formation and additional genetic or epigenetic events are required. Amplification of the PDGF ligand and receptor results in mitogenic stimulation of tumor cells; this stimulation can also be obtained through other genetic and/or epigenetic changes mainly through events that promote progression through the G_1/S checkpoint. The PDGF/PDGFR-mediated signaling cascade may be involved in the initiation of astrocytoma development however abnormal entry in the cell cycle is monitored by precise p53-dependent mechanisms that initiate the apoptosis response. Thus, p53 inactivation is considered usually necessary in conjunction with a promitogenic response.

Crucial to the transition from astrocytoma to anaplastic astrocytoma is the complete disruption of cell cycle control by affecting the various components controlling G_1-S transition. In grade III astrocytomas and glioblastomas, allelic losses of 9p, 13q, and 19q have been reported. *p16* appears to be the tumor-suppressor gene on chromosome 9p, although the pRb gene is on chromosome 13q. A 19q tumor-suppressor gene has not yet been identified. Most frequently *p16* is inactivated (50% of GBMs), usually by homozygous deletions *(76)*. Alternatively, in 20–30% of malignant gliomas, Rb inactivation occurs. CDK4 gene amplification and overexpression may occur in 10–15% of high-grade astrocytomas *(77)*. Deletion of the *p16* gene releases inhibition of CDK4 thereby increases its activity. The same increase may be obtained in CDK gene amplification when an increase in CDK4 may allow the formation of CDK–cyclin complexes. Moreover, higher levels of cyclin D_1 could also result in more CDK–cyclin complexes. These two events, which have a similar effect on cell cycle regulation, are rarely found to coincide.

Taken together, disrupted p53 functional pathway and growth-factor-RTK-Ras signaling cascades are involved in the initiation of astrocytoma formation, whereas pRb-mediated regulation of the cell cycle is important for the progression of astrocytoma development.

4.2. The Genesis of Primary Globlastoma

The kinetics of tumor development in primary and secondary GBMs are dramatically different, although these two subtypes have similar histopathological characteristics and clinical outcomes, as well as similar aberrant genetic pathways *(25,60)*. Primary GBMs arise rapidly without clinical or histological evidence of pre-existing low-grade lesions, which makes it difficult to distinguish between genetic alterations that contribute to the initiation of primary GBMs and those that are associated with the progression of primary GBMs. Certain genetic alterations that have been predominately identified in primary GBMs provide clues for the distinction between the mechanisms that lead to rapid- versus low-progressing glial tumors.

In contrast to secondary GBMs, INK4a mutations are common in primary GBMs (~40%), whereas mutations in TP53 are less frequent (~10%). Furthermore, in these tumors, mutations in TP53 and INK4a are mutually exclusive, although $p16^{INK4a}$ is involved in the TP53-mediated cell cycle-regulatory pathway. This conundrum is probably reconciled by the identification of a second transcript, ARF, from the CDKN2A locus. ARF stabilizes p53 proteins by antagonizing MDM2 (amplification and overexpression of MDM2 are detected in primary GBMs that lack TP53 mutations), which targets p53 for ubiquitin-mediated degradation. So, the p53 pathway is altered in primary GBMs, resulting either from loss of ARF or from upregulation of MDM2. Homozygous deletion of the CDKN2A locus ablates both $p16^{INK4a}$ and $p14^{ARF}$ function, simultaneously dismantling both Rb and p53 pathways. This might explain why primary GBMs manifest so rapidly. In mice, simultaneous disruption of both *Nf1* and *Trp53* genes results in the development of high-grade astrocytomas in certain genetic backgrounds, whereas stepwise loss *Nf1* and *p53* function does not. These mouse studies support the concept that simultaneous loss of two key growth-regulatory pathways in a cell might present more favorable conditions for the development of cancer.

Amplification of the gene that encodes EGFR is found in approx 40% of primary GBMs, but is rare in secondary GBMs. The specific role of the EGFR signaling pathway in primary GBMs is consistent with the observation that EFGR amplification is associated with mutation in the gene that encodes $p16^{INK4a}$ and is mutually exclusive with the *TP53* mutation. Moreover, most tumors with EGFR amplifications (approx 70%) have additional genetic alterations, most of which are intragenic rearrangements that lead to a truncated and constitutively active EGFR. Overexpression of this truncated EGFR confers a growth advantage and tumorigenic properties in glioma cell lines. In vivo, expression of the EGFR mutant in neonatal mouse brains failed to induce tumor formation. Overexpression of this EGFR in the Ink4a-Arf mutant background does, however, lead to the development of glioma-like lesions *(27)*. These data support a model in which the cooperation of EGFR activation and $p16^{INK4a}$-$p14^{ARF}$ deficiency contributes to glioma formation.

Loss of the long arm of chromosome 10 is the most common genetic alteration that is associated with GBMs. Several genetic loci that are associated with these tumors have been identified in this region. Among them, loss of PTEN is found in more than 30% of primary GBMs, but is rare in secondary GBMs (4%). This is confirmed by the observation that enhanced Akt activity has been detected in PTEN-deficient tumors and cell lines from both humans and mice. Furthermore, overexpression of constitutive Akt, as well as oncogenic Ras, in mouse neural stem cells leads to the development of GBMs, supporting the idea that the PI3-K-Akt pathway is pivotal in the etiology of GBMs.

5. CONCLUSION AND FUTURE DIRECTIONS

Molecular genetic studies of gliomas remain an exciting area of current research. During the last decade, dramatic advances have been made in elucidating the molecular genetics regulating normal

CNS development and of brain tumors. Such findings have begun to allow the establishment of glioma mouse models that biologically mimic various glioma subtypes. Disruptions in signaling pathways regulating cell cycle and those mediating growth factor have generally been implicated in gliomagenesis. In secondary GBM, loss of p53 and activation of the growth-factor-RTK signaling pathway initiates tumor formation, whereas disruption of Rb pathway implicates in the tumor progression. Similar genetic pathways are disrupted in primary GBM but through different specific mechanisms. In contrast to secondary GBMs in which TP53 is directly mutated, the alteration of p53 pathway is a result of either loss of p14-ARF or from upregulation of MDM2. EGFR gene amplification and mutations are common in primary GBMs and rare in secondary GBMs, which may explain the rapid growth of primary GBMs. In addition, the loss of PTEN is also found in more than 30% of primary GBMs but rare in secondary GBMs. Combined with the observation of increased Akt activation in GBMs, PTEN loss and PI3-K/Akt pathway is pivotal in the etiology of GBMs.

The details of other pathways and the precise mechanisms by which abnormalities in the cell cycle and growth factor signaling pathways trigger glial transformation remains an important goal for future research. Large-scale genomic or proteomic surveys of tumor specimens should provide starting points for new lines of inquiry into glioma biology. In recent years, several key signaling pathways, including the cytokine-induced JAK/STAT, the bone morphogenic protein (BMP)-induced SMAD, Notch, CBP, and p300, have been implicated in the differentiation of astrocytes *(78)*. Novel insights into glial tumorigenesis will undoubtedly arise from a clearer understanding of the signaling event that underlies glial cell differentiation during normal brain development.

REFERENCES

1. Kleihues P, Louis DN, Scheithauer BW, et al. The WHO classification of tumors of the nervous system. J Neuropathol Exp Neurol 2002;61:215–225; discussion 226–219.
2. Winger MJ, Macdonald DR, Cairncross JG. Supratentorial anaplastic gliomas in adults. The prognostic importance of extent of resection and prior low-grade glioma. J Neurosurg 1989;71:487–493.
3. von Deimling A, von Ammon K, Schoenfeld D, Wiestler OD, Seizinger BR, Louis DN. Subsets of glioblastoma multiforme defined by molecular genetic analysis. Brain Pathol 1993;3:19–26.
4. Kapoor GS, O'Rourke DM. Receptor tyrosine kinase signaling in gliomagenesis: pathobiology and therapeutic approaches. Cancer Biol Ther 2003;2:330–342.
5. Kolibaba KS, Druker BJ. Protein tyrosine kinases and cancer. Biochim Biophys Acta 1997;1333:F217–F248.
6. Hanahan D, Weinberg RA. The hallmarks of cancer. Cell 2000;100:57–70.
7. Hunter T. Signaling—2000 and beyond. Cell 2000;100:113–127.
8. Betsholtz C, Karlsson L, Lindahl P. Developmental roles of platelet-derived growth factors. Bioessays 2001;23:494–507.
9. Bergsten E, Uutela M, Li X, et al. PDGF-D is a specific, protease-activated ligand for the PDGF beta-receptor. Nat Cell Biol 2001;3:512–516.
10. Gilbertson DG, Duff ME, West JW, et al. Platelet-derived growth factor C (PDGF-C), a novel growth factor that binds to PDGF alpha and beta receptor. J Biol Chem 2001;276:27,406–27,414.
11. LaRochelle WJ, Jeffers M, McDonald WF, et al. PDGF-D, a new protease-activated growth factor. Nat Cell Biol 2001;3:517–521.
12. Kazlauskas A, Cooper JA. Autophosphorylation of the PDGF receptor in the kinase insert region regulates interactions with cell proteins. Cell 1989;58:1121–1133.
13. Hermanson M, Funa K, Hartman M, et al. Platelet-derived growth factor and its receptors in human glioma tissue: expression of messenger RNA and protein suggests the presence of autocrine and paracrine loops. Cancer Res 1992;52:3213–3219.
14. Guha A, Dashner K, Black PM, Wagner JA, Stiles CD. Expression of PDGF and PDGF receptors in human astrocytoma operation specimens supports the existence of an autocrine loop. Int J Cancer 1995;60:168–173.
15. Vassbotn FS, Ostman A, Langeland N, et al. Activated platelet-derived growth factor autocrine pathway drives the transformed phenotype of a human glioblastoma cell line. J Cell Physiol 1994;158:381–389.

16. Lokker NA, Sullivan CM, Hollenbach SJ, Israel MA, Giese NA. Platelet-derived growth factor (PDGF) autocrine signaling regulates survival and mitogenic pathways in glioblastoma cells: evidence that the novel PDGF-C and PDGF-D ligands may play a role in the development of brain tumors. Cancer Res 2002;62:3729–3735.
17. Nister M, Libermann TA, Betsholtz C, et al. Expression of messenger RNAs for platelet-derived growth factor and transforming growth factor-alpha and their receptors in human malignant glioma cell lines. Cancer Res 1988;48:3910–3918.
18. Black P, Carroll R, Glowacka D. Expression of platelet-derived growth factor transcripts in medulloblastomas and ependymomas. Pediatr Neurosurg 1996;24:74–78.
19. Andrae J, Molander C, Smits A, Funa K, Nister M. Platelet-derived growth factor-B and -C and active alpha-receptors in medulloblastoma cells. Biochem Biophys Res Commun 2002;296:604–611.
20. Shapiro JR. Genetics of nervous system tumors. Hematol Oncol Clin North Am 2001;15:961–977.
21. von Deimling A, Eibl RH, Ohgaki H, et al. p53 mutations are associated with 17p allelic loss in grade II and grade III astrocytoma. Cancer Res 1992;52:2987–2990.
22. Dai C, Celestino JC, Okada Y, Louis DN, Fuller GN, Holland EC. PDGF autocrine stimulation dedifferentiates cultured astrocytes and induces oligodendrogliomas and oligoastrocytomas from neural progenitors and astrocytes in vivo. Genes Dev 2001;15:1913–1925.
23. Uhrbom L, Hesselager G, Nister M, Westermark B. Induction of brain tumors in mice using a recombinant platelet-derived growth factor B-chain retrovirus. Cancer Res 1998;58:5275–5279.
24. Libermann TA, Nusbaum HR, Razon N, et al. Amplification, enhanced expression and possible rearrangement of EGF receptor gene in primary human brain tumours of glial origin. Nature 1985;313:144–147.
25. Zhu Y, Parada LF. The molecular and genetic basis of neurological tumours. Nat Rev Cancer 2002;2:616–626.
26. Wong AJ, Bigner SH, Bigner DD, Kinzler KW, Hamilton SR, Vogelstein B. Increased expression of the epidermal growth factor receptor gene in malignant gliomas is invariably associated with gene amplification. Proc Natl Acad Sci USA 1987;84:6899–6903.
27. Holland EC, Hively WP, DePinho RA, Varmus HE. A constitutively active epidermal growth factor receptor cooperates with disruption of G1 cell cycle arrest pathways to induce glioma-like lesions in mice. Genes Dev 1998;12:3675–3685.
28. Bachoo RM, Maher EA, Ligon KL, et al. Epidermal growth factor receptor and Ink4a/Arf: convergent mechanisms governing terminal differentiation and transformation along the neural stem cell to astrocyte axis. Cancer Cell 2002;1:269–277.
29. Bos JL. ras oncogenes in human cancer: a review. Cancer Res 1989;49:4682–4689.
30. Guha A, Feldkamp MM, Lau N, Boss G, Pawson A. Proliferation of human malignant astrocytomas is dependent on Ras activation. Oncogene 1997;15:2755–2765.
31. Ding H, Roncari L, Shannon P, et al. Astrocyte-specific expression of activated p21-ras results in malignant astrocytoma formation in a transgenic mouse model of human gliomas. Cancer Res 2001;61:3826–3836.
32. Sonoda Y, Ozawa T, Hirose Y, et al. Formation of intracranial tumors by genetically modified human astrocytes defines four pathways critical in the development of human anaplastic astrocytoma. Cancer Res 2001;61:4956–4960.
33. Simpson L, Parsons R. PTEN: life as a tumor suppressor. Exp Cell Res 2001;264:29–41.
34. Vivanco I, Sawyers CL. The phosphatidylinositol 3-kinase AKT pathway in human cancer. Nat Rev Cancer 2002;2:489–501.
35. Maehama T, Dixon JE. The tumor suppressor, PTEN/MMAC1, dephosphorylates the lipid second messenger, phosphatidylinositol 3,4,5-trisphosphate. J Biol Chem 1998;273:13,375–13,378.
36. Haas-Kogan D, Shalev N, Wong M, Mills G, Yount G, Stokoe D. Protein kinase B (PKB/Akt) activity is elevated in glioblastoma cells due to mutation of the tumor suppressor PTEN/MMAC. Curr Biol 1998;8:1195–1198.
37. Smith JS, Tachibana I, Passe SM, et al. PTEN mutation, EGFR amplification, and outcome in patients with anaplastic astrocytoma and glioblastoma multiforme. J Natl Cancer Inst 2001;93:1246–1256.
38. Samuels Y, Wang Z, Bardelli A, et al. High frequency of mutations of the PIK3CA gene in human cancers. Science 2004;304:554.
39. Holland EC. Gliomagenesis: genetic alterations and mouse models. Nat Rev Genet 2001;2:120–129.
40. Wang H, Zhang W, Huang HJ, Liao WS, Fuller GN. Analysis of the activation status of Akt, NFkappaB, and Stat3 in human diffuse gliomas. Lab Invest 2004;84:941–951.

41. Seoane J, Le HV, Shen L, Anderson SA, Massague J. Integration of Smad and forkhead pathways in the control of neuroepithelial and glioblastoma cell proliferation. Cell 2004;117:211–223.
42. Datta SR, Dudek H, Tao X, et al. Akt phosphorylation of BAD couples survival signals to the cell-intrinsic death machinery. Cell 1997;91:231–241.
43. Cardone MH, Roy N, Stennicke HR, et al. Regulation of cell death protease caspase-9 by phosphorylation. Science 1998;282: 1318–1321.
44. Brunet A, Bonni A, Zigmond MJ, et al. Akt promotes cell survival by phosphorylating and inhibiting a Forkhead transcription factor. Cell 1999;96:857–868.
45. Sherr CJ. Cancer cell cycles. Science 1996;274:1672–1677.
46. Sherr CJ, Roberts JM. CDK inhibitors: positive and negative regulators of G1-phase progression. Genes Dev 1999;13:1501–1512.
47. Weinberg RA. The retinoblastoma protein and cell cycle control. Cell 81:323–330.
48. Hartwell LH, Kastan MB. Cell cycle control and cancer. Science 1994;266:1821–1828.
49. Elledge SJ. Cell cycle checkpoints: preventing an identity crisis. Science 274:1664–1672.
50. Yu JT, Foster RG, Dean DC. Transcriptional repression by RB-E2F and regulation of anchorage-independent survival. Mol Cell Biol 2001;21:3325–3335.
51. Nishikawa R, Furnari FB, Lin H, et al. Loss of P16INK4 expression is frequent in high grade gliomas. Cancer Res 1995;55:1941–1945.
52. Xiao A, Wu H, Pandolfi PP, Louis DN, Van Dyke T. Astrocyte inactivation of the pRb pathway predisposes mice to malignant astrocytoma development that is accelerated by PTEN mutation. Cancer Cell 2002;1:157–168.
53. Holland EC, Hively WP, Gallo V, Varmus HE. Modeling mutations in the G1 arrest pathway in human gliomas: overexpression of CDK4 but not loss of INK4a-ARF induces hyperploidy in cultured mouse astrocytes. Genes Dev 1998;12:3644–3649.
54. Vogelstein B, Lane D, Levine AJ. Surfing the p53 network. Nature 2000;408:307–310.
55. Jung JM, Bruner JM, Ruan S, et al. Increased levels of p21WAF1/ Cip1 in human brain tumors. Oncogene 1995;11:2021–2028.
56. Ohgaki H, Dessen P, Jourde B, et al. Genetic pathways to glioblastoma: a population-based study. Cancer Res 2004;64:6892–6899.
57. Ichimura K, Schmidt EE, Goike HM, Collins VP. Human glioblastomas with no alterations of the CDKN2A (p16INK4A, MTS1) and CDK4 genes have frequent mutations of the retinoblastoma gene. Oncogene 1996;13:1065–1072.
58. Costello JF, Plass C, Arap W, et al. Cyclin-dependent kinase 6 (CDK6) amplification in human gliomas identified using two-dimensional separation of genomic DNA. Cancer Res 1997;57:1250–1254.
59. He J, Reifenberger G, Liu L, Collins VP, James CD. Analysis of glioma cell lines for amplification and overexpression of MDM2. Genes Chromosomes Cancer 1994;11:91–96.
60. Kleihues P, Ohgaki H. Primary and secondary glioblastomas: from concept to clinical diagnosis. Neuro-Oncology 1999;1:44–51.
61. Oliver TG, Wechsler-Reya RJ. Getting at the root and stem of brain tumors. Neuron 2004;42: 885–888.
62. Reynolds BA, Weiss S. Generation of neurons and astrocytes from isolated cells of the adult mammalian central nervous system. Science 1992;255:1707–1710.
63. McKinnon RD, Matsui T, Dubois-Dalcq M, Aaronson SA. FGF modulates the PDGF-driven pathway of oligodendrocyte development. Neuron 1990;5:603–614.
64. Bogler O, Wren D, Barnett SC, Land H, Noble M. Cooperation between two growth factors promotes extended self-renewal and inhibits differentiation of oligodendrocyte-type-2 astrocyte (O-2A) progenitor cells. Proc Natl Acad Sci USA 1990;87:6368–6372.
65. Rajan P, McKay RD. Multiple routes to astrocytic differentiation in the CNS. J Neurosci 1998;18: 3620–3629.
66. Groszer M, Erickson R, Scripture-Adams DD, et al. Negative regulation of neural stem/progenitor cell proliferation by the Pten tumor suppressor gene in vivo. Science 2001;294:2186–2189.
67. Gage FH. Mammalian neural stem cells. Science 2000;287:1433–1438.
68. Temple S. The development of neural stem cells. Nature 2001;414:112–117.
69. Doetsch F, Caille I, Lim DA, Garcia-Verdugo JM, Alvarez-Buylla A. Subventricular zone astrocytes are neural stem cells in the adult mammalian brain. Cell 1999;97:703–716.
70. Noctor SC, Flint AC, Weissman TA, Dammerman RS, Kriegstein AR. Neurons derived from radial glial cells establish radial units in neocortex. Nature 2001;409:714–720.

71. Alvarez-Buylla A, Garcia-Verdugo JM, Tramontin AD. A unified hypothesis on the lineage of neural stem cells. Nat Rev Neurosci 2001;2:287–293.
72. Holland EC, Celestino J, Dai C, Schaefer L, Sawaya RE, Fuller GN. Combined activation of Ras and Akt in neural progenitors induces glioblastoma formation in mice. Nat Genet 2000;25:55–57.
73. Hunter KE, Hatten ME. Radial glial cell transformation to astrocytes is bidirectional: regulation by a diffusible factor in embryonic forebrain. Proc Natl Acad Sci USA 1995;92:2061–2065.
74. Kondo T, Raff M. Oligodendrocyte precursor cells reprogrammed to become multipotential CNS stem cells. Science 2000;289:1754–1757.
75. Ichimura K, Bolin MB, Goike HM, Schmidt EE, Moshref A, Collins VP. Deregulation of the p14ARF/MDM2/p53 pathway is a prerequisite for human astrocytic gliomas with G1-S transition control gene abnormalities. Cancer Res 2000;60:417–424.
76. Walker DG, Duan W, Popovic EA, Kaye AH, Tomlinson FH, Lavin M. Homozygous deletions of the multiple tumor suppressor gene 1 in the progression of human astrocytomas. Cancer Res 1995;55:20–23.
77. Schmidt EE, Ichimura K, Reifenberger G, Collins VP. CDKN2 (p16/MTS1) gene deletion or CDK4 amplification occurs in the majority of glioblastomas. Cancer Res 1994;54:6321–6324.
78. Sauvageot CM, Stiles CD. Molecular mechanisms controlling cortical gliogenesis. Curr Opin Neurobiol 2002;12:244–249.

VII
FUTURE DIRECTIONS

32
Cell Cycle of Encapsulated Cells

Roberto Dal Toso, PhD and Sara Bonisegna, PhD

SUMMARY

Cell therapy is the prevention or treatment of human disease by the administration of cells that have been selected, multiplied, and pharmacologically treated or altered outside the body. The aim of cell therapy is to replace, repair, or enhance the function of damaged tissues or organs. There are, however, two major drawbacks that need to be resolved before cellular allografts or xenografts can achieve clinical acceptance: control of immunorejection and control of implanted cell proliferation. Cell encapsulation by means of organic or biologically derived matrixes, such as alginate, appears to be an avenue by which these goals may be accomplished, and several encouraging results show many potential applications. However, attempts based only on the alginate matrix have not been totally successful. Recently, an inorganic silica-based coating of alginate microspheres has shown to greatly improve long-term stability and cell growth control properties of alginate gels, particularly of human embryonic kidney (HEK) 293 cells engineered with adenovirus, opening new perspectives of research in the field of tissue engineering.

Key Words: Cell cycle; cell transplantation; encapsulation; alginate; Biosil; pancreatic islets; HEK 293.

1. INTRODUCTION

Tissue engineering is the development of biologically based replacement tissues and organs for the repair or restoration of tissue/organ function. Several approaches have been made using autologous tissue-engineered devices. Examples of successful applications for this approach include autologous skin grafts of wounds, parathyroid autotransplantation, of articular cartilage lesions by transplantation of autologous chondrocytes. However, the possible applications of autologous transplantation are rather scarce due to restricted tissue sources. In the early 1960s the utilization of transplantation devices, for the restoration of endocrine, metabolic, and degenerative disorders, was proposed and investigated to allow for the use of a cell and tissue source different from the host body. Indeed, the encapsulation procedure allows for cells to be taken from other organisms of the same species (allografts) or from other species (xenografts). There are many obstacles to a successful long-term and functional implantation, mainly immunological rejection of the grafted cells and biocompatibility of the device surface. To overcome these major challenges, transplantation groups have used either organic or biologically derived polymers and matrices to embed cells in a protective and immune isolating environment. Among the polymers tested so far, alginate has been, and will continue to be, one of the most important scaffolding materials. Alginate crosslinked with Ca^{2+} or Ba^{2+} has been used successfully to encapsulate cells and to maintain their function in tissue culture *(1–3)*. The widespread use of alginate is, however, limited by its tendency to degrade by tissue enzymatic pathways to its two monomeric subunits, mannuronic and guluronic acids *(4–6)*. In an attempt to increase stability,

From: *The Cell Cycle in the Central Nervous System*
Edited by: D. Janigro © Humana Press Inc., Totowa, NJ

the surface of the alginate gels, mainly prepared as microspheres, can also be further coated by other cationic organic polymers such as poly-L-lysine (PLL) which does increase duration, but reduces biocompatibility. The matter is relevant because different shapes and sizes of alginate devices can be easily prepared and the low toxicity of alginate makes it one of the most suitable matrices for cell encapsulation and transplantation. In this chapter we will review some of the relevant data available on the effects of the alginate matrix on the viability and function of the encapsulated cells and on an innovative, silica-based technology to achieve long-term stability and biocompatibility of the microsphere surface: the Biosil process.

2. CELL CYCLE REGULATION BY MATRIX INTERACTIONS

The cell cycle is a highly complex process which, through all the life span, from embryonic development and differentiation to adulthood, is regulated by extracellular signals *(7)*. These may be diffusible factors, such as growth factors (e.g., EGF, PDGF, etc.) growth and differentiation regulating factors (TGFb) *(8)*. The cell cycle is also regulated by cell surface bound adhesion molecules, which interact with analogous molecules of neighboring cells (cadherins, cellular adhesion molecules, selectins), as well as components of the extracellular matrix (integrins interacting with collagen and CD44 interacting with hyaluronic acid) *(9)*. The various types of regulations are sometimes quite complex because there are examples of growth factors, such as basic fibroblast growth factor, that require to be associated to proteoglycans (heparin sulfate) of the extracellular matrix for a high-affinity interaction with its own receptor. The interactions between cell surface molecules with extracellular matrix components determine the various phases of proliferation, migration, and final differentiation during embryonic development and during wound healing. A reduction or loss of interaction with the neighboring cells and extracellular matrix has been suggested as a major hallmark not only for cell tumorigenesis and metastatic transformation but also for chronic inflammatory diseases (e.g., arthritis). Cadherins regulate and are regulated by the intracellular Rho family of guanosine triphosphatase (GTPases) *(10)*, and are frequently downregulated during malignant progression. Integrins are heterodimeric cell surface receptors, connecting a variety of extracellular proteins (i.e., collagen, fibronectin, laminin, tenascin, etc.) with the cytoskeleton. Integrin engagement with components of the extracellular matrix has been shown to activate Rho-GTPases *(11)* which in turn regulate mitogen-activated protein kinases and the cell cycle. In addition, R-RAS and Rap-1-GTPases can regulate the functional status of integrins which make integrins effective signaling proteins linked to cell survival, differentiation, motility, and proliferation. It has also been shown that cell encapsulation in an alginate three-dimensional matrix, either natural or derivatized with collagen mimetic peptides, can deeply modify expression of surface adhesion *(12)* owing to its mechanical properties and stiffness *(13)*. The dependence of attachment on integrins and substrate rigidity suggests that integrins may play a role in sensing the mechanical properties of the matrices to which they are attached. Furthermore, some studies indicate that alginate matrices can reduce the efficacy of bFGF, which strongly binds to the alginate matrix, and enhance intercellular adhesion molecule 1 expression *(14)*.

Hence, the encapsulating material, termed also scaffold or matrix, provides form, size, and sometimes function, for the encapsulated cells or tissue and may also provide topographical or biochemical signals (either naturally released or designed to release) to the embedded cells. Not all cell types, however, prove to be retained by the matrix; and cells that have exited the scaffold undergo a dedifferentiation process *(15,16)*, indicating that a more rigid surface coating may improve the confining properties of the encapsulation device. One further drawback of most of the mentioned biologically derived polymeric substances is their high biodegradability and low biocompatibility which greatly limits the possibility in long-term implant applications.

3. TISSUE-ENGINEERING DEVICES

The first challenge in constructing a biologically based medical device is to match the chemical and mechanical properties of the scaffolding material to the needs of the encapsulated cell

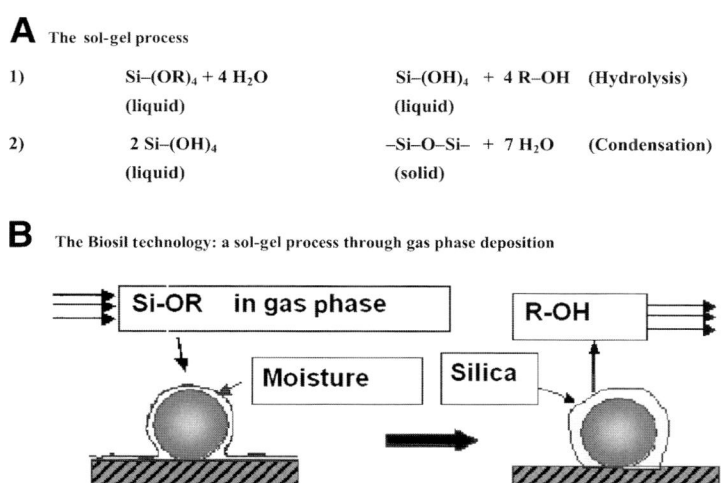

Fig. 1. The Sol-Gel and Biosil processes. (Please *see* companion CD for color version of this figure.)

population and later on to the needs of the developing tissue. Traditional tissue-engineering approaches address either three-dimensional, relatively thick porous systems; for example, an open cell foam or nonporous injectable gels. The major limitations faced in using the former are the difficulties associated with vascularizing a large volume of material in order to support implanted cells, the difficulty of spatially arranging the cells within the scaffold in a defined manner, and also the risks and costs associated with an operative implant procedure. That is, without the infrastructure to supply nutrients and remove waste, clinically viable tissue-engineered devices have been spatially constrained to small volumes. Additionally, large-scale, prefabricated scaffolds require invasive surgery for implantation, are difficult to contour to the defect shape, and have predetermined mechanical properties. Besides alginate *(5)*, attempts to engineer soft tissue (e.g., muscle, fat, and liver) include the use of scaffolds manufactured from natural polymers or synthetic, biodegradable polymers, such as collagen, hyaluronic acid, and chitosan *(17)* gels, and all these materials have therefore emerged as candidates for soft tissue replacement. Injectable materials are advantageous because they require a minimally invasive procedure, they are mechanically similar to soft tissue, and they conform to irregular shapes *(18)*. However, the major limitation of using an injectable gel is that many cells perform poorly when suspended within a large gel due to insufficient nutrient/waste exchange in the central area of the implanted device. Consequently, some researchers have suggested that gel properties could be enhanced by forming smaller, more mobile units of rigid scaffold materials (i.e., microspheres; *see* Figs. 1 and 3). The use of small scaffolds, such as ground demineralized bone matrix *(19)*, alginate beads *(1,20)*, and arginine-glycine-aspartate modified alginate beads *(6)*, lower than 300 μm in diameter in which cells are embedded together with an optimized cell number to volume ratio may contribute in reducing these issues. Important issues for the optimization of alginate properties are the ratio of guluronic to manuronic monomeric subunits *(21)*, and stability *(22)* and absence of mitogenic impurities *(23,24)*. Another essential factor for a successful device is the surface biocompatibility, which determines the intensity of the initial inflammatory response and of the following foreign body reaction of the surrounding tissue hosting the implanted device. These issues are becoming crucial for the survival of the encapsulated cell, because macrophages, overgrown on the device surface, release interleukin-1 and tumor necrosis factor-α that have been shown to be detrimental for the survival of the encapsulated cells and long-term functionality *(25–28)*. Also of concern are ease of sterilizing the material, ease of handling the system in the operating room, and robustness to handling, transport, and implantation.

4. POTENTIAL BIOMEDICAL APPLICATIONS AND CELL LINES

The possibilities of therapeutical applications of cell therapy appear to be enormous and range in a number of pathological situations due to increased efficacy and safety of therapy by localized delivery of therapeutic substances at the target, thus avoiding toxicity with high doses of systemic administration.

One of the first proposed applications has been the encapsulation of insulin-producing cells in semipermeable membranes, which have the potential to provide an effective treatment for insulin-dependent diabetes with little or no immunosuppression of the host *(29,30)*, and has reached clinical human trials *(31)*. The basic idea is to provide a drug delivery device providing an immunoinsulating sheet for the encapsulated cells. In this case the long-term stability of the immunoinsulating layer is the most important feature. Another important area of application is implantation of the central nervous system (CNS) to either to substitute for lacking substances following neurodegenerative or dismetabolic diseases *(32–39)*, for drug delivery and treatment of neurodisorders such as epilepsy *(40,41)* or for the treatment of brain tumors such as multiform glioblastoma *(42–45)*.

Many of these approaches are based on the use of genetically engineered mammalian cell lines that provide the synthesis and release of the substances required; most cell lines, however, need to be restricted in their proliferation and motility capacity for safety purposes. Although many different solutions to this question have been proposed and attempted, to date, long-term stability of the encapsulating alginate device has not been definitely achieved due to the biodegradability of the matrix. Encapsulated primary cell grafts have also been considered as alternatives to the use of tumor derived cell lines. Primary cell grafts, however, as with any other mammalian cell line, face several limitations due to reliable and safe cell sources both for allogenic and xenogenic grafts and to this add tissue availability in adequate amounts for allogenic grafts.

New hope derives from the use of differentiated stem cells that can be derived from adult tissues. Recently, remarkable discoveries that stem cells may be generated in different adult tissues throughout life and may differentiate into a variety of specialized cells have brought into focus their possible therapeutic potential (for review, *see* ref. 46). It is now well established that the vertebrate brain continues to produce new neurons throughout the life span. Moreover, adult stem cells demonstrate surprising pluripotency, for example, stem cells isolated from the bone marrow may differentiate into neurons and glia after transplantation into mouse brain *(47)*. Research on therapeutic application of adult stem cells for treatment of neurological disorders is constrained by ethical considerations, lack of reliable sources and, to date, inability to propagate stem cells in sufficient quantities. Finally, the complete differentiation of stem cells may not always be achieved, with some of the nonterminally differentiated cells retaining an oncogenic potential. The necessity for a biocompatible immunoinsulating and cell confinement device may hence be an obligated complement also for stem cell applications.

Hence, serious obstacles, including a better understanding of immune acceptance issues, remain to be addressed before a clinically applicable therapeutic procedure based on encapsulated cells, regardless of the cell source or type, becomes available as pointed out in a recent commentary *(48)*.

5. THE BIOSIL TECHNOLOGY

Ca^{2+}-alginate gels have several useful features that make them the matrix of choice to microencapsulate single cells and cell aggregates, mainly the absence of toxicity for the embedded cell and the host tissue as well as the ease of handling. The main limitation for a widespread clinical usage is the short-term stability of the matrix and hence of its immunoisolating properties. The addition alginate gel core of an additional coating layer with polycationic organic polymers has improved the durability of the microcapsule, but, as previously mentioned, has generated some problems with the biocompatibility of the implants. An alternative approach to this limit has been pursued by our laboratory with the development of a porous inorganic silica film coating over the alginate gel (for a review, *see* ref. *49*). The coating procedure is based on the sol-gel process. This chemical process takes advantage of the peculiar properties of the organosilan monomer chemistry, which are described in Fig. 1.

In the presence of water, the organically substituted silicon monomers [$Si(OR)_4$], where R is a organic residue such as methanol or ethanol, which in biologically compatible conditions is physically in a liquid phase, hydrolyses to $Si(OH)_4$, and polymerizes leading to an extensive Si-O-Si network as a stable reaction product which is physically a solid (Fig. 1A). Depending on the type of organic residues linked to the silanic monomers and on their percentage in the initial liquid sol phase, the Si-O-Si network of the solid gel phase can be tailored to host pores of various sizes. The sol-gel process is viewed as a soft chemistry process because it can proceed in near-physiological temperature and pH conditions and can thus be used to immobilize and encapsulate a variety of living cells *(50,51)*. The procedure requires the previous mixing of an appropriate amount of the organosilan sol phase with a solution containing the cell to encapsulate and then allowing sufficient time for gelling to occur. When containing living cells, the controlled porosity of the gel matrix allows for the diffusion of essential gas and nutrients from the surrounding environment. However the simple application of the sol-gel process as described has a biological limit in the amount of R-OH, the organic reaction by-product, typically ethanol, which builds up and remains in contact with the encapsulated cells. To cope with this problem, an innovative approach has been devised based on the deposition of the organosilan monomers on the surface to be coated from the gas phase, instead of the direct contact of the surface with the dense liquid phase. Briefly, the desired composition of silan precursors, present in a heated container, is air fluxed to obtain a saturated gas phase, which is then transferred to a reaction chamber where it reaches the wet surface of cells or microspheres. Here the unreacted organosilan monomers partition from the gas phase to the aqueous film covering the cell or device and start the series of reactions described in Fig. 1A, ending in the formation of a thin inorganic layer. The toxic levels of reaction by-products (R-OH) are reduced by the gas carrier flow, as schematically shown in Fig. 1B. The procedure just described is the basis of the Biosil technology and has been applied to immobilize and encapsulate a variety of microbial, plant, and mammalian cells *(51–53)*.

There are several advantages in the use of the Biosil technology over the more conventional sol-gel process: because the amount of monomer precursor interacting with the coating surface from the gas phase is much less than that present in the liquid phase, the thickness of the coating layer can be more easily controlled together with the levels of toxic byproducts. The estimated average thickness of the Biosil film ranges from 100 to 200 nm, which is an extremely thin coating also constituting of a reduced diffusion barrier for permeable molecules. Another important feature of the Biosil-produced membrane is its molecular weight cutoff which can be designed and realized, according to needs, from 10 to more than 300 kDa. For most cell encapsulation and transplantation purposes, a cutoff proximal to 100 kDa appears to be optimal, because it represents an absolute barrier for the diffusion of immunoglobulins and immunocompetent cell migration, and at the same time allow for free diffusion of nutrients, hormones, growth factors, and other signaling molecules *(49)*. This diffusion barrier is obviously effective on molecules directed both towards the inside, as well as the outside of the encapsulated body. The characteristic can be used to confine the motility of undesired agents from the encapsulated cell or microsphere (e.g., viruses or tumor cells), and an example of this possibility is given in the next section.

Biocompatibility has been one of the most relevant issues for implanted materials because it determines the level of inflammatory response from the surrounding tissues and the foreign body response with the formation of an avascular fibrotic scar enveloping and isolating the grafted device. Furthermore, low biocompatibility of the encapsulating surface may also be detrimental for the embedded cells *(26)*. The Biosil-produced films have been extensively tested for the compatibility with blood cell elements (erythrocytes, platelets, and polymorphonucleated cells) and

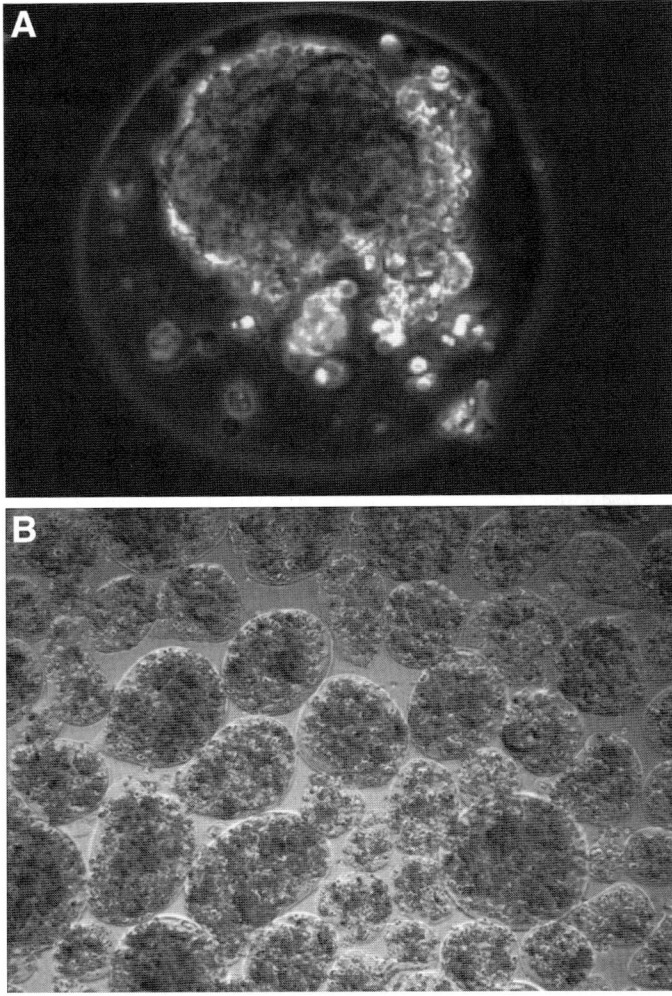

Fig. 2. Encapsulated pancreatic islets. **(A)** Dithizone stained pancreatic islet. **(B)** Microscopic field of islet preparation. (Please *see* companion CD for color version of this figure.)

systems (complement and clotting factors) and has proved to be extremely safe for all hematological parameters tested *(54)*. Furthermore, the general biocompatibility has been evaluated by intraperitoneal injection of Biosil-coated or noncoated alginate beads in both Balb/c and nonobese diabetic (NOD) mice for a period up to 8 wk prior to recovery, with extremely positive results *(49)* because no overgrowth of peritoneal macrophages was observed on Biosil-coated beads.

The Biosil-produced inorganic film thus appears to complement a number of features required to improve alginate gels for clinical application.

6. BIOSIL APPLICATIONS TO DIABETES AND CNS

In a first series of experiments, streptozotocin treated type-1 diabetic dogs were implanted with Biosil-encapsulated islets taken from pig donors. This approach was chosen to maximize the potential for cross-species immunological rejection which typically occurs in xenotransplantation. Nonencapsulated islet cells taken from pigs were transplanted into dogs, they would normally be rejected within 1 or 2 d by acute immune response. Approximately 200,000 microspheres with encapsulated pancreatic islets (*see* Fig. 2A) were surgically injected by portal

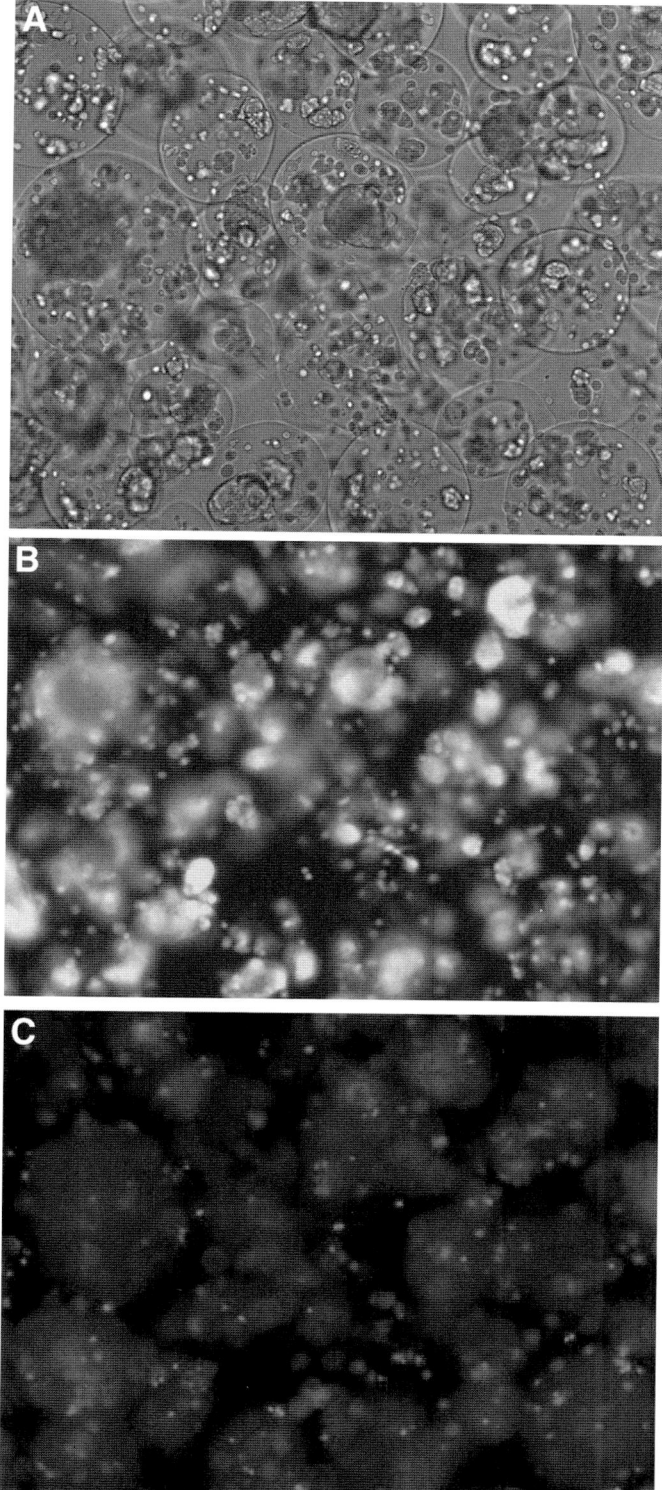

Fig. 3. Viability staining of encapsulated islets with calcein AM (**A**) and ethidium homodimer 1 (**B**). (**C**) Propidium iodide as a marker of dead cells. Cell viability was tested by microscopical fluorescence determination of Calcein and Ethidium homodimer-1 (LIVE/DEAD Viability/Cytotoxicity kit, Molecular Probes). (Please *see* companion CD for color version of this figure.)

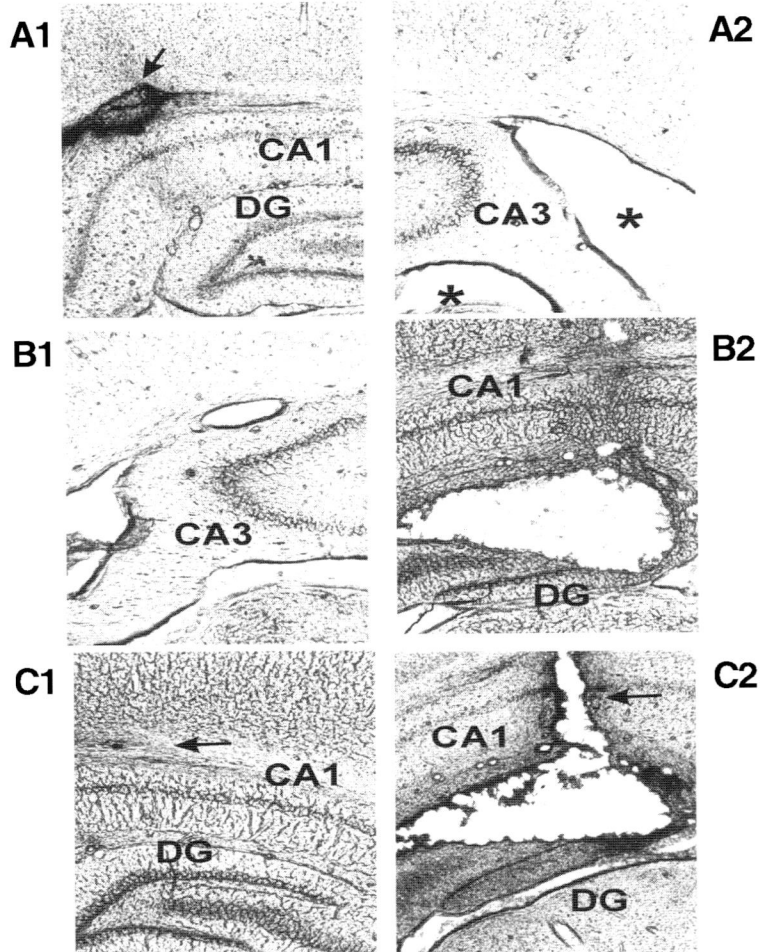

Fig. 4. Histological sections of rat brain injected with HEK 293 cell either Biosil encapsulated (**A1,B1,C1**) or nonencapsulated (**A2,B2,C2**).

vein to the liver of diabetic dogs. Biosil coating prevented rejection in all eight cross-species transplants attempted. It was also found that the encapsulated islet cells were releasing insulin up to 4 wk after transplantation. This is important because indicates a long-lasting viability and function of the transplanted tissue. In addition, following histological evaluation of the implanted tissue, no signs of rejection of the transplanted cells were seen for up to 6 mo *(55,56)*.

In a first attempt to determine the cell confinement capacity of the alginate gel/Biosil system, HEK 293 cells genetically engineered by adenovirus were encapsulated into alginate microspheres coated by the Biosil-produced film *(see* Figs. 2B and 3). The average diameter of microspheres was of 70 μm, and approx 300 were injected in the hippocampal CA1 and dentate gyrus area of Sprague-Dawley rats (250–300 g in weight). Other animals received approximately the same amount of HEK 293 cells without Biosil coating. After 3 d, the animals were sacrificed and the whole rat brain was recovered for histological processing and observation *(see* Fig. 4).

Figure 4 shows bright field images of eosin-hematoxilin-stained brain slices obtained from lesioned animals that received Biosil coated (left panel) or uncoated cells (right panel). Note that in animals where Biosil-coated cells were injected 3 d after injury (A1) needle track damage was still visible as indicated by the arrow. In animals that received a similar treatment but without Biosil coating, large ventricles were visible (A2). Widespread damage was observed in uninjected animals as shown in B2 7 d following injection. Another animal with similar pathology is shown in C2. Also note that the needle track is indicated by an arrow. In animals that received Biosil-coated cells the needle track is still visible (arrow, C1) *(57)*.

7. CONCLUSIONS

Some promising technologies, for example, gene therapy and cell transplantation therapy, have been proposed as devices useful to achieve more specific and effective means to deliver pharmacologically active molecules in clinical applications, particularly to CNS. However, there are protective mechanisms known as the blood–brain barrier, which efficiently shield the CNS from many blood borne molecules. Traditional delivery methods meant to deliver therapeutic agents to the brain are often hampered by this protective mechanism. In attempts to overcome this barrier, cells or viruses have been directly implanted. The results have been modest and when delivery is successful, the intended effects do not last for therapeutically significant periods of time due to the body's immune system response or death of implanted cells. To overcome the latter, the use of immortal or immortalized cell lines has been proposed (e.g., transformed or stem cells). This approach, however, may lead to uncontrolled growth of these cells in the brain. We have tested the effectiveness of siliceous encapsulation in preventing excessive immune response as well as in limiting the improper displacement of the implanted cells after implantation. We believe that a proper combination of encapsulation technologies based on highly purified clinical grade alginate together with silica-based coatings might allow new perspectives for cell transplantation and therapy with the achievement of long-term therapeutical benefits.

REFERENCES

1. Lanza RP, Jackson R, Sullivan A, et al. Xenotransplantation of cells using biodegradable microcapsules. Transplantation 1999;27: 1105–1111.
2. Selden C, Khalil M, Hodgson H. Three dimensional culture upregulates extracellular matrix protein expression in human liver cell lines: a step towards mimicking the liver in vivo? Int J Artif Organs 2000;23:774–781.
3. Hasse C, Bohrer T, Barth P, et al. Parathyroid xenotransplantation without immunosuppression in experimental hypoparathyroidism: long-term in vivo function following microencapsulation with a clinically suitable alginate. World J Surg 2000;24:1361–1366.
4. Atala A, Kim W, Paige KT, Vacanti CA, Retik AB. Endoscopic treatment of vesico ureteral reflux with a chondrocyte-alginate suspension. J Urol 1994;152:641–643.
5. Burg KJ, Holder WD, Jr, Culberson CR, et al. Comparative study of seeding methods for three-dimensional polymeric scaffolds. J Biomed Mater Res 2000;51:642–649.
6. Paige KT, Cima LG, Yaremchuk MJ, Schloo BL, Vacanti JP, Vacanti CA. De novo cartilage generation using calcium alginate-chondrocyte constructs. Plast Reconstr Surg 1996;97:168–178.
7. Nojima H. G1 and S-phase checkpoints, chromosome instability, and cancer. Methods Mol Biol 2004;280:3–49.
8. Li Y, Foster W, Deasy BM, et al. Transforming growth factor-1 induces the differentiation of myogenic cells into fibrotic cells in injured skeletal muscle—a key event in muscle fibrogenesis Am J Pathol 2004;164:1007–1019.
9. Boudreuau NJ, Jones PL. Extracellular matrix and integrin signalling: the shape of things to come. Biochem J 1999;339:481–488.
10. Etienne-Manneville S, Hall A. Rho GTPases in cell biology. Nature 2002;12:629–635.
11. Lee JW, Juliano R. Mitogenic signal transduction by integrin- and growth factor receptor-mediated pathways. Mol Cells 2004;17:188–202.
12. Gigant-Huselstein C, Hubert P, Dumas D, et al. Expression of adhesion molecules and collagen on rat chondrocyte seeded into alginate and hyaluronate based 3D biosystems. Influence of mechanical stresses. Biorheology 2004;41:423–431.
13. Genes NG, Rowley JA, Mooney DJ, Bonassar LJ. Effect of substrate mechanics on chondrocyte adhesion to modified alginate surfaces. Arch Biochem Biophys 2004;422:161–167.

14. Wang L, Geng M, Li J, Guan H, Ding J. Studies of marine sulfated polymannuroguluronate on endothelial cell proliferation and endothelial immunity and related mechanisms. J Pharmacol Sci 2003;92:367–373.
15. Schulze-Tanzil G, Mobasheri A, de Souza P, John T, Shakibaei M. Loss of chondrogenic potential in dedifferentiated chondrocytes correlates with deficient Shc-Erk interaction and apoptosis. Osteoarthr Cartilage 2004;12:448–458.
16. Rokstad AM, Holtan S, Strand B, et al. Microencapsulation of cells producing therapeutic proteins: Optimizing cell growth and secretion. Cell Transplant 2002;11:313–324.
17. Chenite A, Chaput C, Wang D, et al. Novel injectable neutral solutions of chitosan form biodegradable gels in situ. Biomaterials 2000;21:2155–2161.
18. Zimmermann U, Mimietz S, Zimmermann H, et al. Hydrogel-basednon-autologous cell and tissue therapy. Biotechniques 2000;29:564–572.
19. Marijnissen WJ, van Osch GJ, Aigner J, Verwoerd-Verhoef HL, Verhaar JA. Tissue-engineered cartilage using serially passaged articular chondrocytes. Chondrocytes in alginate, combines in vitro with a synthetic (E210) or biologic biodegradable carrier (DBM). Biomaterials 2000;21:571–580.
20. Eiselt P, Yeh J, Latvala RK, Shea LD, Mooney DJ. Porous carriers for biomedical applications based on alginate hydrogels. Biomaterials 2000;21:1921–1927.
21. Wandrey C, Espinosa D, Rehor A, Hunkeler D. Influence of alginate characteristics on the properties of multi-component microcapsules. J Microencapsul 2003;20:597–611.
22. Rokstad AM, Strand B, Rian K, et al. Evaluation of different types of alginate microcapsules as bioreactors for producing endostatin. Cell Transplant 2003;12:351–364.
23. Zimmermann U, Thurmer F, Jork A, et al. A novel class of amitogenic alginate microcapsules for long-term immunoisolated transplantation. Ann NY Acad Sci 2001;944:199–215.
24. Webera M, Steinertb A, Jorka A, et al. Formation of cartilage matrix proteins by BMP-transfected murinemesenchymal stem cells encapsulated in a novel class of alginates. Biomaterials 2002;23: 2003–2013.
25. De Vos P, Van Straaten JFM, Nieuwenhuizen AG, et al. Why do microencapsulated islet grafts fail in the absence of fibrotic overgrowth? Diabetes 1999;48:1381–1388.
26. de Groot M, Schuurs TA, Leuvenink HG, van Schilfgaarde R. Macrophage overgrowth affects neighboring nonovergrown encapsulated islets. J Surg Res 2003;115:235–241.
27. Rokstad AM, Kulseng B, Strand BL, Skjak-Braek G, Espevik T. Transplantation of alginate microcapsules with proliferating cells in mice: capsular overgrowth and survival of encapsulated cells of mice and human origin. Ann NY Acad Sci 2001;944:216–225.
28. Cole DR, Waterfall M, McIntyre M, Baird JD. Microencapsulated islet grafts in the BB/E rat: a possible role for cytokines in graft failure. Diabetologia 1992;35:231–237.
29. Lim F, Sun AM. Microencapsulated islets as bioartificial endocrine pancreas. Science 1980;210:908–910.
30. Sambanis A. Encapsulated islets in diabetes treatment. Diabetes Technol Ther 2003;5:665–668.
31. Soon-Shiong P. Treatment of type I diabetes using encapsulated islets. Adv Drug Deliv Rev 1999;35:259–270.
32. Maysinger D, Berezovskaya O, Fedoroff S. The hematopoietic cytokine colony stimulating factor 1 is also a growth factor in the CNS: (II). Microencapsulated CSF-1 and LM-10 cells as delivery systems. Exp Neurol 1996;141:47–56.
33. Maysinger D, Piccardo P, Liberini P, Jalsenjak I, Cuello C. Encapsulated genetically engineered fibroblasts: release of nerve growth factor and effects in vivo on recovery of cholinergic markers after devascularizing cortical lesions. Neurochem Int 1994;24:495–503.
34. Maysinger D, Jalsenjak I, Cuello AC. Microencapsulated nerve growth factor: effects on the forebrain neurons following devascularizing cortical lesions. Neurosci Lett 1992;140:71–74.
35. Ross CJ, Ralph M, Chang PL. Delivery of recombinant gene products to the central nervous system with nonautologous cells in alginate microcapsules. Hum Gene Ther 1999;10:49–59.
36. Barsoum SC, Milgram W, Mackay W, et al. Delivery of recombinant gene product to canine brain with the use of microencapsulation. J Lab Clin Med 2003;142:399–413.
37. Xue YL, Wang ZF, Zhong DG, et al. Xenotransplantation of microencapsulated bovine chromaffin cells into hemiparkinsonian monkeys. Artif Cells Blood Substit Immobil Biotechnol 2000;28:337–345.
38. Xue Y, Gao J, Xi Z, et al. Microencapsulated bovine chromaffin cell xenografts into hemiparkinsonian rats: a drug-induced rotational behavior and histological changes analysis. Artif Organs 2001;25:131–135.

39. Tobias CA, Dhoot NO, Wheatley MA, Tessler A, Murray M, Fischer I. Grafting of encapsulated BDNF-producing fibroblasts into the injured spinal cord without immune suppression in adult rats. J Neurotrauma 2001;18:287–301.
40. Boison D, Huber A, Padrun V, Deglon N, Aebischer P, Mohler H. Seizure suppression by adenosine-releasing cells is independent of seizure frequency. Epilepsia 2002;43:788–796.
41. Huber A, Padrun V, Deglon N, Aebischer P, Mohler H, Boison D. Grafts of adenosine-releasing cells suppress seizures in kindling epilepsy. Proc Natl Acad Sci USA 2001;98:7611–7616.
42. Visted T, Lund-Johansen M. Progress and challenges for cell encapsulation in brain tumour therapy. Expert Opin Biol Ther 2003;3:551–561.
43. Bjerkvig R, Read TA, Vajkoczy P, et al. Cell therapy using encapsulated cells producing endostatin. Acta Neurochir 2003;88:137–141.
44. Read TA, Sorensen DR, Mahesparan R, et al. Local endostatin treatment of gliomas administered by microencapsulated producer cells. Nat Biotechnol 2001;19:29–34.
45. Joki T, Machluf M, Atala A, et al. Continuous release of endostatin from microencapsulated engineered cells for tumor therapy. Nat Biotechnol 2001;19:35–39.
46. Schaffer DV, Gage FH. Neurogenesis and neuroadaptation. Neuromol Med 2004;5:1–9.
47. Nakano K, Migita M, Mochizuki H, Shimada T. Differentiation of transplanted bone marrow cells in the adult mouse brain. Transplantation 2001;71:1735–1740.
48. Orive G, Hernández RM, Gascón AR, et al. Cell encapsulation: Promise and progress. Nat Med 2003;9:105.
49. Carturan G, Dal Toso R, Bonisegna S, Dal Monte R. Encapsulation of functional cells by sol-gel silica: actual progress and perspectives for cell therapy. J Mater Chem 2004;14:2087–2098.
50. Nassif N, Bouvet O, Rager MN, Roux C, Coradin T, Livage J. Living bacteria in silica gels. Nat Mater 2002;1:42–44.
51. Inama L, Dire S, Carturan G, Cavazza A. Entrapment of viable microorganisms by SiO_2 sol-gel layers on glass surfaces: trapping, catalytic performance and immobilization durability of Saccharomyces cerevisiae. J Biotechnol 1993;30:197–210.
52. Pressi G, Dal Toso R, Dal Monte R, Carturan G. Production of enzymes by plant cells immobilized by Sol-Gel silica. J Sol-Gel Sci Techn 2003;26:1189–1193.
53. Boninsegna S, Dal Toso R, Dal Monte R. Alginate microspheres loaded with animal cell and coated by siliceous layer. J Sol-Gel Sci Tech 2003;26:1151–1157.
54. Boninsegna S, Magagna C, Tommasoli R, Castellani S, Dal Toso R, Bellavite P. Biosil (SiO_2) coated alginate microspheres display high biocompatibility properties, submitted.
55. Grundfest-Broniatowski S, Boninsegna S, Waterman J, et al. Siliceous encapsulation of islet cell xenografts in a canine model of diabetes: a feasibility study. 37th Annual Meeting, 2003, Poster Abstract, Vol 14, Orlando, FL: Pancreas Club, Inc.
56. Grundfest-Broniatowski S, Boninsegna S, Waterman J, et al. SiO_2 encapsulation of islet cell xenografts and allografts, submitted.
57. Garrity-Moses ME, Boninsegna S, Teng Q, et al. Biosil encapsulation of mammalian cells: a novel strategy for cell mediated drug delivery. Program No. 789.1. 2003 Abstract Viewer/Itinerary Planner. Washington, DC: Society for Neuroscience, 2003.

33
Viral Vector Delivery to Dividing Cells

Yoshinaga Saeki, MD, PhD

SUMMARY

Recent progress in stem cell research and developmental neurobiology opens up possibilities of stem cell-based replacement therapies for neurological disorders. Embryonic and adult stem cells of all kinds have been derived from various mammalian species, and extensive efforts have been made to uncover the secrets of "stemness" (self-renewal and multipotency). Viral vector-mediated gene delivery is a powerful technique that would facilitate basic stem cell biology research, as well as the development of stem cell-based therapeutics. In this chapter, five of the most popular viral vector systems are briefly reviewed and potential applications of herpes simplex virus amplicon vectors for stem cell research are discussed.

Key Words: Viral vectors; herpesvirus; adenovirus; AAV; retrovirus; lentivirus; HSV amplicon; iBAC; neural stem cell.

1. INTRODUCTION

In the healthy central nervous system (CNS) at its fully developed stage, the majority of cells are terminally differentiated and not actively dividing. However, in pathological conditions (e.g., stroke, tumors, brain, or spinal cord injuries, and encephalitis), resting astrocytes become activated and start proliferating in response to various cytokines and growth factors (1,2). Brain tumors have been considered to be a unique clinical situation, in which malignant cells are the only rapidly dividing cells in the CNS, and therefore, a number of therapeutic strategies have been devised in order to selectively eliminate actively proliferating cells.

However, recent identification of multipotent neural stem cells (NSCs) in adult mammalian CNS revealed that a small but significant number of cells are continuously dividing and replacing the pre-existing damaged neurons even in physiological conditions (3–5). This has raised a question if one can use and enhance the ability of endogenous NSCs to repair damaged neurons and their impaired networks. Moreover, the recent progress in stem cell biology research now allows us to isolate and propagate embryonic and adult NSCs in culture for extended period of time without losing their multipotency (3,6,7). This certainly has provided neuroscientists with a greater opportunity to study NSCs for their differentiation mechanisms and also to use NSCs as therapeutic resources for treating various neurological disorders. Viral vector-mediated gene transfer is a powerful technique for both basic neuroscience research applications as well as gene therapy applications.

Viruses have evolved through the process of natural selection to provide a variety of sophisticated features that allow for the efficient delivery of genetic information to cells. Thanks to the recent advances in virology and molecular biology, members of virtually any virus family can be used in designing gene transfer vectors. However, becoming clearer is that there are no universally applicable ideal viral vector systems currently available. In this chapter we will define various features of several vectors facilitating identification of specific vectors matching individual application needs. This review will summarize the basic properties of several of the most popular viral vector systems derived from four different classes of virus, and focus mainly on the herpes simplex virus (HSV)-based amplicon system. Further, we will discuss our previous work which has concentrated on the HSV system including its potential applications to dividing cells and our future perspectives on the vector system.

2. CHARACTERISTICS OF VIRAL VECTORS DERIVED FROM FOUR DIFFERENT CLASSES OF VIRUSES

Characteristics of various vector systems derived from four different groups of viruses including adenoassociated virus (AAV), adenovirus (Ad), retrovirus, and herpesvirus are summarized in Table 1. Each system has both advantages and limitations in their appropriate designation to various applications.

2.1. AAV Vectors

AAV is among the smallest viruses characterized by a 4.7-kb single-stranded linear DNA genome (both positive and negative strands) in a nonenveloped capsid with a diameter of approx 20 nm. AAV is a naturally occurring defective virus that requires coinfection of unrelated helper viruses (e.g., Ads and HSVs) in order to undergo productive infection (type of infection that produces progeny virus). In the absence of a helper virus, AAV is able to integrate into the host-cell genome and maintain a latent infection. Superinfection of latently infected cells with a helper virus results in the rescue and replication of the AAV genome. Viral genome integration may occur in a site-specific manner and the best-characterized AAV serotype 2 (AAV2) has been demonstrated to integrate preferentially into adenoassociated virus integration site on human chromosome 19 *(8)*.

In the past several years, AAV has established its position as one of the most popular vector systems for in vivo gene delivery. This is mainly owing to the long-term and efficient transgene expression in various cell types in many organs and tissues (e.g., liver, muscle, lung, and CNS). Recombinant AAV vectors (rAAVs), derived mainly from AAV2 and cloned as a pBR322-based plasmid, have a transgene cassette flanked by inverted terminal repeats (ITRs). ITRs are the only *cis*-acting elements required for rescue, replication, and packaging of the vector sequence into AAV virions. Recently developed helper virus-free packaging methods *(9,10)*, together with affinity and/or ion exchange chromatography-mediated purification methods *(11–14)*, have made the vector system appropriate for large-scale production and clinical grade applications. The advantages of AAV vectors include (1) small particle size, allowing them to spread throughout tissues in vivo, (2) relatively high transduction efficiency in postmitotic cells including neurons, (3) potential psudotyping virions with capsids of different serotypes with at least eight serotypes currently available *(15–17)*, and (4) limited adaptive immune response owing to the lack of dendritic cell transduction.

However, there are limitations associated with AAV vectors. The small transgene capacity of 4.5 kb does not allow for incorporation of large regulatory sequences, genomic transgenes, or large cDNAs. The relatively slow induction of transgene expression, owing to the requirement for single-stranded vector genomes to be converted into double-stranded DNA, may render them less than optimal for certain applications. Further, vector integration into the host genome is extremely inefficient, occurring randomly, if at all owing to the lack of the rep gene. A recent study showed the AAV vector integration often occurs within genes and causes deletions *(18)*. The lack of efficient integration and the slower transduction of the vector limits its usefulness for application in rapidly dividing cells, in which extrachromosomal vector DNA becomes rapidly diluted and lost before transgene expression may begin. Currently, there are a limited number of publications reporting the applications of AAV vectors on neural progenitor or cancer

Table 1
Basic Properties of Commonly Used Viral Vector Systems

Vector system	Transgene size	Integration	Other features
AAV vector	< ~ 4.5 kb	Yes[a]	Multiple vector coats available
			High tissue penetration
			Smallest perticle size (~20 nm in diameter)
			Low immunogenicity and minimal cytotoxicity
Ad vector			Broad host range, both dividing and nondividng Retargetting possible
FG Ad vector		No	Very high titer can be achieved (>10^{12} VPs/mL)
			Some immunogenicity and some cytotoxicity
HD Ad vector (gutless)	~ 30 kb	No	Large transgene capacity
			Low immunogenicity and minimal cytotoxicity
Retroviral vector			Stable transgene expression
			Psudotyping and retargetting possible
MoMLV-based vector	< ~ 8 kb	Yes[b]	No transduction in nondividing cells
Lentiviral vector	< ~ 8–10 kb	Yes[c]	Both dividing and nondividing cells can be transduced
Herpesvirus vector			Broad host range, both dividing and nondividing
Replication-defective vector (recombinant)	< ~ 50 kb[d]	No	Potential immunogenicity and cytotoxicity
			(depending on genotype of its parental mutant)
			Large transgene capacity
HSV amplicon vector	< ~ 150 kb	No	Low immunogenicity and minimal cytotoxicity
			Easy construction and rapid packaging preperation
			Largest transgene capacity

[a]Integration can occur randomly into host genome. Replication is not required. Majority of vector DNA remains episomally.
[b]Integration occurs randomly into host genome. Replication is required.
[c]Integration occurs randomly into host genome. Replication is not required.
[d]This is a theoretical estimate.

cells. Hughes et al. reported that AAV vectors with AAV2, AAV4, and AAV5 capsids transduced mouse primary neural progenitor cells very poorly in culture *(19)*. Surace et al. reported successful transduction of AAV vectors with AAV1, AAV2, and AAV5 capsids on developing mouse retinal progenitor cells in vivo *(20)*. According to their data, vectors with AAV1 or AAV5 virions transduced progenitor cells more efficiently than those with AAV2 virions. Human neural progenitor cells, on the other hand, were reported to be transduced efficiently during both proliferating and differentiated stages *(21)*.

One of the exciting features of AAV vectors is vector-mediated homolog recombination *(22–24)*. Up to 1% of normal human fibroblasts can undergo gene targeting by AAV vectors when infected at extremely high multiplicities of infections (MOIs) typically more than 2000 *(22,23)*. AAV vector targeting in various mammalian stem cells is worthy of further investigation.

2.2. Ad Vectors

Ad is a nonenveloped virus containing a linear, double-stranded genome approx 36 kb long. The 5′ end of each strand is covalently attached to a protein called terminal protein. The viral capsid has an icosahedral structure, the major components of which are the hexon, penton, and fiber proteins. Ad possesses a number of characteristics that makes them attractive tools for gene delivery. Ad is among the most extensively studied viruses and their genome has been well characterized. Recombinant Ad vectors can infect a wide range of dividing and nondividing cells both in vitro and in vivo. In addition, there are well-established techniques that allow for propagation and purification of vector stocks to titers over 10^{12} viral particles (VPs)/mL *(25,26)*. Although most of the currently used Ad vectors are based on serotype 5, there are a number of serotypes available to make chimeric vectors having different tropisms. Moreover, there are well-established methodologies to retarget the vectors to specific cell types by using genetic or chemical modifications of viral virion components *(27,28)*. Ad vectors are generally episomal and their DNA does not integrate into the host genome. There are two classes of nonreplicating Ad vector systems currently available.

2.2.1. First-Generation (E1-Deleted) Ad Vectors

First-generation (FG) Ad vectors are termed so because they were the first Ad-based vectors to be developed. The vector uses most of viral genome and has deletions in the E1 region and other loci (e.g., E3 and E4). These deletions allow for the insertion of transgenes whereas minimizing potential toxicity. They can accommodate up to 8 kb of inserted DNA. Ad vectors can achieve robust transgene expression within a relatively short period of time (1–2 days) after infection lending themselves to successful application in dividing cells. However, the vector DNA does not integrate into the host genome, and the episomal vector DNA can be diluted and eventually lost in dividing cells. Ad vectors can efficiently transduce neural progenitor cells both in vitro and in vivo *(19,29,30)*. However, transgene expression is rather transient and may induce glial differentiation of neural progenitor cells *(19)*. Although defective in viral replication, FG Ad vectors remain capable of inducing cytotoxicity and immunogenicity owing to low-level viral gene expression from the vector backbone and toxic virion component proteins.

2.2.2. Helper-Dependent (Gutless and Gutted) Ad Vectors

Helper-dependent (HD) vectors are a new generation of Ad vectors in which the coding sequence of the viral genome is completely deleted, leaving only the ITRs and packaging signal sequence, which are the only essential *cis*-acting elements required for the viral replication and packaging. The deleted genome is replaced with a foreign DNA fragment that encompasses the expression cassette of interest with or without "stuffer" DNA, which may be required to maintain the optimal genome size (approx 30–35 kb) *(31,32)*. In order to propagate and rescue the ITR-flanked vector genome for use as a recombinant virus, an E1-deleted superinfected helper Ad vector is required. The *Cre-loxP* and *Flp-FRT* site-specific recombination systems have been successfully used to reduce contamination of helper virus in vector preparation *(33–35)*. After further physical separation mediated by a CsCl density gradient, more than 99% purity can be achieved. The HD vector is a very attractive gene transfer system because of its large transgene capacity and lack of viral vector DNA coding sequences. One can introduce multiple-transgene cassettes or a large genomic transgene into the vector, and it is likely that the HD vectors cause less cytotoxicity and immunogenicity in comparison with FG vectors. There is also an attempt to achieve homologous recombination using the large transgene capacity of the HD Ad vectors *(36,37)*.

To achieve the long-term transgene expression, two classes of hybrid vectors have been reported, integrating and nonintegrating. The former incorporates ITRs and the rep gene from AAV (Ad-AAV hybrid vector) *(38,39)* or the inverted repeats and transposase gene from DNA transposones (Ad-transposone hybrid vector) *(40)*. The latter is the Ad–Epstein-Barr virus (EBV) hybrid vector, which can release a circular DNA containing the EBV episomal replicon elements (*EBNA-1* gene and *oriP*) *(41,42)*. These new hybrid vectors require further examination to evaluate their efficacy in neural progenitor cells.

2.3. Retroviral Vectors

Retroviridae encompass a large number of viruses that contain two identical copies of single-stranded RNA molecules per particle and reverse transcriptase. Retroviruses have glycoprotein envelopes and their particle size ranges from 80 to 100 nm in diameter. The size of their genome is typically from 7 to 11 kb in length. The viral RNA is reverse-transcribed into double-stranded DNA in the cytoplasm, and then transported into the nucleus in which it is randomly inserted into host chromosomes. The integrated viral DNA or provirus can produce progeny viruses and be passed on to daughter cells through mitosis. Because of their efficient and stable transduction, retroviral vectors have been widely used in gene therapy research and also in basic research in the areas of developmental biology and neuroscience. In this section, two classes of retroviral vectors, Moloney murine leukemia virus (MoMuLV)-based vectors and lentiviral vectors, will be discussed.

2.3.1. MoMuLV-Based Vectors

MoMuLV-derived vectors are among the most commonly used viral vector systems in gene therapy and basic research applications. The MoMuLV-based vector systems were developed in the early 1980s *(43–45)* and have because gone through a number of modifications and improvements. One of the major advantages of the system is their efficient integration of vector sequences into host chromosome DNA, making stable transgene expression possible even in actively proliferating cells. However, the transgene capacity of this vector system is limited to 8 kb and potential polyadenylation signals and splicing signals need to be avoided in the insert sequences.

Another advantage of the vector system is that envelop glycoproteins can be easily pseudotyped or modified, therefore making the vector capable of transducing a wide variety of cell types including human, primate, rodent, and fish cells *(46–49)*. However, vector integration and transduction can occur in rapidly dividing cells but not in postmitotic cells, because the viral preintegration complex cannot transverse an intact nuclear envelope *(50)*. Although this is a major limitation of the vector system, this unique feature has been used for cell lineage studies and stem cell biology research *(51,52)*. This also led to the MoMuLV vector-mediated therapeutic strategies against brain tumors, in which vector integration and transduction are targeted to highly proliferative tumor cells *(53–55)*.

2.3.2. Lentiviral Vectors

Lentiviruses are a class of retroviruses that cause slowly progressive disorders resulting in chromic degenerative diseases of the nervous, hematopoietic, respiratory, musculoskeletal, and immune systems. Human immunodeficiency virus (HIV) is the best-characterized member of this group. In contrast to MoMuLV and other oncoretroviruses, lentiviruses can infect nondividing cells *(56)*. This unique property of HIV reflects the presence of nuclear localization signals (NLS) on three components of its nucleoprotein complex: matrix (MA), Vpr, and integrase (IN) *(57–60)*. As expected, HIV-derived vectors, pseudotyped with vesicular stomatitis virus G (VSV G) protein, were demonstrated to have all of the advantages of MoMuLV-based conventional retroviral vectors, yet they have the additional ability to transduce nondividing cells *(61,62)*. The biological safety of the vector system has been greatly improved. The third-generation lentiviral vector system is currently being used and is characterized by the elimination of the homology sequences between the vector plasmid and *gag-pol* encoding helper plasmid. All accessory open-reading frames are removed from the *gag-pol* encoding helper plasmid and the *rev* gene is supplied as a separate helper plasmid *in trans*. The enhancer element of 3′-LTR of is deleted to self-inactivate HIV LTR promoter on vector integration *(63,64)*. An increasing number of successful applications of lentiviral vectors on NSCs have been reported *(19,65,66)*.

2.4. Herpesvirus Vectors

Herpesviridae are a family of viruses that have a virion 150–200 nm in diameter consisting of four components: a lipid envelope with surface glycoprotein projections, a tegument of amorphous protein layer, an icosahedral nucleocapsid, and a protein spool on which the DNA is wrapped. The genome consists of a single molecule of linear double-stranded DNA (120–250 kb). To date, more than 100 different herpesviruses have been isolated and partially characterized, all of which share the ability of establishing both lytic and latent infection. A variety of vector systems have been developed from HSV types 1 and 2 as well as other herpesviruses such as pseudorabies virus (PRV) *(67)*, cytomegalovirus (CMV) *(68)*, and EBV *(69–71)*. In this section only HSV type 1 (HSV-1)-based vector systems will be discussed.

HSV-1, one of the eight human herpesviruses so far identified, has been among the most extensively investigated of all viruses. HSV-1 uses heparan sulfate as a primary binding molecule at the cellular surface *(72)* and two recently identified specific cell surface fusion receptors, herpesvirus entry mediator (Hve) A, a member of the tumor necrosis factor receptor family *(73,74)*, and HveC, also called nectin-1 or poliovirus receptor-related protein 1 *(75)*. On entry into the cytoplasm, capsids and some of the tegument proteins are actively transported along microtubules to nuclear pores, and the genome is released into the nucleoplasm. In the nucleoplasm it can circularize to establish latent infection or begin expression of more than 80 viral gene products in a well-organized cascade fashion to undergo lytic infection. The viral DNA does not actively integrate into the host chromosome but remains episomal.

Several features of HSV make it a promising gene transfer vehicle. These include its broad host cell range encompassing both dividing and nondividing cells, its large genome size predictive of its large transgene capacity, and its ability to establish latent infection in sensory neurons.

Two distinct types of vectors, recombinant and amplicon vectors, have been developed based on HSV (reviewed in refs. *76–79*).

2.4.1. Replication Defective Recombinant HSV Vectors

Recombinant HSV vectors retain most of the viral genome but have deletions in one or more viral genes to allow for the insertion of transgenes of interest. This type of vectors can be subdivided into two groups: replication defective and replication conditional or oncolytic vectors, depending on which genes are mutated or replaced. When one or more of the essential immediate-early (IE) genes (*ICP4* and *ICP27*) are deleted, the resulting recombinant viruses cannot replicate except in cells that complement the defective genes *(80–83)*. The lack of viral replication eliminates some viral cytopathogenicity, but the remaining viral gene expression, especially IE genes such as *ICP0*, *ICP4*, *ICP22*, and *ICP27* were shown to be toxic to cells *(83–86)*. To prevent cytotoxicity, a series of vectors has been generated possessing deletions in multiple *IE* genes. A quintuple deletion mutant (d109), null for all five IE genes (*ICP0*, *ICP4*, *ICP22*, *ICP27*, and *ICP47*), was demonstrated to be nontoxic to at least Vero and human embryonic lung (HEL) cells, and the vector genomes were shown to persist for long period of time in infected cells. This suggests that the mutant is in a state very similar to latency after infection *(83)*. However, the replication of this mutant is poor even in fully complementing cells and the vector expresses transgenes at very low levels specifically in the absence of *ICP0* activity.

Chen et al. identified cell type-specific processing of ICP0 protein *(87)*. According to their findings, ICP0 protein accumulates in specific nuclear domains called ND10 in nonneuronal cells, whereas it fails to accumulate in terminally differentiated neurons. This suggested that ICP0-mediated cytotoxicity may not be significant when ICP0-retaining HSV vectors infect neurons. Recent report from Jackson and DeLuca demonstrated that the activity of ICP0 protein in HSV infected cells determines the fate of viral genomes: noncircularization for lytic infection with ICP0 activity, whereas circularization for latency without ICP0 activity *(88)*. The neuron-specific degradation of the ICP0 protein might be responsible for the neuron-specific establishment of HSV latency.

The use of latency-associated transcript (LAT) promoters to drive transgene expression to achieve long-term gene expression has been explored *(89,90)*. The LAT promoter-mediated transgene expression in the context of a quintuple deletion mutant such as d109 is worthy of further investigation. This approach may allow both complete elimination of cytotoxicity and efficient long-term transgene expression. However, additional improvements may be required in order to achieve stable transduction in dividing cells such as progenitor and cancer cells because the vector DNA remains episomal and is not equipped with machineries that allow for active replication and segregation during mitosis.

2.4.2. HSV Amplicon Vectors

HSV amplicon vectors are based on the historical observation of defective HSV mutants after serial passaging of HSV at high MOI *(91)*. Because all the helper functions including viral gene expression and required protein synthesis can be supplied *in trans*, vector sequence are free from viral coding sequences. An "amplicon plasmid," which is a minimal seed unit of an "amplicon vector," consists of a prokaryotic plasmid carrying one or more transgene cassettes of interest and two *cis*-acting, noncoding HSV sequences, an origin of DNA replication (ori) and a DNA cleavage/packaging signal (pac) *(92)*. In the presence of HSV helper functions in HSV permissive cells, the transfected amplicon plasmid DNA undergoes the head-to-tail, rolling-circular-type DNA replication, and the resulting concatemeric DNA of approx 150 kb is then cleaved at the pac signal and incorporated into HSV virions. In other word, the HSV amplicon vector is a plasmid-based expression vector encapsulated in infectious HSV particles. Traditionally, HSV helper functions required for the replication and packaging of amplicon DNA were supplied from superinfected helper viruses, which cannot be physically separated from amplicon vectors. The resulting cytotoxicity caused by the helper viruses has been a critical limitation of this vector system *(93–95)*.

In 1996, Fraefel and his colleagues reported an improved method of amplicon packaging by cotransfecting pac-deleted five overlapping HSV genomic fragments cloned in cosmid vectors as helper DNA *(96)*. This first-generation helper virus-free packaging system reduced the number of helper virus almost completely (vector/helper ratio: $5 \times 10^4/3 \times 10^5$ *[97]*). Challenges remain however, including the relatively complicated packaging procedure, genetic instability of some of the HSV cosmid clones in bacteria, and contamination of a trace amount of replication competent helper virus. In order to overcome these problems, we have devised an improved helper virus-free packaging system (Fig. 1) *(97–99)*. In this current system, we use a bacterial artificial chromosome (BAC)-cloned HSV genome as a helper DNA. This simplified the packaging procedure significantly, increased genetic stability of HSV helper genome in bacteria, and increased the vector titers up to twofold compared with the cosmid-based system. We also included the following three safety features in the system: (1) deletions of all pac signals, (2) deletion of the essential *ICP27* gene from HSV-BAC and ICP27 function is supplied from a separate helper plasmid, and (3) oversizing the HSV-BAC beyond HSV packaging capacity (~178 kb). With these features, we further reduced the risk of helper contamination down to the vector to helper ratio of at least 10^8 *(97)*. The typical amplicon titer obtained with this system is 5×10^6~3×10^7 transducing units (TU)/mL without concentration and 5×10^8~3×10^9 TU/mL after concentration.

The most attractive and unique feature of HSV amplicon is its potentially large transgene capacity. In order to characterize and use this feature, we devised a general strategy in which any genomic DNA library clone in a BAC or P1-based artificial chromosome (PAC) vector may be modified by adding sequences (ori and pac) that will permit its packaging into an infectious HSV virion as an amplicon vector (Fig. 2) *(100)*. This modification can be easily accomplished through a single Cre-mediated recombination in bacteria because of the presence of a *loxP* site in every BAC and PAC clone. With this strategy, we designed an infectious BAC (iBAC) system, and demonstrated that DNA fragments of up to 150 kb in sizes can be efficiently packaged as HSV amplicons and delivered, intact into target cells through HSV infection. We also con-

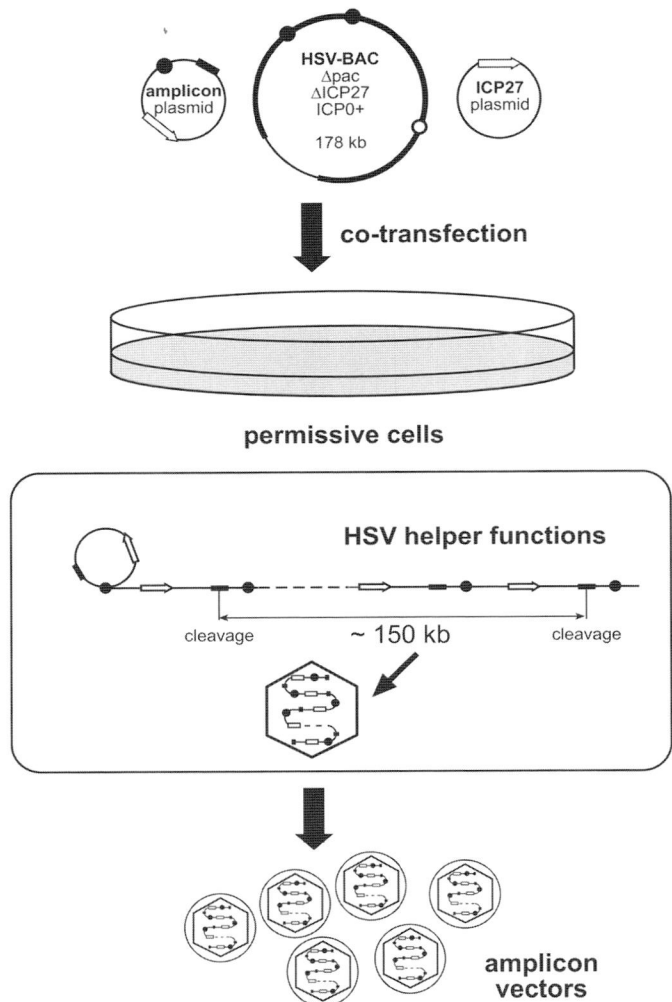

Fig. 1. Schematic diagram of improved helper virus-free HSV packaging method. HSV-permissive cells were cotransfected with amplicon plasmid DNA and two HSV helper plasmids, a pac-deleted, *ICP27*-deleted, oversized HSV-BAC and an *ICP27* complementing plasmid. In the transfected cells, amplicon plasmid undergoes the head-to-tail, rolling-circular-type DNA replication, and the resulting concatemeric DNA of approx 150 kb is then cleaved at the pac site and incorporated into HSV virions. HSV-BAC, on the other hand, expresses all the proteins required for HSV lytic infection and undergoes DNA replication, but itself cannot be packaged because of the lack of the pac signals.

firmed that in the context of HSV amplicon-mediated gene delivery, alternatively spliced variants and transcripts driven from multiple promoters can be expressed from the genomic loci with expression levels physiologically regulated *(101,102)*.

With these technical improvements, HSV amplicon vectors are now capable of offering several unique features: (1) easy and flexible vector constructions and rapid preparation of vectors (typically packaging and concentration of vectors within a week), (2) minimal cytotoxicity and immunogenicity owing to the lack of helper virus contamination and no viral gene expression from vector sequences, (3) large transgene capacity up to 150 kb *(100–102)*, (4) a broad cell

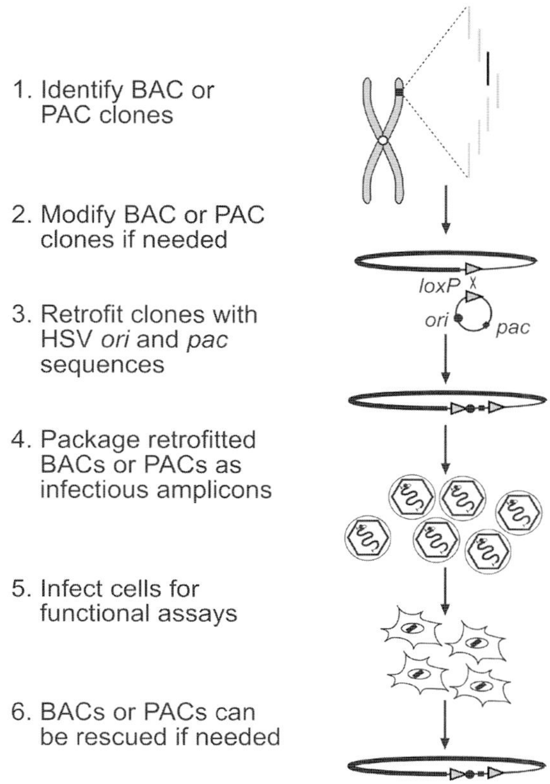

1. Identify BAC or PAC clones

2. Modify BAC or PAC clones if needed

3. Retrofit clones with HSV *ori* and *pac* sequences

4. Package retrofitted BACs or PACs as infectious amplicons

5. Infect cells for functional assays

6. BACs or PACs can be rescued if needed

Fig. 2. An overview of iBAC strategy is described. BAC or PAC clones that cover a gene of interest can be identified either through database searching or experimental screening such as PCR analysis and high-density filter hybridization. The sequence of BAC or PAC clones can be modified in bacteria if required and the modified clones can be further converted into amplicon plasmids by retrofitting ori and pac signals of HSV through Cre-mediated site-specific recombination. The resulting amplicon plasmids can be packaged into infectious HSV virions by the improved helper virus-free packaging system (*see* Fig. 1). The packaged genomic transgene can be efficiently introduced into wide range of mammalian cell types and various functional studies can be performed. Moreover, the packaged BAC- or PAC-based amplicon plasmid can be rescued in bacteria by electroporating high-molecular-weight cellular DNA prepared from the infected cells.

type host range with potential for conversion of resistant cells to infectable cells through overexpression of receptor molecules *(103)*, and (5) vector DNA can be circularized once transported into nucleus.

3. APPLICATIONS OF HSV AMPLICON VECTORS ON DIVIDING CELLS

HSV amplicon vectors, similar to other episomal vectors such as Ad vectors, does not actively integrate into host chromosomal DNA. Therefore, extrachromosomal vector DNA can be easily diluted through cell divisions and eventually be lost in rapidly dividing cells. As shown in Fig. 3, HSV amplicon can efficiently transduce murine primary neural progenitor cells. Typically, strong marker gene expression can be observed within 24 h after infection. However, the transgene expression in dividing cells gradually subsides. Lentiviral vectors typically require 2 d or more to express a marker gene, but almost all progeny cells derived from infected cells are positive for transgene expression. Integrating vectors such as MoMuLV-based and lentiviral vectors have a large advantage over episomal vectors in terms of obtaining stably expressing dividing cells.

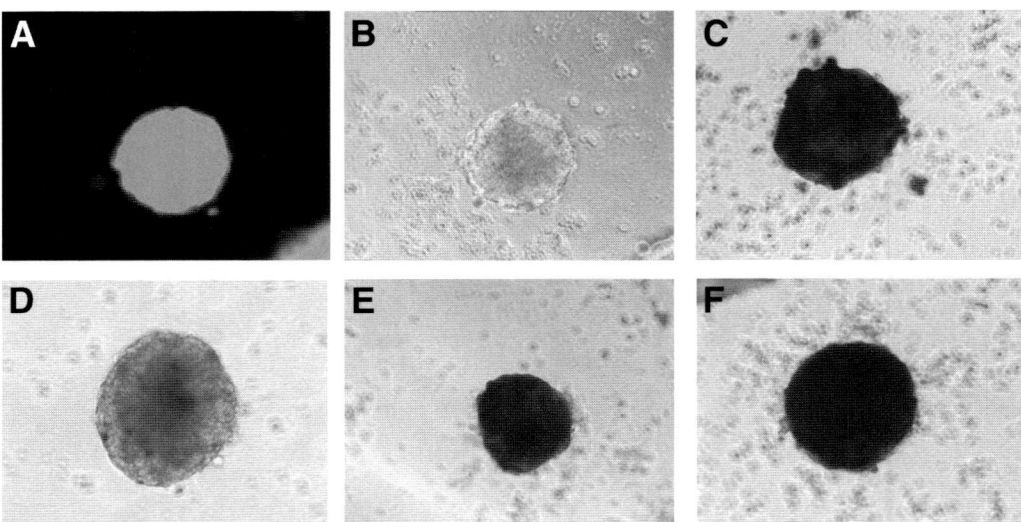

Fig. 3. Efficient transduction of neural progenitor cells using HSV amplicon vectors. Murine fetal (E13.5) neural progenitor cells in neurospheres were infected with a GFP and LacZ-expressing HSV amplicon vector at the MOI of 1 (**A–C**) or a LacZ-expressing FG Ad vector at various MOI (**D**, MOI = 1; **E**, MOI = 10; **F**, MOI = 100). Seventy-two hours after infection, the neurospheres were photographed under a fluorescent microscope ([**A**] GFP, [**B**] phase contrast) or subjected to X-gal staining (**C–F**).

However, because of their relative small transgene capacities, attempts to improve "high-capacity" HSV amplicon systems in order to achieve stable long-term transgene expression are ongoing.

3.1. Integrating HSV Hybrid Amplicon Vectors

The HSV amplicon vector has an indefinite potential to incorporate a variety of DNA elements and genes derived from other viruses and organisms to generate hybrid vectors because of its flexible vector design requirements and large transgene capacity. One example of the hybrid vector system is the HSV/AAV hybrid amplicon vector *(104–107)*. In this system, a transcriptional unit of interest is flanked by a pair of AAV ITRs and the rep 68/78 coding sequence is placed outside of the ITR-flanking cassette. When examined using human cell lines such as human embryonic kidney 293 and Gli36, threefold to tenfold higher stable integration frequencies (up to 10%) were obtained using the rep-containing hybrid amplicons, compared with amplicons without rep. Out of these stable clones obtained with the hybrid amplicons, up to 50% revealed having the transgene insertion at the adenoassociated virus integration site 1. HSV/AAV hybrid amplicon vectors hold future potential in becoming a powerful tool to achieve site-specific integration of a large transgene cassette into human chromosomes even though the frequencies of site-specific integration are not very high using the current system.

One alternative approach is to incorporate DNA elements for DNA transposone as demonstrated with the Ad vector system *(40)*. Another approach would be using a bacteriophage IN system such as phiC31 IN *(108,109)*. This is particularly interesting because the single crossover integration of circularized vector DNA into the chromosomes could result in integration of an entire vector sequence into the host genome.

3.2. Episomally Replicating HSV Hybrid Amplicon Vectors

An alternative approach to achieve long-term stable transduction is to incorporate the DNA elements required for EBV latent replication, *oriP* and *EBNA-1 (110)*, into the HSV amplicon vector.

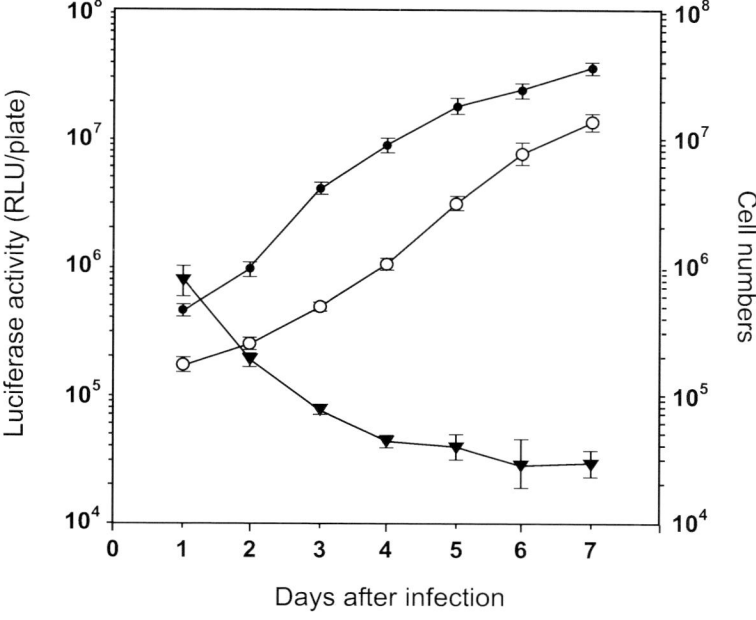

- Cell number per plate (n =6)
- ▼ HSV amplicon vector (n = 3)
- ○ HSV/EBV hybrid amplicon vector (n = 3)

Fig. 4. Stable transgene expression mediated by HSV/EBV hybrid amplicon vectors in actively dividing HeLa cells; 2×10^6 HeLa cells plated in P60 dish were transduced with either an HSV amplicon vector expressing firefly luciferase (Luc) or an HSV/EBV hybrid vector with the identical Luc expression cassette at the MOI of 3 (in triplicate). On the following day, the cells were trypsinized and passaged into seven P100 plates and let them further grow. Every 24 h, the cells on a plate from each infected sample were harvested and examined luciferase activity per plate. Luciferase activities of the conventional HSV amplicon-infected cells dropped very rapidly during the first 3 d, whereas those of HSV/EBV hybrid-infected cells showed a gradual increase, which correlates with the cell proliferation.

This type of vector design was originally reported by Wang and Vos (111) and is often termed the HSV/EBV hybrid amplicon vector. HSV/EBV hybrid amplicons can be packaged as approx 150 kb concatenated linear DNA molecules in HSV virions, which can infect a wide variety of cell types using the HSV infection machineries, become circularized in the nuclei of transduced cells, and replicate and persist as EBV replicons. The HSV/EBV hybrid amplicon allows extended transgene expression in rapidly proliferating cultured cells (Fig. 4). Although the EBV replicon can replicate in human, primate, and canine cells, its replication is not supported in rodent cells (110). However, by introducing any piece of mammalian genomic DNA larger than 10–15 kb, the EBV replicon can replicate using cryptic origins of DNA replication present in the genomic sequences even in rodent cells (112). We have confirmed this phenomenon in the context of HSV/EBV hybrid amplicon vectors. Hybrid amplicon DNA carrying approx 135 kb human low-density lipoprotein receptor (LDLR) genomic loci was successfully retained as a replicating extrachromosomal element in Chinese hamster ovary (CHO) cells for more than 3 mo in culture (101).

The polyoma virus replicon (113) is an alternative autonomously replicating system compatible to rodent cells. We are interested in testing the replicon elements in the context of HSV amplicon vectors.

4. CONCLUSIONS

Recent advances in molecular virology and gene therapy research have led to the development of a wide variety of viral vector systems. Every vector system has merits and limitations inherited from its parental virus. When considering the use of currently available vector systems, it is important to make selections which are application-specific.

In developing and improving existing vector systems, some limitations can be overcome through genetic modifications of virion components or incorporation of genetic components of other viruses or organisms. However, it is the transgene capacity limit that may be the most critical limitation of all, because not much modification or incorporation is possible.

The outstanding, 150-kb packaging capacity, therefore, would make the HSV amplicon vector system a versatile platform for a number of applications. As previously discussed, the amplicon system can adopt features of integrating systems such as AAV, DNA transposones, and bacteriophage INs, or those of the episomal EBV replicon system.

The development of infectious BAC technology now allows for the introduction of the entire genetic loci of up to 150 kb in size into virtually any cell type, taking advantage of HSV infection machineries. This technology, with further improvements, would be a powerful tool for functional genomics studies, drug screening applications, and eventually for gene therapies applicable in the treatment of human disease.

ACKNOWLEDGMENTS

The author thanks Drs Tetsu Yoshida, Hideyuki Okano, Keiro Ikeda, Tomotsugu Ichikawa, Hiroaki Wakimoto, Tatsuya Abe, Kinya Terada, Ryo Inoue, Ken Ishii, Shinji Yamamoto, Hirokazu Kambara, Masayuki Nitta, Shigeru Tanaka, Kazue Kasai, E. Antonio Chiocca, Edyta Tyminski, and Justin Kim for their collaborative insights. This work was supported by NIH R21 NS44514 and ALSA.

REFERENCES

1. Norton WT. Cell reactions following acute brain injury: a review. Neurochem Res 1999;24(2): 213–218.
2. Chen Y, Swanson RA. Astrocytes and brain injury. J Cereb Blood Flow Metab 2003;23(2):137–149.
3. Ray J, Raymon HK, Gage FH. Generation and culturing of precursor cells and neuroblasts from embryonic and adult central nervous system. Methods Enzymol 1995;254:20–37.
4. Weiss S, Dunne C, Hewson J, et al. Multipotent CNS stem cells are present in the adult mammalian spinal cord and ventricular neuroaxis. J Neurosci 1996;16(23):7599–7609.
5. Eriksson PS, Perfilieva E, Bjork-Eriksson T, et al. Neurogenesis in the adult human hippocampus. Nat Med 1998;4(11):1313–1317.
6. Reynolds BA, Weiss S. Generation of neurons and astrocytes from isolated cells of the adult mammalian central nervous system. Science 1992;255(5052):1707–1710.
7. Caldwell MA, He X, Wilkie N, et al. Growth factors regulate the survival and fate of cells derived from human neurospheres. Nat Biotechnol 2001;19(5):475–479.
8. Kotin RM, Linden RM, Berns KI. Characterization of a preferred site on human chromosome 19q for integration of adeno-associated virus DNA by non-homologous recombination. EMBO J 1992; 11(13):5071–5078.
9. Grimm D, Kern A, Rittner K, Kleinschmidt JA. Novel tools for production and purification of recombinant adenoassociated virus vectors. Hum Gene Ther 1998;9(18):2745–2760.
10. Xiao X, Li J, Samulski RJ. Production of high-titer recombinant adeno-associated virus vectors in the absence of helper adenovirus. J Virol 1998;72(3):2224–2232.
11. Summerford C, Samulski RJ. Membrane-associated heparan sulfate proteoglycan is a receptor for adeno-associated virus type 2 virions. J Virol 1998;72(2):1438–1445.
12. Zolotukhin S, Byrne BJ, Mason E, et al. Recombinant adeno-associated virus purification using novel methods improves infectious titer and yield. Gene Ther 1999;6(6):973–985.
13. Kaludov N, Handelman B, Chiorini JA. Scalable purification of adeno-associated virus type 2, 4, or 5 using ion-exchange chromatography. Hum Gene Ther 2002;13(10):1235–1243.

14. Kaludov N, Padron E, Govindasamy L, McKenna R, Chiorini JA, Agbandje-McKenna M. Production, purification and preliminary X-ray crystallographic studies of adeno-associated virus serotype 4. Virology 2003;306(1):1–6.
15. Rutledge EA, Halbert CL, Russell DW. Infectious clones and vectors derived from adeno-associated virus (AAV) serotypes other than AAV type 2. J Virol 1998;72(1):309–319.
16. Rabinowitz JE, Rolling F, Li C, et al. Cross-packaging of a single adeno-associated virus (AAV) type 2 vector genome into multiple AAV serotypes enables transduction with broad specificity. J Virol 2002;76(2):791–801.
17. Gao GP, Alvira MR, Wang L, Calcedo R, Johnston J, Wilson JM. Novel adeno-associated viruses from rhesus monkeys as vectors for human gene therapy. Proc Natl Acad Sci USA 2002;99(18):11,854–11,859.
18. Nakai H, Montini E, Fuess S, Storm TA, Grompe M, Kay MA. AAV serotype 2 vectors preferentially integrate into active genes in mice. Nat Genet 2003;34(3):297–302.
19. Hughes SM, Moussavi-Harami F, Sauter SL, Davidson BL. Viral-mediated gene transfer to mouse primary neural progenitor cells. Mol Ther 2002;5(1):16–24.
20. Surace EM, Auricchio A, Reich SJ, et al. Delivery of adeno-associated virus vectors to the fetal retina: impact of viral capsid proteins on retinal neuronal progenitor transduction. J Virol 2003;77(14):7957–7963.
21. Wu P, Ye Y, Svendsen CN. Transduction of human neural progenitor cells using recombinant adeno-associated viral vectors. Gene Ther 2002;9(4):245–255.
22. Russell DW, Hirata RK. Human gene targeting by viral vectors. Nat Genet 1998;18(4):325–330.
23. Hirata RK, Russell DW. Design and packaging of adeno-associated virus gene targeting vectors. J Virol 2000;74(10):4612–4620.
24. Miller DG, Petek LM, Russell DW. Human gene targeting by adeno-associated virus vectors is enhanced by DNA double-strand breaks. Mol Cell Biol 2003;23(10):3550–3557.
25. Dedieu JF, Vigne E, Torrent C, et al. Long-term gene delivery into the livers of immunocompetent mice with E1/E4-defective adenoviruses. J Virol 1997;71(6):4626–4637.
26. Christ M, Lusky M, Stoeckel F, et al. Gene therapy with recombinant adenovirus vectors: evaluation of the host immune response. Immunol Lett 1997;57(1–3):19–25.
27. Wickham TJ. Targetingadenovirus. Gene Ther 2000;7(2):110–114.
28. Nicklin SA, Baker AH. Tropism-modified adenoviral and adeno-associated viral vectors for gene therapy. Curr Gene Ther 2002; 2(3):273–293.
29. Imai T, Tokunaga A, Yoshida T, et al. The neural RNA-binding protein Musashi1 translationally regulates mammalian numb gene expression by interacting with its mRNA. Mol Cell Biol 2001;21(12):3888–3900.
30. Benraiss A, Chmielnicki E, Lerner K, Roh D, Goldman SA. Adenoviral brain-derived neurotrophic factor induces both neostriatal and olfactory neuronal recruitment from endogenous progenitor cells in the adult forebrain. J Neurosci 2001;21(17):6718–6731.
31. Mitani K, Graham FL, Caskey CT, Kochanek S. Rescue, propagation, and partial purification of a helper virus-dependent adenovirus vector. Proc Natl Acad Sci USA 1995;92(9):3854–3858.
32. Kochanek S, Clemens PR, Mitani K, Chen HH, Chan S, Caskey CT. A new adenoviral vector: Replacement of all viral coding sequences with 28 kb of DNA independently expressing both full-length dystrophin and beta-galactosidase. Proc Natl Acad Sci USA 1996; 93(12):5731–5736.
33. Parks RJ, Chen L, Anton M, Sankar U, Rudnicki MA, Graham FL. A helper-dependent adenovirus vector system: removal of helper virus by Cre-mediated excision of the viral packaging signal. Proc Natl Acad Sci USA 1996;93(24):13,565–13,570.
34. Ng P, Beauchamp C, Evelegh C, Parks R, Graham FL. Development of a FLP/frt system for generating helper-dependent adenoviral vectors. Mol Ther 2001;3(5, Pt 1):809–815.
35. Umana P, Gerdes CA, Stone D, et al. Efficient FLPe recombinase enables scalable production of helper-dependent adenoviral vectors with negligible helper-virus contamination. Nat Biotechnol 2001; 19(6):582–585.
36. Mitani K, Wakamiya M, Hasty P, Graham FL, Bradley A, Caskey CT. Gene targeting in mouse embryonic stem cells with an adenoviral vector. Somat Cell Mol Genet 1995;21(4):221–231.
37. Hirai H, Ogawa S, Kurokawa M, Yazaki Y, Mitani K. Molecular characterization of the genomic breakpoints in a case of t(3;21)(q26;q22). Genes Chromosomes Cancer 1999;26(1):92–96.
38. Fisher KJ, Kelley WM, Burda JF, Wilson JM. A novel adenovirus-adeno-associated virus hybrid vector that displays efficient rescue and delivery of the AAV genome. Hum Gene Ther 1996;7(17): 2079–2087.

39. Goncalves MA, van der Velde I, Knaan-Shanzer S, Valerio D, de Vries AA. Stable transduction of large DNA by high-capacity adeno-associated virus/adenovirus hybrid vectors. Virology 2004; 321(2):287–296.
40. Yant SR, Ehrhardt A, Mikkelsen JG, Meuse L, Pham T, Kay MA. Transposition from a gutless adeno-transposon vector stabilizes transgene expression in vivo. Nat Biotechnol 2002;20(10):999–1005.
41. Tan BT, Wu L, Berk AJ. An adenovirus-Epstein-Barr virus hybrid vector that stably transforms cultured cells with high efficiency. J Virol 1999;73(9):7582–7589.
42. Dorigo O, Gil JS, Gallaher SD, et al. Development of a novel helper-dependent adenovirus-Epstein-Barr virus hybrid system for the stable transformation of mammalian cells. J Virol 2004;78(12): 6556–6566.
43. Mann R, Mulligan RC, Baltimore D. Construction of a retrovirus packaging mutant and its use to produce helper-free defective retrovirus. Cell 1983;33(1):153–159.
44. Cone RD, Mulligan RC. High-efficiency gene transfer into mammalian cells: generation of helper-free recombinant retrovirus with broad mammalian host range. Proc Natl Acad Sci USA 1984; 81(20):6349–6353.
45. Miller AD, Buttimore C. Redesign of retrovirus packaging cell lines to avoid recombination leading to helper virus production. Mol Cell Biol 1986;6(8):2895–2902.
46. Hartley JW, Rowe WP. Naturally occurring murine leukemia viruses in wild mice: characterization of a new "amphotropic" class. J Virol 1976;19(1):19–25.
47. Rasheed S, Gardner MB, Chan E. Amphotropic host range of naturally occuring wild mouse leukemia viruses. J Virol 1976;19(1):13–18.
48. Chen ST, Iida A, Guo L, Friedmann T, Yee JK. Generation of packaging cell lines for pseudotyped retroviral vectors of the G protein of vesicular stomatitis virus by using a modified tetracycline inducible system. Proc Natl Acad Sci USA 1996;93(19):10,057–10,062.
49. Sharma S, Cantwell M, Kipps TJ, Friedmann T. Efficient infection of a human T-cell line and of human primary peripheral blood leukocytes with a pseudotyped retrovirus vector. Proc Natl Acad Sci USA 1996;93(21):11,842–11,847.
50. Roe T, Reynolds TC, Yu G, Brown PO. Integration of murine leukemia virus DNA depends on mitosis. EMBO J 1993;12(5):2099–2108.
51. Geller HM, Dubois-Dalcq M. Antigenic and functional characterization of a rat central nervous system-derived cell line immortalized by a retroviral vector. J Cell Biol 1988;107(5):1977–1986.
52. Price J, Thurlow L. Cell lineage in the rat cerebral cortex: a study using retroviral-mediated gene transfer. Development 1988;104(3): 473–482.
53. Short MP, Choi BC, Lee JK, Malick A, Breakefield XO, Martuza RL. Gene delivery to glioma cells in rat brain by grafting of a retrovirus packaging cell line. J Neurosci Res 1990;27(3):427–439.
54. Ezzeddine ZD, Martuza RL, Platika D, et al. Selective killing of glioma cells in culture and in vivo by retrovirus transfer of the herpes simplex virus thymidine kinase gene. New Biol 1991;3(6):608–614.
55. Oldfield EH, Ram Z, Culver KW, Blaese RM, DeVroom HL, Anderson WF. Gene therapy for the treatment of brain tumors using intra-tumoral transduction with the thymidine kinase gene and intra-venous ganciclovir. Hum Gene Ther 1993;4(1):39–69.
56. Lewis P, Hensel M, Emerman M. Human immunodeficiency virus infection of cells arrested in the cell cycle. Embo J 1992;11(8): 3053–3058.
57. Bukrinsky MI, Haggerty S, Dempsey MP, et al. A nuclear localization signal within HIV-1 matrix protein that governs infection of non-dividing cells. Nature1993;365(6447):666–669.
58. Heinzinger NK, Bukinsky MI, Haggerty SA, et al. The Vpr protein of human immunodeficiency virus type 1 influences nuclear localization of viral nucleic acids in nondividing host cells. Proc Natl Acad Sci USA 1994;91(15):7311–7315.
59. von Schwedler U, Kornbluth RS, Trono D. The nuclear localization signal of the matrix protein of human immunodeficiency virus type 1 allows the establishment of infection in macrophages and quiescent T lymphocytes. Proc Natl Acad Sci USA 1994;91(15):6992–6996.
60. Gallay P, Hope T, Chin D, Trono D. HIV-1 infection of nondividing cells through the recognition of integrase by the importin/karyopherin pathway. Proc Natl Acad Sci USA 1997;94(18):9825–9830.
61. Naldini L, Blomer U, Gallay P, et al. In vivo gene delivery and stable transduction of nondividing cells by a lentiviral vector. Science 1996;272(5259):263–267.
62. Naldini L, Blomer U, Gage FH, Trono D, Verma IM. Efficient transfer, integration, and sustained long-term expression of the transgene in adult rat brains injected with a lentiviral vector. Proc Natl Acad Sci USA 1996;93(21):11,382–11,388.

63. Miyoshi H, Blomer U, Takahashi M, Gage FH, Verma IM. Development of a self-inactivating lentivirus vector. J Virol 1998;72(10):8150–8157.
64. Dull T, Zufferey R, Kelly M, et al. A third-generation lentivirus vector with a conditional packaging system. J Virol 1998;72(11):8463–8471.
65. Englund U, Ericson C, Rosenblad C, et al. The use of a recombinant lentiviral vector for ex vivo gene transfer into the rat CNS. Neuroreport 2000;11(18):3973–3977.
66. Englund U, Fricker-Gates RA, Lundberg C, Bjorklund A, Wictorin K. Transplantation of human neural progenitor cells into the neonatal rat brain: extensive migration and differentiation with long-distance axonal projections. Exp Neurol 2002;173(1):1–21.
67. Boldogkoi Z, Bratincsak A, Fodor I. Evaluation of pseudorabies virus as a gene transfer vector and an oncolytic agent for human tumor cells. Anticancer Res 2002;22(4):2153–2159.
68. Borst EM, Messerle M. Construction of a cytomegalovirus-based amplicon: a vector with a unique transfer capacity. Hum Gene Ther 2003;14(10):959–970.
69. Banerjee S, Livanos E, Vos JM. Therapeutic gene delivery in human B-lymphoblastoid cells by engineered non-transforming infectious Epstein-Barr virus. Nat Med 1995;1(12):1303–1308.
70. Sun TQ, Livanos E, Vos JM. Engineering a mini-herpesvirus as a general strategy to transduce up to 180 kb of functional self-replicating human mini-chromosomes. Gene Ther 1996;3(12):1081–1088.
71. Delecluse HJ, Pich D, Hilsendegen T, Baum C, Hammerschmidt W. A first-generation packaging cell line for Epstein-Barr virus-derived vectors. Proc Natl Acad Sci USA 1999;96(9):5188–5193.
72. Shieh MT, WuDunn D, Montgomery RI, Esko JD, Spear PG. Cell surface receptors for herpes simplex virus are heparan sulfate proteoglycans. J Cell Biol 1992;116(5):1273–1281.
73. Montgomery RI, Warner MS, Lum BJ, Spear PG. Herpes simplex virus-1 entry into cells mediated by a novel member of the TNF/NGF receptor family. Cell 1996;87(3):427–436.
74. Terry-Allison T, Montgomery RI, Whitbeck JC, et al. HveA (herpesvirus entry mediator A), a coreceptor for herpes simplex virus entry, also participates in virus-induced cell fusion. J Virol 1998;72(7):5802–5810.
75. Geraghty RJ, Krummenacher C, Cohen GH, Eisenberg RJ, Spear PG. Entry of alphaherpesviruses mediated by poliovirus receptor-related protein 1 and poliovirus receptor. Science 1998;280(5369):1618–1620.
76. Glorioso JC, Bender MA, Goins WF, Fink DJ, DeLuca N. HSV as a gene transfer vector for the nervous system. Mol Biotechnol 1995;4(1):87–99.
77. Jacobs A, Breakefield XO, Fraefel C. HSV-1-based vectors for gene therapy of neurological diseases and brain tumors: part I. HSV-1 structure, replication and pathogenesis. Neoplasia 1999;1(5):387–401.
78. Jacobs A, Breakefield XO, Fraefel C. HSV-1-based vectors for gene therapy of neurological diseases and brain tumors: part II. Vector systems and applications. Neoplasia 1999;1(5):402–416.
79. Burton EA, Fink DJ, Glorioso JC. Gene delivery using herpes simplex virus vectors. DNA Cell Biol 2002;21(12):915–936.
80. McCarthy AM, McMahan L, Schaffer PA. Herpes simplex virus type 1 ICP27 deletion mutants exhibit altered patterns of transcription and are DNA deficient. J Virol 1989;63(1):18–27.
81. Rice SA, Knipe DM. Genetic evidence for two distinct transactivation functions of the herpes simplex virus alpha protein ICP27. J Virol 1990;64(4):1704–1715.
82. Samaniego LA, Webb AL, DeLuca NA. Functional interactions between herpes simplex virus immediate-early proteins during infection: gene expression as a consequence of ICP27 and different domains of ICP4. J Virol 1995;69(9):5705–5715.
83. Samaniego LA, Neiderhiser L, DeLuca NA. Persistence and expression of the herpes simplex virus genome in the absence of immediate-early proteins. J Virol 1998;72(4):3307–3320.
84. Johnson PA, Wang MJ, Friedmann T. Improved cell survival by the reduction of immediate-early gene expression in replication-defective mutants of herpes simplex virus type 1 but not by mutation of the virion host shutoff function. J Virol 1994;68(10):6347–6362.
85. Wu N, Watkins SC, Schaffer PA, DeLuca NA. Prolonged gene expression and cell survival after infection by a herpes simplex virus mutant defective in the immediate-early genes encoding ICP4, ICP27, and ICP22. J Virol 1996;70(9):6358–6369.
86. Samaniego LA, Wu N, DeLuca NA. The herpes simplex virus immediate-early protein ICP0 affects transcription from the viral genome and infected-cell survival in the absence of ICP4 and ICP27. J Virol 1997;71(6):4614–4625.

87. Chen X, Li J, Mata M, et al. Herpes simplex virus type 1 ICP0 protein does not accumulate in the nucleus of primary neurons in culture. J Virol 2000;74(21): 10,132–10,141.
88. Jackson SA, DeLuca NA. Relationship of herpes simplex virus genome configuration to productive and persistent infections. Proc Natl Acad Sci USA 2003;100(13):7871–7876.
89. Goins WF, Sternberg LR, Croen KD, et al. A novel latency-active promoter is contained within the herpes simplex virus type 1 UL flanking repeats. J Virol 1994;68(4):2239–2252.
90. Goins WF, Lee KA, Cavalcoli JD, et al. Herpes simplex virus type 1 vector-mediated expression of nerve growth factor protects dorsal root ganglion neurons from peroxide toxicity. J Virol 1999;73(1): 519–532.
91. Frenkel N, Locker H, Vlazny DA. Studies of defective herpes simplex viruses. Ann NY Acad Sci 1980;354:347–370.
92. Spaete RR, Frenkel N. The herpes simplex virus amplicon: a new eucaryotic defective-virus cloning-amplifying vector. Cell 1982;30(1): 295–304.
93. Geller AI, Breakefield XO. A defective HSV-1 vector expresses Escherichia coli beta-galactosidase in cultured peripheral neurons. Science 1988;241(4873):1667–1669.
94. Geller AI, Keyomarsi K, Bryan J, Pardee AB. An efficient deletion mutant packaging system for defective herpes simplex virus vectors: potential applications to human gene therapy and neuronal physiology. Proc Natl Acad Sci USA 1990;87(22):8950–8954.
95. Lim F, Hartley D, Starr P, et al. Generation of high-titer defective HSV-1 vectors using an IE 2 deletion mutant and quantitative study of expression in cultured cortical cells. Biotechniques 1996;20(3): 460–469.
96. Fraefel C, Song S, Lim F, et al. Helper virus-free transfer of herpes simplex virus type 1 plasmid vectors into neural cells. J Virol 1996;70(10):7190–7197.
97. Saeki Y, Fraefel C, Ichikawa T, Breakefield XO, Chiocca EA. Improved helper virus-free packaging system for HSV amplicon vectors using an ICP27-deleted, oversized HSV-1 DNA in a bacterial artificial chromosome. Mol Ther 2001;3(4):591–601.
98. Saeki Y, Ichikawa T, Saeki A, et al. Herpes simplex virus type 1 DNA amplified as bacterial artificial chromosome in Escherichia coli: rescue of replication-competent virus progeny and packaging of amplicon vectors. Hum Gene Ther 1998;9(18):2787–2794.
99. Saeki Y, Breakefield XO, Chiocca EA. Improved HSV-1 amplicon packaging system using ICP27-deleted, oversized HSV-1 BAC DNA. Methods Mol Med 2003;76:51–60.
100. Wade-Martins R, Smith ER, Tyminski E, Chiocca EA, Saeki Y. An infectious transfer and expression system for genomic DNA loci in human and mouse cells. Nat Biotechnol 2001;19(11):1067–1070.
101. Wade-Martins R, Saeki Y, Chiocca EA. Infectious delivery of a 135-kb LDLR genomic locus leads to regulated complementation of low-density lipoprotein receptor deficiency in human cells. Mol Ther 2003;7(5, Pt 1):604–612.
102. Inoue R, Moghaddam KA, Ranasinghe M, Saeki Y, Chiocca EA, Wade-Martins R. Infectious delivery of the 132 kb CDKN2A/ CDKN2B genomic DNA region results in correctly spliced gene expression and growth suppression in glioma cells. Gene Ther 2004;11(15):1195–1204.
103. Miller CG, Krummenacher C, Eisenberg RJ, Cohen GH, Fraser NW. Development of a syngenic murine B16 cell line-derived melanoma susceptible to destruction by neuroattenuated HSV-1. Mol Ther 2001;3(2):160–168.
104. Johnston KM, Jacoby D, Pechan PA, et al. HSV/AAV hybrid amplicon vectors extend transgene expression in human glioma cells. Hum Gene Ther 1997;8(3):359–370.
105. Wang Y, Camp SM, Niwano M, et al. Herpes simplex virus type 1/adeno-associated virus rep(+) hybrid amplicon vector improves the stability of transgene expression in human cells by site-specific integration. J Virol 2002;76(14):7150–7162.
106. Heister T, Heid I, Ackermann M, Fraefel C. Herpes simplex virus type 1/adeno-associated virus hybrid vectors mediate site-specific integration at the adeno-associated virus preintegration site, AAVS1, on human chromosome 19. J Virol 2002;76(14):7163–7173.
107. Bakowska JC, Di Maria MV, Camp SM, Wang Y, Allen PD, Breakefield XO. Targeted transgene integration into transgenic mouse fibroblasts carrying the full-length human AAVS1 locus mediated by HSV/AAV rep(+) hybrid amplicon vector. Gene Ther 2003;10(19):1691–1702.
108. Groth AC, Olivares EC, Thyagarajan B, Calos MP. A phage integrase directs efficient site-specific integration in human cells. Proc Natl Acad Sci USA 2000;97(11):5995–6000.
109. Olivares EC, Hollis RP, Chalberg TW, Meuse L, Kay MA, Calos MP. Site-specific genomic integration produces therapeutic Factor IX levels in mice. Nat Biotechnol 2002;20(11):1124–1128.

110. Yates JL, Warren N, Sugden B. Stable replication of plasmids derived from Epstein-Barr virus in various mammalian cells. Nature 1985;313(6005):812–815.
111. Wang S, Vos JM. A hybrid herpesvirus infectious vector based on Epstein-Barr virus and herpes simplex virus type 1 for gene transfer into human cells in vitro and in vivo. J Virol 1996;70(12):8422–8430.
112. Krysan PJ, Calos MP. Epstein-Barr virus-based vectors that replicate in rodent cells. Gene 1993;136(1,2):137–143.
113. Gassmann M, Donoho G, Berg P. Maintenance of an extrachromosomal plasmid vector in mouse embryonic stem cells. Proc Natl Acad Sci USA 1995;92(5):1292–1296.

34
Electrical Stimulation and Angiogenesis
Electrical Signals Have Direct Effects on Endothelial Cells

Min Zhao, MD, PhD

SUMMARY

Electrical stimulation has long been used for various clinical conditions and recently been showed to have profound effects on angiogenesis—new blood vessel formation from pre-existing blood vessels. Electrical signals stimulate angiogenesis and possibly is able to organize blood vessel formation spatially. This chapter will present recent data on the cellular and molecular mechanisms of electrical stimulation-induced angiogenic responses. It will discuss direct effects of electrical stimulation on vascular cells. These include directed cell migration, orientation, elongation, and cell cycle control. Experimental evidence will be presented for vascular endothelial growth factor (VEGF) release and VEGF receptor signaling in electrical stimulation-induced angiogenic responses. The direct effects on endothelial cells, together with indirect effects caused by electric stimulation on other types of cells, such as stimulated VEGF production by muscle cell, may provide potential clinical approaches for angiogenesis control. Such a possibility is emerging, as human trials with electrical stimulation to enhance angiogenesis have shown promising results.

Key Words: Angiogenesis; electric stimulation; endothelium; vascular cells; cell cycle; cell migration; cell alignment; cell signaling.

1. INTRODUCTION

Angiogenesis is the process of new blood vessel formation from existing blood vessels. Control and modification of this process are critically important in many medical and biological events, as this forms "lifeline"—blood circulation for both normal and abnormal tissues and cells during major structural and functional changes. It is one of the most important factors in local homeostasis including cell cycle control in brain as well as in other parts of the body. The interaction between local cell cycle progress and functional cellular activities of a sizable population of cells and blood supply is a dynamic and two-way interactive processes. Blood vessels are the structural basis of blood circulation. Cell growth and activities demand blood supply, hence opening of local microcirculation eventually leads to new blood vessel formation in one way; in the other way, new blood vessel formation controls how active the local cell function can be and also as a population, to what extent the cells can proliferate. Modulation of new blood vessel formation, either to increase the blood supply to ischemic tissues or to inhibit blood supply to undesired neoplasm, such as cancer, offers great hope for treatment of a vast spectrum of diseases *(1)*.

Electric stimulation to enhance angiogenesis has been heralded recently as a potential new chapter in the search for novel approaches to treat ischemia *(2–5)*. This significant enhancement, either by skeletal muscle contraction or via noncontraction-based electric stimulation to the ischemic or nonischemic limbs, has been shown to be mediated by increased expression of vascular endothelial growth factor (VEGF) in the muscle cells *(2,5,6)* (i.e., it is an indirect effect of electric stimulation, via enhanced VEGF production by muscle cells). More recently, we and others have shown that electrical stimulation has direct effects on fundamental cell behaviors important for angiogenesis and stimulates VEGF production directly by endothelial cells *(7–11)*.

This chapter focuses on some basic cell behaviors of angiogenesis *per se* and the direct control (direct effects on vascular cells) with electric stimulation. Blood vessel formation *de novo* during early development and angiogenesis at a later stage are reviewed in Chapters 4, 17 and 18, which give detailed accounts to angiogenesis, related to brain diseases. This chapter will mainly discuss direct effects of electric stimulation on endothelial cells and the molecular mechanisms.

Endothelial cells and other vascular cells are exposed to endogenous electric fields (EFs). This perhaps is the biological basis of why those cells respond to exogenous electrical stimulation. First, a brief account of several examples where endogenous EFs have been experimentally detected is given. Second, some reports of in vivo angiogenesis-induced by electrical stimulation are summarized. Finally, the direct effects of electrical stimulation on vascular cells, endothelial cells in particular, will be described in more detail. The intracellular signaling mechanisms induced by electrical stimulation appear to be very similar or the same as those found for other angiogenic factors, such as shear flow stress and hypoxia. The implication of these findings as promising approaches for angiogenesis control and enhancement will be discussed at the end.

2. ENDOGENOUS EFS

Applied electric stimulation induces significant angiogenesis responses. More importantly, there are endogenous electric potential differences in close relation with the vasculature, with blood flow and in situations in which active angiogenesis occurs, such as in development, wound healing, tissue damage, and abnormal cell proliferation *(12–27)*. This may be significant physiologically as well as clinically.

Application of electric stimulation has the following advantages:
1. It can be applied easily and practically in most situations.
2. It causes biochemical changes locally.
3. It is relatively cheap to use.

We are at the early stages of fully exploiting this technique and basic scientific information will probably lead the way along side in vivo and clinical research.

2.1. Endogenous EFs and Blood Flow

Several types of electric potential difference exist around blood vessel endothelium and may be involved in regulating the development of thromboses *(12–14)*. Zeta potentials for example are created at the vascular lumen side of endothelial surface by blood flow. This type of potential difference ranges from 100 to 400 mV *(13)*. This is the electric potential difference between the shear plane of flood flow to the surface of endothelial cells lining the vessels.

2.2. Endogenous EFs in Development

Endogenous EFs measured directly in development may be important for development control *(15–21)*. For example, a steady direct current (DC) EF of 450–1600 mV/mm has been measured across the wall of the amphibian neural tube during early neuronal development *(18)* and disrupting this perturbs development *(18–21)*. It has not been studied whether this involves some behaviors or proliferation of endothelial cells or vasculature formation in vivo.

2.3. Wounding and Tissue Damage Induce Local EFs

During active tissue construction and reconstruction in wound healing, angiogenesis is a key event. It occurs in the presence of experimentally evident endogenous EFs. How is this generated?

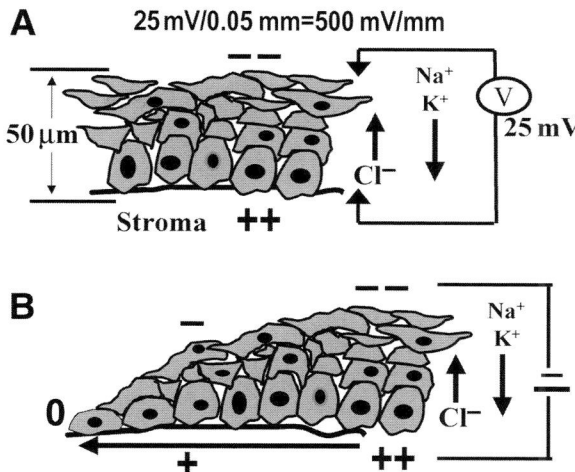

Fig. 1. Endogenous wound EFs. (**A**) Normal polarized corneal epithelial cells establish an electric potential difference of 25–27 mV across the epithelial layer of approx 50 µm distance inside positive. This is maintained by actively pumping sodium and potassium inward to the basal side of the epithelial layer while chloride outward to the apical side. (**B**) Disruption of the epithelial barrier results in a short-circuit current that flows out at the wound. The surrounding nonwounded epithelium continues to pump ions actively, Na^+ and K^+ inward to the basal side and Cl^- outward to the apical side and this serves as a battery keeping current flowing.

Normal polarized epithelial cells separate inside from outside by selectively segregating and actively pumping sodium and potassium inward to the basal side of the epithelial layer, whereas chloride outward to the apical side *(1,6,22–26)*. This results in an electric potential difference of 25–27 mV across the epithelial layer of approx 50 µm distance inside positive in the cornea epithelium for instance (Fig. 1A). Disruption of the epithelial barrier, as happens in injuries or wounding, causes an instantaneous collapse of the transepithelial potential difference and generates a persistent short-circuit current that flows out at the wound (Fig. 1B). The surrounding non-wounded epithelium continues to pump ions actively, Na^+ and K^+ inward to the basal side and Cl^- outward to the apical side, and this serves as a battery, keeping current flowing until the wound heals, the barrier restored and the short-circuit stopped (Fig. 1B). This electric potential difference across epithelial layers has been measured in humans, guinea pigs, rats, and mice (Reid and Zhao, unpublished data) *(25–26)*. We and others have measured consistent outward electrical currents at wounds in cornea and skin in rodent and human skin. The wound-induced electrical fields lasted for hours to days, depending on the wound size. After wounding, a large outward current of 4 µA/cm^2 was measured at the wound edges (Fig. 1B). This gradually increased to 10 µA/cm^2 and persisted at approx 4–8 µA/cm^2. The direction of the current was independent of wound size and the current vector (defined as the flow of positive charge) was directed toward the wound center.

Injured and ischemic tissue also becomes electrically polarized relative to surrounding normal tissue, because of depolarization of cells in the damaged areas and the build-up of extracellular K^+ ions. In the heart, this leads to a flow of injury current, intracellularly through junctionally coupled cells with a return extracellular loop, and is thought to be involved in arrythmogenesis *(27–28)*. The injury currents and hence the extracellular EFs which they produce, vary with the heartbeat. In diastole, damaged extracellular areas become negative relative to undamaged areas and this is reversed in systole, though the currents become more diffuse and are around one-third of their diastolic magnitude *(28)*. Extracellular injury currents, which create directly measured DC EFs of 58 mV/cm, extend more than about 8 mm at the boundary between ischemic and normal tissue *(27)*.

2.4. Cell Proliferation Results in Changes of Membrane Electrical Charges

Rapid and uncontrolled proliferation of cells also causes significant changes in cell surface charge *(29–30)*, and tumors become polarized relative to quiescent surrounding regions. In breast cancer, for example, potential differences between proliferating and nonproliferating regions can be measured at the surface of the skin and are being used diagnostically, because they correlate well with malignancy of the neoplasm *(31)*.

These examples show electrical signals exist around blood vessels, especially when active angiogenesis takes places, such as wound healing, tissue repair, and significant cell proliferation. Angiogenic responses induced by the electrical stimulation will be discussed in the following sections, albeit those induced by externally applied electric stimulation may not necessarily parallel to those happens endogenously.

3. ANGIOGENESIS INDUCED BY ELECTRICAL STIMULATION IN VIVO

Electrical stimulation has long been used to stimulate muscle contraction and it was generally assumed that the activity increase demands in blood supply. Surprisingly, subcontraction electrical stimulation, or electrical stimulation, which does not cause muscle contraction, is also very effective in inducing significant angiogenesis *(2–4)*. Importantly, we and others have recently shown that electrical stimulation has direct effects on angiogenic responses of endothelial cells *(7–11)*. These include increased VEGF production directly by endothelial cells, cell behaviors important for angiogenesis, such as direction cell migration, alignment, and elongation *(7–11)*.

This has significant medical and biological implications, as electrical stimulations can be applied to nonmuscle tissues, as well as muscle tissues, to modulate angiogenic responses; no locomotion is needed to achieve angiogenesis for muscle tissues. Therefore, electrical stimulation may be a novel approach in the management of angiogenesis in vivo. Other advantages using electrical stimulation include easy and well-controlled local applications, which are cheap and easy to use.

Electric stimulation generally includes stimulation with DC electric currents, which can be constant and pulsed with varied frequency, and alternative currents (ACs), with great variety in frequency and strength.

3.1. Electric Stimulation-Induced Blood Vessel Permeability Increase

Using direct observation by fluorescence microscopy, Nannmark et al. showed that applied EFs induced increase in capillary permeability to macromolecules and leukocytes *(32)*. The electrical stimulation they used include 5, 20, or 50 µA of DC or 20 µA AC. Fluorescein isothiocyanate-dextran (molecular weight: 150,000) leaked out after a lag time of 30–160 min. As increase in vascular permeability are associated with angiogenesis *(33)*, this may be one of the initial responses of microvasculature to electric stimulation.

3.2. Electric Stimulation Enhanced Angiogenesis in Skeletal Muscle

It has been know for more than two decades that electric stimulation significantly increased capillary density and blood flow in skeletal muscles *(34)*. This has been further confirmed in various species such as rabbit *(35)*, rat *(36,37)*, and human *(38,39)*. Electrical stimulation has been proved to have a definitive direct effect to increase local capillary density. Local transcriptional and translational increases in VEGF production by muscle cells plays a major role in the angiogenic responses *(36,37)*. This is in addition to the indirect effects mediated by local muscle contractions, as electrical stimulation of the strength, either capable of causing muscle contraction or below the contraction threshold, are effective in enhancing blood flow and capillary density in the stimulated tissue.

When electrical stimulation is applied locally, the local vasculature and endothelial cells are all subjected to the stimulation. It was proposed that electrical stimulation, in addition to its effect on angiogenesis indirectly through inducing VEGF production by other types of cell, can

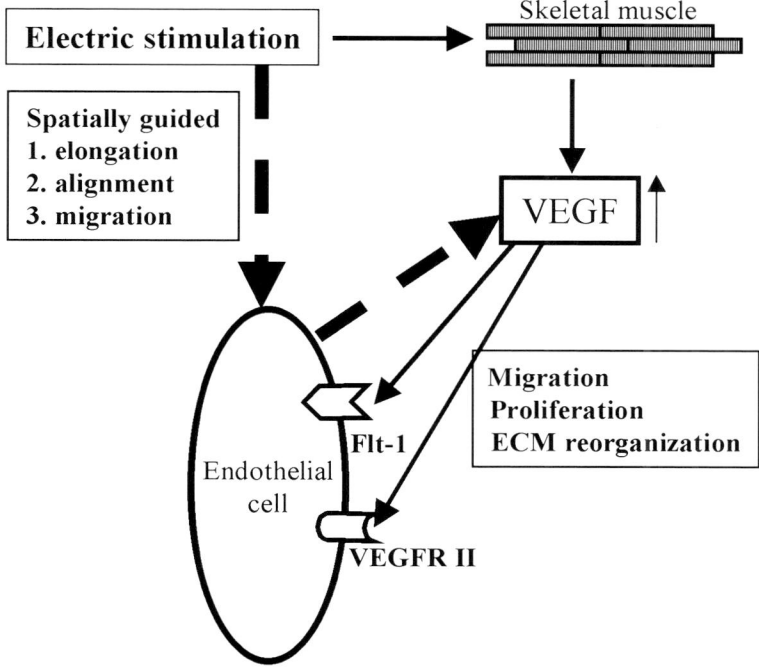

Fig. 2. Effects that electrical stimulation may have on angiogenesis. Electric stimulation acts on endothelial cells via enhanced VEGF production by skeletal muscle (solid line arrow). Electric stimulation can also have a direct effect on endothelial cell shape, alignment, and migration in a spatially controlled manner, and also on VEGF production by endothelial cells themselves (dotted line arrow). This offers potential wide application of electrical stimulation to enhance angiogenesis.

also directly induce angiogenic responses of endothelial cells (Fig. 2). I will present experimental evidence of that in the following sections.

4. ELECTRICAL STIMULATION HAS DIRECT EFFECTS ON ENDOTHELIAL CELLS

Angiogenic induction through indirect effects, that is, enhanced VEGF production by muscle tissues have been reviewed *(3,4)*. The following will mainly discuss direct effects on endothelial cells and other types of vascular cells of electric stimulation *(7–11)*.

4.1. Methodology

4.1.1. Electrical Stimulation

In order to mimic the DC EFs detected endogenously, we use an experimental setup in which well-controlled DC electric signal can be applied *(40)* (Fig. 3). This electrotactic chamber is made with two strips of cover glass glued to the base of a tissue culture dish to form a 22-mm-long × 10-mm-wide trough. Cells are seeded to the base. A roof of cover glass is added and sealed with silicone grease. The final dimensions of this very shallow chamber, through which current was passed, were $22 \times 10 \times 0.2$ mm^3. Agar-salt bridges connect silver/silver chloride electrodes in beakers of physiological saline solution, to the reservoirs of culture medium at the ends of the chambers. This prevents diffusion of electrode products into the cultures. Field strengths were measured directly at the ends of the chamber. This basic setup can be used for various stipulations such as constant or pulsed DC and AC stimulation.

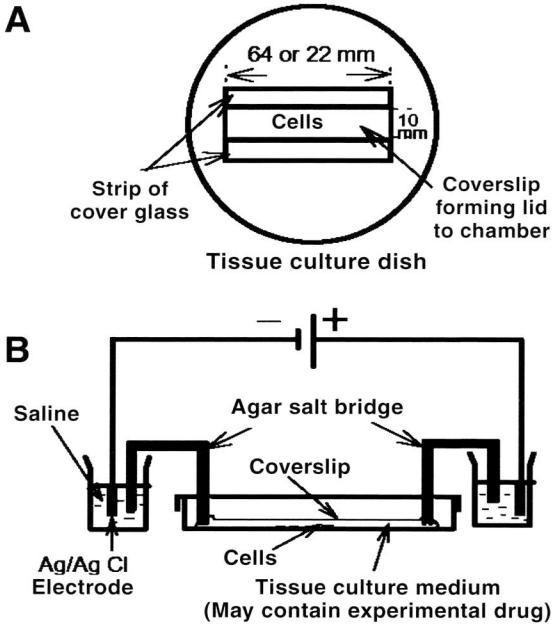

Fig. 3. Experimental setup for study of the effects of electrical stimulation on vascular cells. **(A)** Electrotactic chamber constructed within a tissue culture plastic dish, viewed from above. **(B)** Side-on view includes DC power supply and Ag/AgCl electrodes isolated from the culture chamber using agar-gel salt bridges.

4.1.2. Cell Behavior Analysis

In angiogenesis, endothelial cells must (1) migrate directionally outward from a degraded basement membrane, (2) reorient in concert with their neighbors in three dimensions, and (3) proliferate *(41)*. We have been focusing on those basic endothelial behaviors induced directly by applying DC EFs. We have imposed a striking orientation to blood vessel endothelial cell growth and cell migration, which may have profound implications in controlling angiogenesis. When cultured in EFs, endothelial cells (1) migrated directionally and (2) assumed a most dramatic alignment, perpendicular to the field vector. The strong alignment that these cells showed in two-dimensional cultures may promote tubular structure formation in three-dimensional cultures. This single cue therefore has a remarkable series of effects, which set the scene for the induction of capillary tube formation in three dimensions, because directional cell migration, cell orientation, and cell proliferation are fundamental steps in angiogenesis.

4.1.2.1. DIRECTIONAL MIGRATION

Mean migration rate and directedness were quantified over a defined period of time *(40,42,43)*. The angle that each cell moved with respect to the imposed EF vector was measured. The cosine of this angle (defined as directedness) is 1, for cells moving directly toward the cathode, 0 for cells moving perpendicular to the EF vector, and −1 for cells moving directly toward the anode. Averaging the cosines ($\Sigma\cos\theta)/N$, (where θ is the angle between the field vector and the direction of movement, and N is the total number of cells) yields average directedness of cell movement.

4.1.2.2. PERPENDICULAR ORIENTATION

Cell orientation was quantified as an orientation index (Oi) *(9,40)*. Oi of a cell with respect to the EF was defined as a function of $\cos 2\alpha$, where α is the angle formed by the intersection of

Electrical Stimulation and Angiogenesis

a line drawn through the long axis of each cell with a line drawn perpendicular to the field vector. This Oi varies from −1 to 1. A cell lying parallel to the EF vector has an Oi of −1, and a cell perpendicular to the EF vector, an Oi of 1. A randomly oriented population of cells gives an average Oi (defined by $\Sigma\cos2\alpha/N$) of 0. The angle α was measured using the image analyzer and average Oi for the cell population ($\Sigma\cos2\alpha/N$) was calculated. The significance of this orientation was calculated using Rayleigh's distribution *(40)*. The probability that the population is randomly oriented is given by $P = e^{-(L^2 n)(10^{-4})}$, where $L = [(\Sigma n \sin 2\alpha/n)^2 + (\Sigma n \cos 2\alpha/n)^2]^{1/2}/n$ (0.01), and *n* is the total number of cells. A probability level of 0.001 was used as the limit for significant perpendicular orientation.

4.1.2.3. Cell Elongation

A long:short axis ratio was calculated from the measurements made with the image analyzer. This gives an objective assessment of elongation of the endothelial cells and the effects of an applied EF *(9)*.

4.2. Electrical Signals Directly Induce Angiogenic Responses of Endothelial Cells

Some typical cellular behaviors in direct response to electrical stimulation are discussed in this section. That is not by the release of angiogenic factors of other type of cells or tissues. I will mainly discuss the responses of endothelial cells. Similar responses have been observed for other types of vascular cells, such as vascular fibroblast and smooth muscle cells. However, the heterogeneity of responses of different types of vascular cells does exist and this needs to be taken into account when considering the effects on whole vasculature *(7–11)*.

4.2.1. Small EFs Dramatically Reorient Endothelial Cells

Vascular cells cultured without exposure to the EF had a typical cobble-stone morphology with the long axis of the cell body oriented randomly *(9–11)* (Fig. 4A). By contrast, endothelial cells cultured in DC EFs underwent a striking reorientation, with their long axis coming to lie perpendicular to the vector of the applied EF (Fig. 4B). This remarkable elongation and alignment in an applied EF resembles the response of endothelial cells to fluid shear stress.

Quantitative analysis of the orientation responses of endothelial cells showed both time and voltage dependency. Significant orientation was observed as early as 4 h after the onset of the EF. The threshold field strength inducing perpendicular orientation of the endothelial cells was between 50 and 75 mV/mm. This is low, representing only 0.5–0.75 mV across a cell with a diameter of 10 μm *(7)*.

4.2.2. Small EFs Direct Migration of Endothelial Cells Toward the Anode

Endothelial cells migrated directionally either toward the anode or cathode when cultured in EFs (for details *see* Subheading 4.3.). The directional migration was slow, but steady during the EF exposure. Cells migrated directionally, while elongating and reorientating perpendicularly. Lamellipodial extension toward the anode was marked. Directional migration was obvious at a physiological EF strength of 100 mV/mm (Fig. 4). The threshold of field strength inducing directional migration therefore was less than 100 mV/mm. Cell migration was quantified as previously *(40)* and significant directional migration was evident ($p < 0.0001$). Migration speed, however, remained constant before and after EF exposure, at 1–2 μm/h, which is significantly slower than most other cell types migrating in an EF. Substratum coating and different culture media can significantly decrease or increase the migration rate of vascular cells in response to EFs, but the migration direction remains the same (our unpublished observation).

4.2.3. Small EFs Elongate Endothelial Cells

Most vascular cells elongated dramatically in an EF (Fig. 5B). By contrast, cells cultured with no EF retained a more cobble-stone-like appearance *(9)* (Fig. 5A). Striking cell elongation was induced by a voltage drop of about 0.7–4 mV across a cell of approx 15 μm in diameter. We quantified the

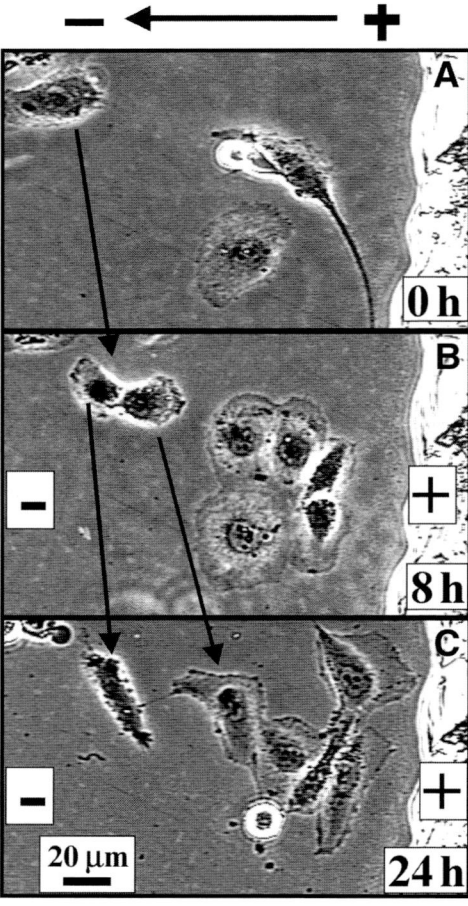

Fig. 4. Directional endothelial cell migration in a small physiological EF. Endothelial cells in culture exposed to an EF of 100 mV/mm migrated directionally, toward the anode. Cells migrated slowly but steadily toward the anode more than 24 h. Movement is evident using the static scratch on the culture dish as a reference (right margin). Note that lamellipodia extended preferentially toward the anode.

elongation of the cells using a long:short axis ratio. A perfectly round cell has a long:short axis ratio of 1. As cells elongate the ratio increases. Control cells (no EF) showed no increase in long:short axis more than 24 h in culture. Elongation responses were both time- and voltage-dependent. The long:short axis ratio of EF exposed cells indicated gradual cell elongation throughout the 24 h experimental period. The voltage dependency of the elongation response was more obvious at later times, with a greater long:short axis ratio for cells cultured at higher EFs. The threshold for EF-induced endothelial cell elongation was between 50 and 75 mV/mm, again 0.5 and 0.75 mV across a cell 10 µm in diameter. The elongation response of endothelial cells was more marked than that seen previously at the same EF strengths, in corneal and lens epithelial cells (9,40).

4.3. Heterogeneity in Electric Stimulation Induced Responses of Vascular Cells

New blood vessel formation is based on the capacity of microvascular endothelial cells to migrate, proliferate, elongate, and organize in three-dimensional tubules (7–14). Small applied EFs may upregulate some growth factor receptors and increase growth factor release (2,44,45). These changes are important for migration of many cell types, including human umbilical vein

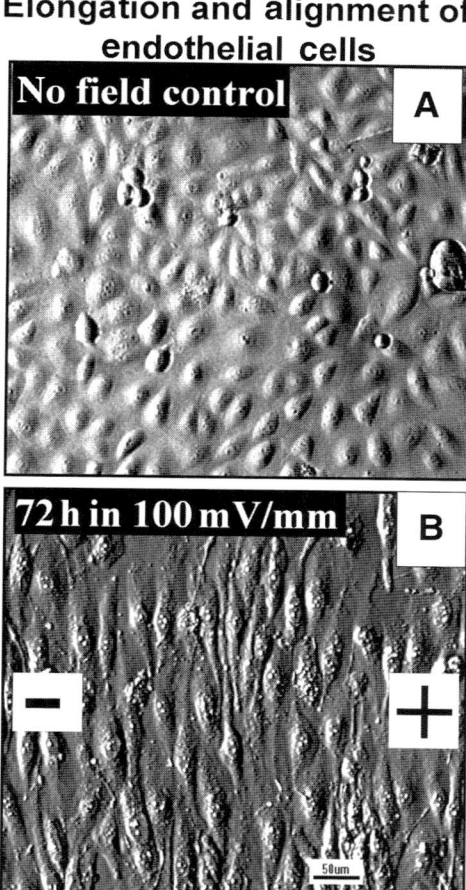

Fig. 5. Cell alignment induced by constant DC EFs. Perpendicular orientation and elongation of endothelial cells in small applied EFs. Control HUVEC cells cultured in same chamber without EFs showed typical cobble-stone morphology and random orientation (**A**). Cells exposed to small applied EFs showed dramatic elongation and perpendicular orientation in EFs (**B**).

endothelial cell (HUVEC) and bovine aortic endothelial cells (BAECs) *(7–10)*. Because angiogenesis occurs mainly in the microvasculature usually at postcapillary venules and not in large blood vessels *(46)*, we compared the effects of a DC EF on a human dermal microvasculature endothelial cell line (HMEC-1) *(47)*, vascular fibroblasts (bovine pulmonary artery fibroblasts [BPAFs]), and vascular smooth muscle cells (murine aortic smooth muscle cells) HUVECs, as described in Subsection 4.2. We show that endothelial cells derived from angiogenic, microvasculature as opposed to macrovascular tissues moved fastest and in the opposite direction in a small DC EF. Different cell types from a common tissue source responded differently to an applied DC EF. This intriguing directional selectivity indicates that a DC electric signal as a directional cue may be able to play a role in the spatial organization of vascular structure.

4.3.1. Micro- and Macrovascular Cells Migrate in Different Direction in EFs

When cultured in EFs, the vascular endothelial cells (HMEC-1 and HUVEC), smooth muscle cells and fibroblast cells showed evident directional migration. Strikingly, HMEC-1 cells migrated toward the cathode, whereas BPAF, mouse aortic smooth muscle cells (MASMCs),

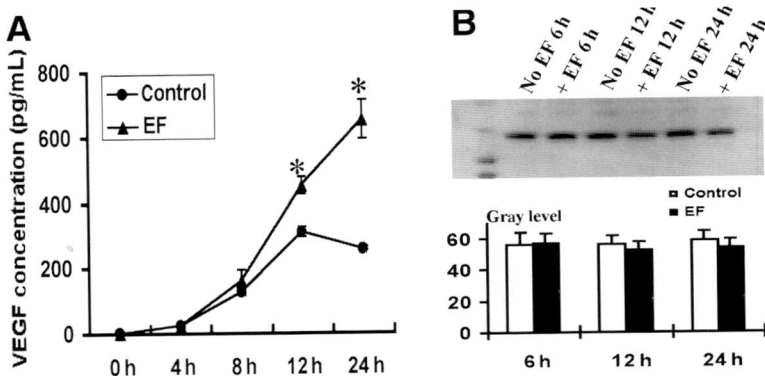

Fig. 6. Exposure to EF increased VEGF secretion (**A**) and did not increase VEGF receptor expression (VEGFR-2) (**B**). HUVEC cells were cultured in DMEM with 10% fetal bovine serum and exposed to EF of 150 mV/mm (for VEGF ELISA) and 200 mV/mm (for VEGF receptor analysis). VEGF in the media were quantified with ELISA and VEGF receptor were detected with western blot. $^*p < 0.01$.

and HUVECs migrated anodally. HMEC-1 and BPAF cells were the most actively migrating cells in the experimental conditions with directional migration obvious within 3 h. Cells extended lamellipodia in the direction of migration within 1 h after the onset of EF. The size and shape of HMEC-1 cells changed more frequently than BPAF cells in EFs *(11)*. The difference in the migration direction in EFs among the four types of vascular cells were confirmed and quantified with detailed analysis of the time-lapse images at different EF strengths.

4.3.2. Micro- and Macrovascular Cells Align and Elongate in EFs

When cultured in EFs, vascular endothelial cells (HMEC-1 and HUVEC), smooth muscle cells and fibroblast cells aligned with their long axis perpendicular to the EF vector. BPAF, MASMC, and HUVEC cells had similar and robust voltage dependency, whereas HMEC-1 cells showed a less dramatic increase in Oi with time in an EF of 200 mV/mm. However, for all the four types of vascular cells, the orientation response had a threshold between 50 and approx 150 mV/mm. As the four types of vascular cells migrated and aligned in an EF, they elongated in response to the EF. By contrast, cells cultured without an applied EF retained a typical nonpolarized morphology.

4.4. Molecular Mechanisms of Cellular Responses

Accumulating data indicates that electrical stimulation induces angiogenic responses through VEGF signaling pathway. More importantly, the focal autocrine release of VEGF by endothelial cells perhaps is critically involved in the initiation of angiogenesis, as a local increase of VEGF in immediate vicinity of endothelial cells. The molecular mechanisms start to crack and we started to understand how those cellular responses are induced molecularly *(2,7,9,10,11)*.

4.4.1. DC EFs Stimulate VEGF Production in Endothelial Cells

VEGF activation is one of the pivotal elements in angiogenic responses. Enhanced angiogenesis by electric stimulation in vivo is mediated through VEGF receptor activation *(2–5)*. Pulsed DC electrical stimulation (10 Hz, 300 μs) induces VEGF expression by muscle cells through increasing VEGF mRNA at a transcriptional level, which started to rise 12 h after EF stimulation and peaked at 24–48 h. This response may play a role in sensing tissue hypoxia following excessive muscle activation *(6)*.

To test whether DC EF-induced endothelial cell orientation might involve VEGF signaling, we quantified levels of VEGF. EF exposure (200 mV/mm; the same as that measured at skin

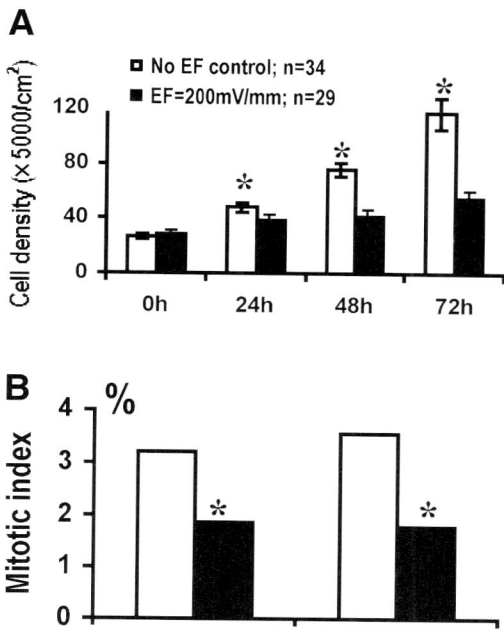

Fig. 7. Exposure to EF inhibits endothelial cell cycle progress. Cell density (**A**) and mitotic index ([**B**], number of mitotic spindles in 100 cells) of HUVEC cells are both significantly lower for cultures exposed to a DC EF of 200 mV/mm. $^*p < 0.01$ compared with control without EF exposure, student's t-test, or χ^2-test. Data are means ± SEM; n is number of microscopic fields assessed from at least two independent experiments.

wounds) significantly enhanced levels of VEGF released into the culture medium. Marked elevation of VEGF (VEGF 165) in the culture medium was observed as early as 5 min after onset of the EF, this was reduced at 1 and 2 h, rose again at 4 h, and reached a high level by 24 h (Fig. 6A). A constant DC EF induced significant VEGF production by endothelial cells within 5 min after the onset of an EF and showed a first peak after 30 min (data not shown in Fig. 6A). This probably is too early to result from VEGF transcription *de novo* and indicates that the EF must increase VEGF secretion. By contrast, the later elevation of VEGF at 24 h in a DC EF could be owing to new VEGF synthesis. This difference in VEGF production might be as a result of the differences in stimulation modalities (pulsed DC vs constant DCs) and cell types (muscle cells versus endothelial cells). The early rise of VEGF within the immediate vicinity of endothelial cells is necessary and causes the initiation of the behavioral responses of endothelial cells. VEGF receptor (VEGFR II) expression by the endothelial cells remains unchanged after EF stimulation during 24 h of constant DC electrical stimulation (Fig. 6B). These results suggest that electrical stimulation induces VEGF expression directly by endothelial cells as well as other type of cells.

4.4.2. VEGF Receptor, Phosphatidylinositol 3-Kinase/Akt, and Rho Kinase Signaling in EF-Induced Responses

Our experiments suggest that the proximal element, which transduces the EF and induced the preangiogenic responses is the VEGF receptor with downstream signaling elements involving Phosphatidylinositol 3-Kinase/Akt, Rho/Rho associated kinase (ROCK), and the F-actin cytoskeleton *(9)* (Fig. 7). Our finding that Rho-p160ROCK is involved in EF-induced cell alignment is in keeping with a similar role in cell alignment in shear stress *(48)*. Apart from its involvement in Ca^{2+} regulation, p160ROCK or Rho-kinase can enhance the phosphorylation of myosin light chain *(49)*. A small DC EF may activate the Rho-p160ROCK pathway, which in

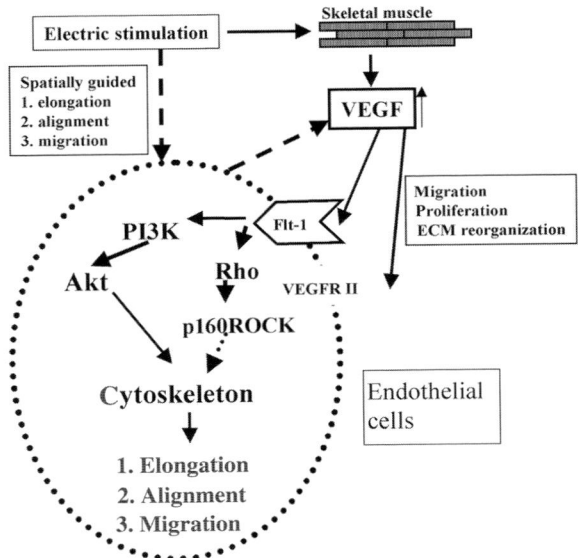

Fig. 8. Signaling mechanisms of EF-induced angiogenic responses of endothelial cells.

turn modulates myosin light chain phosphorylation, or intracellular Ca^{2+} to regulate the cell alignment.

Thus, phosphatidylinositol 3-kinase/Akt and Rho/Rock are two major pathways which act separately and downstream of EF-VEGF receptor activation to induce cell elongation and orientation. The integrin αvβ3 interacts with VEGF receptor signaling during mechanotransduction in endothelial cells responding to shear stress (50), but this is not involved in DC EF-induced orientation. This is interesting, given the common downstream signaling elements shared by cells orienting to shear stress and to a small EF (51). Thus EFs use one of the most important signaling strategies to modulate endothelial cell behaviors. There are parallels here with directional migration of epithelial cells, which is stimulated by a chemical gradient of epidermal growth factor (EGF). EF-induced epithelial cell migration is initiated by epidermal growth faltor receptor signaling (9,52,53,54).

5. ELECTRIC CONTROL OF CELL CYCLE OF ENDOTHELIAL CELLS

Proliferation of vascular endothelial cells (VECs) is a key event in angiogenesis (1,55,56). The target of many angiogenesis inhibitors is the VECs, with inhibitors that selectively affect a number of endothelial cell functions acquired during angiogenesis (57–59). Steady EFs of 200 mV/mm inhibited HUVEC cell proliferation (Fig. 8). However, physiological electrical fields of half that strength (50–100 mV/mm) did not interfere with cell cycle progression. This inhibition is through cell cycle block at the G_1–S phase transition, which involved down regulation of cyclin EF and upregulation of p27kip1. Modulation of proteins, which regulate cell cycle progression by a physiological EF, is a novel mechanisms of regulating endothelial proliferation with electric stimulation. The possible involvement of apoptosis has been excluded by the experiments to show that Z-VAD-FMK (an apoptosis inhibitor) did not prevent the EF-induced reduction of cell proliferation, nor did field-exposed cells show significant caspase-3 staining (indicating apoptosis).

The inhibition of applied EFs on proliferation of endothelial cells at higher magnitude of DC EFs should be particularly noted, when the stimulation is intended to enhance angiogenesis. However, whether this reflects what will happen in vivo has yet to be demonstrated. It should be noted that whereas lower field strength had no significant effects on HUVEC cell proliferation, they do have significant effects on directed migration, orientation, and elongation (*see* Section 4). These indicate that electric signals could exert differential effects on migration and proliferation of endothelial cells. This shows the possible complex roles of electric stimulation in controlling of angiogenesis.

6. CONCLUSIONS

Electrical stimulation is an effective way to induce angiogenesis in vivo. Clinical trials show a promising result of using this practical treatment to enhance angiogeneses in human subjects with external subsensory or subcontractile stimulation *(38,39)*. These are generally believed to be mediated by enhanced VEGF production of muscle cells. Recent research has shown that electrical stimulation induces directional migration, orientation, and elongation responses of vascular endothelial cells, fibroblast cells, and smooth muscle cells. Those responses appear to be mediated by VEGF receptor signaling without the presence of other type of cells. This may be significant, as this indicates electric stimulation can be applied to other noncontractile tissues to enhance or regulate angiogenesis. Distinct heterogeneity in the responses existed among different types of the vascular cells. This may have important physiological and clinical implications in areas in which electric stimulation is used to promote angiogenesis or vasculature remodeling. Combination of the advances in both basic research and clinical trials is expected to optimize the strategy and lead to effective electrical stimulation in angiogenesis control and induction.

ACKNOWLEDGMENTS

M.Z. held a Wellcome Trust University Award (058551). Research in author's laboratory was supported by the Wellcome Trust (068012), the British Heart Foundation (FS/2000056, PG/99191), and Royal Society.

REFERENCES

1. Folkman J. Angiogenesis in cancer, vascular, rheumatoid and other disease. Nat Med 1995;1:27–31.
2. Kanno S, Oda N, Abe M, et al. Establishment of a simple and practical procedure applicable to therapeutic angiogenesis. Circulation 1999;99:2682–2687.
3. Patterson C, Runge MS. Therapeutic Angiogenesis The New Electrophysiology? Circulation 1999;99:2614–2616.
4. Cuevas P, Asin-Cardiel E. Electromagnetic therapeutic angiogenesis: the next step. Neurol Res 2000;22:349–350.
5. Linderman JR, Kloehn MR, Greene AS. Development of an implantable muscle stimulator: measurement of stimulated angiogenesis and poststimulus vessel regression. Microcirculation 2000;7:119–128.
6. Hang J, Kong L, Gu JW, Adair TH. VEGF gene expression is upregulated in electrically stimulated rat skeletal muscle. Am J Physiol 269:H1827–H1831.
7. Zhao M, Morgan P, Barker JA, Yin Y, Forrester JV, McCaig CD. Alignment of endothelial cells in a physiological electric field: involvement of MAP kinase and PI3 kinase signalling. J Physiol 2001;528:85P.
8. Zhang L, Zhao M, Forrester JV, McCaig CD. Directed migration of cultured bovine retinal capillary endothelial cells in a small physiological electric fields. J Physiol 2000;528:12P–13P.
9. Zhao M, Bai H, Wang E, Forrester JV, McCaig CD. Electrical stimulation directly induces pre-angiogenic responses in vascular endothelial cells by signaling through VEGF receptors. J Cell Sci 2004;117(Pt 3):397–405.
10. Li X, Kolega J. Effects of direct current electric fields on cell migration and actin filament distribution in bovine vascular endothelial cells. J Vascular Res 2002;39:391–404.
11. Bai H, Wang E, Forrester JV, McCaig CD, Zhao M. DC electric fields induce distinct preangiogenic responses in microvascular and macrovascular cells. Arterioscler Thromb Vasc Biol 2004;24:1–6.

12. Sawyer PN, Pate JW. Bio-electric phenomena as an etiologic factor in intravascular thrombosis. Am J Physiol 1953;175:103–107.
13. Sawyer PN, Himmelfarb E, Lustrin I, Ziskind H. Measurement of streaming potentials of mammalian blood vessels, aorta and vena cava, in vivo. Biophys J 1966;6:641–651.
14. Harshaw DH, Jr, Ziskind H, Mazlen R, Sawyer PN. Electrical potential difference across blood vessels. Circ Res 1962;11:360–363.
15. Robinson KR, Messerli MA. Electric embryos: the embryonic epithelium as a generator of developmental information. In: McCaig CD (ed.) Nerve Growth and Nerve Guidance. London: Portland Press, 1996, pp 131–150.
16. McCaig CD, Zhao M. Physiological electrical fields modify cell behaviour. Bioessays 1997;19:819–826.
17. Jaffe LF, Stern CD. Strong electrical currents leave the primitive streak of chick embryos. Science 1979;206(4418):569–571.
18. Borgens RB, Shi R. Uncoupling histogenesis from morphogenesis in the vertebrate embryo system by collapse of the transneural tube potential. Dev Dyn 1995;203:456–467.
19. Nuccitelli R. Endogenous ionic currents and DC electric fields in multicellular animal tissues. Bioelectromagnetics 1992;(Suppl 1):147–157.
20. Hotary KB, Robinson KR. Evidence of a role for endogenous electrical fields in chick embryo development. Development 1992;114:985–996.
21. Hotary KB, Robinson KR. Endogenous electrical currents and voltage gradients in Xenopus embryos and the consequences of their disruption. Dev Biol 1994;166:789–800.
22. Jaffe LF, Vanable JW, Jr. Electric fields and wound healing. Clin Dermatol 1984;2:34–44.
23. Nuccitelli R. A role for endogenous electric fields in wound healing. Curr Top Dev Biol 2003;58:1–26.
24. Barker AT, Jaffe LF, Vanable JW, Jr. The glabrous epidermis of cavies contains a powerful battery. Am J Physiol 1982;242:R358–R366.
25. Foulds IS, Barker AT. Human skin battery potentials and their possible role in wound healing. Br J Dermatol 1983;109:515–22.
26. Nuccitelli R. Endogenous electric fields in embryos during development, regeneration and wound healing. Radiat Prot Dosimetry 2003;106(4):375–383.
27. Kleber AG, Janse MJ, van Capelle FJ, Durrer D. Mechanism and time course of S-T and T-Q segment changes during acute regional myocardial ischemia in the pig heart determined by extracellular and intracellular recordings. Circ Res 1978;42:603–613.
28. Coronel R, Wilms-Schopman FJG, Opthof T, van Capelle FJL, Janse MJ. Injury current and gradient of diastolic stimulation threshold, TQ potential and extracellular potassium concentration during acute regional ischemia in the isolated perfused pig heart. Circ Res 1991;68:1241–1249.
29. Brent TP, Forrester JA. Changes in surface charge of HeLa cells during the cell cycle. Nature 1976;215:92–93.
30. Elul R, Brons J, Kravitz K. Surface charge modifications associated with proliferation and differentiation in neuroblastoma cultures. Nature 1975;258(5536):616–617.
31. Cuzick J, Holland R, Barth V, et al. Electropotential measurements as a new diagnostic modality for breast cancer. Lancet 1998;352(9125):359–363.
32. Nannmark U, Buch F, Albrektsson T. Vascular reactions during electrical stimulation. Vital microscopy of the hamster cheek pouch and the rabbit tibia. Acta Orthop Scand 1985;56:52–56.
33. Dvorak HF, Brown LF, Detmar M, Dvorak AM. Vascular permeability factor/vascular endothelial growth factor, microvascular hyperpermeability, and angiogenesis. Am J Pathol 1995;146(5):1029–1039.
34. Brown MD, Cotter MA, Hudlicka O, Vrbova G. The effects of different patterns of muscle activity on capillary density, mechanical properties and structure of slow and fast rabbit muscles. Pflugers Arch 1976;361(3):241–250.
35. Hudlicka O, Tyler KR. The effect of long-term high-frequency stimulation on capillary density and fibre types in rabbit fast muscles. J Physiol 1984;353:435–445.
36. Kwong WH, Vrbova G. Effects of low-frequency electrical stimulation on fast and slow muscles of the rat. Pflugers Arch 1981;391:200–207.
37. Brutsaert TD, Gavin TP, Fu Z, et al. Regional differences in expression of VEGF mRNA in rat gastrocnemius following 1 hr exercise or electrical stimulation. BMC Physiol 2002;2(1):8.
38. Clover AJ, McCarthy MJ, Hodgkinson K, Bell PR, Brindle NP. Noninvasive augmentation of microvessel number in patients with peripheral vascular disease. J Vasc Surg 2003;38:1309–1312.

39. Peters EJ, Armstrong DG, Wunderlich RP, Bosma J, Stacpoole-Shea S, Lavery LA. The benefit of electrical stimulation to enhance perfusion in persons with diabetes mellitus. J Foot Ankle Surg 1998;37(5):396–400; discussion 447–448.
40. Zhao M, Agius-Fernandez A, Forrester JV, McCaig CD. Orientation and directed migration of cultured corneal epithelial cells in small electric fields are serum dependent. J Cell Sci 1996;109: 1405–1414.
41. Risau W. Mechanisms of angiogenesis. Nature 1997;386(6626):671–674.
42. Erickson CA, Nuccitelli R. Embryonic fibroblast motility and orientation can be influenced by physiological electric fields. J Cell Biol 1984;98:296–307.
43. Gruler H, Nuccitelli R. Neural crest cell galvanotaxis: new data and a novel approach to the analysis of both galvanotaxis and chemotaxis. Cell Motil Cytoskeleton 1991;19(2):121–133.
44. Zhao M, Dick A, Forrester JV, McCaig CD. Electric field directed cell motility involves upregulated expression and asymmetric redistribution of the EGF receptors and is enhanced by fibronectin and lamin. Mol Bio Cell 1999;10:1259–1276.
45. Fitzsimmons RJ, Strong DD, Mohan S, Baylink D. Low-amplitude, low-frequency electric field-stimulated bone proliferation may in part be mediated by increased IGF-II release. J Cell Physiol 1992;150:84–89.
46. Klagsbrun M, Folkman J. Angiogenesis. In: Sporn MB, Roberts AB (ed.) Handbook of experimental Pharmacology. Berlin:Springer, 1990, pp 549–586.
47. Ades EW, Candal FJ, Swerlick RA, et al. HMEC-1: establishment of an immortalized human microvascular endothelial cell line. J Invest Dermatol 1992;99:683–690.
48. Li S, Chen BP, Azuma N, et al. Distinct roles for the small GTPases CDC42 and Rho in endothelial responses to shear stress. J Clin Invest 1999;103:1141–1150.
49. Kimura K, Ito M, Amano M, et al. Regulation of myosin phosphatase by Rho and Rho-associated kinase (Rho-kinase). Science 1996;273:245–248.
50. Wang DI, Gotlieb AI. Fibroblast growth factor 2 enhances early stages of in vitro endothelial repair by microfilament bundle reorganization and cell elongation. Exp Mol Pathol 1999;66:179–190.
51. Chen KD, Li YS, Kim M, et al. Mechanotransduction in response to shear stress. Roles of receptor tyrosine kinases, integrins, and Shc. J Biol Chem 1999;274:18,393–18,400.
52. Zhao M, Dick A, Forrester JV, McCaig CD. Electric field directed cell motility involves upregulated expression and asymmetric redistribution of the EGF receptors and is enhanced by fibronectin and lamin. Mol Biol Cell 1999;10:1259–1276.
53. Fang KS, Ionides E, Oster G, Nuccitelli R, Isseroff RR. Epidermal growth factor receptor relocalization and kinase activity are necessary for directional migration of keratinocytes in DC electric fields. J Cell Sci 1999;112:1967–1978.
54. Zhao M, Pu J, Forrester JV, McCaig CD. Membrane lipids, EGF receptors, and intracellular signals colocalize and are polarized in epithelial cells moving directionally in a physiological electric field. FASEB J 2002;16:857–859.
55. Wang E, Yin Y, Zhao M, Forrester JV, McCaig CD. Physiological electric fields control the G1/S phase cell cycle checkpoint to inhibit endothelial cell proliferation. FASEB J 2003;17(3):458–460.
56. Buschmann I, Schaper W. Arteriogenesis versus angiogenesis: two mechanisms of vessel growth. News Physiol Sci 1999;14:121–125
57. Matsubara T, Saura R, Hirohata K, Ziff M. Inhibition of human endothelial cell proliferation in vitro and neovascularization in vivo by D-penicillamine. J Clin Invest 1989;83:158–167
58. Buckley CD. Science, medicine, and the future. Treatment of rheumatoid arthritis. BMJ 1997; 315(7102):236–238.
59. Szekanecz Z, Szegedi G, Koch AE. Angiogenesis in rheumatoid arthritis: pathogenic and clinical significance. J Invest Med 1998;46:27–41.

35
Developmental and Potential Therapeutic Aspects of Mammalian Neural Stem Cells

L. Bai, MD, PhD, **S. L. Gerson,** MD, **and R. H. Miller,** PhD

SUMMARY

Advances in our understanding of the biology of stem cells, including neural, hematopoietic and mesenchymal stem cells, have opened new avenues for cell-based therapeutic approaches to replace damaged or lost neurons. The mature central nervous system (CNS) has traditionally been considered an unfavourable environment for the regeneration of damaged axons or the generation of new neurons. The recent realization that neural stem cells exist not only in the developing mammalian nervous system but also in the adult CNS of most mammals—including humans—raises the possibility of using either endogenous or transplanted neural stem cells for therapy. Although the location of neural stem cells in the adult CNS is relatively restricted, the restorative potential of stem cells is enhanced by their ability to migrate throughout the CNS. Many aspects of the biology of neural stem cells, including migrational control, proliferative, and differentiative cues are still poorly understood. In order to effectively utilize neural stem cells, either after transplantation or by promoting the mobilization of endogenous cells, will require the ability to manipulate the molecular cues that mediate commitment to particular cell lineages and their differentiation to functional cell types within the settings of neural injury or disease. If neural stem cells can be guided to replace lost neuronal or glial cell populations in the damaged adult CNS, they may offer a highly effective therapy for a wide range of degenerative diseases and insults to the adult human nervous system.

Key Words: Neural stem cells; development; repair; central nervous system; cell therapy; neurogenesis.

1. INTRODUCTION

The vertebrate nervous system is unique in terms of cellular diversity as well as the complexity and specificity of the cellular interactions that mediate its function. This is particularly true for the brain and spinal cord that comprise the central nervous system (CNS). Given the level of cellular complexity of the CNS and the apparent failure of the tissue to functionally recover after injury or insult it was generally accepted that there was little or no cell turnover in the adult brain and spinal cord. Studies over the last two decades have however, clearly demonstrated that the adult CNS retains stem cells as well as more restricted precursors that have the potential for extensive cellular replacement and current studies are exploring ways of harnessing this potential for functional restoration. Here we discuss sources for neural cell replacement and their potential for cell-based therapies.

The adult mammalian CNS is composed primarily of three major differentiated cells types—neurons, astrocytes, and oligodendrocytes. Neurons, a highly diverse cell population, are characterized by unique electrical properties and morphologies. They are the functional unit of the CNS, responsible for forming specific connections and acting as the conduit for information flow in the nervous system. Astrocytes and oligodendrocytes are the primary glial cells of the CNS. Astrocytes, a diverse population of cells, all of which share the characteristic of expression of a particular intermediate filament protein known as glial fibrillary acidic protein (GFAP) sometime during their development, provide both trophic and structural support to neurons. By contrast, the best-characterized role for oligodendrocytes is the generation of myelin the fatty insulation that surrounds axons and facilitates rapid electrical conduction. In addition to these mature cell types, the adult CNS contains microglial cells, derived from peripheral tissue and a number of different precursor cells including stem cells. The term neural stem cell (NSC) is used to describe cells that (a) can generate neural tissue or are derived from the nervous system, (b) have the ability to self-renew, and (c) can give rise to cells other than themselves through asymmetric cell division (1). NSCs are more frequent in the developing mammalian nervous system but are retained in the adult nervous system of all mammalian organisms, including humans. Developmentally, cell genesis occurs primarily in two waves—a prenatal wave that generates most of the neurons and an early postnatal wave that generates most of the astrocytes and oligodendrocytes. Once this second wave of proliferation is completed, the CNS has traditionally been considered mitotically quiescent that may reflect the inability of the CNS to replace nonfunctional tissue lost owing to disease or injury.

A major focus in brain repair has traditionally been on keeping neurons alive following injury and promoting their ability to extend processes and re-establish functional cell connections. The recent realization that new neurons continue to be generated, perhaps from NSCs into adulthood in many species, suggests that cell replacement therapies might also be highly effective. These findings have led to intense interest in the identification and characterization of NSCs and neural progenitor cells both for basic developmental biology studies and for therapeutic applications to the damaged brain. Transplantation of NSCs or their derivatives into a host, recruitment of new neurons from endogenous neural precursors, induction of the proliferation and differentiation of endogenous stem cells *in situ* by pharmacological manipulations and hematopoietic stem cell (HSC) gene therapy are potential treatments for many neurodegenerative diseases and brain injuries. Possible targets for such therapy include Parkinson's disease, Alzheimer's disease, and genetic disorders that damage the brain, such as Batten's, Gaucher's, and Tay-Sach's diseases, brain ischemia, spinal cord injury, and even for the therapy of human cancer. The glial cells derived from NSCs, moreover, could lead to axonal remyelination of neurons in diseases like multiple sclerosis. Continued progress in stem cell research will therefore provide a new future for brain repair.

2. STEM CELLS IN THE CNS

In general, pluripotent stem cells have two important characteristics that distinguish them from other cell types. First, they are unspecialized cells that renew themselves for long periods through cell division. Second, under certain physiological or experimental conditions, they can be induced to generate cells with special functions such as the beating cells of the heart muscle or the insulin-producing cells of the pancreas. These features make stem cells important elements in embryonic development and in adult tissue for maintaining cell number following injury, disease, or natural cell turnover. During development, in the very early embryo or blastocyst, a small group of about 30 cells called the inner cell mass gives rise to the hundreds of highly specialized cells needed to make up an adult organism. In the developing fetus, stem cells in developing tissues give rise to the multiple specialized cell types that make up the heart, lung, skin, and other tissues. In some adult tissues, such as bone marrow (BM), muscle, and gut, discrete populations of adult stem cells generate replacements for cells that are lost through normal wear and tear, injury, or disease. Traditionally, stem cells are more common and better characterized from tissues that exhibit a great deal of cell turnover. By contrast, tissues that have little turnover such as the CNS are likely to contain fewer stem cells.

3. CHARACTERIZATION OF NSCs IN VITRO

With the development of effective methods for neural cell culture the analyses of the biochemical and proliferative characteristics of neural precursors in embryonic and early postnatal tissue became possible. Several types of precursors were identified in primary cultures but none exhibited stem cell features. In the hematopoietic system the definitive demonstration of a stem cell is its ability to reconstitute the tissue after lethal irradiation; however, the critical nature of the cytoarchitecture of the brain and spinal cord makes such studies impossible in the CNS. Furthermore, the lack of specific markers for the unambiguous characterization of NSCs results in their identification being based on functional features such as self-renewal and the ability to generate a large number of progeny. Using growth factor stimulation and cell selection approaches Reynolds and Weiss succeeded in isolating cells both from the embryonic and adult CNS (2–4) that demonstrated extensive proliferative capacity and the capability to generate multiple cell types both of which are characteristics of stem cells.

A critical component of the isolation of putative NSCs is their positive selection from a heterogeneous primary culture by growing the cells under conditions in which committed progenitor cells and differentiated mature cells are rapidly eliminated by cell death, whereas undifferentiated NSCs survive and proliferate (5). In suspension cultures in the presence of growth factors such as epidermal growth factor (EGF) or basic fibroblast growth factor (bFGF) NSCs proliferate and form small clonal clusters of cells by 2–3 d. The clusters continue to expand and by day 7 the clusters, called neurospheres, typically measure 100–200 µm in diameter, and comprise approx 10,000 cells. These spheres can be successfully passaged by dissociation and reculture. The secondary neurospheres so generated begin to proliferate within the first 24 h and form new clonally derived clusters that can be repassaged 7 d later. This procedure results in a relatively consistent arithmetic increase in cell number. Differentiation of neurosphere-derived cells is induced by mitogens removal and plating the progeny on an adhesive substrate either as intact clusters or dissociated cells. After several days virtually all progeny will differentiate into neurons, astrocytes, and oligodendrocytes (6–8). Even in suspension cultures not all neurosphere cells retain stem cell features, in fact many cells undergo spontaneous differentiation. Thus, a neurosphere is a mixture of NSCs, progenitor cells, and even differentiated neurons and glia, depending on the size of the neurospheres and its time in culture (9). To maintain significant numbers of stem cells in a neurospheres requires repeated dissociation and subcloning, whereas continued subcloning can result in the formation of stable NSC cell lines (5,10).

The initial characterization of NSCs has facilitated studies of their biology. For example, the isolation of neurosphere-producing cells is not restricted to the rodent CNS. NSCs can be cultured from the CNS of different mammalian species at many stages of development (11,12). Further, the ability to generate neurospheres is not restricted to the developing CNS. Rather, self-renewing NSCs with a high-proliferative capacity can be isolated from distinct regions of the CNS throughout adulthood (13). The molecular characteristics of NSCs are not well established. As in other systems, the expression of CN133 and CD45 antigens has been used as markers for the preliminary characterization of potential mammalian CNS stem cell subsets. A distinct subset of human fetal CNS cells with the phenotype $CD133^+$, $5E12^+$, $CD34^-$, $CD45^-$, and $CD24^{-/lo}$ has demonstrated the ability to form spheres in culture, initiate secondary sphere formation and differentiate into neurons and astrocytes. A similar approach was taken in the purification of stem cells from the adult murine periventricular region in which NSCs were identified as $nestin^+$ PNA^- $CD24^-$ (14) and were purified to up to 80% (100-fold increase in NSC frequency) by flow cytometry.

The ability of neurospheres forming cells to give rise to different populations of neural cells has been tested both in vitro and after transplantation into the intact CNS. In culture it is clear that all three major classes of neural cells, neurons, astrocytes, and oligodendrocytes, can be obtained from neurospheres and that the bias of cell type generated can be altered through modification of the environment in which the cells are grown. In transplantation studies, in vitro expanded cells migrate widely throughout the developing CNS and can give rise to both neurons and glia. Whether such cells functionally integrate into the CNS in a manner analogous to normal cells has yet to be defined. Under pathological conditions, transplantation of NSCs has been demonstrated to result in improved functional responses, although whether such improvements reflect the direct integration of transplanted cells or more pleiotropic effects on host cells remains to be determined *(15)*. Several outstanding issues with respect to the biology of NSCs remain to be resolved. For example, it is unclear whether NSCs isolated from the adult CNS are equivalent to those isolated during development. Furthermore, there is some evidence that the properties of the NSCs isolated from different regions of the developing CNS may be different, however, the extent and the nature of such differences are currently unclear.

Whereas such in vitro studies provide compelling evidence for the presence of multipotent cells in the vertebrate CNS, they provide little insight into whether such cells contribute significantly to development or repair processes in the adult CNS. However, in vivo studies provide substantial evidence for the continued presence of proliferative cells in particular regions of the adult CNS.

4. CHARACTERIZATION OF NSCs IN VIVO

The structural complexity of the adult vertebrate CNS has significantly impaired the detection and characterization of NSCs compared with that in models systems such as Drosophilia *(16)*. Although the identification of "stem cells" has been difficult, the concept that the adult CNS retains populations of proliferating cells in distinct locations has been known for many years. For example, the subependymal region of the rodent forebrain was shown to contain proliferating cells in the 1960s *(17)*. This was initially believed to be species-specific, but has subsequently been demonstrated in multiple species. What cell types are generated from proliferating cells in the ventricular lining has recently been the subject of intense investigation. In the adult mammalian brain, the genesis of new neurons has been consistently documented in the subgranular layer of the dentate gyrus of the hippocampus and the subventricular zone (SVZ) of the lateral ventricles *(18)*. From the SVZ, newly generated neurons reach their final destination in the olfactory bulb, (OB) *(2)* after long-distance migration through a well-defined path called the rostral migratory stream (RMS) *(19,20)*. It seems likely that the proliferating cells of the SVZ are equivalent to the neurospheres forming cells in vitro. Consistent with this idea, the SVZ is the region of adult brain with the highest neurogenetic rate, and the isolated NSCs have a characterized ability to give rise to nonneural cells *(3,21)*. Based on the proliferative index within the SVZ it has been estimated that there is a complete turnover of the resident proliferating cell population every 1228 d with about 30,000 new neuronal precursors (neuroblasts) being produced every day and migrating to the OB *(22)*.

The precise nature of the proliferating cells in the SVZ is becoming clear. Two main cell types are found in the SVZ: migratory, proliferating neuroblasts and astrocytes. The latter form a network throughout the whole SVZ and are suggested to be organized into channels termed glial tubes oriented along the anteroposterior axis. In these glial tubes, NSC-generated neuroblasts undergo rostral forward, tangential migration in tightly associated chains of cells *(23,24)*. The neuroblasts ultimately reach the distal tip of the OB, and separate and leave the tubes, shifting their migration pattern from tangential to radial. Cells within the glial tubes express GFAP but may not be fully differentiated astrocytes because they also contain the cytoskeletal proteins vimentin and nestin *(25,26)*. The putative immature nature of these cells suggests a role for glial tubes in influencing the migration/guidance of neuroblasts in the SVZ reminiscent of the role of radial glia in the radial neuroblast migration in the developing cortex.

The highly restricted localization of neuron-producing putative NSCs in the adult CNS suggests that local environmental cues contribute to their retention during development. Thus, particular regions of the CNS may provide an environment that is particularly permissive for ongoing neurogenesis. The nature of such environmental cues is unknown. Frequently, proliferating clusters of neural cells in the adult hippocampus are found in close proximity to blood vessels, raising the possibility that blood-derived factors regulate neurogenesis. Consistent with that idea factors that promote endothelial cell proliferation also increase neurogenesis in the mammalian forebrain, possibly indicative of an important relationship between these two processes *(27)*. Neurogenesis is also influenced by other neural cells. Astrocytes from the adult hippocampus promote the proliferation of neural stem and instruct them to adopt a neuronal fate *(28)*. By contrast, astrocytes from non-neurogenic regions, such as the adult spinal cord, do not promote neurogenesis, suggesting that the local characteristics of astrocyte population may play a role in the creation of a neurogenic environment. The influence of astrocytes on neurogenesis may extend beyond the initial stages because they are known to influence the maturation and synapse formation of the newly generated neurons *(14,29)*. The neurogenic activities of astrocytes is not specific to hippocampus; in the neurogenic niche that generates the adult-born olfactory neurons, SVZ astrocytes have similar effects on the proliferation and neuronal differentiation of neural stem/progenitor cells as their hippocampal counterparts *(30)*. Not only do astrocytes promote neuronal development, but distinct GFAP+ cells of the astrocyte lineage may be NSC capable of generating neurons. For example, ablation of rapidly proliferating SVZ cells by administration of subdural, cytotoxic doses of antimitotic drugs results in residual astroglia, which appear to be competent to regenerate the entire SVZ. In this model a subset of SVZ astrocytes named type B-cells is thought to represent relatively quiescent stem cells that normally proliferate at a low rate and generate neuronal precursors (type A-cells), the C-cell (or D-cell) in the hippocampus *(31)*. The direct correlate of NSCs in these SVZ cells is not established but it is likely that they originate from type B-cells that retain radial glia-like features within specific regions of the adult brain *(32)*. The presence of NSCs is not restricted to neurogenic regions. NSCs can be isolated from non-neurogenic periventricular regions, in which the mature parenchyma is directly in contact with the ependymal monolayer, such as the fourth ventricle or the spinal cord. These findings raise the possibility that stem-like cells may exist throughout the whole adult CNS, but that only those that reside in particular regions manifest normal "stem cell-like" behavior, whereas those in other regions remain dormant *(33,34)*.

The cellular characterization of neurogenic microenvironments will likely provide insights into the molecular cues regulating NSC biology. Several factors have been implicated in NSC biology, including the regulation of proliferation of adult NSCs by flial growth factor-2, EGF *(35)*, vascular endothelial growth factor (VEGF) *(36)*, and brain-derived neurotrophic factor (BDNF) *(37)*, and fate choice by adult NSCs in response to bone morphogenetic protein (BMP) *(38)* or Wnt signaling to instruct NSCs to adopt a neuronal fate and survival of newly generated neurons. Within the SVZ intracerebral administration of growth factors such as EGF, FGF, or transforming growth factor-α (TGF-α) *(39)*, increases proliferation of cells and the fate of the progeny can change depending on the type of factors. In particular, EGF results in an increased production of cells in the SVZ, in a diversion of their migration pattern from tangential to radial, and in the eventual generation of cells of the glial lineage rather than neurons *(40,41)*. In all situations the definitive identification of NSCs depends on its characteristics in vitro, including the ability to proliferate, self-maintain, and produce large quantities of differentiated progeny and respond to injury *(42)*. One concern with this model of localized NSCs is the ability to culture neurospheres from regions of the adult brain not to normally undergo self-renewal *(43)*, raising the possibility that neurosphere-forming stem cells can arise in culture from precursor or differentiated cells.

5. NSC PLASTICITY AND TRANSDIFFERENTIATION

5.1. Neuroectodermal-Derived NSCs

Recent studies suggest that stem cells isolated from distinct tissues including brain retain significant plasticity. Plasticity or transdifferentiation is the ability to differentiate into multiple cell types. For example HSCs have been suggested to differentiate into three major types of neural cells (neurons, oligodendrocytes, and astrocytes), skeletal muscle cells, cardiac muscle cells, and liver cells. BM stromal cells may differentiate into cardiac muscle cells and skeletal muscle cells. NSCs may differentiate into blood cells and skeletal muscle cells *(21)*. This concept of NSCs plasticity and of their dependence on environmental cues is strengthened by transplantation and manipulation/recruitment studies in vivo. Both cultured and freshly isolated NSCs are influenced by environmental cues. For example, transplantation of hippocampal stem cells into the hippocampus results in the generation of the neuronal types normally found within this region; however, when transplanted to new CNS locations, the cells adopt some of the characteristics appropriate to the new environment *(44)*, including the formation of glia in nonneurogenic CNS regions. These observations support the notion that NSCs possess a rather broad developmental potential and functional repertoire whose expression is strongly influenced by extracellular cues.

5.2. Nonneurodermal-Derived NSCs

The distinction between neural and non-NSCs is not well defined. NSCs appear to be able to cells normally derived from germ layers other than the neuroectoderm. For example, adult murine CNS stem cells contribute to the formation of chimeric chick and mouse embryos *(45)*, whereas BM cells appear pluripotential and generate many mesoderm derivatives, including skeletal *(46–47)*. Similarly, muscle precursors have been reported to generate hematopoietic cells *(48–49)* although the original cells undergoing conversion may be hematopoietic in origin *(50)*. The cell products derived from stem cells are not always restricted to the same germinal layer. Mesoderm-derived mesenchymal stem cells (MSCs) have been reported to generate astrocytes and neurons both in vivo *(51)* and in vitro *(52)*. In vitro MSCs are heterogeneous and the cell fates they assume depend on environmental cue such that they generate smooth muscle cells, adipocytes, chondrocytes and even neurons depending on the extra cellular matrix, and signal molecules in their environment. The conversion of neural cells into blood derivatives and vice versa is currently controversial. NSCs have been proposed to contribute to hematopoiesis when transplanted into the tail vein of irradiated mice *(21)*. NSC derivatives integrate into many of the host hematopoietic tissues including spleen, thymus, BM, and gave rise to various types of blood progenitors. These progenitors did eventually differentiate into a wide range of blood cells including megacaryocytes, granulocytes, macrophages, and B- and T-lymphocytes *(53)* suggesting NSCs may rescue lethally irradiated animals. Potential complications of the interpretation of these studies include the contamination of initial cell populations and the possibility of cell fusion events *(54)* whereas other studies have failed to confirm the transdifferentiation of neural cells into mesenchymal derivatives *(55)*. It may be that transdifferentiation is an uncommon phenomenon that occurs only under very peculiar circumstances, however, if such mechanisms could be identified, stem cells from healthy tissue might be induced to repopulate and repair diseased tissue. For therapeutic purposes identifying the potential of stem cells is more critical than defining their normal functions *(56)*.

6. CELLULAR THERAPY FOR NEURAL REPAIR

Despite the anatomical protection provided to the CNS by the skull and vertebral column, it is still vulnerable to a variety of injuries as well as a number of neurodegenerative diseases. Most neurodegenerative diseases are progressive in their course and few effective therapies exists to delay disease progression or to promote significant recovery. Likewise, functional loss as a result of CNS injury is rarely fully reversible. With the emerging understanding of the biology of stem cells, there is now considerable hope that neural or nonneural-derived stem cells may be effective in promoting brain repair. Several different aspects of cellular therapy are currently being explored.

7. INDUCTION OF ENDOGENOUS NEUROGENESIS FOR REPAIR

Without some form of intervention the effect of endogenous precursors in brain repair is extremely limited. This likely reflects the limited number of NSCs or the unfavorable microenvironment of the injured adult brain that seldom allows effective neurogenesis and gliogenesis. Manipulation of endogenous NSCs through exogenous administration of bioactive molecules is limited, but initial studies suggest it might be highly effective. The injection of bFGF and EGF into the lateral ventricle of normal animals, expands SVZ progenitor population, leading prominently to neurons after bFGF stimulation and astrocytes after EGF stimulation in the OB *(41)*. Infusion of TGF-α in animals with a selective lesion of the dopaminergic nigrostriatal system induces massive proliferation of forebrain stem cells, followed by migration of both glial and neural progenitors toward injection side *(57)*. This treatment resulted in increased numbers of differentiated neurons in the striatum and functional improvement. These findings indicate that lesion itself is not sufficient to "activate" endogenous precursors and further supports the hypothesis that sensitivity of endogenous NSCs to microenvironmental signals occurs during a temporal window provided by pathological conditions and/or by administration of exogenous molecules.

One attractive set of candidate molecules for NSC activation are the neurotrophins. Neurotrophin-like nerve growth factor (NGF), brain-devived neurotrophic factor, and glial cell line-derived neurotrophic factors (GDNF) are essential regulators of nervous tissue development. For example, nerve growth factor is the growth factor for differentiation and survival of sympathetic, sensory and central cholinergic neurons. Neurotrophins also influence adult NSCs and have been demonstrated to influence multipotent cells from the human SVZ and other regions of the CNS *(48,58,59)*. Moreover, the multiple roles of neurotrophins as mediators in cell cycle regulation, cell survival, and differentiation during development make them potential candidates for the physiological regulation of NSC proliferation and differentiation in adult brain and possible targets for exogenous regulation of such processes in brain repair *(60)*. The recruitment of endogenous stem cells may also involve the incorporation of other stem cell types delivered by systemic rather than intracranial injection. For example, BM-derived stem cells have been reported to enter the CNS and contribute to the neuronal populations although the frequency and mechanism of their integration are unclear *(61,62)*. Potentially, either BM cells or umbilical cord blood cells could be delivered by intravenous injection.

8. NSCs' TRANSPLANTATION FOR REPAIR

One possibility to enhance repair of the injured adult CNS is to increase the number of NSCs through transplantation. For instance, Parkinson's disease is a common neurodegenerative disorder caused by a progressive degeneration and loss of dopamine-producing neurons, which leads to tremor, rigidity, and hypokinesia. The successful generation of an unlimited supply of dopamine neurons could make transplantation available for patients with Parkinson's disease in the future, however, the limited number of functionally active cells surviving transplantation and developing synapses suggests that postmitotic cells are not the best source for transplantation. Embryonic neural cells, although possibly more efficient, are unreliable for large-scale use owing to logistic and ethical problems related to material collection. One currently attractive notion is to use engineered multipotent stem cells to create a potentially unlimited source of neurons expressing a defined neurochemical phenotype *(63)*, a direction that will likely be applied for other degenerative diseases, such as stroke, dementia, and demyelinating diseases. The sources of NSCs remain uncertain. Viable NSCs may be recovered from postmortem human CNS tissue *(64)*, which require in vitro expansion and differentiation before transplantation into non-neurogenic regions of the CNS *(65)*. Alternatively, the induction of suitable regulatory signals at the time of transplantation either by the cells themselves or adjuvant therapy might facilitate the development of specific neural populations from endogenous cells *(66)*.

9. HSC THERAPY

Stem cell transplants are routinely used to treat patients with cancers and other disorders of the blood and immune systems. Recently, HSCs appear to be able to form other kinds of cells, such as muscle, blood vessels, and bone as well as neural cells. There are currently three sources of HSC: the BM, the mobilized peripheral blood, and the umbilical cord blood. BM is the classical source of HSC and has the advantage of more rapid hematopoietic recovery after transplantation. It has been known for some time that peripheral blood contains twice as many HSCs as BM, engraft more quickly, and respond well to cytokines such as granulocyte colony stimulating factor (GCSFs); and recently, the majority of autologous (in which the donor and recipient are the same) and allogenic (in which the donor and recipient are different) "BM" transplants have been white blood cells drawn from peripheral circulation. In the late 1980s and early 1990s, human umbilical cord blood (HUCB) and placenta were identified as a source of HSCs. HUCB has a number of advantages as a stem cell source. It is easily accessible in unlimited supply without jeopardizing the mother or infant. It is easily preserved and has robust viability *(67)*, lower incidence of graft vs host disease (GVHD) *(68,69)*, and less ethical controversies than other sources. The immaturity of cord blood cells has been postulated as the reason for this low rejection rate because they are still of primitive ontogeny and exposed to limited immunological challenge *(70)*.

Like other stem cells HUCBs appear to be able to generate neural cells under appropriate stimulation. Exposure to EGF and bFGF increased proliferation and expressions of the neuronal markers Musashi-1, β-tubulin III, and GFAP *(71)*. The nature of the responding cells has not been defined, however, removal of $CD34^+$ and $CD45^+$ cells did not inhibit neurogenesis *(72)* and the residual clone-forming cells expressed nestin and were not capable of producing hematopoietic cells. Development of differentiated neurons and glia required additional neural-derived stimulation. Collectively, these studies demonstrate that HUCB cells may be directed in culture to neural phenotypes but a subpopulation of NSCs have not yet been identified from the heterogeneous cord blood cell population. Identification of such cells might be more therapeutically beneficial then the heterogeneous population of HUCB cells.

Whether HUCB-derived cells will generate neural cells in the intact CNS is currently less clear. Mononuclear fractions of HUCB cells can improve motor behavior deficits observed in a rat spinal cord injury model and HUCB cells can migrate specifically to the injured spinal cord *(73)*. Further intravenous delivery of HUCB to the superoxide dismutase-1 (SOD-1) animal model of amyotrophic lateral sclerosis (ALS) characterized by a rapid degeneration of the motor neurons in the spinal cord, motor cortex, and brain stem nuclei resulted in increased life-span *(74)*. Results showed a delay in the disease progression and longer life-span when compared with control mice, with HUCB-derived cells in the parenchyma of the brain, spinal cord, spleen, kidneys, liver, lungs, and heart of host. Cells within the parenchyma of the brain-stained positive for neural markers like β-tubulin III and GFAP, and it was the first demonstration that HUCB cells can migrate to the diseased brain and spinal cord without a traumatic injury. These studies suggest a potential for HUCB systemic infusion in other neurodegenerative models, such as Parkinson's disease, Alzheimer's disease, or even MS *(75)*.

10. MSCs THERAPY

MSCs have a number of advantages for therapeutic interventions in the CNS. They are a relatively accessible stem cell source for which there is extensive clinical experience. Considerable evidence suggest that if MSCs are placed in a new environment, they may have the capability to differentiate into a wide range of mature cell types *(76)*, consistent with their stromal cell origin *(77)*. MSCs are a minor fraction (0.0001%) of the total nucleated cells in marrow but can be enriched in vitro through growth or antibody selections using reagents such as Stro-1, SB-10 (CD166) *(78)*, SH2 (CD105) *(79)*, SH3, and SH4 (CD73) *(80)*. Several studies suggest that MSCs have the capacity to interact with neural tissue and possibly generate neural derivatives. When implanted in the brain, MSCs survive and migrate broadly. Some cells may differentiate into astroglia, but specific treatment in vitro, such as using condition medium to enrich, instruct, or select for neural lineage cells, may generate neurons *(81,82)*. For example, growth of MSCs in B104-conditioned medium resulted in neurons *(83)*, although what molecular signals that regulate the switch is unknown.

In transplantation studies MSCs appear to have a broad range of potential therapeutic applications. They have been suggested to be effective not only for the repair of heart disease *(84)*, bone disease *(85)*, and arthritis but also for the repair for neurodegenerative diseases. For example, implantation of stem cells from BM into the sites of degeneration is reported to have a beneficial functional effect *(76)* in models of traumatic brain injury. This benefit could be owing to several factors, including the implanted cells acting as protective barrier to the damaged neurons, or secreting factors that increase the brain's plasticity or resistance to disease. The functional improvement does not seem to depend on the replacement of the lost neurons but may be owing to increased expression levels of neurotrophic factors suggesting a positive influence regardless of actual cellular transdifferentiation. Similarly, although MSCs appear to improve functional responses in some demyelinating diseases, they are not necessarily the result of transdifferentiation of the stem cell, but rather generation of a less inhibitory environment. Consistent with this notion *(86)*, in a rodent model of cerebral ischemia MSCs engrafted and resulted in functional improvement but did not have the expected morphology despite staining with markers for oligodendrocytes, astrocytes, and neurons. The characterization of the molecular interactions between MSCs and NSCs will provide critical insights into the bases of the proported efficacy of MSCs in neural repair. A potentially exciting approach to enhance the therapeutic capability of transplanted stem cells is to engineer desirable properties into the cells before transplantation. Although not yet widely used in the therapy of neural injuries or tumors, this approach has gained substantial credence in hematopoeitic stem cell therapy *(87–89)*.

11. CHALLENGES OF TRANSLATING ANIMAL STUDIES TO HUMAN THERAPIES

The majority of information about adult neurogenesis and the capacity of the adult brain to accept new cells into existing neuronal circuitry are primarily derived from rodent models. Interspecies differences may compromise the smooth transition of this information to clinical application. For example, the human SVZ appears to have significantly less neuronal differentiation than is seen in rodents *(90)*. Why SVZ cells fail to become neurons in human is unclear. It may be they are not competent to differentiate into neurons although this is unlikely because they are capable of generating neurons in vitro. Alternatively, commitment to the neuronal lineage is actively suppressed in the human CNS possibly to protect established circuitry. If so these signals would have to be counteracted to achieve neuronal differentiation of grafted or recruited NSCs. The final possibility is that failure of neuronal differentiation reflects the lack of necessary induction signals in mature human systems indeed. Downregulation of neuronal induction signals in the mature brain would be more efficient than maintaining expression of inhibitory signals. This scenario is also more favorable to therapeutic strategies; for example, turning on appropriate signals may be easier to achieve than widespread blocking of suppression signals. These inhibitory mechanisms are not mutually exclusive and multiple approaches including regulation of the balances between glial suppression and neuronal induction will likely be important *(66)*.

12. CONSIDERATIONS BEFORE USING STEM CELL THERAPIES FOR NEUROLOGICAL DISEASE AND INJURY

One consideration in the development of therapeutic strategies for CNS repair is the nature of the insult. Focal damage can result from mechanical injury or ischemic insults to particular CNS regions whereas more general traumatic brain injury or global ischemia arising from cardiac arrest or increased intracranial pressure results in widespread, damage. Neurodegenerative diseases can also be relatively localized, such as Parkinson's disease, or widespread, such as Alzheimer's disease. In the case of MS, the pathology can be both focal and wide-ranging over time. In addition, there are a number of developmental or perinatal injuries, such as cerebral palsy or perinatal hypoxia, which produce widespread impairment of CNS function and for which therapies have yet to be developed.

Although considerable progress has been made in developing new techniques for the culture of human stem cells, the successful clinical application of these cells is presently limited by our understanding of both the intrinsic and extrinsic regulators of stem cell proliferation as well as those factors controling cell lineage determination and differentiation *(91)*. Progress in the field has been further limited by the lack of suitable for the identification and selection of NSCs from heterogeneous populations of precursor cells. In transplantation studies several criteria have to be addressed. For example, it is unclear whether similar strategies will need to be adopted depending on whether NSCs are required to deliver missing neurotransmitters, grow or trophic factors, to establish local or long-distance connections, to remyelinate central axons, or to replace extensive neural tissue loss with multiple neural phenotypes. Moreover, successful therapy requires a full understanding of the biology of the host tissue, whether it has adapted to chronic disease or is suffering from acute damage. Likewise, the host reaction (integration or rejection) of transplanted cells and the level of inflammation will modulate outcome. Unfortunately, unlike other tissues, cellular reconstruction of the CNS will not simply follow the introduction of new and healthy cells *(92)*.

Each distinct neuropathology represents a unique challenge for the potential use of stem cells. It seems likely that therapies directed toward focal disease pathology or injury will have a higher probability of success because in addition to a smaller anatomical area, focal injuries provide a more defined repair strategy. For example, not all neurons are equivalent and the more defined the need for a particular neuronal phenotype, the more likely that appropriate lineage determination signals can be presented to NSCs to achieve the desired cell type. In some ways a more daunting problem is the functional integration of the new neurons into the damaged circuitry as well as preventing ongoing degeneration in neurodegenerative diseases, and concerns that disease pathology may generate an unfavorable or toxic environment. Finally, studies in rodents suggest the aging brain is less capable of supporting endogeneous neurogenesis *(93)* and because many individuals suffering from neurodegenerative disease are older patients this could compromise the success of any therapies. Additional studies on age-related changes in the human brain will likely illuminate directions to enhance the success of stem cell therapies for older patients.

13. CONCLUSIONS

Rapid progress in understanding the biology of NSCs has generated enthusiasm for the development of therapeutic strategies using these cells. There remain, however, substantial gaps in our understanding of the biology of the developing and adult CNS. In the adult brain neurogenesis is highly spatially restricted and the signals that distinguish neurogenic brain regions from non-neurogenic area are not fully defined. Recent evidence of widely distributed neural progenitor cells that can differentiate into neurons under appropriate culture conditions or within neurogenic regions of the brain suggests that it may be possible to recruit these local cells to repopulate injured regions. The CNS stem cells are now beginning to be recognized as a population of precursors that differentiate into neurons and glial cells. It is important to define stem cells from the CNS accurately in molecular levels. Embryonic stem cells can differentiate to specific neuron and glial types through defined intermediates that are similar to the cellular precursors that normally occur in brain development, and this area represents an important avenue for future studies. There is convincing evidence that the differentiated progeny of embryonic stem cells and CNS stem cells show the expected functions of neurons and glia. Advances in several distinct areas such as (a) cell cycle control, (b) the control of cell fate, and (c) early steps in neural differentiation are required before widespread application of NSC therapies to a variety of neurodegenerative disorders can be accomplished. Therefore, whereas the presence of stem cells in the mammalian brain opens the door to new therapeutic avenues aimed at replacing lost or damaged CNS cells and although there is reason to be optimistic, it will likely be some time before the use of NSCs moves from the research laboratory to become a safe and effective clinical therapy for neurological disease or injury.

REFERENCES

1. Gage FH. Mammalian neural stem cells. Science 2000;287:1433–1438.
2. Reynolds B, Stevens PA, Adamson JK, et al. Effects of clearfelling on stream and soil water aluminium chemistry in three UK forests. Environ Pollut 1992;77:157–165.
3. Reynolds BA, Tetzlaff W, Weiss S. A multipotent EGF-responsive striatal embryonic progenitor cell produces neurons and astrocytes. J Neurosci 1992;12:4565–4574.
4. Reynolds BA, Weiss S. Clonal and population analyses demonstrate that an EGF-responsive mammalian embryonic CNS precursor is a stem cell. Dev Biol 1996;175:1–13.
5. Bottai D, Fiocco R, Gelain F, et al. Neural stem cells in the adult nervous system. J Hematother Stem Cell Res 2003;12:655–670.
6. Doetsch F, Caille I, Lim DA, et al. Subventricular zone astrocytes are neural stem cells in the adult mammalian brain. Cell 1999;97:703–716.
7. Johansson CB, Momma S, Clarke DL, et al. Identification of a neural stem cell in the adult mammalian central nervous system. Cell 1999;96:25–34.
8. Gage FH. Structural plasticity: cause, result, or correlate of depression. Biol Psychiatry 2000; 48:713,714.
9. Suslov ON, Kukekov VG, Ignatova TN, et al. Neural stem cell heterogeneity demonstrated by molecular phenotyping of clonal neurospheres. Proc Natl Acad Sci USA 2002;99:14,506–14,511.
10. Li L, Yoo H, Becker FF, et al. Identification of a brain- and reproductive-organs-specific gene responsive to DNA damage and retinoic acid. Biochem Biophys Res Commun 1995;206:764–774.
11. Temple S. The development of neural stem cells. Nature 2001;414:112–117.
12. Gritti A, Vescovi AL, Galli R. Adult neural stem cells: plasticity and developmental potential. J Physiol Paris 2002;96:81–90.
13. Arsenijevic Y. Mammalian neural stem-cell renewal: nature versus nurture. Mol Neurobiol 2003;27:73–98.
14. Toda H, Takahashi J, Mizoguchi A, et al. Neurons generated from adult rat hippocampal stem cells form functional glutamatergic and GABAergic synapses in vitro. Exp Neurol 2000;165:66–76.
15. McDonald JW, Liu XZ, Qu Y, et al. Transplanted embryonic stem cells survive, differentiate and promote recovery in injured rat spinal cord. Nat Med 1999;5:1410–1412.
16. Doe CQ, Fuerstenberg S, Peng CY. Neural stem cells: from fly to vertebrates. J Neurobiol 1998;36:111–127.
17. Altman J, Bayer SA. Development of the diencephalon in the rat. III. Ontogeny of the specialized ventricular linings of the hypothalamic third ventricle. J Comp Neurol 1978;182:995–1015.
18. van Praag H, Schinder AF, Christie BR, et al. Functional neurogenesis in the adult hippocampus. Nature 2002;415:1030–1034.
19. Luskin MB. Restricted proliferation and migration of postnatally generated neurons derived from the forebrain subventricular zone. Neuron 1993;11:173–189.
20. Lois C, Alvarez-Buylla A. Long-distance neuronal migration in the adult mammalian brain. Science 1994;264:1145–1148.
21. Bjornson CR, Rietze RL, Reynolds BA, et al. Turning brain into blood: a hematopoietic fate adopted by adult neural stem cells in vivo. Science 1999;283:534–537.
22. Craig CG, D'sa R, Morshead CM, et al. Migrational analysis of the constitutively proliferating subependyma population in adult mouse forebrain. Neuroscience 1999;93:1197–1206.
23. Lois C, Garcia-Verdugo JM, Alvarez-Buylla A. Chain migration of neuronal precursors. Science 1996;271:978–981.
24. Jankovski A, Sotelo C. Subventricular zone-olfactory bulb migratory pathway in the adult mouse: cellular composition and specificity as determined by heterochronic and heterotopic transplantation. J Comp Neurol 1996;371:376–396.
25. Peretto P, Merighi A, Fasolo A, et al. Glial tubes in the rostral migratory stream of the adult rat. Brain Res Bull 1997;42:9–21.
26. Peretto P, Merighi A, Fasolo A, et al. The subependymal layer in rodents: a site of structural plasticity and cell migration in the adult mammalian brain. Brain Res Bull 1999;49:221–243.
27. Palmer TD, Willhoite AR, Gage FH. Vascular niche for adult hippocampal neurogenesis. J Comp Neurol 2000;425:479–494.
28. Song H, Li X, Zhu C, et al. Glomerulosclerosis in adriamycin-induced nephrosis is accelerated by a lipid-rich diet. Pediatr Nephrol 2000;15:196–200.

29. Song HJ, Stevens CF, Gage FH. Neural stem cells from adult hippocampus develop essential properties of functional CNS neurons. Nat Neurosci 2002;5:438–445.
30. Lim DA, Alvarez-Buylla A. Interaction between astrocytes and adult subventricular zone precursors stimulates neurogenesis. Proc Natl Acad Sci USA 1999;96:7526–7531.
31. Seri B, Garcia-Verdugo JM, McEwen BS, et al. Astrocytes give rise to new neurons in the adult mammalian hippocampus. J Neurosci 2001;21:7153–7160.
32. Alvarez-Buylla A, Garcia-Verdugo JM, Tramontin AD. A unified hypothesis on the lineage of neural stem cells. Nat Rev Neurosci 2001;2:287–293.
33. Weiss S, Dunne C, Hewson J, et al. Multipotent CNS stem cells are present in the adult mammalian spinal cord and ventricular neuroaxis. J Neurosci 1996;16:7599–7609.
34. Palmer TD, Markakis EA, Willhoite AR, et al. Fibroblast growth factor-2 activates a latent neurogenic program in neural stem cells from diverse regions of the adult CNS. J Neurosci 1999;19:8487–8497.
35. Taupin P, Ray J, Fischer WH, et al. FGF-2-responsive neural stem cell proliferation requires CCg, a novel autocrine/paracrine cofactor. Neuron 2000;28:385–397.
36. Jin K, Mao XO, Sun Y, et al. Heparin-binding epidermal growth factor-like growth factor: hypoxia-inducible expression in vitro and stimulation of neurogenesis in vitro and in vivo. J Neurosci 2002;22:5365–5373.
37. Benraiss A, Chmielnicki E, Lerner K, et al. Adenoviral brain-derived neurotrophic factor induces both neostriatal and olfactory neuronal recruitment from endogenous progenitor cells in the adult forebrain. J Neurosci 2001;21:6718–6731.
38. Molne M, Studer L, Tabar V, et al. Early cortical precursors do not undergo LIF-mediated astrocytic differentiation. J Neurosci Res 2000;59:301–311.
39. Tropepe V, Craig CG, Morshead CM, et al. Transforming growth factor-alpha null and senescent mice show decreased neural progenitor cell proliferation in the forebrain subependyma. J Neurosci 1997;17:7850–7859.
40. Craig CG, Tropepe V, Morshead CM, et al. In vivo growth factor expansion of endogenous subependymal neural precursor cell populations in the adult mouse brain. J Neurosci 1996;16:2649–2658.
41. Kuhn HG, Winkler J, Kempermann G, et al. Epidermal growth factor and fibroblast growth factor-2 have different effects on neural progenitors in the adult rat brain. J Neurosci 1997;17:5820–5829.
42. Nakatsuji Y, Miller RH. Selective cell-cycle arrest and induction of apoptosis in proliferating neural cells by ganglioside GM3. Exp Neurol 2001;168:290–299.
43. Shihabuddin LS, Horner PJ, Ray J, et al. Adult spinal cord stem cells generate neurons after transplantation in the adult dentate gyrus. J Neurosci 2000;20:8727–8735.
44. Gage FH, Coates PW, Palmer TD, et al. Survival and differentiation of adult neuronal progenitor cells transplanted to the adult brain. Proc Natl Acad Sci USA 1995;92:11,879–11,883.
45. Clarke DL, Johansson CB, Wilbertz J, et al. Generalized potential of adult neural stem cells. Science 2000;288:1660–1663.
46. Jiang Y, Vaessen B, Lenvik T, et al. Multipotent progenitor cells can be isolated from postnatal murine bone marrow, muscle, and brain. Exp Hematol 2002;30:896–904.
47. Ferrari G, Cusella-De Angelis G, Coletta M, et al. Muscle regeneration by bone marrow-derived myogenic progenitors. Science 1998;279:1528–1530.
48. Johansson CB, Svensson M, Wallstedt L, et al. Neural stem cells in the adult human brain. Exp Cell Res 1999;253:733–736.
49. Bhagavati S, Xu W. Isolation and enrichment of skeletal muscle progenitor cells from mouse bone marrow. Biochem Biophys Res Commun 2004;318:119–124.
50. Issarachai S, Priestley GV, Nakamoto B, et al. Cells with hemopoietic potential residing in muscle are itinerant bone marrow-derived cells. Exp Hematol 2002;30:366–373.
51. Kopen GC, Prockop DJ, Phinney DG. Marrow stromal cells migrate throughout forebrain and cerebellum, and they differentiate into astrocytes after injection into neonatal mouse brains. Proc Natl Acad Sci USA 1999;96:10,711–10,716.
52. Sanchez-Ramos J, Song S, Cardozo-Pelaez F, et al. Adult bone marrow stromal cells differentiate into neural cells in vitro. Exp Neurol 2000;164:247–256.
53. Shih CC, Wu YW, Lin WC. Ameliorative effects of Anoectochilus formosanus extract on osteopenia in ovariectomized rats. J Ethnopharmacol 2001;77:233–238.

54. Castro O, Sandler SG, Houston-Yu P, et al. Predicting the effect of transfusing only phenotype-matched RBCs to patients with sickle cell disease: theoretical and practical implications. Transfusion 2002;42:684–690.
55. Morshead CM, Benveniste P, Iscove NN, et al. Hematopoietic competence is a rare property of neural stem cells that may depend on genetic and epigenetic alterations. Nat Med 2002;8:268–273.
56. Anderson RN, Miniño AM, Fingerhut LA, et al. Deaths: injuries, 2001. Natl Vital Stat Rep 2004;52:1–86.
57. James Fallon, Steve Reid, Richard Kinyamu, et al. In vivo induction of massive proliferation, directed migration, and differentiation of neural cells in the adult mammalian brain. Proc Natl Acad Sci USA 2000;97:14,686–14,691.
58. Kirschenbaum B, Nedergaard M, Preuss A, et al. In vitro neuronal production and differentiation by precursor cells derived from the adult human forebrain. Cereb Cortex 1994;4:576–589.
59. Nunes MC, Roy NS, Keyoung HM, et al. Identification and isolation of multipotential neural progenitor cells from the subcortical white matter of the adult human brain. Nat Med 2003;9:439–447.
60. Calza L, Giuliani A, Fernandez M, et al. Neural stem cells and cholinergic neurons: regulation by immunolesion and treatment with mitogens, retinoic acid, and nerve growth factor. Proc Natl Acad Sci USA 2003;100:7325–7330.
61. Mezey E, Nagy A, Szalayova I, et al. Comment on "Failure of bone marrow cells to transdifferentiate into neural cells in vivo". Science 2003;299:1184.
62. Wagers AJ, Sherwood RI, Christensen JL, et al. Little evidence for developmental plasticity of adult hematopoietic stem cells. Science 2002;297:2256–2259.
63. Arenas E. Stem cells in the treatment of Parkinson's disease. Brain Res Bull 2002;57:795–808.
64. Laywell ED, Kukekov VG, Steindler DA. Multipotent neurospheres can be derived from forebrain subependymal zone and spinal cord of adult mice after protracted postmortem intervals. Exp Neurol 1999;156:430–433.
65. Suhonen JO, Peterson DA, Ray J, et al. Differentiation of adult hippocampus-derived progenitors into olfactory neurons in vivo. Nature 1996;383:624–627.
66. Chmielnicki E, Benraiss A, Economides AN, et al. Adenovirally expressed noggin and brain-derived neurotrophic factor cooperate to induce new medium spiny neurons from resident progenitor cells in the adult striatal ventricular zone. J Neurosci 2004;24:2133–2142.
67. Hows JM, Howard MR, Downie T, et al. Unrelated bone marrow donor transplantation (UD-BMT): interim results of the IMUST study. International Unrelated Search and Transplant. Leukemia 1992;6(Suppl 4):163.
68. Madrigal JA, Cohen SB, Gluckman E, et al. Does cord blood transplantation result in lower graft-versus-host disease? It takes more than two to tango. Hum Immunol 1997;56:1–5.
69. Wagner JE. Umbilical cord blood stem cell transplantation. Am J Pediatr Hematol Oncol 1993;15:169–174.
70. Vaziri H, Dragowska W, Allsopp RC, et al. Evidence for a mitotic clock in human hematopoietic stem cells: loss of telomeric DNA with age. Proc Natl Acad Sci USA 1994;91:9857–9860.
71. Bicknese AR, Goodwin HS, Quinn CO, et al. Human umbilical cord blood cells can be induced to express markers for neurons and glia. Cell Transplant 2002;11:261–264.
72. Buzanska L, Machaj EK, Zablocka B, et al. Human cord blood-derived cells attain neuronal and glial features in vitro. J Cell Sci 2002;115:2131–2138.
73. Saporta S, Kim JJ, Willing AE, et al. Human umbilical cord blood stem cells infusion in spinal cord injury: engraftment and beneficial influence on behavior. J Hematother Stem Cell Res 2003;12:271–278.
74. Garbuzova-Davis S, Willing AE, Zigova T, et al. Intravenous administration of human umbilical cord blood cells in a mouse model of amyotrophic lateral sclerosis: distribution, migration, and differentiation. J Hematother Stem Cell Res 2003;12:255–270.
75. Bron D, De Bruyn C, Lagneaux L, et al. Hematopoietic stem cells: source, indications and perspectives. Bull Mem Acad R Med Belg 2002;157:135–145; discussion 145,146.
76. Akiyama Y, Radtke C, Honmou O, et al. Remyelination of the spinal cord following intravenous delivery of bone marrow cells. Glia 2002;39:229–236.
77. Bonilla S, Alarcon P, Villaverde R, et al. Haematopoietic progenitor cells from adult bone marrow differentiate into cells that express oligodendroglial antigens in the neonatal mouse brain. Eur J Neurosci 2002;15:575–582.

78. Bruder SP, Ricalton NS, Boynton RE, et al. Mesenchymal stem cell surface antigen SB-10 corresponds to activated leukocyte cell adhesion molecule and is involved in osteogenic differentiation. J Bone Miner Res 1998;13:655–663.
79. Barry FP, Boynton RE, Haynesworth S, et al. The monoclonal antibody SH-2, raised against human mesenchymal stem cells, recognizes an epitope on endoglin (CD105). Biochem Biophys Res Commun 1999;265:134–139.
80. Barry F, Boynton R, Murphy M, et al. The SH-3 and SH-4 antibodies recognize distinct epitopes on CD73 from human mesenchymal stem cells. Biochem Biophys Res Commun 2001;289: 519–524.
81. Lu P, Blesch A, Tuszynski MH. Induction of bone marrow stromal cells to neurons: differentiation, transdifferentiation, or artifact? J Neurosci Res 2004;77:174–191.
82. Black IB, Woodbury D. Adult rat and human bone marrow stromal stem cells differentiate into neurons. Blood Cells Mol Dis 2001;27:632–636.
83. Kang SK, Jun ES, Bae YC, et al. Interactions between human adipose stromal cells and mouse neural stem cells in vitro. Brain Res Dev Brain Res 2003;145:141–149.
84. Deb A, Wang S, Skelding KA, et al. Bone marrow-derived cardiomyocytes are present in adult human heart: A study of gender-mismatched bone marrow transplantation patients. Circulation 2003;107:1247–1249.
85. Quarto R, Mastrogiacomo M, Cancedda R, et al. Repair of large bone defects with the use of autologous bone marrow stromal cells. N Engl J Med 2001;344:385–386.
86. Nagai N, Zhao BQ, Suzuki Y, et al. Tissue-type plasminogen activator has paradoxical roles in focal cerebral ischemic injury by thrombotic middle cerebral artery occlusion with mild or severe photochemical damage in mice. J Cereb Blood Flow Metab 2002;22:648–651.
87. Gerson SL. MGMT: its role in cancer aetiology and cancer therapeutics. Nat Rev Cancer 2004;4:296–307.
88. Koc ON, Phillips WP, Jr, Lee K, et al. Role of DNA repair in resistance to drugs that alkylate O6 of guanine. Cancer Treat Res 1996;87:123–146.
89. Lee K, Gerson SL, Maitra B, et al. G156A MGMT-transduced human mesenchymal stem cells can be selectively enriched by O6-benzylguanine and BCNU. J Hematother Stem Cell Res 2001;10:691–701.
90. Sanai N, Tramontin AD, Quinones-Hinojosa A, et al. Unique astrocyte ribbon in adult human brain contains neural stem cells but lacks chain migration. Nature 2004;427:740–744.
91. Ostenfeld T, Svendsen CN. Recent advances in stem cell neurobiology. Adv Tech Stand Neurosurg 2003;28:3–89.
92. Calza L, Fernandez M, Giuliani A, et al. Stem cells and nervous tissue repair: from in vitro to in vivo. Prog Brain Res 2004;146:75–91.
93. Hallbergson AF, Gnatenco C, Peterson DA. Neurogenesis and brain injury: managing a renewable resource for repair. J Clin Invest 2003;112:1128–1133.

36
Mammalian Sir2 Proteins
A Role in Epilepsy and Ischemia

Barbara Aumayr, BS and Damir Janigro, PhD

SUMMARY

The family of sirtuins is a relatively uninvestigated group of proteins that is gaining more and more attention with every new features being discovered. It involves almost all key cell regulatory systems especially in those guarantying cellular integrity and it seems very plausible that silent information regulator (Sir)2 is the missing link between metabolic activity and cell proliferation. Most importantly, Sir2 apparently keeps the balance between apoptosis and proliferation under stress. Thus, this chapter will provide the most relevant information available about mammalian Sir2 proteins and an attempt to establish their relevance in central nervous system disease.

Key Words: Sirtuins; SIRT1; Sir2; HDAC3; FOXO; PARP-1; nicotinamide; p53; epilepsy; ischemia.

1. INTRODUCTION

Reversible protein acetylation is emerging as a critical post-translational modification important to many biological processes. Though pioneering studies focused on the role of histone acetylation in transcriptional control, more recent findings have generalized the concept of reversible protein acetylation to many nonhistone proteins *(1,2)*. Eighteen different human histone deacetylases (HDACs) have been classified and are grouped into three classes based on their sequence homology to *Saccharomyces cerevisiae* HDACs: reduced potassium dependency 3 (RPD3) (class I), histone deacetylose1 (HDA1) (class II), and silent information regulator (Sir)2 (class III) *(3)*. Only recently, increasing effort has been put into characterizing Sir2 and its implications in biological processes. Most work has been carried out in *S. cerevisiae*. Today it is accepted that Sir2 is a nicotinamide adenine dinucleotide (NAD^+)-dependent histone deacetylase that regulates gene silencing, cell cycle, DNA-damage repair, and life span *(4)*. However, the picture of Sir2 in the human cell is by far incomplete, and the data available so far only underscores the need for further investigations of this remarkable group of proteins. This chapter will first provide a concise description of Sir2 functions in different organisms and then focus on the role of human sirtuins and then discuss its possible implications for pathogenesis of seizure disorders and ischemia.

2. THE FAMILY OF SIRTUINS

The *Sir2* gene family of proteins is conserved from bacteria to humans *(5)*. The most intensively studied member is the yeast Sir2 protein, which has been found to regulate heterochromatin silencing at the mating type loci, telomeres, and ribosomal DNA repeats *(6,7)*. Moreover, Sir2 modulates life span in both *S. cerevisiae (8)* and *Caenorhabditis elegans (9)*. In *Drosophila melanogaster*, Sir2 controls segmentation, heterochromatin silencing, as well as sex determination

From: *The Cell Cycle in the Central Nervous System*
Edited by: D. Janigro © Humana Press Inc., Totowa, NJ

via direct interaction with members of the hairy/enhancer of Split bHLH repressor family *(10,11)*. Surprisingly, the functional role of mammalian Sir2 proteins differs from the yeast Sir2 functions, which are all in some way associated with chromatin structure. In mammals, Sir2 is capable of directly deacetylating other target proteins (i.e., p53, Forkhead transcription regulators, etc.) in addition to histones. The mouse Sir2 homolog, SIR2-α, plays a role in the growth and maturation of the embryo and in gametogenesis in both sexes *(12)*. Unlike the Sir2p in yeast, the human gene is also found in the cytoplasm and the over expression does not cause chromosome loss or cell toxicity *(13)*. There are seven human Sir2 homologs (SIRT1–7), of which sirtuin 1 (SIRT1) is the most closely related to the yeast Sir2p as well regarding the mouse SIR2α homolog *(14)*. Table 1 gives a concise summary of human Sir2-like proteins regarding their predicted cellular localization, size, orthologs, and postulated function if available.

3. CLASSIFICATION OF SIRTUINS

The family of sirtuins can be grouped into four distinct classes, designated here as classes I–IV. A fifth class of sirtuin has been discovered in Gram-positive bacteria and *Thermotoga maritima*. In prokaryotes only members of classes II and III are present, *S. cerevisiae* has five members of class I sirtuins and *D. melanogaster* as well as *C. elegans* have sirtuins from classes I, II, and IV. The seven human sirtuin genes include members of all four classes: SIRT1, SIRT2, and SIRT3 are class I, SIRT4 is class II, SIRT5 is class III, and SIRT6 and SIRT7 are class IV. The main branch leading toward Sir2 is designated as class I. The yeast Sir2 and proteins such as Hst1, SIRT1, C.ele1, and D.mel1 belong to a subgroup called class Ia. The yeast Hst2 and proteins that include SIRT2, SIRT3, and D.mel2 are members of the subgroup class Ib. In summary, currently available data suggest that class I, II, and IV sirtuins could be present in all metazoan organisms as well as in some protists, whereas most prokaryotic sirtuin gene sequences are of class III type, the most ancient class of sirtuins *(15)*.

4. STRUCTURE AND FUNCTION

Each member of the family of sirtuins is characterized by a conserved core domain that is in some instances attached to additional *N*- or *C*-terminal sequences. The NAD^+-dependency of Sir2-like enzymes distinguishes them from the class I and II HDACs, which employ a zinc-catalyzing mechanism *(17)*. In contrary to conservative HDACs, which deacetylate their substrates by simple hydrolysis, Sir2-like proteins employ a complex chemistry in order to deacetylate acetyl-lysine residues. This unique reaction consumes NAD^+, and releases *O*-acetyl-adenosine diphosphate (ADP)-ribose as well as the deacetylated product. One can think of it as a two-step process: In the first step, two substrates, an acetylated protein and NAD^+, interact with the active site of the enzyme. Subsequent conformational changes enable a nucleophilic attack on the Cl^- of the nicotinamide (NAM) ribose resulting in the cleavage of the *N*-pyridinium glycoside bond of NAD^+ with concomitant release of NAM, which in turn generates a high-energy ADP-ribose intermediate. This first step is reversible in the presence of excess NAM:

$$NAD^+ + Sir2 + acetyl\text{-}lysine \Leftrightarrow NAM + Sir2\sim ADP\text{-}ribose\ acetyl\text{-}lysine$$

The second step is irreversible in nature and involves the reaction of high-energy ADP-ribose intermediates with the ε-amino group of the lysine residue. A novel compound, acetyl-ADP-ribose is created when ensuing transfer of the acetyl group to the high-energy intermediate takes place. Finally, the deacetylated protein and acetyl-ADP-ribose are released from the enzyme's active site:

$$Sir2\sim ADP\text{-}ribose \bullet acetyl\text{-}lysine \Leftrightarrow lysine + Sir2 + acetyl\text{-}ADP\text{-}ribose$$

The formation of *O*-acetyl ADP-ribose (OAAR) is closely linked to NAD^+-deacetylase activity (NDAC), which raises the possibility that OAAR might act as a second messenger, acting only focal in the cell at the site in which SIRT proteins interact with their acetylated substrates and thus *de novo* generation of OAAR occurs. This theory is supported by the fact that OAAR,

Table 1
Classification of Mammalian Sir2-Related Proteins Regarding Their Substrates

Gene name	NCBI locus link ID[a]	Human chromosomal location	Protein product(s), conceptual translation	PSORT protein prediction (results of the k-NN prediction)[b]	Comments	Substrate proteins
SIRT1	23411	10q21.3	Isoform 1 (747 aa), isoform 2 (555 aa)	SIRT1 (747 aa): 69.9% nuclear, 13% mitochondrial, 8.7% plasma membrane, 4.3% vesicles of the secretory system, 4.3% cytoplasmic	Known as Sir1-α in mice; ortholog of the *Drosophila* SIRT gene located at 34A7-8; ortholog of the *C. elegans* SIRT gene found on chromosome IV; probable ortholog identified in zebrafish as EST (expressed sequence tag)	H3, H4, p53, FOXO3a, Hes/Hey
SIRT2	22933	19q13.1	Isoform 1 (389 aa), isoform 2 (352 aa)	SIRT2 (352 aa): 30.4% cytoplasmic, 21.7% mitochondrial, 17.4% extracellular (including cell wall), nuclear, 4.3% vacuolar, 4.3% cytoplasmic, 4.3% endoplasmic reticulum	Known as SIR2L2 in mice; ortholog of the *Drosophila* SIRT gene found on 3R92E; no apparent ortholog in *C. elegans*	α-tubulin
SIRT3	23410	11p15.5	Isoform A (257 aa), isoform B (346 aa)	SIRT3 (399 aa): 44.4% endoplasmic reticulum,[c] 22.2% Golgi, 11.1% mitochondrial, 11.1% nuclear, 11.1% cytoplasmic	Known as SIR2L3 in mice; no apparent ortholog in *Drosophila* or *C. elegans*	
SIRT4	23409	12q	314 aa	SIRT4 (314 aa): 33.3% endoplasmic reticulum, 22.2% vacuolar, 22.2% plasma membrane, 11.1% Golgi, 11.1% mitochondrial	*Drosophila* ortholog found at 65E-6 on chromosome 3; in *C. elegans* there are two closely linked apparent orthologs	
SIRT5	23408	6pter-p25.1	Isoform 1 (310 aa), isoform 2 (299 aa)	SIRT5 (310 aa): 62.5% mitochondrial, 8.7% cytoplasmic, 8.7% nuclear, 4.3% vacuolar, 4.3% extracellular (including cell wall), 4.3% plasma membrane, 4.3% peroxisomal	No apparent orthologs in either *Drosophila* or *C. elegans*	
SIRT6	51548	19p13.3	Isoform 1 (355 aa), isoform 2 (237 aa)	SIRT6 (355 aa): 78.3% nuclear, 13.0% cytoskeletal, 4.3% Golgi, 4.3% mitochondrial	*Drosophila* ortholog found at 3R 85F-86A; apparent *C. elegans* ortholog found on chromosome 1	
SIRT7	51547	17q25	400 aa	SIRT7 (400 aa): 56.5% nuclear, 26.1% cytoplasmic, 13.0% mitochondrial, 4.3% cytoskeletal	No apparent orthologs in either *Drosophila* or *C. elegans*	

[a]LocusLink ID is available from the electronic repository at the National Center for Biotechnology Information (NCBI), available at http://www.ncbi.nlm.nih.gov/LocusLink.
[b]PSORT prediction is a summation of results of the k nearest neighbor (k-NN) prediction according to the predictive software accessible at http://psort.nibb.ac.jp/form2.html.
[c]Converse to this prediction, SIRT3 has been found to be imported into the mitochondrial matrix where it becomes active on cleavage. It is a soluble mitochondrial matrix protein.
(Adapted from ref. *15*.)

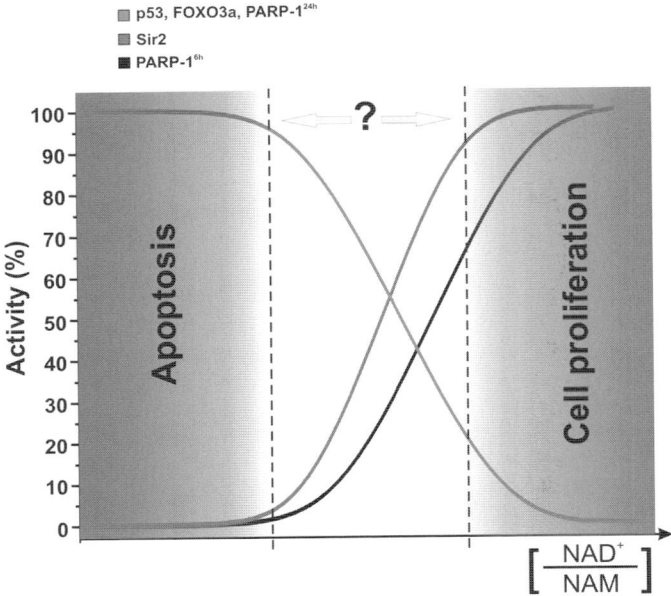

Fig. 1. This scheme roughly depicts the individual dependency of Sir2 and PARP (6 or 24 h after activation) on [NAD+/NAM] and the resulting activity of p53 and FOXO3a (only at early time-point). Whereas it is clear that low levels of NAM in addition to high levels of NAD+ lead to cell proliferation and that the converse state results in apoptosis, it is not known when the apoptotic response shifts to cell proliferation. (Please *see* companion CD for color version of this figure.)

when injected in starfish oocytes or blastomeres, resulted in the delay or complete blockade of the cell cycle during development *(18)*.

Sir2-like enzymes are regulated in function either via altered free NAD+ levels or via the inhibitory noncompetitive effect of NAM on the deacetylation reaction. However, the relative inhibitory contribution of NAM is likely to be more substantial than the small increases in activity predicted with noncatastrophic alterations in NAD+ or NAD+/ NADH ratios. Bitterman et al. proposed a model by which NAM inhibits deacetylation by binding to a conserved pocket adjacent to NAD+, thereby blocking NAD+ hydrolysis. The nature of NAM inhibition, indicates that it cannot be alleviated by increased levels of NAD+. Besides, in vitro experiments have clearly depicted that the concentration of NAM required for significant SIRT1 and Sir2 inhibition is within physiological concentrations (IC_{50} <50 μM), which have been reported to range from 11 to 400 μM. Taken together, it can be assumed that NAM levels, being physiologically relevant, are the main regulating factor of Sir2 function. However, taken into consideration the NAD+ dependency of SIRT1 it might be most accurate to relate SIRT1 activity to the quotient of NAD+ and NAM concentration, respectively ([NAD+]/[NAM]) (Fig. 1 provides with a scheme of the relative activities of proteins involved in dependency of [NAD+]/[NAM]) *(19–23)*.

5. SIRTUIN 1

SIRT1, predominantly nuclear in localization, is the best-investigated member of the human family of sirtuins and suggested to play an essential role in life-span extension (on caloric restriction), oxidative stress response (PARP; [poly(ADP-ribose) polymerase]), as well as in regulation of forkhead transcription factors (FOXOs) and p53 (*see* Fig. 2).

SIRT1 has also been shown to regulate fat mobilization in white adipocytes, by binding to and repressing genes controlled by the fat regulator peroxisome proliferators-activated receptor-γ

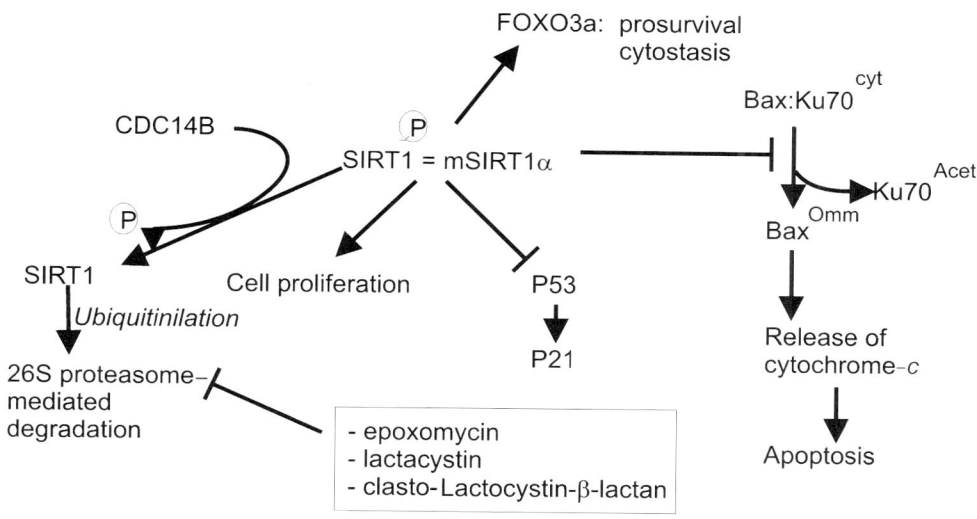

Fig. 2. General features of SIRT1. Phosphorylation prevents SIRT1 from degradation, thus CDC14B counteracts SIRT1 by dephosphorylating it and allowing for ubiquitinylation and ensuing degradation by the 26S proteasome, which can be inhibited by several cytostatica (i.e., epoxomycin, lactacystin, and clasto-lactocystin-β-lactan). SIRT1 further inhibits p53-mediated apoptosis via deacetylation, and inhibits Bax-mediated apoptosis, by keeping Ku70 in a deacetylated state, which retains Bax in the cytoplasm instead of relocating to the outer mitochondrial membrane and triggering apopotis. Besides many more important functions, SIRT1 shifts the effects of FOXO3a to survival and cytostasis. Taken together, SIRT1 activation results in cytostasis and ensuing cell proliferation.

(PPAR-γ), including genes that mediate fat storage. Picard et al. suggest that SIRT1 represses PPAR-γ by docking with its cofactors nuclear receptor corepressor (NCoR) and silencing mediator of retinoid and thyroid hormone receptors (SMRT). In differentiated fat cells SIRT1 upregulation induces lipolysis and loss of fat, suggesting that SIRT1 may provide a molecular link between caloric restriction and life extension in mammals, because a reduction in fat is sufficient to extend murine life span *(24)*. Besides, Takata and Ishikawa uncovered that SIRT1 associates with the human hairy related basic helix loop helix (bHLH) repressors hairy and enhancer of split (HES1) and hairy/enhancer-of-split-related with YRPW motif 2 (HEY2) both in vivo and in vitro and is involved in HES1- and HEY2-mediated repression *(25)*. SIRT1 was also found to regulate skeletal muscle gene expression and differentiation concomitant with changes of the redox state. Fulco et al. argue that in the course of muscle differentiation the decreasing ratio of NAD^+ to NADH impacts SIRT1 activity and thus builds a positive feed back loop resulting in enhanced differentiation when differentiation was initiated. Myocytes with reduced levels of SIRT1 were observed to differentiate prematurely, whereas SIRT1 overexpression retards muscle differentiation. Evidence was provided that SIRT1 forms a complex with the acetyltransferase p300/CBP-associated factor (PCAF) and MyoD to carry out these functions. Moreover, taking into account that administration of lactate decreases the ratio of NAD^+ to NADH, it seems reasonable that SIRT1 might be contributing to myocyte differentiation in response to exercise. Generally speaking, SIRT1 is suggested to act as a sensor of redox state in response to exercise, food intake, and starvation *(26,27)*.

6. SIRT1 AND p53

Among its other functions SIRT1 has been shown to be involved in DNA damage responses *(28–30)*. In mammalian cells, p53 is the primary mediator of these. p53 is a short-lived protein and its activity is maintained at low levels under normal circumstances. On DNA damage p53 is protected from rapid degradation (normally executed by mouse double minute 2 homolog [MDM2]-dependent proteasomal degradation) and gains transcriptional activation functions mostly owing to post-translational modifications, including ribosylation, phosphorylation, and acetylation. Importantly, acetylation is a major contributor to p53 stabilization and consequent target gene activation *(31–33)*.

CBP/p300, a protein possessing histone acetyltransferase activity (HAT), is a transcriptional coactivator of p53 and potentiates its transcriptional activity, through enhanced DNA binding, coactivator recruitment and stabilization, as well as its biological function in vivo *(34–36)*. In humans, acetylation of p53 occurs at multiple lysine residues, including K320, K373, and K382, all situated at the C-terminal end of the protein. Phosphorylation of p53 is believed to further enhance p300-mediated acetylation of p53, which in turn leads to stabilization of p53 by inhibiting MDM2-mediated ubiquitination. Thus, in response to DNA damage and oxidative stresses, p53, on stabilization, exerts proapoptotic and cytostatic effects. Mice engineered to have high levels of p53 activity were found to have a short life-span in addition to being less susceptible to tumor formation *(37)*. Modulating p53 level and activity thus seems to be a key link between tumor suppression and longevity. Interestingly, p53-dependent apoptosis is also implicated in monitoring spontaneous DNA damage during neurogenesis *(38–40)*.

As indicated above, MDM2 is the predominant negative regulator of p53 under normal conditions, whereas SIRT1 has no obvious effect unless the cell is stressed, whereupon SIRT1 (as well as mSir2α) interacts with and specifically deacetylates K382 of p53 *(41–44)*. Thus, on oxidative stress or DNA damage, SIRT1-mediated deacetylation of p53 mitigates p53-mediated apoptosis and transcriptional activation by NAD^+-dependent and NAM-sensitive deacetylation. Furthermore, whereas Sir2α specifically inhibits p53-ediated apoptosis it is incapable of averting Fas-mediated cell death. Therefore, SIRT1 was proposed to be critical for cell survival, when MDM2 is severely attenuated by various stress impacts, enabling the induction of p53 activity for DNA repair before committing to apoptosis. Luo et al. propose a speculative model in which p53 and Sir2α interaction leads to p53 deacetylation, ensuing recruitment of the p53-Sir2α complex to the target promoter (e.g., of the p21 gene) and subsequent transcriptional repression, which may act both via decreasing p53 transactivation capability or through Sir2α-mediated histone deacetylation at the target promoter region. Besides SIRT1 deacetylation, p53 is also deacetylated by HDAC1, which is recruited to p53 by metastasis-associated protein 2 (PID/MTA2), a p53 interacting protein. Because many DNA-damaging drugs (e.g., etoposide) are very effective antitumor drugs in cancer therapy, it is possible that combining DNA-damaging drugs with HDAC-1 inhibitors and SIRT1 inhibitors might have synergistic effects in cancer therapy by maximally activating p53 *(45–47)*.

7. SIRT1 AND THE REGULATION OF FOXOs

A link between SIRT1 and FOXOs has recently been elucidated. It was shown that SIRT1 binds to and deaceltylates FOXO3 in mammalian cells and thus represses the ability of FOXO3 to activate its target genes. Further data suggest that SIRT1 also represses FOXO1 and FOXO4.

There are four different FOXOs: FOXO1 Forkhead in rhabdonyosarcoma (FKHR), FOXO3 (FKHRL1), FOXO4 acute-lymphocytic-leukemia-1 fused gene from chromosome X[AFX], and FOXO6, which is less related to the other FOXOs. The nonphosphorylated forms of the FOXO proteins localize to the nucleus of mammalian cells, in which they function as transcription factors. However, phosphorylation of FOXO by v-akt murine thymoma viral oncogene (Akt)/PKB on two or three threonin or serine residues, promotes interaction with 14-3-3 proteins, resulting in exclusion of the FOXO protein from the nucleus and consequent inhibition of its interaction with target genes. FOXO3 has been shown to be acetylated at five different lysine residues and phosphorylated at eight different serine or threonine residues. The deacetylase activity of SIRT1 is suspected to downregulate activated forkhead by either destabilizing the protein, decreasing its DNA binding activity, or changing protein/protein interactions. Besides, SIRT1 may also repress FOXO target genes by deacetylating histones at those loci. The exact mechanisms remain

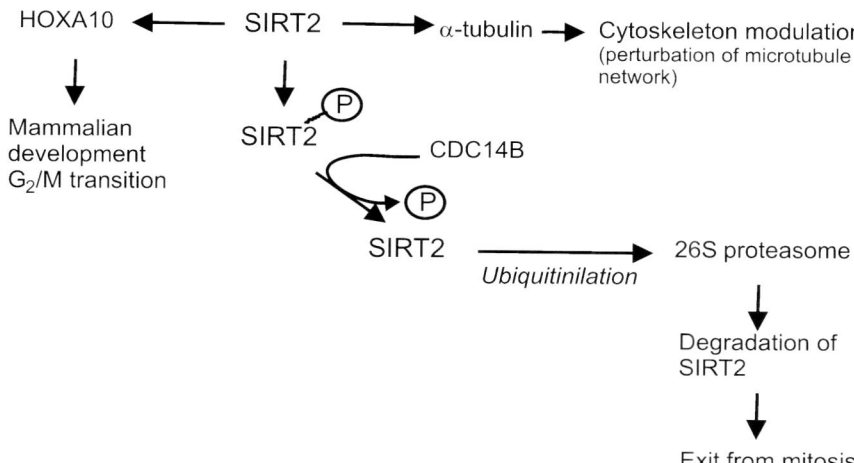

Fig. 3. Functions and degradation of SIRT2. SIRT2 results in the activation of HOXA10 which is involved in mammalian development and G_2/M transition, and furthermore in the deacetylation of α-tubulin, leading to perturbation of the microtubule network and therefore modulation of the cytoskeleton. Similar to SIRT1, SIRT2 becomes ubiquitinylated on dephosphorylation by CDC14B, and thereafter degraded via the 26S proteasome, which results in exit from mitosis.

to be further elucidated. However, as mentioned above, there is evidence that SIRT1 binds to and deacetylates all FOXOs in mammalian cells (except FOXO6), and represses their ability to bind to their target genes by sequestering them in the cytoplasm. Interestingly, the interaction of SIRT1 with FOXO3 appears to occur only in the presence of stress stimuli. Furthermore, SIRT1 was shown to bind to and deacetylate p300/CBP, a transcriptional coactivator for FOXOs *(48)*.

Depending on the activation signal, FOXO can regulate apoptosis (via Fas ligand and BIM) *(49–51)*, cell cycle arrest (via p27KIP) *(52–56)*, differentiation *(57–59)*, or the activation of genes that control repair of damaged DNA (GADD45) *(60,61)* and oxidative stress resistance via ROS detoxification (Mn catalase and dismutase) *(62–66)*.

For example, in *C. elegans* longevity and stress-resistant phenotypes are dependent on the FOXO abnormal DAuer Formation family member (DAF)-16, implicating FOXOs in the regulation of stress-resistance-related genes such as that for the cytosolic catalase (CTL-1) in *C. elegans* and GADD45 in *Mus musculus*, respectively. GADD45 expression was found to be increased on promoter activation by FOXOs (a FOXO-binding motif occurs at position –507 5′ of the GADD45 promoter) and occurred in response to oxidative stress in a FOXO-dependent but p53-independent pathway. On inhibition of AKT, SIRT1 contributes to the increased expression of GADD45. GADD45 mediates G_2 arrest by interacting with and inhibiting the kinase activity of Cdc2. Given that GADD45 is involved in the cellular response to DNA damage (including activation of cell cycle checkpoint and DNA repair, thereby contributing to maintenance of genomic integrity), and given that FKHRL1 (FOXO3) is abundant in postmitotic neurons in the adult mouse brain, these findings suggest that GADD45 may serve to protect postmitotic neurons from oxidative stress on FOXO3-dependent induction. This theory is further supported by the fact that the activation of FOXO proteins contributes to cellular resistance to oxidative stress.

However, SIRT1 does not regulate all FOXO target genes in the same manner. Although SIRT 1 appears to contribute to increased expression of the stress response gene GADD45 in response to inhibition of the AKT pathway and although it potentiates FOXO ability to induce cell cycle arrest, possibly allowing more time for cells to detoxify ROS and repair damaged DNA, its influence on bcl-2 interacting mediator of cell death (BIM) expression is unclear. Generally speaking, SIRT1 mitigates FOXO's proapoptotic and metabolic effects and enhances FOXO-mediated survival (cell cycle arrest, DNA repair, and oxidative stress resistance). A scheme including SIRT1 influence on FOXO is presented in Fig. 3.

8. SIRT1 AND PARP-1: A CONVERGING PATHWAY?

The physiological functions of PARP-1, an abundant nuclear protein in most eukaryote tissues, comprise a large array of cellular activities, including safeguarding genomic integrity by limiting sister chromatid exchange (SCE), regulating gene expression and facilitating DNA repair *(67,68)*. It is one of the main consumers of cellular NAD^+ and accounts for more than 90% of cellular poly-ADP-ribose production, generating an equal amount of NAM as a byproduct. Though being inactive under normal conditions, on oxidative stress PARP is activated by DNA breaks or single-stranded DNA. Besides its NAD^+ dependency, PARP shares other similarities with SIRT1: PARP is also inhibited by NAM and PARP poly-ADP-ribosylation activity also targets histones and p53. It furthermore appears that p53's DNA binding activity is down-regulated in vitro by poly-ADP-ribosylation. Thus, it seems plausible that there is a linkage between the two ancient pathways that mediate broad biological activities in multicellular organisms *(69)*. Interestingly, it was found that NAM inhibition of PARP occurred only at early time-points (6 h) after DNA damage but enhanced PARP activity later on (24 h) *(70,71)*.

Sir2 is significantly more sensitive to inhibition by NAM than PARP. NAD^+ depletion occurring along with PARP activity was found to be associated with increased DNA-fragmentation, suggesting the initiation of a death cascade involving PARP, NAD^+ depletion, and NAM accumulation, resulting in Sir2 and PARP inhibition and consequent cell death. Taking all the findings together, one might propose a model in which PARP is recruited to DNA breaks after injury (e.g., caused by oxidative stress) and the ensuing NAM accumulation is primarily inhibiting SIRT1, which if not entirely inhibited leads to cytostasis instead of apoptosis via its actions on FOXO and p53. However, this occurs only as long as NAM levels remain within a certain range. Furthermore, PARP activity mitigates p53 activity to prevent from apoptosis. When severe cell damage occurs, NAM levels are expected to remain high or even increase throughout cellular response to injury. As indicated above, NAM increases PARP activity at late time points. This then results in a tremendous increase in NAM, total NAD^+ depletion, which is translated into complete SIRT1 inhibition along with complete p53 activation, ultimately exerting its proapoptotic functions. However, it remains unclear how PARP activity is increased by NAM after 24 h conversely to the initial inhibition. A scheme of PARP action on SIRT1 on oxidative stress is given in Fig. 3. Importantly though, depletion of PARP activity by NAM is associated with genomic instability and increased risk of neoplastic growth in some experimental models, perhaps owing to concomitant loss of SIRT1 activity *(72)*.

9. SIRTUIN 2

SIRT2 proteins, belonging to class Ib sirtuins, bear a NDAC (NAD^+-dependent deacetylase) activity that is specifically dependent on NAD^+, resistant to sodium butyrate and sensitive to NAM. Human SIRT2 is most closely related to yeast Hst2p, both of which are of cytoplasmic localization. However, substrate specificities of these two proteins have diverged significantly. In addition to deacetylating histone-H3 peptide acetylated on lysine-14, SIRT2 is also capable of deacetylating an acetylated α-tubulin peptide, an ability that Hst2p clearly lacks. Besides, SIRT2 is the only member of the sirtuin family that has been shown to specifically deacetylate α-tubulin. Further investigations unveiled that endogenous SIRT2 and HDAC6 colocalize along the microtubule network, suggesting that they are part of a single multiprotein complex although differing functionally. Both proteins are capable of deacetylating tubulin heterodimers and microtubules comprising of purified tubulin proteins, but only SIRT2 and not HDAC6 deacetylates tubulin (dimers or microtubules) reassembled from cell lysates. Thus, a selective HDAC6 inhibitor might be present in cellular lysate microtubules. However, inhibition of either SIRT2 or HDAC6 alone (via siRNA) is sufficient to induce hyperacetylation of tubulin, indicating an interdependent mechanism of function. As tubulin acetylation is implicated in regulation of cell shape, intracellular transport, cell motility, and cell division it will be of future interest to address the role of SIRT2 in tubulin deacetylation also in the concept of cental nervous system (CNS) diseases with special regards to cytoskeleton modulation in glioma pathogenesis.

Hiratsuka et al. discovered that SIRT2, which is located at 19q13.2 (a region known to be frequently deleted in human gliomas), was dramatically diminished in 12 out of 17 gliomas and glioma cell lines. Further analysis showed that ectopic expression of SIRT2 in glioma cell lines led to the perturbation of the microtubule network and caused a remarkable reduction in the number of stable clones expressing SIRT2 as compared with that of a control vector in colony formation assays. These findings suggest that SIRT2 may act as a tumor suppressor gene in human gliomas possibly through the regulation of microtubule network and may serve as a novel molecular marker for gliomas *(73)*.

Even more importantly, Dryden et al. addressed the role of SIRT2 NAD-dependent deacetylase activity in control of the mitotic exit in cell cycle. They discovered that SIRT2 is upregulated during mitosis and that overexpression of SIRT2 results in a prolonged mitotic phase. It was further found that especially at the G_2/M transition but generally in late G_2 phase, M phase, and during cytokinesis phosphorylation of SIRT2 occurs, which apparently prevents SIRT2 from degradation. These findings fit in a model in which CDC14B mediates ubiquitination and subsequent degradation of SIRT2 proteins via the 26S proteasome by dephosphorylating SIRT2, thus leading to the exit from mitosis. This model is supported by the fact that CDC14B leads to a decrease in SIRT2 protein and those inhibitors of the 26S proteasome (i.e., lactacystin, clastolactocystin-β-lactan, and epoxomycin) result in the stabilization of SIRT2. A schematic view of this model is given in Fig. 4. SIRT2 proteins also appear to be modified by an additional post-translational mechanism other than phosphorylation, whose nature is currently under investigation *(74,75)*.

It has further been shown that SIRT2 interacts with the homeobox transcription factor HOXA10 raising the intriguing possibility that SIRT2 plays a role in mammalian development *(76)*.

10. SIRTUIN 3

The human SIRT3 is the last member of class Ib sirtuins to be discussed. It is conserved in prokaryotes and eukaryotes. The *SIRT3* gene (www.ncbi.nlm.nih.gov/omim: MIM604481; contig NT_035113) is situated at the telomeric terminal on 11p15.5 chromosome. Interestingly, SIRT3 is not an imprinted gene (i.e., a gene that shows silencing of one parental allele) although being located in a domain of imprinted genes potentially associated with longevity (v-Ha-ras Harvey rat sarcoma viral oncogene homologs, insulin-like growth factor 2, proinsulin, and tyrosine hydroxylase). Nevertheless, evidence has been given that SIRT3 is involved in aging *(77,78)*. Despite prior analysis of amino acid motifs for protein sorting and trafficking by using P-SORT-II algorithms (based on the *k*-nearest neighbor classifier) predicted SIRT3 to be localized at the endoplasmatic reticulum, recent studies revealed that SIRT3 is located at the mitochondrial matrix. Mitochondrial import of SIRT3 is dependent on NH_2-terminal amphipatic α-helix rich in basic residues. Concisely, an approx 44 kDa form hSIRT3p is prevalent in its inactive state in the cytoplasm and is then transported into mitochondria in which it becomes active on NH_2-terminal cleavage. This R99/100-dependent cleavage is mediated by matrix processing peptidase (MPP) and yields the NAD^+ dependent 28 kDa active form of hSIRT3, which is a soluble mitochondrial matrix protein *(79)*.

Given that SIRT3 (mitochondrial in localization) is ubiquitously expressed, particularly in metabolically active tissue and that SIRT3 has a NAD^+ dependent protein deacetylase activity it might represent an important sensor for metabolic activity modulating silencing according to cellular energy levels. It has been suggested that on NAD^+ increase a metabolic pathway is triggered which leads to the activation of SIRT3, resulting in the deacetylation of specific targets. However, the identity and function of these targets remain to be unveiled *(80)*. Conversely, an alternative hypothesis argues that stable NAD^+ levels present in mitochondria allow hSIRT3 to be constitutively active, ensuring the constitutive deacetylation of one or several mitochondrial proteins *(81,82)*. Thus, certain conditions leading to a sudden NAD^+ decrease might strongly impair the activity of hSIRT3. These involve the opening of the highly conducting mitochondrial permeability transition pore (mtPTP) in the inner mitochondrial membrane, triggered by

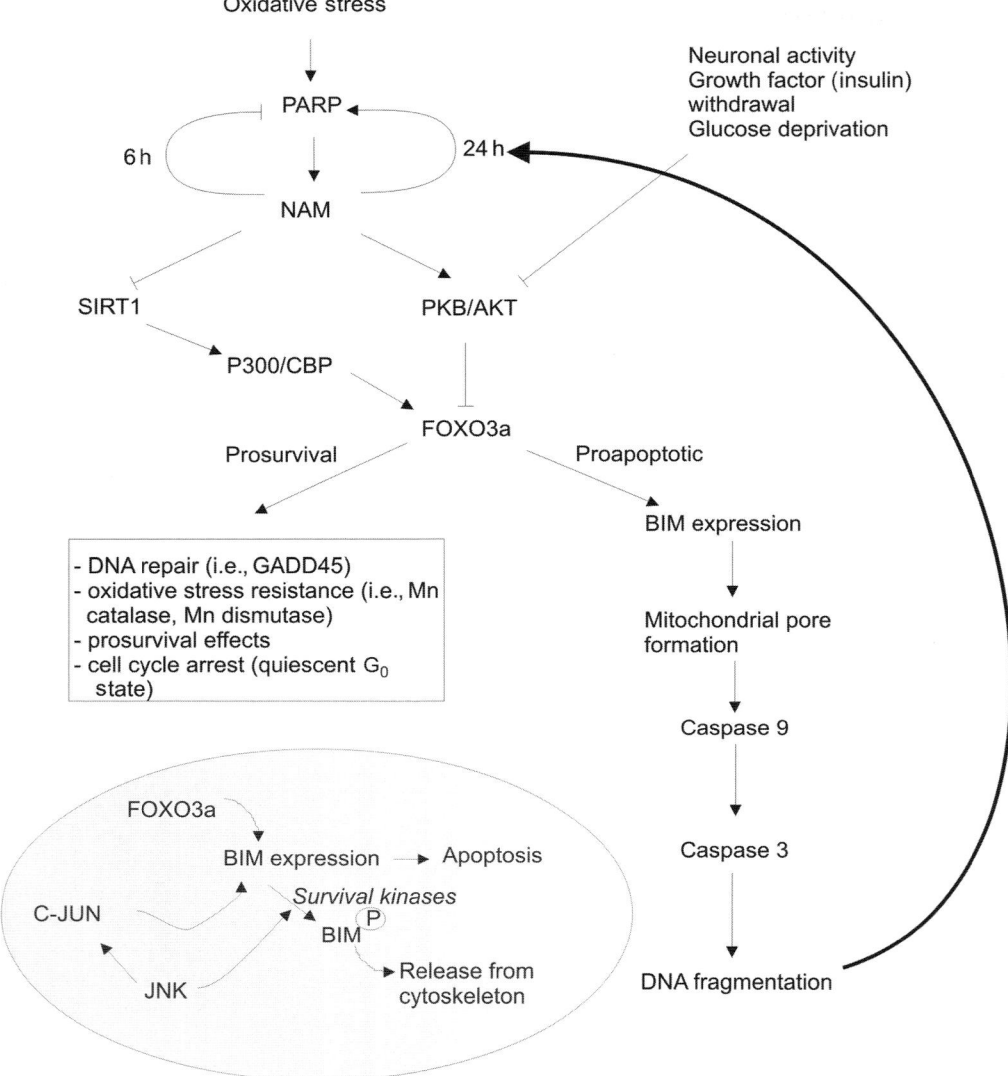

Fig. 4. Influences on the Sir2 pathway. Importantly, oxidative stress results in the activation of PARP, being strongly associated with decreasing NAD^+ and increased NAM levels upon activation. This results in the inhibition of Sir2, which therefore, depending on the grade of inhibition, is impaired in carrying out its prosurvival effects (i.e., DNA repair, induction of cell cycle arrest, inhibition of FOXO-mediated apoptosis, etc.). It is thus assumed that minor oxidative stress results in PARP activation and minor inhibition SIRT1, whose response still exists but becomes increasingly mitigated the more PARP is activated—leading to cell cycle arrest allowing for DNA repair. This also explains that if major damage has occurred by oxidative stress, NAM levels from the very beginning are very high—SIRT-mediated effects are entirely mitigated. FOXO3a can exert its proapoptotic function resulting in DNA-fragmentation therefore further enhancing PARP activity ending in apoptosis. Also if damage cannot be fixed within 24 h, the negative feedback loop of NAM and PARP switches to a positive feedback loop resulting in apoptosis as well. Interestingly, FOXO-activation can be counteracted by growth factor withdrawal, neuronal activity, or glucose deprivation (caloric restriction) via the PKB/AKT pathway, mitigating apoptosis. (Please *see* companion CD for color version of this figure.)

either reactive oxygen species or an increase in Ca^{2+} levels, which results in NAD^+ efflux and ensuing NAD^+ hydrolyzation by NAD glycohydrolases (NADases) in the intermembrane space. Importantly, this reaction generates ADP-ribose and NAM, the strongest inhibitor of Sir2-like enzymes *(83–87)*. In summary this very probable model suggests that matrix NAD^+ depletion, increased NAD^+ hydrolysis accompanied by NAM formation, and inhibition of hSIRT3 import into the matrix would constitute important regulatory mechanisms for SIRT3 activity that under normal conditions (steady NAD^+ levels) protects the cell against apoptosis.

It would also be formally possible that SIRT3 is involved in mitochondrial ADP-ribosylation, taken into account that ADP-ribosyl transfer has been reported for some sirtuins *(88)*. Besides, the deacetylation of a substrate by Sir2-like enzymes is coupled to the formation of O-acetyl-ADP-ribose, a novel metabolic agent, whose intrinsic metabolic activity further expands the number of metabolic processes possibly controlled by SIRT3 *(89)*. In the end the exact function of SIRT3 remains to be elucidated. However, its unique cellular localization gives important clues to its functional targets and strengthens the need for further investigations of this possible link between cellular energy levels and silencing mechanisms.

11. NAM AND NAD^+ METABOLISM IN THE LIGHT OF EPILEPTIC, ISCHEMIC, AND OTHER CNS DISORDERS

Mitochondrial oxidative phosphorylation provides the major source of adenosine triphosphate in cortical neurons. During seizures mitochondria depolarize, so there is the risk of increased production of free radicals. Furthermore, sustained epileptic seizures change the redox potential and reduce ATP content, leading to collapse in energy production and supply in brain. Because of the increase in blood flow during seizures, the elevated oxygen supply consequently further enhances the risk of reactive oxygen species (ROS) generation. Thus, oxidative stress is a main feature of epileptic disorders, as well as it occurs in the course of ischemic disorders.

Under these circumstances special attention should be drawn to complex I whose primary functions include oxidizing NADH in the mitochondrial matrix, reducing ubiquinone to ubiquinol, and pumping protons across the inner mitochondrial membrane to drive ATP synthesis. Complex I is markedly more susceptible to oxidative stress than other respiratory chain complexes, is suggested to play a role in apoptosis, and its selective loss contributes to neurodrodegenerative diseases such as Parkinson's disease, Huntington's disease, and hippocampal sclerosis. An animal model closely resembling status epilepticus of temporal lobe origin provided evidence that the activity of NADH-cytochrome-c oxidoreductase (NCCR) (complexes I and III) in the hippocampus underwent a significant decrease during prolonged (180 min) kainic acid (KA)-induced status epilepticus, whereas succinate cytochrome-c reductase (SCCR) (complexes II and III) and cytochrome-c oxidase (CCO) (complex IV) remained essentially unchanged. These data suggest that prolonged epileptic seizures result in the dysfunction of complex I (in the mitochondrial electron-transport chain of the hippocampus, resulting in incomplete mitochondrial electron transport and therefore in decreased ATP production.

Furthermore, it was shown that elevated NAM levels occur in response to ultraviolet (UV)-B but only at ultraviolet-B levels high enough to cause oxidative stress, only providing more evidence for elevated NAM levels resulting from oxidative stress *(90)*. Taken together, epileptic neurons and neurons affected by ischemia are subject to oxidative stress, that ultimately elevates NAM levels, and additionally display decreased ATP generation. Conversely, elevated ATP levels would be required either for repair of ischemia-caused damage or for replenishing the energy needed for the excessive firing of epileptic neurons. Neuronal death then might be mitigated by the effects of SIRT1.

Oxidative stress ultimately results in the activation of PARP. However, as indicated above, excessive activity of PARP might be detrimental to cellular function. Augmented activation of

PARP leads to a rapid depletion of NAD$^+$, its sole substrate, accumulation of NAM, inhibiting SIRT1, and furthermore lowers ATP production. Ensuing consumption of ATP by the cell, in an effort to regenerate cellular NAD$^+$ levels, results in a cellular energy crisis that precipitates apoptosis. Interestingly, NAM might offer cytoprotection through a series of pathways that involve both maintenance of PARP integrity and preservation of cellular energy reserves. NAM prevents the degradation of PARP and allows for DNA repair by direct inhibition of caspase-3-like activity *(91–93)*. As discussed above, NAM can also inhibit the activity of PARP within a 6-h time period following the onset of focal cerebral ischemia to prevent excessive energy depletion *(94,95)*. Nevertheless, NAM appears to function as a double-edged sword that can also have detrimental effects not only because PARP activity and energy depletion become significantly increased over a 24 h period as a result of NAM administration, but also because NAM inhibits Sir2 and along with it all its prosurvival effects *(96–98)* (*see also* Figs. 1 and 3).

In summary, there are ample data on Sir2 already available only underlining the importance of investigating this protein. However, future work will have to provide with essential clues on Sir2 regulation including precise NAM levels under various stages of disease and unveil some clues for proper disease treatment.

REFERENCES

1. Sterner DE, Berger SL. Acetylation of histones and transcription-related factors. Microbiol Mol Biol Rev 2000;64:435–459.
2. Fu M, Wang C, Zhang X, Pestell RG. Acetylation of nuclear receptors in cellular growth and apoptosis. Biochem Pharmacol 2004;68:1199–1208.
3. Schwer B, North BJ, Frye RA, Ott M, Verdin E. The human silent information regulator (Sir)2 homologue hSIRT3 is a mitochondrial nicotinamide adenine dinucleotide-dependent deacetylase. J Cell Biol 2002;158:647–657.
4. Dryden SC, Nahhas FA, Nowak JE, Goustin AS, Tainsky MA. Role for human SIRT2 NAD-dependent deacetylase activity in control of mitotic exit in the cell cycle. Mol Cell Biol 2003;23:3173–3185.
5. Frye RA. Phylogenetic classification of prokaryotic and eukaryotic Sir2-like proteins. Biochem Biophys Res Commun 2000;273:793–798.
6. Rine J, Herskowitz I. Four genes responsible for a position effect on expression from HML and HMR in Saccharomyces cerevisiae. Genetics 1987;116:9–22.
7. Guarente L. Sir2 links chromatin silencing, metabolism, and aging. Genes Dev 2000;14:1021–1026.
8. Kaeberlein M, McVey M, Guarente L. The SIR2/3/4 complex and SIR2 alone promote longevity in *Saccharomyces cerevisiae* by two different mechanisms. Genes Dev 1999;13:2570–2580.
9. Tissenbaum HA, Guarente L. Increased dosage of a sir-2 gene extends lifespan in Caenorhabditis elegans. Nature 2001;410:227–230.
10. Rosenberg MI, Parkhurst SM. Drosophila Sir2 is required for heterochromatic silencing and by euchromatic Hairy/E(Spl) bHLH repressors in segmentation and sex determination. Cell 2002;109:447–458.
11. Takata T, Ishikawa F. Human Sir2-related protein SIRT1 associates with the bHLH repressors HES1 and HEY2 and is involved in HES1- and HEY2-mediated transcriptional repression. Biochem Biophys Res Commun 2003;301:250–257.
12. McBurney MW, Yang X, Jardine K, Bieman M, Th'ng J, Lemieux M. The absence of SIR2alpha protein has no effect on global gene silencing in mouse embryonic stem cells. Mol Cancer Res 2003;1:402–409.
13. Afshar G, Murnane JP. Characterization of a human gene with sequence homology to *Saccharomyces cerevisiae* SIR2. Gene 1999;234:161–168.
14. McBurney MW, Yang X, Jardine K, Bieman M, Th'ng J, Lemieux M. The absence of SIR2alpha protein has no effect on global gene silencing in mouse embryonic stem cells. Mol Cancer Res 2003;1:402–409.
15. Frye RA. Phylogenetic classification of prokaryotic and eukaryotic Sir2-like proteins. Biochem Biophys Res Commun 2000;273:793–798.
16. Dryden SC, Nahhas FA, Nowak JE, Goustin AS, Tainsky MA. Role for human SIRT2 NAD-dependent deacetylase activity in control of mitotic exit in the cell cycle. Mol Cell Biol 2003;23:3173–3185.

17. Finnin MS, Donigian JR, Cohen A, et al. Structures of a histone deacetylase homologue bound to the TSA and SAHA inhibitors. Nature 1999;401:188–193.
18. Dryden SC, Nahhas FA, Nowak JE, Goustin AS, Tainsky MA. Role for human SIRT2 NAD-dependent deacetylase activity in control of mitotic exit in the cell cycle. Mol Cell Biol 2003;23:3173–3185.
19. Tanner KG, Landry J, Sternglanz R, Denu JM. Silent information regulator 2 family of NAD-dependent histone/protein deacetylases generates a unique product, 1-O-acetyl-ADP-ribose. Proc Natl Acad Sci USA 2000;97:14,178–14,182.
20. Lin SJ, Defossez PA, Guarente L. Requirement of NAD and SIR2 for life-span extension by calorie restriction in *Saccharomyces cerevisiae*. Science 2000;289:2126–2128.
21. Di Lisa F, Ziegler M. Pathophysiological relevance of mitochondria in NAD(+) metabolism. FEBS Lett 2001;492:4–8.
22. Anderson RM, Bitterman KJ, Wood JG, Medvedik O, Sinclair DA. Nicotinamide and PNC1 govern lifespan extension by calorie restriction in *Saccharomyces cerevisiae*. Nature 2003;423: 181–185.
23. Bitterman KJ, Anderson RM, Cohen HY, Latorre-Esteves M, Sinclair DA. Inhibition of silencing and accelerated aging by nicotinamide, a putative negative regulator of yeast sir2 and human SIRT1. J Biol Chem 2002;277:45,099–45,107.
24. Picard F, Kurtev M, Chung N, et al. Sirt1 promotes fat mobilization in white adipocytes by repressing PPAR-gamma. Nature 2004;429:771–776.
25. Takata T, Ishikawa F. Human Sir2-related protein SIRT1 associates with the bHLH repressors HES1 and HEY2 and is involved in HES1- and HEY2-mediated transcriptional repression. Biochem Biophys Res Commun 2003;301:250–257.
26. Fulco M, Schiltz RL, Iezzi S, et al. Sir2 regulates skeletal muscle differentiation as a potential sensor of the redox state. Mol Cell 2003;12:51–62.
27. Alcendor RR, Kirshenbaum LA, Imai S, Vatner SF, Sadoshima J. Silent information regulator 2alpha, a longevity factor and class III histone deacetylase, is an essential endogenous apoptosis inhibitor in cardiac myocytes. Circ Res 2004;95:971–980.
28. Cheng HL, Mostoslavsky R, Saito S, et al. Developmental defects and p53 hyperacetylation in Sir2 homolog (SIRT1)-deficient mice. Proc Natl Acad Sci USA 2003;100:10,794–10,799.
29. Smith J. Human Sir2 and the "silencing" of p53 activity. Trends Cell Biol 2002;12:404–406.
30. Vaziri H, Dessain SK, Ng EE, et al. hSIR2(SIRT1) functions as an NAD-dependent p53 deacetylase. Cell 2001;107:149–159.
31. Gu W, Luo J, Brooks CL, Nikolaev AY, Li M. Dynamics of the p53 acetylation pathway. Novartis Found Symp 2004;259:197–205.
32. Ito A, Lai CH, Zhao X, et al. p300/CBP-mediated p53 acetylation is commonly induced by p53-activating agents and inhibited by MDM2. EMBO J 2001;20:1331–1340.
33. Appella E, Anderson CW. Post-translational modifications and activation of p53 by genotoxic stresses. Eur J Biochem 2001;268:2764–2772.
34. Avantaggiati ML, Ogryzko V, Gardner K, Giordano A, Levine AS, Kelly K. Recruitment of p300/CBP in p53-dependent signal pathways. Cell 1997;89:1175–1184.
35. Gu W, Shi XL, Roeder RG. Synergistic activation of transcription by CBP and p53. Nature 1997;387:819–823.
36. Lill NL, Grossman SR, Ginsberg D, DeCaprio J, Livingston DM. Binding and modulation of p53 by p300/CBP coactivators. Nature 1997;387:823–827.
37. Tyner SD, Venkatachalam S, Choi J, et al. p53 mutant mice that display early ageing-associated phenotypes. Nature 2002;415:45–53.
38. Frank KM, Sharpless NE, Gao Y, et al. DNA ligase IV deficiency in mice leads to defective neurogenesis and embryonic lethality via the p53 pathway. Mol Cell 2000;5:993–1002.
39. Gu Y, Sekiguchi J, Gao Y, et al. Defective embryonic neurogenesis in Ku-deficient but not DNA-dependent protein kinase catalytic subunit-deficient mice. Proc Natl Acad Sci USA 2000;97:2668–2673.
40. Gao Y, Sun Y, Frank KM, et al. A critical role for DNA end-joining proteins in both lymphogenesis and neurogenesis. Cell 1998;95:891–902.
41. Langley E, Pearson M, Faretta M, et al. Human SIR2 deacetylates p53 and antagonizes PML/p53-induced cellular senescence. EMBO J 2002;21:2383–2396.
42. Luo J, Nikolaev AY, Imai S, et al. Negative control of p53 by Sir2alpha promotes cell survival under stress. Cell 2001;107:137–148.
43. Vaziri H, Dessain SK, Ng EE, et al. hSIR2(SIRT1) functions as an NAD-dependent p53 deacetylase. Cell 2001;107:149–159.

44. Luo J, Nikolaev AY, Imai S, et al. Negative control of p53 by Sir2alpha promotes cell survival under stress. Cell 2001;107:137–148.
45. Luo J, Nikolaev AY, Imai S, et al. Negative control of p53 by Sir2alpha promotes cell survival under stress. Cell 2001;107:137–148.
46. Blandino G, Levine AJ, Oren M. Mutant p53 gain of function:differential effects of different p53 mutants on resistance of cultured cells to chemotherapy. Oncogene 1999;18:477–485.
47. Smith J. Human Sir2 and the "silencing" of p53 activity. Trends Cell Biol 2002;12:404–406.
48. Motta MC, Divecha N, Lemieux M, et al. Mammalian SIRT1 represses forkhead transcription factors. Cell 2004;116:551–563.
49. Brunet A, Bonni A, Zigmond MJ, et al. Akt promotes cell survival by phosphorylating and inhibiting a Forkhead transcription factor. Cell 1999;96:857–868.
50. Nakamura N, Ramaswamy S, Vazquez F, Signoretti S, Loda M, Sellers WR. Forkhead transcription factors are critical effectors of cell death and cell cycle arrest downstream of PTEN. Mol Cell Biol 2000;20:8969–8982.
51. Burgering BM, Kops GJ. Cell cycle and death control:long live Forkheads. Trends Biochem Sci 2002;27:352–360.
52. Dijkers PF, Medema RH, Pals C, et al. Forkhead transcription factor FKHR-L1 modulates cytokine-dependent transcriptional regulation of p27(KIP1). Mol Cell Biol 2000;20:9138–9148.
53. Schmidt M, Fernandez dM, van der HA, et al. Cell cycle inhibition by FoxO forkhead transcription factors involves downregulation of cyclin D. Mol Cell Biol 2002;22:7842–7852.
54. Burgering BM, Kops GJ. Cell cycle and death control:long live Forkheads. Trends Biochem Sci 2002;27:352–360.
55. Stahl M, Dijkers PF, Kops GJ, et al. The forkhead transcription factor FoxO regulates transcription of p27Kip1 and Bim in response to IL-2. J Immunol 2002;168:5024–5031.
56. Kops GJ, Medema RH, Glassford J, et al. Control of cell cycle exit and entry by protein kinase B-regulated forkhead transcription factors. Mol Cell Biol 2002;22:2025–2036.
57. Hribal ML, Nakae J, Kitamura T, Shutter JR, Accili D. Regulation of insulin-like growth factor-dependent myoblast differentiation by Foxo forkhead transcription factors. J Cell Biol 2003;162:535–541.
58. Bois PR, Grosveld GC. FKHR (FOXO1a) is required for myotube fusion of primary mouse myoblasts. EMBO J 2003;22:1147–1157.
59. Nakae J, Kitamura T, Kitamura Y, Biggs WH, III, Arden KC, Accili D. The forkhead transcription factor Foxo1 regulates adipocyte differentiation. Dev Cell 2003;4:119–129.
60. Furukawa-Hibi Y, Yoshida-Araki K, Ohta T, Ikeda K, Motoyama N. FOXO forkhead transcription factors induce G(2)-M checkpoint in response to oxidative stress. J Biol Chem 2002;277:26,729–26,732.
61. Tran H, Brunet A, Grenier JM, et al. DNA repair pathway stimulated by the forkhead transcription factor FOXO3a through the Gadd45 protein. Science 2002;296:530–534.
62. Nemoto S, Finkel T. Redox regulation of forkhead proteins through a p66shc-dependent signaling pathway. Science 2002;295:2450–2452.
63. Stahl M, Dijkers PF, Kops GJ, et al. The forkhead transcription factor FoxO regulates transcription of p27Kip1 and Bim in response to IL-2. J Immunol 2002;168:5024–5031.
64. Murphy CT, McCarroll SA, Bargmann CI, et al. Genes that act downstream of DAF-16 to influence the lifespan of Caenorhabditis elegans. Nature 2003;424:277–283.
65. Nemoto S, Finkel T. Redox regulation of forkhead proteins through a p66shc-dependent signaling pathway. Science 2002;295:2450–2452.
66. Furukawa-Hibi Y, Yoshida-Araki K, Ohta T, Ikeda K, Motoyama N. FOXO forkhead transcription factors induce G(2)-M checkpoint in response to oxidative stress. J Biol Chem 2002;277:26,729–26,732.
67. de Murcia G, Menissier-de Murcia J, Schreiber V. Poly(ADP-ribose) polymerase:molecular biological aspects. Bioessays 1991;13:455–462.
68. de Murcia JM, Niedergang C, Trucco C, et al. Requirement of poly(ADP-ribose) polymerase in recovery from DNA damage in mice and in cells. Proc Natl Acad Sci USA 1997;94:7303–7307.
69. Zhang J. Are poly(ADP-ribosyl)ation by PARP-1 and deacetylation by Sir2 linked? Bioessays 2003;25:808–814.
70. Yang J, Klaidman LK, Nalbandian A, et al. The effects of nicotinamide on energy metabolism following transient focal cerebral ischemia in Wistar rats. Neurosci Lett 2002;333:91–94.
71. Klaidman LK, Mukherjee SK, Adams JD, Jr. Oxidative changes in brain pyridine nucleotides and neuroprotection using nicotinamide. Biochim Biophys Acta 2001;1525:136–148.

72. Hageman GJ, Stierum RH. Niacin, poly(ADP-ribose) polymerase-1 and genomic stability. Mutat Res 2001;475:45–56.
73. Hiratsuka M, Inoue T, Toda T, et al. Proteomics-based identification of differentially expressed genes in human gliomas:down-regulation of SIRT2 gene. Biochem Biophys Res Commun 2003;309: 558–566.
74. Dryden SC, Nahhas FA, Nowak JE, Goustin AS, Tainsky MA. Role for human SIRT2 NAD-dependent deacetylase activity in control of mitotic exit in the cell cycle. Mol Cell Biol 2003;23:3173–3185.
75. North BJ, Marshall BL, Borra MT, Denu JM, Verdin E. The human Sir2 ortholog, SIRT2, is an NAD+-dependent tubulin deacetylase. Mol Cell 2003;11:437–444.
76. Bae NS, Swanson MJ, Vassilev A, Howard BH. Human histone deacetylase SIRT2 interacts with the homeobox transcription factor HOXA10. J Biochem (Tokyo) 2004;135:695–700.
77. Onyango P, Celic I, McCaffery JM, Boeke JD, Feinberg AP. SIRT3, a human SIR2 homologue, is an NAD-dependent deacetylase localized to mitochondria. Proc Natl Acad Sci USA 2002;99: 13,653–13,658.
78. Rose G, Dato S, Altomare K, et al. Variability of the SIRT3 gene, human silent information regulator Sir2 homologue, and survivorship in the elderly. Exp Gerontol 2003;38:1065–1070.
79. Schwer B, North BJ, Frye RA, Ott M, Verdin E. The human silent information regulator (Sir)2 homologue hSIRT3 is a mitochondrial nicotinamide adenine dinucleotide-dependent deacetylase. J Cell Biol 2002;158:647–657.
80. Onyango P, Celic I, McCaffery JM, Boeke JD, Feinberg AP. SIRT3, a human SIR2 homologue, is an NAD-dependent deacetylase localized to mitochondria. Proc Natl Acad Sci USA 2002;99: 13,653–13,658.
81. Di Lisa F, Ziegler M. Pathophysiological relevance of mitochondria in NAD(+) metabolism. FEBS Lett 2001;492:4–8.
82. Schwer B, North BJ, Frye RA, Ott M, Verdin E. The human silent information regulator (Sir)2 homologue hSIRT3 is a mitochondrial nicotinamide adenine dinucleotide-dependent deacetylase. J Cell Biol 2002;158:647–657.
83. Bernardi P. Mitochondrial transport of cations:channels, exchangers, and permeability transition. Physiol Rev 1999;79:1127–1155.
84. Ziegler M, Jorcke D, Schweiger M. Identification of bovine liver mitochondrial NAD$^+$ glycohydrolase as ADP-ribosyl cyclase. Biochem J 1997;326(Pt 2):401–405.
85. Bernardi P. Mitochondrial transport of cations:channels, exchangers, and permeability transition. Physiol Rev 1999;79:1127–1155.
86. Bernardi P, Scorrano L, Colonna R, Petronilli V, Di Lisa F. Mitochondria and cell death. Mechanistic aspects and methodological issues. Eur J Biochem 1999;264:687–701.
87. Vinogradov A, Scarpa A, Chance B. Calcium and pyridine nucleotide interaction in mitochondrial membranes. Arch Biochem Biophys 1972;152:646–654.
88. Onyango P, Celic I, McCaffery JM, Boeke JD, Feinberg AP. SIRT3, a human SIR2 homologue, is an NAD-dependent deacetylase localized to mitochondria. Proc Natl Acad Sci USA 2002;99: 13,653–13,658.
89. Borra MT, O'Neill FJ, Jackson MD, et al. Conserved enzymatic production and biological effect of O-acetyl-ADP-ribose by silent information regulator 2-like NAD+-dependent deacetylases. J Biol Chem 2002;277:12,632–12,641.
90. Landry J, Sternglanz R. Enzymatic assays for NAD-dependent deacetylase activities. Methods 2003;31:33–39.
91. Chong ZZ, Lin SH, Maiese K. Nicotinamide modulates mitochondrial membrane potential and cysteine protease activity during cerebral vascular endothelial cell injury. J Vasc Res 2002;39:131–147.
92. Chong ZZ, Lin SH, Maiese K. The NAD$^+$ precursor nicotinamide governs neuronal survival during oxidative stress through protein kinase B coupled to FOXO3a and mitochondrial membrane potential. J Cereb Blood Flow Metab 2004;24:728–743.
93. Lin SH, Vincent A, Shaw T, Maynard KI, Maiese K. Prevention of nitric oxide-induced neuronal injury through the modulation of independent pathways of programmed cell death. J Cereb Blood Flow Metab 2000;20:1380–1391.
94. Yang J, Klaidman LK, Nalbandian A, et al. The effects of nicotinamide on energy metabolism following transient focal cerebral ischemia in Wistar rats. Neurosci Lett 2002;333:91–94.
95. Klaidman LK, Mukherjee SK, Adams JD, Jr. Oxidative changes in brain pyridine nucleotides and neuroprotection using nicotinamide. Biochim Biophys Acta 2001;1525:136–148.

96. Yang J, Klaidman LK, Nalbandian A, et al. The effects of nicotinamide on energy metabolism following transient focal cerebral ischemia in Wistar rats. Neurosci Lett 2002;333:91–94.
97. Yang J, Klaidman LK, Chang ML, et al. Nicotinamide therapy protects against both necrosis and apoptosis in a stroke model. Pharmacol Biochem Behav 2002;73:901–910.
98. Klaidman LK, Mukherjee SK, Adams JD, Jr. Oxidative changes in brain pyridine nucleotides and neuroprotection using nicotinamide. Biochim Biophys Acta 2001;1525:136–148.

Index

A

A549 cell cultures, 428
Aβ protein precursor (AβPP) in Alzheimer's disease, 299
ABC transporters in MDR, 373–379
ABCG2. *See* breast cancer resistance protein (BCRP) in MDR
Acridine orange in apoptosis detection, 153
Actin in the ECM, 211
Actinomycin-D and apoptosis, 111, 152
Acute lymphoblastic leukemia, 433
Acyclophostin, effect on ER, 106
Addiction and neurogenesis, 331, 332, 339, 343–346
Adenoassociated virus (AAV) vectors, 478–479
Adenophostin-A, effect on ER, 106
Adenovirus, 153, 397, 403, 479–481
Adriamycin, resistance to, 380
AG3340, 405
Age and neurogenesis, 73
Agyria described, 51
AIDS, BRCP in, 378
Akt pathway
    in apoptosis, 74–75, 534
    in cell cycle regulation, 210, 211, 273, 336
    electrical stimulation of, 504–506
    in gliomagenesis, 396–398, 436–437, 450–453, 459
    induction of, 389, 392
    in neuron survival, 267
    oxidative stress in, 287
    in targeted therapy, 404
    and VEGF regulation, 236, 247
Alginate, 465–469
ALSin gene in ALS, 249
Alzheimer's disease
    about, 195–196, 299–301
    amyloid precursor protein (APP) in, 195, 196, 288, 299, 301–304, 362–364
    and apoptosis, 111, 179
    CDK1 expression in, 284, 285
    cell cycle reentry in, 283, 300–304
    cell therapy for, 512, 518
    CRND regulation in, 288–289

DNA damage in, 288
DNA replication in, 284, 290
factors influencing, 361
inflammation in, 286
neurogenesis and, 76, 340, 359–364
neuronal apoptosis in, 170–172
oxidative stress in, 287
therapy
    transplantation, 201
    VEGF, 256, 257
AMPA (α-amino-3-hydroxy-5-methyl-isoxazole-4-propionic acid) receptors in epileptogenicity, 313
Amphetamine and neurogenesis, 344
Amphiregulins, 123, 265
Amyloid-β (Aβ) protein
    in Alzheimer's disease, 363–365
    described, 299, 301–303
    and NSAIDs, 286
Amyloid precursor protein (APP) in Alzheimer's disease, 195, 196, 288, 299, 301–304, 362–364
Amyloidosis, Aβ expression in, 301
Amyotrophic lateral sclerosis (ALS)
    cell cycle reentry in, 283
    cell cycle regulators in, 171
    DNA damage in, 288
    inflammation in, 286
    treatment of, 252–254, 518
    VEGF in, 245, 249–252
Anaphase described, 4
Anaplastic astrocytomas, 430, 434–435, 437
Angiogenesis
    about, 31, 33–34, 222, 320, 495
    angiopoietins in, 35, 37, 38, 222, 325–326
    augmentation of, 227
    in the brain, 33–34
    and electrical activity, 495–496, 498–499, 507
    and endothelial cells, 33, 37–39, 506
    EPCs in, 37, 221–222, 225, 226
    FGF-2 in, 211
    HSCs in, 36–37
    hypoxia and, 37, 38, 212
    Id transcription factors and, 226
    inhibition of, 38–39

in muscle cells, skeletal, 498–499
and nerve regeneration, 255–256
PlGF ligand in, 236, 246
post-ischemic, 240–241
postnatal, 36–38
poststroke, 241
regulation of, 34–39, 209, 235, 245, 506
(*See also* vascular endothelial growth factor [VEGF])
in skeletal muscle, 498–499
in the telencephalon, 33–34
TGF-β in, 34, 35
VEGF in, 31, 32, 34–35, 37–38, 222, 223, 225, 496, 498, 499, 504–506
vs vasculogenesis, 225
in wound-healing, 37, 497
Angiopoietins
in angiogenesis, 35, 37, 38, 222, 325–326
in cerebral ischemia, 239–240
in targeted therapy, 405
Angiostatin in targeted therapy, 405, 406
Animal models, astrocytic tumors, 441–442
Annexin V in apoptosis detection, 148–150
Anthracyclines, resistance to, 374
Antiangiogenesis therapy, 405–406
Anticonvulsant drugs, resistance to, 373, 379–380
Antidepressants and neurogenesis, 332, 341–342
Antiepileptic drugs, resistance to, 373, 379–380
Antigen presenting cells (APCs) in targeted therapy, 403
Antioxidants and hippocampal neurogenesis, 345
Antipsychotics and neurogenesis, 346
Apaf-1
in apoptosis, 74, 110, 144, 145
and HDAC, 287
Apolipoprotein E (ApoE) in Alzheimer's disease, 299, 304
Apoptosis. *See also* cells, senescence
about, 71, 73, 74, 105, 143–145, 178, 310
AIF in, 75, 144, 145
Apaf-1 in, 74, 110, 144, 145
ataxia telangiectasia mutated protein in, 145
Bak gene in, 105, 108
Bax protein in, 105, 108, 144, 145
Bcl2 protein in, 105, 108, 110, 111, 144, 145
Bid protein in, 110
Ca^{2+} channel in, 105–108, 112, 144, 145
calcimycin and, 111
calreticulin and, 106, 107
caspases in, 74, 75, 105, 109–111, 144–146
CDK1 expression in, 285
ceramide and, 111
Cl^- channel in, 109, 111
clofilium and, 112

clotrimazole and, 111
cytochrome-c in, 74, 75, 106, 107, 109, 144, 145
detection of, 14, 145–155 (*See also individual method by name*)
diazoxide and, 113
in the diseased brain, 73–75, 251–252
ERK pathway in, 74
etoposide and, 106, 107
Fas ligand in, 74, 112, 144, 145
glutamate in, 107, 110
growth factors in, 145
Huntington's disease and, 179
hypoxia-induced, 112, 255
induction of, 167, 179
ischemia and, 179
JNK kinase in, 75
K^+ channel, 105, 108–113
levocromakalim and, 113
mitochondria in, 109, 146
models
monkey, 111
mouse, 73–75, 107, 108, 111
rat, 111
Na^+ channel in, 108, 109, 113
neural tube in, 73
neuronal, 74, 169–172, 178, 198
p38 transcription factor in, 75, 112
and p53 expression, 373, 380
p53 transcription factor in, 145, 153
phosphorylation in, 110, 112
pinacidil and, 113
PKC in, 112
regulation of, 64, 323
RTKs in, 112
in Schwann cells, 255, 256
in smooth muscle cells, 111
staurosporine and, 106, 110, 111
stress kinases in, 74–75
TEA and, 111, 112
TNF-α in, 74, 111, 144, 145, 153
transient cerebral ischemia and, 179
UVB and, 107, 110–111
valinomycin and, 108
zinc in, 113
Apoptosis-inducing factor (AIF) in apoptosis, 75, 144, 145
ARF pathway in glioblastoma, 458
Arteriogenesis described, 31
ARX gene in lissencephaly, 51
Astrocytes. *See also* GFAP cells
and brain ECs, 234, 235, 322–323
characterization of, 82, 164, 512
differentiation of, 456
EGFR in, 266–268, 270, 450

genesis of, 15, 24–26, 125, 126–127, 178, 390
MDR in, 381, 382
MRP in, 378, 379
and neurogenesis, 515
P-gp expression in, 375, 379
reactive, 169, 249
role of, 168–169
transformation of, 433, 440–442
and VEGF, 249, 252
Astrocytomas
anaplastic, 430, 434–435, 437
in the cerebellum, 434
classification of, 433–436
development of, 455–458
diffuse, 434
EGFR in, 272, 392, 450
ErbB signaling in, 391–392
malignant, 437
PDGF/PDGFR in, 390, 450, 451
PET imaging of, 424–427, 429–431
pilocytic, 434
progression of, 37, 38, 396
RTK signaling in, 389, 395, 450
SEGA, 434, 436
therapy, 406
Ataxia telangiectasia mutated protein in apoptosis, 145
Atherosclerosis, 213
ATP-binding cassette (ABC) transporters in MDR, 373–379
Avastin (bevacizumab) in targeted therapy, 405
AVMs described, 38

B

Bach 1 and HO-1, 213
Bacterial endotoxin, 382, 401
BAD (BCL2 antagonist of cell death), 285, 437, 453
Bak gene in apoptosis, 105, 108
Balloon cells in MCDs, 313, 315
Band heterotopia described, 47, 48, 51–52
Basement membrane (BM)
described, 233–234
disruption of, 235–239
Batimastat (BB-94) in targeted therapy, 405
Batten's disease, 512
Bax protein
in apoptosis, 105, 108, 144, 145
detection of, 148, 152, 154
Bay12-9566 in targeted therapy, 405
Bcl-xL and STAT, 399
Bcl2 protein
in apoptosis, 105, 108, 110, 111, 144, 145
detection of, 148, 154
in MCDs, 313, 315
and STAT, 399
and thrombospondin, 211

BDNF. *See* brain-derived neurotrophic factor (BDNF)
Behavior and neurogenesis, 361
Benzodiazepines and GABA, 96
Bergmann glia cells, 27
Bevacizumab (Avastin) in targeted therapy, 405
Bicuculline and GABA signaling, 99, 101
Bid protein in apoptosis, 110
Biglycan, expression of, 210
Bilateral perysylvian PMG, 50
Biocompatibility in cell therapy, 466, 467, 469–470
Biosil process, 467–473
Bipolar disorder and neurogenesis, 340
BLBP (*FABP7*) expression, 62, 63, 66
Blood flow and electrical stimulation, 496, 498
Blood–brain barrier (BBB)
about, 221, 226, 233, 322
and ALS treatment, 253
cell transport across, 238–239, 241, 321
drug transport across, 373–379, 401, 419–423, 427
repair of, 227
in spinal cord ischemia, 255, 256
in stroke, 254
vascular differentiation in, 319–323, 326
Bmi1 gene in cell cycle exit, 129–132
Bone marrow as cell source, 223, 226
Bone morphogenetic proteins (BMPs)
in astrocyte differentiation, 433
in astrogenesis, 26
in neurogenesis, 24, 515
in NSC renewal, 121, 123, 132
VEGF and, 252
Bouin's fixative in mitosis counts, 4
Bovine models, MDR, 378
Brain
angiogenesis in, 33–34
blood supply to, 233
cells, visualizing, 334–339
development, 43, 59–60, 131, 165
EGFR in, 266, 268–270
VEGF in, 247
diseased
abnormalities in, 331, 338, 347, 363
apoptosis in, 73–75, 251–252
neurogenesis in, 75–76
endothelial cells, invasion by, 34
environment, regulation of, 168–169
injury
gene/protein expressions in, 239–241
ischemic (*See* stroke)
and neovascularization, 226, 227
and the NV unit, 234, 236–239, 241
TBIs, 200, 283
malformations in, 45–52
neogenesis in, 13–16, 164

repair
    in Alzheimer's disease, 363–364
    and EGFR, 270–271
structure/function and drugs of abuse, 344, 345
tumors (*See* gliomas)
VEGF delivery to, 253
Brain angiogenesis inhibitor 2 (BAI2) in cerebral ischemia, 239
Brain-derived neurotrophic factor (BDNF)
    for ALS, 252, 253
    GABA and, 101
    in neurogenesis, 72, 73, 515
Breast cancer resistance protein (BCRP) in MDR, 377–379
Breast carcinoma
    integrin signaling in, 88
    metastatic mitotic activity in, 6
    PI3-K/Akt signaling in, 396
    therapy, 406
    therapy, targeted, 401
Brefeldin-A, effect on ER, 106
Bromodeoxyuridine (BrdU) labeling
    about, 6–8, 14, 72, 76
    in Alzheimer's disease, 196
    cerebellum, 282
    in CRND, 283–284
    in epilepsy, 198–199, 313, 314
    in neurogenesis studies, 333, 335, 360
    in neuronal migration, 269
    in NSC renewal, 123

C

C-11 thymidine labeling, 420–427
VE-Cadherin in vasculogenesis, 36, 223, 236
Cadherins, 36, 223, 236, 466
Caffeine, effect on ER, 106
Cajal-Retzius neurons, 44
Calcimycin and apoptosis, 111
Calcium (Ca^{2+}) channel
    in apoptosis, 105–108, 112, 144, 145
    in cell cycling, 81–90, 101
Calmodulin in EGFR binding, 265
Calreticulin and apoptosis, 106, 107
CaM kinase II in cell cycle regulation, 85
Canaries, neurogenesis in, 73
Canine models, cell therapy, 470–472
Capillary hemangioblastomas, EGFR in, 272
Capillary tube formation, reproduction of, 235
Carbamazepine, resistance to, 379
Cardiac myocytes, repolarization of, 84
Caspases
    in apoptosis, 74, 75, 105, 109–111, 144–146
    in astrocyte transformation, 437, 453
    detection of, 151–152
    and HDAC, 287
    and STAT, 399
    and Thrombospondin, 211
    and VEGFs, 248
β-Catenin and presenilins, 289
Caveolae described, 87
Caveolin-1 in cell signaling, 87, 88
CBP/p300 protein, 530, 531
CCg cofactor in NPC proliferation, 16
CCI-779 in targeted therapy, 404
CCMs described, 38
CCPs. *See* cell cycle-associated proteins (CCPs)
CD44 in cell cycle regulation, 466
CD68 antibodies in cell proliferation evaluation, 9
CD133 in MCDs, 313, 315
Cdc42 pathway
    in Alzheimer's disease, 300
    in gliomagenesis, 451
CDKN2A locus, 439–440, 454–456, 458
CDKs. *See* cyclin-dependent kinases (CDKs)
Cell cycle
    aberrations in gliomagenesis, 453–454
    about, 3–4, 163–168, 172, 177–179, 323
    assessment methodologies, 6–8, 150
    exit from, 60–65
    ion channels in, 81–88
    in neurons, 281–283
    and the NV, 233–241
    progression
        in Alzheimer's disease, 300–304
        in vitro, 284
        in vivo, 283–284
    proliferation, retinoic acid in, 59–68
    reentry
        in Alzheimer's disease, 283, 300–304
        as disease mechanism, 169–172, 179, 184–185, 323–325
        in Down's syndrome, 283
        mechanisms of, 284–289
        mouse models, 281–283
        and neuronal loss, 289–291
        in Niemann-Pick syndrome type C, 283
        in Parkinson's disease, 283, 323
        in Pick's disease, 283, 323
        in stroke, 283
        in TBIs, 283
    regulation (*See also* amyloid-β[Aβ] protein; cyclin-dependent kinases [CDKs]; p53 transcription factor; retinoblastoma [Rb] protein)
        Akt pathway, 210, 211, 273, 336

bFGF in, 208–211, 466
CaM kinase II in, 85
CD44 in, 466
cell cycle-dependent receptor (CDE) in, 209
checkpoints in, 165–167, 184
chimeric receptor (CHR) in, 209
in the CNS, 164–165, 177–179
cyclin A in, 165, 168, 207–209, 212, 323
cyclin B in, 170, 171, 283, 324
cyclin D, 165, 167, 170–172, 207, 212, 283, 323, 324, 453, 454
cyclin E in, 168, 170, 171, 207, 208, 212, 323, 325
E2F family transcription factors, 168, 171, 177, 208, 209, 282, 324, 438–439, 454
EAG (ether-a-go-go) family in, 84, 90, 111
ECM in, 207–212, 325
endothelial cells, 207–209, 213
ERG channels in, 84
ERK pathway, 210
extracellular matrix (ECM), 207–212, 325, 466
FGFs, 60, 66, 67, 325, 456
GFS, 325–326
growth factors in, 209, 302, 466
GTPases, 466
hERG channels in, 84, 85, 87
HO-1/2 gene in, 213
integrins in, 466
interferon-γ, 208–209
Ki-67 antibody, 170, 172
MAPK pathway, 210, 212
Na$^+$ channel, 89
p14 transcription factor, 454, 455
p15 transcription factor in, 165, 323
p16 transcription factor, 165, 395, 454
p21 transcription factor, 165, 167–168, 177–178, 208, 323, 395
p27 transcription factor, 165, 168, 178, 208, 323–325
pathways, schematic, 324, 438
PCNA, 282
phosphorylation, 207
PI3-K pathway, 210, 211, 273
PKC, 208
Ras pathway, 166
SV40 Tag in, 282
TGF-β, 208–210, 324, 453
TNF-α, 208–209
VEGF, 208, 211, 325–326
volume homeostasis in, 83–85
in the telencephalon, 309–311
Cell cycle-associated proteins (CCPs)

in the cerebellum, 283
in neurological disorders, 281–283, 287
and neuronal death, 290
regulation of, 284–285
Cell cycle-dependent receptor (CDE) in cell cycle regulation, 209
Cell cycle-related neuronal death (CRND)
about, 281–283
DNA replication in, 284, 290–291
regulation of, 288–289
Cells
balloon, 313, 315
electrical stimulation, response to, 504–506
encapsulation of (*See* encapsulation procedure)
grafts, primary, 468
morphology changes in apoptosis detection, 146, 147, 150
nonproliferative, characterization of, 164
proliferation, 341, 419, 498
evaluation of, 8–9
Ki-67 antibody, 9, 14
MIB-1 antibody in, 9, 10, 166
molecular markers, 17, 334–339
PCNA, 9, 14, 197
retrovirus in, 14, 119, 334
and glutamate, 342
growth factors, 342
haloperidol, 342, 346
RVD in, 85
senescence, 130–132, 310 (*See also* apoptosis)
therapy, 465–466, 516
tracking methods, limitations of, 225
visualizing, 334–339
Central nervous system (CNS)
cell cycle regulation in, 164–165, 177–179
cell types in, 168, 511–512
development of, 82, 247, 517
neurogenesis in, 13–17
neuronal migration in, 43–45
P-gp expression in, 375
self-repair in, 200
stem cells in, 23, 512
tumors of, 183–184
vasculature, 319
VEGF in, 245, 247
Central neurocytomas, EGFR in, 272
Ceramide and apoptosis, 111
Cerebellar granule
HB-EGF in, 266
HDAC in, 287
Cerebellum
astrocytomas in, 434
Bmi1 expression in, 129
BrdU labeling of, 282

CCPs in, 283
CDKs in, 165
cell cycle proteins in, 283
development of, 282
GABA in, 99
neurogenesis in, 27, 266
Cerebral amyloid angiopathy, Ab expression in, 301
Cerebral edema, 241
Cerebrospinal fluid (CSF), drug transport within, 373, 374
Cerebrovasculature, embryogenesis of, 221–222
Cerebrovasculogenesis
bFGF in, 223
cytokines in, 223, 224, 227, 236
in gliomas, 226
mouse models, 222, 225, 226, 235, 247
rodent models, 225, 227
thrombomodulin in, 222
UEA-1 binding in, 223
Cetuximab for gliomas, 401
Charybdotoxin and apoptosis, 111
Checkpoints in cell cycle regulation, 165–167, 184
Chemotherapy in tumor control, 39, 373–375
Chimeric receptor (CHR) in cell cycle regulation, 209
Chinese hamster models
apoptosis, 111
EAG channel, 84
MDR, 375
viral vectors, 487
Chlorine (Cl-) channel
in apoptosis, 109, 111
in cell cycling, 81–90
Choroid plexus, MRP expression in, 378
Chromatin condensation in apoptosis detection, 147–149
Chromosomes in prophase, 3–4
CHX10 transcription factor, 129
CI1033 in targeted therapy, 404
Ciliary margin zone, RSCs in, 124–125
Cisplatin, resistance to, 375
CLCA2 channels, integrin activity in, 88
Clofilium and apoptosis, 112
Clotrimazole and apoptosis, 111
CNTF
for ALS, 252, 253
in astrocyte differentiation, 433, 456
cell stimulation by, 123, 126
Cocaine and neurogenesis, 344
Cognition and neurogenesis, 360–363. *See also* learning
Colon carcinomas, therapy, 406
Colorectal cancer, 401, 435, 453

Cortex
genesis of, 309–310, 315–316
HB-EGF in, 266
neurogenesis in, 76
origin of, 24
Cortical dysplasia, 381
Corticosterone
opiates and, 343
and stress, 339
COUP-TF1(NR2F1) gene in neural cell development, 64, 66
CREB pathway in Alzheimer's disease, 304
Creutzfeld–Jacob disease, Aβ expression in, 301
Cyclin A
in Alzheimer's disease, 300
in cell cycle regulation, 165, 168, 207–209, 212, 323
Cyclin B
in Alzheimer's disease, 300, 301
in cell cycle regulation, 170, 171, 283, 324
in neurodegenerative diseases, 285
phosphorylation of, 336
Cyclin C, 300
Cyclin D
in Alzheimer's disease, 300, 301
in astrocytic tumors, 438, 457
β-catenin and, 289
in cell cycle regulation, 165, 167, 170–172, 207, 212, 283, 323, 324, 453, 454
in glioblastoma multiform, 392, 395
in neurodegenerative diseases, 285, 286
and STAT, 399
Cyclin-dependent kinases (CDKs)
in ALS, 171, 286
in Alzheimer's disease, 172, 283, 300–304
in astrocytic tumors, 438, 457
CDK1 in cell detection/visualization, 334, 336
in cell cycle regulation, 207–209, 212, 273, 284–285, 324–325, 453, 454
in the cerebellum, 165
and CRND, 290
in glioblastoma multiform, 392, 395
inhibition of (CDKIs), 168, 301
in mitosis, 3, 164–165, 323
regulation of, 177, 178
in tumorigenesis, 184
Cyclin E
in Alzheimer's disease, 300, 301
in cell cycle regulation, 168, 170, 171, 207, 208, 212, 323, 325
and HDAC, 287
Cyclins
in cell cycle regulation, 164, 170, 177, 207, 323
in NSC proliferation, 124, 130
Cyclopamine in NSC proliferation, 124

Cystatin-C
  in NPC proliferation, 16
  and NSC proliferation, 122
Cytochrome-c
  in apoptosis, 74, 75, 106, 107, 109, 144, 145
  detection of, 152
Cytokines
  in cell signaling, 87, 208–209, 373
  in cerebrovasculogenesis, 223, 224, 227, 236
  in EGFR activation, 270, 271
  in epilepsy, 382
  in immunotherapy, 403
  in MMP induction, 237

D

D54 human glioma cells, 154
Dacron grafts, 222
Dantrolene, effect on ER, 106
DAOY medulloblastoma cell lines, 405
DCX gene
  in cell detection/visualization, 334, 337
  in lissencephaly, 51, 52
  as molecular marker, 73, 196
  in neurogenesis, 76
Decorin, expression of, 210
Delta pathway in NSC proliferation, 126
Deltex and Notch expression, 132
Dementia, frontotemporal, 170. *See also*
    Alzheimer's disease
Dendritic cells in targeted therapy, 403
Dentate gyrus (DG)
  cell function in, 199
  EGFR in, 266, 268
  and epileptic seizures, 198, 200
  neurogenesis in, 13–16, 45, 72, 124, 179,
    196, 310–311, 360
Depression and neurogenesis, 200, 331, 340,
    341, 361, 362
Dexamethasone and VEGF expression, 38
Diabetes
  cell therapy for, 468, 470–473
  and vascular injury, 212–213
  VEGF and, 254–255
Diazoxide and apoptosis, 113
Diffuse astrocytomas, 434
DNA
  damage in inflammation, 287–288
  fragmentation
    in apoptosis detection, 145, 146, 149,
      152–154
    resistance to, 111
  repair, 532, 534
  replication

  in Alzheimer's disease, 284, 290
  in CRND, 284, 290–291
  in neurodegeneration, 284
  PET imaging of, 419–421
DNA polymerase α
  in cell proliferation evaluation, 8–9
  in neuronal death, 290
DNA topoisomerase-II α in cell proliferation
    evaluation, 9
Doublecortin. *See DCX* gene
Down's syndrome
  Aβ expression in, 301
  CDK1 expression in, 284–285
  cell cycle reentry in, 283
  neuronal apoptosis in, 170, 171
Doxorubicin, resistance to, 375, 381

E

E2F family transcription factors
  in cell cycle regulation, 168, 171, 177, 208,
    209, 282, 324, 438–439, 454
  in cell detection/visualization, 334
  in cell senescence, 131–132
  and CRND, 289, 290
  and HDAC, 287
  and VEGF, 249
EAG (ether-a-go-go) family in cell cycle
    regulation, 84, 90, 111
EBV as vector, 487
Ecstasy and neurogenesis, 344
EEG pattern in FCD, 48
Efflux mechanisms in MDR, 373, 374
EGF. *See* epidermal growth factor (EGF)
EGFR. *See* epidermal growth factor receptor
    (EGFR)
EKB-569 in targeted therapy, 404
EKLF gene in neural cell development, 64, 66
Electrical activity
  and angiogenesis, 495–496, 498–499, 507
  in CNS development, 82
  endogenous, 496–498
  and endothelial cells, 499–506
  in neurogenesis, 25, 164, 313–314
  in the SVZ, 98
  in tumor development, 185–186
Electrical stimulation
  Akt pathway, 504–506
  blood flow and, 496, 498
  cellular response to, 504–506
  endothelial cells, 499–506
  muscle cells, smooth, 503–504
  PI3-K pathway, 504–506
  Rho kinase pathway, 504–506
EMD 121974 in targeted therapy, 405

EMX2 gene
  expression profiling of, 63–64, 66, 68
  in NSC renewal, 128–129
  in PMG/schizencephaly, 51
Encapsulation procedure, 465–468, 473; *See also* Biosil process; Sol-Gel process
Endoplasmic reticulum (ER) as Ca^{2+} storage pool, 106–107
Endostatin in targeted therapy, 405, 406
Endothelial cells
  angiogenesis in, 33
  in angiogenesis regulation, 37–39, 506
  biological processes in, 214
  brain invasion by, 34
  cell cycle regulation in, 207–209, 213
  cerebral, about, 233–234
  circulating, 222–223
  differentiation of, 319–320, 326–327
  electrical stimulation of, 499–506
  function of, 207
  and hypoxia, 234–236
  neurogenesis in, 164
  and NSC renewal, 126
  P-gp expression in, 375
  turnover rate of, 222
  vasculogenesis in, 31–32
  and VEGF, 34–36, 246, 252
Endothelial nitric oxide synthetase (eNOS) in EPCs, 223, 225
Endothelial progenitor cells (EPCs)
  in angiogenesis/vasculogenesis, 37, 221–222, 225, 226
  characterization of, 223–225
  origins of, 31
  sources of, 222–223
Environment and neurogenesis, 72
Enzymes in the cell cycle, 130–131, 321, 323
Ependymomas. *See also* gliomas
  EGFR expression in, 272, 392
  PDGF/PDGFR in, 391, 451
Ephrins, 36, 125, 325
Epidermal growth factor (EGF)
  in Alzheimer's disease, 302, 304
  in astrocyte differentiation, 433, 456
  and C-cells, 268
  in cell signaling, 86, 87
  in EGFR binding, 265
  in gliomagenesis, 452, 458
  and HSCs, 518
  in neurogenesis, 75–76, 515, 517
  and NPCs, 16
  and NSC renewal, 120–124, 128, 132
  and NSCs, 252, 513
  in potassium channel modulation, 183
  and Thrombospondin, 211

Epidermal growth factor receptor (EGFR)
  about, 265–266
  in astrocytomas, 435
  in brain development, 266, 268–270
  in brain repair, 270–271
  in cell signaling, 389, 391–392, 396
  developing brain, expression in, 266
  in gliomagenesis, 436, 452, 458
  in gliomas, 272–273, 398, 399
  in immunotherapy, 403, 404
  STAT activation by, 399, 400
Epilepsy
  development of, 168, 179, 311–313
  focal (*See* focal epilepsy, development of)
  and glioma prognosis, 185
  ion channels in, 82
  limbic, 315
  MDR in, 373–375, 379–380
  NAM/NAD+ metabolism in, 535–536
  neurogenesis, postnatal, 197–200
  neuronal viability and p53, 381–382
  and NPC proliferation, 313–315
  and PXAs, 435
  temporal lobe, 170, 198, 199, 313–314, 323, 379
Epithelial cells, 497
ErbB family
  in cell signaling, 391–392, 396
  in gliomagenesis, 451
  in targeted therapy, 401
Erbitux for gliomas, 401
ERG channels in cell cycle regulation, 84
ERK pathway. *See also* MAPK pathway
  in Alzheimer's disease, 300, 304
  in apoptosis, 74
  in cell cycle regulation, 210
  induction of, 395, 396, 398
  ion channels in, 86, 87
  in VEGF regulation, 247–248
Estrogen and vascular injury, 212
Ethanol and neurogenesis, 343, 345
Etoposide, 106, 107, 375
Exercise and neurogenesis, 72, 249, 332, 361
Experimental allergic encephalomyelitis (EAE), VEGF in, 256
Expression profiling described, 60
Extracellular matrix (ECM)
  in cell cycle regulation, 207–212, 325, 466
  hormones and, 212–213
  integrins in, 211, 236–237, 321, 325
  in malignant gliomas, 405
  and NV remodeling, 233, 235–239
  production of, 321
  proteolysis of, 237–238

F

FAK (focal adhesion kinase) pathway in cell signaling, 87, 88, 396
Familial ALS. *See* Amyotrophic lateral sclerosis (ALS)
Familial amyloid polyneuropathy, Aβ expression in, 301
Farnesyl transferase inhibitors (FTIs) in Ras inhibition, 404–405
Fas ligand in apoptosis, 74, 112, 144, 145
Felbamate, resistance to, 379
Fetal cells in transplantation therapy, 201
FG Ad vector, 479, 480
FGF. *See* fibroblast growth factors (FGFs)
Fibroblast growth factors (FGFs)
    bFGF
        in Alzheimer's disease, 302
        in cell cycle regulation, 208–211, 466
        in cerebrovasculogenesis, 223
        and HSCs, 518
        and NSCs, 513, 517
        and VEGF, 252
    in cell cycle regulation, 60, 66, 67, 325, 456
    FGF-2
        in angiogenesis, 211
        and EGFR, 271
        and NPCs, 16
        and NSCs, 252, 267
        regulation of, 210–211
        and VEGF, 249
    in neurogenesis, 75–76
    and NSCs, 120–126, 128, 132
    in vasculogenesis, 31, 35–36
Fibronectin in endothelial growth, 210–212, 325
Fibrosarcoma, therapy, 406
Fixatives/fixation in mitosis counts, 4, 5
FK-506, effect on ER, 106
Flavopiridol, residence to, 378
FLICA assay in apoptosis detection, 151–152
Flk-1
    in vasculogenesis, 31–32, 34, 37, 223, 236
    and VEGF, 245
FLN1 gene in PNH, 49–50
Flow cytometry
    about, 8
    in apoptosis detection, 148–150, 152, 153, 155
Flt-1
    in vasculogenesis, 32, 34, 37, 236
    and VEGF, 245
FLT ([F-18]-3'-deoxy-3'-fluorothymidine) labeling, 420–422, 427–431
Fluorochromes in apoptosis detection, 152, 154
Flux constant, calculation of, 428
Focal brain ischemia. *See* stroke
Focal cortical dysplasia (FCD) described, 46, 48–49
Focal epilepsy, development of, 45, 48, 50
Food restriction and neurogenesis, 364
Forkhead transcription factor family, activation of, 437
FOXO proteins, 528, 530–531, 534
FRAP/mTOR kinase inhibition in targeted therapy, 404
Free radicals. *See* reactive oxygen species (ROS)
Fukutin gene in lissencephaly, 51

G

G protein-coupled receptors (GPCRs) in EGFR binding, 266
$G_0/G_1/G_2$ phase of mitosis described, 3, 164
GABA (γ-aminobutyric acid)
    in the cerebellum, 99
    in FCD, 49
    mechanism of action, 100–101
    in neuronal cell migration, 45, 95–97, 100, 101
    receptor activation, 97–100
    signaling, paracrine, 97–99
GATA family genes in neural cell development, 64, 66, 222
Gaucher's disease, 512
Gel electrophoresis, limitations of, 147
Gene delivery, viral vectors, 477–485, 488
Gene expression
    activation of, 65–68
    in astrocytomas, 391–392
    in glial cells, radial, 62, 129
    in MDR, 373–375, 379–381
    phases of in neurogenesis, 60–65
Gene transfer
    in immunotherapy, 403
    VEGF, 255
Gerbil models, methamphetamine, 344
Germ cell state, regulation of, 60, 124
GFAP cells
    and cocaine, 344
    described, 95, 96, 166, 512
    detection of, 337, 518
    GABA receptor activation in, 96–100
    in gliomas, 456
    in MCDs, 313, 315
    in neurogenesis, 514, 515
    TGF-α in, 393
GIRK channels, integrin activity in, 88
GLAST (*SCL1A8*) expression, 62, 66
Glial cell-derived neurotrophic factor (GDNF)
    for ALS, 252, 253

Glial cells, radial
    described, 311, 456
    formation of, 26–27
    gene expression in, 62, 129
    in neuronal migration, 44, 45, 101
Glioblastoma multiform (GBM)
    about, 435, 450, 453, 455, 458–459
    CDKN2A mutations in, 440
    cell lines derived from, 440, 441
    cell proliferation in, 10, 37, 38
    development of, 434, 436–438, 456–458
    ErbB signaling in, 391–392, 396–398, 452
    JNK pathway in, 395–396
    mouse models, 437, 451, 452
    p53 mutations in, 440, 455
    PDGF/PDGFR in, 390
    PI3-K/Akt signaling in, 396, 404, 437
    Ras activity in, 452
    therapy, targeted, 401, 404, 405
Glioblastomas
    Bmi1 expression in, 130
    EGF/EGFR in, 272, 273, 452
    p53 expression in, 455
    PDGF/PDGFR in, 391, 451
Gliogenesis. *See also* astrocytes; neurogenesis; oligodendrocytes, genesis of
    about, 15, 163–164
    and EGFR expression, 122
    and Notch expression, 126
    in the spinal cord, 16, 17, 24
Gliomagenesis
    about, 391, 396–398, 449–450
    cellular origin of, 455–456
    molecular pathways of, 450–455
Gliomas
    about, 183, 389–393, 406
    anaplastic, recurrent, 425
    cell proliferation in, 9, 10, 37, 38
    cerebrovasculogenesis in, 226
    classification of, 449
    detection of, 148–150, 152
    EGFR in, 272–273
    genesis (*See* gliomagenesis)
    glypican-1 in, 211
    ion channel activity in, 89
    malignant, 405
    MAPK pathway in, 393–396
    mortality, decreasing, 185
    p53 tumor suppressor protein in, 153
    PET imaging of, 419–420, 424–431
    and PLC-γ, 398, 399
    proliferation, PET detection of, 419–430
    right frontal, recurrent, 424
    small molecule therapy for, 404–405
    STAT pathway in, 398–400
    therapy, 143, 154, 155
        targeted, 400–406
Gliosarcoma, 429
Gliosis
    mitotic activity in, 5
    reactive, 164, 166, 168–169
Glucocorticoids and stress, 339, 345
Glucose, 213, 248
Glucuronide conjugates, resistance to, 375
Glutamate
    in apoptosis, 107, 110
    and cell proliferation, 342
    and EGFR, 267
    in gliomas, 90
    receptors in epilepsy, 313
    toxicity, 248
Glutathione conjugates, resistance to, 375
Glypican-1 in glioma, 211
GM-CSF (granulocyte-macrophage colony-stimulating factor) in immunotherapy, 403
Gp130 and NSC renewal, 123–124, 132
Graft vs host disease, 518
Growth and differentiation factor (GDF) 3, 60
Growth factors
    in apoptosis, 145
    in astrocyte transformation, 436–438
    in cell cycle regulation, 209, 302, 466
    and cell proliferation, 342
    receptors and ion channels, 86–88
Gsh2 transcription factor in radial glial cells, 27
Guanosine triphosphatases (GTPases)
    in cell cycle regulation, 466
    in gliomagenesis, 450–452
    and integrins, 87, 88
    in NSC renewal, 130
    PDGF/PDGFR and, 391

H

H19 gene in neural cell development, 64
Haloperidol and cell proliferation, 342, 346
Hamster models. *See* Chinese hamster models
Harlequin (*Hq*) mice, 287, 288
HBGAM (heparin-binding growth-associated molecule), 125
HD Ad vector, 479–481
HDACs. *See* histone deacetylases (HDACs)
HDM2 in p53 regulation, 439, 440
Heart tissue, polarization in, 497
HEK 293 cells, 472
HeLa cells, ion channels in, 82, 85
Hemangioendothelioma, therapy, 406

Hematopoietic stem cells (HSCs)
  in angiogenesis/vasculogenesis, 36–37
  in cell therapy, 512, 513, 517–518
  origins of, 31, 515
  renewal of, 129
  source of, 223
Heparin, 106, 252
Heparin-binding EGF (HB-EGF), 265, 266, 268, 270–271
Heparin sulfate PGs in FGF-2 regulation, 210–211
Hepatocyte growth factor (HGF), 125
Herceptin (trastuzumab) for gliomas, 401
HERG channels in cell cycle regulation, 84, 85, 87
Heroin and neurogenesis, 343–344
Herpes simplex virus, 403, 479, 482–485
Hes-1 gene in NSC proliferation, 126, 127
Hes-5 gene in NSC proliferation, 126
High-power field variation in microscopy, 4–5
Hippocampus
  and addiction disorders, 343–346
  in Alzheimer's disease, 359–365
  in epilepsy, 170, 197–199
  HB-EGF in, 266
  and mood disorders, 340, 346
  neurogenesis in, 13–15, 17, 45, 72, 196, 249, 331–333
  NSCs in, 122–124
  and stress, 339, 340
  VEGF protection of, 248
Histone deacetylases (HDACs)
  classification of, 525
  mechanism of action, 526
  oxidative stress and, 287
Histones in cell detection/visualization, 336
HO-1/2 gene in cell cycle regulation, 213
Homeodomain transcription factors, 128–130
Hormones
  and the ECM, 212–213
  in EGFR activation, 270, 271
  and neurogenesis, 361
HOX gene expression, 63, 66–68
HSV amplicon vectors, 479, 483–487
HUCB-derived stem cells, 518
Human telomerase (hTERT), 392, 452
Huntington's disease
  about, 196–197, 340
  and apoptosis, 179
  transplantation therapy for, 201
  VEGF therapy for, 257
HUVEC cells, 503–506
Hyaluronic acid, 466
Hybridomas, 124
8-Hydroxydeoxyguanosine (8-OHdG), 288

Hypercholesterolemia and vascular injury, 212–213
Hypoxia
  and angiogenesis, 37, 38, 212
  and apoptosis, 112, 255
  in cerebral ischemia, 239
  mouse models, 236
  in neurological diseases, 246–247
  and neuronal death, 289, 290
  and the NV unit, 233–236
  and VEGF, 246–248, 251
Hypoxia-inducible transcription factors (HIFs)
  in EC activation, 234–236
  and VEGF, 246–247, 250

I

IBAC strategy, 485
Id transcription factors and neoangiogenesis, 226
Idoxifene and vascular injury, 212
IGF (insulin-like growth factor) family
  for ALS, 252, 253
  in Alzheimer's disease, 302
  in NSC renewal, 121–123, 132
  in potassium channel modulation, 183
  role of in adult brain, 212
  and Thrombospondin, 211
Image analysis systems in cell proliferation evaluation, 9
IMC-C225 for gliomas, 401
Immunoglobulins, 36, 66, 401
Immunohistochemical methodologies
  in cell detection/visualization, 335
  in CRND, 283–284
  described, 8–10, 153
Immunotherapy, 401–404
In situ histone hybridization described, 8
Infection and circulating EC's, 222
Inflammation
  DNA damage in, 287–288
  in neurodegenerative diseases, 286, 301, 373, 382
ING4 tumor suppressor protein, 440
INK4A pathway, 439–440, 454–456, 458
Inositol, effect on ER, 106
Insulin, 255, 472
Integrins
  in breast carcinoma, 88
  in cell cycle regulation, 466
  in the ECM, 211, 236–237, 321, 325
  GTPases and, 87, 88
  in malignant gliomas, 405
  neutrophils and, 87
  receptors, ion channels in, 87–88
  in vasculogenesis, 36, 236
Interferon-γ, 208–209, 403

Interleukin-1 (IL-1), 382
Interleukin-2 (IL-2), 403
Interleukin-4 (IL-4), 403
Interleukin-6 (IL-6), 323, 382
Interleukin-12 (IL-12), 403
Internal ribosome entry sites (IRESs) in VEGF induction, 246
Interobserver variability in cell proliferation evaluation, 9–10
Ion channels. *See also individual channel by name*
   in cell cycling, 81–85
   integrin receptors in, 87–88
   and signaling pathways, 86–88
   and tumor growth, 89–90
Iressa (ZD-1839) in targeted therapy, 404
Irinotecan (CPT-11) in targeted therapy, 401
Ischemia
   and apoptosis, 179
   brain (*See* stroke)
   cell cycles in, 163, 170–171
   cerebral, cell interactions in, 236–239 (*See also* transient cerebral ischemia)
   and EGFR, 271
   gene/protein expressions in, 239–241
   hindlimb, 255
   NAM/NAD+ metabolism in, 535–536
   polarization in, 497
   resistance to, 74, 75, 112
   semaphorins and, 241
   treatment, 227, 512
   and VEGF, 251
Ischemic peripheral neuropathy, 255
Ivermectin, resistance to, 375

J

Jagged pathway in NSC proliferation, 126
JAK pathway in gliomas, 452
JNK kinase, 75, 86, 395

K

KDR pathway, 236, 245
Kennedy's disease. *See* spinal and bulbar muscular atrophy (SBMA)
Ki-67 antibody
   in Alzheimer's disease, 300, 302
   in cell cycle regulation, 170, 172
   in cell detection/visualization, 334–336
   in cell proliferation evaluation, 9, 14
Kit ligand and EPCs, 225

L

Laminins in cerebral ischemia, 239
Lamotrigine, resistance to, 379
Laser scanning cytometry (LSC) in apoptosis detection, 150, 152–155
Lateral ganglionic eminence (LGE) in neuronal migration, 44, 45
Learning. *See also* cognition and neurogenesis
   and neurogenesis, 72, 360, 361, 364
   and stress, 339
Lentivirus
   and ALS, 254
   as vector, 479, 481, 485
Leukemia-inducing factor (LIF)
   in astrocyte differentiation, 433
   and NSCs, 120, 123–124
   VEGF and, 252, 253
Levocromakalim and apoptosis, 113
Lewy body disease, 323
Li-Fraumeni syndrome, 433
LIF. *See* leukemia-inducing factor (LIF)
Light scatter in apoptosis detection, 146, 148–149
LIS gene in lissencephaly, 51, 52
Lissencephaly described, 47, 48, 51–52
Lithium and cell proliferation, 341
Lithium and neurogenesis, 201
Locomotion described, 44
Lou Gehrig's disease. *See* Amyotrophic lateral sclerosis (ALS)
Luciferase (Luc) in apoptosis detection, 154
Lung carcinomas, therapy, 406
Lurcher mice, 282, 287, 323
LY294002 in cell signaling inhibition, 393, 404
Lymphocyte proliferation, ion channels in, 85

M

M14 melanoma cells, 395
MAb 806 in targeted therapy, 401
MAb Y10 in targeted therapy, 401
Macaque cortex, neurogenesis in, 72
Magnetic resonance imaging (MRI) in tumor imaging, 419, 420, 424, 427
Magnetic resonance spectroscopy (MRS), 420
Malformations of cortical development (MCDs), 43, 45–52, 311–313
MAO inhibitors and cell proliferation, 341
MAP2 microtubule markers and VEGF, 248
MAPK pathway. *See also* ERK pathway
   in Alzheimer's disease, 300, 302–304
   in cell cycle regulation, 210, 212
   and EGFR, 272
   GABA in, 101
   in gliomas, 393–396
   and glucose, 213
   induction of, 389, 394, 396, 450, 452
   inhibition of in targeted therapy, 404–405
   ion channels in, 86, 87
   in neuron survival, 267, 271
   in NSC proliferation, 121, 123–125, 132
   oxidative stress in, 287
Marijuana. *See* tetrahydrocannabinoid (THC) and neurogenesis

Marimastat in targeted therapy, 405
Matrix metalloproteinases (MMPs)
    in ECM proteolysis, 237, 239
    and EGFR, 270
    in malignant gliomas, 405, 438
    in vasculogenesis, 36, 323
MDM2 protein, 458, 530
Medial ganglionic eminence (MGE) in neuronal migration, 44, 45
Medulloblastomas
    Bmi1 expression in, 130
    EGFR expression in, 392, 393
    PDGF/PDGFR in, 391, 451
    and STAT, 399
MEIS1/2 gene in neural cell development, 64
MEK pathway
    in Alzheimer's disease, 304
    induction of, 389, 393, 394, 450, 452
    in VEGF regulation, 247–248
Melanoma, therapy, 406
Memory
    in Alzheimer's disease, 360, 362, 364
    and heroin, 344
    and neurogenesis, 200, 333, 338
    radiation and, 72
    and stress, 339
Meningiomas, 5, 272
Mercury fixatives in mitosis counts, 4
Mesenchymal stem cells (MSCs)
    in neovascularization, 227
    in neurogenesis, 516
    in therapy, 518–519
MEST (PEG1) gene in neural cell development, 64
C-MET in NSC proliferation, 125
Metallospondin in targeted therapy, 405
Metaphase described, 4
Methamphetamine and neurogenesis, 344
Methotrexate, resistance to, 375, 378
MIB-1 antibody in cell proliferation evaluation, 9, 10, 166
Microglia cells
    MRP in, 378, 379
    and NSAIDs, 286
    origin of, 27
Microscopy, high-power field variation in, 4–5
Midkine (*MDK*) gene expression, 63, 66
Migration. *See* neuronal migration
Miller-Dieker syndrome, 51
Miller glia cells, 27
Mineralocorticoid and stress, 339
Minocycline in ALS, 286
Mitochondria
    in apoptosis, 109, 146
    ATP production in, 535
    as Ca^{2+} storage pool, 106–107
    changes, studying, 148, 150, 152, 422

Mitosis
    counts, 4–6
    and development, 82
    ion channels in, 85
    NSCs, 126
    and opiates, 345
    perceptions of, 163
    phases of, 3–4, 164, 323
    regulation of, 130, 131
Mitoxantrone, residence to, 378
Molecular markers. *See also* individual marker by name
    Alzheimer's disease, 302
    in apoptosis detection, 146, 152
    in cell proliferation evaluation, 17, 334–339
    DCX gene, 73, 196
    Huntington's disease, 196–197
MoMuLV-based vector, 479, 481, 485
Monkey models, apoptosis, 111
Monoclonal antibodies (MAbs) in targeted glioma therapy, 401
Monocytes, peripheral blood, 223–225, 227
Mood disorders and neurogenesis, 340–342
Morphine and neurogenesis, 343–344
Mossy fiber (MF) sprouting, 198, 199
Motor neuron degeneration. *See also* Neurodegeneration
    HIFs in, 247
    and VEGF, 250–252
Mouse models
    ALS, 249–254, 286
    Alzheimer's disease, 196, 362, 363
    anaplastic astroctyomas, 437
    apoptosis, 73–75, 107, 108, 111
    cell cycle reentry, 281–283
    cerebrovasculogenesis, 222, 225, 226, 235, 247
    DNA damage, 288
    EGFR, 268, 272, 392
    GABA signaling, 97
    GBM, 437, 451, 452
    Huntington's disease, 196
    hypoxia, 236
    IGF-1, 212
    MAb therapy, 401
    MDR, 375, 378–381
    neurogenesis, 333, 337–339, 360
    NSC renewal, 122–132
    potassium channels, 180
    PTEN expression, 398, 456
    retinoblastoma, 289
    small molecule therapy, 405
    spinal cord ischemia, 255
    TGF-α, 266, 393
    volume homeostasis, 85
MTOR, activation of, 437

Müller cells
  characterization of, 82
  in NSC renewal, 126, 132
Multidrug resistance-associated protein 1 (MRP1)
  in MDR, 373, 375–379
  and p53 expression, 381
  substrates of, 382
Multidrug resistance (MDR)
  described, 373–374
  in epilepsy, 379–380
  p53 in, 373, 380–381
  transporters, 373–379
Multiple-hit hypothesis, 289–290
Multiple myeloma, Aβ expression in, 301
Multiple sclerosis
  transplantation therapy for, 201, 518
  VEGF therapy for, 256–257
Multipolar migration described, 44
Musashi family, 128, 518
Muscimol and GABA, 101
Muscle cells
  skeletal, angiogenesis in, 498–499
  smooth
    apoptosis in, 111
    electrical stimulation of, 503–504
    ion channels in, 85
    vascular, 212
  stem cells in, 512
Mutations
  in astrocytomas
    p53, 434, 435, 457
    PDGF, 434, 435
    PTEN tumor suppressor, 435, 437, 457
    retinoblastoma (Rb) protein, 435, 457
  CDKN2A, 440
  in cell regulatory pathways, 454
  in oligodendrocytes, 456
  in oncogenesis, 449, 453–454
  p53
    in astrocytomas, 434, 435, 457
    GBM, 440, 455
  PIK3CA, 453
  SOD1, 171, 249–253, 286, 518
Myc gene, 167, 399
C-*Myc* in Alzheimer's disease, 300, 303
N-*Myc* in NSC proliferation, 124, 132
Myocardial infarction, etiology, 222

N

NANOG in germ cell regulation, 60, 66
Nasopharyngeal carcinoma, 424
NCAM-1 gene, expression profiling of, 63, 65, 66
Necrosis in the cell cycle, 178

Nerve growth factor (NGF)
  in Alzheimer's disease, 302, 304
  in cell signaling, 86
  and VEGF, 255
Nestin
  cell differentiation, assessment of, 337–338
  in EGFR expression, 266
  in NSC proliferation, 130, 132
  in radial glial cells, 27, 313, 315
Netrin-1 and oligodendrocyte formation, 25–26
Neural progenitor cells (NPCs)
  about, 15–16, 23
  in Alzheimer's disease, 196
  development of, 178, 309–310
  in disease therapy, 201
  and EGFR, 268–270
  epilepsy and, 313–315
  GABA receptor activation in, 96–100
  PDGF/PDGFR in, 451
  proliferation, 16, 313–315
  radiation damage to, 72
  VEGF in, 248–249
Neural stem cells (NSCs)
  in cell therapy, 477, 511–512, 517, 520
  characterization of
    in vitro, 513–514
    in vivo, 514–515
  CHX10 transcription factor, 129
  in the CNS, 23, 195
  described, 13, 15–16, 119
  development of, 16–17, 120–121, 132, 178
  differentiation of, 433, 455–456, 515–516
  and EGFR, 268–270
  FGFs and, 252, 267, 513, 517
  Notch pathway, 121, 126–128, 132
  proliferation, cyclins in, 124, 130
  proliferation, MAPK pathway, 121, 123–125, 132
  proliferation, PI3-K pathway, 123, 124, 132
  renewal of
    about, 119–120
    cell cycle machinery and, 131–132
    control, intrinsic, 128–131
    epigenetic factors in, 122–126
  STAT pathway, 132
  transforming growth factor-α, 123, 515
  in transplantation therapy, 201–202
  and VEGF, 247, 249, 252
Neural tube, 32, 73
Neuroblastoma, EGFR in, 272
Neurodegeneration
  in Alzheimer's disease, 299–301
  CCPs in, 283
  DNA replication in, 284

features of, 285–289
inflammation in, 286, 301, 373, 382
motor neuron, 247, 250–252
neuronal loss in, 285–286
oxidative stress in, 286–287
stresses, secondary, 290
and VEGF, 250
Neurofibrillary tangles (NFTs) described, 299
Neurofibromatosis 1/2, 433
Neurogenesis
about, 23–24, 59–60, 67–68, 71–72, 74, 163–164, 178–179, 309–310, 512
age and, 73
in Alzheimer's disease, 76, 340, 359–364
amphetamine and, 344
antidepressants and, 332, 341–342
antioxidants and, 345
antipsychotics and, 346
astrocytes and, 515
BDNF in, 72, 73, 515
behavior and, 361
BMPs in, 24, 515
and brain function, 333
BrdU labeling of, 333, 335, 360
in the cerebellum, 27, 266
characterizing, 14, 515
in the CNS, 13–17
cocaine and, 344
cognition and, 360–363
in the cortex, 76
DCX gene in, 76
in the dentate gyrus, 13–16, 45, 72, 124, 179, 196, 310–311, 360
depression and, 200, 331, 340, 341, 361, 362
in the diseased brain, 75–76, 200, 201
ecstasy and, 344
EGF in, 75–76, 515, 517
and EGFR, 268–270
electrical activity in, 25, 164, 313–314
endogenous, induction of, 516–517
in endothelial cells, 164
environment and, 72
ethanol and, 343, 345
exercise and, 72, 249, 332, 361
factors influencing, 72, 361
FGFs in, 75–76
food restriction and, 364
function of, 72
GABA limitation of, 99
gene expression, phases of, 60–65
GFAP cells in, 514, 515
heroin and, 343–344
in the hippocampus, 13–15, 17, 45, 72, 196, 249, 331–333

hormones and, 361
inhibition of, 518
learning and, 72, 360, 361, 364
lithium and, 201
markers for, 72–73
memory and, 200, 333, 338
models
canaries, 73
mouse, 333, 337–339, 360
primate, 310–311, 316, 333, 340, 360
rat, 75–76, 333, 340, 360
rodent, 269, 310–311, 514
songbirds, 73, 126
zebra finches, 73
mood disorders and, 340–342
morphine and, 343–344
MSCs in, 516
neurotrophic factors in, 517
nicotine and, 343–345
in the olfactory bulb, 13–15, 45, 332
opiates and, 343–344, 345
in Parkinson's disease, 200, 252, 340
PDGF in, 25
postnatal, 195, 310–313, 316
in epilepsy, 197–199
hippocampal, 13–15, 17, 45, 72, 196, 249, 331–333, 359–365
opium-induced, 343–344
and radiation, 199
and rostral migration, 332, 514
schizophrenia and, 331, 332, 339, 346–347
serotonin in, 72, 342
sonic hedgehog (SHH) in, 24–26
in spinal cord, 13, 15, 24
stimulants and, 343, 344
stress and, 362
stroke and, 75–76, 200
in the subgranular zone, 14, 15, 71, 72, 75–76, 332
in substantia nigra, 15, 197, 271
in the subventricular zone, 14, 15, 17, 45, 71, 72, 75–76, 179, 196, 310–311
in TBIs, 200
TGF-β in, 24
THC and, 343, 345
thymidine labeling of, 332–334
and VEGF, 72, 247, 248–249, 252
in the visual cortex, 72
Neurological diseases
about, 195–197
cell function in, 199–200
hypoxia in, 246–247
therapy for, 200–201
Neuron-restrictive silencing factor (NRSF), regulation of, 65

Neuronal migration. *See also* rostral migration
   BrdU labeling of, 269
   in the CNS, 43–45
   described, 43, 72, 514
   failures in, 49
   and GABA, 45, 95–97, 100, 101
   glial cells, radial, 44, 45, 101
   LGE in, 44, 45
   mechanisms of, 43–45, 209, 268, 269, 500, 503–504
   MGE in, 44, 45
   and VEGF, 249
Neurons
   about, 511–512
   apoptosis in, 74, 169–172, 178, 198
   Cajal-Retzius, 44
   cell cycle in, 281–283
   CRND (*See* cell cycle-related neuronal death [CRND])
   degenerating, CCPs in, 284–285
   and EGFR, 267
   genesis of (*See* neurogenesis)
   hypoxia in, 289, 290
   loss of and cell cycle reentry, 289–291
   MAPK pathway and, 267, 271
   migration (*See* neuronal migration)
   NER activity in, 288
   regeneration of, 255–256, 516
   survival, Akt pathway in, 267
   and VEGF, 247–248
   viability of in epilepsy, 381–382
Neuropilins, expression of, 72, 236, 246, 247
Neurotransmitters in EGFR activation, 271
Neurotrophic factors, 302, 517
Neurovascular (NV) unit
   about, 233, 241–242
   brain injury alterations to, 234, 236–239
   in hypoxia, 233–236
Neutrophils, 87, 237–238
NF-κB pathway
   and Akt, 453
   in apoptosis detection, 153
   and atherosclerosis, 213
   in cell signaling, 87, 209, 382
   oxidative stress in, 287
NF1 (neurofibromatosis type-I) expression in gliomas, 392, 395
Nicotinamide (NAM) and sirtuin family proteins, 526, 528, 530, 532, 534–536
Nicotine and neurogenesis, 343–345
Niemann-Pick syndrome type C, 283
NIH3T3 cells, 391, 395
NMDA (*N*-methyl-D-asparate) receptors in epileptogenicity, 313, 361

Noggin in NSC renewal, 123
Notch pathway
   in Alzheimer's disease, 289, 303
   in gliogenesis, 24
   in NSC renewal, 121, 126–128, 132
   in vascular differentiation, 321
Novartis (PKI166) in targeted therapy, 404
NPCs. *See* neural progenitor cells (NPCs)
NRTIs (nucleoside reverse transcriptase inhibitors), resistance to, 378
NSAIDs (nonsteroidal anti-inflammatory drugs) and Alzheimer's disease risk, 286
NSCs. *See* neural stem cells (NSCs)
NT2 cells, gene expression in, 60–68
Nucleostemin in NSC renewal, 130
Numb and Notch expression, 127, 128, 132
Numblike and Notch expression, 127, 128

O

Occludin in BBB degradation, 239
N-OCT3 (*POU3F2, brn2*) gene in neural cell development, 64, 67
OCT4 (*POU5F1*) in germ cell regulation, 60, 66, 67
Olfactory bulb (OB)
   cell function in, 199
   neurogenesis in, 13–15, 45, 332
Olig2 transcription factor in radial glial cells, 27
Oligoastrocytomas
   classification of, 434
   EGFR in, 392
   PDGF/PDGFR in, 390, 451
Oligodendrocyte precursor cells (OPCs)
   in astrogenesis, 26
   in gliogenesis, 24, 25, 456
Oligodendrocytes
   about, 512
   genesis of, 15, 24–26, 65, 126–127, 178, 390
   mutations in, 456
Oligodendrogliomas
   EGFR in, 272, 392
   origins of, 455
   PDGF/PDGFR in, 390, 451
   PET imaging of, 424–427
Oncogenes in tumor development, 165–167
Oncogenesis described, 449
Oocytes, ion channels in, 85, 88
OPCs. *See* oligodendrocyte precursor cells (OPCs)
Opiates and neurogenesis, 343–344, 345
Optic nerve, oligodendrocyte formation in, 25–26
OSI-774 (Tarceva) in targeted therapy, 404
Osteonectin. *See* SPARC
Ovarian carcinomas, therapy, 406
Oxidative stress in neurodegenerative diseases, 286–287, 303, 304

Index

P

P-glycoprotein (P-gp)
  in MDR, 373–376
  substrates of, 378, 379, 381, 382
P14 transcription factor
  and CDKN2A deletions, 458
  in cell cycle regulation, 454, 455
  in cell senescence, 131
  mutations in astrocytomas, 435, 438, 440
P15 transcription factor in cell cycle regulation, 165, 323
P16 transcription factor
  in Alzheimer's disease, 300, 302, 304
  and CDKN2A deletions, 458
  in cell cycle regulation, 165, 395, 454
  in cell senescence, 131
  in glioblastoma multiform, 392
P19 transcription factor, 132, 392
P21 transcription factor
  in Alzheimer's disease, 300–304
  in cell cycle regulation, 165, 167–168, 177–178, 208, 323, 395
  in cell senescence, 131
  and dm-LDL, 213
  and glucose, 213
  induction of, 455
  and STAT, 399
P27 transcription factor
  in Alzheimer's disease, 300, 303
  in cell cycle regulation, 165, 168, 178, 208, 323–325
  in cell senescence, 131
  and glucose, 213
  and PTEN, 397
P34 transcription factor, 300, 336
P38 transcription factor
  in Alzheimer's disease, 300, 304
  in apoptosis, 75, 112
P53 transcription factor
  and Akt, 453
  in Alzheimer's disease, 303
  in apoptosis, 145, 153
  in cell cycle regulation, 165, 167, 177, 323, 438–440, 453–455
  in cell senescence, 131
  in glioblastoma multiform, 392, 458
  in MDR, 373, 380–382
  mutations in astrocytomas, 434, 435, 457
  oxidative stress and, 287
  PDGF/PDGFR and, 390, 451
  and SIRT1, 529–530
  and sirtuin proteins, 528
P105 protein, 9, 300

P107 in Alzheimer's disease, 300
PACAP (pituitary adenylate cyclase-activating polypeptide), 125
Pachygyria described, 51
Pancreatic islets, encapsulated, 470
Parkin expression, 171
Parkinson-amyotrophic lateral sclerosis of Guam, neuronal apoptosis in, 170
Parkinson's disease
  about, 197
  cell cycle reentry in, 283, 323
  cell cycle regulators in, 170, 171
  cell therapy for, 512, 518
  inflammation in, 286
  neurogenesis in, 200, 252, 340
  oxidative stress in, 287
  VEGF therapy for, 256, 257
PARP
  in apoptosis detection, 146, 151, 154–155
  dependency of, 528
  and sirtuin family proteins, 532, 534, 536
Patched (ptc) receptor in NSC proliferation, 124
PAX6 gene
  expression profiling of, 63–64, 66–68
  in NSC renewal, 129
  in radial glial cells, 27
PBX1 gene expression, 66
PDGF. *See* platelet-derived growth factor (PDGF)
PEG1,3,10 gene in neural cell development, 64
PEG3 gene in neural cell development, 64
PEG10 gene in neural cell development, 64
Pericytes
  described, 234
  regulation of, 235, 321–322
Peripheral blood monocytes, 223–225, 227
Periventricular nodular heterotopia (PNH) described, 46, 48–50
Perlecan in FGF signaling, 210–211
PH in mitosis counts, 4, 153
Phenobarbital, resistance to, 379
Phenytoin, resistance to, 379
Phospholipase C (PLC)-γ, 391, 398, 399
Phosphorylation
  in apoptosis, 110, 112
  in cell cycle regulation, 207
  of ion channels, 86, 88
  of VEGF, 246
PI3-K pathway
  in astrocyte transformation, 437
  in cell cycle regulation, 210, 211, 273
  and EGFR, 396, 397
  electrical stimulation of, 504–506

in GBM, 396, 404, 437
in gliomagenesis, 396–398, 450–453, 459
induction of, 389, 391, 392
in NSC proliferation, 123, 124, 132
oxidative stress in, 287
in targeted therapy, 404
and VEGF regulation, 236, 247
Pick's disease
CDK1 expression in, 285
cell cycle reentry in, 283, 323
neuronal apoptosis in, 170, 171
PIK3CA mutations in gliomagenesis, 453
Pilocarpine in SE induction, 315
Pilocytic astrocytomas, 434
Pinacidil and apoptosis, 113
Pituitary tumors, 433
Pixel values, measurement of, 150
PKI166 (Novartis) in targeted therapy, 404
Plasma membrane changes in apoptosis detection, 146–148, 152
Plasminogen activator (PA) in ECM proteolysis, 237
Platelet-derived growth factor (PDGF)
in Alzheimer's disease, 302
in cell signaling, 86, 389–391
in EGFR binding, 267
in gliomagenesis, 436–437, 450–451, 456, 457
mutations in astrocytomas, 434, 435
in neurogenesis, 25
STAT activation by, 400
in targeted therapy, 404, 405
in vasculogenesis, 32, 34, 36
Platelet-derived growth factor receptor (PDGFR)
in gliomagenesis, 436–437, 450–451, 457
Pleomorphic xanthoastrocytoma (PXA), 434, 435
Plexin, expression of, 72
PlGF ligand in angiogenesis, 236, 246
PMN domains in gliogenesis, 24–26
PNET/MBs, TGF-α in, 393
Polymicrogyria (PMG) described, 46, 48, 50–51
Polyoma virus replicon, 487
Polysialylated adhesion molecules in AD, 76, 196
Positron emission tomography (PET) in glioma detection, 419–430
Potassium (K^+) channel
in apoptosis, 105, 108–113
in cell cycling, 81–90, 169, 177, 179–180, 186–187, 248
homeostasis, regulation of, 108
modulation of, 182–183, 185–186
in tumorigenesis, 184
Presenilins
in Alzheimer's disease, 195, 196, 299, 303–304, 362
as CRND regulators, 288–289
in memory/neurogenesis relationships, 338

Primate models
cerebral ischemia, 237
neurogenesis, 310–311, 316, 333, 340, 360
Proliferating cell nuclear antigen (PCNA)
in Alzheimer's disease, 283, 300, 302
in cell cycle regulation, 282
in cell detection/visualization, 334, 335, 340
in cell proliferation evaluation, 9, 14, 197
Prophase, chromosomes in, 3–4
Propidium iodide in apoptosis detection, 149–150, 152
Propiomelanocortin in cell differentiation assessment, 338–339
Prostate cancer, therapy, 406
Protein acetylation, reversible, 525
Protein kinase C (PKC)
in apoptosis, 112
in astrocyte transformation, 436–438
in cell cycle regulation, 208
induction of, 398
and mTOR activity, 436
in potassium channel modulation, 183, 186
Proteoglycans (PGs) in endothelial growth, 210
Proteomics analysis, cerebral ischemia, 239, 240
Prox1 gene in cell cycle exit, 129
Psychiatric disorders, brain structure abnormalities in, 331, 338, 347, 363
PTB (phosphotyrosine-binding) domains in EGFR binding, 265
PTEN tumor suppressor
in cell senescence, 132
and EGFR, 392, 456, 458
and mTOR activity, 436
mutations in astrocytomas, 435, 437, 457
PI3-K/Akt signaling and, 396–398, 404, 451–453, 459
Purkinje cells
HB-EGF in, 266
loss of, 287
SV40 Tag in, 282

R

R115777 in targeted therapy, 405
Rabbit models, ischemic peripheral neuropathy, 255
Rac-1 in cell signaling, 88
Rac-Rho pathway in gliomagenesis, 452
Radial unit hypothesis, 312
Radiation
and astrocyte transformation, 433
and memory, 72
neurogenesis and, 199
Radionecrosis, 424, 426, 428, 430
Raf pathway, induction of, 389, 393, 394, 450, 452
RalGEFs in glioma, 395

Rapamycin in targeted therapy, 404
Ras pathway
   in Alzheimer's disease, 300, 302, 304
   in cell cycle regulation, 166
   and EGFR, 272, 392
   in gliomagenesis, 436–437, 450–452
   induction of, 389, 391, 393–395, 451
   inhibition of in targeted therapy, 404–405
Rat C6 cells, 442
Rat models
   apoptosis, 111
   astrocytic tumors, 442
   CDK1 expression, 336
   cerebral ischemia, 239, 241
   EAG channel, 84
   epilepsy, 200, 313–315, 379, 382
   ethanol, 345
   gliosarcoma, 406
   idoxifene, 212
   immunotherapy, 403
   ion channels, 86
   MDR, 378, 379
   neurogenesis, 75–76, 333, 340, 360
   nicotine, 344–345
   Notch expression, 126
   NSC proliferation, 122, 123, 129
   Parkinson's disease, 197, 256
   stimulants, 344
   VEGF, 248–249, 255
Rat1 fibroblasts, 395
RC2 antigen in radial glial cells, 27
Reactive gliosis, 82
Reactive oxygen species (ROS)
   in diabetes, 213
   in neurodegenerative diseases, 286, 304
Receptor tyrosine kinases (RTKs)
   about, 265, 389–390
   in apoptosis, 112
   in astrocytomas, 389, 395, 450
   in cell signaling, 86, 88, 183, 325–326, 393, 394
   in glioma therapy, 400–406
   in gliomagenesis, 436–438, 450–453, 459
   and mTOR activity, 436
   in NSC renewal, 122–123
   small molecule therapy and, 404–405
   in vasculogenesis, 31, 35, 236
Reelin gene in lissencephaly, 51
Regulatory volume decrease (RVD) in cell proliferation, 85
Renal cancer, therapy, 406
Repamycin, effect on ER, 106
Replication-defective vector, 479, 482–483
Resistance to therapy, 374–382
REST, regulation of, 65, 66

Resting (G0) phase of mitosis described, 3
Resting membrane potential (RMP) in tumor development, 182–184
Retinal stem cells (RSCs)
   described, 119, 122, 125
   proliferation of, 129
Retinoblastoma (Rb) protein
   in ALS, 286
   in Alzheimer's disease, 300, 303
   in cell cycle regulation, 165, 168, 171, 177, 208, 282, 289, 324, 438–439, 453, 454
   in cell detection/visualization, 334–335
   in cell senescence, 131, 132
   in glioblastoma multiform, 392, 459
   and HDAC, 287
   inhibition of, 455, 458
   mutations in astrocytomas, 435, 457
   in tumorigenesis, 184
Retinoic acid in cell cycle proliferation, 59–68
Retrovirus
   in cell proliferation evaluation, 14, 119, 334
   as cell therapy vector, 479, 481
Reward pathway in drug abuse/addiction, 342
RGDs in targeted therapy, 405
Rho kinase pathway, electrical stimulation of, 504–506
Riluzole for ALS, 252
Rodent models. *See also* mouse models; rat models
   CCP regulation, 284–285
   cerebrovasculogenesis, 225, 227
   CNS repair, 200, 270
   MDR, 375, 378–381
   neurogenesis, 269, 310–311, 514
   oxidative stress, 287
   volume homeostasis, 85
Rodent repairadox, 288
Rostral migration
   described, 45, 72, 95, 310
   and EGFR, 269
   neurogenesis and, 332, 514
RTKs. *See* receptor tyrosine kinases (RTKs)
Rx gene, 129
Ryanodine, effect on ER, 106, 107

S

S phase of mitosis
   described, 3, 164
   evaluating, 6–8, 14
   progression, requirements of, 290
SCH66336 in targeted therapy, 405
Schizencephaly described, 47, 48, 50–51
Schizophrenia and neurogenesis, 331, 332, 339, 346–347

Schwann cells
    apoptosis, prevention of, 255, 256
    in CNS development, 247
    and Notch expression, 126
Seizures, kainite-induced, 170, 198, 199, 266. *See also* epilepsy
E-Selectin in vasculogenesis, 36
Selectins, 36, 466
Selective serotonin reuptake inhibitors and cell proliferation, 341
Semaphorins, expression of, 72, 246, 247
Semaphorins and ischemia, 241
SERCA pump in Ca^{2+} movement, 106, 107, 108
Serotonin in neurogenesis, 72, 342
SH-SY5Y neuroblastoma cells, 248
Sickle cell anemia, etiology, 222
Signaling pathways and ion channels, 86–88
Sir2 protein, 525–526, 528
SIRT1 protein, 526–532
SIRT2 protein, 527, 531–534
SIRT3 protein, 527, 533, 535
SIRT4 protein, 527
SIRT5 protein, 527
SIRT6 protein, 527
SIRT7 protein, 527
Sirtuin family proteins
    about, 525–526
    classification of, 526, 527
    nicotinamide (NAM) and, 526, 528, 530, 532, 534–536
    PARP and, 532, 534, 536
    structure/function of, 526–528
Six3 gene, 129
Skeletal muscle, angiogenesis in, 498–499
SKMG-3 glioblastoma cell lines, 441
Small molecule therapy for gliomas, 404–405
Smooth muscle cells
    apoptosis in, 111
    electrical stimulation of, 503–504
    ion channels in, 85
    vascular, 212
Smoothened (smo) receptor in NSC proliferation, 124
Sodium (Na^+) channel
    in apoptosis, 108, 109, 113
    in cell cycle regulation, 89
Sodium phenobarbital, resistance to, 379
Sol-Gel process, 467, 469
Somal translocation described, 44
Somites in vasculogenesis, 32
Songbirds, neurogenesis in, 73, 126
Sonic hedgehog (SHH)
    in neurogenesis, 24–26
    in NSC proliferation, 121, 123–124, 126

SOX2 gene expression, 67
Sox2 transcription factor, 129
Sox8 transcription factor, 24
Sox9 transcription factor, 24
Sox10 transcription factor, 24
Sp1 transcription factor and p53, 381
SPARC (secreted protein acidic and rich in cysteine) described, 237–238
Spinal and bulbar muscular atrophy (SBMA), 250
Spinal cord
    in ALS, 251–252
    gliogenesis in, 16
    ischemia and VEGF, 255–256
    neurogenesis in, 13, 15, 24
    NSCs in, 17, 122, 129
    VEGF delivery to, 253
Src family
    in astrocyte transformation, 437, 450–451
    in cell signaling, 86, 265, 391, 393
Staggerer mice, 282, 287, 323
Standard uptake value (SUV), calculation of, 428
STAT pathway
    in gliomas, 398–400, 452
    ion channels in, 86
    in NSC renewal, 132
    and vascular injury, 213
Status epilepticus (SE), 198–200, 313–315
Staurosporine and apoptosis, 106, 110, 111
Stella/pgc7/dppa3 in germ cell regulation, 60
Stem cells. *See also* neural stem cells (NSCs); retinal stem cells (RSCs)
    bone marrow-derived, 223
    in cell therapy, 468, 519–520
    in the CNS, 512
    hippocampal, 337
    HUCB-derived, 518
    mesenchymal (*See* mesenchymal stem cells [MSCs])
    recruitment signaling in, 164
Steroids in EGFR activation, 270
Stimulants and neurogenesis, 343, 344
Stress
    disorders of, 339–342
    kinases in apoptosis, 74–75
    and neurogenesis, 362
Stress-activated protein kinases (SAPKs), 393, 395–396
Stroke
    angiogenesis following, 241
    in apoptosis, 74, 170–171
    CDK1 expression in, 284
    cell cycle reentry in, 283
    DNA damage in, 288
    ECM degradation in, 237

ion channels in, 82
ischemic, 112–113, 237, 239, 360
and neurogenesis, 75–76, 200
treatment, 227
and VEGF upregulation, 37
Subcortical heterotopia described, 49
Subependymal giant cell astrocytoma (SEGA), 434, 436
Subependymal zone (SEZ). See subventricular zone (SVZ)
Subgranular zone (SGZ)
cell detection/visualization in, 334–339
neurogenesis in, 14, 15, 71, 72, 75–76, 332
opiates in, 343–344
Substantia nigra, neurogenesis in, 15, 197, 271
Subventricular zone (SVZ)
CNTF, cell stimulation by, 123
described, 95
EGFR in, 266, 268, 269
GABA signaling in, 97–99
Huntington's disease in, 197
neurogenesis in, 14, 15, 17, 45, 71, 72, 75–76, 179, 196, 310–311
stress and, 339
Sulfate conjugates, resistance to, 375
Superoxide dismutase 1 (SOD1) mutations in ALS, 171, 249–253, 286, 518
SV40 Tag in cell cycle regulation, 282
Synapsin III and GABA, 98
Synaptogenesis, 227
Syndecan-4 in FGF signaling, 211

T

T-cell leukemia homeobox (TLX) described, 130
τ Protein in Alzheimer's disease, 299–303
T98G glioblastoma cells, 396
Tarceva (OSI-774) in targeted therapy, 404
TAU gene in ALS, 249
Taxol and CDK1, 285
Tay-Sach's disease, 512
TdR kinase 1/2 (TK1/2) expression, 422, 427
TEA and apoptosis, 111, 112
Telencephalon
angiogenesis in, 33–34
cell cycle in, 309–311
neural migration in, 45
NSCs in, 124
oligodendrocytes in, 25
radial glial cells in, 27
Telomerase in NSC proliferation, 130–131
Telophase described, 4
Temperature in mitosis counts, 4
Temporal lobe epilepsy (TLE), 170, 198, 199

Terminal deoxynucleotidyltransferase-mediated dUTP-biotin nick-end labeling described, 14
Tert gene transfer and cell senescence, 130
Tetracycline, TGF-α induction by, 393
Tetrahydrocannabinoid (THC) and neurogenesis, 343, 345
Tetrodotoxin and oligodendrocyte formation, 25
TFAP2A gene expression, 63, 67
TGIF gene in neural cell development, 64
Thalamus
astrocytoma in, 434
cyclin D regulation in, 170
EGFR in, 266, 268
HB-EGF in, 266
neuronal migration in, 44
Thalidomide in tumor control, 39
Thapsigargin, effect on ER, 106, 108
Therapy
adenovirus in, 153
for ALS, 252–254, 518
ischemia, 227, 512
neurodegenerative diseases, 200–201
resistance to, 374–382
stroke, 227
studies, evaluation of, 519
Therapy, targeted
AG3340 in, 405
Akt pathway, 404
angiopoietins in, 405
angiostatin in, 405, 406
antigen presenting cells (APCs) in, 403
Avastin (bevacizumab) in, 405
bacterial endotoxin in, 401
Batimastat (BB-94) in, 405
Bay12-9566 in, 405
breast carcinoma, 401
CCI-779 in, 404
CI1033 in, 404
dendritic cells in, 403
EKB-569 in, 404
EMD 121974 in, 405
endostatin in, 405, 406
FRAP/mTOR kinase inhibition in, 404
GBM, 401, 404, 405
gliomas, 400–406
IgG in, 401
Iressa (ZD-1839) in, 404
irinotecan (CPT-11) in, 401
MAb 806 in, 401
MAb Y10 in, 401
MAbs in, 401
MAPK pathway, inhibition of, 404–405
Marimastat in, 405
metallospondin in, 405

Novartis (PKI166), 404
OSI-774 (Tarceva) in, 404
PDGF in, 404, 405
PI3-K pathway, 404
R115777, 405
rapamycin in, 404
Ras pathway, inhibition of, 404–405
RGDs in, 405
SCH66336 in, 405
thrombospondins in, 405
TIMPs in, 405, 422
Thrombomodulin in cerebrovasculogenesis, 222
Thrombospondins
   in cerebral ischemia, 239–241
   in EC proliferation, 211
   in targeted therapy, 405
Thymidine labeling
   C-11, 420–427
   described, 6–8, 72
   in neurogenesis studies, 332–334
Tie family receptors in vasculogenesis, 35, 38, 222, 223, 325–326
Tight junctions, 239, 321, 327, 373
TIMPs. *See* tissue inhibitors of metalloproteinases (TIMPs)
Tissue engineering, 465–468
Tissue inhibitors of metalloproteinases (TIMPs)
   regulation of, 438
   in targeted therapy, 405, 422
Tissue staining, 4, 5, 8
TNF-α. *See* tumor necrosis factor-α (TNF-α)
TOAD-64 (TUC-4), 196, 199
Topiramate, resistance to, 379
Topotecan, resistance to, 375, 378
Transcranial magnetic stimulation (TMS), 342
Transcription factors in gene expression, 66; *See also individual factor by name*
Transforming growth factor-α (TGF-α)
   in cell signaling, 389
   in EGFR binding, 265–268, 270–271, 392–393
   in gliomagenesis, 450
   in NSC renewal, 123, 515
Transforming growth factor-β (TGF-β)
   in Alzheimer's disease, 302
   in angiogenesis, 34, 35
   in cell cycle regulation, 208–210, 324, 453
   in cerebral ischemia, 239
   in germ cell regulation, 60, 63, 66
   in immunotherapy, 403
   in neurogenesis, 24
Transient cerebral ischemia
   and apoptosis, 179
   gene/protein expression in, 239–241
   VEGF and, 255

Transplantation therapy, 201, 225, 226, 465–466, 512, 517
Trastuzumab (herceptin) for gliomas, 401
Traumatic brain injury (TBIs), 200, 283
Tuberous sclerosis, 433
TUJ1 microtubule markers, 248, 313
Tumor necrosis factor-α (TNF-α)
   in apoptosis, 74, 111, 144, 145, 153
   in cell cycle regulation, 208–209
   in cell signaling, 87
   in epilepsy, 382
   in immunotherapy, 403
Tumors
   astrocytic
      animal models of, 441–442
      cyclin D in, 438, 457
   brain (*See* gliomas)
   CDKs in, 438, 457
   cell migration in, 398
   chemotherapy in, 39, 373–375
   classification of, 449
   CNS, 183–184
   development
      EGFR in, 272–273, 392–393
      electrical activity, 185–186
      electrical activity in, 185–186
      oncogenes in, 165–167
      potassium channel in, 184–185
      SAPKs, 393, 395–396
   growth
      factors affecting, 143, 209
      and ion channels, 89–90
      ion channels in, 89–90
   MDR in, 373–379
   MRI imaging, 419, 420, 424, 427
   PET imaging, 419–420
   pituitary, 433
   RTK signaling in, 389, 395, 450
TUNEL assay, 147, 152, 282
Tunicamycin, effect on ER, 106
Turcot's syndrome, 433
Tyrosine kinase receptors. *See* receptor tyrosine kinases (RTKs)

U

U2OS tumor cell line, necleostemin in, 130
U87MG glioblastoma cell lines, 393, 397, 405, 441
U251MG glioblastoma cell lines, 441
U373MG glioblastoma cell lines, 441
U1242MG cell line, 393
Ulex europaeus agglutinin-1 (UEA-1) binding in cerebrovasculogenesis, 223
UVB and apoptosis, 107, 110–111

V

Vaccination for gliomas, 403
Valinomycin and apoptosis, 108
Valproic acid, resistance to, 379
Vascular endothelial growth factor (VEGF)
    about, 236, 245–246, 257
    in angiogenesis/vasculogenesis, 31, 32, 34–35, 37–38, 222, 223, 225, 496, 498, 499, 504–506
    in cell cycle regulation, 208, 211, 325–326
    in cerebral ischemia, 239–241
    EGF/EGFR and, 396
    hypoxia and, 246–247, 250
    in malignant gliomas, 405
    neural cells, effects on, 247–252
    in neurogenesis, 72, 247
    PDGF/PDGFR and, 391
    therapeutic potential of, 252–257
Vascular system
    about, 245
    differentiation in, 326–327
    genesis (*See* vasculogenesis)
    injury, response to, 209, 212–213, 225
Vasculogenesis
    about, 31–33, 222, 319
    augmentation of, 227
    postnatal, 36–38, 222
    VEGF and, 31, 32, 34–35, 37–38, 222, 223, 225
    vs angiogenesis, 225
Vector-mediated gene delivery, 477–485, 488
VEGF. *See* Vascular endothelial growth factor (VEGF)
Ventriculomegaly, 241
Versican, expression of, 210
Vimentin in radial glial cells, 27, 313
Vinblastine, resistance to, 375
Vinca alkaloids, resistance to, 374, 375, 381
Vincristine, resistance to, 381
Viral vector-mediated gene delivery, 477–485, 488
Visual cortex, neurogenesis in, 72
Volume homeostasis in cell cycle regulation, 83–85
Von Hippel-Lindau syndrome, 433
Von Willebrand factor (vWF) in EPCs, 223, 227

W

Wortmannin in cell signaling inhibition, 393, 404
Wound-healing
    angiogenesis in, 37, 497
    ion channels in, 82

X

X-ray computed tomography (CT) in tumor imaging, 419
Xestospongin, effect on ER, 106
XLIS gene in lissencephaly, 51, 52

Z

ZD-1839 (Iressa) in targeted therapy, 404
Zebra finches, neurogenesis in, 73
Zinc
    in apoptosis, 113
    and GABA, 96
Zonnula occludens, 239, 321, 327, 373